Applied Aquatic Ecosystem Concepts

Gerald L. Mackie

KENDALL/HUNT PUBLISHING COMPANY
4050 Westmark Drive Dubuque, Iowa 52002

Other art work provided by Gerald Mackie.

ISBN 0-7872-7490-9

Printed in the United States of America
10 9 8 7 6 5 4 3 2 1

Preface

I consider myself one of the most fortunate people in the world with respect to my job and personal life. I work (do research) with water, I teach "water" (aquatic biology) and I play with water (I have a cottage where I work and play!). I teach an introductory course on aquatic environments, biology of running waters and biology of polluted waters. Nearly all the students are able to regurgitate at the ends of the courses the plethora of limnological concepts, but fewer are able to apply them in explaining water events (floods, waves, flotation, fog, etc.) or water uses (adding salt to boil potatoes, make artificial ice, water treatment, etc.) that we experience or observe on nearly a daily basis. This is particularly true of people who own cottages.

There are several excellent limnological texts, each describing limnological concepts in a unique and interesting way. Many provide numerous references that provide even more detail than given in the texts. However, very few *apply* the concepts to explain the events and uses of water. This text summarizes the limnological concepts and then demonstrates or illustrates their use. Lake morphometric formulae (area, volume, retention time, fetch, wave height) are not only given, they are applied; nutrient cycles and dissolved gases are not only described, they are applied to explaining and measuring degrees of eutrophication; algae, macrophytes, invertebrates and fish are not only described, they are applied to the biological assessment of water quality; and biodiversity is not only described, but applied to the strengths and weaknesses of endangered and exotic species. Emphasis is given on integrating the fundamental concepts of the physical and chemical attributes of water and applying them to the physical, chemical and biological assessment of water quality.

Finally, I am happy to acknowledge the literature and suggestions about the application of limnological concepts that I received from numerous colleagues, graduate students and undergraduate students. I particularly would like to thank Josephine (Jo) Archbold for her contribution to the pesticide section in Chapter 10, for preparing the glossary and for helping to mark words in the text for the index. Susan Reynolds also researched and contributed to many of the definitions for the glossary. All illustrations are drawings and photographs made by the author, except for a few kindly provided by Jonathan Witt. Chad Boyko, Cara Smith and Susan Reynolds all helped with formatting the illustrations for the text. Jo, Susan, Andrew Wannan and Dave Zanatta assisted with the proof reading of the manuscript.

Gerald L. Mackie

TABLE OF CONTENTS

CHAPTER 1

UNIQUE PROPERTIES OF WATER and THE HYDROLOGICAL CYCLE

Why Read This Chapter?

There are myriad events in our lives that occur only because water has very unique physical properties. This chapter examines eight unique of them and relates their importance to aquatic organisms and to human's every day use of water. All life needs water. For example, consider the numerous every-day events (e.g. boiling potatoes, making artificial ice, swimming, etc.) in our lives as being a result of, in one way or another, the unique physical properties of water. You will discover:

☞ 1. Four ways to make freshwater from salt water
☞ 2. The molecular structure of water
☞ 3. Why aquatic organisms are exposed to less temperature extremes than are terrestrial organisms
☞ 4. Why coastal climates are more moderate than inland climates
☞ 5. Why lake and pond temperatures lag behind that of air temperatures
☞ 6. Why fog is more common in the fall than in the spring
☞ 7. Why a household aquarium can provide a better indication of average room temperature than can a wall thermometer
☞ 8. Why snow flakes are hexagonal
☞ 9. Why pipes burst when the water freezes
☞ 10. Why ice floats
☞ 11. Why the bottom of deep lakes is a permanent refrigerator
☞ 12. The difference between overflow, interflow and underflow
☞ 13. How salt affects the molecular structure of water
☞ 14. Why potatoes cook faster when salt is added to water
☞ 15. Why seawater is an antifreeze
☞ 16. How to make artificial ice
☞ 17. Why, after swimming, freshwater evapourates more quickly from the skin than does seawater
☞ 18. Why aquatic organisms need to spend less energy to stay afloat in winter than in

summer

☞ 19. Why it is easier to float in sea water than in freshwater
☞ 20. Why pulling a boat out of water is easier than lifting it out
☞ 21. Why we can clean and reuse water polluted with sewage and other contaminants
☞ 22. Why you can apply electricity to water
☞ 23. Why and how to use water as a levelling medium
☞ 24. How Christmas trees can be fed a constant water supply
☞ 25. How little freshwater our freshwater supplies are
☞ 26. What the sources of water are for streams and lakes
☞ 27. What pathways water use to get to streams and lakes
☞ 28. What happens to subsurface water

Introduction

Water covers 75% of the earth's surface; about 96% of this is found in the ocean. The remaining 4% is in polar ice caps (~2%), groundwater (~1.8%), freshwater lakes and rivers (<0.020%) and in atmospheric vapour (<0.001%). Only 0.017% of the earth's water is fresh and available at the surface for use or consumption by humans. Of the groundwater, 0.62% is available, mostly within the top 200 m of the surface of the earth.

Even though a mere 0.62% of the global freshwater is available, some of it has undesirable natural substances, is polluted, or is otherwise unfit for human consumption. Because fresh water is a vital substance for every living organism and its supply is being limited even further by human wastefulness and contamination, there is increasing pressure on the human population to stop contaminating our water supplies and find new sources. Today, the availability of potable water (water suitable for drinking) determines the number of people who can inhabit any particular geographical area. There are many areas in North America areas are located close to the oceans. Because of this, the marine resource is being used to make fresh water from salt water
1☞ using a variety ***desalination*** processes. The most common method is ***distillation*** where sea water is first boiled and then the vapour is condensed. About three quarters of the world's desalinated water is produced with this method. ***Freezing*** is another method, but it is very costly. In this process ice crystals exclude salt as they form and the ice can then be captured and melted for use as fresh water. An increasingly popular process is ***reverse osmosis desalination***, where sea water is forced against a semi-permeable membrane at high pressure. Fresh water seeps through the membrane's pores while the salts remain behind. About 25% of today's desalinated water is produced in this way. The method uses less energy per unit of freshwater than does freezing, however the membranes are fragile and costly.

> How to make fresh water from salt water

The most recent technology, although it is somewhat low-tech in nature, relies on ***evapouration*** and ***condensation***. For example, huge plastic bubbles are placed over large pools of seawater, allowing the seawater to evapourate inside the bubble. The outer surface of the bubble is kept cooler than the inside so that the evapourated water condenses on the walls of the plastic bubble. The condensed fresh water on the inner walls of the bubble is then collected in troughs and distributed to the consumer.

A bonding relationship!

Molecular Structure of Water

2 ☞

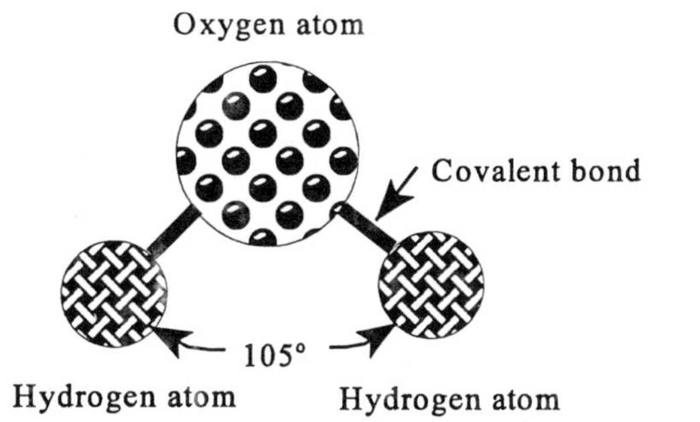

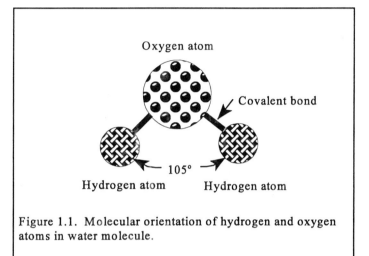

Figure 1.1. Molecular orientation of hydrogen and oxygen atoms in water molecule.

Figures 1.1 and 1.2 show the structure of a water molecule. A water molecule is formed by two hydrogen atoms and one oxygen atom. In the water molecule, electrons are shared between two hydrogen atoms and one oxygen atom. Because the bonds are formed by shared pairs of electrons the bonding is known as ***covalent bonding***. The angle formed by the two hydrogen atoms and the central oxygen atom is 105° (Figure 1.1).

Each water molecule can be thought of as having a positive end and a negative end because positively charged ions, called protons, at the centre of the hydrogen atom, are left partially exposed as the electrons are bonded closely to the oxygen atom. In essence, the water molecules act like magnets, the hydrogen atom being positive and attracting oxygen atoms with a negative charge.

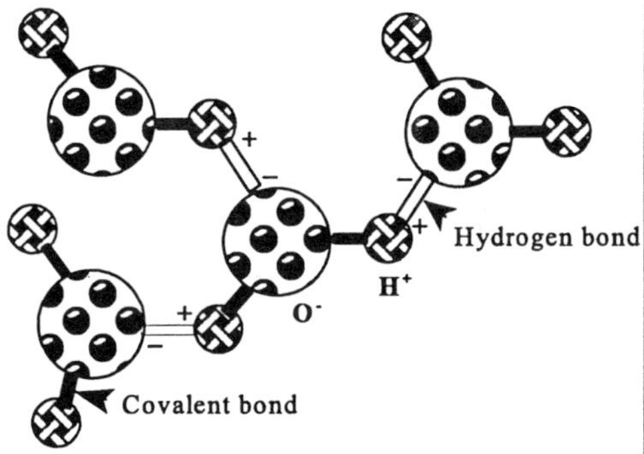

Figure 1.2. Covalent and hydrogen bonds of a water molecule.

When a hydrogen atom in one water molecule is attracted to the oxygen atom of another water molecule, an electrical bond, called a ***hydrogen bond***, is formed (Fig. 1.2). The hydrogen bonding is fairly loose and enables water to flow. It is the hydrogen bonding that greatly affects the physical properties of water. The following describes how these physical properties affect our lives and why water is such a unique and important resource.

Unique Physical Properties of Water

(I) Heat Relations

Temperature is among the most important factors in the aquatic environment because it greatly influences the well being of organisms, the rates of chemical reactions and the physical properties of water in general. Actually, it is the heat relations and not temperature per se that affect the well being of organisms. There are three heat relations: ***the latent heat of vapourization; specific heat; and the latent heat of fusion.***

> Talk about temperature extremes - why we as humans have more to worry about than do aquatic organisms!

(i) Latent Heat of Vapourization - pure water has a very high latent heat of vaporization, about 585 calories per gram of waters at 20°C. "Latent" refers to heat that does not cause a change in temperature but does produce a change of state, in this case from the liquid state to the vapour, gaseous, or steam state (Fig. 1.3). Because of the large number of hydrogen bonds a lot of heat is required to break the bonds to create vapour. Water is remarkably resistant to heating and because of the latent heat of vaporization, it is more stable in thermal qualities than is the atmosphere. Hence, aquatic organisms are subject to less temperature variations than are terrestrial organisms.
3 ☞ For example, terrestrial organisms commonly experience about 10°C variation in a 24 h period, but aquatic organisms usually experience only 1-2 °C in the same period. The latent heat of
4 ☞ vapourization also explains why coastal climates are more moderate than inland climates, and why rooms and climates with high humidity have less variable temperatures than do rooms and climates with drier air.

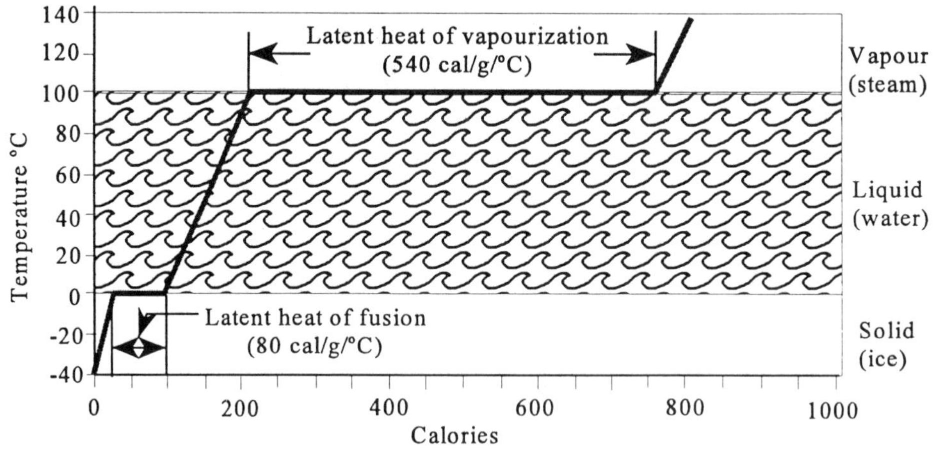

Figure 1.3. Latent heats and changes of state of water. The latent heat of melting (or fusion) (80 cal/g/°C) is much less than latent heat of vapourization (540 cal) because only a few bonds must be broken to convert 1 g of ice to water while all remaining bonds must be broken to convert 1 g of water to steam.

(ii) Specific Heat - water also has a remarkable capacity to hold large amounts of heat with only a relatively small change in temperature. Pure water has a specific heat of one gram calorie. Very few other solvents have a higher specific heat. By definition, *specific heat is the amount of heat required to raise the temperature of 1 gram of water 1 °C.*

Specific heat also has important implications to aquatic organisms, as well as to organisms on land. The high specific heat of water acts as a buffer against wide
5 ☞ fluctuations in temperature and hence moderates terrestrial climates near large bodies of water. The slow rate of cooling and warming of lakes is attributable to the high specific heat of water. Therefore, lake and pond temperatures lag well behind those of the atmosphere. As a result, aquatic organisms are subjected to less rapid changes in temperature than are terrestrial organisms. It also explains why fog is more common in the fall than in the spring. When water is warmer than the air, water evapourates from the surface (of lakes, ponds, wet ground etc.) and, because it is lighter, rises and condenses in
6 ☞ the cooler air above, creating fog. In the spring, lake and pond (and wet ground) temperatures rise slower than air temperatures so that the air immediately above the lake surface is cooled and sinks.

> Why fog is more common in fall than in spring

7 ☞ The specific heat property also implies that you can use the temperature of water in an aquarium as a more reliable mean temperature of a room (with the aquarium) than taking several readings from a wall thermometer. The daily room temperature may vary 2 -5 °C, but the daily temperature of water in the aquarium will vary only 0.1 - 0.2 °C.

> How to tell the average temperature of a room.

(iii) *Latent Heat of Fusion* - The heat required to change ice to water with no change in temperature is known as the latent heat of fusion. For freshwater, 80 calories of heat are required to melt one gram of ice at 0°C. This represents about 15% of the heat necessary to separate hydrogen bonding in the vapourization process because the heat generated increases the motion of hydrogen molecules. This motion, through collision of molecules, helps collapse the lattice network of ice and induces melting.

Figure 1.4 shows the lattice network of ice. Notice
8 ☞ the hexagonal arrangement of the atoms at the molecular level. The hexagonal pattern explains the six-sided structure of snow flakes (Fig.1.5). As water cools to 0°C the crystal lattice forms as the angle of the hydrogen atoms in water expand from about 105° to slightly more than 109°. This expansion represents about 9% of the original water volume and

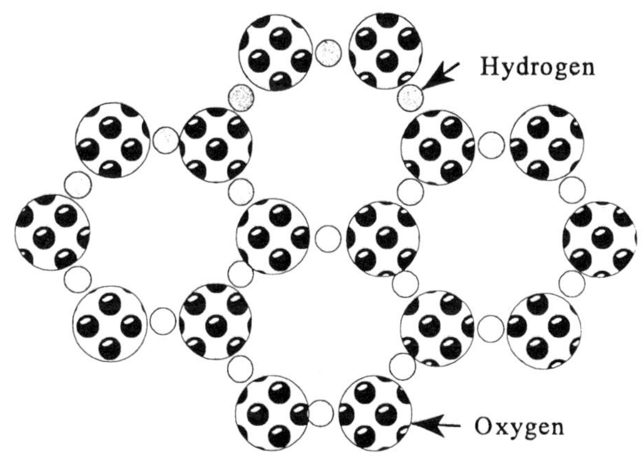

Figure 1.4. Molecular structure of ice.

9 ☞ explains why frozen pipes burst in the winter time. Because the molecules are more widely spaced in ice than in water, ice is less dense than liquid water and floats. Ice at 0°C weighs only 0.917 grams per cubic centimeter whereas liquid water at 0°C weighs 0.999 grams per cubic centimeter. This property is important to aquatic organisms because they are protected during the winter months by ice and because water freezes from the surface downward. Since ice is less dense than water, the lower portion of the lake rarely freezes and organisms at or near the bottom of a lake are rarely exposed to temperatures less than 4°C, due to the density-temperature relations of water.

Why pipes burst in the winter.

Some aquatic organisms are never exposed to freezing temperatures - and they rarely have to worry about being hot either!

Figure 1.5. Hexagonal structure of a snowflake.

(II) Density

Density is affected by variations in ***temperature, pressure*** and ***dissolved material***.

10 ☞ ***(i) Temperature*** - like most liquids, water becomes heavier with cooling, as shown in Figure 1.6, and it becomes lighter with heating. Water is unique in that it reaches its maximum density at 3.98°C (we usually round this off to 4°C). Other liquids are most dense at their freezing point, 0°C, but as water cools below 3.98°C, it again becomes lighter as the crystalline °C°C lattice of ice develops (Figure 1.4). But even without ice, water at temperatures below 3.98 °C will always be near the surface (unless mixed by wind). Hence, deep water organisms in lakes are rarely exposed to temperatures less than (or even more than, as shown later) 4°C. This is the average temperature of most refrigerators.

How to keep your "root beer" cold when the camp refrigerator is out of order!

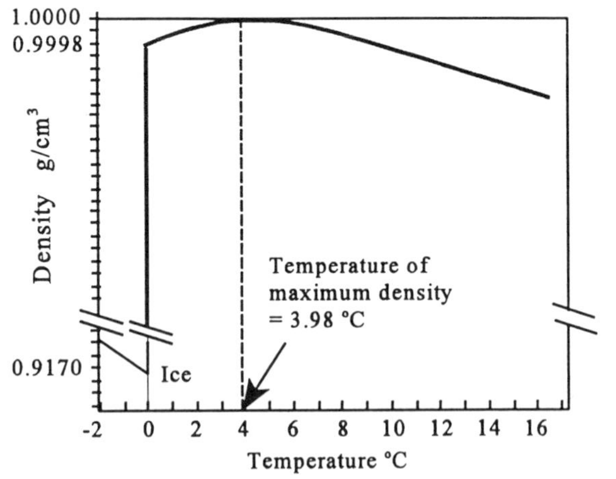

Figure 1.6. Effect of temperature on density of pure water.

11 ☞ The changes in density as the temperature changes ultimately have a great effect on mixing processes in

lakes. Figure 1.6 shows that the density differences are rather small but when you examine the difference per degree lowering, the density difference decreases greatly below 1 °C and is lightest (~0.9170 g/cm^3) just before it freezes at 0 °C. The ***density difference per degree lowering*** is significant when lakes thermally stratify because it offers ***thermal resistance to mixing*** between stratified layers. The greater the differences in temperature between stratified layers, the greater the thermal resistance to mixing.

(ii) Pressure - density is also affected by pressure. Pure water that is subject to a pressure of one atmosphere is considered to have a density of unity, 1.0000. At ten atmospheres the density of pure water increases to about 1.0005, and at twenty atmospheres it increases to about 1.0010. In other words, the density of freshwater increases with increasing pressure at a rate of approximately 0.0005 per increase of 10 atmospheres.

(iii) Dissolved Materials - all fresh water contains some substances in solution, but never as much as seawater. Dissolved materials will increase the density of water in proportion to the concentration and the specific gravity of the solute. Evapouration will increase the density simply by concentrating the dissolved material.
12 ☞ Alternatively, precipitation in the form of rain or snow, which is usually very low in dissolved materials, will dilute the solutes causing the density to decrease. The density of spring melt water flowing into lakes from parent streams is usually lower than that of the lake, resulting in ***overflow*** (light water on top of heavy water, Fig.1.7a); during the summer, stream temperatures are often cooler than the surface temperatures of lakes and will cause ***underflow*** (cooler, heavier water flowing under warmer, lighter water, Fig. 1.7b). If stream temperatures are similar to surface lake temperatures, water enters the lake as ***interflow*** (Fig. 1.7c).

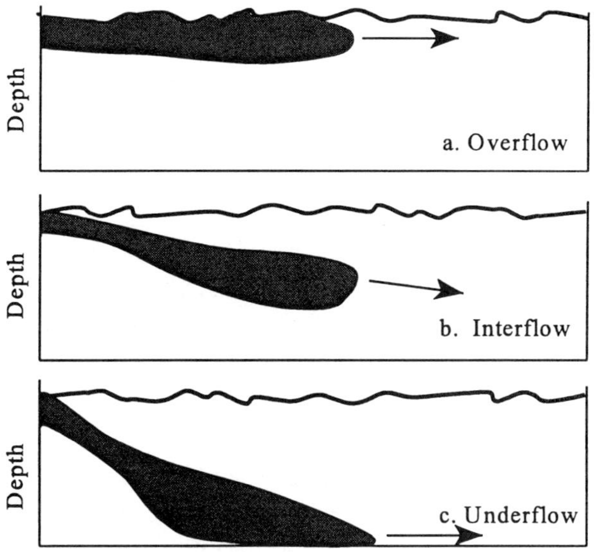

Figure 1.7. Three types of flow patterns that develop when different density layers meet

Variations in density also occur because of the amounts of suspended materials. All fresh waters contain some suspended material, such as aquatic organisms (algae, plankton), debris, and silt, and if present in sufficient quantities, can increase the weight or density of the water.

When salt, such as sodium chloride, is added to water the ions of sodium and chloride absorb water molecules and become hydrated (Fig. 1.8). Salt dissolves in water because the polarity of the molecule keeps positive ions separated from negative ions. As a result, sodium ions are surrounded by water molecules with the negatively charged
13 ☞ portion (i.e. O^-) attracted to the positive (i.e. Na^+) ion; chloride ions are surrounded by water molecules with their positively charged portion (i.e. H^+) attracted to the negative ion (i.e. Cl^-). Hence, the chloride ion bonds with four hydrogen ions and the sodium ion bonds with the oxygen ions (Fig. 1.8). In order to freeze or boil saline water, both the hydrogen bonding between hydrogen and oxygen ions and the additional hydrogen

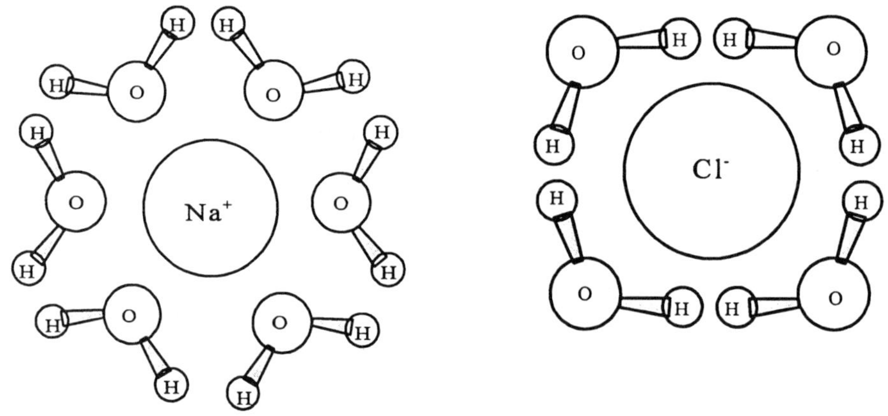

Figure 1.8. Orientation of water molecules about sodium and chloride ions in sea water.

bonding between the chloride and hydrogen ions must be overcome. The additional bonding effectively reduces the normal physical properties of water. There are five important changes in the physical properties of water when salt is added:

1) The ***specific heat*** decreases with increasing salinity. That is, less heat is required to raise the temperature of seawater 1 °C than to raise the temperature of freshwater by the same amount. The specific heat also increases with increasing temperature in saline waters (e.g. sea water). Hence, as the temperature of the
14 ☞ water increases it becomes harder to remove the last few water molecules from the hydrated ions. Therefore, the boiling point of water is increased with increasing salinity. This explains why we add salt to water when boiling potatoes. It takes longer to boil the water because the salt increases the boiling point, but the water gets hotter and therefore cooks the potatoes faster than water without salt.

2) The density increases nearly linearly with increasing salinity (Fig. 1.9). Pure water has a maximum density at 3.98 °C but the addition of salt lowers the temperature of maximum density (Fig. 1.9). At salinities greater than 24.7 ppt, the maximum density occurs at a temperature below the normal freezing point,

0 °C. Therefore as ocean water is cooled, it grows denser and, since cooling is from the surface down, the surface water becomes heavier and sinks by ***convection***. The underlying warmer and less dense water rises to the surface to be cooled in a similar process. This process continues until the whole water mass is

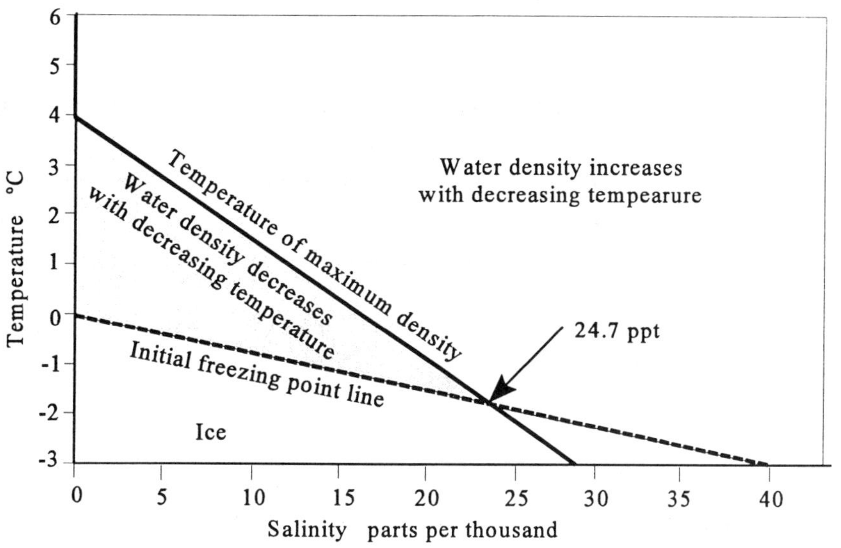

Figure 1.9. Effect of salinity on temperature of maximum density and freezing point

Why the ocean never freezes!

at the same temperature or until the entire mass freezes. Continued cooling in any ocean with a salinity >25 ppt (most are ~35 ppt) would result in the ice sinking and the whole water mass freezing. The oceans do not freeze because of the huge volumes of water and global currents (e.g. due to ***Coriolis forces*** resulting from rotation of the earth on its axis - see Chapter 4) that move water from warmer climates to colder climates. At salinities less than 24.7 ppt, the temperature of maximum density is reached before the freezing point is reached. With continued cooling the water will get less dense, freezes and floats on the surface.

15 ☞

The anti-freeze use of salt water, or how to make artificial ice!

16 ☞

3) Figure 1.9 also shows that the freezing point is lowered with increasing salinity. This ***anti-freeze effect*** of salt is due to the hydrogen bonding between the hydrogen ions and the chloride ions in salt water. Ice in sea water forms at a lower temperature than ice in freshwater. It is this principle that is used to create ice in arenas. Brine is run through pipes that lie within the concrete floor of the arena. The temperature of the brine is lowered to less than 0 °C, usually somewhere around -1 °C and -2 °C, and the freshwater, when it is laid on the surface by a Zamboni, freezes almost immediately at 0 °C. Jet ice in some newer arenas is made by using distilled or deionized water instead of tap water because the latter has more solutes.

17 ☞

Drying off is easier in fresh water than in salt water.

4) The ***vapour pressure*** is also lowered with increasing salinity because the salt makes the water molecules less available for evapouration. Vapour pressure is a measure of how easily water molecules escape from the liquid state to the vapour state. Sea water evapourates more slowly than freshwater because of the additional hydrogen bonding between chloride and hydrogen ions. Anyone who has bathed in both sea water and freshwater has probably noticed how quickly water evapourates from the skin after swimming in freshwater compared to sea water.

5) ***Osmotic pressure*** is the pressure exerted on a biological membrane by differences in salinity on each side of the membrane. Water will flow (creating osmotic pressure) from the low-salinity side to the high-salinity side. Increasing salinity causes increases in osmotic pressure. This is an important physiological factor in transmitting water into and out of cells. The direction of flow depends on whether the osmotic pressure of the internal medium of the organism is greater or lower than the external aquatic medium; flow is always toward the more concentrated medium.

Of the five variables affected by increasing salinity, four of them, the increase in boiling point, the freezing point depression, the vapour pressure elevation and the increase in osmotic pressure, are often referred to as ***colligative properties***. Freshwater does not have colligative properties.

(III) Viscosity

18 ☞

It's harder to float in the summer than in the winter?!

By viscosity we mean ***mobility***. Water is a mobile liquid and has internal friction, called ***viscosity***, that varies inversely with temperature. The resistance of water to flow is due to the energy contained in hydrogen bonding. With increased temperature, bonding is weakened and flow is eased. This means that the mixing and stirring of water from wind action is more complete and requires less work at higher temperatures than at lower temperatures. Viscosity also affects aquatic organisms, especially fish; it is easier to stay afloat in cold, viscous water than it is in warm, more fluid water.

(IV) Buoyancy

19 ☞

It's much easier to keep afloat in sea water than in fresh water !

Buoyancy is the direct outcome of density and varies with the same factors. The ***Law of Archimedes*** states that a body in water is buoyed up by a force equal to the weight of the water displaced. The greater the density, the greater the buoyant force. The denser the water, the higher a floating object will rise, explaining why its easier to swim in salt water than in freshwater. Ships passing from freshwater into seawater float high on the surface. Hence, ships with a light cargo load must carry ballast to weight them down, making the ship more stable by floating lower in the water. When the ship reaches a body of freshwater, it must unload the ballast so that the ship does not float too low in

> How zebra mussels reached the Great Lakes!

the water. This procedure explains how zebra mussels were introduced into the Great Lakes; a ship from Europe took on freshwater (containing zebra mussels) as ballast to cross the Atlantic Ocean, then discharged the ballast water (with zebra mussels) in western Lake Erie before sailing up the rather shallow Detroit River to Lake St. Clair and thence up another shallow river, St. Clair River, into Lake Huron. If the ballast water was not released, the ship would have struck bottom on its way up the Detroit and St. Clair rivers. Since cold water is more dense (and more buoyant) than warm water, organisms and objects (including ships) are more buoyant in winter than in summer, and swimmers have to work harder to remain afloat in fresh water than in salt water.

(V) Surface Tension

> Water has a surface skin capable of supporting needles and razor blades.

Hydrogen bonds allow water molecules to stick to each other, a property known as ***cohesion***. Cohesion gives water an unusually high surface tension, or a "***surface skin***" capable of supporting needles, razor blades, and organisms. ***Adhesion*** is the ability of water to stick to other materials; it allows water to adhere to solids and to make them wet. Without hydrogen bonding, water molecules would dissipate. Adhesion is the tendency of the liquid to cling to the surface by means of bonds established between the hydrogen atoms of the water and the oxygen atoms of the object. A force is exerted toward the water, because there is only one layer of oxygen atoms at the surface of the object and several layers of hydrogen atoms below. To test this, try pulling a wide board from and parallel to the surface of the water; it is much easier to tilt the board first, breaking the bonds at one end of the board, and then along the length of the board as it is pulled up at
20 ☞ an angle. This explains why it is easier to pull a boat out of water than to try and lift it straight up and out. Without surface tension, water would stick to everything, much like syrup.

(VI) Solvent Action

21 ☞

> Why sewage treatment plants are effective.

There are three types of solvent action. One type is ***inert*** because water exercises practically no chemical reaction on most of the substances it dissolves, and organisms can receive some solutes in essentially an unmodified form. Also, water as a solvent can be used over and over again. It is this principle that allows sewage treatment plants to strip organic and inorganic wastes out of water so that the water can be reused. It also explains why sea water can be desalinated to produce fresh water.

Another type of solvent action is ***ionization***. Water is unparalleled in its capacity to split or ionize salts. Hence, some organisms can extract specific ions without taking in a myriad other ions of little use to them.

Thirdly, water has a very high ***charge of separation***. As a result, an extremely
22 ☞ high electric charge is needed to separate hydrogen from oxygen. Hence, we are able to use electro-shocking to control zebra mussel settlement in industrial facilities. In fact,

How to control zebra mussels - "zap" them.

low levels (8-10 volts/inch) of alternating current are being used to control zebra mussels on concrete and steel surfaces in some hydro-generating plants.

(VII) Transparency

Pure water is transparent and since many organisms, especially plants, rely on deep penetration of light waves, water clarity is a vital physical characteristic for the normal functioning of aquatic ecosystems. SCUBA divers exploit this feature as well. Most natural waters are somewhat coloured (but transparent) and contain suspended materials (that decrease transparency) that selectively absorb certain light waves. Nevertheless, most natural waters are transparent enough to allow substantial amounts of photosynthesis by algae, the main source of oxygen for aquatic organisms in lakes and rivers. Later, we will examine in detail the selective absorption of light waves in water.

(VIII) Liquid Nature

Finally, and perhaps the most obvious unique feature of water, is its liquid nature. There are very few inorganic substances that occur naturally in the liquid state. Mercury is the only other one. Water, an inorganic substance, is unique simply in that it is liquid in nature. It can be used as a levelling medium. Simply fill a hose with water and stretch it across any
23 ☞ variable surface. Since water finds its own level it can be used to level two stations almost perfectly (Fig. 1.10). As long as both ends of the hose are open and unobstructed, gravity pushes equally on water at each end. Otherwise a vacuum is created and the water level is then controlled by the negative pressure created at the obstructed end of the hose. Only when air is allowed to enter will the water flow, by gravity, toward the unobstructed end.

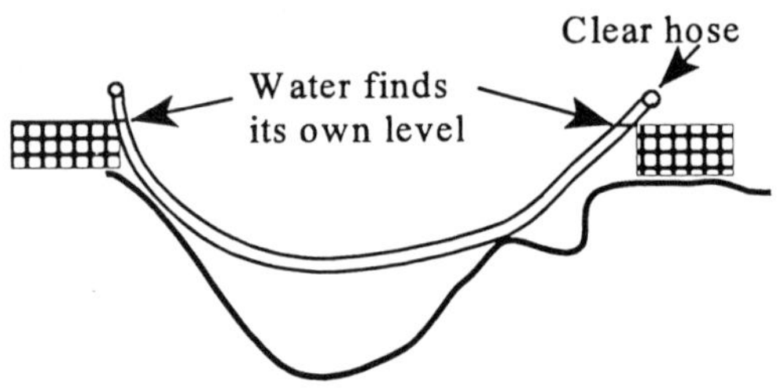

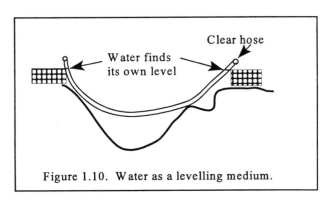

Figure 1.10. Water as a levelling medium.

The same principle is used for devices created for automatic watering of Christmas trees. To make your own, simply fill a large bottle (a 2 L plastic soft drink bottle will do) with water. Obtain a length of plastic hose (5 mm maximum inside
24 ☞ diameter) long enough to reach from the Christmas tree to the bottle. Next, get a cork or rubber stopper large enough to fit securely in the opening of the bottle. In the cork, drill a hole large enough to receive the outside diameter of the hose. Place one end of the hose in the cork and insert the cork in the bottle. Fill the Christmas tree stand about three quarters full. Place the other end of the hose about half way down the stand. Invert the bottle and place it in a stand, ensuring that the hose is not kinked. Some imagination is

needed to make the stand, but two clay or plastic planting pots, one large enough to hold the inverted bottle and one slightly larger to act as the stand will do. Each pot must have a hole in the bottom centre. Invert the larger pot and place the smaller one on top (Fig. 1.11). Water will flow until it has replaced the air in the hose and then stop. Water will flow again after the water level in the Christmas tree stand falls below the opening of the hose.

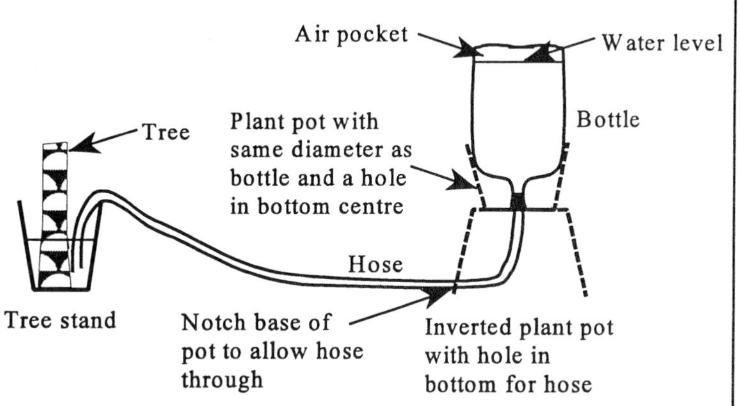

Figure 1.11. Applying gravity principles for automatic watering of Christmas trees. Water will flow when level in tree stand falls below opening of hose. Air enters the hose, rises to top of bottle, allowing water to flow by gravity until hose opening is covered with water.

The Hydrological Cycle

Table 1.1 shows the distribution of water on earth. The largest pool of water is in the ocean, accounting for all but 4% of the total. Less than 1% of the total amount of water is available for human use and, as discussed earlier, some of this is tainted by
25 ☞ human negligence. The largest freshwater source is in precipitation, 87% of which is supplied from the oceans as evapouration. However, only about 7% reaches land, as clouds, the remaining 80% falls back on the oceans. The only other major source of water is the continents which contribute about 13% through evapouration from lakes, rivers and land and evapo-transpiration from plants. Figure 1.12 shows the global water
26 ☞ balance, the cycling of water and the major pathways and processes involved in the hydrological cycle.

The hydrological cycle can be divided into gains and losses for a watershed and for the body of water (e.g. lake) itself (Fig. 1.13).

Gains:

Precipitation:

Precipitation is the primary source of water for both the watershed and the lake basin, but
27 ☞ lakes may also receive water from groundwater seepage, spring seepage and runoff. Cold water fish, such as trout, often congregate around these seepage areas, particularly during the summer when the water temperature rises in upper water strata.

Table 1.1. Distribution of water on earth

Source	Volume km^3	% of total
Oceans	1,322,000,000	96.0
Polar ice	29,200,000	2.1
Ground water	25,000,000	1.8
Freshwater lakes	125,000	0.01
Saline lakes	104,000	0.01
Soil and subsoil	65,000	<0.01
Atmospheric vapour	14,000	<0.001
Rivers	1,200	<0.0001

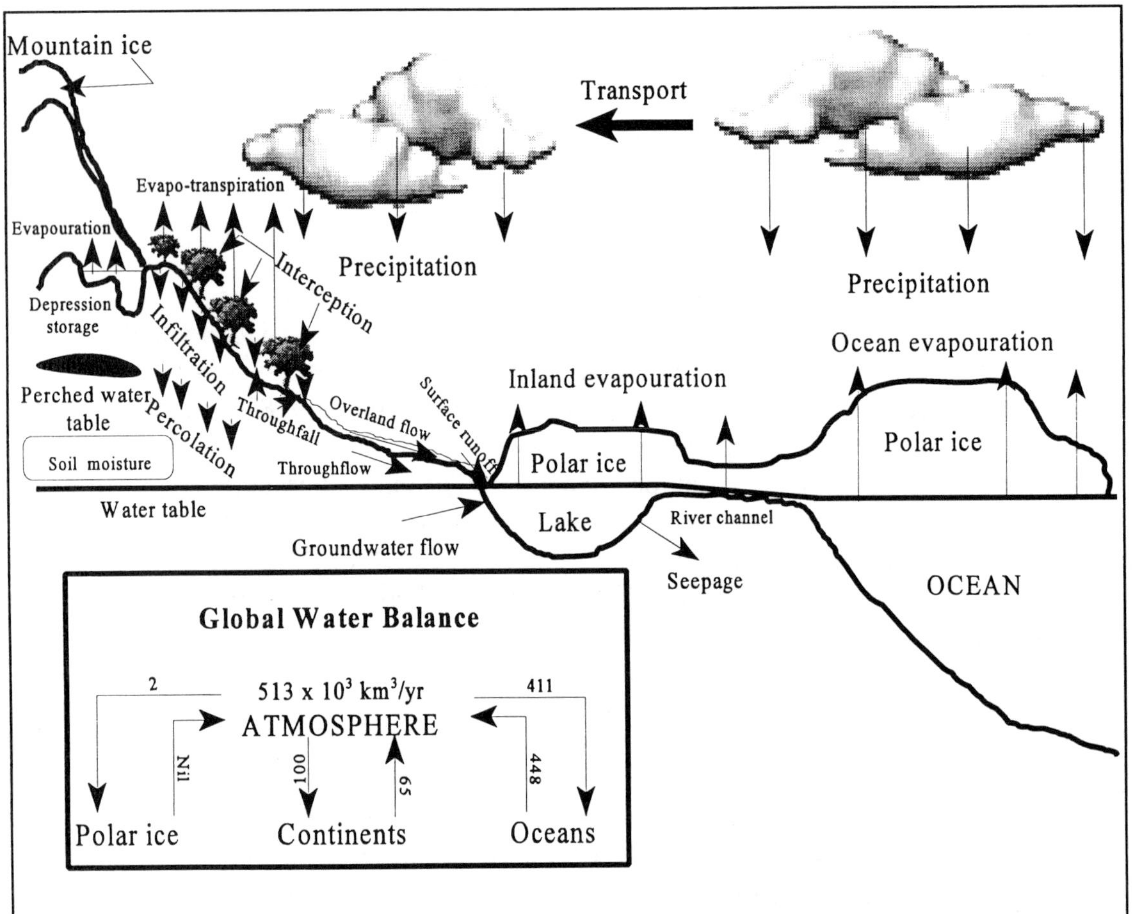

Figure 1.12. The hydrological cycle and global water balance.

Figure 1.13. Hydrological regions and the major pathways of the hydrological cycle in an exorheic region and, occasionally endorheic regions

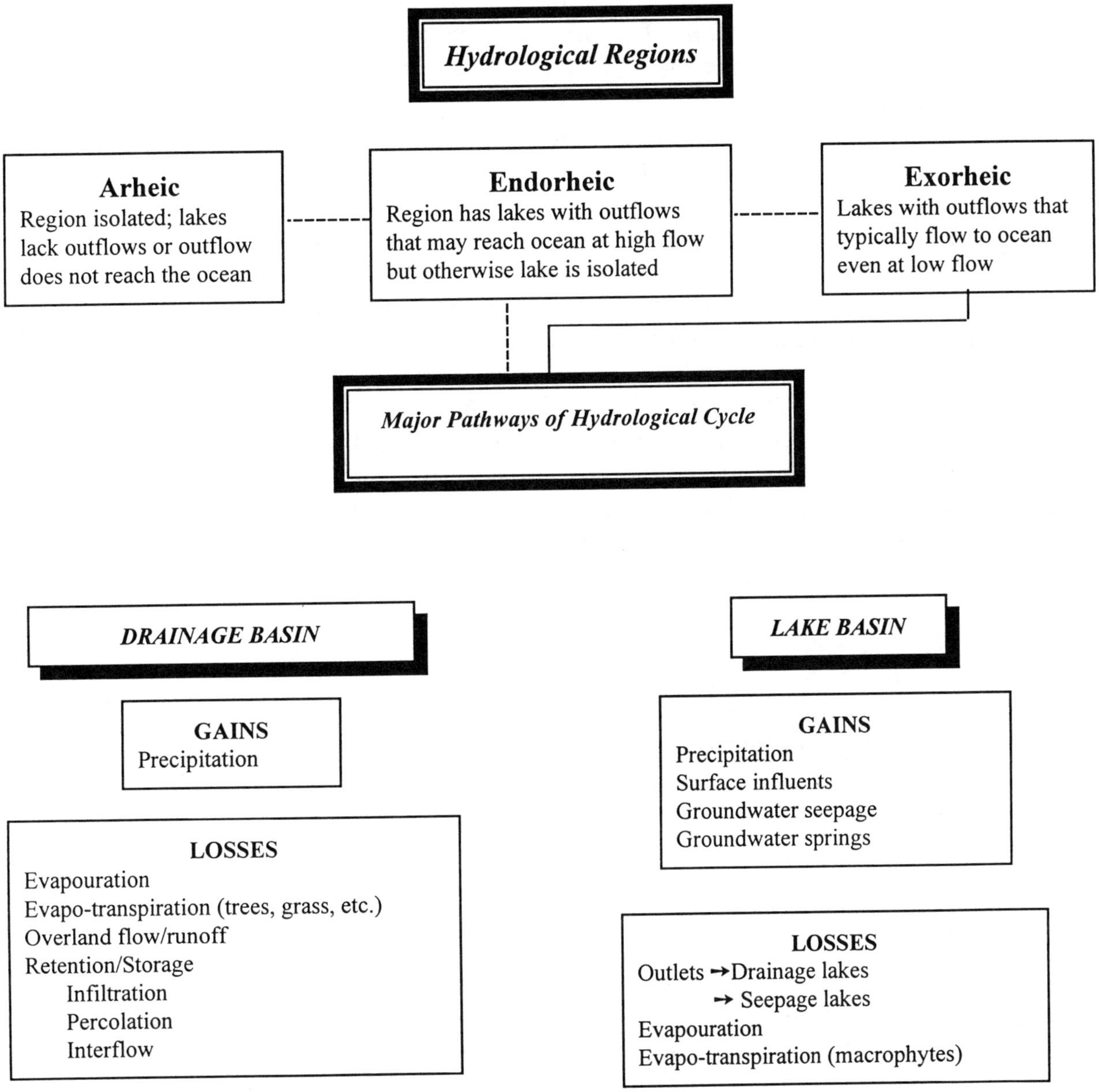

Losses:

Drainage Basin Losses:

Losses may occur by several processes after water falls from the clouds to earth.
28 ☞ ***Evapouration*** from the ground and ***evapo-transpiration*** from vegetation are two major pathways for returning water to the atmosphere as vapour. The rates of evapouration and evapo-transpiration increase with decreasing humidity.

A variety of ***runoff and flow processes*** contribute to major losses of water from the drainage basin. If the soil is already saturated with water, any additional supplies will runoff into rivers and lakes. If the soil is unsaturated, water will enter the soil in a process known as ***infiltration*** and once in the soil, is distributed downward by gravity through a process called ***percolation***. The infiltration and percolation rate is a function of the soil type, particle size and wetness. Water infiltrates and percolates faster in coarse sand and gravel than in mud and fine sand. When the intensity of precipitation exceeds the infiltration rate of the soil, water begins to accumulate on the ground surface and eventually runs off as ***overland flow***. No surface runoff occurs if the rate of infiltration into the soil exceeds precipitation intensity. However, infiltration rates of soils decrease with prolonged precipitation events, to the point that the infiltration rate is exceeded by precipitation intensity, leading again to surface runoff. Overland runoff due to a higher precipitation intensity than infiltration rate is known as the ***Horton model*** of runoff. The processes and dynamics of infiltration and percolation and their role in stream-flow generation are described in Chapter 2.

A second model deals with ***interflow*** or ***throughflow*** of water by lateral movement through soil layers. Since the permeability of soil tends to decrease with depth., the velocity of water moving downward also decreases with depth. Throughflow rates are dictated by, (i) the precipitation intensity, (ii) the infiltration rate, (iii) the percolation rate through soil layers, (iv) the permeability of the soil layers and (iv) the degree of slope of the land. These processes are also described in Chapter 2.

Lake Basin Water Losses:

Evapouration from the surface also accounts for significant losses from the lake basin. Evapo-transpiration from emergent macrophytes also contributes to some water loss. However, the largest amounts of water are lost daily from outlets in drainage lakes and/or from groundwater seepage. Seepage lakes lose most of their water through groundwater seepage.

Hydrological Regions

The hydrological cycle varies among hydrological regions. ***Arheic regions*** are those in which lakes have no outflow, or the outflow is minor and water does not reach the ocean. In many of these lakes, water losses occur through groundwater seepage, as well as evapouration and evapo-transpiration; losses through outflows are usually minor. Many arheic region lakes are in polar and desert areas and, while the gains (e.g. precipitation, runoff, underground sources) are similar to lakes in other regions, the amounts are significantly less. ***Endorheic regions*** are those which most lakes have outflows but the flows are so low that the water usually does not reach the ocean. Most processes of the hydrological cycle apply to endoreic regions, but usually only during wet seasons. Occasionally, lakes in arheic regions have flow rates high to resemble lakes in endoreic regions. ***Exorheic regions*** are regions containing lakes with outflows that reach the ocean through one or more tributaries. All pathways and processes of the hydrological cycle usually apply to watersheds of exorheic regions. Occasionally, lakes in endorheic regions have flow rates high enough, for example during spring runoff, that some of the water reaches the ocean and they resemble lakes in exorheic regions (dashed line, Fig. 1.13). Conversely, a few lakes in exorheic regions may have temporary outflows and do not reach the ocean during the summer months and resemble lakes typical of the endorheic region. Many meromictic lakes (see Chapter 4) in the exorheic region have small outflows and the water often does not reach the ocean.

CHAPTER 2

PARENT STREAMS, THE MOTHERS OF LAKES

Why Read This Chapter?

Lakes are not merely big holes filled with water. They have character, they have unique qualities, and they have life. Their character, their quality of life, and their degree of vitality depend largely on the streams that feed them. Discover:

☞ 1. The importance of topographic maps
☞ 2. How to use a topographic map
☞ 3. How to find all the inflows (parent streams) and outflows of a lake
☞ 4. The importance of stream "order" size and how to measure it
☞ 5. When floating docks are better than stationary ones
☞ 6. How to determine the size and shape of a watershed
☞ 7. How the watershed determines the fisheries potential of a lake
☞ 8. Twelve basin characteristics and how to measure them
☞ 9. Nine channel characteristics and how to measure them, including:
☞ 10. Channel slope
☞ 11. Water velocity
☞ 12. Discharge
☞ 13. The value of stream hydrographs
☞ 14. The generation of stream flow
☞ 15. The longitudinal zonation and terminology of streams
☞ 16. The vertical zonation and terminology of streams
☞ 17. The river continuum concept (RCC)
☞ 18. The importance of wetlands and differences between swamps, marshes, bogs, and fens
☞ 19. How to determine the frequency and size of flood events using hyetographs
☞ 20. How to determine surface areas of odd shapes, like watersheds, cross-sections and lakes

Introduction

Nearly all lakes have one or more inflows (***parent streams***). There are usually
1 ☞ several inflows, including groundwater supplies, and one outflow (and/or groundwater seepage). Watershed or drainage basin shapes and sizes can be determined from ***topographic maps.*** These are maps that show the ***topography*** (hills and valleys) of the land in an area. They come in different scales, 1:25,000, 1:50,000, 1:100,000 and 1:250,000. 1:50,000 is the most common used map, which means that 1 unit on the map equals 50,000 units on the ground. For example, 1 cm = 50,000 cm, or 500 m (since 1 m = 100 cm), or 0.50 km (since 1000 m = 1 km) on the ground; or if feet or miles are preferred, 1 inch on the map = 50,000 inches, or 50,000/12 = 416.67 ft, or 416.67/3 = 138.9 yards or 416.67/5,280 = 0.79 miles on the ground. Work in metric units; it is much easier to divide by 10, or 100, or 1000 than it is by 3, 12 or 5,280!

> Maps that not only show you how to get to your destination, but the hills and valleys that you must go through.

Topographic maps must be purchased at a map store (corner service stations will not do, in most cases at least. Stores that sell government documents will probably have them. All topographic maps have a name and map number, like Milnet, No. 41-I/15. Many stores will take orders over the telephone or by mail order. If so, be ready to provide the name and number, or at least the district or county name that the lake is located within. Alternatively, call the nearest Ministry of Natural Resources or Ministry of Environment office and ask them for information needed to obtain a topographic map.

The Topographic Map and How to Use it

Unroll or unfold the map and hold the corners down with some weights. The longitude of a site can be determined from the numbers given on bottom or top axes and
2 ☞ the latitude from the numbers given on the left or right axes. The longitudes and latitudes are written in black numbers and given in degrees (e.g. 45°), minutes (e.g. 55') and seconds (e.g. 42"). Degrees and minutes are given at each corner of the topographic map. Alternating black and white bars are used to divide the longitudes and latitudes into 1' (minute) intervals, with numbers provided at every 5' interval. Seconds (") must be estimated from the position of a site relative to a 1' interval (e.g. a site located at one-third of 1' interval is 60 "/3 = 20 ").

Alternatively, a "grid reference" can be used to identify a specific location. The location is based on its position within the "One Thousand Metre Universal Transverse Mercator Grid". Every topographic map gives a grid zone designation, usually in the right margin of the map. The map also gives an example of how to estimate a grid reference location. Before determining the grid location, find the grid "zone number" of the map. The zone number is given in the right margin of the topographic map. Golf Lake is in Zone 17T (hypothetically). "Easting" and "northing" is then used to produce a grid reference number composed of 6 digits, the first 3 for easting and the last 3 for northing. For easting, read the number (2 digits) on the grid line immediately to the left of the point of interest. Then estimate the tenths of a 1000 m square (3rd digit) from this

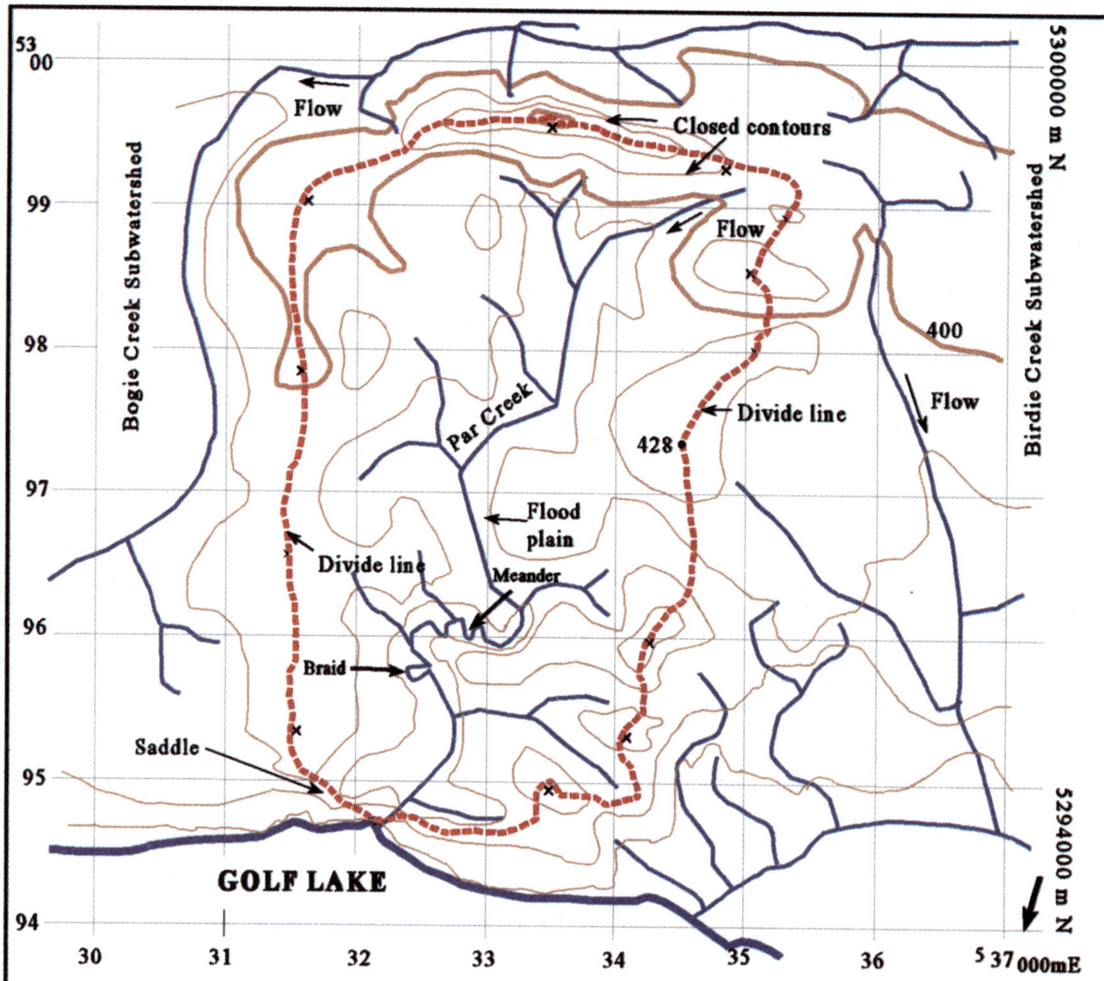

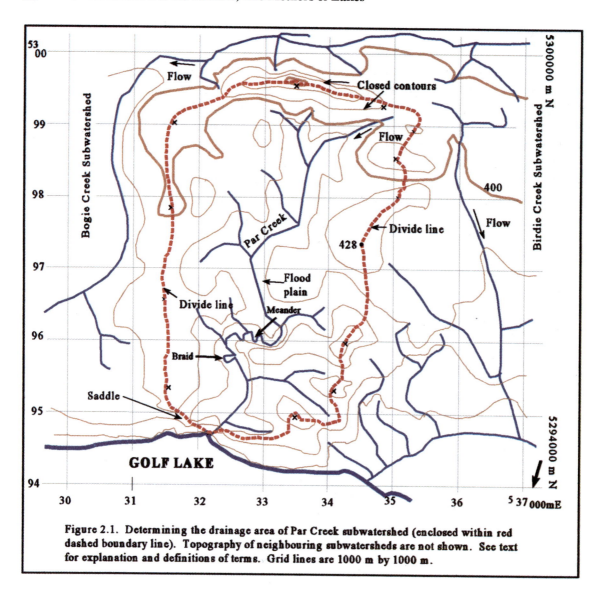

Figure 2.1. Determining the drainage area of Par Creek subwatershed (enclosed within red dashed boundary line). Topography of neighbouring subwatersheds are not shown. See text for explanation and definitions of terms. Grid lines are 1000 m by 1000 m.

line eastward to the point. For northing, read the number (2 digits) on the grid line immediately below the point, then estimate the tenths of a 1000 m square (3rd digit) from this line northward to the point. As an example, the grid reference for the 428 m high point in Fig. 2.1 is 345974 (easting = 345; northing = 974) in Zone 17T. The 5 of 345 represents 5/10 of 1000 m E of grid 34 and the 4 of 974 represents 4/10 of 1000 m N of grid 97.

The grid numbers actually represent, in thousands of metres, a portion of the distance from the ***mercator*** (0° Longitude, 0° Latitude). To determine the actual distance, first find the "prefix values" for the east and north distances. They may be found in any corner of the map or in the right margin. The values represent the distance in a 100,000 m square. In Fig. 2.1, the values are 5 for easting and 52 for northing. Hence, the actual distance of the highest point in the watershed from the mercator is 534500 m east and

AMSL is a standard acronym meaning Above Mean Sea Level

Where to find fish!

The total equals the sum of its parts, but some parts are better than others!

Where to build a cottage, without getting flooded, without losing your dock, without mildew growing every where, without constant shore erosion, and still have drinking water and a great location!

5242800 m north (± 100 m each direction).

With the foregoing basic information, locate your lake on the map. Note the 1000 m x 1000 m grids. Each grid is 2 cm x 2 cm, so 1 cm = 500 m (i.e. map scale is 1:50,000). Around the lake are brown contour lines, all of which have a common elevation interval. Modern topographic maps are metric with 10 m contour intervals, shown in light brown (Fig. 2.1). Contour lines at 50 m intervals are shown with a heavy brown line (Fig. 2.1). Older maps are marked in feet, usually at 50 foot intervals. Large lakes often have an elevation marked on the lake, with a "±" indicating that the lake level may change due to spring floods, precipitation, summer-time evapouration, dam (if present) discharge regulation, etc.. The "±" number is the mean elevation of the lake, or the elevation ***Above Mean Sea Level*** (AMSL). If an AMSL ± number is not provided, it must be estimated from the contour lines adjacent to the lake. The lake level will be that

3 ☞ shown between the contour line nearest the lake at the inflow and the contour line nearest the lake at the outflow. Inflows are always at higher elevations than outflows. Often, one of the contour lines around the lake appears to touch the water's edge; this contour elevation can be taken as the AMSL ±. Golf Lake, in Fig. 2.1, would have an AMSL of 390 m±.

Why is it important to know the lake elevation and how to determine inflows and outflow? Inflows are favourite feeding areas of many fish because the inflows provide a constant supply of food items and usually a fresh supply of oxygenated water. Most rivers have turbulent flow which mixes oxygen into the water, and lakes, especially those that have some kind of oxygen stress in the summer will often have fish congregating at the inflows.

Outflows are usually great fishing spots because the water from the lake is compressed into a narrow channel. This means that any food in the outflow is continually concentrated into a small area making it easier for the fish to find and capture the food. Also, most lakes have only one outflow and several inflows. Where do fish congregate? At the single outflow that is carrying food produced in the lake, including any remaining in the inflow sources, or to a single inflow that is carrying only a portion of the food for the lake?

Parent streams are an important source of food materials for a lake. The greater the number of parent streams, the greater the contribution. Also, the greater the discharge, the greater the amount of food. The quality of water and the size and productivity of the stream, the geochemistry of the stream bed, and many other factors vary among parent streams, but they all contribute to the lake's physical, chemical and biological characteristics.

There are other reasons it is important to know how to read a topographic map. Suppose one wishes to build a cottage on a lake; where do you build it? Most of us would do a slow run around the lake looking for the perfect spot. We all have our criteria for the perfect spot, usually a nice beach, a southern- or eastern-facing slope (to take full advantage of the sun since southern- and eastern-facing slopes are warmer and have more sunlight than northern- or western-facing slopes), a mix of trees, like some cedars, pine, spruce, birch and poplar (if in the northern region), or maples, willows, oaks and cedars (if in the southern region). But what about flood events? Is the dock going to disappear every spring? All these questions can be answered from topographic maps.

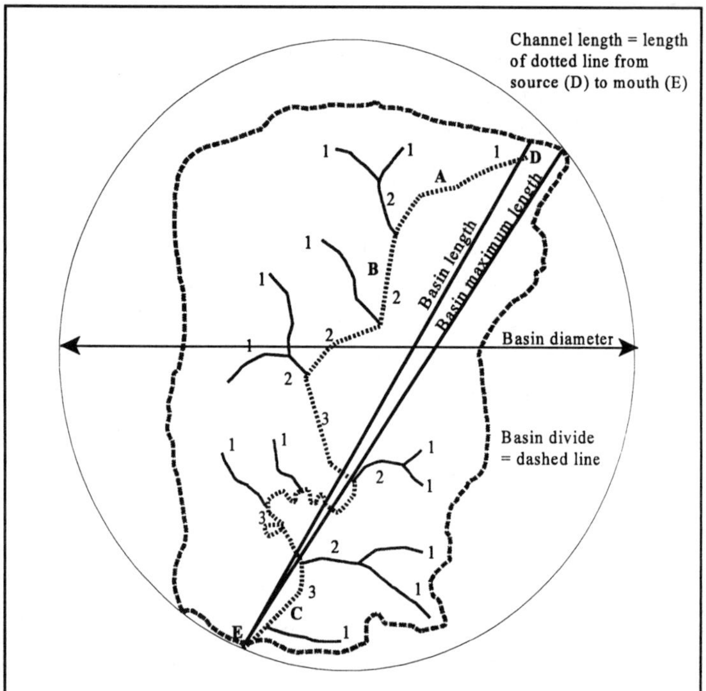

Figure 2.2. Determining stream order and some basin characteristics of Par Creek (from Fig. 2.1). See text for explanation of letters A, B, C and definitions of terms.

Stream Order

4 ☞

Putting some *order* into this confusion.

Before describing how to determine the shape and size of watersheds, it is prudent to first examine the classification of stream size. Scientists refer to stream size as ***order***, a first order stream being a stream without tributaries. They represent source streams. The more first order streams, the more sources. Two first order streams combine to make a second order stream, and two second order streams combine to make a third order stream, and so on. This method of stream ordering is known as the ***Strahler method*** because Strahler (1966)[1] developed the concept from principles described by Horton (1945)[2]. Elongated watersheds tend to have numerous first and second order streams.

Figure 2.2 shows the ordering process using the Strahler method. The basic rule to remember is that streams do not increase in order size until two streams of equal stream order size combine. Hence, if a first order stream combines with a third order stream, the order size stays the same as the largest order size (Order 3); a stream order of 4 results only when two third order streams meet.

Shreve (1966)[3] argued that the Strahler method did not accurately represent changes in physical characteristics of streams, such as increases in width and discharge,

---

[1]Strahler,, A. N., 1964. Quantitative geomorphology of drainage basins and channel networks. In Chow, V. T. (Ed). Handbook of applied hydrology. McGraw-Hill, NY. Sections 4-11.

[2]Horton, R. E. 1945. Erosional development of streams: quantitative physiography factors. Bull. Geol. Soc. Amer. 56: 275-370.

[3]Shreve, R. L. 1966. Statistical law of stream numbers. J. Geol. 74: 17-37.

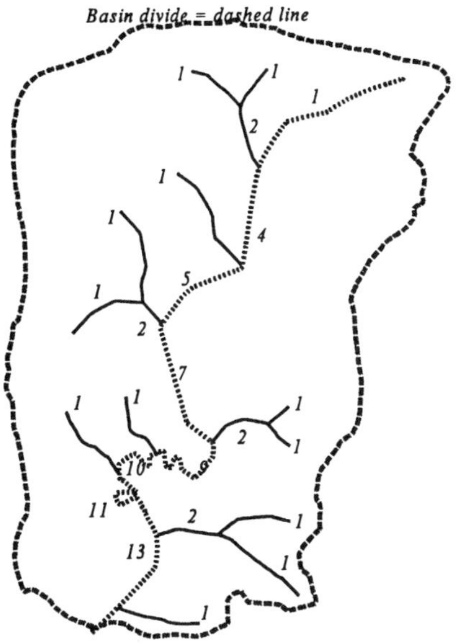

Figure 2.3. Determining stream order using Shreve's method.

because the entry of a stream segment of a given order into one of a higher order, does not increase the order of the larger stream. Shreve (1966) overcame this by summation of stream orders. In ***Shreve's method***, the order of each stream link indicates the number of first order tributaries discharging into it. Hence, a stream of order 2 entering a stream of order 3 increases to order 5, and two streams of orders 5 and 2 coming together create a stream of order 7, and so on (Fig. 2.3). Nevertheless, the Strahler method has become the method of choice and is used in the river continuum concept, discussed later in this chapter.

There are several stream features associated with Strahler's stream ordering concept. Many of these were developed in the original work of Horton (1945):

1. The number of streams is inversely related to stream order size. For example, there is 3 to 4 times more streams of order 1 than of order 2 and 3 to 4 times more streams of order 2 than of order 3, and so on (Fig.2.4, note log scale on y-axis; modified from Allan (1995)[4]).
2. The length of a stream is directly related to its size (Fig. 2.5, note log scale on y-axis; figure is modified from Allan (1995)). First order streams are about half the length of most second order streams; fourth order streams are about 10 times longer than first order streams.
3. The width and depth of a stream are directly related to stream order size. The erosion capacity of flowing waters causes increases in width and depth with increasing distance downstream. In general, the width tends to increase as the square root of the mean discharge, or:

$$\text{Width} = pQ^{0.5}$$

where, p is a function that varies for each stream, Q is discharge (m^3/sec) and 0.5 is function that is more-or-less constant for all streams. Depth increases much less rapidly.
4. The cross-sectional area of a stream is directly related to its order size.

---

[4]Allan, J. D. 1995. Stream ecology, structure and function of running waters. Chapman & Hall, New York, NY.

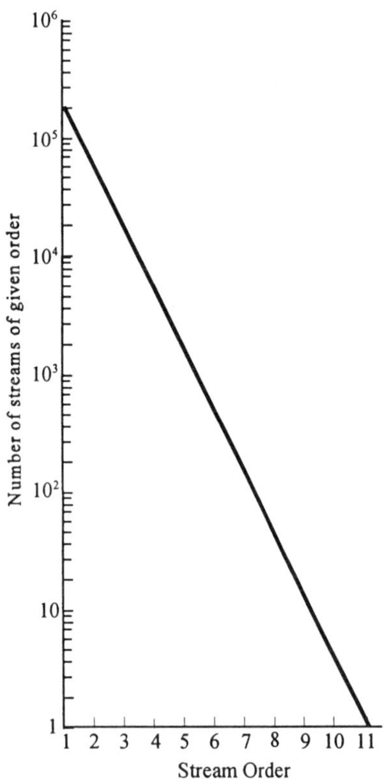

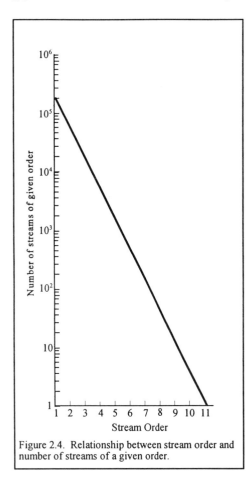

Figure 2.4. Relationship between stream order and number of streams of a given order.

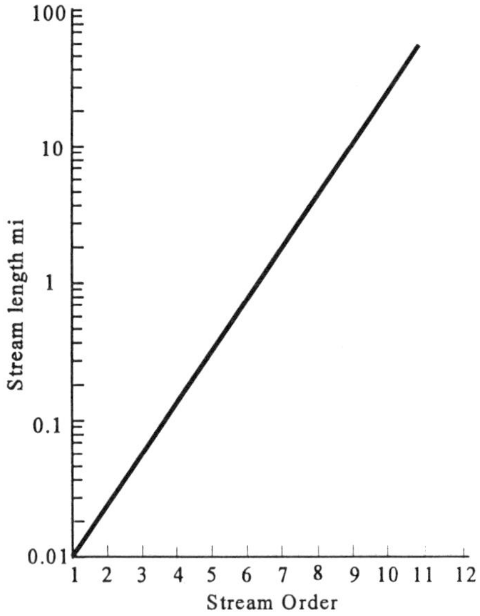

Figure 2.5. Relationship between stream order and stream length.

This is a direct result of increasing depth and width of channel with distance downstream.

5. The slope of a stream is more-or-less inversely related to stream order size. Steeper slopes are usually seen in headwater areas, gentler slopes in downstream areas, near the mouth.

6. Current velocity (speed of flow, e.g. cm/sec) is *directly* related to stream order size. This may seem surprising because slope decreases with increasing distance downstream; one would think that gentle slopes have slower currents. However, slope does not appear to control velocity. In fact, current speed is known to increase over a constant slope. Such factors as damming by the build up of rocks and debris raises the water level, elevates discharge and alters current speeds along the length of a channel. It is shown later (see Manning equation) that bottom roughness greatly offsets the effects of reductions in slope. Channel depth tends to increase and substrates generally get finer with increasing distance downstream, hence resistance decreases longitudinally.

7. Discharge (volume flow rate, e.g. m^3/sec) is directly related to stream order size, a direct result of the above (3 to 6) relationships.

4 ☞ Stream order size can be important when fishing for trout. Trout become more and more uncommon as stream order size increases. They are very common in stream

orders 1-3 but become increasingly hard to find in stream orders 4 and higher. In the Strahler method, the largest order recorded is 12, but with numerous large bays or coves, large rivers begin to emulate lakes in limnological characteristics. Trout are absent in
5 ☞ large rivers where the water is warm and stagnant.

Floating docks are often better than stationary docks.

Floods result from an accumulation of water as it flows with increasing distance downstream; the longer and more numerous the channels, the greater the flood potential and the longer the period that floods occur. Figure 2.6 shows the amount of water accumulating at three points on Par Creek in Fig. 2.2. After a precipitation event, high water first appears at "A", then a short time afterward at "B", and finally at "C". Note that discharge increases with distance downstream because more source streams contribute to the flood event at each site. The bottom line? Expect greater flood potential in watersheds that have a lot of first order streams! Or, if everything else is equal, expect larger floods on lakes that are further down a watershed than on lakes that are higher up the watershed. Obviously, lakes that are near the bottom of a watershed will receive more runoff than lakes that are higher up on the same number of first order streams or if your lake is near the bottom of a watershed.

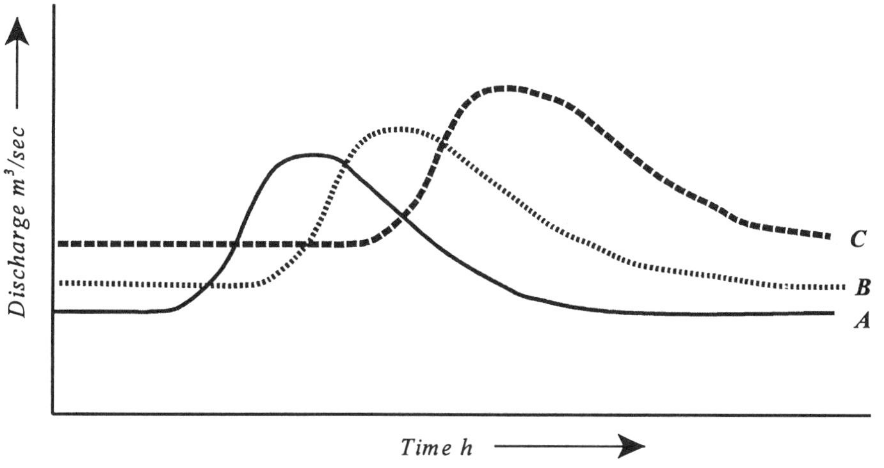

Figure 2.6. Changes in discharge level over time and distance (A, B, C, as shown in Fig. 2.2) downstream

So, how is the high water mark determined? Look at the shrubs and trees on the shoreline. Visit the site in the early summer and look for drift debris hanging from branches of shrubs and trees on the shores of any inflow. Look at water stains and erosion on concrete walls of bridges that cross the stream. Measure the distance from the water level to the height of the debris. Remember, the lowest water level is during the warmest month, usually July or August in Ontario. It is not unusual to find high water marks greater than 1 m (3.3 ft), but this will depend on the streams location in the watershed. Remember, there will be much more runoff near the bottom of the watershed because of runoff from all lakes and streams above. The amount and frequency of potential flooding can be determined from procedures described below (see "Predicting Floods").

Determining Watershed Shape and Area

The size, shape and drainage pattern of a watershed affects the size and duration of spate (flood) events. Figure 2.7 shows two common shapes of drainage basins and the most common types of drainage networks associated with each. Most drainage basins are **elongate** in shape and have either a ***dendritic*** (Fig. 2.7a), ***rectangular*** (Fig. 2.7b), ***parallel*** (Fig. 2.7c) or ***trellised*** (Fig. 2.7d) network of streams. Because of the large number of small stream orders (e.g. 1 and 2), discharge tends to be high at the mouth of each

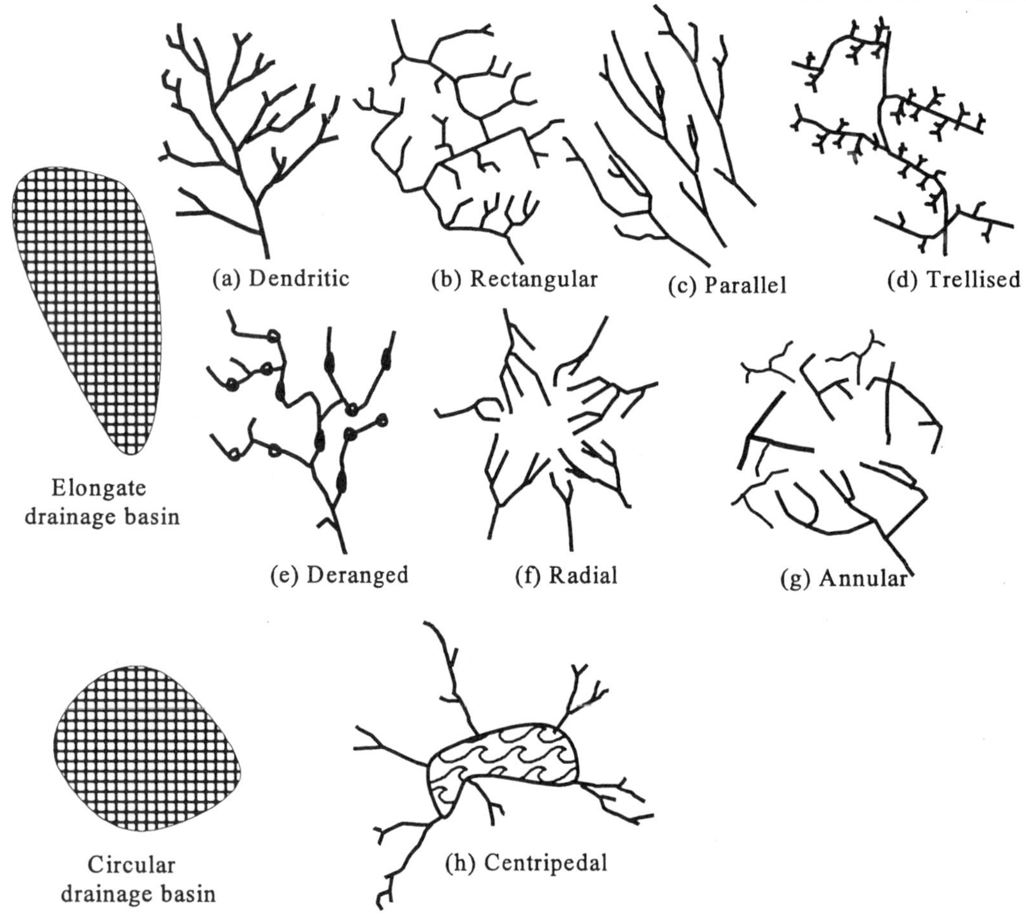

Figure 2.7. Various types of drainage networks associated with elongate and circular drainage basins.

network. Networks with a lot of small stream orders (e.g. dendritic, rectangular, trellised) will have greater discharges and larger spate events at its mouth than those with few small order streams. A ***deranged*** network (Fig. 2.7e) also leads to elongated watersheds, but the small numbers of first order streams and the insertion of several small lakes along the network results in small discharges at the mouth. ***Radial*** and ***angular*** drainage networks (Figs. 2.6 f, g) often result in several adjacent elongated subwatersheds, each of which has a low discharge because each shares runoff at the top of the watershed.

Circular watersheds tend to have a ***centripetal*** stream network (Fig. 2.7 h). While discharge at the mouth of each centripedal stream is small, lake discharge levels are usually elevated above those in elongated watersheds. Each drainage system in the centripetal network is usually of the dendritic type.

The size and shape of the drainage basin can only be determined from topographic maps. Fig. 2.1 is a simulation of a topographic map showing a hypothetical watershed with a dendritic drainage network. By following some simple principles, the boundary of

6 ☞

the watershed can be easily determined. **The key** *is to locate and delimit all neighbouring* ***subwatersheds*** *within a* ***watershed***. That is, each lake or river has a large watershed that is composed of several small subwatersheds. Consider hypothetical Golf Lake in Fig. 2.1 for example. Golf Lake has three subwatersheds, Par Creek, Bogie Creek (above Par!) and Eagle Creek (below Par!). The elevations are shown entirely for Par Creek subwatershed, but only partially for the neighbouring two subwatersheds. The eastern boundary of Bogie Creek subwatershed forms the western boundary of Par Creek subwatershed; the western boundary of Eagle Creek subwatershed forms the eastern boundary of the Par Creek subwatershed. The following describes how the subwatershed boundary can be determined:

> The difference between a watershed and a subwatershed

> Applying the principles.

- First, mark an "x" at the highest elevations between adjacent watersheds. The highest elevations are the smallest "***closed contours***", as in Fig. 2.1.
- Then draw drainage boundaries (***divide lines***) perpendicular to the contour lines of the closed contours (Fig. 2.1, 2.8). The line should split the closed contour in half, either through the long axis or the short, depending on the overland flow (Fig. 2.8).

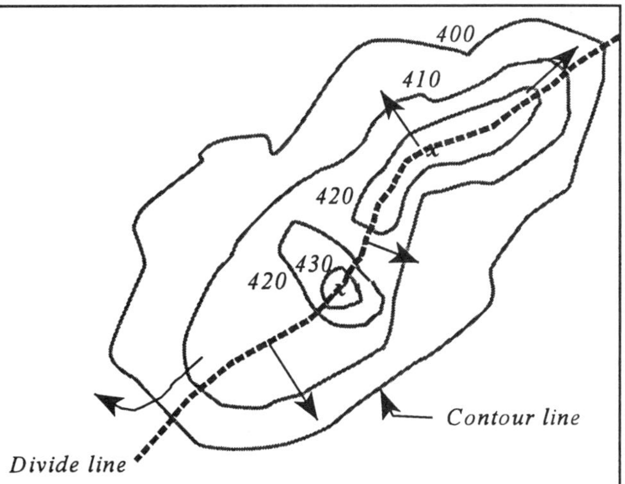

Figure 2.8. Locating the divide line through closed contours. Arrows show direction of runoff. Numbers are elevations in meters. See text for explanation.

Overland flow and shallow subsurface water flows perpendicular to contour lines.

- Draw divide lines through the centre of ***saddles*** (Fig. 2.1, 2.9).
- The lowest elevation in the watershed is the stream itself. The basin, or trough, containing the stream water is the ***stream channel***. The space, or flat area, between the contours on each side of the stream channel represents the ***flood plain*** (Fig. 2.1).
- Finally, the base of the watershed is at the outflow, or mouth, of the river.

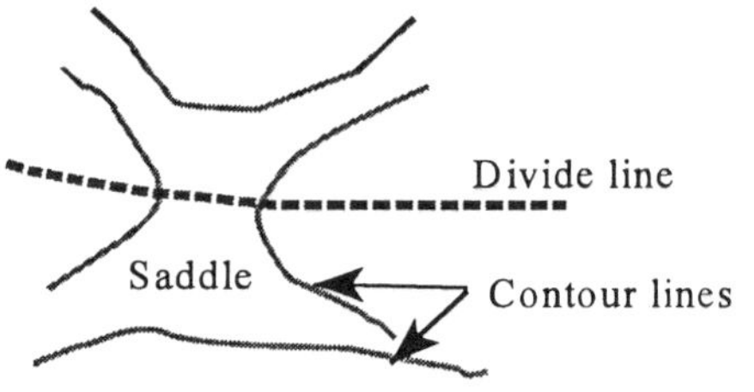

Figure 2.9. Locating a divide line through a saddle.

Figure 2.10. Water flows ***to*** the Vs of contour lines, or ***from*** the opening of the Vs.

The importance of knowing the influence of the watershed on the fishing and recreational potential of your lake!

The direction of flow in streams can be determined in two ways. First, since water flows from high elevations to low elevations, find two contours that cross the river and then determine the elevations of the contours. Water flows from the higher contour to the lower one. Second, water flows ***from*** the opening of the Vs of contour lines that cross the stream (Fig. 2.1, 2.10).

What is Golf Lake's watershed area? One of the two methods described at the end of this chapter can be used to determine watershed area. As a guestimate, the 1 km^2 grids can also be used. According to Fig. 2.1, about 15 squares (i.e. 15 km^2) are included within the watershed boundary. Using methods described later, the actual = 14.72 km^2 . The watershed areas of ***oligotrophic lakes*** (low nutrients levels, low productivity, but good lake trout lakes) usually range between two and ten times the lake area. ***Eutrophic lakes*** (high nutrient levels, high productivity, poor lake trout lakes) typically have watershed area to lake area ratios above 50, but the shape of the lake's bottom (e.g. deep with steep sides vs shallow with gentle sloping sides) and the geochemistry of the watershed's soils and bedrock also affect the lake's trophic status. More will be said about trophic status (i.e. nutrient level) in the following chapters. The terms are introduced here because watersheds play a key role in determining the trophic state of lakes, and in their recreational and fishing potential.

The Watershed and its Influence on a Lake

7 ☞ The chemistry of water is determined largely by the bedrock and soils that the water flows over and through before it enters the stream and lake. For example, groundwater which flows over limestone (calcium carbonate) will contain more calcium

and carbonate than groundwater that flows over granite or quartz (mostly silica). Most parts of northern (e.g. Sudbury, Sault Ste. Marie) and central Ontario (e.g. Muskoka district) have granitic bedrock. Most parts of southern Ontario have limestone bedrock. Lakes fed by streams flowing over limestone will have higher calcium carbonate levels than streams flowing over granite. Streams and lakes with high calcium carbonate content have much greater fisheries and body contact recreational potential than do streams and lakes with just silica (glass fragments!!). Lakes lying over granitic bedrock have so few chemical constituents (ions) that they are often referred to as "beakers". Details of physical and chemical reactions and the contribution of different bedrock types to ions in the water are described in Chapter 5.

The productivity of a stream depends on whether the constituents were produced *within* (i.e. ***autochthonous*** (pronounced aw-tock-thon-ous) production) or *outside* (i.e. ***allochthonous*** (pronounced a-lock-thon-ous) production) the stream. Algae (and their products) produced within the stream is an example of autochthonous material. Material supplied by trees (e.g. leaf litter, twigs) and ***riparian vegetation*** (e.g. grass and shrubs along side the stream), by the atmosphere (e.g. fallout or material in precipitation), and by groundwater are examples of allochthonous materials. But both the autochthonous and allochthonous materials carried by parent streams into lakes are allochthonous materials for the lake. Autochthonous materials of lakes include the algae and large plants that are produced within the lake.

Some Drainage Basin Features That Influence Water Quality

Water quality not only depends on the chemistry of soils and bedrock in the watershed, but on the amount of water sources (e.g. number of first order streams) and the
8 ☞ retention time of the water (e.g. basin slope, total length of all stream constituents, size of the watershed). Twelve drainage basin features can be measured to provide these kinds of information. Alphabetically, they are:

Basin area (A_B): The drainage area enclosed within the divide line for the watershed. The method for determining basin area was described earlier.

Basin diameter (D_B): The diameter of the smallest circle that will encompass the entire basin. The basin diameter of Par Creek subwatershed is 5.5 km (Fig. 2.2).

Basin length (L_B): Straight-line distance from the outlet to the point on the basin divide used to determine the main channel length, L_c. For Par Creek subwatershed, the L_c = 5.1 km (Fig. 2.2).

Basin perimeter (P_B): The length of the divide line that defines the boundary of the basin. For Par Creek subwatershed, the P_B is 15.82 km. See compactness ratio for the importance of the basin perimeter.

Basin shape (SH_B): A measure of the shape of the basin calculated as the ratio of its length (L_B) to its average width (W_B), or:

$$SH_B = (L_B)^2/A_B$$

The shape of the basin is a function of the drainage network pattern. The basin shape for Par Creek subwatershed is $(5.1)^2/14.72 = 1.77$.

Basin slope (S_B): Average land slope estimated as rise/run, the rise being the change in altitude between contours nearest a grid intersection and the run being the straight-line distance between the two measurements. Several estimates are made from the top of the basin to the base of the basin and a mean is calculated from them. The S_B for Par Creek subwatershed varies from 0.004 near its mouth, to 0.067 near the headwaters, and has a mean of 0.022. The greater the slope, the greater the erosion capacity of the stream. Gentle slopes increase the contact time between the water and the sediments, allowing for greater chemical exchange reactions between the two media.

Basin width (W_B): Average width of the basin determined by dividing the drainage basin area, A_B, by the basin length, L_B, or:

$$W_B = A_B/L_B$$

For Par Creek subwatershed, the W_B is 14.72/5.01 = 2.94 km. Along with basin length, the basin width determines the drainage area. The larger the drainage area, the greater the potential for different sources of ions to enrich the chemical composition of the stream.

Bifurcation ratio (BR): The ratio of the number of stream segments in an order relative to the number of stream segments in the next highest order, or:

$$BR = n_i/n_{i+1}$$

where, n_i = number of stream segments in order i and n_{i+1} is the number of stream segments in the next highest order.

For Par Creek subwatershed (Fig. 2.1), the BR for first and second order streams is 13/4 = 3.25. Most dendritic drainage networks have BR values between 3 and 5 for first order streams. The greater the number of first order streams, the greater the numbers of sources of ions from the soils and bedrock.

Compactness ratio (CR_B): The ratio of the perimeter of the basin to the circumference of a circle of equal area, or:

$$CRB = P_B/2\sqrt{(\pi A_B)}$$

A CR_B value of 1 indicates a circular watershed; since $A = \pi r^2$ and $P_B = \pi$ x diameter = π x 2r, hence $CR_B = 2\pi r/2\sqrt{(\pi^2 r^2)} = 2\pi r/2\pi r = 1$. The compactness ratio for Par Creek subwatershed is $15.82/2(3.14 \times 14.72)^{1/2} = 1.16$. The longer the basin perimeter, the greater the compactness ratio and the greater the numbers of tributaries contributing to the drainage network.

Drainage density (DD): The ratio of the total length of all stream segments to the drainage basin area, or:

$$DD = \sum L/A_B \text{ km}$$

where, L = length of all stream segments (km), A = drainage area (km^2).

The greater the drainage density, the greater the stream discharge. This affects erosional processes, especially during spring spate events. The bases of drainage basins with high DD values are highly prone to flooding during spring thaw and after precipitation events. For Par Creek subwatershed, the DD = 15.5/14.72 = 1.05.

Elongation ratio (ER): The ratio of the diameter of a circle having the same drainage area to the length of the longest axis of the drainage basin, or:

$$ER = d_B / l$$

where, l = longest length of basin, d = diameter of a circle with same area as watershed = 2 x radius (r), $r = \sqrt{(A/\pi)}$, or:

$$d = 2\sqrt{(A_B/\pi)}$$

where A_B = drainage basin area. Hence, ER =

$$ER = (2\sqrt{(A_B/\pi)})/l$$

Watersheds with elongated basins (smaller ER values) tend to have large numbers of low order streams (e.g. stream orders 1 and 2) relative to stream orders 3 and 4. Many first and second order streams enter at stream orders 3 to 6. Circular watersheds ($ER \approx 1$) have many first and second order streams that enter at stream order 3, and stream orders higher than 4 become increasingly rare. For Par Creek subwatershed,

$$ER = (2\sqrt{(14.7/3.14)})/5.4 = 0.80$$

Sinuosity ratio (P): The ratio of the main channel length (L_C) to the basin length (L_B), or:

$$P = L_C/L_B$$

The sinuosity ratio for Par Creek subwatershed is 8.3/5.1 = 1.62. High values indicate a meandering or braided river (see channel characteristics below). A sinuous stream has greater surface area for water to flow over and increases the amount of exchange reactions between water and sediments.

Some Channel Characteristics

Nearly all channel features are measured for determining water velocity and discharge. The most basic of the channel features are its length, width, depth and slope.
9 ☞ Channel cross-sectional area can be calculated from the channel width and depth data. Water velocity can be measured directly or estimated from an equation (called Manning equation, discussed later) that incorporates channel slope, cross-sectional area, bottom features. The most common channel characteristics measured are described below, in alphabetical order.

Channel length (L_C): For the main channel, it is the length of the channel from its mouth to its source nearest the basin divide. The main channel length for Par Creek is 8.3 km (Fig. 2.2). Lengths can be determined for any tributary, from its source to its mouth.

Channel length is directly related to the area drained for both small and large rivers. On average, the relationship is defined as:

$$L_C = 1.4\ A^{0.6}$$

The relationship is a power curve (e.g. $Y = aX^b$), with a y-intercept ("a") value of 1.4 and a slope ("b") of 0.6.

Channel slope (S_C): A measure of the gradient (rise/run) of the channel. It can be calculated two ways: (i) determine the length of channel between two contour lines for
10 ☞ several segments of the channel. Divide the change in altitude by the length of channel to estimate the slope of each segment, then determine the mean slope. For example, suppose the channel lengths and changes in altitude for 3 segments of a stream are 4, 9, 6 km, with a change in altitude of 10 m for each segment. The slopes are 10/4000 = 0.0025, 10/9000 = 0.001 and 10/6000 = 0.002, yielding a mean slope value of 0.0018. (ii) The second method is used to estimate the slope of the main channel. Compute the difference in stream-bed altitude at point 10% (E_{10}) and 85% (E_{85}) of the distance along the main channel from the outlet to the basin divide and divide by the length of channel between, allowing for a 25% error. That is:

$$S_C = (E_{85} - E_{10})/0.75L_C$$

Discharge features

In order to measure discharge it is first necessary to measure the ***cross-sectional area*** and ***water velocity*** of a stream. Cross-sectional area is simply the product of the ***mean depth*** (m) and ***width*** (m). Each of these variables is described below.

Cross-sectional area can be determined two ways. The more accurate method of the two is to first measure the depth at equal intervals (if a narrow channel) or at 0.5 to 1.0 m intervals across the channel. Plot, on ordinary graph paper, the depth at each interval across the stream. The resulting figure is an inverted cross-sectional profile of the stream, with the water surface laying on the x-axis (Fig. 2.11, from data in Table 2.1). The ***channel width*** is the length of the water line on the x-axis. The ***mean channel depth*** for the transect is the sum of the depths divided by the number of measurements plus 1, or:

$$z = (z_1 + z_2 + z_3 \ldots\ldots + z_n)/(N + 1)$$

The cross-sectional area is the area enclosed within the plot. The area can be determined by either method (squared paper or paper weight vs area) outlined at the end of this chapter. Alternatively, and much simpler, is to estimate the area by multiplying the mean depth by its width. For the data in Table 2.1, the area is 2.65 m^2.

Table. 2.1. Depth measurements across a stream 10 m wide

Stream width (m)	Depth (cm)	Stream width (m)	Depth (cm)
0	0	6	43
1	10	7	72
2	12	8	50
3	15	9	25
4	30	10	0
5	35		

Current velocity - Current is the downstream water movement in the stream channel. It is the current that erodes the channel and determines its path and the nature of the
11☞ sediments. Gravel, stones, rubble and rip-rap (large rocks) are associated with fast currents and form ***riffles*** or ***rapids***. Fine sediments occur in slow currents; the slower the current, the finer the sediment (Table 2.2). Particle size can be determined by direct measurement, from settling velocities or by sieving. Table 2.2 relates particle size to current velocities. Occasionally, particle size is measured on a Wentworth scale

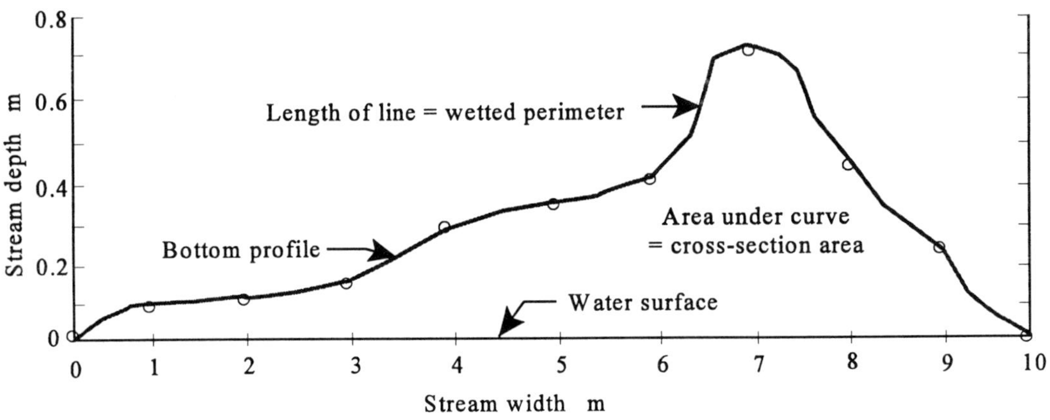

Figure 2.11. Determining cross-sectional area of a stream, using data in Table 2.2.

classification in Phi (φ) units. The Phi unit is a $-\log_2$ transformation of the smallest particle diameter within a size category. For example, the Phi value for silt (<0.063 mm) is ≥5, for fine sand (0.125-0.25 mm) it is 3, for small pebbles (16-32 mm) it is -4, and for boulder it is ≤-8.

Table 2.2. Relationship between current velocity and sediment particle size. Modified from Allan (1995, cited earlier) and Wetzel and Likens (1995)[5].

Velocity (cm/sec)	Particle size (mm)	General description
<3	<0.063	Silt
3-20	0.063-0.125	Mud, very fine sand
20-40	0.125-0.5	Fine to medium fine sand
40-60	0.5-8.0	Coarse sand to medium gravel
60-120	8.0-64.0	Coarse gravel to large pebbles
120-200	128-256	Small to large cobble
>200	>256	Boulders

To estimate current velocity, place an orange, or any float with neutral buoyancy, in the water and measure the time (in seconds) it takes to travel a known distance (10 m is

[5]Wetzel, R. G. and G. E. Likens. 1995. Limnological analyses. Springer-Verlag, New York, NY.

appropriate). Divide the distance travelled by the time to estimate velocity in m/sec. For example, suppose it takes 15 seconds to flow 10 m; velocity = 10/15 = 0.67 m/sec. If the water is flowing too quickly to measure the time accurately, use a 20-m distance and divide 20 by the time. Any object, like an orange, that has neutral buoyancy makes an excellent float. With half the object above the water surface and half below the water surface, winds do not affect its motion as much as objects without neutral buoyancy, like empty plastic bottles or cans. At least 5 velocity measurements should be made to derive a mean value. Of the five measurements, one is taken near each shore, one in the middle and one between the middle and each shore measurement. Even though five measurements are taken, the mean velocity measured in this manner is probably an underestimate. Figure 2.11 shows that water velocity tends to increase from the surface down to about 1/3 the depth and then decrease (to near zero at the mud-water interface in slow-flowing waters). Hence, current meters should be used to more accurately measure water velocity. As implied by Fig. 2.12, mean velocity should be measured at a depth about 0.6 of the distance from the water surface to the stream bed.

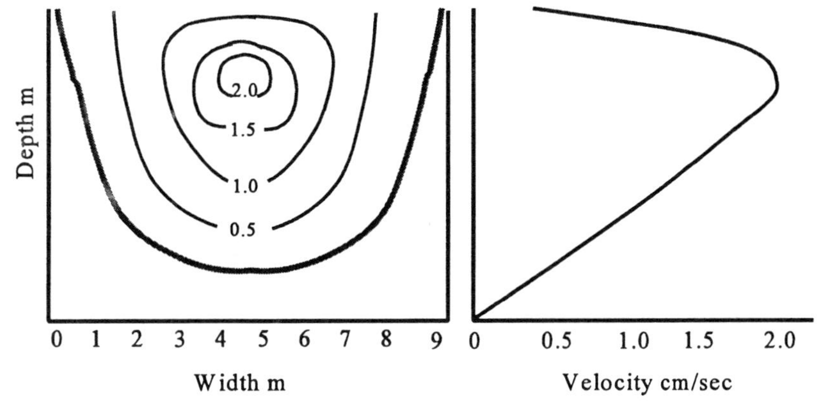

Figure 2.12. Idealized water velocity (cm/sec) in a channel cross section (left) and in a vertical profile athe the midpoint of the cross section in a stream with laminar flow on the substrate (right) .

Current is the single most important factor affecting stream organisms. Many species adapt to and are confined to a specific range of velocities. Chapter 8 examines several strategies and adaptations of invertebrate organisms to flowing waters.

Discharge (Q) is the volume of water passing through a cross-sectional area of the
12☞ stream channel per unit of time (e.g. m^3/sec), or:

$$Q=A_C v$$

Since the cross-sectional area = mean depth (m) times the width (m) of channel, discharge can also be expressed as:

$$Q=\bar{z}W_C v$$

Discharge can also be determined from the Manning Equation:

$$Q = A_C(1/n)R^{2/3}S^{1/2} \text{ (metric)}$$

where, n is a ***roughness coefficient*** that varies with the size and shape of substrate, the nature of bed forms (logs, etc.) and vegetation characteristics; R is the ***hydraulic radius*** (cross-sectional area divided by ***wetted perimeter***, the wetted perimeter being the length, from shore to shore, of wetted bottom (Fig. 2.11); and S is the slope of the channel. The roughness coefficient of most natural streams varies between 0.025 for smooth-bottom streams to 0.075 for rough-bottom streams. Table 2.3 gives "n" values for a range of bottom features.

Table 2.3. Roughness coefficient values for natural streams with various bottom features.

Stream Bottom Features	Roughness value
Smooth and straight	0.025 - 0.033
Smooth and winding	0.034 - 0.044
Rough, with weeds and stones	0.045 - 0.060
Winding, with weeds and pools	0.05
Very weedy, deep pools	0.075 - 0.150
With heavy brush and timber	0.1

In the Manning equation, $(1/n)R^{2/3}S^{1/2} = v$, current velocity. The equation allows an estimate of discharge simply by determining only the average depth and width of stream; the two variables permit measures of cross-sectional area (A_C) and wetted perimeter. The wetted perimeter is needed to estimate the hydraulic radius, R. The slope, S, can be determined from a topographic map and "n" can be estimated from Table 2.3. Alternatively, if discharge, current velocity and area are known, the slope can be estimated from:

$$S = (Q/A(1/n)R^{2/3})^2$$

Hydrographs - are plots of discharge against time. Fig. 2.13 shows the common features
13☞ of a hydrograph. Regular flow is known as ***base flow***. During a precipitation event, the water first penetrates and then infiltrates the soil until it is saturated with water. At this point, water runs off and enters the stream. Some of the soil water also enters the stream as through flow. At this point, the water level in the stream begins to rise. The rise in water level appears as a ***rising limb*** on the hydrograph. A peak in discharge occurs after a certain time, or ***lag period*** that varies among catchments and seasons. Some catchments produce hydrograph peaks over a short time interval, and these streams are termed ***flashy streams***. Others exhibit low hydrograph peaks after a long lag period, and these are termed ***sluggish streams***. Soils that are already saturated with water release runoff sooner than those that are dry, and the lag period is reduced. Spring floods occur very quickly because the soil is frozen and the snow-melt runs off immediately into the stream. After the peak is reached (and runoff and through flow subsides), the water level in the streams falls and is recorded in the hydrograph as a ***falling limb***. Discharge levels continue to fall until ***base flow*** is reached.

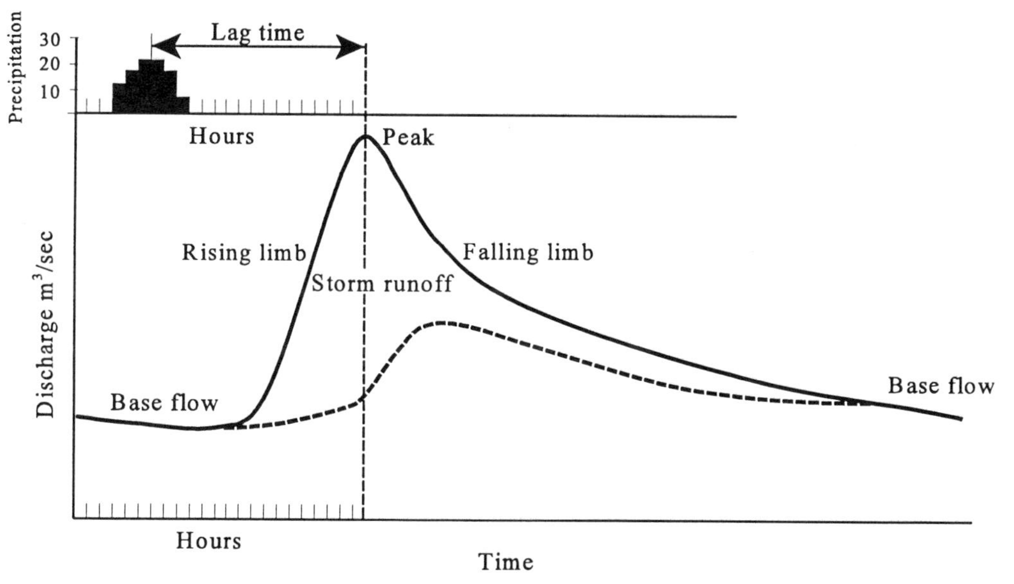

Figure 2.13. Stream hydrograph formed from a single storm event.

Several factors contribute to the form of a hydrograph (Fig. 2.14). Many of them are permanent and change very little over time, such as channel characteristics, the drainage network and basin characteristics. Other factors are more transient in nature, like storm characteristics, interception and surface storage, infiltration rate,

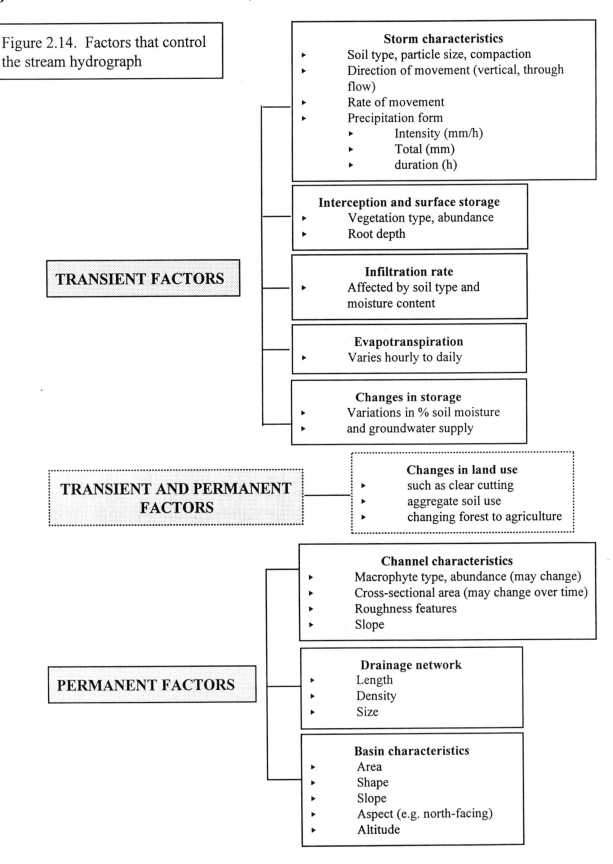

Figure 2.14. Factors that control the stream hydrograph

TRANSIENT FACTORS

Storm characteristics
- Soil type, particle size, compaction
- Direction of movement (vertical, through flow)
- Rate of movement
- Precipitation form
  - Intensity (mm/h)
  - Total (mm)
  - duration (h)

Interception and surface storage
- Vegetation type, abundance
- Root depth

Infiltration rate
- Affected by soil type and moisture content

Evapotranspiration
- Varies hourly to daily

Changes in storage
- Variations in % soil moisture
- and groundwater supply

TRANSIENT AND PERMANENT FACTORS

Changes in land use
- such as clear cutting
- aggregate soil use
- changing forest to agriculture

PERMANENT FACTORS

Channel characteristics
- Macrophyte type, abundance (may change)
- Cross-sectional area (may change over time)
- Roughness features
- Slope

Drainage network
- Length
- Density
- Size

Basin characteristics
- Area
- Shape
- Slope
- Aspect (e.g. north-facing)
- Altitude

evapo-transpiration and changes in amounts of soil and groundwater storage. Some factors, like land use, can be either transient or permanent.

Each of the permanent and transient factors are governed by subsets of other factors. Fig. 2.14, adapted from Whitton (1975)[6], summarizes the more important factors that control the stream hydrograph. Note that some channel characteristics (e.g. macrophytes, cross-sectional area) can be either permanent or transient. ***Macrophyte*** (large aquatic plants) abundance and diversity tends to change over time. Streams become wider and shallower over a period of time and the cross-sectional area is subsequently altered.

Sinuosity: Nearly all flowing waters follow a sinuous (S-shaped) course. The degree of sinuosity is generally determined from a sinuosity index, SI, that relates the distance travelled by water (A) to its straight, downvalley distance (B, Fig. 2.15), or:

SI = Channel distance/Downvalley distance

Small channels tend to have small curves, large channels tend to have large curves. Streams with SI values exceeding 1.5 are called ***meanders*** (Figs. 2.1, 2.15). SI values may range between unity in simple, well-defined channels to 4 in highly meandering channels.

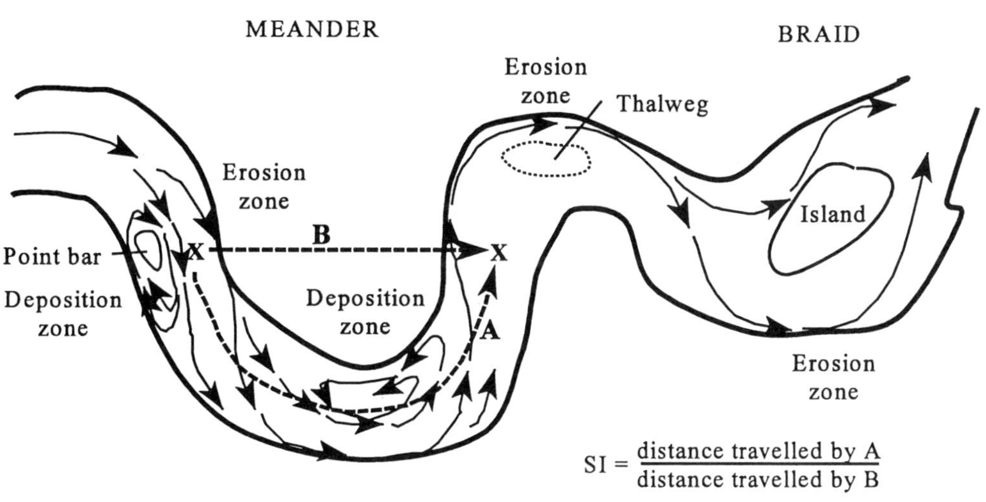

Figure 2.15. Meandering and braided reaches showing the line of maximum velocity throughout and separations of flow that produce areas of deposition and erosion..

[6]Whiton, B. A. 1975. River ecology. University of California Press, Los Angeles, CA.

Flow through a meander causes predictable erosion and deposition patterns (Fig. 2.15). Maximum velocity and greatest erosion occurs on the outer side of each bend, creating a deep hole, or ***thalweg***. As water flows around a bend, the water level is raised at the outside of the bend and causes a helical flow of water towards the opposite bank (Fig. 2.15). A back eddy results and deposits sediments eroded from the bed and bank receiving high water velocities. This explains why ***point bars*** develop in depositional zones in a downstream direction.

Braided or ***reticulate streams*** (Figs. 2.1, 2.15) are composed of a series of anastomosing channels separated by islands. They are more commonly seen in streams with glacier debris, moraine gravel or loose, readily moveable material. Any obstruction, like a group of large stones or logs, tends to divide the stream and pile up material on the downstream side. The resulting diverted streams on each side erode their beds and the obstruction emerges as an island which grows in a downstream direction. Braids usually occur in meandering streams with SI > 2.

Stream-flow Generation

The generation of stream flow is initiated with precipitation. If the precipitation is snow, its contribution to stream flow is delayed (as ***snow storage***) until the snow pack melts (Fig. 2.16). Some rain can fall and be stored as ice within the snow pack. In the spring, the snow melts, contributes to ***surface storage*** and enters the stream channel as ***overland runoff***. Stream-flow generation by overland runoff will continue until the soil begins to thaw. Once the surface depressions are depleted of water and the soil loses moisture, rain water can ***infiltrate*** (penetrate) the soil surface layer and contribute to ***soil water storage***. The ***Horton model*** of runoff states that overland runoff will occur as long as the infiltration rate is exceeded by the precipitation rate. Two components of infiltration are recognized. First, there is a ***diffusion component*** during which unsaturated pore spaces are filled with water through ***capillary action*** from the surface downwards, in a random fashion. The diffusion rate is initially rapid but slows as the wetting front (the division between the saturated and unsaturated pore spaces) moves downwards through the soil. A ***transmission component*** describes the ***gravitational flow*** of water through the pore system of the soil. It remains relatively constant in amount during a storm event.

The snow-melt runoff is essentially carbonic acid, low in dissolved materials, and has a pH of ~5.6, similar to that of water when it is in equilibrium with carbon dioxide. In fact, the pH of early spring runoff is usually much lower (e.g. < 4.0) because the hydrogen ions migrate down the snow pack as the snow melts from the surface downward. Hydrogen ions accumulate at the base of the snow pack and will cause a severe ***spring pH depression*** if the receiving waters do not have an alkalinity high enough to buffer the huge H^+ input. In addition, any silt in the snow pack, and silted added by the erosion of soil as the water runs over its surface, is carried into the stream creating a high silt load. The temperature of the runoff is highly variable, initially being

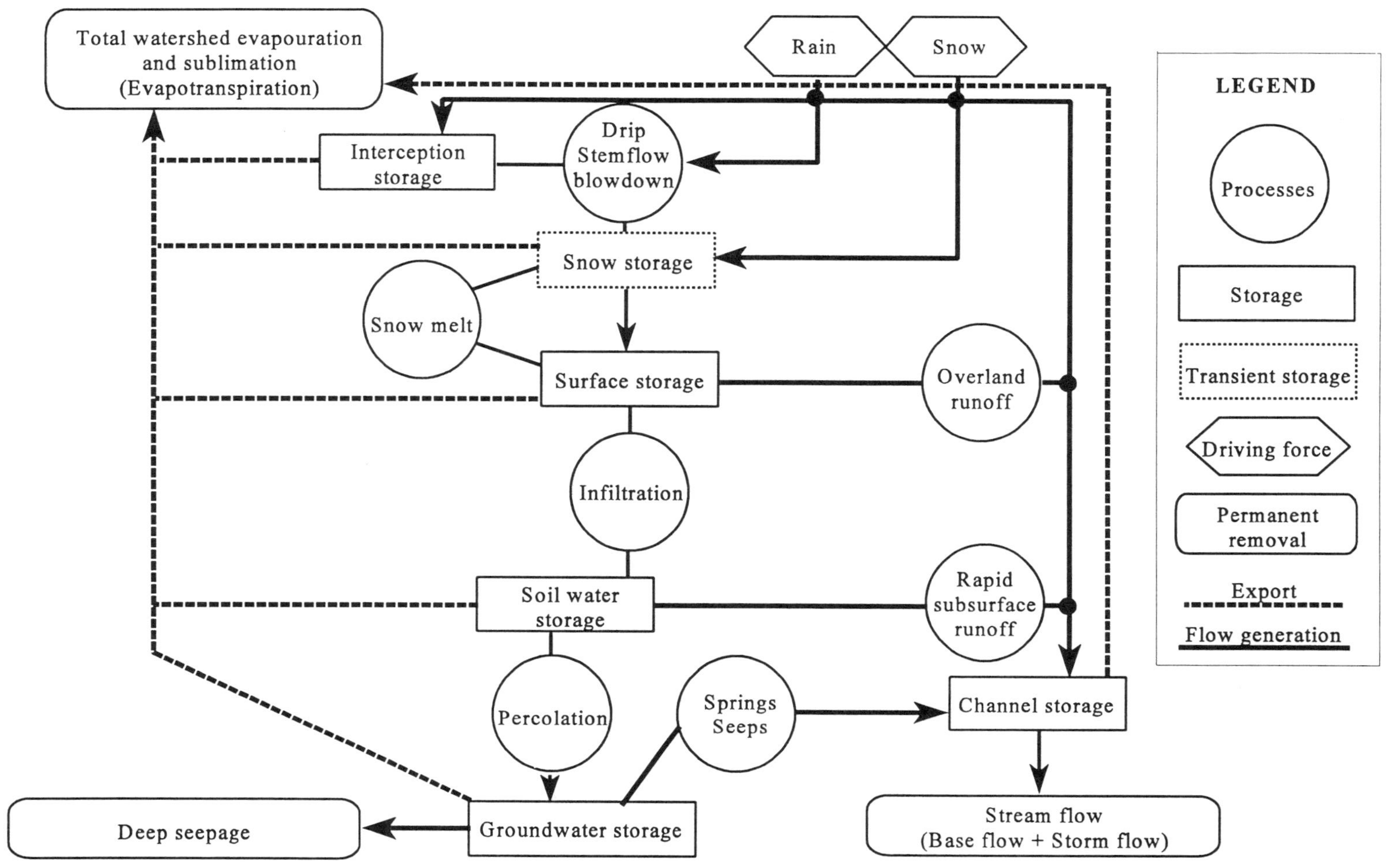

Figure 2.16. Stream flow generation. Legends for symbols and lines are shown at right.

close to 0 °C when the snow melts, and then rises as the air temperature rises.

If the pore spaces in the upper soil surface layers are partially filled with water, or melt water is still being generated, water will enter the stream channel as ***subsurface runoff***, or ***throughflow***. The subsurface runoff has much different physical and chemical characteristics than overland runoff. Because subsurface water is in intimate contact with soil particles, it has a much higher level of dissolved materials, a higher pH level and lower silt loads than does overland runoff. Also, because underground water is insulated from the sun, its temperatures are less variable than surface water.

Throughflow will occur as long as the infiltration rate or the permeability of the soil exceeds the precipitation rate (Fig. 2.17A). Otherwise, water ***percolates*** downward through the soil until it reaches the ***groundwater storage*** pool. The infiltration rate depends on the permeability of the soil. Hence, the maximum infiltration rate equals the permeability of the soil. Generally, the upper layers are more permeable than the lower layers due to increasing compactness of soil particles with increasing depth. The amount of throughflow can be estimated by dividing the soil profile into layers, as shown in Fig. 2.17 and then determining the permeability of each layer (in mm/h). For simplicity, rainfall intensity is assumed to equal the infiltration rate and also the percolation rate in the top layer of soil. Throughflow occurs only if the percolation rate through a layer exceeds the permeability of the layer below (Fig. 2.17B). The rate of throughflow depends on the slope of the ground; if the ground is level, the water will tend to back up towards the surface and form a pool of water. Water is deflected laterally only on slopes, hence the greater the slope, the greater the rate of throughflow, especially during periods of high precipitation intensity (Fig. 2.17A).

Throughflow is less prominent during periods of low precipitation intensity because the percolation rate of each soil layer is less than the permeability of the layer below. Two phases of water movement are recognized. The first is ***saturated flow*** (i.e. percolation), whereby water movement continues downward as long as sufficient water is supplied at the surface and no barriers are present to limit downward movement. When the water supply at the ground surface from precipitation decreases or is limited, saturated flow soon ceases. Any continued downward movement becomes entirely capillary and ***unsaturated flow*** takes place. Under such situations, tension differences exist between the moisture films coating the soil particles. Water now begins to move from regions of low tension (thick moisture films, high moisture content) to regions of high tension (thin films, low moisture content). Water movement can be either upwards, downwards or lateral, depending on the tension gradients that are developed. Unsaturated flow continues until an equilibrium is reached, with moisture contents being nearly equal throughout the volume of the soil profile. However, equilibrium is probably rarely attained because atmospheric and environmental conditions are continually changing.

Some of the groundwater enters the stream channel through ***springs*** and ***seeps***, but otherwise, it is not available to stream-flow generation because most of it ends up in ***deep storage***. The groundwater supply contributes to the ***baseflow*** of hydrographs. ***Stormflow*** represents the peak in discharge in the hydrograph. In contrast to the omnipresent groundwater supply, stormflow is erratic and varies rapidly in response to the overland

and subsurface runoff.

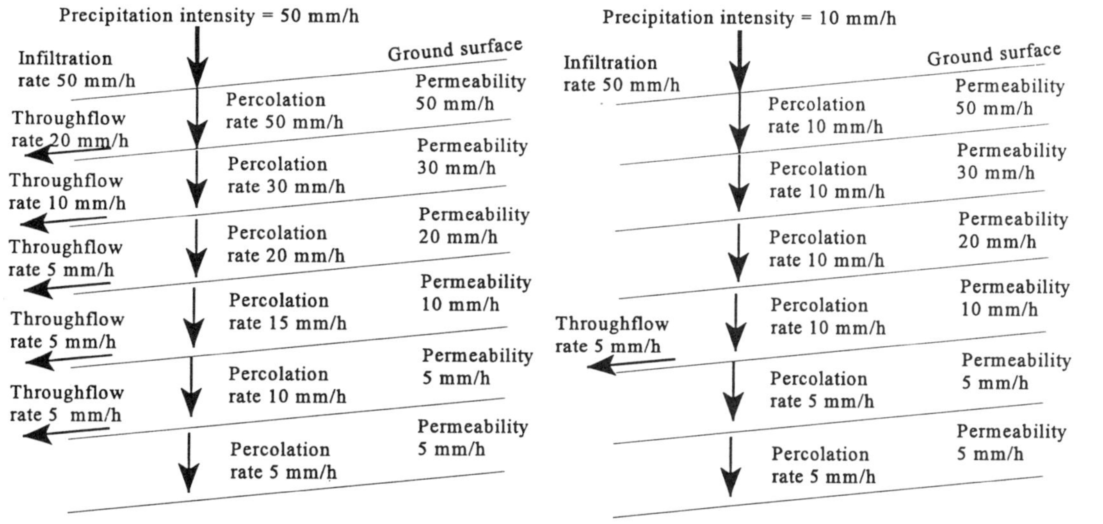

A. Throughflow model with heavy rainfall **B. Throughflow model with light rainfall**

Figure 2.17. Changes in rates of throughflow with depth of soil under heavy and light precipitation.

So far, the discussion has centered around the supply of precipitation directly to the ground. However, much precipitation falls on the foliage of trees and either drips or is blown off to the ground surface. Some water on the foliage evapourates into the atmosphere, along with some water in surface and subsurface storage (Fig. 2.16). The roots of trees and other vegetation take up some of the water in surface storage and remove it from the stream-flow generating processes. Some moisture within the plant foliage is released by evapo-transpiration. Evapo-transpiration also occurs from leaves of emergent macrophytes within the stream, as discussed in the hydrological cycle in Chapter 1.

Longitudinal Zonation of Streams

All streams begin in the headwater region as ***springs*** or ***seeps***. Limnologists have grouped the springs into a ***crenobiont division***. Within the crenobiont division are two
14☞ zones. The ***boil zone,*** where water bubbles up from underground sources, is characterized by low temperature and oxygen content. The boil zone contains biota called ***eucrenon***. The ***brook zone, or spring,*** follows the boil zone; it is characterized by slow flow rate, cool temperatures, rising oxygen levels, and has biota called ***hypocrenon***. If the spring discharges directly to a channel, the spring is called a ***rheocrene***. If the spring discharges into a pond and then to a channel, the spring is called a ***limnocrene***.

Finally, if the spring discharges into a marsh and then to the channel, the spring is called a ***helocrene***

Following the crenobiont divisions are the ***stream divisions***. Behind the brook zone, and the first of the stream divisions, is the ***rhithron***. It has three subdivisions, the ***epirhithron*** (nearest brook zone), the ***hyporhithron*** (farthest from brook zone), and the ***metarhithron*** (between epi- and hyporhithron). The rhithron is characterized by cool waters with monthly mean stream temperatures < 20 C, high oxygen levels, and fast to turbulent flow.

The rhithron is followed by the ***potamon***, which also has three divisions, the ***epipotamon***, ***metapotamon***, and the ***hypopotamon***, in the same downstream sequence described above. The hypopotamon is the largest region, has brackish water and is near the sea. The potamon is characterized by monthly mean temperatures > 20 C, common oxygen deficits, slow flow, sandy bottoms, and has eurytherms (tolerate a wide temperature range) and/or warm water stenotherms (tolerate a narrow but warm temperature range).

Vertical Stream Zonation

Streams also have a set of terminology for describing vertical zonation. The lowest zone is the ***phreatic zone***, or the groundwater area. Between the streambed and
15 the phreatic zone is the ***hyporheic zone***, or the upper 10 cm of substrate. The ***psammon*** is the wetted edge of the stream and moves up and down the "beach" with changes in water level. The ***hygropetric zone***, occasionally referred to as the ***madicolous habitat***, is the thin sheet of water on the surface of a rock. These specialized habitats are discussed in Chapter 10.

River Continuum Concept

The river continuum concept (RCC) states that physical, chemical and biological attributes of a stream change as a continuum of gradients as stream order size increases
16 from its headwaters to its mouth (Vannote *et al* 1980)[7]. The following chapters examine the changes in physical and chemical attributes of water, primary production, algal and macrophyte diversity, benthic invertebrate diversity in form and function, especially with respect to functional feeding groups, and fish diversity in form and function. Only the changes in stream morphology are described below.

17 As described above, most streams arise as springs or seeps. Many springs are seasonal, containing water in the spring and then drying up until a storm event. As water flows down hill and picks up momentum, it erodes a path to form the brook zone, which

---

[7]Vannote, R. L., G. W. Minshall, K. W. CumminsJ. R. Sidel and C. E. Cushing. 1980. The river continuum concept. Can. J. Fish. Aq. Sci. 37: 130-137.

usually is a permanent aquatic segment of a stream. The RCC begins in the brook zone, at stream order 1, the smallest tributary. To simplify the understanding of the dynamics involved, the RCC examines the continuum in three segments, stream orders 1 to 3, 4 to 6, and ≥ 7 (Fig. 2.18). Stream orders 1 to 3 are characterized by a heavy canopy of trees that shade the stream, keeping the stream cool, and supply a considerable load of ***coarse particulate organic matter (CPOM)*** in the form of leaf litter and twigs. The turbulent water maintains high oxygen levels and an erosional substrate water.

Stream orders 4 to 6 are wider, more exposed to light and are characterized by the presence of both riffles and pools. The water is warmer than upstream, but the riffle zones maintain high oxygen levels. Most of the leaf litter from upstream has been

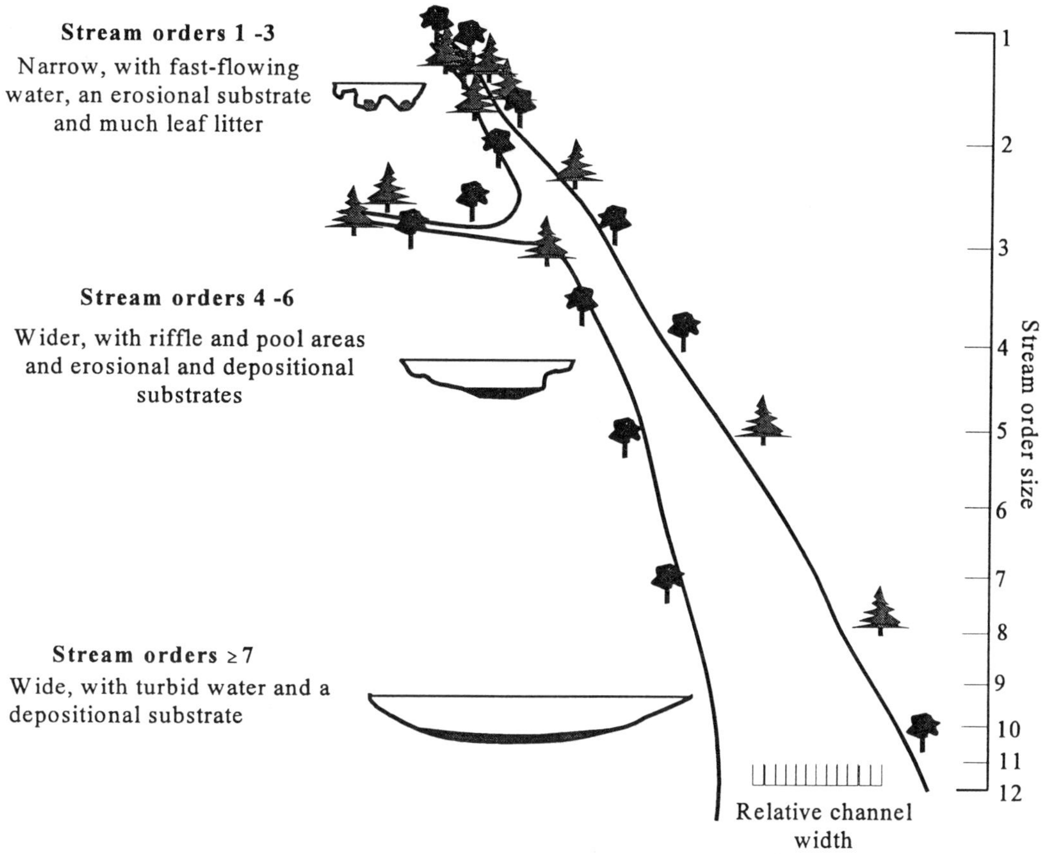

Figure 2.18. Some morphological attributes of a river systemaccording to the river continuum concept.

processed into ***fine particulate organic matter*** (***FPOM***).

Current velocity is greatly reduced in stream orders greater than 7, producing depositional substrates dominated by clay, mud or fine sand. The river increasingly takes on lake characteristics as stream order size approaches 10 to 12.

The Importance of Wetlands

Wetlands are the biological filters of streams before they enter lakes. Not all streams flow through a wetland, but those that do are purified before they enter a lake. Wetlands include ***marshes***, ***swamps***, ***bogs*** and ***fens***. Table 2.4 summarizes some of the basic differences among the four types of wetlands.

Marshes are flooded areas (i.e. they have an external source of water) that have ***emergent*** and ***submergent macrophytes***. Emergent macrophytes are aquatic plants that rise above the surface of the water, like cattails (*Typha*), bullrushes (*Scirpus*), arrowheads (*Sagittaria*), pampas grass (*Phragmites*), and purple loosestrife (*Lythrum salicaria*). The open water areas often have white (*Nymphaea*) and yellow (*Nuphar*) water lilies, with submergent plants below, such as pond weeds (*Potamogeton* spp.). Most marshes are like small ponds or lakes, except in lakes the macrophytes are replaced by ***phytoplankton*** (floating algae) as the most important primary producers of oxygen. In marshes, attached algae (called ***periphyton***) contribute most of the food and oxygen for aquatic organisms (mostly snails, freshwater shrimp, and minnows). The periphyton are found attached to the stems and leaves of the macrophytes.

In the fall most of the macrophytes die and fall to the bottom where they accumulate as organic matter and peat. Some may be removed by floods, but most decompose rather slowly and accumulate. For this reason, marshes are characterized by ***anoxia*** (lack of oxygen) or at least low levels of oxygen. If it was not for the external source of water (usually overflow from streams and rivers or groundwater) and submerged macrophytes to replenish the oxygen supply, marshes would have little or no aquatic life. Most marshes have fairly short hydraulic retention times, usually 3-7 days in the summer time.

Table 2.4. Summary of features of four types of wetlands

Wetland Type	Typical Emergent Plants	Submerged/Floating Macrophytes	pH	Water Source	Relative [Organic]
Marsh	Cattails, rushes, reeds	Lilies, pondweeds, milfoil	5.1-7.0	Stream(s)	High
Swamp	Trees (willows, cedars, cypress)	Water lilies, duckweed, water hyacinths, water lettuce	3.0-7.0	Stream(s)	Least
Bog	*Sphagnum*	*Sphagnum*	3.6-4.7	Precipitation	Highest
Fen	Mosses, sedges, grasses	Horsetails, cotton grass	5.1-7.6	Groundwater	High

Swamps are flooded areas that have trees, like willows (*Salix* spp), maples (*Acer* spp.) and cedars (*Thuja* spp.). Macrophytes may be present in open, sunlit areas. Swamps are like marshes in that they have an external source of water but they do not accumulate large amounts of peat or other organic materials. Swamps also are characterized by anoxia and organisms depend on submerged macrophytes and flood waters to replenish oxygen supplies. Many swamps have ***ephemeral pools*** and the organisms that live within them have adapted to a brief aquatic stage. Many produce eggs that are resistant to ***desiccation*** and have very rapid growth rates so as to complete their life cycle in the short period of time that water is present. Many ephemeral ponds are ***vernal pools,*** the water being present in the swamp only during the spring time. Most swamps have very short ***hydraulic retention (resident) times***, from a few hours to 1-2 days during the summer period.

Bogs are dominated by moss (*Sphagnum*) and the water has an acidic pH (~3.5-5) due to ***humic acids***. In the northern parts of North America, bogs are called ***muskeg***; Europeans call them ***mires*** or ***moors***. Most bogs consist entirely of living and dead mounds of *Sphagnum*, with open water in the depressions. Enormous amounts of peat accumulate on the bottom, with a characteristic "rotten egg" smell due to the production of hydrogen sulphide gas by sulphur bacteria. Only the wary dare try to wade into *Sphagnum* bogs because of the thick organic ooze in which waders often get stuck. The depth of the ooze seems endless, lending the expression "false bottom" to many bogs. Bogs have much longer hydraulic residence times than any other wetland type, ranging from two to four months.

Fens have a higher pH (~5-8) than bogs and have a variety of aquatic vegetation. They share characteristics of bogs and marshes, having a mineral-rich groundwater source, but they have a more neutral pH than bogs. The water is usually clearer and not as brown as that in bogs. Mosses, sedges and willows are common plants of fens. Hydraulic residence times are also more similar to marshes (4-8 days) than to bogs.

Wetlands are biological filters, acting as sinks or accumulators for silt and other suspended particles and soluble inorganic nutrients (e.g. nitrates and phosphates), and filters and detoxifiers of many pesticides. However, they are also sources for dissolved and particulate organic matter. Only wetlands that contain significant amounts of algae and periphyton will remove nutrients.

Perhaps the most significant role of wetlands is to transform substances from inorganic nutrients to soluble and particulate organic compounds. Nitrates and ammonia are taken up by plants and transformed into microbial materials or organic matter. Nitrogen and sulphur bacteria strip nitrogen and sulphur from the water, this removal being an important mechanism for reducing eutrophication potential of water leaving wetlands before flowing into lakes. *Sphagnum* acts as an ion-exchange resin, with calcium ions replacing hydrogen ions in the filamentous appendages of the plant. They filter out excessive nutrients, and even some contaminants, before the water enters the lake. If you have wetlands around your lake, and there's a good chance you do, ***protect them***; they are vital to maintaining the water quality of the lake.

Wetlands can also be classified according to the sources of water. Wetlands that are adjacent to lakes and obtain their water from rising water levels of the lake surface are called ***lacustrine wetlands***. Only the edge of the lacustrine wetland adjacent to the lake shore is wet most of the year. The rest of the wetland is temporary aquatic habitat (Fig. 2.19a). ***Palustrine wetlands*** are topographical depressions that fill with water but the

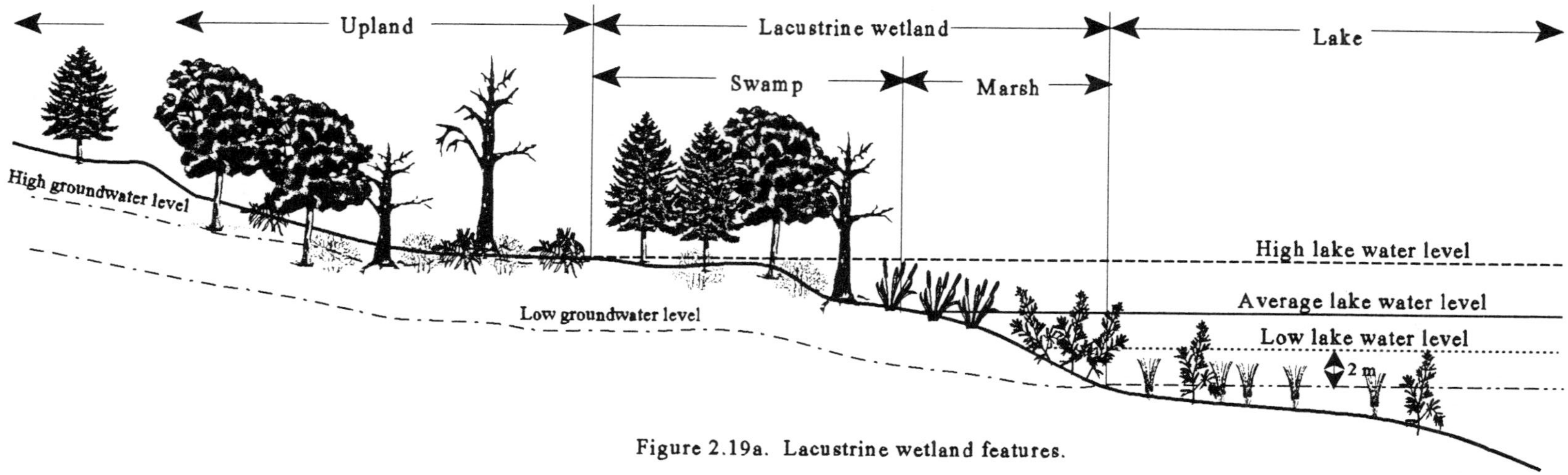

Figure 2.19a. Lacustrine wetland features.

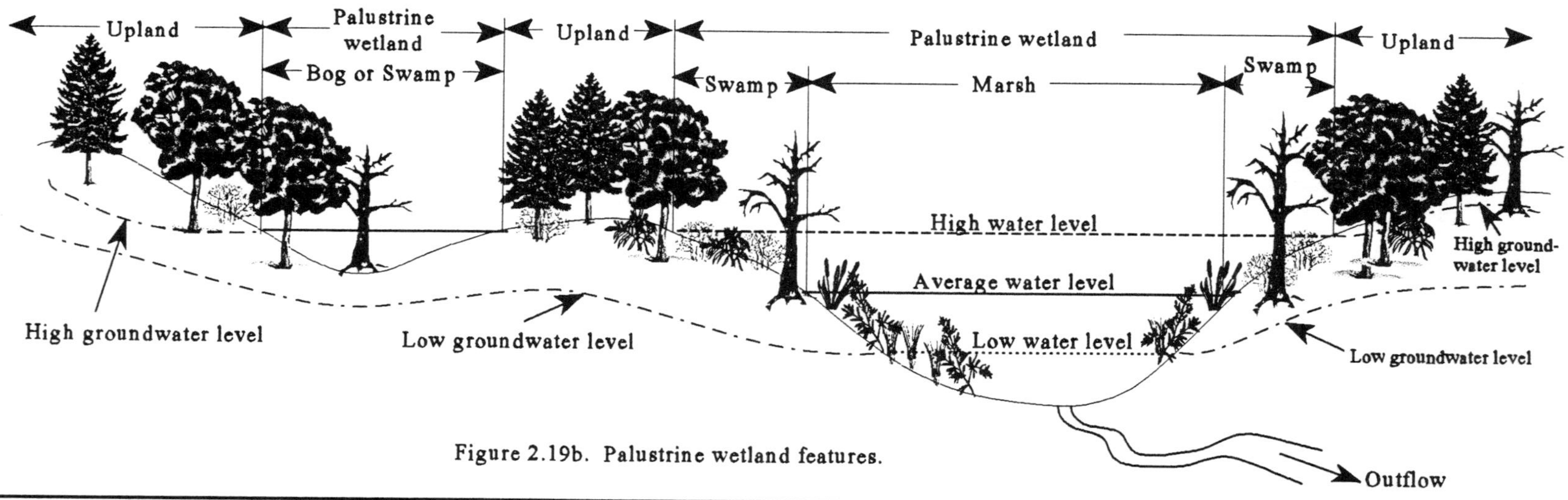

Figure 2.19b. Palustrine wetland features.

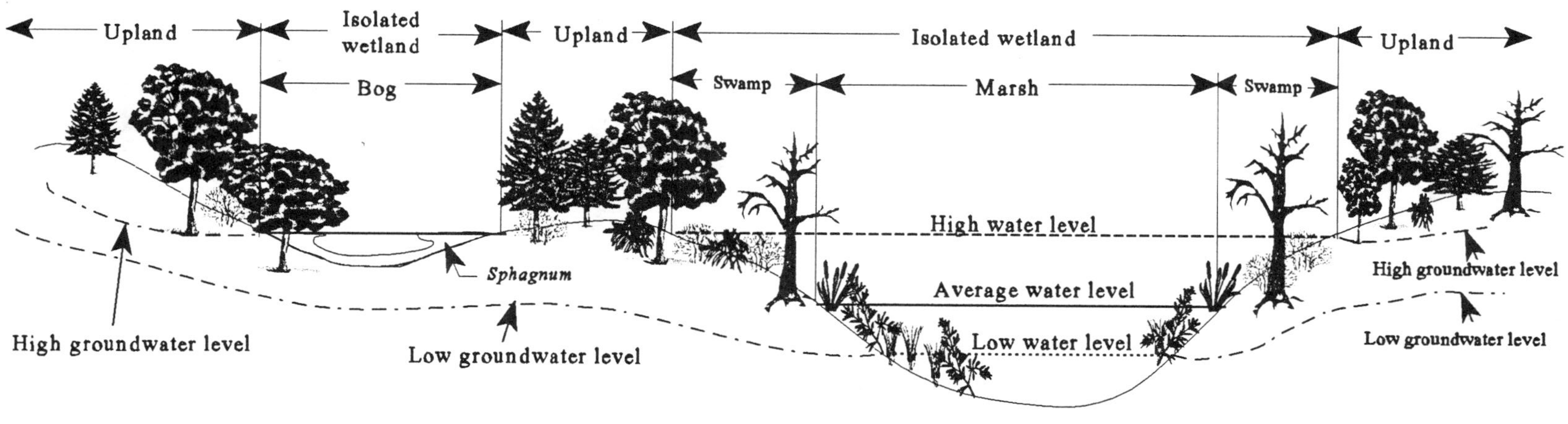

Figure 2.19c. Isolated (no outflows) wetland features.

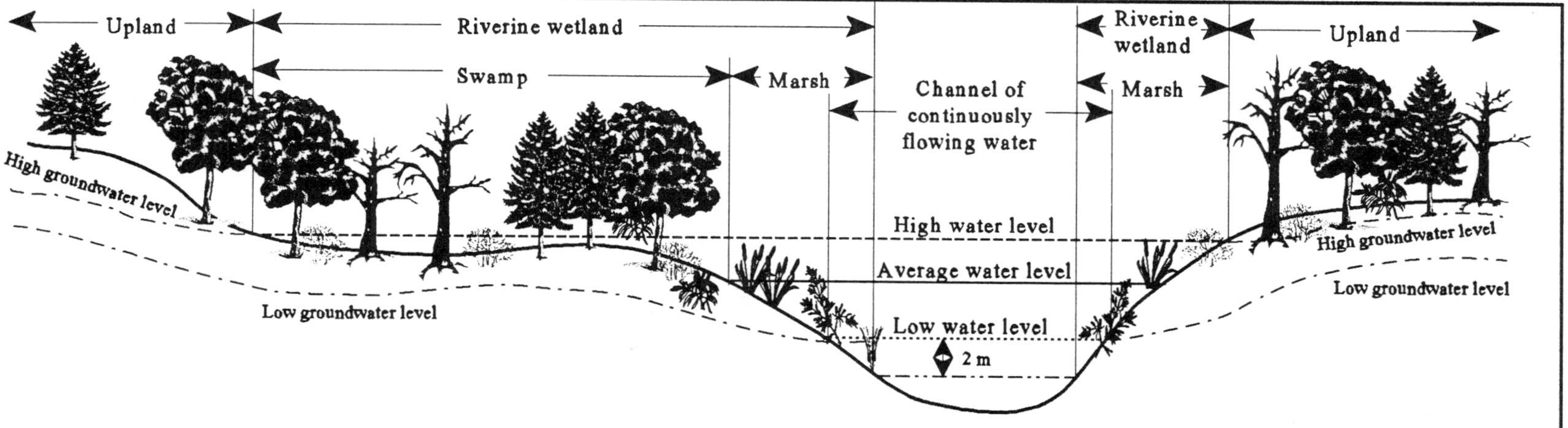

Figure 2.19d. Riverine wetland features.

marsh has an outflow (Fig. 2.19b). Bogs and/or swamps may also be part of a palustrine wetland. In general, palustrine wetlands are permanent bodies of water, although in dry seasons, water levels may fall below the level needed to maintain an outflow. ***Isolated wetlands*** have similar features to palustrine wetlands, except there are no outflows (Fig. 2.19c). Otherwise, marshes, bogs and swamps are common components of isolated wetlands. A fourth type of wetland is the ***riverine wetland*** (Fig. 2.19d). Like lacustrine wetlands, they arise from high water levels. Swamps are common in depressions located more elevated parts of adjacent lands.

Floodplains/Riparian Zones

Floodplains are the vegetated (usually) areas at the edge of streams. Some have grasses, others have shrubs or trees. The landward edge of the foodplain is usually taken as twice the distance of the highest-water mark. The highest water mark can be very high, often 2-4 m, or 7-10 ft, above the mean water level! The edge adjacent to the stream is called the ***riparian zone***. It is usually taken as the vegetated area 30 m from the water's edge plus another 30 m to allow for wind-fallen trees/shrubs, or a total of 60 m from the water's edge.

Riparian zones are "***buffer strips***" that absorb (or buffer) chemical and physical extremes. For example, they absorb, or moderate (buffer) impacts of excessive nutrients (e.g. phosphates and nitrates), pesticides and silty runoff before they enter the stream. Some riparian zones have trees that shade the stream and prevent water temperatures from rising to excessive levels. The roots of plants in the riparian zone also absorb or strip nutrients and contaminants from runoff before they enter the stream.

Riparian zones also filter the water. As water flows through vegetation, the current is slowed and suspended particles tend to settle. Hence, riparian zones also act as efficient water clarifiers by filtering suspended particles out of the water column.

Predicting floods

When will floods occur and how high will it be? Scientists can predict, fairly
18☞ accurately, the size of a flood event from one year to the next. Scientists have a great deal
more difficulty predicting the period and extent of the highest of high water events.

A ***hyetograph*** is used to predict the "lowest of lows" and the "highest of highs". Hyetographs (Figure 2.20) are graphs that plot precipitation intensity (e.g. mm/hr) on the y-axis against discharge (e.g. m^3/sec) on the x-axis. Usually curves are drawn for yearly averages, 10 year averages and 100 year averages. Some flood level can be expected every year. Somewhat higher discharges can be expected every 10 years, and abnormally high discharges can be expected every 100 years. Restated, there is a chance of 1 in 10 (= 0.1% probability) that we will experience a flood of level "x", or a chance of 1 in 100 (= 0.01% probability) that we will experience a flood of several levels greater than "x". The size of "x" is determined from numerous scientific measurements over several years. Typically, one estimates the probability of a "1-in-N-year" flood event of a given size or larger.

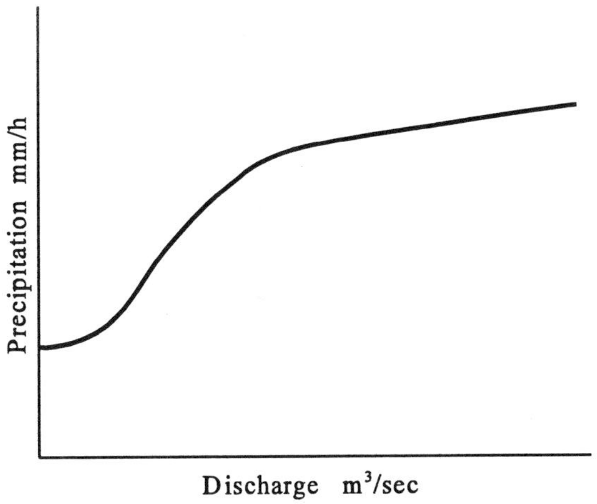

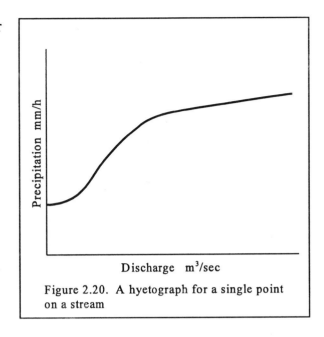

Figure 2.20. A hyetograph for a single point on a stream

Flood probability (P) and the average recurrence or return period (Tr) are inversely related:

$$P = 1/Tr$$

Tr is the average number of years in which a given event is equalled or exceeded. The most common flood frequency analysis is performed on a data set consisting of the single highest flow each year, and is plotted on probability paper or is fitted to any one of several probability distributions. Annual maximum floods are based on peaks in the flood hydrograph rather than on average daily discharge because peak flow may be under-estimated for small rivers in which peak flow may pass in hours. The recurrence interval for an individual flood is calculated as:

$$Tr = (n + 1)/ m$$

where n = number of years of observation and m = rank or order assigned to each member of the data set. The largest event is assigned 1. From the analysis of mass hyetographs of storm events one can calculate the maximum intensity of precipitation over time periods of as little as 5 minutes to 24 hours or longer. Use of these data, along with estimated return periods, one can make a series of depth-duration, or intensity-duration graphs (Fig. 2.21) for any point where detailed precipitation information has been collected.

Cottagers would especially be interested in flood-frequency curves which give the probability of annual maximum discharges for a particular stream (Fig. 2.22). The bankfull flood is estimated using T = 1.5 years. The probability or recurrence of more extreme events (e.g. 10- and 50-year floods) can be read from the graph.

The average high-water mark in a stream can be easily determined by measuring the height of debris hanging from the branches of shrubs in the riparian zone; 1-2 metres is probably an average height! This can be tripled or quadrupled for a 1 in 100 year flood event in larger streams!

Hyetographs and intensity-duration graphs are not only useful for predicting the size and frequency of flood events, but it is useful for determining how high up on the shore buildings should be placed. At least consider the 1 in 10 year flood event; most people cannot afford the time or money to replace a building every 10 years! Floating docks and boat houses suddenly become more attractive than fixed docks and boat houses! But, the location of a lake in a watershed is also important. Go back to the topographic map and determine where a lake is in the watershed. The lower the lake is in a watershed, the greater the flood potential. Consider the factors that affect the hydrograph, like the aspect (north-facing vs south-facing) of the slope. North-facing slopes tend to get more snow but runoff tends to be more gradual than south-facing slopes.

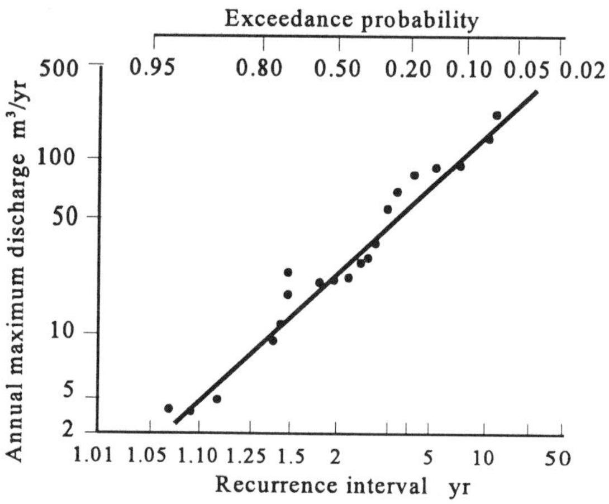

Figure 2.22. Hypothetical flood-frequency analysis for a stream based on annual peak events for a 20-yr period.

Methods for Estimating Areas of Odd Shapes

There are formulae for estimating areas of definite shapes, like rectangles or
19☞ squares (A = length x width), circles (A = 3.14 x (radius)^2), triangles (A = (height x base)/2)), ellipses (A = 3.14 x length x width), and rhomboids (A = ½ diagonal a x diagonal b). Specialized devices, like ***planimeters*** and computerized digitizers, accurately measure areas of odd-shapes, but they are expensive. If precision is not required, here are two methods that can be used to measure areas of any shape, including cross-sectional areas of streams, drainage areas and surface areas of lakes. One method, the ***squared paper method*** requires the use of graph paper with squares about 1-2 mm per side. The other, the ***paper weight method***, requires the use of a microbalance, which any high school or university with a chemistry, physics, or biology lab will have. The microbalance is a balance that weighs to four or five decimal places in grams. The paper weight method is much easier and faster to use than the squared paper method.

Squared Paper Method:

1. Obtain a sheet of squared graph paper. Trace the outline of the area (e.g. watershed area, lake area, cross-sectional area) on to the graph paper. It is important to always keep careful record of the scale being used, especially if the tracing has been enlarged.
2. Count the number of full squares within the area and record the number.
3. Next, count the total number of triangles of half-square size at the margin of the area, divide the total by 2 and record the number of full squares made up by half triangles. Colour in all triangles after they have been counted to make sure that triangles are not double counted.
4. Repeat for triangles of 1/3 the area of a full square; divide by 3 to determine number of full squares made up by triangles of 1/3 square area.
5. Repeat the process for triangles of 1/4, 1/5, etc. until all areas have been accounted for.
6. Once the total number of squares has been determined, by summing all the squares made up by full sized squares and different sized triangles, multiply the number by the scale size. For example, to determine the cross-sectional area of a stream with depths and width shown in Table 2.1, plot the data on graph paper. The grid is not shown in Fig. 2.11, but the image will be similar, with an inverted cross-sectional view of the stream, the surface of the stream being on the x-axis. Note that the scale on the y-axis (0 to 0.8 m in 0.1 m increments) is different than that on the x-axis (1 to 10 m in 1 m increments). I traced the curve onto graph paper with a grid size of 0.05 m (depth) by 0.2 m (width), hence, a unit area was 0.05 m x 0.2 m = 0.01 m^2. I counted 265 squares, including triangles, so the cross-sectional area is 265 x 0.01 = 2.65 m^2).

Paper Weight Method:

The following method is great for determining areas of any shape. It is routinely used to determine areas of rock surfaces to which zebra mussels are attached. The procedure is simple; wrap aluminum foil around that part of the rock to which the mussels are attached, pressing the foil into crevices and cutting off any "folds" that result from the crimping process. Then weigh the piece of foil. The area was determined from a standard curve (line) relating known foil areas to its foil weight, using the same procedures described below. Again keep careful records of the change in scale if it the tracing was enlarged for any reason.

1. Trace the outline of the area onto a piece of blank paper.
2. Cut out the tracing and weigh the tracing on a microbalance. Record the weight in milligrams or grams (= 1000 mg).
3. Using the same sheet of paper from which the tracing was removed, cut out pieces of paper that are 1, 2, 4, 8, 16, 32, 64, 128, 256, 512 cm^2 (i.e. pieces measuring

1x1, 1x2, 2x2, 2x4, 4x4, 8x8, 8x16, 16x16, 16 cm x 32 cm). Larger pieces may have to be cut, depending on the size of the area being determined.

4. Weigh each piece and plot weight in mg vs size (cm^2) on squared paper, with area on the x-axis and weight on the y-axis. A straight line that runs through the origin should result (Figure 2.23).
5. Determine the area of the tracing by relating the weight of the tracing in step 2 to the weight-area line in step 4.
6. Account for the scale used for the tracing. For example, if 1 cm = 248 m, 1 cm^2 = (248 x 248) = 61,504 m^2. Therefore, the actual area is 26 x 61,504 = 1,599,104 m^2 = 1.60 km^2.
7. Most photocopy paper weighs about the same. Hence, if a tracing is photocopied, and the area is cut out of the copy, the area can be approximated by using the following relationship that is based on the mean of about five brands of photocopy paper:

$$\text{Area of paper (cm}^2\text{)} = [(\text{paper weight} + 0.013)/7.394]$$

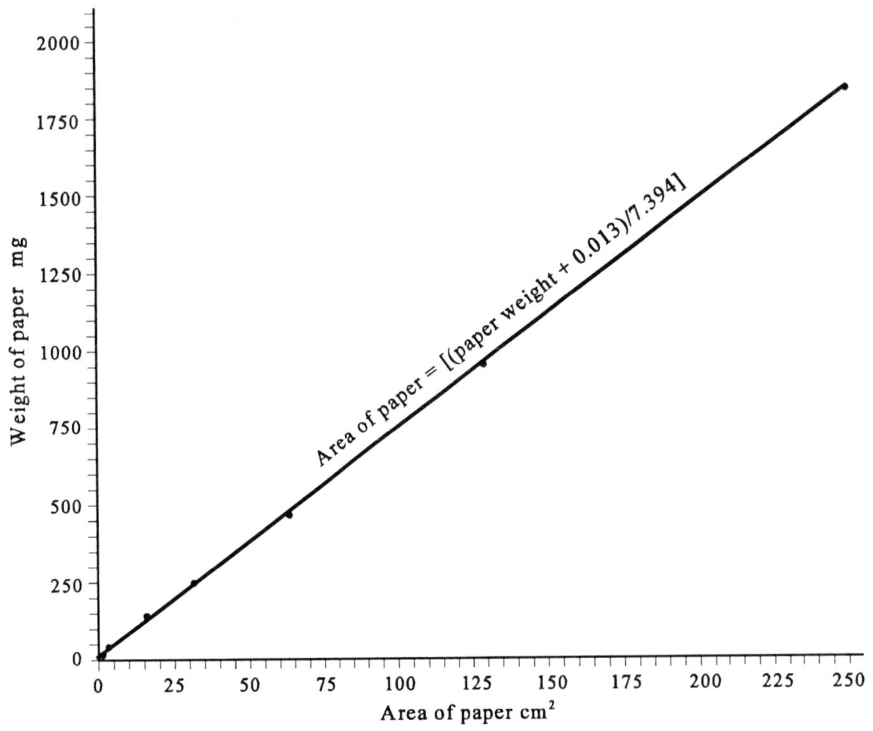

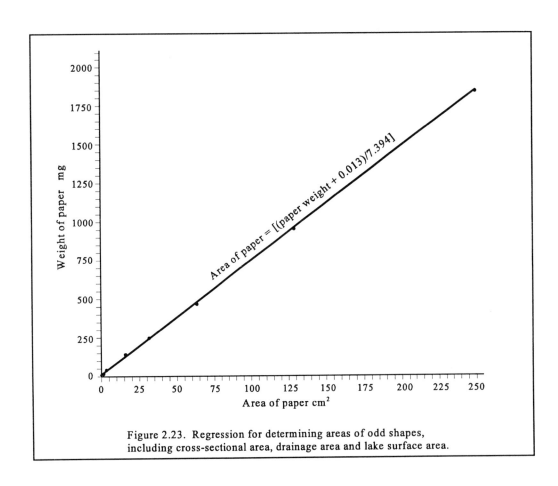

Figure 2.23. Regression for determining areas of odd shapes, including cross-sectional area, drainage area and lake surface area.

The tracing of the cross-sectional area in Fig. 2.11 was photocopied, area cut out of the copy and its weight found to be 122.43 mg. The area = (122.43 + 0.013)/7.394 = 16.56 cm^2. But 1 cm^2 = 0.16 m^2 for the scale used. Hence, the actual cross-sectional area is 16.56 x 0.16 = 2.65, which agrees with the squared paper method.

CHAPTER 3

LAKES, THEIR ORIGIN AND DESIGN

Why Read This Chapter?

Lakes are perhaps the most familiar units of aquatic ecosystems, yet they contain less than 0.020% of the earths water supply. They are also known as a ***lentic*** (standing) body of water, ponds and pools being other types of lentic systems. Lentic systems come in different sizes, shapes, and depths, ranging from puddles and small ponds to huge, deep lakes, like Lake Baikal in Siberia. Lake Baikal is the deepest lake in the world (1,632 m or 2,611 ft!)) and contains about 20% of the world's freshwater resources, a bit less than that found in all of the Great Lakes! All lakes have distinct features, each being unique in their design (otherwise known as ***morphometry***) and physical, chemical and biological features. This chapter introduces some general concepts about lakes - their origins, important watershed characteristics and some commonly measured lake morphometrics.

In order to fully understand the functional processes of lakes, it is essential to know how to derive some of the lake's basic design characteristics, like mean and maximum depth, surface area, volume, maximum width and length, fetch, shoreline length, shoreline development, and retention time. A thorough understanding of these features, integrated with knowledge of the lake's origin, its watershed characteristics, and its inputs and outputs in the hydrological cycle, will go a long way to understanding dynamic processes within a lake. Several, if not most, of these concepts play vital roles in determining a lake's fisheries and recreational potential. To help illustrate the different concepts, a hypothetical lake, called Golf Lake (introduced in Chapter 2) is used.

By the end of this chapter you will know:

☞ 1. The basic features of lakes, ponds, and other lentic bodies of water
☞ 2. The zonation of lakes
☞ 3. How to identify and list different types of lake origins and their forces, including:
  - ☞ a Glacial forces
  - ☞ b Tectonic forces
  - ☞ c Volcanic forces
  - ☞ d Extra-terrestrial

☞ e Solution forces
☞ f Fluviatile forces
☞ g Aeolian forces
☞ h Landslide forces
☞ i Shoreline forces
☞ j Impoundment forces

☞ 4. The importance and methods of measuring lake morphometric features, including:

☞ a Lake surface area
☞ b Island area
☞ b Maximum length
☞ d Maximum width
☞ e Mean width
☞ f Mean length
☞ g Mean length to width ratio
☞ h Shoreline length
☞ i Shoreline development
☞ j Maximum fetch
☞ k Bathymetry
☞ l Volume
☞ m Maximum depth
☞ n Mean depth
☞ o Relative depth
☞ p Hydraulic residence time
☞ q Hypsographic and volume curves
☞ r Volume development

☞ 5. Know what cryptodepressions are and how to measure them.

Lentic (Standing Water) Ecosystems

1☞ Lentic ecosystems include lakes, ponds and puddles. ***Lakes*** are generally thought of as large bodies of water which have sufficient depths and waves to create a beach zone, or a wave swept zone. ***Ponds*** are smaller bodies of water with less depth and waves cannot be generated, or at least will have insufficient force, to maintain a wave swept beach zone. Ponds that are temporary and filled with water in the spring are called ***spring ponds*** or ***vernal ponds***; if filled only after a large precipitation event, they are called ***ephemeral ponds***. ***Estuaries*** are places where the river meets the sea and share some river features with ocean features.

2☞ Figure 3.1 shows some important features of lakes. Most lakes have two distinct zones, a ***pelagic zone*** or an open water zone, and a littoral zone, or a shore reaching zone with ***macrophytes*** or ***hydrophytes*** (large aquatic plants). The pelagic zone consists of a ***photic*** or ***euphotic zone***, which extends from the lake surface down to where the light is about 1% of that of the surface (= the ***1% incident light level*** zone. At this depth the rate of photosynthesis (or production of oxygen) equals the rate of respiration (or

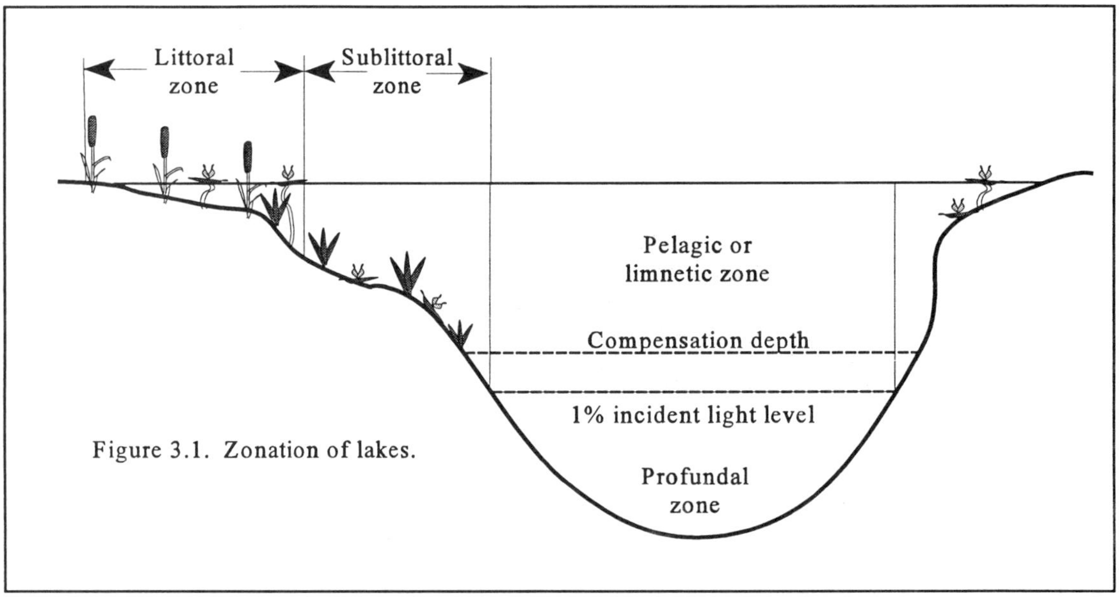

Figure 3.1. Zonation of lakes.

oxygen consumption). The lower limit of the photic zone is usually referred to as the ***compensation depth***. The portion of the lake where incident light level is less than 1% (i.e. total darkness) is known as the ***aphotic zone***. Strictly speaking, the pelagic zone is used for the open water zones of oceans and seas, and the ***limnetic zone*** is used for the open water zone of fresh waters.

The littoral zone has been the subject of numerous definitions and re-definitions. For some people, the littoral zone is that zone with aquatic vegetation, including emergent macrophytes, floating leaf macrophytes, and submerged macrophytes, and the ***sublittoral zone*** is the transition zone between the littoral zone and then the deep-water, plantless zone known as the ***profundal zone***. However, the transition zone is often very difficult to define. We will take a simpler approach; the littoral zone is defined here as that zone inhabited by ***emergent*** and ***floating leaved aquatic macrophytes***; the sublittoral zone is the zone dominated by ***submerged aquatic plants***. The profundal zone is the deep, plantless part of the lake, where there is insufficient light to support growth of plants, and the bottom is usually muddy and very soft.

Lake Origins

First, a brief review of the relative sizes (surface areas) of inland waters, most of which are freshwater lakes. The Black Sea is the largest inland body of water, but it is a
3☞ saline body connected to the Mediterranean Sea. The Caspian Sea is another large inland body of water, saline, but much smaller than the Black Sea. The Aral Sea (formerly southwest USSR), also saline, was once the fourth largest inland body of water in the world (the third being Lake Superior, North America) but is shrinking in surface area, being only about 60% of its former size, due to diversions of water for crop irrigation.

Even if all the Great Lakes are combined into one body of water, the total surface area (244,160 km^2) ranks third, behind the Black Sea and the Caspian Sea. However, the Great Lakes is the largest body of ***freshwater*** in the world. In fact, Lake Superior alone ranks as the largest (82,100 km^2) freshwater lake in the world. If the Aral Sea continues to shrink from its current 64,000 km^2, Lakes Huron (59,600 km^2) and Michigan (57,800 km^2) will soon rank as the fourth and fifth largest inland bodies of water in the world.

Most lakes can be assigned to one of ten different origins. The ***geomorphology*** (shapes determined by geological processes) of lakes is intimately related to physical, chemical and biological events within the drainage basin, and all play a major role in the lake's metabolism. For example, the rates of inputs of nutrients to a lake, the volume of influx, the nature of the drainage basin, the thermal stratification patterns, and the distribution of dissolved gases are all, in some way, related to the geomorphology of the lake.

Most lakes have had some kind of catastrophic origin, that is, they were formed by some kind of catastrophic force, including glacial, tectonic and volcanic events. Other forces, like fluviatile (riverine), landslide, damming and wind-blown events are not catastrophic but are physical enough to form lakes.

Lakes of a common origin tend to be localized into lake districts in which large numbers of lake basins are concentrated. More recently, humans have built large numbers of dams that have created impoundments, or reservoirs, but most are less than 20 m in depth. Table 3.1 summarizes the different origins and kinds of lakes found within each.

Glacial Forces

Glacial activity is probably the most important lake-creating force over the last few millennia. As the glaciers melted and retreated, they gouged huge cavities in the
3a☞ earth, with the ice melt filling the cavities. The best known example is the Great Lakes. Because of the massive sizes of glaciers, one usually finds thousands of lakes in huge areas that have glacial origin. Most lakes formed by glacial scouring have an elongated shape and, because the ice retreated in one direction, have their long axes oriented in the same direction. The orientation of glacial lakes is most evident on topographic maps of scale 1:100,000 1:250,000 or larger.

Several types of lakes have been formed by ice scour. In mountainous areas freezing and thawing action has produced ***ampitheater-shaped basins*** (Fig. 3.2). Lakes with an ampitheater-like formation are referred to as ***cirques*** (from french, meaning semicircle or amphitheater). Such lakes evolve in rocky cliff basins where the slow downhill movement of ice and the continual freezing and thawing activity erodes and fractures the rock. Cirque lakes are usually deepest near the cliff and shallowest near the outlet. Often a chain of cirque lakes will occur as the river flows down the mountain side, much like a string of beads, to form ***paternoster lakes***. ***Fjord lakes*** and ***piedmont lakes*** are formed in mountainous regions, the former being along glaciated coast lines and the latter at the foot of mountains in inland alpine areas. Because of the proximity of fjord lakes to oceans and seas they often have a salinity higher than piedmont lakes.

Kettle lakes form when blocks of ice are trapped in glacial debris called ***moraine***. As the ice melts, the moraine is deposited around the outside, and the melt water fills the depression. Some lakes formed this way have steep sides and are surrounded by high hills, resulting in ***meromictic lakes***, or lakes that only partially mix (see Chapter 5). In other cases, moraine was deposited and impounded the melt water as the ice retreated to create ***moraine impoundments*** or ***moraine lakes***. ***Cryogenic,*** or ***thaw lakes***, occur when frozen ground or permafrost thaws, leaving the melt water to fill the depression. Finally, the mere weight of some glaciers was sufficient to create depressions in the earth's surface, the depression being filled with melt water to create ***glacial hydraulic lakes***.

Table 3.1. A summary of the different kinds of lakes and their origins.

Glacial Forces	Volcanic Forces	**Fluviatile Forces**
Existing Glaciers	Within volcanic craters	Levee lakes (Pools)
Pre-existing Glaciers	Calderas, Marrs	Oxbow lakes
Inground moraine, e.g. Kettles, Pits	Within lava	Deltaic lakes
Morainic impoundment	Coulee Lakes	Eversion
Hydraulic forces	**Extra-terrestrial Forces (Astroblemes)**	- Pothole Lakes
Glacial Scour - Fjords	Meteoric crater lakes	**Aeolian Forces**
Cirques, Paternosters	**Solution (Chemical) Forces**	Aeolian Lakes
Piedmont lakes	$CaCO_3$	**Shoreline Forces**
Cryogenic or Thaw lakes	- Karsts, Dolines, Sinks, Swallow holes	Shoreline Lakes
Tectonic Forces	NaCL	Landslide Forces
Faulting - e.g. Grabens	- Salt lakes	Rock, mud, etc. slides
Epeirogeny - e.g. Uplifts	$CaSO_4$	**Impoundments**
Rift lakes	- Sulphur lakes (pools)	Human made
Earthquakes	Sandstone	Made by other animals
	- False karsts (pipes)	

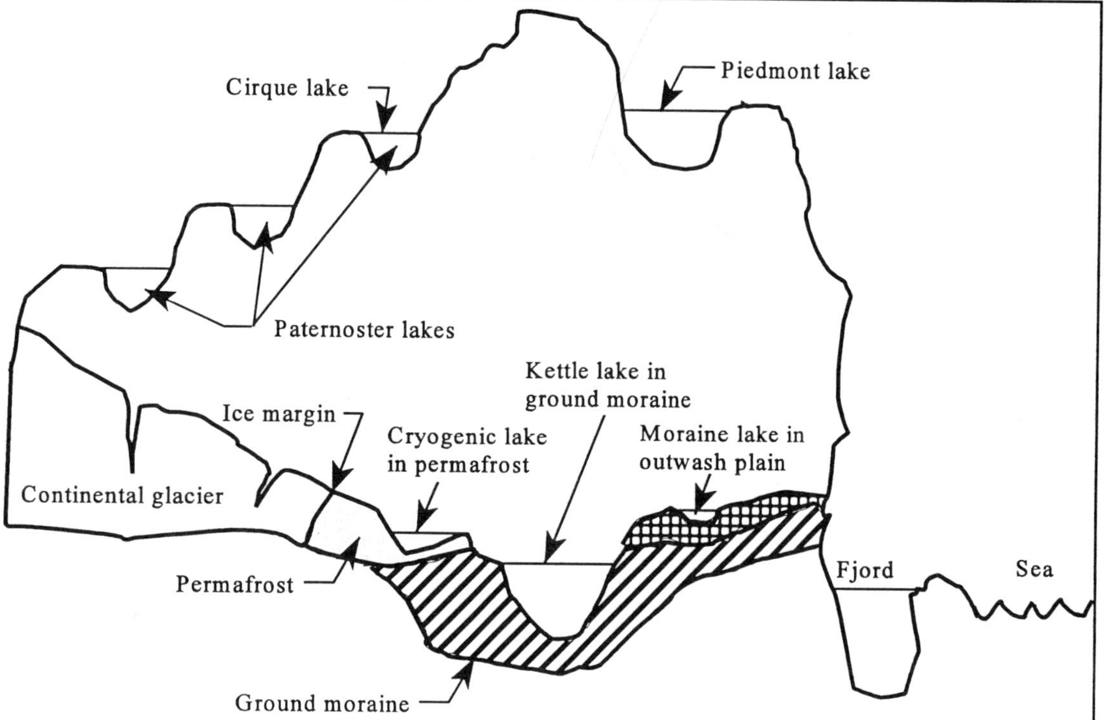

Figure 3.2. Different types of galcial lakes.

Tectonic Forces

Tectonically-formed lakes are lakes created by warping, faulting, fracturing,
3b☞ buckling, folding, thrusting or quaking of the earth's shell. ***Graben lakes*** are the most common type of tectonic basin. They are formed as a result of large depressed areas located between adjacent faults, where the fault block is untilted and forms a flat bottom in a trough (Fig. 3.3a). Lake Tahoe, California, is a graben lake. ***Rift lakes*** occur with a single fault event, the depression brought about by tilting (Fig. 3.3b). Lake Baikal was formed by one great fault event. Most tectonic lakes are very deep and tend to have a rectangular shape to their surface area. Lake Tanganyika in Africa is also a graben lake.

The Caspian Sea and the Sea of Aral also are tectonic lakes but were separated by formation of uplifting of portions of the sea floor during the miocene period. Lake Okeechobee, Florida, was formed in the same way. ***Uplifting***, also called ***epeirogeny***, of inland areas can alter or even reverse existing drainages and form lakes in upstream sections.

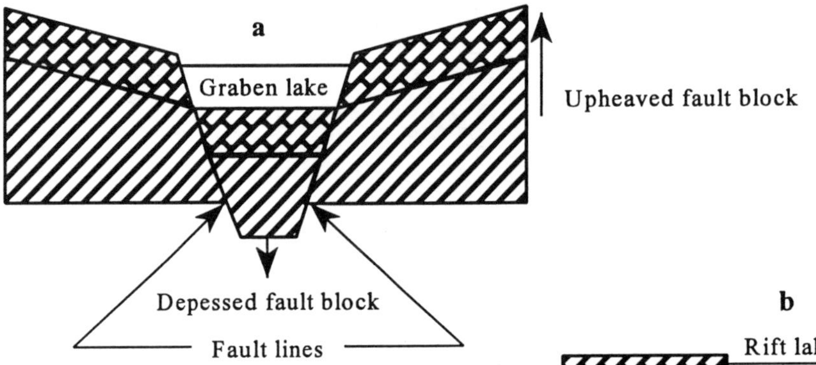

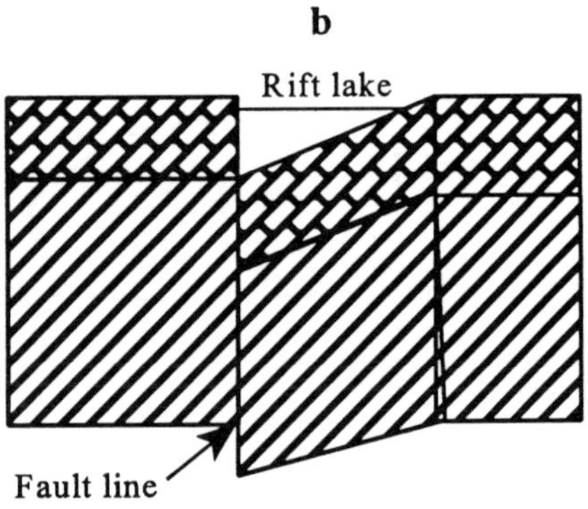

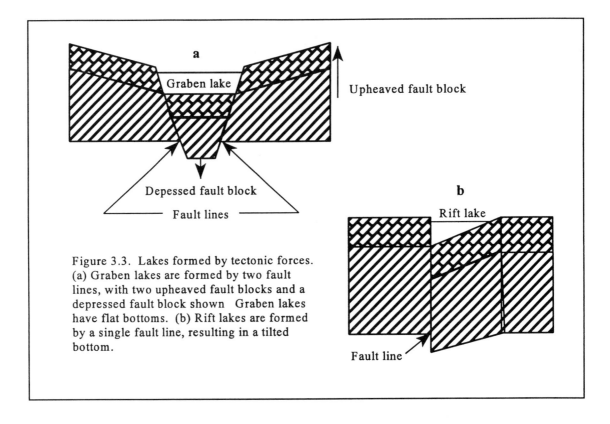

Figure 3.3. Lakes formed by tectonic forces. (a) Graben lakes are formed by two fault lines, with two upheaved fault blocks and a depressed fault block shown Graben lakes have flat bottoms. (b) Rift lakes are formed by a single fault line, resulting in a tilted bottom.

Volcanic Forces

When volcanoes erupt, materials are rejected and depressions and cavities, called ***craters***, are formed. Since the craters are undrained, they fill with water and eventually
3c☞ form a lake (Fig. 3.4). Because most ***crater lakes*** are associated with basaltic minerals, which are low in nutrients and dissolved minerals, the water tends to be relatively unproductive. Crater Lake, Oregon, with a depth of 600 m, is perhaps the most famous crater lake. Depressions with diameters less than 2 km are termed ***maars***. They are nearly circular in shape and can be extremely deep in relation to their small surface area. ***Calderas*** are basins formed by the subsistence of the low rim or roof around the volcanic cone after the magma has been ejected. Crater Lake in Oregon is a spectacular caldera which was formed by the collapse of the centre of the volcanic cone. Lava flowing from volcanic activity can form ***coulee lakes*** at the base of the volcanoes and where lava can form walls to contain pools or basins of water. Lava flowing across a river may also dam the river and form a small lake upstream.

Extra-terrestrial Forces

3d☞ These are not UFO forces! The extra-terrestrial forces are meteors that impact with earth to create ***astroblemes*** (***"star scars"***). The impact creates a huge depression

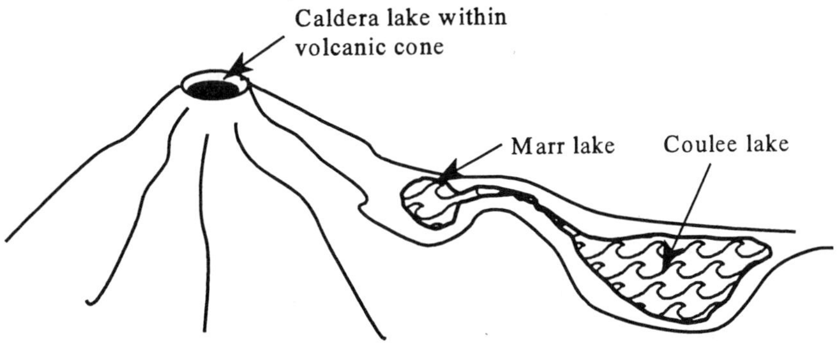

Figure 3.4. Volcanic lakes. Marrs are nearly circular and result from lava coming in contact with ground water. Calderas are partially emptied magmatic chambers with a rim or roof around it. Coulees are formed by lava flows that form depressions at the base of the volcanoe and fill with w ater.

that fills with water to create ***meteoric crater lakes***. There are a few meteoric crater lakes in North America but most have dried up. The most famous of the extinct lakes is Coon Butte Crater Lake in Colorado Plateau, Arizona. It was once believed to be of volcanic origin but recent evidence indicates that it was formed by a meteor. The basin contains 30 m of ***lacustrine sediments***. It, like most other meteoric lakes, are nearly circular in outline and have their maximum depth near their centre. Of meteoric lakes still in existence are Gilmour Lake and Tecumseh Lake located in Algonquin Park, Ontario, Canada. Both are part of the Brent Crater, formed by a giant meteor 450 mybp (million years before present). The meteor pulvarized the overlying granitic rock and exposed underlying limestone that resulted in the formation of ***hardwater lakes*** (see Chapter 5) that are otherwise relatively uncommon in the Canadian shield.

Solution Forces

3e☞ ***Solution lakes*** are created in areas where there are deposits of soluble rock that slowly dissolve by purcolating water. Rocks dominated by rock salt (sodium chloride or halite), limestone (calcium carbonate), or gypsum (calcium sulphate) in particular, contribute to the formation of solution lakes. Solution basins are usually very circular and conically-shaped. They are commonly referred to as ***sinks***, ***dolines***, or ***swallow holes***. Lakes formed within limestone bedrock are specifically referred to as ***karsts***, many of which are common in areas with hot springs. The hot water increases the solubility of calcium carbonate, making the water very hard and alkaline (see Chapter 5). Most karsts have a very unique fauna and ecology and are protected by local conservation authorities. However, the conservation of karsts is often compromised by their appeal to bathers as "***natural hot tubs***". Solution lakes that form within sandstone are called ***false karsts*** or ***pipes***.

Fluviatile Forces

3f☞ These are forces generated by the activities of flowing waters. Perhaps the most common type of fluviatile lake is the ***plunge pool lake***, formed by the erosional forces of water at the base of a waterfall. Trout and minnows often congregate in plunge pools to feed on organisms retained by eddies as they pass from upstream to downstream sites. The erosional action of river eddies is known as ***eversion*** and can form small ***pothole lakes*** within the river. Rivers entering quiet waters of a lake often deposit large amounts of sediments to form ***deltas***. Within many deltas are long narrow lakes called ***deltaic lakes***. Deltaic lakes formed adjacent to oceans receive saltwater at high tides and are brackish. ***Oxbow lakes*** form on large meandering rivers when one of the loops is cut off by deposition of silt. The deposition of suspended loads at the margins of rivers can also isolate the mainstream from an adjacent (see chapter 5) levee lake. ***Levees*** are banks of eroded material, the suspended load being deposited along the margin after a flood and as the water velocity decreases. Levee lakes and oxbow lakes are refilled when flood-waters overtop their banks. They can also be flushed and destroyed by spring ***spates*** (floods events).

Photosynthetic activities of blue-green algae and other plants can produce massive precipitation of calcium carbonates in calcareous waters. Over time these deposits can form barriers and isolate small lakes within the river, with their margins rising above the water level, like the ***atolls***, a type of coral reef in oceans.

Aeolian Forces

3g☞ Aeolian forces involve the actions of wind, tornadoes, and hurricanes. Wind action operates in arid regions to create lake basins by piling up pieces of eroded or broken rocks or by re-distribution of sand, resulting in the formation of ***dune lakes***. Most dune lakes are temporary, their permanency depending on fluctuations in climate. Wind action can also topple trees and pile up debris and soil across rivers, which over time accumulates enough organic and inorganic materials to effectively dam the river and create a small ***aeolian lake***. Such lakes are usually shallow, the deepest part being at the dam. Beavers will often enhance the size of the aeolian dam to create an even larger impoundment. Depending on the typography, aeolian lakes can have a fairly large surface area, especially if the flood plain is low and large, or very small surface areas in parts of drainages with little or no flood plain. In most cases, aeolian lakes have a large length to width ratio.

Landslide Forces

3h☞ Large quantities of unconsolidated material often slide into floors of stream valleys and create dams that form lakes, often of very large size. The landslides may be in the form of rock falls, mudflows or ice slides and even flows of large amounts of peat, but these are usually found in glaciated mountains. Dams formed by mud or very light materials are susceptible to rapid erosion and at some point in time may cause a disastrous flood downstream.

Shoreline Forces

3i☞ Activities of waves on the shoreline of some lakes can create fairly shallow, usually temporary, ***shoreline lakes*** or ***shoreline pools*** They are typically found adjacent to large bodies of water, such as seas or large lakes, that possess some irregularity or indentation along the shoreline. ***Sand bars*** often form across bays to form a ***coastal lake***.

Impoundments

3j☞ Humanity has had, and continues to have, an enormous impact on the development of dams and artificial impoundments to meet the growing demand for hydroelectric power (e.g. James Bay Project, Quebec) and flood control (e.g. Mississippi River). In Africa (e.g. Lake Kariba), and North America, many dams are built for a variety of purposes, including flood control, fisheries, low flow augmentation, irrigation, and recreation. Most, if not all, have a ***dendritic pattern***, each dendrite representing one of several major inflows. The deepest area is always near the dam.

Impoundments or pools can be created by the damming action of plant materials, accumulated either by natural riverine events or by activities of animals, particularly beavers. We have already discussed (under Fluviatile Forces) some of the kinds of lakes formed by natural riverine events, and also have alluded to the activities of beavers (see Aeolian Forces). Beavers are masters at making dams large enough to create ponds or lakes with surface areas of one to two square kilometers. Beaver ponds are favourite breeding areas of many minnow species. However, ***humic materials*** leached from the wood and leaf debris, left by the beavers, turn the water brown and oxygen levels slowly drop over time. Only the more thermally-tolerant fish species remain, and those that do often congregate at the dam near areas of higher flow rates and turbulence that usually result in higher dissolved oxygen levels. The impact of artificial impoundments on physics, chemistry and biology of streams is discussed in Chapter 10.

Some Lake Features: Their Importance And How To Measure Them

Lake morphometry is the dimensional characteristics of a lake and plays a key role in determining a lake's recreational and fisheries potential. In many cases, the calculation of one variable depends on the measurement or estimation of another. The morphometrics are discussed in order of their importance or need for estimating other more complex variables.

Lake Morphometry

This section examines the different measurements used for determining the size and shape of lakes. Many of the morphometric measurements correlate well with faunal
4☞ and floral productivity of lakes, discussed in later chapters.

The first measurements required are lake surface morphometrics, including surface area, maximum length and maximum width. They are determined from a tracing of the lake's outline from an aerial photograph or from a topographic map, usually of 1 - 25,000 scale. If a 1:25,000 scale topographic map, or an aerial photograph, is not available, use a 1:50,000 scale topographic map and enlarge it using a photocopier with enlargement capabilities. Most photocopiers will magnify an image at least 1.40 times, but one can keep enlarging each enlargement until the map is 8 to 10 times larger than that shown on the 1:50,000 scale map. Try to enlarge the lake to a size large enough to draw depth contours at 1-5 m intervals and still have it fit a standard 8.5 x 11 inch sheet of paper. Make sure some grid lines are shown on the photocopy because they provide information on the scale and true or magnetic north orientation of the lake. Remember, the grid lines are separated by 2 cm, or 1 km on a 1:50,000 scale map. Hence, if a lake on a 1:50,000 topographic map is enlarged 8 times, the scale should be 1 cm = 0.5 km/8 = 0.0625 km.

Figure 3.5 is an enlargement of Golf Lake, a hypothetical lake from a 1:50,000 scale topographic map. The scale is shown at bottom centre. The magnetic north and true north lines are determined with reference to the grid north lines using the declination shown on the topographic map. From the enlargement one can determine lake area, island areas (not to be included in the lake area), maximum length, maximum and mean width, mean length, mean length to width ratio, shoreline length, shoreline development, and maximum fetch (the maximum distance that wind can blow without interruption by land).

Lake Surface Morphometrics

Lake Area: There are several methods available to determine surface area. An older, but still popular method, is the use of a ***planimeter***. It is a tool designed for deriving areas on
4a☞ flat surfaces by tracing the perimeter, or shoreline, of the image. As the tracing is made, readings on a dial increase accordingly so that the length of the perimeter and the area of the image are determined by integration. Planimeters are calibrated in either square

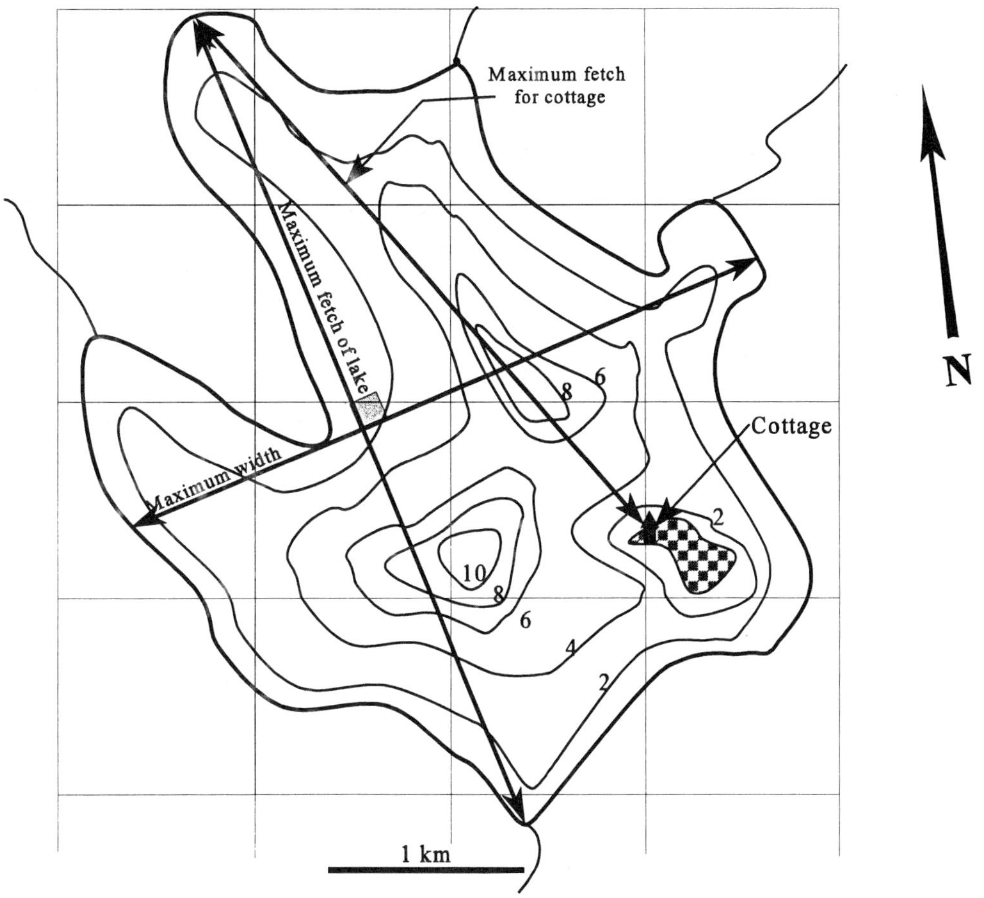

Figure 3.5. Some morphometric features of a lake. Maximum lake and cottage fetches assume wind is blowing in direction of arrows. Depth contours are in 2-meters intervals

inches or square centimeters. Most planimeters have easy-to-follow instructions making operation of the instrument a relatively simple task. However, they are fairly expensive.

Alternatively, one can rely on computer technology and use a ***digitizer***. With proper software, one can make areal estimates for any shape accurately and quickly. In this method, a ***mouse*** with a pointer is used to trace the outline of the image. The mouse button is pressed several times as the outline is traced. The more often the button is pressed, the more accurate the estimate of the area enclosed, or the length of line being traced. The computer program integrates the data and converts it into area (e.g. lake area) or length of line traced (e.g. shoreline length). However, digitizers and digitizing programs are also fairly expensive.

Two other methods for measuring lake areas from topographic maps are described in Chapter 2 under the heading, "Measuring Areas". Surface area is designated by A_0. Island areas are not included in lake area measurements. For Golf Lake (Fig. 3.5), the surface area is 8.30 km^2, using either the squared paper or the paper weight method. For the paper weight method, the area of paper (cm^2) = [(paper weight + 0.013)/7.394] (see

Chapter 2). The lake cutout (without the island) weighed 434.10 mg. Hence, the area of the cutout is (434.10 + 0.013)/7.394 = 58.71 cm^2. According to the scale of Fig. 3.5, 1 km^2 = 7.023 cm^2 (~1.75 times enlargement of a 1:50,000 scale map in which 1 km^2 = 4 cm^2). Therefore, the lake surface area (without the island) is 58.71/7.023 = 8.36 km^2, which is very close to the squared paper method.

The area of a lake by itself is more than a mere descriptor. It is important for estimating other lake morphometrics needed for determining the recreational and fisheries potential of a lake. These morphometrics (also called ***metrics*** when formulae are used) include mean depth and shoreline development.

Island Area (A_{island}): Island area is determined in much the same way as lake surface area. If the paper weight or squared paper method is used, it is important to enlarge the
4b☞ lake several times in order to accurately measure the areas of small islands. The island in Golf Lake (Fig. 3.5) has a surface area of A_{island} = 0.0933596 km^2 = 93,360 m^2.

Island area may be important for its real estate value, but it also indirectly affects the productivity of the lake by adding to its shoreline development.

Maximum Length (l_{max}): This is the maximum straight-line distance across the water from one shoreline to the next, regardless of whether it passes through an island or one or
4c☞ more points of the lake. It is determined from a topographic map using a ruler and adjusting the measurement for the scale of the map used. The maximum length of Golf Lake in Fig. 3.5 is 4.38 km (scale: 1 cm = 0.388 km).

The maximum length is important for determining the maximum potential of wave height and wave velocity in a lake, as discussed in Chapter 4. Lake length also affects the erosion potential on the windward shores and safety on the water.

Maximum Width (B_{max}): The maximum width of a lake is determined by measuring, at
4d☞ right angles to the axis of the maximum length, a line that connects the greatest distance between two points on opposite shores. Golf Lake (Fig. 3.5) has a maximum length of 3.51 km.

The maximum width affects the same factors as maximum length, but to a lesser degree because lake widths are usually shorter than lake lengths.

Mean Width ($\overline{B}$): The mean width is determined by dividing the surface area (A_0) of the
4e☞ lake by its maximum length l_{max}. The mean width of Golf Lake (Fig. 3.5) is 8.36 km^2 / 4.38 km = 1.91 km.

Mean Length ($\overline{l}$): Just as the mean width is determined by dividing the lake's surface
4f☞ area by its maximum length, the mean length is obtained by dividing the same surface area by the maximum width. The mean length of Golf Lake is 8.36 / 3.51 = 2.38 km.

4g☞ ***Mean Length to Width Ratio:*** This metric gives an indication of a lake's relative length.

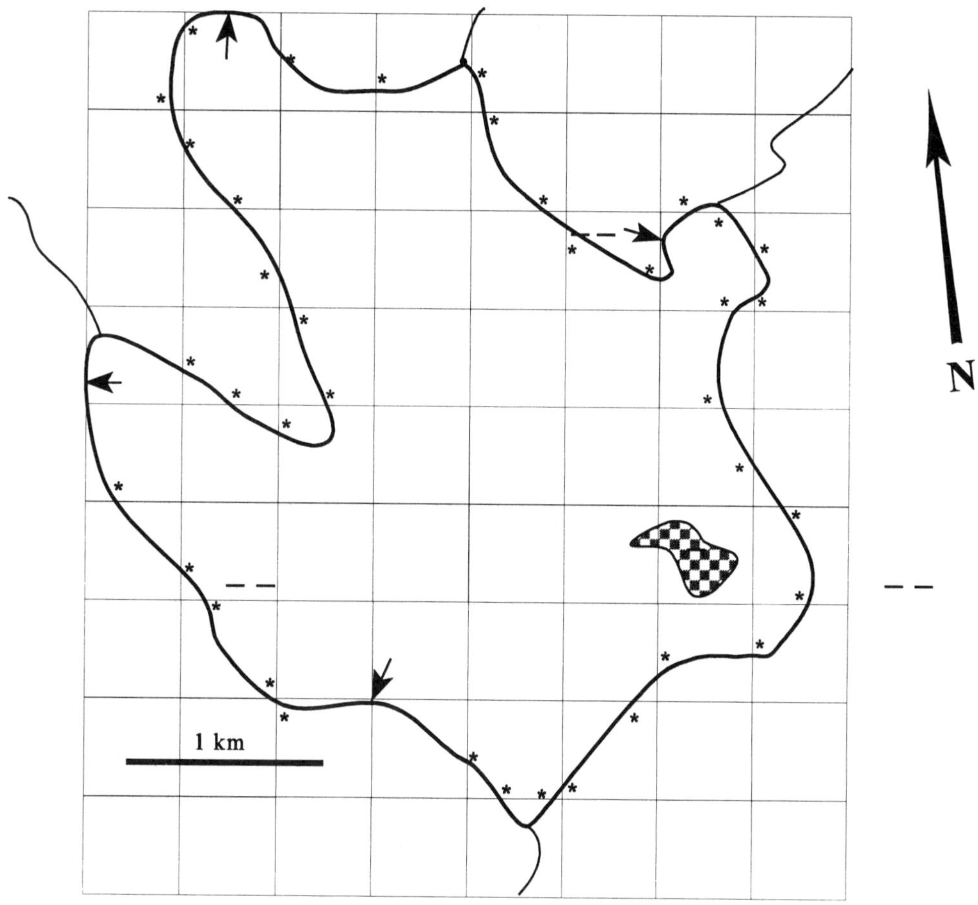

Figure 3.6. Grid intersection method for measuring shoreline length. Arrows are tangents, asterisks are grid line crossings.

Most fluviatile lakes have an elongated axis, with L/B ratios ranging from 3:1 to 4:1. The average L/B ratio for most lakes is between 1 and 1.5. The L/B ratio for Golf Lake is 2.38/1.91 = 1.25.

4h☞ ***Shoreline Length (L):*** This is the length of shoreline around the perimeter of the lake at zero depth. Lakes with many bays and inlets will have longer shorelines than those without them. The shoreline length may be determined by: (i) wrapping a string or thin wire around the perimeter of a lake traced from a topographic map; (ii) a ***cartometer*** which is specifically designed for measuring lengths of shorelines; (iii) an inexpensive ($5 - $10) road map measuring devices, such as a ***meilograph*** (a form of cartometer), or ***map measurer***; (iv) by using the grid intersection technique described below. Using the meilograph, the shoreline length of Golf Lake (Fig. 3.5) is 15.8 km.

Since much of the aquatic vegetation is in shallow water near shore, the length of shoreline greatly affects the productivity of a lake. Lakes with several bays and inlets are generally more productive (have more weeds and fish) than lakes without them.

Measuring Shoreline Length: The grid intersection technique is a fast and fairly accurate method for measuring lengths of irregular lines. Figure 3.6 is a copy of Fig. 3.5, with only the shoreline shown. It is recommended that the long axis of the lake be drawn diagonal to the grid pattern. Count the times that the shoreline crosses the horizontal and vertical grid lines (= 38 asterisks on Fig. 3.6). Add to this the number of tangents divided by 2 (a tangent is any line parallel to and touching a grid line = 4 arrows in Fig. 3.6). The sum of intersections + tangents/2 is multiplied by 0.785 and the grid scale. For Golf Lake, the values are (38 + 4/2) (0.785) (0.5 km grid) = 15.7 km, a value very close to that obtained by the meilograph above.

Shoreline Development (D_L): The shoreline development index describes the "roundness" of a lake or the degree of dissectedness of the shore line. Lakes with values
4i☞ near 1 are circular and have little shoreline development. The shoreline development index is calculated as:

$$D_L = \frac{L}{2\sqrt{\pi A_0}}$$

where, L = shoreline length, A_0 = lake surface area, $\pi = 3.1415$

Rounded lakes have D_L values near 1. A perfectly circular lake has a $D_L = 1$ because, for a circle, L = perimeter = π x diameter = $2\pi r$; $A = \pi r^2$; hence $D_L = 2\pi r / 2\sqrt{(\pi)(\pi r^2)} = 2\pi r/2\pi r = 1$.

Maximum Fetch (F_{max}): The fetch is the distance that wind can blow without interruption by land or large objects. The maximum fetch is often the same as the
4j☞ maximum length of the lake, provided there is no land to interfere with the wind as it blows along the longest axis. Obviously, wind direction is the critical factor. Hence, when the wind is blowing, the fetch is the straight-line distance, from shore to shore, parallel to the wind.

All cottages have a maximum fetch. Figure 3.5 shows maximum fetch directions for Golf Lake and for a cottage on the island. The maximum fetch for the lake is the same as its maximum length (4.39 km). The maximum fetch for the cottage is 3.45 km, when the wind blows from the north, northwest (Fig. 3.5).

Subsurface Lake Morphometrics

Bathymetry: Among the most important lake descriptors is its bathymetry, or its depth
4k☞ profile. A ***bathymetric map*** shows the depth contour (***isobath***) characteristics of a lake. Subsurface contours must be established and plotted on the bathymetric map in order to determine lake volume and retention time.

A lake must be ***sounded***, or its depths measured using a ***depth sounder*** (e.g. fish finder) or a sounding line marked off in meters, in order to draw a bathymetric map. Usually several transects are drawn across the lake and depths are recorded at regular intervals on each transect. Figure 3.5 is based on the depth data shown in Fig. 3.5. The depth contours were drawn at 2-m intervals, as shown in Fig. 3.5, based on the depths recorded in Fig. 3.7. From the bathymetric map can be measured lake volume, mean and maximum depth and retention time.

Lake Volume (V): The volume of water in a lake is determined from data in the bathymetric map. Total lake volume is determined by summing the volume of water in

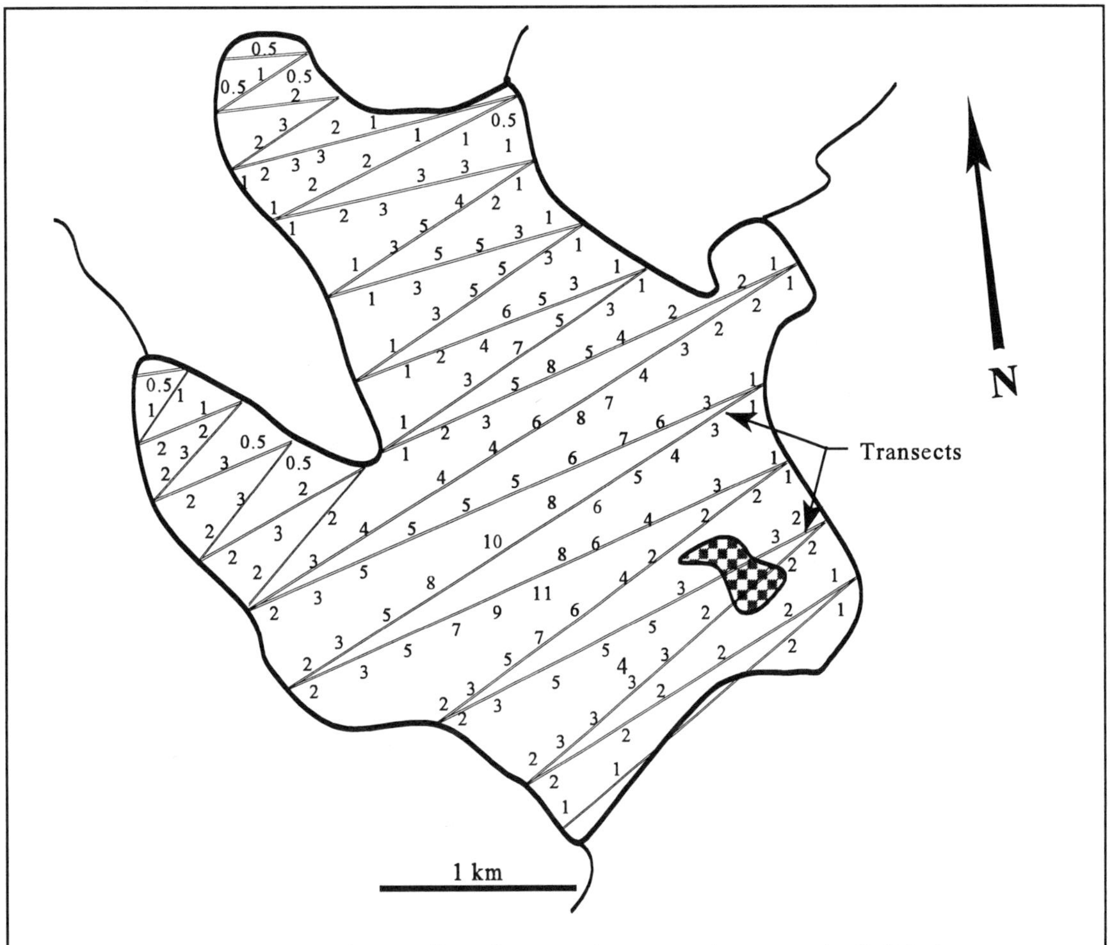

Figure 3.7. Depth recordings (in meters) on several transects in Golf Lake.

41☞ each stratum, as defined by the following formula:

$$V=\frac{(b-a)}{3}\sum\left(A_a+A_b+\sqrt{(A_aA_b)}\right)$$

where, b and a are depths

and A is surface area of each stratum. Volume may be determined in either cubic meters or cubic kilometers, BUT KEEP THE SAME UNITS THROUGHOUT (i.e depth and area in m and m^2 OR depth and area in km and km^2).

For Golf Lake the surface areas of each depth stratum (shown as subscripts) are:

$$A_0 = 8.36 \text{ km}^2$$
$$A_2 = 5.62 \text{ km}^2$$
$$A_4 = 2.52 \text{ km}^2$$
$$A_6 = 0.87 \text{ km}^2$$
$$A_8 = 0.38 \text{ km}^2$$
$$A_{10} = 0.064 \text{ km}^2$$

Hence, V = [(2-0)/3000] [8.36+5.62 + √((8.36)(5.62))] + [(4-2)/3000] [5.62+2.52 + √((5.62)(2.52))] + [(6-4)/3000] [2.52+0.87 + √((2.52)(0.87)) + [(8-6)/3000] [0.87+0.38 + √((0.87)(0.38))] + [(10-8)/3000] [0.38+0.064 + √((0.38)(0.064))] km^3 = (0.0139 + 0.0079 + 0.0032 + 0.0012 + 0.0004) = 0.0266 km^3. Note that the depth is divided by 1000 to convert from m to km.

An easier method for computing V is to use a hypsographic curve, discussed in detail below. When depth (m) is plotted against area (m^2), the area under the resulting curve represents the volume (m^3). A plot of volume against depth (depth-volume curve) allows one to interpolate volumes for depth contours that are not shown on bathymetric maps. For example, the volume of Golf Lake at 5 m is about 49,000 m^3 (Fig. 3.8).

4m☞ ***Maximum Depth (z_{max}):*** The maximum depth is the deepest part (hole) of the lake. Golf Lake's maximum depth is 11 m (Fig. 3.7).

Mean Depth ($\bar{z}$): The mean depth is calculated as:

4n☞

$$\bar{z} = \frac{V}{A}$$

where, V = lake volume and A = lake area. The mean depth for Golf Lake is 0.0266 km^3/8.36 km^2 = 0.0032 km or 3.2 m.

Mean depth is known to correlate well with several biological variables, including primary production, benthic biomass and fish production, all of which are discussed in later chapters.

4o☞ ***Relative Depth (z_r):*** Relative depth is the ratio of the maximum depth (m) to the average diameter of the lake surface (m^2), expressed as a percentage, or:

$$z_r = 100\ z_{max}/d_{A0} = 100\ z_{max}/2\sqrt{(A_0/3.14)} = 88.6\ z_{max}/\sqrt{A_0}$$

where z_{max} = maximum depth and d = diameter of circle whose area = lake surface area, or $2\sqrt{(A_0/3.14)}$. For Golf Lake the relative depth is 974.6/√ 8,360,000 = 0.34%. Most lakes have relative depths less than 2%. Deep lakes with small surface areas have relative depths > 4%. Crater lakes have among the highest z_r values, many ranging between 7% and 75%, but the record to date is the Hawaiian crater lake, Kauhako, with an A_0 value of 0.35 ha and a z_{max} value of 250 m, or a z_r value of 374%!

Hydraulic Residence Time (RT): The time required for a lake to fill or empty once is
4p☞ known as the hydraulic residence time or ***hydraulic retention time***. It is estimated by:

$$RT = V/Q_{outflow}$$

where V = lake volume and $Q_{outflow}$ = discharge of the outflow(s). Most lakes have hydraulic residence times of several days to a few years.

Hypsographic and Volume Curves: The ***hypsographic curve*** is a ***depth-area curve*** that graphically shows the relationship between the surface area of a lake and its depth. The
4q☞ area plotted may be the actual area or the percentage of area overlying a given depth (Fig. 3.8). If the actual area is used, the area under the curve is the total volume of the lake. The depth-volume curve is closely related to the hypsographic curve and represent lake volume relative to depth. The ***depth-volume curve*** is an easy way to determine how much water should be released or stored within a reservoir during periods of flood control. Figure 3.8 shows three types of curves using depth, area and volume data calculated earlier.

Volume Development (D_v): This metric examines basin shape by comparing it to an inverted cone with a height (h) equal to z_{max} and a base (A) equal to the lake's surface
4r☞ area. A lake that is perfectly cone-shaped has a $D_v = 1$. A lake with a relatively larger volume has a $D_v > 1$ and a lake with a relatively smaller volume has a $D_v < 1$ (Fig. 3.9). Since the volume of a cone is 1/3 hA, and volume = Az, the ratio of the actual volume to the volume of the theoretical cone is:

$$D_v = A(z) / (1/3\ A(z_{max}))$$
$$D_v = 3\ z / z_{max}$$

The average D_v for most lakes is near 1.25. The D_v for Golf Lake is 3(3.2/11) = 0.87, which is atypical but being a hypothetical lake, it is not surprising. Glacial lakes on the Precambrian shield have D_v near 1.23. The values suggest that most lakes are U-shaped. Generally, oligotrophic lakes have U-shaped profiles, eutrophic lakes have V-shaped or saucer-shaped profiles and D_v values < 1. However, it is dangerous to equate D_v with lake trophic status because factors other than depth, such as soil types, affect productivity.

5☞ ***Cryptodepressions (z_c):*** Some lakes have their maximum depth below sea level. The portion of the lake below sea level is termed a ***cryptodepression***. Cayuga Lake, New

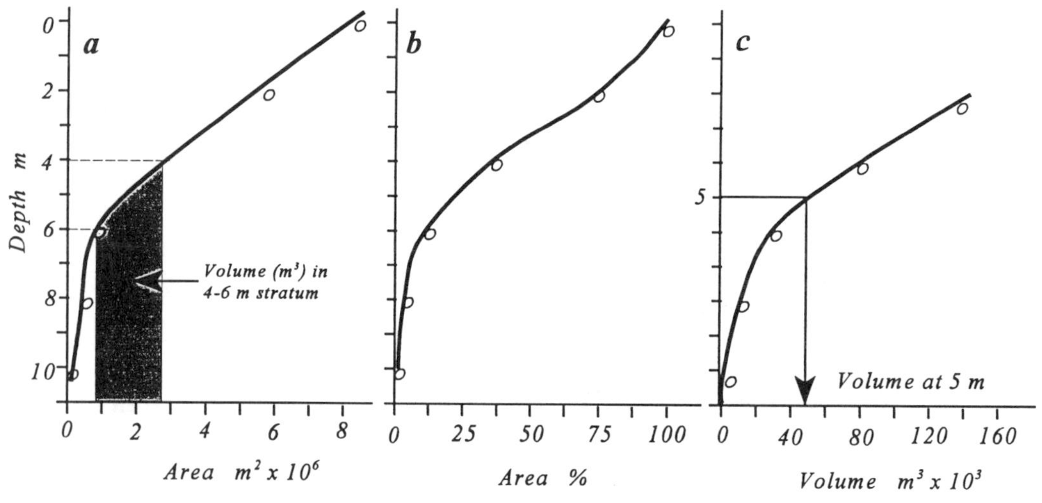

Figure 3.8. Hypsographic curves (a and b) using depth and absolute area (a) and depth and percent area (b), and a depth-volume curve (c). All data are from Golf Lake, as described in text.

York, lies 116 m above sea level and has a maximum depth of 133 m; by subtraction, it has a z_c of 17 m. The saline Dead Sea is entirely below sea level (surface is at -399 m AMSL), meaning its entire basin is a cryptodepression.

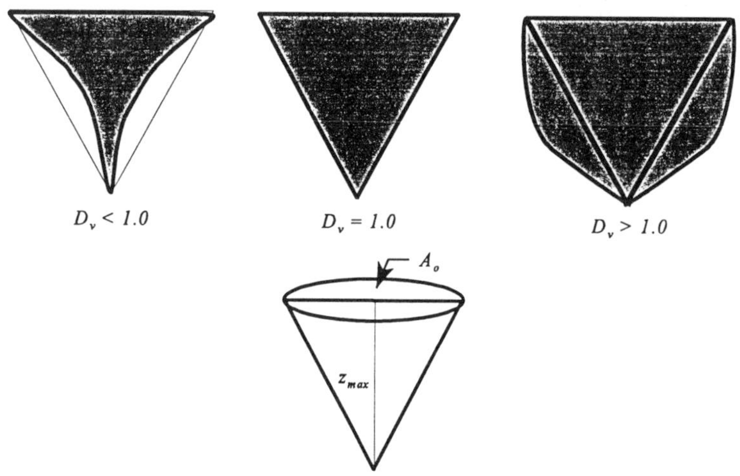

Figure 3.9. Volume development shapes relative to D_v values

CHAPTER 4

KNOW A LAKE'S VITAL SIGNS

Why Read This Chapter?

Every lake has its vital signs, or physical attributes, like wave height, that give it some measure of roughness or gentleness, of its ability to support a large diversity of life and of its utility for fishing, swimming, sailing, boating, diving, or just plain relaxation. In humans, vital signs include the heartbeat, temperature and respiration. In lakes, vital signs include its wave height and velocity, its currents, its temperature (or thermal) profile and its ability to transmit light (or water clarity). These can be measured, or at least estimated, fairly easily and will inspire confidence in one's use of the lake. With these measurements one will know when and where to avoid its potential to sink water crafts, drown even the best of swimmers, or alter or destroy shorelines, break walls or boathouses.

The chapter begins by examining the radiant energy, light wave characteristics, and the dynamics of light waves in the visible part of the light spectrum as they penetrate the atmosphere and the water column. It examines the role of light in heating the lake and its influence on the annual temperature cycle that causes lakes to stratify in the summer, with warmer water on top and colder water at the bottom, and then mix in the spring and fall. The chapter also examines the development of seasonal stratification and how to use this information for finding different species of fish in different parts of the lake at different times of the year. Some lakes mix at all depths in the spring and/or fall, others are poorly mixed, and some do not mix at all. One can classify a lake into one of six mixing types and examine the importance of lake classification to the lake's fisheries and recreational potential.

Light waves impart a colour to water, but the kinds of materials dissolved and suspended in the water will modify the apparent colour. A lake's apparent colour can be more than just a beauty mark; it can inform us about the relative health of a lake. In fact, one can learn a lot about the light characteristics of a lake with a simple device called a ***Secchi disc***, and it can be made quickly and cheaply. The role of light in providing energy for the growth of aquatic plants is examined in Chapter 6.

Most of a lake's vital signs can be estimated from measurements that were described in Chapter 3 for determining a lake's morphometric features. These measurements include the lake's mean and maximum length, its maximum fetch and its mean and maximum depth. These metrics and others are used in this chapter for describing some of a lake's vital signs.

By the end of this chapter you will be able to:

☞ 1. Understand what happens to light as it penetrates the atmosphere
☞ 2. Explain how winds are created and what determines their patterns
☞ 3. Recognize different light wave characteristics
☞ 4. Explain why the sky is blue
☞ 5. Explain what happens to light at the lake's surface
☞ 6. Explain why submerged objects appear to be bent from our angle of direct view
☞ 7. Explain what happens to light in the water column
☞ 8. Understand why lakes have different apparent colours and how you can use colour to assess the health of your lake
☞ 9. Understand differences between true colour and apparent colour
☞ 10. Determine the water clarity of your lake using a Secchi disc
☞ 11. Recognize the different parts of the seasonal thermal stratification cycle
☞ 12. Determine the summer thermal profile of a lake
☞ 13. Determine the relative location of fish whose distribution is determined by temperature
☞ 14. Begin to evaluate the trophic (enrichment) status of a lake
☞ 15 Recognize the importance of water currents and the different types of currents
☞ 16 Differentiate between laminar and turbulent flow
☞ 17. Recognize convection currents
☞ 18. Recognize the influence of Hadley, Ferrel and Polar air cells on wind and weather patterns for a lake
☞ 19. Recognize different types of surface waves in a lake, including travelling surface waves called ripples and capillary waves
☞ 20. Estimate maximum wave height at a beach
☞ 21. Differentiate between plunging and spilling breakers
☞ 22. Understand Coriolis forces and know why water flows clockwise or counter clockwise down a sink and toilet
☞ 23. Explain the formation of Langmuir streaks and currents in a lake
☞ 24. Recognize whole lake standing wave movements, including:
    (a) Surface seiches
    (b) Internal seiches
    (c) Kelvin waves
    (d) Poincare′ waves
    (e) Ekman Spiral and Transport
☞ 25. Recognize thermal bars
☞ 26. Recognize different flow patterns of influent water

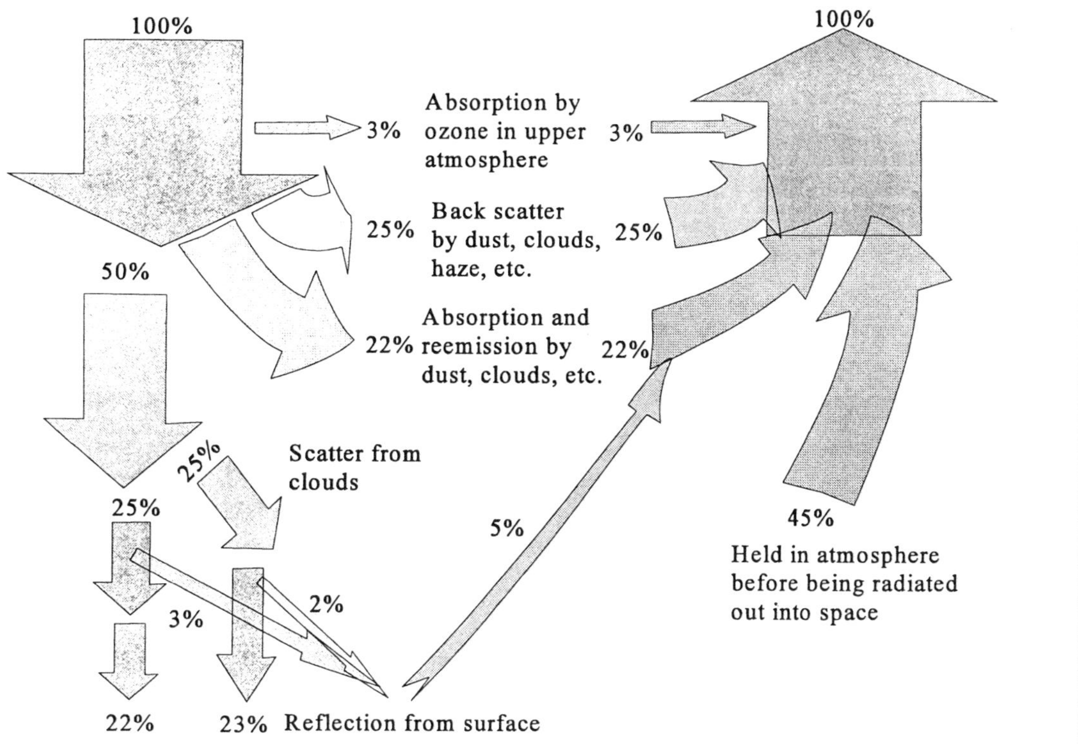

Figure 4.1. Pathways of short wave solar radiation showing reflection, scattering, absorption and transmission of light as it passes and returns through the atmosphere.

Light Energy and its Role in Heating Lakes

Light Dynamics in the Atmosphere

We will first examine the characteristics of radiant energy and the distribution of light and heat in the atmosphere. The ***solar constant*** quantifies the amount of energy
1☞ emitted from the sun. It varies somewhat but is presently about 1.94 calories per square centimetre per minute. This is an enormous quantity of energy and if it all reached the plants on land and in the water, the plants would be "fried". However, much of this energy does not reach the photosynthesizing organisms because the quality and quantity of light is significantly altered as it passes through the atmosphere and water. In fact, more than half the energy is lost before it reaches the surface of the lake (Fig. 4.1). This loss occurs because of ***scattering***, ***reflection***, and ***absorption*** by molecules of carbon dioxide, ozone and water. The remainder is ***transmitted*** to the lake surface.

It is the solar radiation that provides the heat that contributes to the world's wind
2☞ patterns. The current patterns are ultimately created by four forces; ***solar radiation***, ***wind***, ***gravity***, and ***geostrophic effects***. The geostrophic effects are due to the rotation of the earth. We will examine the effects of the earth's rotation on wind patterns later in this

chapter, but for now it is important to know that solar energy is a precursor for substantial mixing powers in lakes and oceans.

There are three important characteristics of light. One is its ***intensity***, which is the number of ***quanta*** or ***photons*** passing through a unit of given area. The second is its
3☞ ***wavelength***, usually abbreviated by lamda (λ). Wavelength defines the colour of the light beam. The third characteristic is ***frequency***, or the penetration power of the light wave. The frequency, ν, of a light beam can be calculated simply by dividing the ***speed of light***, C, by its wavelength, λ, or:

$$\nu = C/\lambda$$

Since we know that the speed of light is 3 X 10^8 m/sec, the frequency is simply:

$$\nu = 3 \text{ X } 10^8/\lambda, \text{ the units being cycles/sec}$$

Figure 4.2 shows the visible part of the wave spectrum to be near 400-700 nanometres (nm). This is also the range of ***photosynthetically active radiation***, or ***PAR***, which is the light that is useable by most aquatic plants. The figure also shows an inverse
4☞ relationship between wavelength and frequency. Ultraviolet (UV) light, at the left end of the light spectrum, has the shortest wavelength and highest frequency and infrared light, at the right end of the spectrum, has the longest wavelength and shortest frequency. Light with short wavelengths and high frequencies is scattered more than light with long wavelengths and short frequencies. The amount of scattering is a function of the fourth

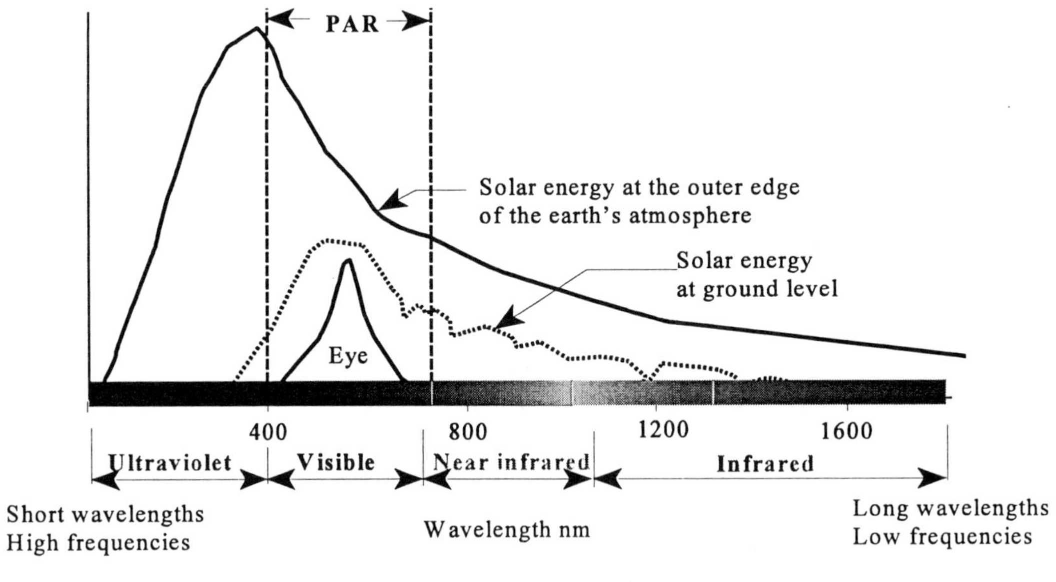

Figure 4.2. Solar energy at the outer edge of the earth's atmosphere and at ground level. Portion of spectrum visible to naked eye ranges from about 400 to 700 nm, as does PAR.

power of the frequency of light (i.e. v^4) so blue and ultraviolet light are scattered the most. Skies appear blue because blue and UV light has high frequencies and are scattered more than red or infrared light with low frequencies.

As light passes through the atmosphere there is selective absorption of wavelengths by gasses and particles of water, carbon dioxide, and ozone. The increase in CO_2, particularly in industrialized areas, has caused an increase in absorption of infrared light. It is this phenomenon that will contribute to global warming over the next century. If this occurs, dramatic changes in seasonal temperatures, sea level fluctuations and precipitation intensity and frequency will be observed. Since ozone gas absorbs UV light, there is scientific evidence that ozone concentrations in the upper atmosphere will decline due to reaction with atmospheric pollutants, such as fluorocarbons.

Light Dynamics at the Lake Surface (Albedo)

The amount of light and the quality of light impinging on the lake surface
5☞ depends on several factors. First, the quality of the air, which includes the amount of clouds, dust, and fog, will determine which wavelengths and how much will be absorbed or scattered, before it reaches the lake surface. The angle of incidence of the sun to the lake's surface is also a major factor, larger angles contributing to greater reflection of light off a lake surface, and less absorption, than smaller angles. The light that is reflected from the surface is known as ***albedo***. Albedo also varies with seasons. The earth orbits the sun and moves further away from the sun in its orbital path during the winter months. The greater the angle of incidence of the sun on the lake surface, the greater the albedo. Albedo is least during the summer when the earth, in its orbit, is closest to the sun. Latitude also affects albedo. Since the earth is tilted on its axis (23° 47'), locations on the globe that are tilted toward the sun will get more heat than locations tilted away from the sun .

There are daily variations as well. The greatest intensity of light occurs when the sun is directly overhead, or at its ***zenith*** position. Although a zenith sun provides the greatest penetration of light into a lake, the smoothness of the lake's surface is also important, more light being reflected from a rough surface than from a smooth surface. Any light that does impinge on a lake's surface is of two kinds, ***direct light*** and ***indirect light***, or ***diffuse light***. Indirect light is light that has been scattered and reflected off particles in the atmosphere, off mountains, and off clouds. Indirect radiation accounts for about 20% of the total light impinging on a lake's surface. About half of the solar energy coming from space is absorbed at the earth's surface and is converted into heat. This light energy is short wave (and high frequency) energy. Heat leaves the earth as infrared, or long wave, radiation. Since energy cannot be created or destroyed, the amount of energy arriving at the earth's surface must equal the amount of energy leaving the earth's surface. This is depicted in Fig. 4.2, where 51% of the energy is absorbed by water and land and 51% leaves the surface as long wave radiation.

Dynamics of Light in the Water Column

As light impinges on the surface of the water it is ***refracted***, ***transmitted***,
scattered, and ***absorbed***. Light that is absorbed and moves from one medium into
another with different optical densities (e.g. from air into water) is bent by ***refraction***.
6☞ This explains why submerged objects appear to be bent from the angle of direct view.
The angle of refraction is higher for shorter wavelengths than for longer wavelengths.
Hence, blue light is bent more than red light. Just as in the atmosphere, light is scattered
and absorbed or refracted by particles in the water. In fact, if the water did not have any
suspended particles, we would not see any light and water would appear black.
Remember, it is the suspended particles that scatter the light passing through the water
and air. Most lakes will appear blue for the same reasons that the sky appears blue; that
is, the amount of scattering in the water is also a function of the fourth power of the
frequency of light.

Light intensity decreases exponentially with depth of water. The decrease in light
intensity can be expressed by the ***extinction coefficient***, or the fraction of light absorbed
7☞ per metre of water. The higher the extinction coefficient, the lower the transmission of
light, or the less transparent the water. Algae constitute some of the particles in the
water column and absorb light as it penetrates the water. Only about half of the
underwater spectrum of transmitted light can be used for photosynthesis. This fraction is
known as PAR, discussed earlier. The fraction of PAR absorbed by algae can vary
greatly, from 2 - 60%. Each species of alga has an optimal PAR range. Some species of
algae occur near the surface of the water, others just below the surface and use those parts
of the light spectrum that provide optimal photosynthesis for the species.

Seasonal variations in PAR also occur. There is less PAR in autumn and winter than in summer, but lesser amounts of light are sufficient for photosynthesis by algae that predominate during the winter months. Indeed, some light intensities can inhibit photosynthesis. The most damaging wavelengths are called ***UVD***. They occur between 290 and 320 nm and can destroy DNA and ***chloroplasts*** (packets of chlorophyll). In general, not many species of algae are found at the surface of the water because of the damaging wavelengths present. As light penetrates lake water, there is selective absorption, this absorption being most pronounced at the red end of the light spectrum. That is, the reds and infrareds are absorbed first and, therefore, penetrate the least. In deep lake water, only blue and violet light remains at 100 m. For pure water, the order of increasing extinction, or decreasing light transmission, is blue, green, UV, red and infrared (i.e. blue transmits the most, infrared the least). However, this order changes if there is dissolved or suspended materials in the water, as in most lakes (Fig. 4.3). With increasing amounts of dissolved and suspended materials, there is decreasing amounts of blues and greens and increasing amounts of UVs. Black results when all the other colours are absorbed (i.e. filtered out)

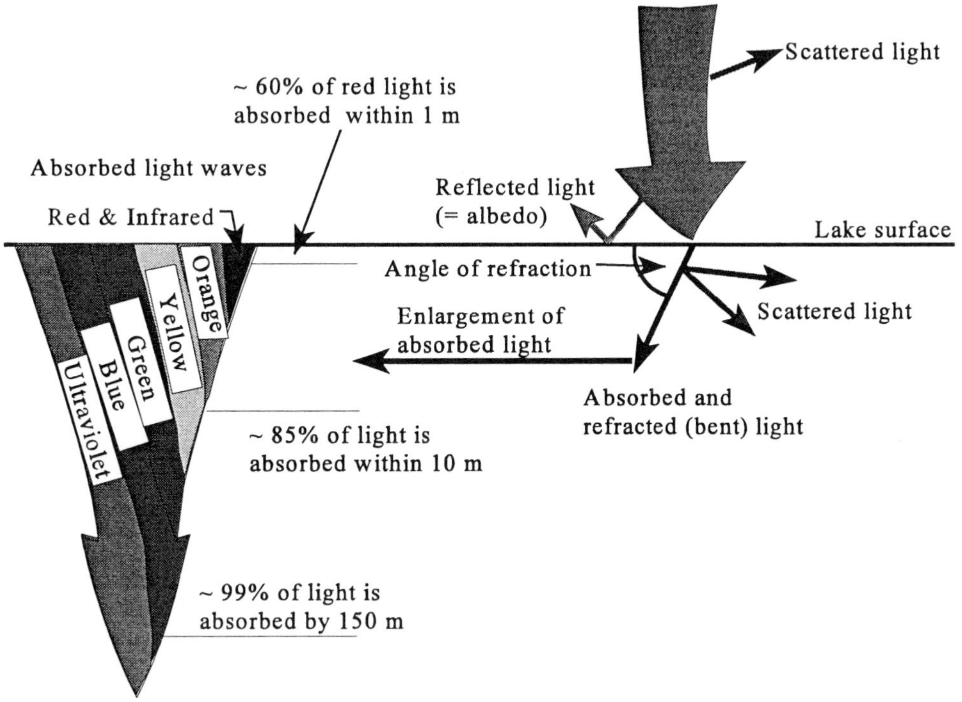

Figure 4.3. Effects of light waves as they pass through the atmosphere, and penetrate the water surface. Selective absorption of light in clear water is shown at left., with decay of waves at the red and yellow end of the spectrum and the violet and blue waves penetrating the deepest. Depth is not drawn to scale

Colour

The quality of material suspended in the water affects the transmission of certain
8☞ light waves. For example, lakes with an abundance of blue-green algae have a green or turquoise hew. Blue-green algae absorb light of a complimentary colour, that is, red and orange, and transmit green and blue-green colours which are reflected upwards from the bottom, or scattered by any suspended particles, and a blue-green colour will be perceived by the viewer. Ultra-clear lakes should appear bluish because there are few particles to absorb the light; also recall that light waves at the left end of the spectrum (which includes blue) is scattered the most because of their high frequencies. They also penetrate the deepest in clear lakes. This explains why blue light is the dominant colour scattered back to the surface in transparent lakes. Lakes which contain considerable amounts of reddish brown organic material absorb blue light and transmit red light. Hence, enriched lakes appear brownish to the viewer.

Since the colour of the lake bottom often affects our perception of the water's colour, much of the colour perceived is ***"apparent colour"***. For deep lakes, apparent colour can be a useful way to assess the ***"trophic status"***, or enrichment status of a lake, discussed in detail in Chapter 5. In general, black or bluish lakes indicate clear, oligotrophic (poorly enriched) lakes (i.e. they are good candidates for supporting lake trout populations!); brownish lakes indicate eutrophic (enriched) lakes (i.e. they are poor lakes for a lake trout fisheries, but great for perch, bass, and catfish).

True colour is determined on a water sample that previously has been filtered or centrifuged to remove suspended particles (which impart the apparent colour). The water sample is then compared to a series of colour standards. The true colour eliminates the
9 effects of the colour of the bottom of a lake, the colour of the sky, the colour of the suspended particles and other factors. Unfortunately, true colour has not been correlated with trophic status of a lake.

Secchi Depth

The ***Secchi disc*** is, arguably, the most useful device in freshwater studies of water
10 quality. It is a disc, usually 20 cm in diameter, with black and white quadrats (Fig. 4.4). Some scientists use solid white discs, but the most common is the black and white quadrat disc. An eye bolt is fastened to the centre of the disc and a rope, or line of some sort, that is clearly marked off in metres, or half metres, is attached to it. The discs are easy to make. The black and white quadrats may be painted on, but some cottage associations supply a decal, at little or no cost, that can be stuck to the top surface of the disc.

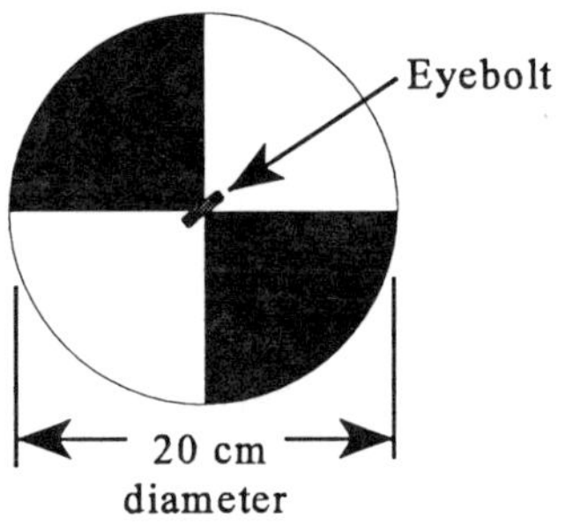

Figurer 4.4. Secchi disc.

The use of the Secchi disc is quite simple. Merely lower the disc into the water until the white quadrats disappear from view. Note the depth on the rope. Lower the disc a bit further, then raise the disc until it re-appears. Note the depth again. The depth at which it disappeared will probably be different from the depth at which it re-appeared. Repeat the procedure until the depth at which the disc disappears is the same as the depth at which it re-appears. This is the ***Secchi depth***.

The Secchi depth is a measure of the relative depth of light penetration. It is a "relative" measurement because the readings usually vary from one individual to the next, from one day to the next, and even from one hour to the next. Factors such as cloud cover, surface water conditions (smooth vs rough), and whether you take the measurement from the shaded side or the sunlit side of the boat will affect the measurement. Normally, the measurement is taken at a similar time of day from the

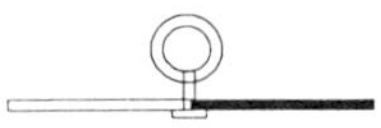

shaded side of the boat. Yet, in spite of all the these variables, there is an excellent correlation between Secchi depth and trophic status of a lake. In fact, the Secchi depth is used in later chapters to predict several physical, chemical and biological characteristics of a lake. ***The Secchi disc symbol, shown in the left margin, will appear in the margin every time it is related to some measure of trophic status.***

It is time to elaborate further on the term, "trophic status". Lakes are classified as either ***oligotrophic***, ***mesotrophic***, or ***eutrophic***, depending on the amount of nutrients contained. Each is a trophic state. Oligotrophic lakes have little or few nutrients ("oligo" meaning poor; "troph" meaning food or nutrients, hence poor nutrient levels). They are generally deep (usually U-shaped basins), cold lakes with close to 100% oxygen saturation at all depths. Most northern glacial lakes are oligotrophic and support healthy lake trout populations. Eutrophic lakes are highly enriched lakes ("eu" meaning true; "troph" meaning food or nutrients, hence truly enriched). Such lakes are often shallow (usually v-shaped basins) and warm and have anoxic (no oxygen) waters in their deepest portion. Fish are restricted to the upper, warm water layers where there is oxygen present. Only fish that can tolerate lower oxygen levels and warmer temperatures are found in these lakes. Mesotrophic lakes are those that are moderately enriched, between oligotrophy and eutrophy. Most lakes are probably mesotrophic. They support a diversity of fish populations, like oligotrophic lakes, but the salmonid (e.g. trout) populations are small or absent because oxygen levels are low in the deepest portions of the lake. Other physical, chemical and biological characteristics will be applied to the three trophic levels in most of the remaining chapters.

Now, for the first application of the Secchi disc to trophic status, scientists have found that:

Secchi depths greater than 5 m are characteristic of **oligotrophic lakes**
Secchi depths less than 2 m are common for **eutrophic lakes**
Secchi depths between 2 and 5 m are characteristic of **mesotrophic lakes**

Table 4.1 gives some morphometric characteristics of three trophic states.

Table 4.1. Some diagnostic features of lakes of three trophic states. Only those features which have been discussed in the first three chapters, and to this point in this chapter, are listed.

Features	**Oligotrophic Lakes Secchi depth > 5 m**	**Mesotrophic Lakes Secchi depth 2- 5 m**	**Eutrophic Lakes Secchi depth < 2 m**
Lake basin shape	U-shaped	V-shaped	V-shaped
Depth	Very deep	Moderately shallow	Usually shallow
Colour	Black or blue	greenish or brownish	green or brown

The Annual Temperature Cycle of Lakes

Winter Thermal Stratification

Several concepts from Chapter 1 need to be addressed in order to understand the
11 annual temperature cycle. Perhaps the most important is the relationship between density and temperature; remember, water gets heavier as it is cooled, until 4 °C at which point

fresh water is the most dense. Water gets lighter with cooling below 4 °C, until at 0 °C when ice forms and floats. This results in a low density surface layer (under the ice) at 0 °C and sits on top of a dense (actually the densest) layer near 4 °C (Fig. 4.5a). Since cold water (0 °C) sits on top of warm(er) water (4 °C), the condition is known as a ***temperature inversion***, that is a condition different (inverted) from the normal one where warm water usually sits upon cold water. Temperature inversion is the typical pattern in winter months.

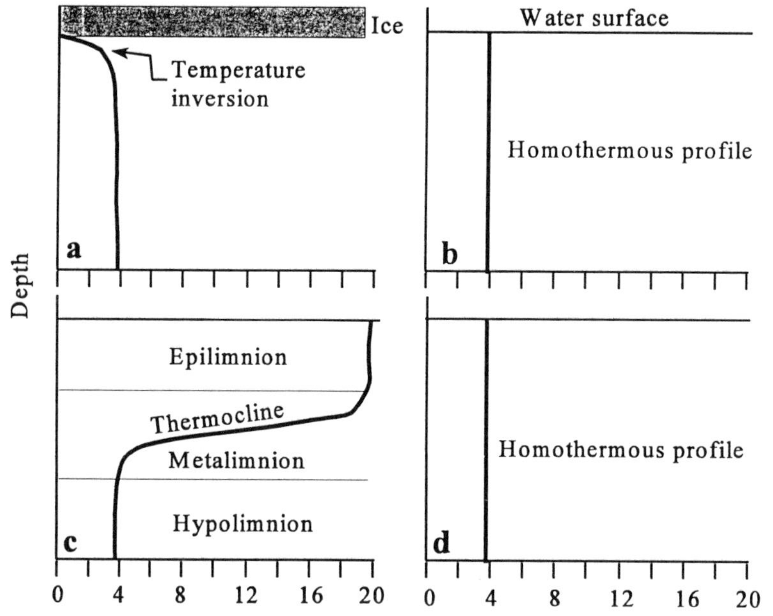

Figure 4.5. Annual temperature cycle showing, (a) winter thermal profile and temperature inversion, (b) spring turnover, (c) summer thermal stratification and (d) fall turnover.

Spring Turnover

When the earth, in its orbit, comes closer to the sun, the length of daylight hours begins to increase and the amount of solar radiation and heating at the earth's surface increases. The sun's rays melt the ice from the surface down, but the water temperature during ice melt remains close to 0 °C. When all the ice has melted, the surface layer of water is close to 0 °C. With increased heating, the surface water temperature gradually increases above 0 °C and becomes heavier until it reaches 4 °C. At this point the entire water column is ***homothermous*** (one temperature at all depths, Fig. 4.5b) and the lake is in its ***spring turnover*** state. With sufficient wind forces, water mixes throughout all depths.

Summer Thermal Stratification

The sun gradually comes to lie directly overhead (known as the Zenith position) at which time the surface layers of the lake are rapidly heated and become much warmer and lighter than the water below. Because water is heated from the surface down, only the upper one to two metres can be heated by the red and infrared light waves. The clearer the water, the deeper the red and infrared light can penetrate and the deeper the heating process can occur. But, as described earlier, nearly all the red light is absorbed in the first meter of water. The heated atmosphere generates winds which are needed to mix

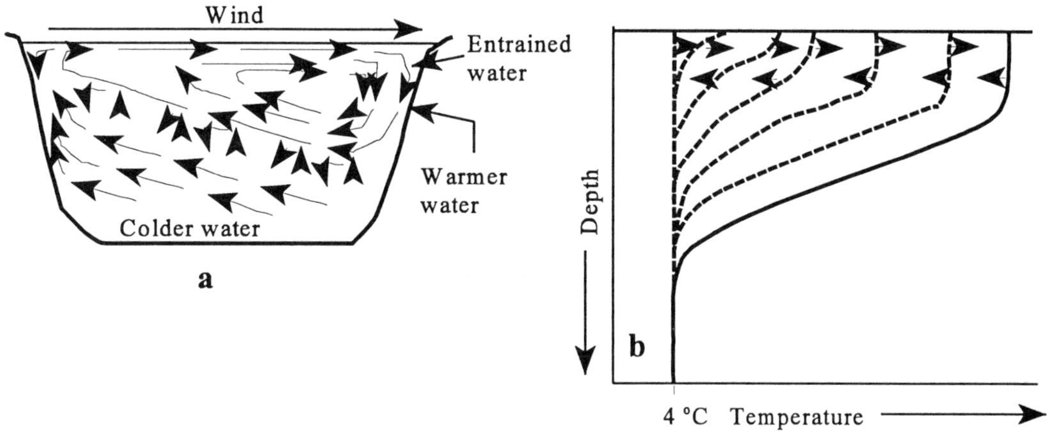

Figure 4.6. Role of wind in mixing warm water with cold water (a) and the progression of the thermocline downward over time as wind blows warm water downward (b). In (b), right arrows show formation of thermocline in summer, left arrows show decay of thermocline in the fall as water cools from the surface down.

the warmer, lighter water of the surface layers with the cooler, heavier water below. The stronger and more frequent the winds, the deeper the warm water can be mixed with the colder water below. The mixing action is strongest at the windward end of the lake, where water is piled up and warm water becomes entrained (mixed) with the cooler water (Fig. 4.6). The entrained, warm water flows under the surface layer, against the direction of the wind, to the leeward end of the lake. The warmer water rises to the surface as it flows across the lake to be reheated and re-circulated. Once the warm water is mixed below, it rises by convection because the warm water is lighter than the cold water beneath it. But at the leeward end, cold water is forced to the surface by the pressure of wind on the windward end of the lake (Fig. 4.6). The heating and mixing processes continue and eventually result in the formation of three thermal layers in the lake; an upper, warm layer called the ***epilimnion***, a lower, cold layer known as the ***hypolimnion*** and an intermediate layer known as the ***metalimnion*** (Figure 4.5c). The metalimnion is the zone of rapid change in temperature, more than 1 °C per metre advance in the water column. The ***thermocline*** refers to the temperature gradient. In some lakes, the metalimnion may be extremely thin, as thin as 8-10 mm and the thermal gradient is very steep. Such a thin metalimnion can occur only in lakes where there is sufficient continual heating to set up a thermal gradient that is very resistant to mixing. The thinner the metalimnion, the greater the ***thermal resistance to mixing***. The thermal resistance to mixing is sufficient to prevent mixing of water between the epilimnion and the hypolimnion. As a result, the water in the cold hypolimnion becomes isolated from the warm epilmnion during the summer months.

The temperature of the hypolimnion of deep lakes is usually close to 4 °C. Warmer hypolimnia occur only if the entire lake is mixed during the spring months and the water temperature of the entire water mass can rise to > 4 °C. This often happens in

Lake Erie. In such large lakes, the winds can be so intense during the spring months that the entire water column is mixed and heated. When the surface layers are heated more rapidly than the lower layers, the epilimnion begins to form and the hypolimnion become isolated at 6-7 °C, or whatever the temperature was before the waters began to stratify. Thermal stratification during the summer is referred to as ***direct stratification*** because warm water lies on top of the cold water.

Fall turnover

The air gets colder as the earth in its orbit moves further from the sun, cooling the surface water and making it more dense. The denser water becomes heavier and sinks, gradually eroding the epilimnion and the thermocline (Fig. 4.6b, left arrows). The thickness of the metalimnion increases as the epilimnion is eroded. Eventually, the stratified thermal layers disappear, the entire destratification process occurring within the epilimnion and the metalimnion. The hypolimnial waters do not become part of the destratification process until the water above it approaches 4 °C. At this point, the lake is in its fall turnover period and water mixes at all depths.

Lakes with both a spring and fall turnover are called ***dimictic*** (two mixes). Some lakes do no freeze and the fall turnover period extends right through to the spring. In these lakes the water temperatures do not fall below 4 °C and are called ***warm monomictic*** (one mix) lakes. Lake Superior is a warm monomictic lake. Lake Ontario may be dimictic one year and warm monomictic another. Most lakes in tropical climates are warm monomictic lakes. Lakes in polar regions that are permanently frozen and the water never mixes are ***amictic*** (no mix). The ice on some northern lakes may thaw but water temperatures never exceed 4 °C and are called ***cold monomictic*** lakes. Tropical lakes that are permanently stratified and rarely mix are ***oligomictic*** (poorly mixed). Most lakes at high altitudes near the equator are poorly stratified because they are constantly mixing; such lakes are ***polymictic***.

Determining the Summer Thermal Profile of a Lake

Temperature meters are used to measure water temperature. The meters come with a detachable thermocouple, or temperature probe. However, to determine thermal profiles, a long thermocouple line (with thermocouple on the end) is needed. The length
12☞ of the thermocouple line needed depends on the maximum depth of the lake or the depth at the top of the hypolimnion, whichever is less, but usually a 15-m line is sufficient. Once the 4°C hypolimnion is penetrated, the water temperature remains constant. The top of the hypolimnion in most lakes is rarely lower than 15 m (~50 ft) from the surface. The thermocouple line is lowered into the lake and the temperature is recorded at 1-metre intervals. Some of the newest (and more expensive!) models of fish finders also have the ability to detect the depth of the thermocline.

Alternatively, and a much less expensive method, is to measure the temperature of water obtained from different depths with a small water sampler (Fig. 4.7). The samplers are available in 1-, 2-, 5- and 10-litre sizes. The sampler is a P.V.C (polyvinyl chlorine) cylinder with hinged doors at the top and bottom. A rope is attached to a handle so that the cylinder can be lowered to any depth. As the cylinder is lowered, the water forces the lower door to open into the cylinder and the top door to open out of the cylinder. When the cylinder is hauled back up, the water pressure pushes the hinged doors closed on their seals and water is trapped inside. By lifting the top door, a thermometer can be placed inside the cylinder to measure the water temperature. This should be done as soon as possible, before the water temperature changes. A valve at the bottom of the sampler also allows one to drain the water from the cylinder.

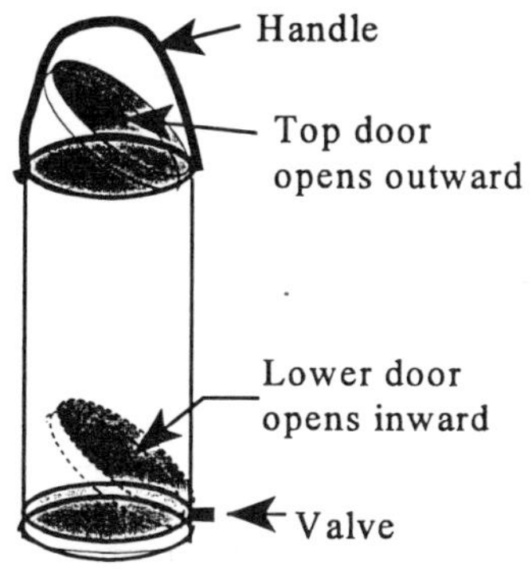

Figure 4.7. A student water sampler.

The Importance of Knowing the Summer Thermal Profile of a Lake

So, why is the shape of the summer thermal profile important? Canadian
13 scientists, like Dr. D. Rawson, in the 1940s and 1950s, discovered that the volume of the hypolimnion and the depth of the thermocline were related to the trout fisheries potential of the lake. Lakes with a large hypolimnial volume and a high thermocline had an abundance of lake trout and whitefish; lakes with a small hypolimnial volume and a deep thermocline had no trout or whitefish. Salmonids, like lake trout, and closely related whitefish, need cold, well oxygenated water. The larger the hypolimnion, the colder (usually close to 4 °C) the water and the more lake trout and whitefish that are likely to be present. A high thermocline provides for a smaller volume of epilimnial water but a larger volume of hypolimnial water. The relative volume of the hypolimnion and the depth of the thermocline are useful criteria for assessing the trophic status of a lake; oligotrophic lakes tend to have a large hypolimnial volume with a water temperature near
14 4 °C and a high thermocline; eutrophic lakes have little or no hypolimnial volume, the temperature being above 4 °C when a hypolimnion was present, and a deep thermocline. These features can be added to the list of criteria developed above for the depth. Table 4.2 incorporates all the criteria discussed so far.

Water - A Medium In Perpetual Motion

Water is never at rest. There are vertical and horizontal currents that continuously move water and materials from one shore to the other and from one layer to the next. First, we will examine water motions on a small scale, such as molecular diffusion, eddy

Table 4.2 Some correlates between Secchi depth and lake morphometry, colour and thermal characteristics.

Features	**Oligotrophic Lakes Secchi depth > 5 m**	**Mesotrophic Lakes Secchi depth 2 - 5 m**	**Eutrophic Lakes Secchi depth < 2 m**
Lake basin shape	U-shaped	V-shaped	V-shaped
Depth	Very deep	Moderately shallow	Usually shallow
Colour	Black or blue	greenish or brownish	green or brown
Hypolimnial temperature	4 °C	about 4 °C, or slightly higher	usually > 4 °C
Hypolimnial volume	Large	Moderately small	Small to absent
Thermocline depth	High	Moderately low	Low to absent

motion, and turbulence and laminar flow. On a larger scale, and involving movements in the entire lake, are ***standing waves***, ***progressive waves***, and ***seiches*** (see-saw motions). Also examined are the effects caused by the rotation of the earth, called ***Coriolis forces***, that influence water movements, called ***geostrophic currents***, in oceans and large lakes. Geostrophic currents also affect weather patterns.

Water movements are vital for the distribution of the different kinds of energy, nutrients, dissolved gases, and even organisms. Lake currents can vary from near 0
15☞ cm/sec to more than 500 cm/sec, the latter typical of breaking waves on large lakes and oceans. Water currents may be either ***laminar*** or ***turbulent***. ***Laminar flow*** occurs when water particles slide smoothly past each other. ***Turbulent flow*** is random, chaotic movement of water particles around each other or any object in the water and is sufficient to create eddies. Flow often proceeds from laminar to turbulent, the onset of turbulence being predicted from the ***Reynolds Number, R_e***. R_e *is the ratio of inertial forces to viscous forces, or the stirring energy divided by the viscosity of the water*. The stirring energy can be estimated by multiplying the water current velocity by the depth or thickness of the water layer of interest. The viscosity of the water at different temperatures is determined from tables in most physics books and all editions of
16☞ "Handbook of Chemistry and Physics", Published by The Chemical Rubber Publishing Company, Cleveland, Ohio. The viscosity of water decreases with increasing temperature. This means that cold water has more ***viscous drag*** than warm water and animals must work harder to swim in cold water than in warm water, or cold water flows more slowly than warm water. The unit of viscosity is centipoise (cP). Water has a viscosity of 1.79 cP at 0 °C, 1.52 cP at 5 °C, 1.31 cP at 10 °C, 1.14 cP at 15 °C, 1.01 at 20 °C, and 0.89 at 25 °C. The viscosity of water is exactly 1.0000 cP at 20.20 °C. A large R_e value implies high water velocity, a thick water layer, a low viscosity, or a combination of the three. Laminar flow exists when the Reynolds Number is less than 500; turbulent flow exists at Reynolds Numbers greater than 2000. Laminar flow will occur in lakes at depths less than 5 m and where water current velocity is less than 1 cm/sec. Most water

motions in lakes are turbulent. If laminar flow is present, it is probably in transition to a turbulent state.

Small currents, such as those generated by ***molecular diffusion*** and ***convection***, have two basic types of flow, ***mean flow*** and ***turbulent flow***. Mean flow has a specific direction and moves materials around the lake by ***advection***, or mass transport of water, typically by horizontal currents. Turbulence is random in direction and creates random, chaotic diffusion and spreading of patches of water in the form of horizontal eddies. Horizontal eddy diffusion is common near the surface of lakes and in mixed layers, such as in the metalimnion of lakes. Vertical eddy diffusion occurs in both the thermocline and the hypolimnion. Molecular diffusion occurs when water is heated and in ionic solutions. Thermal diffusion occurs in lakes where there are differences in temperature between two layers of water strata.

Convection currents are most pronounced in two layers of different density, such as that existing between the epilimnion and the hypolimnion. Convection can be
17☞ demonstrated by making one or two loops of a garden hose in an aquarium containing cool water (e.g. 10°C) and then running hot water through the hose (the hose must drain into a sink or floor drain, not into the aquarium). Soon afterwards, upward convection at the surface of the hose can be seen where aquarium water bathing the hose is heated and rises to the surface in swirls and eddies. Downward convection can be demonstrated just as easily by running cold water through the hose. The warm water bathing the surface of the hose is made colder and sinks in swirls, creating downward convection. The greater the difference in temperature of water in the hose and in the aquarium, the easier it is to see convection currents.

Wind, Weather, and Climate

The oceans cover approximately 75% of the earth's surface and account for about 98% of the water on the globe. Because the oceans are so massive, weather and climate will affect the intensity and direction of winds over both oceans and lakes. The intensity and direction of winds in turn greatly influence the intensity and direction of currents in the oceans and lakes. Inland, changing weather patterns can alter winds which in turn can modify the intensity and direction of currents in lakes.

Weather is the state of the atmosphere at a specific time and place. ***Climate*** is a long-term average of weather in a specific area. Forecasters are able to predict climate more accurately than weather because the factors that influence climate operate over larger areas and over longer periods of time. ***Wind*** is the mass movement of air and its patterns are dictated by variations in solar heating with latitude and season, as well as by the rotation of the earth. The inclination of the earth causes the change of seasons. Because the earth is tilted 23°27' on its rotational axis, relative to the plane of its orbit around the sun, there is uneven solar heating at different latitudes. For example, at mid-latitudes in the northern hemisphere there is about three times more solar energy per day in July than in December.

Convection currents are created by differences in temperatures between the equator and the polar regions. Considering the effects of uneven solar heating alone, we would expect the cool air to fall at the North Pole, flow southward along the surface of the earth towards the equator, where the air is heated, then rise into the atmosphere and flow back to the polar regions. But this is not what happens. The global circulation of air is also controlled by another factor, the ***geostrophic force*** caused by the rotation of the earth. The eastward (counter-clockwise) rotation of the earth on its axis deflects the air and water away from its initial course. This deflection is called the ***Coriolis effect.*** In the northern hemisphere the water and air are deflected about 15° to the right of their expected paths, creating clockwise currents. In the southern hemisphere they are deflected to the left, creating counter-clockwise currents. The Coriolis force is discussed further below.

The Coriolis force greatly affects the atmospheric circulation pattern, as shown in Fig. 4.8. Due to the Coriolis force, three atmospheric cells are created above and below
18☞ the equator. ***Hadley cells*** occur between 0° and 30° latitude and they create the trade winds (***northeasterly trades*** above the equator and ***southeasterly trades*** below the equator). Between 30° and the polar region are the ***Ferrell cells*** (northern and southern

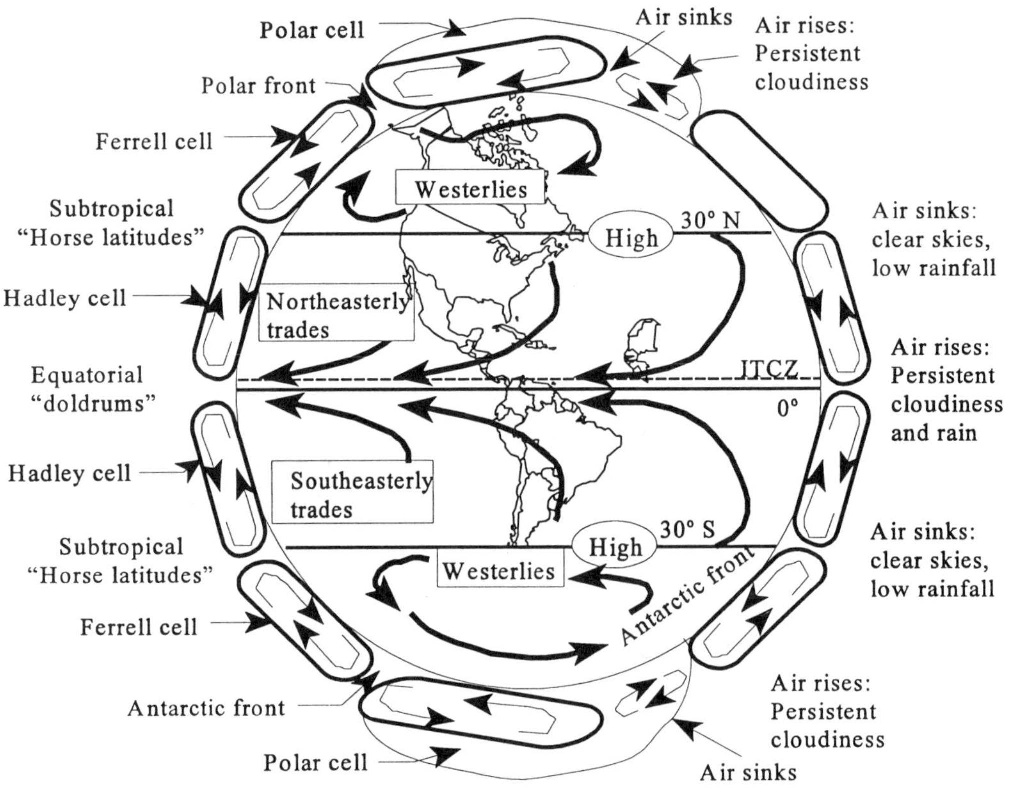

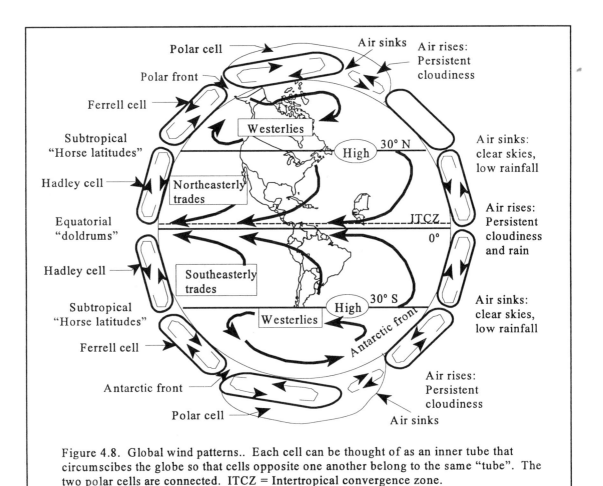

Figure 4.8. Global wind patterns.. Each cell can be thought of as an inner tube that circumscibes the globe so that cells opposite one another belong to the same "tube". The two polar cells are connected. ITCZ = Intertropical convergence zone.

hemispheres) that create the north and south westerlies. ***Polar cells*** occur at the polar regions and they create the polar fronts. Air rises at the equator between the two Hadley cells and then falls as it approaches the Ferrel cells at the ***Horse latitudes*** (Fig. 4.8). In the northern hemisphere, the Coriolis force creates a clockwise circulation of the winds in the Hadley cell, a counter-clockwise circulation of winds in the Ferrel cell and a clockwise circulation of air in the Polar cell. Conversely, in the southern hemisphere, the air mass flows counter-clockwise in the Hadley cell, clockwise in the Ferrel cell, and counter-clockwise in the Polar cell.

Knowing the wind patterns, one may determine its effects on a lake at any latitude. Remember, the predominant wind patterns on any lake is determined by the wind patterns at the latitude the lake is on. For example, the Great Lakes are near 45° latitude and the predominant flow of wind and water should be counter-clockwise (below the Ferrel cell).

Wind-Generated Waves

Wind will generate two kinds of currents or motions. One is ***periodic*** wave-like motion and the other is ***aperiodic***. Each type of wave has a characteristic ***length***, ***height***, ***period***, and ***frequency*** (Fig. 4.9). The wave length is measured between two adjacent crests or troughs. Wave height is the vertical distance between a crest and a trough. Wave amplitude is half the wave height. Wave period is the time required for the passage of two crests or two troughs past a point. The inverse of wave period is known as the wave frequency. Wave length and period are determined mainly by the ***fetch***, or the distance that wind can blow over water without interruption by land.

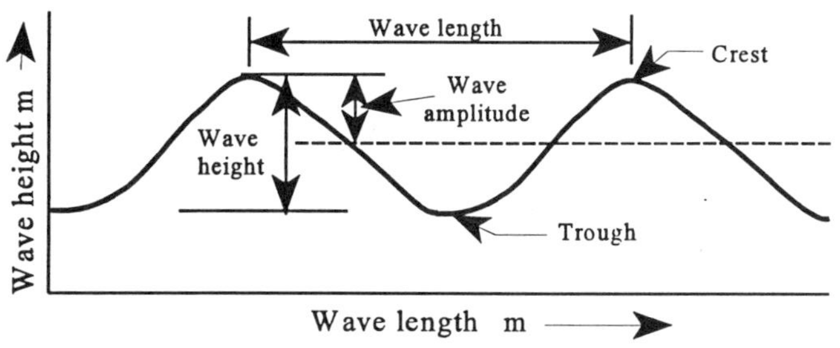

Figure 4.9. Travelling wave characteristics.

Water Movements at the Surface of the Lake

These include ***travelling surface waves*** (also called ***progressive waves***), ***surface currents*** and ***Langmuir circulation*** and ***Langmuir streaks***. Travelling surface waves are
19☞ the waves visible on the surface of lakes. They are created by the friction of the wind blowing over the lake surface. The wind sets the water surface into motion, producing ***wind-drift***. Although travelling surface waves are the most conspicuous water movements, they are of little limnological importance because their energy is confined primarily to the surface layer of the lake. Only the shore areas of lakes are significantly

affected by travelling surface waves. As wind blows over the surface, it causes water particles to move in a circular path, or orbit, such as that shown in Fig. 4.10. The cycloid path has very little significant horizontal motion. This can be demonstrated by placing a cork in the water and comparing its vertical movement to its horizontal movement; the cork has much greater vertical movement than horizontal movement, the horizontal displacement being very slight. At the lake surface, the diameter of the cycloid path of particles is equal to the wave height so the vertical movement of the cork represents the cycloid diameter. However, the vertical oscillation is attenuated rapidly with depth. *The decrease in vertical motion with increasing depth can be approximated by a halving of the cycloid diameter for every depth increase of* $\lambda/9$. For example, if a surface wave has a wave length of 18 m and a height of 1 m, the cycloid diameter (1 m) would be halved for every depth increase of 18/9 (= $\lambda/9$), or 2 m. Hence, the vertical oscillation at a depth of 2 m would be 1/2 x 1 m = 0.5 m; at 4 m it would be 1/2 x 0.50 m = 0.25 m; at 6 m it would be 1/2 x 0.25 m = 0.125 m, and so on.

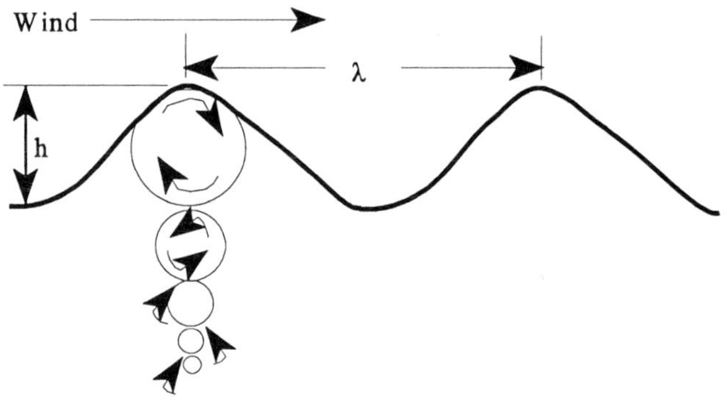

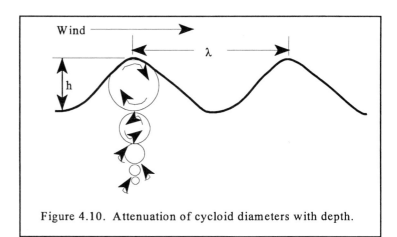

Figure 4.10. Attenuation of cycloid diameters with depth.

Short surface waves with a wave length greater than 2 x π = 6.28 cm are referred to as ***gravity waves***. Waves with a length less than this are known as ***ripples*** or ***capillary waves***. The height of the highest wave, h_{max}, observed on a lake can be approximated from the relation:

20 ☞

$$h_{max} = 0.105 \sqrt{\text{fetch}}\ \text{(cm)}$$

Note that fetch is in centimetres. For example, suppose a cottage is at the end of the lake, the greatest distance from the farthest shore (no islands in between) being 4 km (i.e. = fetch); the maximum wave height that one can expect on the cottage shore is 0.105 $\sqrt{400{,}000}$ cm = 66.4 cm (= 26.1 in or 2.2 ft!).

Table 4.3 provides wave height estimates for fetches at 0.1 km intervals. Use of the table or formula is restricted to ***deep water waves***, or ***short waves***, where the wave length is much less (usually 20 times) than the water depth. When the wave length is greater than 20 times the water depth the wave is transformed into a ***shallow water wave***, or a ***long wave*** and the cycloid movements are transformed into to-and-fro sloshing motions that extend to the bottom of the lake. The waves pick up sediments off the bottom and the water becomes turbid and abrasive. As the wave enters shallow-water

areas, the velocity decreases as the square root of depth decreases, and there is a corresponding reduction in wave length. Eventually water collapses over the front of the wave and becomes a ***breaker***. Breaker waves may be ***plunging***
21 ☞ ***breakers*** or ***spilling breakers*** (Fig. 4.11). In a plunging breaker the forward face of the wave becomes convex and the crest curls over and collapses underneath to complete a vortex. In a spilling breaker the crest collapses forward, spilling downward over the front of the wave as shown in Fig. 4.11.

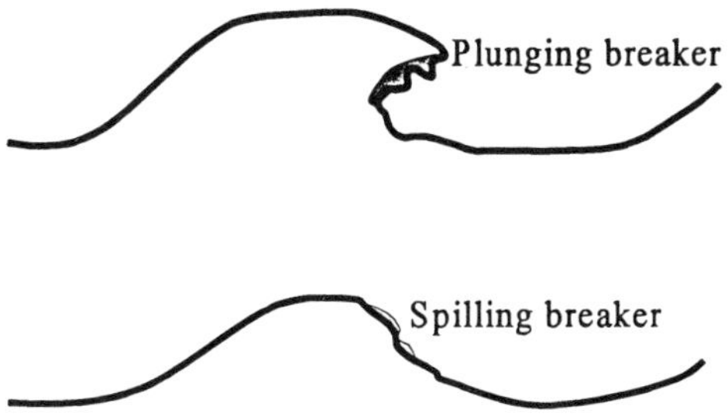

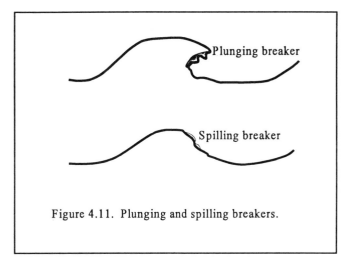

Figure 4.11. Plunging and spilling breakers.

Surface Currents

Surface currents are non-periodic water movements generated by one or more external forces such as wind, change of atmospheric pressure, density gradients caused by differential heating or by diffusion of dissolved materials from the sediments, and geostrophic forces caused by the counter clockwise rotation of the earth.

A Closer Look at the Coriolis Force

The most common geostrophic force is the Coriolis force (named after a french scientist, Gaspard Coriolis). Its effects are more pronounced in oceans than in lakes, but they can be seen in moderate to large-sized lakes as well (and perhaps even in your home, as described below!). Due to the Coriolis force, objects moving horizontally, such as surface water currents, are deflected to the right (relative to the direction of the wind) in the northern hemisphere (between 0° and 30° latitude) and to the left in the southern hemisphere (between 0° and 30° latitude). This sets up large gyres that flow clockwise in oceans and large lakes in the northern hemisphere and counter-clockwise in oceans and large lakes in the southern hemisphere.

Figure 4.12 demonstrates the Coriolis effect using the city of Guelph in the northern latitude and Quito at the equator. Both cities are in the same longitude, but Guelph has a smaller orbit around the earth than does Quito because the earth diameter is
22 ☞ smaller. Hence, Quito must travel faster around the equator to come back to the same position in 24 hours than does Guelph. In fact Quito moves at 1,660 km/hr whereas Guelph would move at 1,279 km/hr. Suppose we fire two rockets, one from Quito toward Guelph and one from Guelph toward Quito. The rocket from Quito will land off

Table 4.3. Theoretical maximum wave heights (metres) for lakes with fetches up to 30 km.

Even Km	Fractions of a kilometre									
	0	0.1	0.2	0.3	0.4	0.5	0.6	0.7	0.8	0.9
1	0.33	0.34	0.36	0.37	0.39	0.40	0.42	0.43	0.44	0.45
2	0.47	0.48	0.49	0.50	0.51	0.52	0.53	0.54	0.55	0.56
3	0.57	0.58	0.59	0.60	0.61	0.62	0.63	0.63	0.64	0.65
4	0.66	0.67	0.68	0.68	0.69	0.70	0.71	0.72	0.72	0.73
5	0.74	0.75	0.75	0.76	0.77	0.77	0.78	0.79	0.8	0.80
6	0.81	0.82	0.82	0.83	0.84	0.84	0.85	0.85	0.86	0.87
7	0.87	0.88	0.89	0.89	0.90	0.90	0.91	0.92	0.92	0.93
8	0.93	0.94	0.95	0.95	0.96	0.96	0.97	0.97	0.98	0.99
9	0.99	1.00	1.00	1.01	1.01	1.02	1.02	1.03	1.03	1.04
10	1.05	1.05	1.06	1.06	1.07	1.07	1.08	1.08	1.09	1.09
11	1.10	1.10	1.11	1.11	1.12	1.12	1.13	1.13	1.14	1.14
12	1.15	1.15	1.16	1.16	1.16	1.17	1.17	1.18	1.18	1.19
13	1.19	1.20	1.20	1.21	1.21	1.22	1.22	1.22	1.23	1.23
14	1.24	1.24	1.25	1.25	1.26	1.26	1.26	1.27	1.27	1.28
15	1.28	1.29	1.29	1.29	1.30	1.30	1.31	1.31	1.32	1.32
16	1.33	1.33	1.34	1.34	1.34	1.35	1.35	1.36	1.36	1.37
17	1.36	1.37	1.37	1.38	1.38	1.38	1.39	1.39	1.40	1.40
18	1.40	1.41	1.41	1.42	1.42	1.42	1.43	1.43	1.44	1.44
19	1.44	1.45	1.45	1.45	1.46	1.46	1.47	1.47	1.47	1.48
20	1.48	1.48	1.49	1.49	1.5	1.50	1.50	1.51	1.51	1.51
21	1.52	1.52	1.52	1.53	1.53	1.54	1.54	1.54	1.55	1.55
22	1.55	1.56	1.56	1.56	1.57	1.57	1.57	1.58	1.58	1.58
23	1.59	1.59	1.59	1.60	1.60	1.61	1.61	1.61	1.62	1.62
24	1.62	1.63	1.63	1.63	1.64	1.64	1.64	1.65	1.65	1.65
25	1.66	1.66	1.66	1.67	1.67	1.67	1.68	1.68	1.68	1.69
26	1.69	1.69	1.7	1.70	1.70	1.70	1.71	1.71	1.71	1.72
27	1.72	1.72	1.73	1.73	1.73	1.74	1.74	1.74	1.75	1.75
28	1.75	1.76	1.76	1.76	1.76	1.77	1.77	1.77	1.78	1.78

29	1.78	1.79	1.79	1.79	1.8	1.80	1.80	1.81	1.81	1.81
30	1.81	1.82	1.82	1.82	1.83	1.83	1.83	1.84	1.84	1.84

course (to the right, as viewed from Quito) because Guelph has not moved as much in its rotation of the earth as has Quito. This is the Coriolis effect. The rocket from Guelph would also miss Quito, again to the right (as viewed from Guelph). The Coriolis deflection is approximately 15° so the distance that the rocket would miss is roughly 400 km.

The flow patterns of gyres of water draining down sinks and toilets in your home is in part due to the Coriolis force created by the counter clockwise rotation of the earth. The gyres flow in a clockwise direction in sinks and toilets of the northern hemisphere,
23☞ between 0° and 30° latitude (below the north Hadley cell), and a counter clockwise direction in the southern hemisphere, between 0° and 30° latitude (below the south Hadley cell). In a house at 45° latitude, below a Ferrel cell, which has counter clockwise currents, water should flow counter clockwise in toilets and sinks. Which direction does the gyre move in sinks and toilets on the equator? Theoretically, straight down, without a gyre!

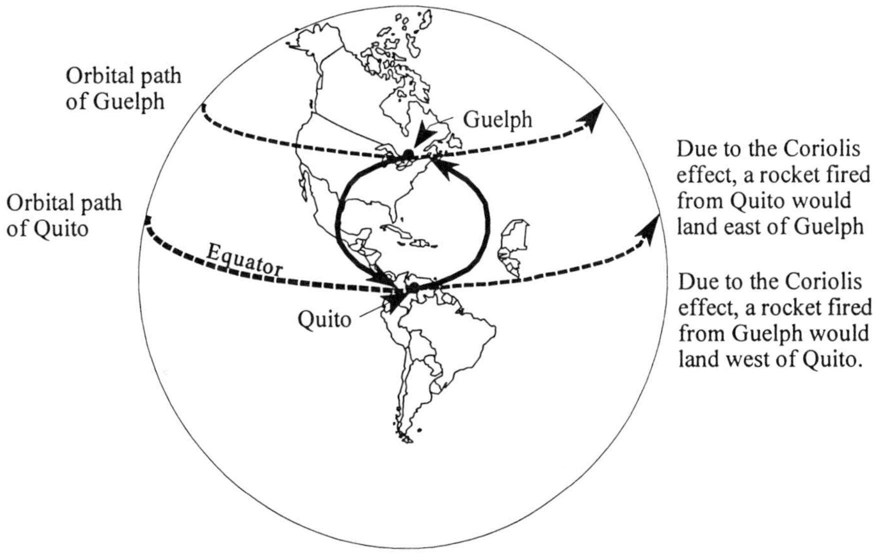

Figure 4.12. Effect of the Coriolis force on deflecting objects to the right in the northern hemisphere and to the left in the southern hemisphere.

Langmuir Circulation and Streaks

Under certain circumstances motions associated with turbulent transport can be organized into vertical, helical currents in the upper layers of lakes, such as that shown in Fig. 4.13. Convection from this vertical motion generates streaks, known as ***Langmuir streaks***, that are oriented in the same direction as the wind. The Langmuir streaks coincide with lines of surface ***convergence*** and downward movements of water particles. Suspended particles accumulate in the streaks making it denser than the water on either side. Upwelling occurs at the zones of ***divergence*** where the water particles move apart. The streaks are most commonly observed at very low wind speeds, between 2 and 5 cm/sec (0.045 - 0.11 mi/h). They are rarely observed at wind speeds above 7 cm/sec (0.16 mi/h).

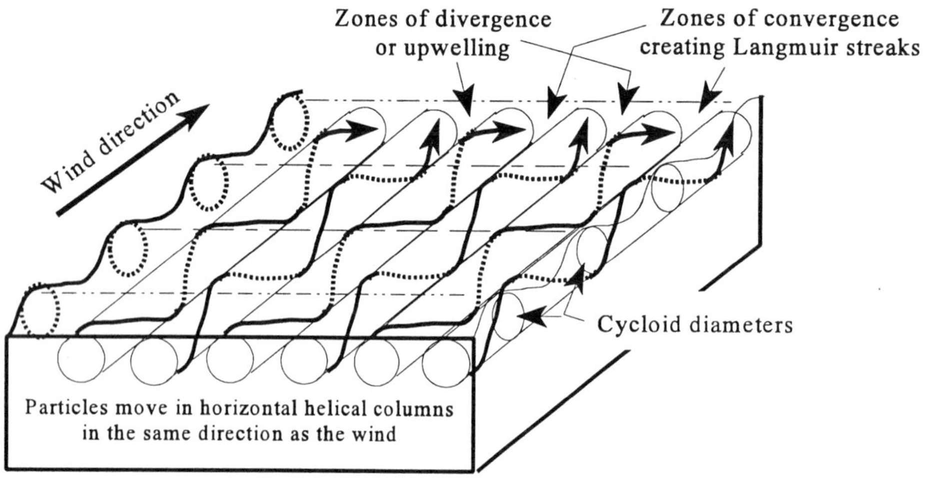

Figure 4.13. Diagramatic representation of movement of particles in horizontal helical columns to create Langmuir streaks at zones of convergence.

The mechanisms for generating Langmuir circulation involve wind, which sets the water particles into motion in their cycloid paths, and the Coriolis force, which converts the cycloid paths into horizontal helices. Because of friction, adjacent cycloid paths rotate in opposite directions, causing the helices to rotate in opposite directions. This creates alternating zones of convergence and divergence (Fig. 4-13). Downward velocities at convergences are about three times the upward velocities at divergences. These currents are sufficient to greatly influence the distribution of zooplankton and phytoplankton suspended in the surface waters of lakes, usually resulting in long, wide horizontal patches of plankton.

Water Movements Inside a Lake

While the movements occur internally, the forces that initiate them are largely external. Only the most common kinds of internal water movements are examined below.

Seiches

Both ***surface seiches*** and ***internal seiches*** are ***long standing waves***. These are waves with rhythmic motion and they have wavelengths of the same order as the length
24a☞ of the lake basin. The water surface at opposite ends of the lake oscillates up and down, like a teeter totter, about a line of no vertical motion at the ***node*** (Fig. 4.14). The to-and-fro horizontal motion is maximum at the node and minimum at the ***antinodes*** (ends of the lake). The term, ***seiche***, refers to the periodic exposure and drying period of shallow littoral zones.

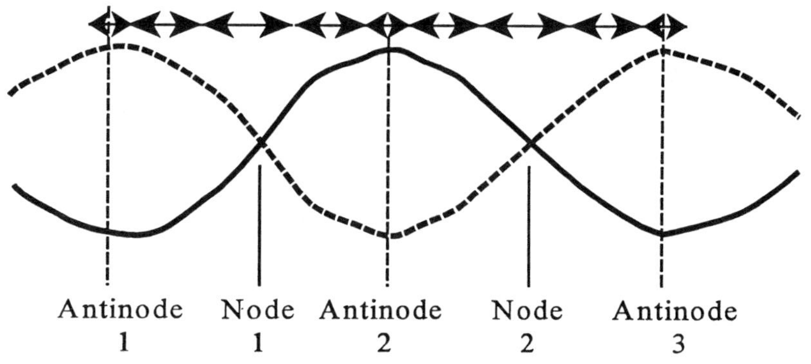

Figure 4.14. Characteristics of a long standing wave, with the relative magnitude of horizontal particle trajectories shown for one cycle.

The most common cause of seiches is wind-induced tilting of the surface waters and of the thermocline. This occurs by metalimnetic entrainment in the stratified layer of a lake. (Fig. 4.15). The wind blows across the surface of the lake, piling the water up at the windward end. This causes the warm water in the epilimnion to be piled up at one end of the lake. However, while this warm water piles up at one end, the cold water from the hypolimnion is forced up to replace the water at the other end of the lake. This sets up a tilted thermocline where cold water passes diagonally from the leeside of the lake down to the lower end of the epilimnion on the windward side (Fig. 4.15). When the wind stops blowing, gravity takes over and the lake begins to rock back and forth with a predictable periodicty. The water rises and falls (the drying events), usually less than a metre, at the ends of the lake. Most lakes have a seiche with a single node at the centre where the water rocks back and forth without changing height. Some large lakes may have more than one node. The period of a uninodal surface oscillation (in a near rectangular basin) is approximated by:

$$t = 2l/\sqrt{gz}$$

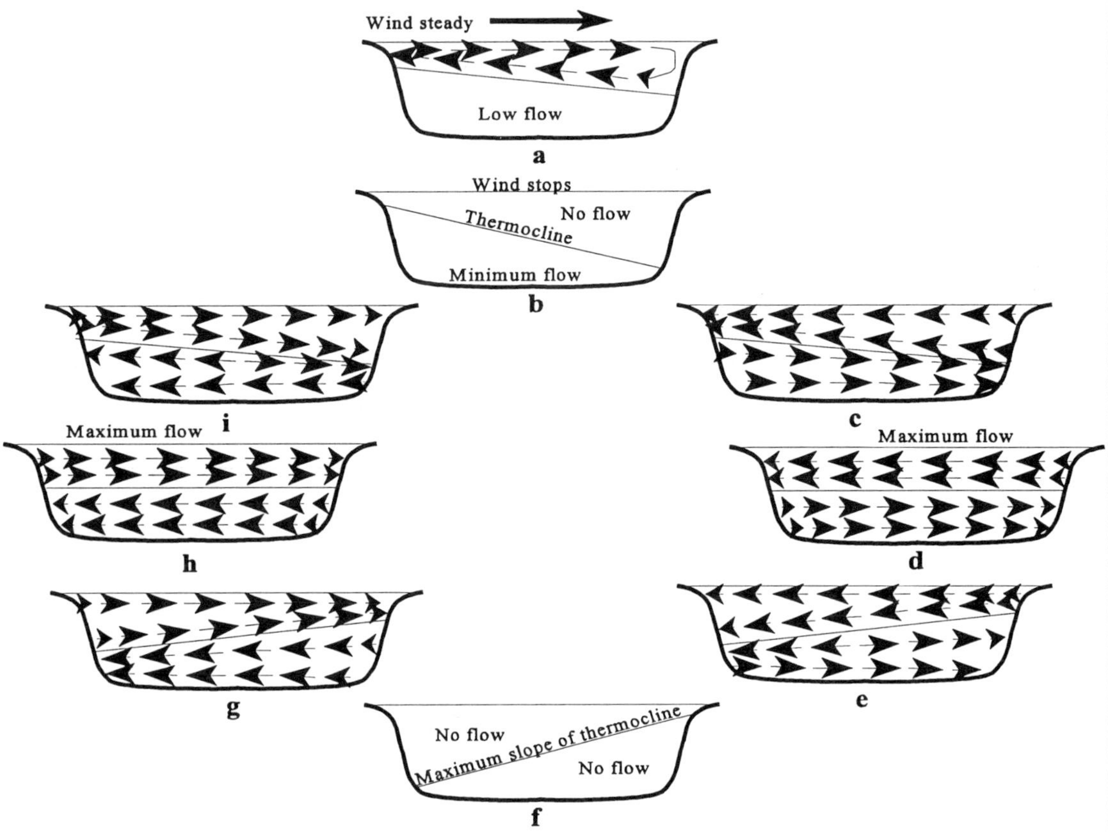

Figure 4.15. Wind stress (a) and its role in the development of an internal seiche after the wind stops (b). The thermocline becomes tilted (b) and water begins to swing back to equilibrium (c). Water attains its maximum flow when the thermocline returns to level (d), but momentum continues to move the water to the other end of the lake (e) until its maximum slope is attained (f). A reverse flow (g) occurs when the thermocline begins to swing back to equilibrium. At equilibrium (h), maximum flow is again attained but in the opposite direction to (d). Again, momentum carries the water back to its original state (i), and the process continues until the oscillating standing wave is completely dampened.

where, t = the period in seconds
l = the length of the lake basin at the surface (in cm)
z = the mean depth of the basin (in cm)
g = acceleration due to gravity = 980.7 cm/sec^2.

Once a surface seiche is set into motion, friction and gravity will dampen the oscillation and eventually the water mass returns to equilibrium until the next wind occurs (Fig. 4.15). The magnitude of the dampening process varies with water depth and complexity of the basin shape. In Lake Erie, surface seiche amplitudes can exceed 2 m and have a period of about 14 hours. Such high amplitudes can have destructive effects on the shore line. While seiches occur in most lakes, their amplitudes are so small that

24b☞

they are almost undetectable.

Internal Seiches and Kelvin Waves

Internal seiches involve layers of different densities and are of much greater limnological importance than are surface seiches. Because two density layers are involved in the oscillation, the period for the oscillation is more complex than that for surface seiches. In large lakes, internal seiches can be affected by the Coriolis force which sets up a ***Kelvin wave***. The Coriolis force induces a counter-clockwise rotation in the northern hemisphere (on the plain of the internal seiche) and a clockwise rotation in
24c☞ the southern hemisphere. In Kelvin waves, the thermocline not only rocks back and forth, but it tilts and rotates to the side in a back and forth fashion. This sets up wave amplitudes that increase exponentially toward the shore. The greatest effect of Kelvin waves is at the perimeter of the plane of the thermocline where water may leak from the epilimnion into the hypolimnion as the wave rotates around the shores of the lake. This is an effective way of bringing nutrients from the shore zone into the middle of the lake.

Poincaré Waves

Coriolis forces also produce ***Poincaré waves***. In these waves the wave amplitude
24d☞ does not decrease exponentially away from the shore as in Kelvin waves. Poincaré waves occur in the open water of large lakes and are transformed into Kelvin waves when encountering shore boundaries. Poincaré waves undulate in a standing wave pattern both across and along the channel or basin producing a cellular pattern of wave-associated currents.

Ekman Spirals

Ekman spirals are common in oceans and large lakes (> 100 m deep) that have large gyres. The top-most layer of water in oceans and large lakes between 0° and 30°
24e☞ latitudes in the northern hemisphere flows at about 45° to the right of the wind direction due to the Coriolis force. The layer of water immediately below the upper layer moves at an angle to the right of the overlying water. But the angle is less than 45° because of friction. The same happens in the layers below, each layer sliding horizontally over the one beneah it, and each layer moving at an angle slightly to the right of the one above it (Fig. 4.16). The resulting pattern creates an Ekman spiral, named after its Swedish discoverer. Actually, the water does not spiral downward in a whirlpool-like motion as depicted in Fig. 4.16. The spiral is only a way of conceptualizing the horizontal movements in a layered water column, each layer moving in a slightly different horizontal direction. However, the most significant effect of the Ekman spiral is that water at some depth will be flowing in the opposite direction from the surface current. The net movement of water in the upper 100 m of water is known as Ekman transport, the theoretically direction of which is 90° to the right of the wind direction. In fact, the 90° direction of Ekman transport is likely rarely achieved; it is probably closer to 45° because of the interaction of the Coriolis force and the force of gravity which forces particles to fall down the gyre.

The clockwise motion of water in the gyre of the northern hemisphere tends to pile the water up in a mound to form a low hill, the height of which is less than 1 m in oceans and even less in large lakes. The centre of the hill is offset to the west (in the northern hemisphere) due to the Coriolis force. Water coming down the right side of the hill due to gravity is deflected and carried westward as it flows along the side of the hill. The westward moving water is balanced then between the Coriolis force, which turns water to the right, and gravity, which tends to turn water to the left. Hence water in a gyre moves along the outside edge of an ocean or lake basin. The fastest currents in the gyres are those that flow westward (in the northern hemisphere), due to westward intensification created by the Coriolis force.

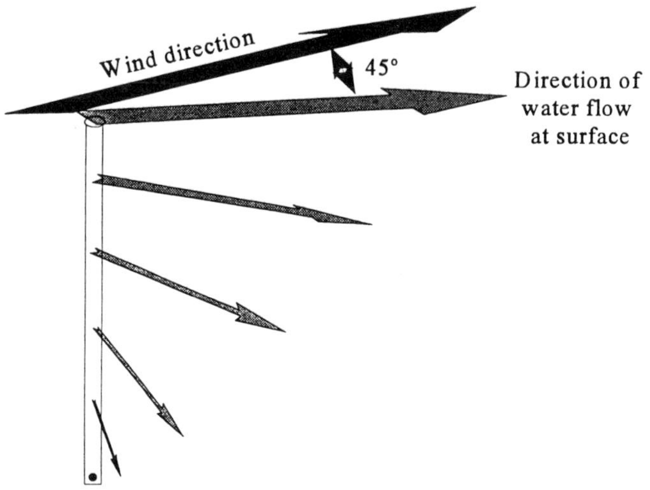

Figure 4.16. Idealized diagram of an Ekman spiral showing surface water deflected about 45° to the right (in northern hemisphere) of the wind direction. Net flow decreases and moves further to the right with increasing depth. Flow is nil when at right angles to the wind direction.

Thermal Bars and Their Effect on Circulation

Thermal bars are a feature of all lakes, but are more often seen in larger lakes. Thermal bars are the narrow transition zone consisting of a nearly vertical 4°C isotherm
25☞ that develops between the open water and stratified water in the shallow beach waters. The density gradient created by rapid heating of the shallow waters combines with the coriolis force to induce a counter clockwise shore current, inside the thermal bar.

Overflow, Underflow and Interflow Currents

These are currents created by river influents and are especially common in artificial impoundments and reservoirs, as discussed in Chapter 10. Rivers entering lakes
26☞ may create the same kinds of flow patterns. Overflow occurs when the density of the inflowing river water is less than the density of the receiving water, resulting in a surface layer of lighter water. Underflow occurs when the density of the river water is greater than that of the lake, so that the water flows close to the sediment, near the bottom of the lake. Interflow occurs when the density of the river and the lake are similar, so that the water from the river flows into the middle of the lake. The currents generated are from molecular diffusion and thermal diffusion.

CHAPTER 5

KNOWING A LAKE'S HEALTHY SIGNS

Why Read This Chapter?

The health of a lake is measured by its water quality. The quality of water in a lake depends to a large extent on the chemistry of the water that supplies the lake. Every lake has its own watershed, and the chemistry of the soil and bedrock in the watershed will largely (but not entirely) dictate the chemistry of the water flowing through it and over it. The chemistry may change over time, the rate and amount of change depending on several factors, such as type and amount of bedrock, climate, and the kinds of weathering processes, but the changes are somewhat predictable, as long as humans do not interfere with the process. In essence, if chemistry of the watershed is known, one should be able to predict the chemistry of a lake; or if the chemistry of a lake is known, one should be able to predict the chemistry of the watershed that supplies the lake with water.

In fact, even if humans do influence the health of the lake, enough is known about the effects of *most* contaminants and stressors that one can predict approximately the effects on the ecosystem. But this will be discussed in detail in Chapters 11 and 12 when aquatic toxicity and water quality assessment techniques are examined.

The trophic status of a lake is, essentially, the metabolic health of the lake. Water chemistry plays a significant role in determining a lake's metabolic health, or trophic status. This chapter examines the role of major ions (e.g. calcium, magnesium, and iron), nutrients (e.g. phosphorus, nitrogen, carbon, and sulphur), and gases (e.g. oxygen, carbon dioxide, and nitrogen) in influencing the trophic status of a lake. At least six more criteria (in addition to those from previous chapters) will be provided to help assess the trophic status of a lake.

In summary, by the end of this chapter you will:

☞ 1. Know the chemical units commonly used in aquatic sciences
☞ 2. Know the major sources of ions in freshwater
☞ 3. Know the major cations (+'ve ions) and anions (-'ve ions) in freshwater
☞ 4. Understand the importance of bedrock types in contributing certain ions
☞ 5. Know the importance of calcium to lake productivity
☞ 6. Know difference between allochthonous and autochthonous materials

☞ 7. Know the difference between and the importance of total dissolved solids and total suspended solids in water

☞ 8. Know the importance of conductivity and understand its relationship with total dissolved solids

☞ 9. Know the relationship between conductivity and salinity

☞ 10. Know the sources of oxygen in water

☞ 11. Know the factors that affect the solubility of oxygen in water

☞ 12. Be able to estimate the per cent saturation of oxygen in a lake

☞ 13. Know how oxygen is produced in photosynthesis

☞ 14. Be able to assess the trophic status of a lake using the oxygen depth profile

☞ 15 Know how to recognize and explain unusual oxygen depth profiles

☞ 16 Be able to explain winter fish kills

☞ 17. Be able to estimate the actual oxygen deficit in a lake

☞ 18. Be able to estimate the absolute oxygen deficit in a lake

☞ 19. Be able to estimate the relative areal actual oxygen deficit and use it to assess the trophic status of a lake

☞ 20. Be able to measure the dissolved oxygen content in a lake

☞ 21. Know the sources of and importance of nitrogen in a lake

☞ 22. Know the nitrogen cycle and the useable and toxic forms of nitrogen

☞ 23. Be able to assess the trophic status of a lake using the total nitrogen value

☞ 24. know how nitrate and ammonia levels change seasonally in a lake

☞ 25. Be able to assess the trophic status of a lake using the nitrate-, nitrite-, or ammonia-nitrogen depth profile

☞ 26. Be able to assess the trophic status of a lake using the carbon dioxide depth profile

☞ 27. Know four states of carbon dioxide in a lake

☞ 28. Know how carbon dioxide makes bicarbonates to help neutralize lake acidity

☞ 29. Know why some plants and rocks have thick layers of chalk

☞ 30. Know what water hardness means

☞ 31. Know why kettles and water heaters form scale

☞ 32. Know why it is difficult to get lather from soap and detergents from some water

☞ 33. Know why bath tub rings form

☞ 34. Know what alkalinity means

☞ 35. Know the importance of alkalinity

☞ 36. Know what redox potential means and its importance

☞ 37. Be able to use redox potential to determine trophic status of a lake

☞ 38. Know the importance of detritus and seston in a lake

☞ 39. Know the carbon cycle and its importance as a source of nutrients

☞ 40. Know the phosphorous cycle and its importance as a source of nutrients

☞ 41. Know how phosphorous and plant chlorophyll are related

☞ 42. Know how to use phosphorous and chlorophyll for assessing trophic status

☞ 43. Know how to use phosphorous loading rate for assessing trophic status
☞ 44. Know how to use the phosphorous depth profile for assessing trophic status
☞ 45. Know the sulphur cycle and its importance in affecting nutrient cycles
☞ 46. Know how to use sulphur depth profiles to determine trophic status
☞ 47. Know the silica cycle and its importance in lakes
☞ 48. Know the iron cycle and its importance in lakes

A Quick Primer on Chemical Units

1☞ Chemistry can be intimidating if one does not understand some of the basic principles for expressing concentrations of ions. So we'll start with a quick primer at how to express chemical results. The metric system is usually used. The most common unit of concentration used is ***ppm*** (parts per million), or 1 mg of solute per kg of solution. 1,000 ppm = 1 ppt (one part per thousand) or 0.1%; 10,000 ppm = 10 ppt = 1%. Parts per thousand can also be abbreviated as ‰. ***Mg/L*** (milligrams per litre) is 1 mg of solute in 1 litre of water. In most cases mg/L and ppm are the same, especially if the solvent is distilled or de-ionized water. Differences in values between mg/L and ppm will occur only if the density of a solution exceeds 7,000 ppm (0.7%). This rarely occurs in fresh water so ppm and mg/L are usually considered equal. If, however, the concentration does exceed 7,000 ppm, correct for mg/L by dividing mg/L by the density or the specific gravity of the solution.

Gas concentrations are often expressed as mmol/L (millimoles per litre), which is the weight of the gas in mg divided by its molecular weight. For example, 12 mg of CO_2/L equals 12/44, or 0.272 mmol/L, where 44 = molecular weight of CO_2 (C = 12, O = 2 x 16 = 32). For ions, ***chemical equivalent*** is preferred because it takes into account the ***combining weight*** or ***equivalent weight*** of the ion (= ***ionic weight*** divided by ***ionic charge***) needed to examine ionic balances. Ionic concentrations are also expressed as meq/L (mil-equivalents per litre). Mil-equivalents/L is mg of ion per litre of water divided by its equivalent weight. For example, calcium with an atomic weight of 40.08 and a 2^+ charge (= Ca^{2+}), has an equivalent weight of 40.08/2 = 20.04. The equivalent weight of 1 g of Ca^{2+} is 1/20.04, or 0.0499 or 0.0499 x 1000 = 49.9 mil-equivalents. Mil-equivalents/L is often used to express the alkalinity of water, discussed later in this chapter. *Non-ionized elements (e.g. iron, silica, dissolved gases) cannot be reported in mil-equivalents.* For these, *mg/L or ppm must be used.* Table 5.1 lists several major ions and gives the mil-equivalent and mg value for each.

Ionic Composition of Fresh Water

2☞ There are two major sources of ions for inland fresh waters. The first is the soil and weathered rocks, collectively known as the ***edaphic*** source. Edaphic ions are released into the watershed by several mechanisms. The mechanisms include simple solution of ions such as carbonates, oxidation and reduction reactions, complexing or

Table 5.1. Converting dominant cations (positive charge) and anions (negative charge) of fresh waters to their mil-equivalent and milligram values.

Ions	Atomic +weight	Charge	To obtain meq weight, multiply mg value by:	To obtain mg value, multiply meq value by:
Calcium	40.08	2+	0.04990	20.04
Magnesium	24.32	2+	0.08224	12.16
Sodium	22.99	1+	0.04350	22.99
Potassium	39.10	1+	0.02558	39.10
Bicarbonates	61.0	1-	0.01639	61.02
Carbonates	60.0	2-	0.03333	30.01
Sulphate	96.06	2-	0.02082	48.03
Chloride	35.45	1-	0.02820	35.46

chelation reactions, and by the action of hydrogen ions, particularly in carbonic acid which constitutes atmospheric precipitation. Several kinds of physical and chemical weathering processes also contribute to the mechanisms for releasing edaphic ions into lakes.

The atmosphere is the second source of ions. It receives ions from the oceans, dust, volcanic explosions, and industrial and domestic pollution. All are present in rain and snow, and dry fallout. Dust incorporates most of the edaphic ions. Volcanic ash includes products of the earth's magma, such as chlorine, carbon dioxide, sulphur, hydrogen, fluorine, nitrogen, nickel, iron, magnesium, and water vapour. Industrial and domestic pollution contribute organic and inorganic contaminants, as well as sulphur, which is quickly oxidized to form sulphur dioxide and sulphuric acid. Nitrogen is also released and is quickly oxidized in the atmosphere to form nitrous dioxide and nitric acid. Water in equilibrium with carbon dioxide has a pH of about 5.6, but the acidification of precipitation by sulphur and nitrogen may depress the pH to 4.0, or lower.

Table 5.2 gives the concentrations of major ***cations*** (positively charged ions) and ***anions*** (negatively charged ions) in fresh waters. Concentrations are given for
3☞ very soft and very hard water as well as for an average North American fresh water system. Notice the large variations in concentrations of cations and anions between soft water and hard water systems. Bedrock in the Guelph area and most southern Ontario localities is dominated by sedimentary rocks consisting mostly of limestone and dolomite. Such rocks contribute high concentrations of calcium and magnesium cations and carbonate/ bicarbonate anions. At the other extreme are areas, like Sudbury, Ontario, dominated by granitic bedrock which contributes little or no calcium or

Table 5.2. Composition of surface waters in watersheds dominated by quartz (soft water lakes) or limestone (hard water lakes) in relation to average surface waters of North America. The areas are listed from very hard (Guelph area) to very soft (central Ontario) All values are mg/L. The North American data are from Livingstone (1963)[1].

Ions	North America	Guelph Area	Great Lakes	Eastern Ontario	Central Ontario
Calcium (Ca^{++})	20.9	84.4	25.0	13.6	4.4
Magnesium (Mg^{++})	5.0	18.8	5.2	3.1	1.1
Sodium (Na^{+})	8.8	15.4	7.5	2.0	1.1
Potassium (K^{+})	1.5	1.8	1.3	1.0	0.5
Carbonate (CO_3^{--})/ Bicarbonate (HCO_3^{-})	0/70.0	2.5/90.6	1.5/75.5	0/30.5	0/4.9
Sulphate (SO_4^{--})	20.5	69.5	22.5	10.0	7.8
Chloride (Cl^{-})	8.2	30.8	10.6	1.8	0.7
Nitrate (NO_3^{-})	0.91	0.45	0.80	0.10	0.12
Silica (SiO_2)	9.3	8.1	10.0	10.6	11.7
Iron (Fe)	0.9	1.0	0.8	0.45	0.10

bicarbonate ions. However, some granitic bedrock has calcareous outcrops and waters have intermediate levels of calcium and bicarbonates.

Climate, geology, geography, topography, biotic activity and time contribute to variation in chemical composition of fresh waters. Each watershed is unique in its interaction of these factors such that the chemical composition of a lake is diagnostic of the chemical composition of the soils and bedrock in its watershed.

Rock Classification

Rocks are classified into three main groups on the basis of their origin.

*1. **Igneous rocks*** are formed by the cooling and crystallization of molten rock magma. Igneous rock is termed ***intrusive*** if the magma has cooled and crystallized below the surface, or ***extrusive*** if the magma has cooled on the rock surface. The intrusive rocks

---

[1]Livingstone, D. A. 1963. Chemical composition of rivers and lakes. *In:* D. G. Frey (ed.), Limnology in North America. University of Wisconsin Press, Madison, WI. Pp. 559-574.

are generally formed within dome-like bodies called ***basoliths***. Extrusive rocks that were poured out as flat sheets of lava on the earth's surface are termed ***lava flows***. Igneous rocks include granite, an intrusive rock composed principally of quartz (SiO_2) and feldspars (e.g. orthoclase ($KAlSi_3O_8$) and plagioclase ($NaAlSi_3O_8$, $CaAl_2Si_2O_8$)); syenite, an intrusive rock composed mostly of orthoclase and plagioclase feldspars; mica, a mineral that has hydrous forms of potassium, magnesium, iron and/or aluminum silicates; and basalt, an extrusive rock formed of magma that contains 40-70% ***ferromagnesian*** (iron and magnesium) minerals. The ***structure*** (outward shape and form) of igneous rocks is generally ***massive*** (not bounded by crystal faces) and ***jointed*** (several hair-line cracks).

2. ***Sedimentary rocks*** are formed by deposition of mineral and rock material by wind, water, ice, etc. The material is derived from weathering and erosion processes on surface rocks. The rocks may be igneous, sedimentary or metamorphic. Examples include: sandstone, a compressed sand consisting of quartz and feldspar sand; limestone, composed mainly of calcite ($CaCO_3$) or dolomite ($CaMg(CO_3)_2$); shale, a consolidated clay; gypsum ($CaSO_4.2H_2O$); and halite (sodium chloride, NaCl, laid down by excessive evapouration processes of seas). Coal is also a sedimentary mineral composed of organic material. Sedimentary rocks are characterized by their ***bedded*** or ***stratified structure*** (layered mineral constituents).

3. ***Metamorphic rocks*** are formed from igneous and sedimentary rocks by heat, pressure, and solutions within the earth's crust. Metamorphic rocks are characterized by their ***foliated*** or ***schistose structure*** (mineral constituents are aligned) owing to recrystallization under the effects of pressure and heat on the source rocks. The altered rock structure frequently attains a new mineral assemblage as well. Common metamorphic rocks include: ***gneiss***, composed of crystalline grains of quartz, feldspar and ferromagnesian
minerals oriented in a banded fashion; ***slate***, a rock that splits easily into thin sheets; ***quartzite***, a recrystallization of quartz sandstone; and ***marble***, a recrystallization of limestone.

In most lakes, cation concentrations tend to occur in the proportions of calcium (Ca^{++}) > magnesium (Mg^{++}) ≥ sodium (Na^{+}) > potassium (K^{+}) and anions in the
4☞ proportions of bicarbonates ($HCO3^{-}_{3}$) and carbonates > sulphates > chlorides. Table 5.3 summarizes the bedrock types that contribute to the different cation and anion pools. Notice that quartz contributes very few ions, silica being the main byproduct.

The adsorption and release rates of ions to the runoff and percolating water depends upon: (i) ionic concentrations and proportions in the interstitial (soil spaces) water; (ii) the availability of cations; (iii) the number of cation exchange sites on a soil particle; and (iv) the volume of water bathing the soil particles. Soils and bedrock containing limestone or dolomite release Ca^{++} and Mg^{++} to the runoff during rainfall events, mainly by exchange reactions with hydrogen ions in the carbonic acid weathering process (discussed in detail later). As the soil water becomes saturated with

Table 5.3. Bedrock/soil types and their cationic and anionic mineral contributions.

Bedrock	Cations	Anions, Free Radicals
Apatite ($Ca_3(PO_4)_2$ or $Ca_5(PO_4)_2$)	Ca	PO_4
Calcite (limestone, $CaCO_3$)	Ca	CO_3, HCO_3
Dolomite ($Ca.MgCO_3$)	Ca, Mg	CO_3, HCO_3
Gypsum ($CaSO_4$)	Ca, S	SO_4
Pyrite (FeS_2)	Fe, S	SO_4
Quartz (SiO_2)	Si	O_2
Halite or Rock Salt (NaCl)	Na	Cl
Mirabilite ($Na_2SO_{4.}\ 10H_2O$)	Na, S	SO_4
Kaolinite (Clay) ($xAlSi_3O_8$)	Si, Al, x = Ca, K, Na	O_8
Feldspars ($xAlSi_3O_8$)	Si, Al, x = Ca, K, Na	O_8

H^+, the Ca^{++} and Mg^{++} are released until all exchange sites are filled with H^+. At this point, the soils have no more capacity to contribute calcium and magnesium to the runoff. Any additional rainfall will contribute H^+ to the runoff, causing severe pH depressions (e.g. lake acidification, see Chapter 13).

Because calcium is required by most organisms, lakes with high calcium levels tend to be more productive than lakes with low calcium levels. However, caution has
5 to be exercised when relating trophic status to calcium levels because many hard water lakes are oligotrophic. Nevertheless, the relationship holds true for most lakes, even the Great Lakes where the least productive (i.e. most oligotrophic, Secchi depth = 6-8 m) is Lake Superior, with a calcium level of 10 mg/L. Lake Erie is the most productive (mesotrophic, Secchi depth = 3-5 m) with a calcium level of 35 mg/L. The ranking of the Great Lakes in relation to calcium level and Secchi depth (SD), from most oligotrophic to most eutrophic, is: Lake Superior ([Ca] = ~10 mg/L, SD = 6-8 m); Lake Huron ([Ca] = ~15 mg/L, SD = 5-6 m); Lake Ontario ([Ca] = 40, SD = 5-7 m); Lake Michigan ([Ca] = ~38, SD = 3-5 m); Lake Erie ([Ca] = ~38 mg/L, SD = 1-3 m (increased to 7.5 m in 1992 during zebra mussel invasion).

Sodium, potassium and chlorides are contributed largely by sedimentary deposits rich in marine salts. Feldspars are a group of aluminum silicate minerals that make up about 60% of the outer 15 km (9 mi) of the earth's crust. They contribute not only aluminum and silicates, but potassium, sodium, and some calcium. Iron, manganese, sulphur, nitrogen, phosphorous, and carbon compounds are contributed by
6 soils and bedrock (contributions from sources outside the lake are called ***allochthonous*** in nature) containing these elements. However, oxidation and reduction reactions involved in microbial decomposition processes greatly affect the rate of supply to the receiving waters. The cycling of these nutrients within the lake (contributions from

sources within the lake are called ***autochthonous*** in nature) maintains their availability to organisms. The sulphur, nitrogen, phosphorous, and carbon cycles are discussed in detail later.

Total Dissolved Solids and Conductivity

Most of the cations and anions are dissolved in solution. The total concentration of cations and anions in water is expressed in mg/L as ***total dissolved***
7☞ ***solids*** (abbreviated ***TDS***). TDS is determined by taking a known quantity of water (usually 100 ml), filtering it to remove all the ***suspended solids*** (also called ***seston***), then evapourating the filtrate at 105°C in a tared (pre-weighed) container until a constant weight is obtained. The materials left on the bottom of the beaker are solutes that precipitated out as the water evapourated. The amount of water needed for a TDS measurement depends on the concentration of dissolved solids; the lower the TDS level, the more water needed to detect the small amounts present.

The suspended solids retained by the filter paper can be dried, at 60°C, and weighed to provide a ***total suspended solids*** (***TSS***) estimate, in mg/L. Both the TDS and TSS fractions contain inorganic and organic residues. The organic fraction can be removed simply by burning or ***ashing*** in a furnace at 500-550°C. The material remaining after ashing at 500-550°C is the inorganic fraction (e.g. Ca, Mg, Fe), and the loss in weight is the organic fraction. Glass-fibre filter papers must be used when ashing for organic and inorganic weights of TDS and TSS.

A much easier method for measuring TDS is to determine the ***specific conductance***, or ***conductivity***, of the water sample and multiply by 0.65, as in Equation 5.1:

8☞

$$\text{TDS} = \text{conductivity} \times 0.65 \qquad \text{Equation 5.1}$$

A conductivity meter is used to measure the flow of electrons (cations and anions) through the water. The specific conductance is measured between two electrodes, each with 1 cm^2 surface area and separated by 1 cm. Specific conductance is the reciprocal of resistance. Since resistance is measured in ***ohms,*** and conductivity is the reciprocal of resistance, specific conductance is expressed as ***mhos*** (the inverse of ohms). The more recent unit of expression is ***siemens*** (abbreviated S). One siemen equals one mho. Specific conductance is typically expressed as μS/cm (or S/cm if S is multiplied by 1000). Temperature affects the flow of electrons, and therefore conductivity measurements. However, most meters automatically compensate for the effect of temperature on conductivity readings.

The relationship, TDS (mg/L) = conductivity (μS/cm) x 0.65, is reasonably accurate for fresh waters in the normal range of conductivity (10-500 μS/cm). The
9☞ relationship begins to break down with the error increasing above 1,000 μS/cm. The levels of total dissolved solids of waters in the low salinity (1% or 10 ppt) range should

be determined by titration for ***chlorinity*** instead of measurement by conductivity. Chlorinity is the amount of chloride ions in water. It is the standard unit of measurement for salinity in sea water, where

$$\text{Salinity} = 0.03 + 1.805\ (\text{Chlorinity}) \qquad \text{Equation 5.2}$$

The chloride content of sea water is directly proportional to the salinity. However, in freshwater, the chloride content is not proportional to the salinity (or conductivity) because of the variations in dilutions of various ions. The ***constancy of proportions*** of one ion in relation to others in the oceans contrasts greatly with the inconstancy of ionic proportions in fresh waters, as shown in Table 5.3.

The concentrations of cations and anions in the total dissolved solids are determined with standard methods. The book, "Standard Methods for the Examination of Water and Wastewater", published by the American Public Health Association is the standard reference. Some methods are rather inexpensive, simple colorimetric titrations that almost every one can perform. Others require elaborate (and expensive) methods and equipment, such as flame spectroscopy and inductively coupled plasma (ICP) spectroscopy.

Composition Of Dissolved Gases

The gases of importance in freshwater ecosystems are oxygen, carbon dioxide, and nitrogen. It is the oxygen that is dissolved in water that is measured, not the oxygen that is attached to the hydrogen atom in the water molecule. Dissolved oxygen is vital to the survival of nearly all aquatic organisms, anaerobic bacteria being an exception.

Carbon dioxide can exist in several forms in freshwater, as ***free dissolved*** CO_2, ***half-bound*** CO_2, ***bound*** CO_2, and ***hydrated*** CO_2. The free dissolved form is required by photosynthetic organisms. All forms are important in the chemical equilibrium relationships of carbonates and bicarbonates that affect the alkalinity, pH and buffering capacity of natural waters. Other important carbon sources are dissolved organic carbon (DOC) and detrital, or particulate organic carbon (POC).

Nitrogen is not essential for life, except to micro-organisms capable of cracking the covalent triple bonds of the nitrogen molecule. Nevertheless, it is the most abundant gas in the atmosphere (78%) and contributes to the formation of nitrate, a limiting nutrient for many algae. The spatial (e.g. depth variations) and temporal variations (e.g. seasonal variations) are described for each gas. The levels of dissolved oxygen and total nitrogen, and their depth profiles are diagnostic of the trophic status of a lake.

Dissolved Oxygen

There are two primary sources of oxygen in aquatic systems, the atmosphere
10☞ and photosynthetic production by algae and macrophytes.

Atmospheric Oxygen

The rate of diffusion of oxygen from the atmosphere into water varies directly with the pressure difference across the air-water interface. The greater the differences in partial pressures, the faster the diffusion rate of oxygen from the atmosphere into water. Both Henry's Law and Dalton's Law govern the solubility of gases. ***Henry's***
11☞ ***Law** states that the concentration of a solution of a gas which has reached equilibrium is proportional to the partial pressure at which the gas is supplied.* ***Dalton's Law*** *states that the pressure of each component of gas in a mixture is proportional to its concentration in the mixture, the total pressure of all gases being equal to the sum of its components*. The partial pressures and the solubilities at ambient temperatures determine how much of each gas can be dissolved in water. Air contains about 21% oxygen by volume, or approximately 300 mg/L of air. Oxygen is nearly twice as soluble as nitrogen at a given temperature but its partial pressure (21% in atmosphere) is only a quarter that of nitrogen (78% in atmosphere). Hence, the ratio of both gases dissolved in water is roughly one part oxygen to two parts nitrogen (2 x 1/4). Carbon dioxide is about 33 times more soluble than oxygen in water and has a partial pressure about 1/700 ((0.0014% in atmosphere) that of oxygen. Because of the high solubility of carbon dioxide in water, carbon dioxide is relatively (i.e. compared to oxygen) more abundant in water than in air.

Fresh water at 0°C and mean sea level contains about 14.6 mg/L but oxygen solubility decreases with increasing temperatures (Table 5.4). Because of the effect of temperature on the solubility of oxygen the concentration of oxygen in water is often
11☞ expressed as ***percent saturation***. To determine % saturation, divide the amount of oxygen (mg/L) dissolved in the water by the amount of oxygen that the water can hold at the observed temperature. For example, suppose one observes 13 mg/L at 0°C. The water should have 14.62 mg/L (Table 5.4), so % saturation is 13/14.62 x 100 = 88.9% at mean sea level.

Altitude (and therefore partial pressure) also affects the solubility of oxygen. The solubility of oxygen increases with an increase in altitude. To correct for the effect of altitude on dissolved oxygen saturation, one must multiply the % saturation
11☞ value by the appropriate factor given in Table 5.4. For example, suppose the dissolved oxygen content for water at 5 °C and 500 m above sea level is 10 mg/L Table 5.4 shows the saturation value at sea level to be 12.8 mg/L; therefore the % saturation at sea level would be 10/12.8 x 100 = 78.13%. However, at 500 m, where there is less oxygen in the atmosphere than at sea level, the% saturation would be 1.06 times
12☞ greater (from Table 5.4), or 1.06 x 78.13 = 82.8%. This is the same as multiplying the measured value of 10 mg/L by 1.06 and then dividing by 12.8 and multiplying by 100. Alternatively, one may use a nomogram, such as that shown in Fig. 5.1. To use

Table 5.4 Solubility of oxygen from an atmosphere containing 20.9 percent oxygen by volume. See text for explanation of altitude factor.

Temp.	Oxygen from normal atmosphere		Correction factors for oxygen saturation at various altitudes			
			Altitude		Pressure	
Temp.	mg per L	ml per L	ft	m	mm	Factor
0	14.62	10.23	0	0	760	1.00
1	14.23	9.96	330	100	750	1.01
2	13.84	9.69	655	200	741	1.03
3	13.48	9.44	980	300	732	1.04
4	13.13	9.19	1310	400	723	1.05
5	12.80	8.96	1640	500	714	1.06
6	12.48	8.74	1970	600	705	1.08
7	12.17	8.52	2300	700	696	1.09
8	11.87	8.31	2630	800	687	1.11
9	11.59	8.11	2950	900	679	1.12
10	11.33	7.93	3280	1000	671	1.13
11	11.08	7.75	3610	1100	663	1.15
12	10.83	7.58	3940	1200	655	1.16
13	10.60	7.42	4270	1300	647	1.17
14	10.37	7.26	4600	1400	639	1.19
15	10.15	7.11	4930	1500	631	1.20
16	9.95	6.96	5250	1600	623	1.22
17	9.74	6.82	5580	1700	615	1.24
18	9.54	6.68	5910	1800	608	1.25
19	9.35	6.55	6240	1900	601	1.26
20	9.17	6.42	6560	2000	594	1.28
21	8.99	6.29	6900	2100	587	1.30
22	8.83	6.18	7220	2200	580	1.31
23	8.68	6.08	7550	2300	573	1.33
24	8.53	5.97	7880	2400	566	1.34
25	8.38	5.86	8200	2500	560	1.36
26	8.22	5.75				
27	8.07	5.64				
28	7.92	5.54				
29	7.77	5.44				
30	7.63	5.34				

the nomogram, place a ruler on the temperature and oxygen concentration (mg/L) axes so that the edge lines up with the temperature and oxygen concentration (mg/L) values observed for the water. Read percent saturation on the diagonal line. Correct for the effect of pressure and altitude by multiplying by the appropriate factor given in the inset Table 5.4. Note that lakes at high elevations (where there is less oxygen per unit volume of air) contain less dissolved gas per unit volume than do lakes with the same temperature at sea level.

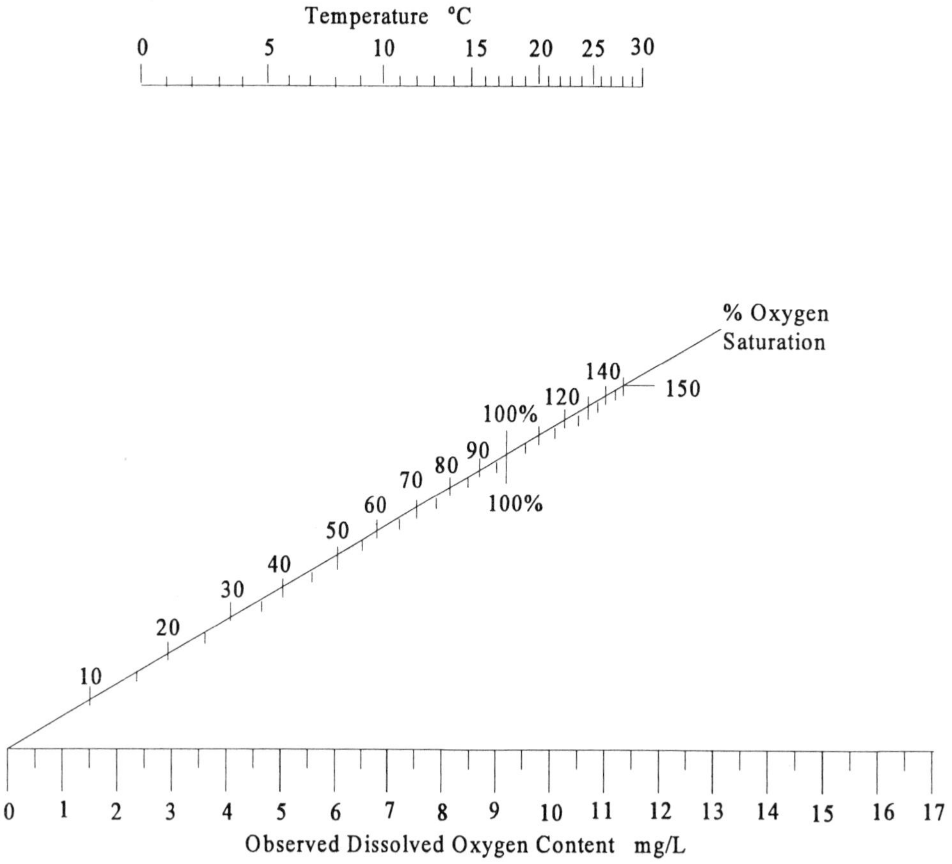

Figure 5.1.. Nomogram for estimating percent oxygen saturation when dissolved oxygen content and temperature are know. See text for explanation

Physical factors also affect the diffusion rate of oxygen across the air-water interface. Oxygen diffuses more rapidly into rough, or choppy, water than into a calm surface because of the increased surface area produced by the ripples or waves. Streams tend to be naturally aerated because of the turbulent flow of water.

Photosynthetic Production of Oxygen

The amount of oxygen produced by photosynthesis depends on the abundance of algae in the water. The reaction involved in photosynthesis is:

13 $$nCO_2 + nH_2O \rightleftharpoons (CH_2O)_n + nO_2. \qquad \text{Equation 5.3}$$

where n is 3 (pyruvate), 6 (glucose), or 12 (sucrose). The relative concentrations of CO_2 and O_2 depend on whether the process moves to the left or to the right. It follows from this, and Dalton's Law, that when the concentration of carbon dioxide is high, the concentration of oxygen in the water will be low. Hence, an inverse relationship exists between the amount of O_2 and the amount of CO_2 dissolved in the water.

Since photosynthesis requires light, diurnal fluctuations in dissolved oxygen concentration occur, with an increase during the day and a decrease during the evening. Conversely, the concentration of dissolved (free) CO_2 in the water decreases during the day as it is being used for photosynthesis and increases at night due to respiration by aquatic organisms, including plants. Lakes with high algal biomasses often have partial pressures greater than that provided by the atmosphere, resulting in super saturation, or concentrations exceeding 100%.

It follows from the previous discussions that lake productivity should affect the dissolved oxygen - depth profile. Oligotrophic lakes have close to 100% oxygen
14 saturation at all depths (Figure 5.2a). The concentration of oxygen in the epilimnion is determined by the amount of photosynthesis occurring there as well as by the amount diffused from the atmosphere. The amount of oxygen in the hypolimnion remains high because of its high solubility at 4 °C and because of the low organic matter content. With meagre amounts of organic material, metabolic and respiratory demands of bacteria are reduced and oxygen levels remain high. Oxygen profiles which exhibit 75% to 100% saturation at all depths are called ***orthograde*** (Fig. 5.2a).

Eutrophic lakes have an abundance of nutrients which leads to high primary productivity. While the epilimnion has 100% oxygen saturation, or even super saturation, it also has large amounts of organic material composed largely of moribund algae that sinks and settles on the bottom. Bacteria convert the organic material into carbon dioxide and in the process use up any oxygen present. The oxygen supply cannot be replenished because the water is thermally stratified. The stratified water column effectively isolates the hypolimnion from any renewable sources of oxygen. By the end of the summer the hypolimnion is anoxic. Lakes that exhibit high saturation values in the epilimnion and low values in the hypolimnion have ***clinograde oxygen profiles*** (Fig. 5.2b). Both eutrophic and mesotrophic lakes have clinograde oxygen profiles, but eutrophic lakes have 0 to 25% and mesotrophic lakes have 25 to 75% oxygen saturation in the hypolimnion.

At least two variations in oxygen profiles occur. **The *hypolimnetic oxygen minimum profile***, also known as the ***negative heterograde oxygen curve***, exhibits
15 minimum concentrations of oxygen in the metalimnion (Fig. 5.2c). Such profiles arise when organic material accumulates in the metalimnion which, through oxidation and

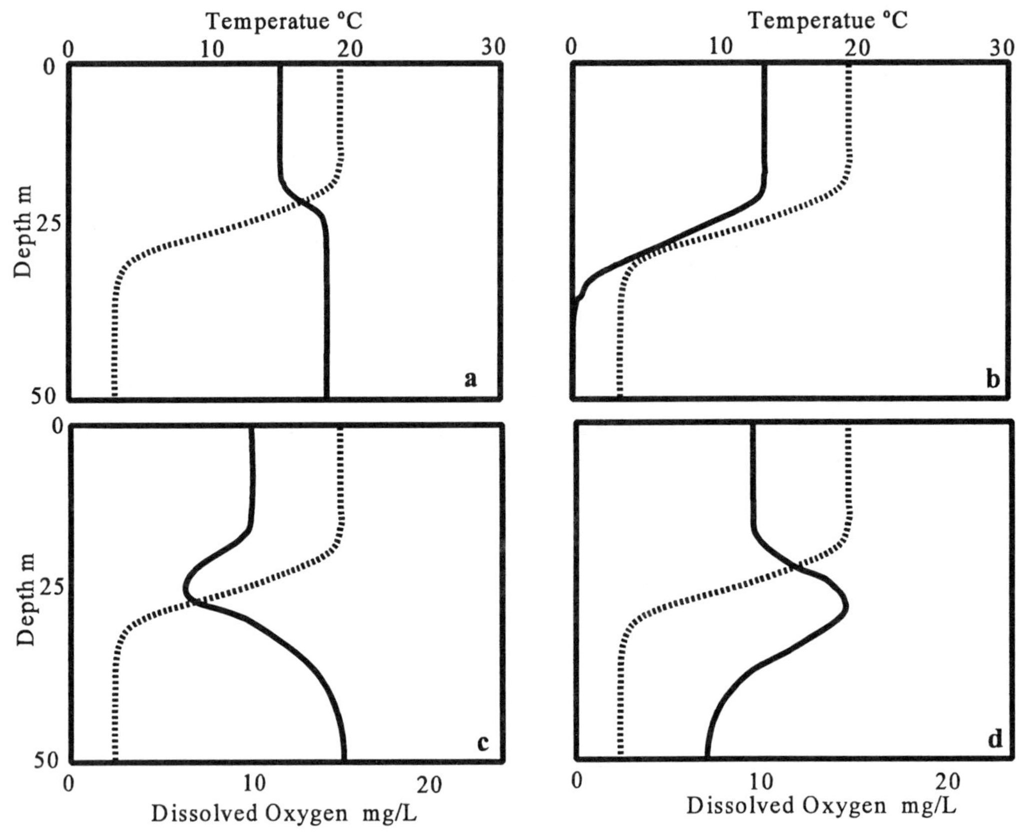

Figure 5.2. Common summer vertical oxygen profiles in lakes. (a) Orthograde. (b) Clinograde. (c) Negative heterograde. (d) Positive heterograde.

metabolism by bacteria, cause a decrease in the oxygen content. Respiration by metalimnetic aggregations of zooplankton have also been attributed to the decrease in oxygen levels. A third mechanism relates to the flow of water across the lake in the lower layers of the epilimnion and adjacent layers of the metalimnion. As flowing, epilimnial water approaches the shoreline, bacteria in the sediments use oxygen to break down the organic material. The water in its return flow in adjacent layers of the metalimnion has reduced levels of oxygen that ultimately creates a metalimnetic oxygen minimum.

Lakes may also exhibit maximum oxygen concentrations in the metalimnion. Such a profile is known as a ***metalimnetic oxygen maximum*** or ***a positive heterograde oxygen curve*** (Fig. 5.2 d). The oxygen maximum is usually attributed to an
15☞ accumulation of a filamentous, blue-green algae, *Oscillatoria* in the metalimnion. Apparently, the species present are cold stenotherms, that is they are able to tolerate water over a narrow range of cold temperatures.

The shapes of the oxygen depth profiles change on a daily and seasonal basis. The daily, or ***diel*** (day plus night), changes reflect the high oxygen concentrations produced during the day and the lower oxygen concentrations due to respiration in the evening. The evening period is known as the ***nocturnal*** part of the diel cycle and the daytime period is known as ***diurnal*** part of the diel cycle. The diel variations in oxygen concentrations are greater in eutrophic lakes than in oligotrophic lakes because of the different levels of productivity that greatly affect photosynthesis and respiration.

Seasonal variations in oxygen depth profiles occur primarily because of two factors, the mixing of lake waters at spring and fall turnover and the increasing solubility of oxygen with decreasing temperatures. Oligotrophic lakes have higher oxygen concentrations in the hypolimnion than do eutrophic lakes because of oxygen consumption by bacteria in the sediments. During the winter, ice restricts light needed
16 for photosynthesis and it prevents the diffusion of oxygen from the atmosphere into the water. Highly eutrophic lakes typically develop ***anoxia*** where oxygen levels fall to zero at all depths. Anoxia is usually followed by the formation of hydrogen sulphide which results in "***winter fish kills***".

Oxygen Deficits

Because of large decreases in oxygen concentration that occur in the hypolimnion, especially in highly eutrophic lakes, it is convenient to express the loss in oxygen as an ***oxygen deficit***. The *oxygen deficit in a lake is the amount of oxygen required to reach the saturation level.* All calculations must account for the atmospheric pressure at the surface of the lake by using the correction factors in Table 5.4.

There are two kinds of oxygen deficits, actual and absolute, depending upon the saturation temperature used in the calculation. The ***actual oxygen deficit*** is the difference
17 in oxygen concentration between that observed and the *saturation value of the same quantity of water (i.e. same depth and place) at the observed temperature*. For example, suppose one observes 8.0 mg O_2/L at 10 °C in the upper part of the hypolimnion; the actual oxygen deficit is 8.0 - 11.33 (the saturation value for 10 °C) = -3.33 mg/L. The main criticism of actual oxygen deficits is it assumes that the water was saturated at the observed temperature during spring turnover.

The ***absolute oxygen deficit*** is the difference in oxygen concentration between that observed and the *saturation value at spring turnover*. For example, suppose one
18 observes 8.0 mg O_2/L at 10 °C; the absolute oxygen deficit is 8.0 - 13.13 (the saturation value at 4 °C) = -5.13 mg/L. The absolute oxygen deficit suffers the same criticism as the actual oxygen deficit in that it assumes saturation at 4 °C, which is not necessarily true.

The most meaningful oxygen deficit measurement is the ***relative hypolimnial oxygen deficit*** because it accounts for oxygen deficits in layers below the hypolimnetic surface. Essentially, one measures the dissolved oxygen contents of several vertical
19 columns of water at the beginning and at the end of the stratification period. Each column has a cross-sectional area of 1 cm^2. The height (depth) of a column is usually 1 to 2 m (100 to 200 cm). The difference in oxygen contents for each layer (column)

between the beginning and end of the stratified period is summed to give a ***relative areal hypolimnial oxygen deficit***, or **RAHOD**. ***Oligotrophic lakes have RAHOD values of < 0.017 mg O_2 lost cm^{-2} day^{-1} and eutrophic lakes have > 0.033 mg O_2 lost cm^{-2} day^{-1}***. The values reflect the amount of respiration by hypolimnetic bacteria in decomposing all moribund phytoplankton that grew in the epilimnion during the stratified period.

Table 5.6 provides dissolved oxygen data for a 92-day period between June 1 and September 1 in a lake 20 m deep. The hypolimnion has a surface area of 1.5×10^6 m^2 (= 15×10^9 cm^2).

The RAHOD is calculated in the following manner:
(1) Multiply the volume of each layer by the observed dissolved oxygen value for each of June 1 and September 1. Note that volume is in m^3, but we need units in litres, so
19 multiply by 10^3 (i.e. $10^6 \times 10^3 = 10^9$ L). (2) Subtract the two values between the two periods for each layer of water. Since oxygen is lost in each layer between June and September, each difference is a deficit; express the differences in mg O_2; (3) Sum the deficit values in mg O_2. (4) Divide the total hypolimnial oxygen deficit value by the area of the hypolimnion to give the areal hypolimnial oxygen deficit in mg/cm^2 for the 92 day period. Finally, (5) divide the mg/cm^2 by 92 days to express RAHOD in mg/cm^2/day. The following table shows the calculations for the data in Table 5.6:

(1) calculate amount of oxygen (mg O_2/L) in each layer (m^3) for each period:					
mg O_2/L	10-12 m	12-14 m	14-16 m	16-18 m	18-20 m
	$6 \times 10^9 \times 9.87$	$4 \times 10^9 \times 10.5$	$2 \times 10^9 \times 10.8$	$1 \times 10^9 \times 11.13$	$0.5 \times 10^9 \times 11.13$
	$6 \times 10^9 \times 8.13$	$4 \times 10^9 \times 8.13$	$2 \times 10^9 \times 8.13$	$1 \times 10^9 \times 8.13$	$0.5 \times 10^9 \times 8.13$
(2) take the difference in total oxygen content between periods for each layer:					
Difference = mg O_2	$6 \times 10^9 \times 1.74$	$4 \times 10^9 \times 2.37$	$2 \times 10^9 \times 2.67$	$1 \times 10^9 \times 3.00$	$0.5 \times 10^9 \times 3.00$
Total mg O_2 deficit / layer	10.44×10^9	9.48×10^9	5.34×10^9	3×10^9	1.5×10^9
(3) Total hypolimnial deficit = sum of deficits in all depths = 29.76×10^9 mg (or 29.76 metric tons)					
(4) Areal hypolimnial oxygen deficit = 29.76×10^9 mg/ 15×10^9 cm^2 = 1.98 mg/cm^2/92 days					
(5) Relative Aerial Hypolimnial Oxygen Deficit = 1.98 mg/cm^2/92 days = **0.022 mg/cm^2/day**					

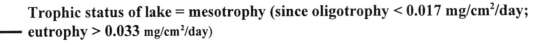

Trophic status of lake = mesotrophy (since oligotrophy < 0.017 mg/cm^2/day; eutrophy > 0.033 mg/cm^2/day)

Table 5.6. Some temperature and oxygen values for a hypothetical lake with a hypolimnion between the 10-m and 20-m depths, each layer (column) of the hypolimnion having the volumes shown. The oxygen saturation values are taken from a nomogram for a lake at sea level. The surface area of the hypolimnion is 15 x 10^5 m^2 (= 15 x 10^9 cm^2).

Parameters	Depths (m)				
	10 - 12	12 - 14	14 - 16	16 - 18	18 - 20
Volume m^3 (multiply values shown by 1000 for volume in litres)	6 x 10^6	4 x 10^6	2 x 10^6	1 x 10^6	0.5 x 10^6
June 1, Temperature, [O_2] mg/L	8 °C, 9.87	6 °C, 10.5	5 °C, 10.8	4 °C, 11.13	4 °C, 11.13
September 1, Temperature, [O_2] mg/L	4 °C, 8.13	4 °C, 8.13	4 °C, 8.13	4 °C, 8.13	4 °C, 8.13
Saturation values: 8°C, 100% [O_2] = 11.87 mg/L; 6°C, 100% [O_2] = 12.48 mg/L; 5°C, 100% (from nomogram) [O_2] = 12.80 mg/L; 4°C, 100% [O_2] = 13.13 mg/L					

Methods for Measuring Dissolved Oxygen Content

There are two basic methods, one being a chemical test, the other electronic.
20☞ The chemical method, called the ***Winkler titrametric method,*** involves a titration with sodium thiosulphate on a water sample that has had its oxygen chemically fixed in a glass-stoppered bottle. A 100-ml glass-stoppered bottle is filled with water. To the sample is added 1 ml of manganous sulphate and 1 ml of a mixture of potassium hydroxide with potassium iodide. The glass stopper is applied and the sample is shaken. Any dissolved oxygen in the sample oxidizes the manganous hydroxide (from the reaction of $MnSO_4$ with KOH) to manganic hydroxide, which forms a precipitate. The bottle is set aside for a few minutes to allow the precipitate to settle in the lower half of the Winkler bottle. The glass stopper is removed and 1-2 ml of concentrated sulphuric acid is added to dissolve the precipitate and convert manganic hydroxide to manganic sulphate. During the reaction, there is replacement of iodine (as potassium iodide) by sulphate and a clear yellow solution results. The 100-ml sample is then titrated with 0.1N sodium thiosulphate until all free iodine has combined into sodium iodide, producing a clear solution. Addition of starch aids detection of the end point because starch turns iodine solutions dark blue which disappears when the end point is reached during the titration process. Knowing; (i) the normality of the titrant, (ii) the volume of titrant needed to change the colour from a dark blue to clear, and (iii) the volume of the sample titrated, allows one to estimate the amount (mg/L) of dissolved oxygen in the sample.

Chemical methods have been replaced by electronic devices which employ a measurement of electronic potential across a membrane on an oxygen probe. The amount of electronic potential that develops in a chemical solution within the probe depends on the amount of oxygen in the water. The probes are attached to the end of a cable that can be purchased in lengths from 3 to 30 m. Use of electronic devices speeds up the measurement of dissolved oxygen, and with 30 m cables, one is able to measure the dissolved oxygen concentration *in situ.*

Nitrogen

Nitrogen is the most abundant gas in the atmosphere, comprising a total of 78% of the volume. Approximately 50% of the dissolved gas in freshwater is nitrogen.
21☞ The upper layers of water are usually saturated with nitrogen. The primary source of nitrogen is the atmosphere, with some nitrogen provided through the nitrogen cycle by algae and bacteria. Many organisms require nitrogen to build protein but they cannot use the free nitrogen in the atmosphere and water directly. The nitrogen must first be fixed into usable forms by specialized bacteria. The major forms of nitrogen available to autotrophs (bacteria, fungi, algae) are ammonium and nitrate. Nitrogen can be a growth-limiting factor to plants in some aquatic systems, especially those that are extremely oligotrophic or extremely eutrophic.

Before examining the nitrogen cycle and nitrogen as a nutrient, we should examine some of the key components of the nutrient cycle. Figure 5.3 shows some common pathways in many nutrient cycles, like nitrogen, carbon and phosphorous. The key components are the autochthonous and allochthonous sources, utilization by autotrophs (bacteria, fungi, algae, etc.), and losses to the sediments and occasionally the atmosphere. The ***allochthonous sources*** are the atmosphere, edaphic supplies (includes soil elements in runoff and groundwater), and influents (parent streams). The ***autochthonous sources*** are everything produced within the lake, including re-mobilization of elements from the sediments, metabolic and excretory products from organisms, chemical transformations of soluble and insoluble compounds that occur in the water, the sediments, or at the sediment water interface, leaching of elements from living and moribund organisms, etc. Allochthonous sources predominate for some elements, like phosphorous and sulphur, most of which ends up on the bottom of the lake, unavailable to autotrophs. Allochthonous sources contribute to the nutrient pools for carbon, nitrogen and silica, but autochthonous sources are able to recycle and maintain the nutrient supply. ***Biologically active compounds*** are any elements or compounds that can be used and assimilated (digested) by autotrophs. For some elements (e.g. phosphorous and sulphur), the lake bottom is a sink and the elements accumulate in the sediments.

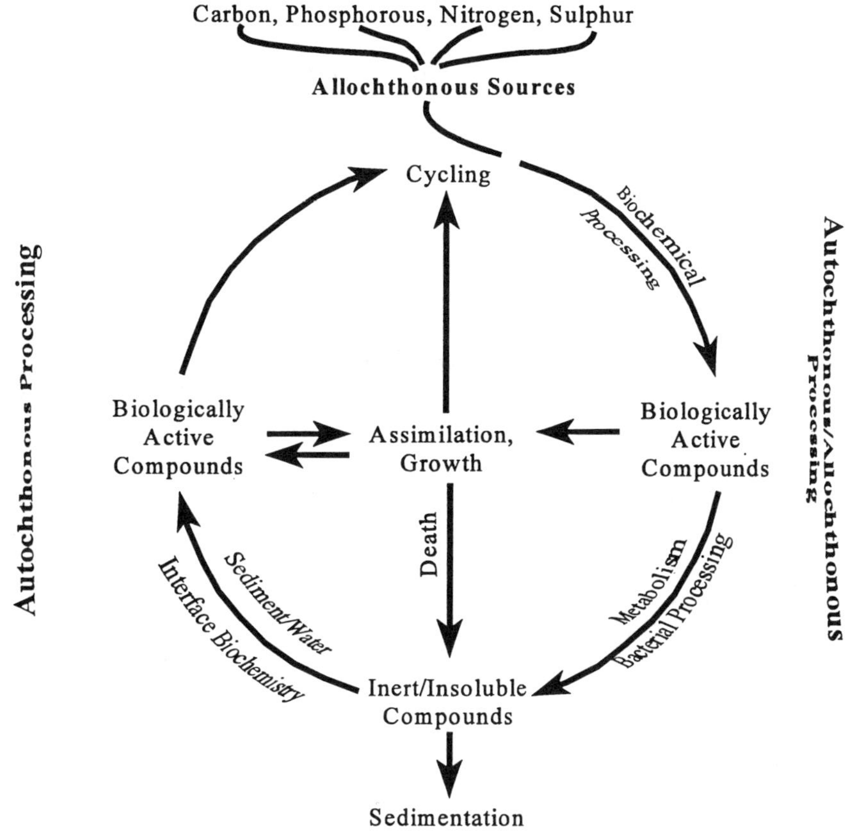

Figure 5.3. Common pathways in biogeochemical cycles

Nitrogen Cycle

The different processes that occur within the nitrogen cycle are illustrated in
22 ☞ Fig. 5.4. Proteins are added to the autochthonous supply of nitrogen upon death and decay of organisms. The excretory products (including ammonia) of organisms contribute to the ammonium pool. The ammonium is present mainly as the dissociated ion, NH_4^+. The ammonium ion is useable by plants, but ammonia (NH_3) is very toxic. The proteins of moribund organisms are broken down to amino acids which are then reduced to ammonium by bacterial metabolism. Under conditions of anoxia, the decay process may occur by putrefaction. The ammonium released by this ***ammonification process*** is in addition to the ammonia contributed in the excreta of organisms. Most of the ammonia (NH_3) dissolves immediately in water to form ammonium hydroxide (NH_4OH) which dissociates to give ammonium (NH_4^+) and hydroxyl ions, as illustrated in the following reaction:

$$NH_3 + H_2O \rightleftharpoons NH_4OH \rightleftharpoons NH_4^+ + OH^- \qquad \text{Equation 5.4}$$

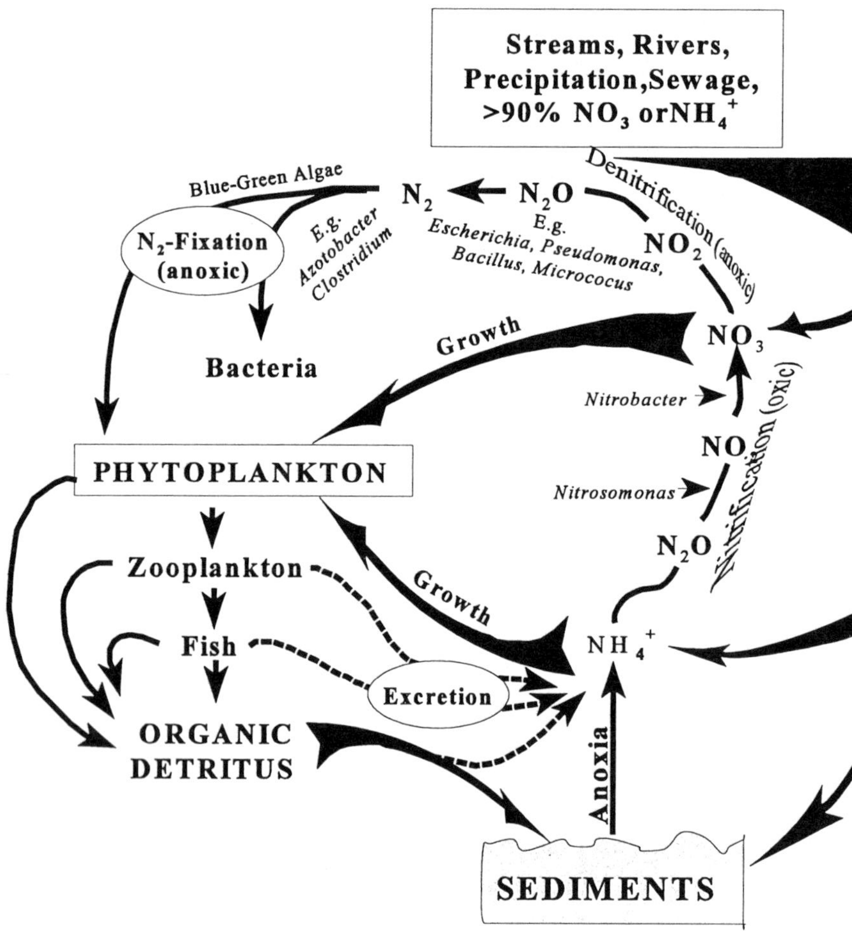

Figure 5.4. The nitrogen cycle

Undissociated ammonium hydroxide is also toxic to organisms, the amount and toxicity of which increases with increasing pH and temperature and decreasing dissolved oxygen content. Ammonification is one of the first steps in the recycling of nitrogen in the nitrogen cycle.

The next step is ***nitrification***, the conversion of ammonium or organic nitrogen into nitrate. Actually, nitrification is often tightly coupled with ***denitrification*** which is discussed next. Nitrification is an **aerobic bacterial process** where ammonium is first oxidized to nitrite by *Nitrosomonas* and then the nitrite is oxidized to nitrate by *Nitrobacter*, as illustrated in the following reactions:

$$NH_4^+ \rightarrow N_2O \rightarrow NO_2 \rightarrow NO_3 \qquad \text{Equation 5.5}$$

The nitrate is then available for use by the autotrophs.

In eutrophic lakes, nitrate produced in the oxygenated epilimnion can move down to the anoxic hypolimnion where it is denitrified. ***Denitrification*** is an **anaerobic bacterial process** where nitrates are reduced to release N_2 gas as the end product, as shown in the following reaction:

$$NO_3 \rightarrow NO_2 \rightarrow N_2O \rightarrow N_2 \qquad \text{Equation 5.6}$$

Most denitrification occurs in the sediments and hypolimnion of eutrophic lakes. Common denitrifying bacteria are *Escherichia, Bacillus, Pseudomonas*, and *Micrococcus*.

The nitrogen (N_2) produced in the denitrification process is available for ***biological nitrogen fixation***. Only a few genera of blue-green algae, such as *Anabaena, Nostoc*, and *Gloeotrichia*, and bacteria, such as *Clostridium* and *Azotobacter,* have the ability to fix nitrogen. Although nitrogen fixation is an anoxic process, most nitrogen fixation occurs in the epilimnion, especially by blue-green algae which require light to photosynthesize. The oxygen produced by algae is prevented from diffusing into the blue-green algal cells by specialized structures called ***heterocysts***. The heterocysts not only consume the oxygen but prevents nitrogen from entering the cells. Nitrogen fixation can greatly accelerate eutrophication even when nitrate-nitrogen and ammonium-nitrogen are in low supply.

The various forms of nitrogen (NH_4 -N, NO_2 -N, NO_3 -N) are not only available from recycled organic materials (i.e. autochthonous nitrogen sources) described above, but from outside sources (i.e. allochthonous nitrogen sources). These allochthonous sources include parent streams and rivers, precipitation, atmospheric fallout, and sewage. More than 90% of the biologically available nitrogen can be contributed from these sources.

The concentrations and depth profiles for the different forms of nitrogen can be useful criteria for assessing trophic status of a lake. However, beware of seasonal differences due to spring and fall turnovers, utilization of nutrient forms of nitrogen (nitrate and ammonium), and ice formation during the winter which restricts photosynthesis and diffusion of oxygen into the water. Because nitrates and ammonia are used by plants, the best time to measure total nitrogen is during the spring turnover period. Table 5.7 gives trophic status criteria for total nitrogen. Figure 5.5 shows typical seasonal variations in (a) nitrate and (b) ammonia in the epilimnion of lakes of different trophic status. Notice that the greatest variations in levels of nitrate and ammonia occur in eutrophic lakes, the least variation being in oligotrophic lakes.

Oxidized forms of nitrogen (i.e. nitrate) can exist only at depths where oxygen is present. Because nitrate is a nutrient, expect its levels to be reduced in the epilimnion during the summer when algae are growing and using it for tissue growth. Oligotrophic lakes have high nitrate levels all year in the hypolimnion (Fig. 5.6A). Reduced forms of nitrogen (i.e. nitrite and ammonia) can exist in large quantities only under low oxygen or anoxic conditions, such as occurs in eutrophic lakes. Hence,

Table 5.7. Criteria for assessing trophic status of lakes using spring total nitrogen levels.

23

Criterion	Oligotrophy Secchi > 5m	Mesotrophy Secchi 2-5 m	Eutrophy Secchi < 2 m
Total nitrogen μg/L	< 300	300-600	> 600

24

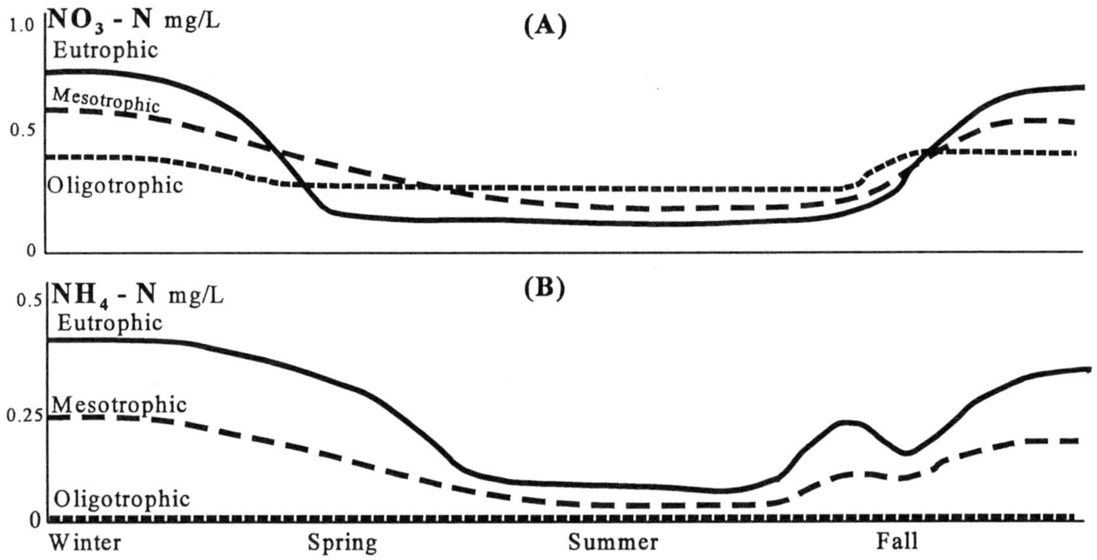

Figure 5.5. Seasonal variations in nitrate nitrogen (A) and ammonium nitrogen (B) in oligotrophic, mesotrophic and eutrophic lakes.

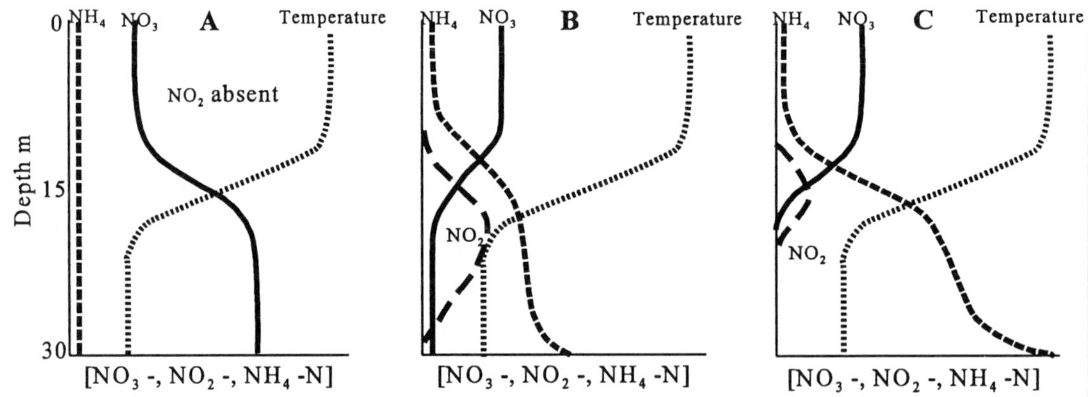

Figure 5.6. Summer depth profilesfor nitrate, nitrite and ammonium ion in oligotrophic (A), mesotrophic (B) and eutrophic (C) lakes.

lakes with higher summer concentrations of nitrite and ammonia in the hypolimnion than in the epilimnion, as shown in Figure 5.6C, are eutrophic. Mesotrophic lakes have intermediate levels and depth profiles of all forms of nitrogen (Figure 5.6B).

Carbon Dioxide

A primary source of carbon dioxide in water is the atmosphere. Additional carbon dioxide is produced from decomposition of organic material by bacteria. Much of this occurs in the sediments and at the sediment-water interface. Some carbon dioxide is contributed by the respiratory activities of all aquatic organisms, including plants, but the major contribution of CO_2 comes from the respiration of bacteria and their metabolism of organic material.

Although carbon dioxide is less concentrated in the atmosphere (<0.1% by volume) than is oxygen (~21% by volume), carbon dioxide is more abundant in water because it is about thirty-three times more soluble than oxygen. The carbon dioxide that we measure in water is ***free carbon dioxide***. The solubility of free CO_2 varies inversely with temperature and salinity and directly with partial pressure, but pH also affects its solubility (Fig. 5.7). This is discussed below, but suffice it to say for now that the solubility of free CO_2 increases with decreasing pH (or increasing hydrogen ion concentration).

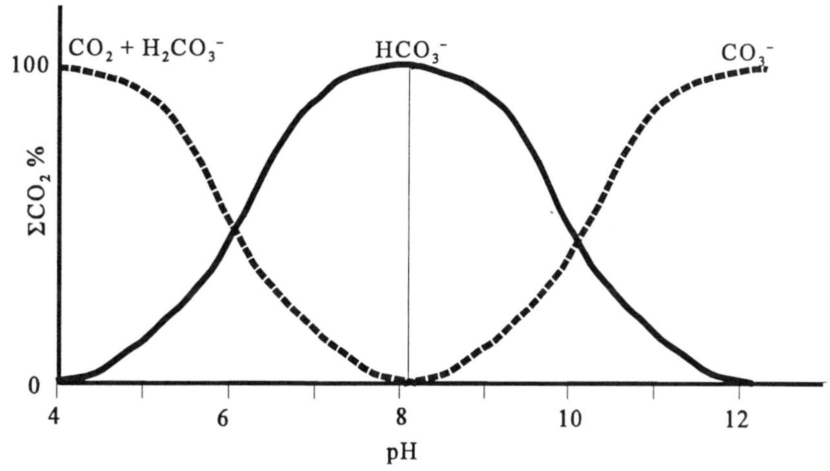

Figure 5.7. Relative proportions of three forms of carbon dioxide, free CO_2, bicarbonate and monocarbonate.

The depth profiles for free CO_2 are nearly the mirror image of those for dissolved oxygen. Hence, dissolved oxygen concentrations are high when the free CO_2 concentrations are low. A sharp increase in free CO_2
26 occurs at the sediment water interface in eutrophic lakes. The free CO_2 profiles that appear in lakes with different trophic levels are shown in Fig. 5.8. The processes that contribute to the shapes of CO_2 and dissolved oxygen profiles are complementary. In

the epilimnion, the photosynthetic activities of algae use carbon dioxide and produce oxygen, creating an inverse relationship between the concentrations of the two dissolved gases. In the hypolimnion, especially in eutrophic lakes where the sediments contain large amounts of organic material, bacteria use oxygen during the metabolism and break down of organic material, releasing CO_2 and elevating the free CO_2 levels at the sediment-water interface. As a result, oligotrophic lakes have low free CO_2 levels in the hypolimnion and an ***inverted orthograde profile*** such as that shown in Figure 5.8a, while eutrophic lakes have high free CO_2 levels in the hypolimnion and an ***inverted clinograde profile*** such as that shown in Fig. 5.8b.

The Carbon Dioxide-Bicarbonate-Carbonate Equilibrium

As mentioned earlier, pH affects the solubility of carbon dioxide. Figure 5.7 shows the effect of pH on the concentrations of free CO_2, bicarbonates, and carbonates in aquatic ecosystems. Carbon dioxide exists in fresh water in four forms: (i) ***free***
27☞ ***CO_2***, the dissolved state that is available to plants; (ii) ***hydrated CO_2***, or undissociated carbonic acid; (iii) ***half-bound CO_2***, or bicarbonates; and (iv) ***bound CO_2***, or monocarbonates. The free dissolved form is required by photosynthetic organisms and, with the remaining forms, is important in the chemical equilibrium relationships of carbonates and bicarbonates that affect the pH, alkalinity and buffering capacity of the water in the following manner. Equation 5.7 shows that carbon dioxide combines readily with water in the atmosphere (i.e. precipitation) to form carbonic acid which has a pH of about 5.6.

$$\underset{\text{dissolved gas}}{CO_2} + \underset{\text{water}}{H_2O} \leftrightharpoons \underset{\text{undissociated carbonic acid}}{H_2CO_3} \qquad \text{Equation 5.7}$$

Carbonic acid weakly dissociates into bicarbonate and monocarbonate ions, as shown in Equation 5.8.

28☞

$$\underset{\text{undissociated carbonic acid}}{H_2CO_3} \leftrightharpoons \underset{\text{bicarbonate}}{HCO_3^-} + H^+ \leftrightharpoons \underset{\text{monocarbonate}}{CO_3^{2-}} + 2\,H^+ \qquad \text{Equation 5.8}$$

Since most North American freshwater lakes lie within a pH range of 6 - 9, most of the CO_2 is present as bicarbonate (Fig. 5.7). The monocarbonate ion can exist only in water with a pH greater than 8.2. Carbon dioxide in lakes with pH > 12 is present as monocarbonate ions. Such lakes will also have free hydroxide ions (OH^-), but they are rare in North America. Hence, plants in lakes with pH > 8 potentially have an inexhaustible supply of CO_2 for photosynthesis.

The free CO_2 necessary to maintain bicarbonate in solution is called ***equilibrium CO_2***. As carbonic acid flows through the soil and comes into contact with

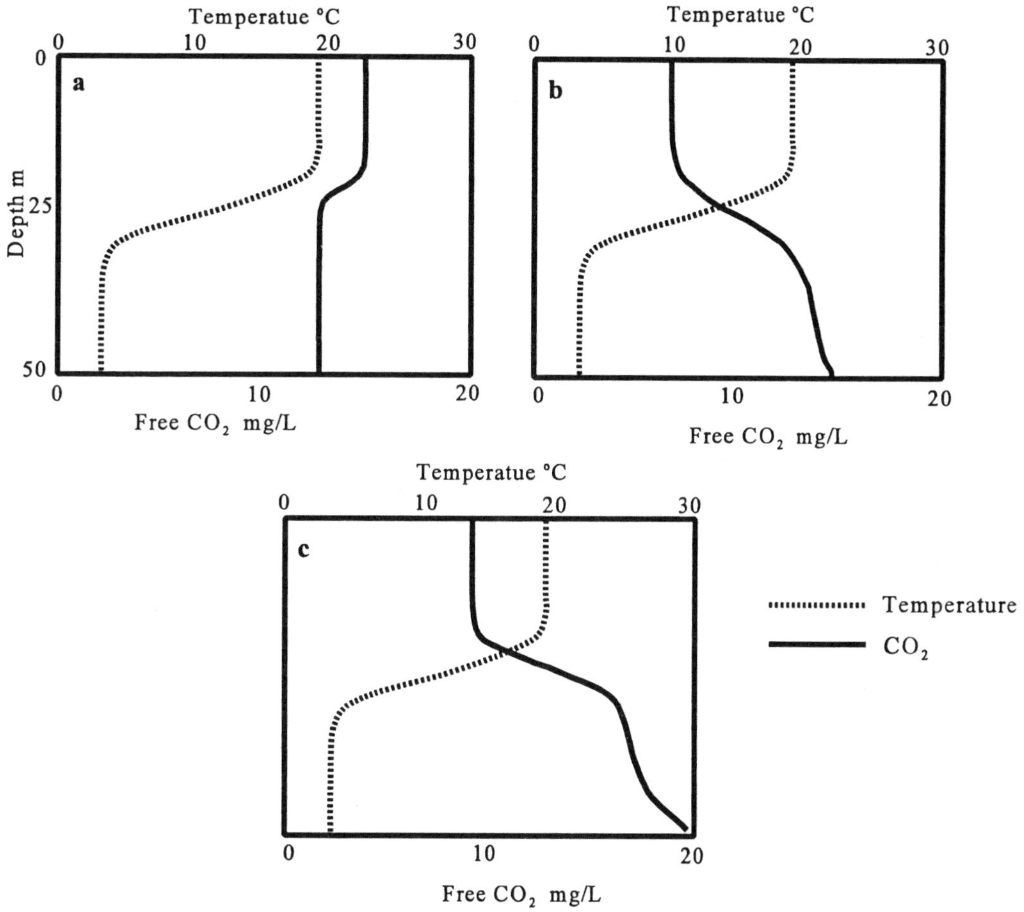

Figure 5.8. Free carbon dioxide depth profiles in (a) oligotrophic, (b) mesotrophic and (c) eutrophic lakes.

limestone, it reacts with the calcium carbonate to put calcium and bicarbonate ions into solution, as shown in Equation 5.9.

$$H_2CO_3 + CaCO_3 \rightarrow Ca^{2+} + 2HCO_3^- \qquad \text{Equation 5.9}$$

Calcium and bicarbonate ions react to form soluble calcium bicarbonate, with equilibrium CO_2 maintaining the reaction (Equation 5.10) in the pH range of 6 - 9. At pH levels greater than 8.2, any CO_2 removed from the water by photosynthesis will cause calcium carbonates to precipitate out, producing a floc that causes ***lake whitening***
29☞ which greatly reduces lake transparency. Calcium carbonate will also precipitate out as chalk, or ***marl***, on leaves of submersed plants in high pH lakes. In hard water streams, attached algae deposit calcium carbonate on rocks as ***travertine***. The travertine can also be produced by turbulence, where carbon dioxide is displaced by oxygen in the turbulent zone. The reaction that produces lake whitening, marl and travertine is shown in Equation 5.11.

$$\underset{}{Ca^{2+}} + 2HCO_3^- \leftrightharpoons \underset{\text{calcium bicarbonate}}{Ca(HCO_3)_2} \qquad \text{Equation 5.10}$$

$$\underset{\text{calcium bicarbonate}}{Ca(HCO_3)_2} \leftrightharpoons \underset{\substack{\text{calcium carbonate}\\ \text{Also called marl}\\ \text{or travertine}}}{CaCO_3\downarrow} + H_2O + \underset{\substack{\text{used by}\\ \text{plants}}}{CO_2} \qquad \text{Equation 5.11}$$

Any additional CO_2, such as that introduced from dehydration of carbonic acid that causes dissolution of $CaCO_3$, is called ***aggressive CO_2***. This can also occur during periods of darkness when respiration rates of plants and animals exceed photosynthetic rates of plants, or when other nutrients are limiting. However, the reaction is essentially irreversible since at pHs greater than 8.3, any additional CO_2 produced by dehydration of carbonic acid is quickly taken up by plants.

In essence, reactions involving addition of CO_2 are the same as reactions involving addition of hydrogen ions. Any CO_2 that is added combines with hydrogen ions to form carbonic acid which helps dissolve monocarbonates, such as calcium carbonate, to produce bicarbonates. The reactions (Equations 5.7-5.11) will continue to the left until all of the carbonates and bicarbonates have been reduced to carbonic acid and, eventually, carbon dioxide. Lakes, such as those in the Muskoka district of Ontario, which have little or no calcium carbonate dissolved in the water, will have a pH less than 8.2 (see Fig. 5.7). Even if the bicarbonate content is low, which is characteristic of lakes with a pH near 7, the pH level in the lake will drop very quickly if any hydrogen ions are introduced, such as those from acidic precipitation, because there is little or no bicarbonates to absorb, or buffer the hydrogen ions added. Such lakes acidify very quickly.

Water Hardness

The hardness of water is contributed by the cations of calcium, magnesium, and to some extent, iron. Total hardness is usually taken as the sum of the concentrations
30☞ of calcium and magnesium and expressed as mg $CaCO_3$/L. Hardness is determined by titration with EDTA (ethylenediaminetetra-acetic acid) which chelates calcium and magnesium, the titrant being standardized to give results equivalent to mg of $CaCO_3$ per litre of water. Hence, if water contained compounds of $CaCO_3$, $Ca(HCO_3)_2$, $MgCO_3$, $Mg(HCO_3)_2$, $CaSO_4$, $MgSO_4$, $CaCl_2$, and $MgCl_2$, the total hardness would be the sum of the concentrations of the calcium combined with monocarbonates (CO_3), bicarbonates (HCO_3), sulphates (SO_4), and chlorides (Cl). Of the compounds that contribute to total hardness, only those combined with monocarbonates and bicarbonates will contribute to total alkalinity, as discussed below. The total concentration of cations of Ca, Mg and Fe combined with carbonates and bicarbonates is known as ***temporary hardness*** because the hardness can be removed by boiling the water and converting
31☞ carbonates and bicarbonates to carbon dioxide, leaving calcium uncombined in solution or deposited on the walls of containers as "***scale***". The total concentration of cations of

Ca, Mg, (and Fe) combined with non-carbonates (e.g. SO_4, Cl, PO_4, etc.) is known as ***permanent hardness*** because the hardness cannot be removed by boiling. If only the concentration of calcium combined with carbonates and non-carbonates is considered, the hardness is expressed as ***calcium hardness***. Because nearly 100% of the total hardness is due to cations of calcium and magnesium in most surface waters (iron usually contributes less than 1% of the total hardness), an estimate of the ***magnesium hardness*** can be made by subtracting the calcium hardness from the total hardness. ***Hence, the total hardness can be estimated in two ways: (1) by summing the temporary and permanent hardness; (2) by summing the calcium and magnesium hardnesses.***

The concept of hardness originated in the early days to express the soap-
32 consuming power of water. Hard water (> 120 mg $CaCO_3$/L) causes soap to precipitate, leaving a ***scale*** on the surface of the vessel. Forming a lather with soap or shampoo is difficult when using hard water because the cations of calcium and magnesium cause the soap to precipitate. Scale (= $CaCO_3$) develops on the walls of kettles or any boiling vessel (e.g. dishwashers, car radiators) when hard water is used. Boiling removes the temporary hardness, leaving only permanent hardness in the water. Soft waters have < 60 mg $CaCO_3$/L; medium hard waters have 60-120 mg $CaCO_3$/L; hard waters have 120-240 mg $CaCO_3$/L; very hard waters have > 240 mg $CaCO_3$/L. Some well water has over 1000 mg $CaCO_3$/L!

> One can "guestimate" the range of water hardness by the soap test; if you can form a lather easily, the water is soft; if it is difficult to form a lather, the water is hard.

Hard water should be "softened" or hot-water appliances, such as water heaters, kettles etc. will develop thick layers of scale, rendering them useless. Water softeners "soften" the water by replacing cations of calcium with either sodium or potassium. The "ring-around-the-tub" is a common sight in hard-water areas. The bathtub ring is
33 calcium and/or magnesium that has precipitated the soap. This does not occur with soft water.

Alkalinity

Alkalinity is a measure of the ability of water to resist a change in pH. Water resists changes in pH due to the presence of anions of carbonates, bicarbonates (and occasionally hydroxides). The ***total alkalinity*** of water is the sum of the concentrations of bicarbonate (HCO_3) and carbonate (CO_3) anions (and hydroxide ions if present).
34 This contrasts with total hardness which is the total concentration of cations of calcium and magnesium (and iron if present at high levels). Since temporary hardness measures the amounts of anions of bicarbonates and carbonates, it is essentially the same as total alkalinity. If we use the same compounds in the water sample as used to discuss hardness above (i.e. $CaCO_3$, $Ca(HCO_3)_2$, $MgCO_3$, $Mg(HCO_3)_2$, $CaSO_4$, $MgSO_4$, $CaCl_2$, and $MgCl_2$), the compounds which contribute to alkalinity are $CaCO_3$, $Ca(HCO_3)_2$, $MgCO_3$ and $Mg(HCO_3)_2$ because only they have anions of carbonate and bicarbonate; the remaining compounds do not contribute to total alkalinity.

Alkalinity is usually expressed in terms of equivalents of bicarbonate or carbonate. or mg of $CaCO_3$ per litre of water. Total alkalinity is the sum of ***carbonate alkalinity*** and ***bicarbonate alkalinity***. To measure total alkalinity, one must determine the carbonate alkalinity first and then the bicarbonate alkalinity. Carbonate alkalinity need not be measured if the pH is 8.2 or less because carbonate alkalinity can only exist in waters with higher pHs (Fig. 5.7). If the pH is 8.3 or greater, carbonate alkalinity is determined by adding an indicator, such as phenolphthalein, which changes colour at pH 8.3. The sample is then titrated with a weak acid, such as 0.02 N H_2SO_4, until the colour changes. Because phenolphthalein is the most commonly used indicator at this step, carbonate alkalinity is also known as ***phenolphthalein alkalinity***. At this point (pH 8.2), all the carbon dioxide is present as bicarbonates (Fig. 5.7). Bicarbonates can be reduced to carbonic acid by adding an acid until pH 4.4 (Fig. 5.7). Hence, indicators that change colour at pH 4.4, such as methyl orange and bromthymol green methyl red, can be used to detect the change. The usual procedure is to take exactly 100 ml of water, add two or three drops of phenolphthalein, which turns the water sample pink, then titrate with 0.02 N H_2SO_4 until the water sample turns clear; note the millilitres of acid titrant added. Since 1 ml of 0.02 N H_2SO_4 corresponds to 1 mg $CaCO_3$, and a 100 ml of water sample was used, multiply the millilitres of acid titrant by 10 to give carbonate alkalinity in mg $CaCO_3$ /L. At this point, all the monocarbonates have been converted to bicarbonates, where each carbonate ion takes up one hydrogen ion to become a bicarbonate ion. Now add methyl orange indicator and continue titrating until the colour changes from orange to blue. Note the millilitres of acid titrant added; multiply the millilitres of acid added by 10 to determine the bicarbonate alkalinity. At this point all the bicarbonates have been converted to carbonic acid and carbon dioxide. Simply sum the carbonate alkalinity and the bicarbonate alkalinity to determine total alkalinity. Alternatively, sum the millilitres of acid titrant added and multiply by 10.

If the pH of the water is greater than 12, such as occurs in alkali lakes or soda lakes in some arid regions where sodium carbonate (Na_2CO_3) and bicarbonate ($NaHCO_3$) abound, the total alkalinity will also be represented by hydroxyl (OH^-) ions. In this case, the hydroxide radical is equivalent to a normal carbonate radical. The reactions that occur with addition of 0.02 N H_2SO_4 are:

Hydroxides are converted to water:

$$H_2SO_4 + Ca(OH)_2 \rightarrow CaSO_4 + H_2O \tag{5.12}$$

Carbonates are converted to bicarbonates:

$$H_2SO_4 + 2CaCO_3 \rightarrow CaSO_4 + Ca(HCO_3)_2 \tag{5.13}$$

Bicarbonates are converted to carbonic acid:

$$H_2SO_4 + Ca(HCO_3)_2 \rightarrow CaSO_4 + 2H_2CO_3 \tag{5.14}$$

Buffer Systems

Buffer solutions are those that resist changes in hydrogen ion concentration

when acids or bases, are added. Fresh water with pH < 7 (acidic) becomes a buffer when an alkali is added; at pH > 7 (alkaline) fresh water becomes a buffer when an acid is added. Because the major source of hydrogen ions is the atmosphere, and precipitation has a pH near 5.6, most lakes must buffer additions of acid. Lakes that are on the acidic side of neutrality have very little buffering capacity, while lakes on
35 the alkaline side of neutrality have more buffering capacity. Clearly, lakes that lie over limestone ($CaCO_3$) or dolomitic ($Ca.Mg.CO_3$) bedrock have an unlimited buffering capacity. Lakes that lie over granitic bedrock have little or no buffering capacity. Lakes that are acidifying are doing so because they have little or no buffering capacity. For example, lakes in the Guelph region, where dolomitic rock prevails, will never acidify even though they are receiving the same acidic inputs as lakes in the Sudbury area where granitic rocks predominate.

The reactions shown in equations 5.13 to 5.14 represent the buffering reactions of lake water containing some dissolved monocarbonates ($CaCO_3$), the sulphuric acid (H_2SO_4) being acid precipitation present over most of Canada (and any other industrialized country). Lakes and rivers containing at least some bicarbonates have some ability to buffer the additions of acid, where strong acids, like sulphuric acid, are converted to much weaker acids like, carbonic acid. Chapter 13 provides additional details on the chemistry and effects of acid precipitation.

Redox Potential

The redox potential, or ***reduction/oxidation potential***, indicated with the symbol ***Eh***, is the electrical voltage that occurs between two electrodes, one made of hydrogen and the other made of platinum. The change in oxidation state of many
36 metallic ions is defined by the redox potential. At neutral pH and 25°C oxygenated lake water has a redox potential of about +500 millivolts. Under these conditions most common metals, such as iron, are stable in their most oxidized forms. Hence, iron will be present as ferric iron (Fe^{3+}), not the reduced form, ferrous iron (Fe^{2+}). Similarly, nitrates rather than nitrite or ammonia, which are reduced forms of nitrate, will be present. As the oxygen concentration falls, the redox potential drops and when it reaches about 250 millivolts ferric iron is converted to ferrous iron. Below +250 millivolts, Fe^{2+} is favoured. At 0 millivolts hydrogen sulphide (H_2S) is favoured over sulphate and sulphur.

The greatest changes in redox potential occur at the sediment-water interface where a thin oxygenated zone exists. Since redox potential is affected greatly by dissolved oxygen content, changes in redox potential are much more dramatic in eutrophic lakes than in oligotrophic lakes. This information is also useful for assessing
37 trophic status of a lake. For example, since oligotrophic lakes have orthograde oxygen profiles, and Eh is directly related to dissolved oxygen levels, one can expect to see an orthograde shape to the Eh depth profile as well. Eutrophic lakes will have a clinograde-shaped Eh depth profile and mesotrophic lakes will have an intermediate

shape for their Eh depth profiles.

Detritus Pool and Seston

So far we have talked about the sources of the inorganic pool of carbon, that is carbon dioxide, bicarbonates and carbonates. Organic sources of carbon are collectively called the ***detritus pool*** and consists of ***dissolved organic material*** (***DOM***) and ***particulate organic material*** (***POM***). POM is defined on the basis of what is retained on a filter with a maximum pore size of 0.45 μm. Usually 0.45 μm is rounded off to 0.5 μm to separate POM from DOM. POM can be further categorized as ***CPOM***, or ***coarse particulate organic matter***, which has a diameter greater than 1 mm (1000 μm), and fine FPOM, or ***fine particulate organic matter,*** which includes material less than 1 mm in diameter. CPOM and FPOM are acronyms commonly used in stream studies involving metabolism of organic material.

DOM includes the soluble metabolites and excreta of organisms. POM includes living and dead particulates. The term, ***detritus*** refers to the dead plants and
38☞ corpses or cast-off parts of organisms that once were living in the water or on the bottom of the lake or river. ***Seston*** includes detritus and inorganic particulate material. Both detritus and seston are an important source of nutrients to a variety of organisms.

The origins of seston are several, but can be divided broadly into autochthonous and allochthonous material. Within each of these components is a dissolved fraction and a particulate fraction. The particulate fraction can be removed with fine-pore filters, the filter paper containing the particulate fraction, the filtrate containing the dissolved fraction. The DOM, includes the excreta of all aquatic organisms, especially zooplankton, any metabolic byproducts that are released as soluble organic compounds during bacterial decay of plant and animal material, and dissolved organic nitrogen and phosphorus.

DOM can be converted to POM by two main mechanisms. The first is called ***agitation aggregation*** whereby tiny plate-like aggregates of organic composition form when air is bubbled through filtered water. Aggregates occur because some compounds, such as soluble fatty acids, proteins and humic materials, readily adhere to bubble surfaces. The second mechanism involves bacteria and algae. Bacteria divide and re-divide incorporating DOM into their cells and daughter cells. Some blue-green algae, or Cyanobacteria, are also able to convert DOM to POM.

Organic material is ultimately broken down to carbon dioxide by bacteria. All organic materials, and the organisms and processes involved in taking up and returning carbon dioxide to the atmosphere, make up the carbon cycle (Fig. 5.9). The main sources of carbon in the cycle are from CO_2 in the atmosphere, which autotrophs (algae
39☞ and macrophytes) use to produce oxygen, and allochthonous materials (CO_2 , DOC, POC, CPOM and FPOM, etc.) from parent streams. There are also autochthonous materials produced within the lake. All materials are utilized and processed by bacteria

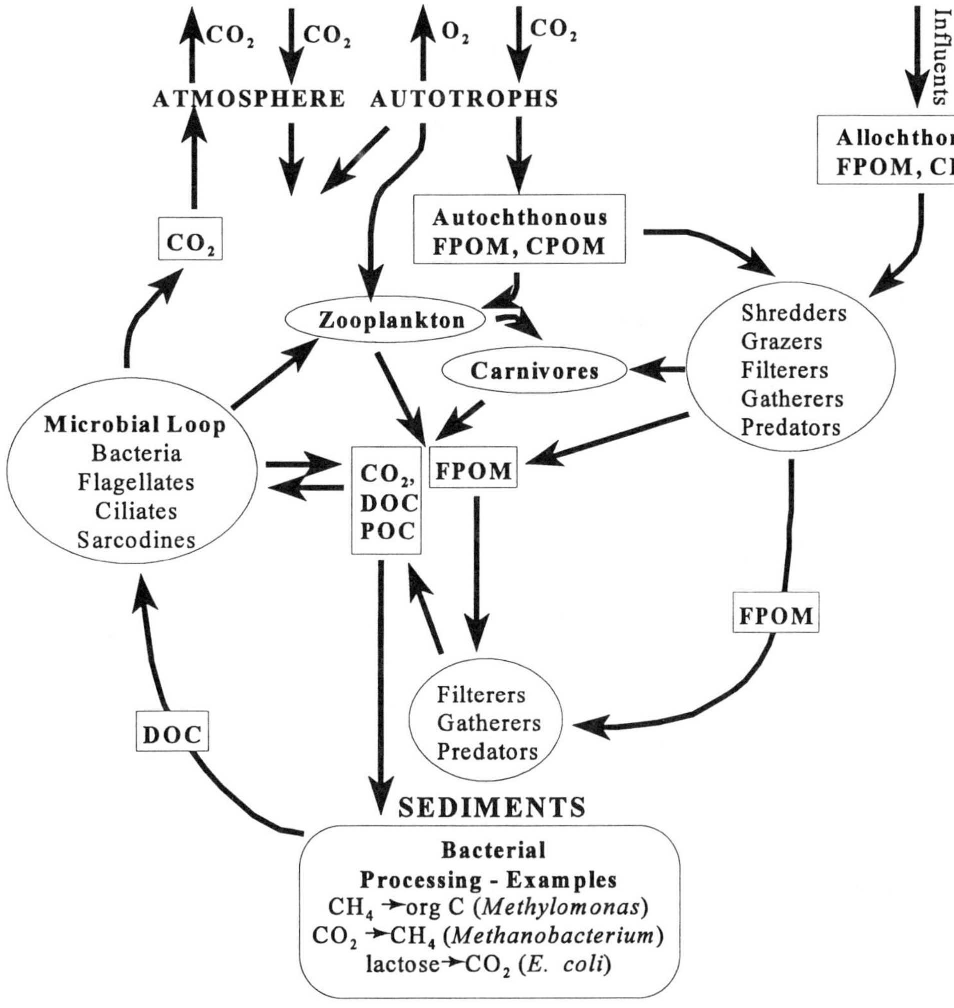

Figure 5.9. The carbon cycle in deep lakes

and phytoplankton which make up the microbial loop (see Chapter 6, zooplankton, for details), zooplankton and a few macro-invertebrates

Allochthonous sources tend to be more ***refractory*** (resistant to decomposition) than autochthonous sources. Autochthonous sources include algae, macrophytes and animal tissues (e.g. molted skins or "exuviae") and are more ***labile*** (easily broken down to CO_2 and carbon). Figure 5.9, shows that allochthonous organic matter is processed by organisms of different functional feeding groups where ***shredders*** convert coarse particulate organic matter (CPOM), like leaves from trees, into fine particulate organic matter (FPOM); ***grazers*** feed on algae and also contribute to the FPOM pool; ***gatherers,*** like oligochaete worms and amphipod and isopod crustaceans, collect and feed on the ditritus (FPOM), while the ***filter feeders*** (e.g. all bivalves and many insects) filter FPOM from the overlying water; the ***predators*** are diverse and specialized for feeding on any of the functional groups, including other predators. The assemblage of functional feeding groups in the profundal zone is different from that in the littoral zone; shredders and grazers are absent because the sediments consist mainly of FPOM (hence no shredders) and there is no light for algal growth (hence no grazers).

Sediment material itself contributes to the carbon pool. Bacteria such as *Methanobacterium,* convert carbon dioxide to methane gas in anoxic lakes; *Methanomonas* converts methane back to carbon dioxide; and *Escherichia coli* converts lactose in plants to carbon dioxide. All are a major part of the bacterial assemblage in sediments. The bacteria themselves are part of the POM and are fed upon by filter feeding invertebrates and zooplankton. POM gathers on the bottom only when the rate of sedimentation exceeds the rate of conversion back to carbon dioxide. This carbon dioxide is either recycled as autochthonous carbon or is exported, with some DOC and POC, out the outflow

Other Nutrients

The main nutrients for autotrophs are carbon, phosphorous and nitrogen. Other important elements, to autotrophs and heterotrophs, are sulphur, because it forms an integral part of proteins (e.g. sulfhydryl chains), iron and manganese, because they are important in forming blood and vitamins, and calcium and silica, because they are the most common structural body components (e.g. bones and exoskeletons). All are essential elements. Some elements are essential to some organisms and non-essential to others. For example, copper is required to form a blood pigment called hemocyanin in many molluscs and crustaceans, and zinc is an important element in many vitamins. However, only trace amounts of these metals are required and both can become toxic when present in even small amounts to most organisms, including those for which the metal is essential.

Nitrogen, carbon and calcium have already been discussed. Phosphorous, sulphur, iron and manganese and silica are discussed below.

Phosphorous

The phosphorous cycle is shown in Fig. 5.10. Of the several sources of phosphorous, only one, bedrock, is natural. Apatite ($Ca_3(PO_4)_2$ or $Ca_5(PO_4)_2$) and
40☞ ferric phosphate ($Fe_3(PO_4)_2$) are the most common phosphorous-bearing rocks. Phosphorous is eroded from the rock and supplied, mostly as insoluble particulates, to a lake by parents streams and rivers. Some phosphorous is also supplied by aerial deposition. Nearly all (> 90%) loadings end up on the bottom of the lake, largely because of clays which have a high phosphorous-holding capacity. Less than 10% enters lakes as soluble phosphate.

All other sources of phosphorous are anthropogenic in origin, such as detergents in sewage, but most (> 90%) of it is present in soluble form and very little (< 10%) is insoluble. Because algae utilize soluble phosphate (e.g. orthophosphate) as a nutrient, anthropogenic sources are the main culprit causing algae blooms, and hence eutrophication (enrichment), in lakes. Fortunately, this was discovered in the mid 1970s and enrichment with anthropogenic phosphorous was greatly curtailed by reducing phosphorous levels in detergents and improving phosphorous removal methods in sewage treatment facilities. Eutrophication rates in the Great Lakes, especially Lake Erie and Lake Michigan, have slowed greatly and the future of sport and commercial fisheries and recreation is not nearly as bleak as it was in the 1970s. However, because most lakes are sinks for phosphorous, there is still a large amount in lake sediments. The mechanism for mobilizing phosphorous is complex but mobilization rates increase with decreasing dissolved oxygen levels. Hence, anoxic sediments of eutrophic lakes are more likely to release phosphorous and maintain eutrophication rates than are mesotrophic or oligotrophic lakes. This ***internal loading*** of phosphorous explains why many shallow lakes remain eutrophic long after all anthropogenic sources have been removed. In deep lakes, internal loading may be confined to the hypolimnion, with only small amounts of phosphorous being leaked to the epilimnion where primary production can occur.

Redox conditions and pH at the sediment-water interface control, in part, the mobilization rate of phosphorous from sediments to water because many insoluble compounds become more soluble as the pH and redox potential drops. For example, at redox potentials less than +250 mV, insoluble ferric phosphate ($Fe^{3+}PO_4$) is reduced to the more soluble ferrous phosphate ($Fe^{2+}(PO_4)_2$), which dissociates into free phosphate and iron according to the following reactions:

$$Fe^{3+}PO_4 \rightleftharpoons Fe^{2+}PO_4 \rightleftharpoons 3Fe^{2+} + 2PO_4^{3-} \qquad (5\text{-}15)$$

Similarly, phosphorous can be mobilized when it is adsorbed onto compounds whose solubilities change with decreasing pH and Eh. Hence, any phosphate adsorbed onto insoluble ferric hydroxide can be set free (mobilized) when ferric hydroxide is reduced to soluble ferrous hydroxide:

$$Fe^{3+}(OH)_3 + 2PO_4^{3-} \rightleftharpoons Fe^{2+}(OH)_2 + PO_4^{3-} \qquad (5\text{-}16)$$

Such reactions are not balanced stoichiometrically because of the sorption processes.

The amount of precipitated phosphorous is different from sorbed phosphate even though both are controlled by pH and Eh. Phosphate can be sorbed directly onto hydrous iron, aluminum oxides (in clays), calcite particles and apatite, but the kind and amount of metallic phosphorous compounds varies with pH. One mineral may buffer the release rate of phosphorous from other minerals. Biological activity of benthic invertebrates may alter the release rate of phosphorous by altering sediment profiles (e.g. worms and insect larvae mix upper and lower layers of sediments) and excreting soluble phosphorous compounds into the pore waters.

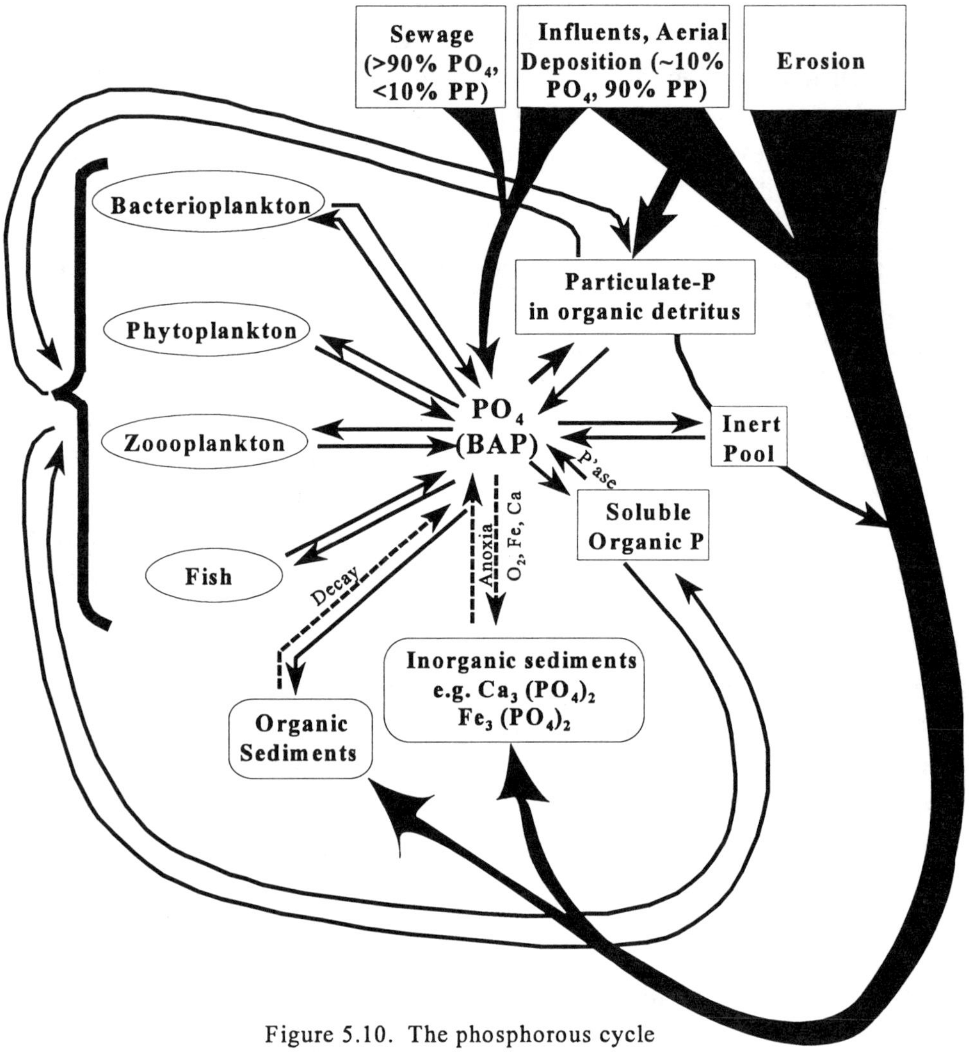

Figure 5.10. The phosphorous cycle

Once phosphorous is mobilized, some or all of it, depending on the degree of thermal stratification and resistance to mixing, may reach the epilimnion where it can be utilized by phytoplankton for growth. The phosphorous available to algae is known as ***BAP***, or ***biologically active phosphate***. Some phytoplankton produce ***alkaline phosphatase***, an esterase that cleaves the bond between phosphate and the molecule to which it is attached. The esterase is produced when there is a phosphorous deficiency in the water.

Macrophytes also play an important role on phosphorous cycling. Uptake of phosphorous by macrophytes occurs through the leaves if there is a rich supply of phosphate, but most (~ 85%) is taken up by the roots from the interstitial sediment pore water. The phosphorous is recycled back to the plants when they die and decay each year, but the amount released varies among species. The retained phosphorous is known as ***refractory phosphorous*** and remains in the sediments as unavailable phosphorous.

All aquatic organisms release phosphorous from their bodies, either in excreta or when they die and decay. Phosphorous in animals is obtained by eating plants or algae. Hence, grazing and herbivorous species play an important role in regulating the amount of chlorophyll and phosphorous in water. Unlike the nitrogen and sulphur cycles, which are driven by bacterial process, the phosphorous cycle is mainly a chemical one, controlled mostly by pH, oxygen levels and Eh at the mud-water interface.

Because phosphorous is limiting, nearly all (~90%) of the variation in chlorophyll *a* levels can be explained by variations in spring total phosphorous (P_T)
41 levels (see Fig. 6.17). The relationship is used to predict trophic status of lakes because it is levels of chlorophyll *a* and phosphorous that really determine the amount of eutrophication that can occur. Eutrophic lakes characteristically have spring P_T levels exceeding 30 μg/L and average summer chlorophyll *a* levels above 5 μg/L. Table 5-8 gives ranges in values of phosphorous and chlorophyll *a* for lakes of different the trophic states. The application of the phosphorous-chlorophyll relationship to determining trophic status of lakes is described further in Chapters 6 and 13.

The P_T-chlorophyll relationship is also useful for predicting the effects of the
42 watershed on phytoplankton blooms. When phosphorous is the growth-limiting nutrient, the amount of P_T entering the lake each year is related to the phytoplankton biomass. A ***phosphorous loading curve*** can be drawn to relate phosphorous loading (on *y* axis) to some lake-related parameter, such as depth (on *x* axis). Depth is usually used because it is related to retention time which indicates the amount of internal loading to expect from the sediments. Shallow lakes have a rapid turnover of inflowing
43 nutrients and much of the sediment is exposed to mixing by wave action. Deep lakes have only a small fraction of their sediments in the epilimnion; the rest is in the hypolimnion and most of the phosphorous mobilized in the hypolimnion is unavailable to the epilimnetic waters. Phosphorous loading curves are drawn only for lakes in which phosphorous is the limiting nutrient. They cannot be used for lakes that are nitrogen limited. Chapter 13 examines the use of the phosphorous loading curve to assess the fisheries and recreational potential of lakes.

Table 5.8. Values for spring total phosphorous and average summer chlorophyll a levels in lakes of three trophic states.

Trophic State	Total Phosphorous μg/L	Chlorophyll *a* μg/L
Oligotrophic	< 10	< 2
Mesotrophic	10-30	2-5
Eutrophic	> 30	> 5

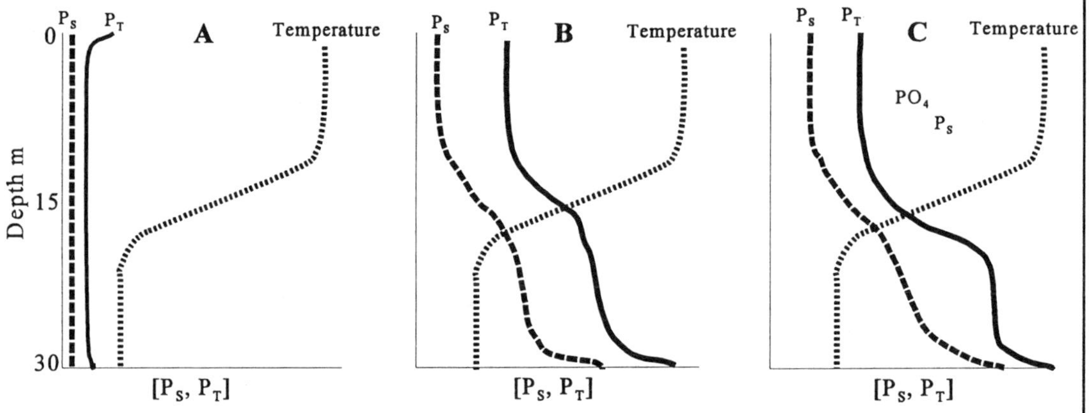

Figure 5.11. Summer depth profiles for soluble phosphorous (P_S) and total phosphorous (P_T) in oligotrophic (A), mesotrophic (B) and eutrophic (C) lakes.

An easy way to assess trophic status of a lake based on phosphorous levels is to examine the shape of the total phosphorous depth profile (Fig. 5.11). Knowing that
44 phosphorous is mobilized under anoxia, one can expect an increase in phosphorous levels in the hypolimnion of eutrophic lakes (Fig. 5.11C) and little or no phosphorous present in the hypolimnion of oligotrophic lakes (Fig. 5.11A). Mesotrophic lakes have an intermediate total phosphorous depth profile (Fig. 5.11B).

Sulphur

Sulphur is important in protein structure and plays a role in cell division. It is usually very abundant in water (10-20 mg/L as sulphate) and therefore rarely limits growth or distribution of organisms. The sulphur cycle, shown in Fig. 5.12, is driven
45 mainly by bacteria. The major sources of sulphur are the atmosphere, which contributes hydrogen sulphate ($H_2SO_4^{2-}$), groundwater and bedrock, such as gypsum ($CaSO_4$) and pyrite (FeS_2), which contribute elemental sulphur and sulphate. Sulphur emissions are a major cause of acid precipitation. It is highest in industrialized areas and comes from the burning of fossil fuels and smelting of sulphur-rich ores. Lakes have a wide range of pH values, due to sulphate and nitrate ions from anthropogenic sources, the lowest pH values being localized to industrialized areas. However, remember that buffering capacity of the water dictates the pH level of the water.

Sulphate (SO_4^{2-}) is the most oxidized form of sulphur and is the compound available to autotrophs. Heterotrophs (organisms which consume other organisms or detritus) obtain their sulphur through the food chain. Sulphur is returned to water when autotrophs and heterotrophs die. Under anoxic conditions, sulphur is returned as hydrogen sulphide by bacterial decomposition and putrefaction of proteins. Under oxic

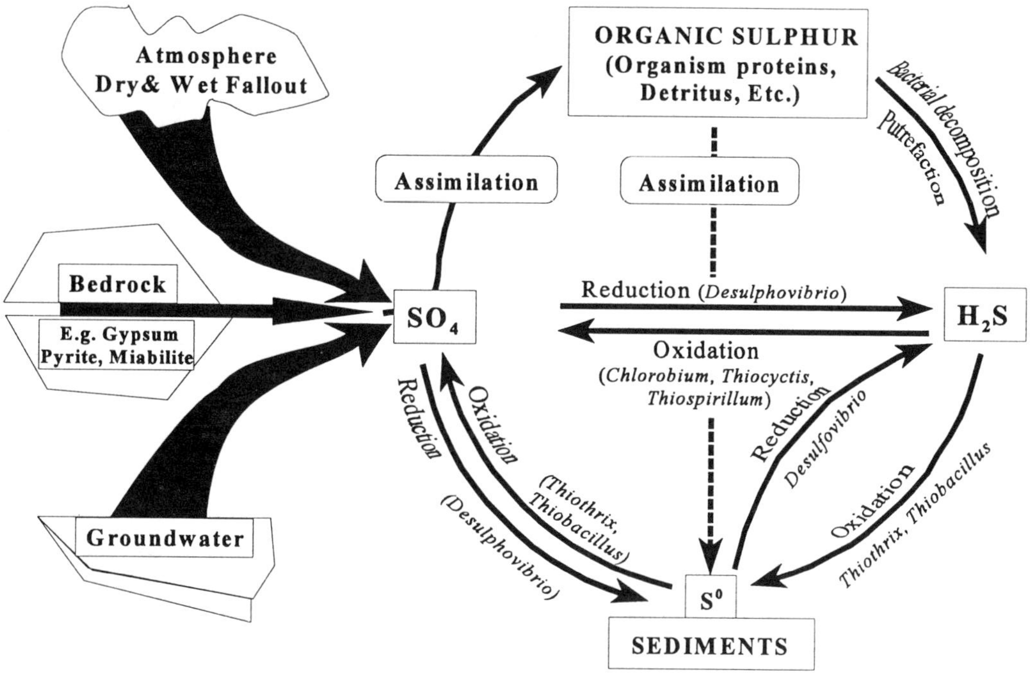

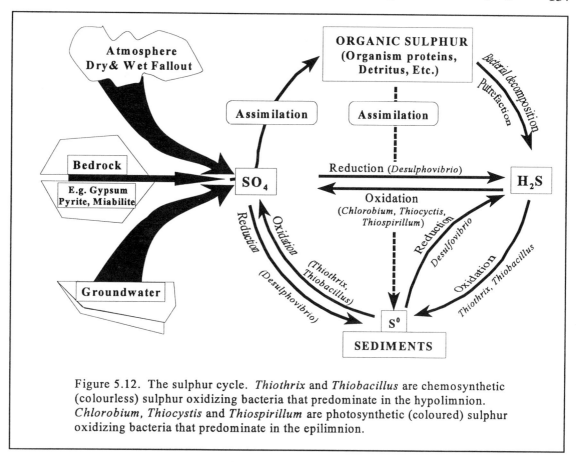

Figure 5.12. The sulphur cycle. *Thiothrix* and *Thiobacillus* are chemosynthetic (colourless) sulphur oxidizing bacteria that predominate in the hypolimnion. *Chlorobium, Thiocystis* and *Thiospirillum* are photosynthetic (coloured) sulphur oxidizing bacteria that predominate in the epilimnion.

conditions, hydrogen sulphide is quickly oxidized to sulphate (Fig.5.12). Eutrophic lakes, with their anoxic sediments, often release hydrogen sulphide gas (H_2S), which gives a "rotten egg smell". As H_2S rises through the water column, it is oxidized to SO_4^{2-} when mixed with oxygenated water. In highly anoxic sediments, H_2S combines with iron to produce ferrous sulphide (FeS) which precipitates out as a black ooze.

Figure 5.12 shows two basic bacterial processes in the sulphur cycle: (1) ***sulphate and sulphur reduction***; (2) sulphur oxidation. *Desulfovibrio* is an example of ***sulphate reducing bacteria***. ***Sulphur oxidizing bacteria*** are of two types; ***chemosynthetic or colourless bacteria*** and photosynthetic or coloured bacteria. The chemosynthetic bacteria deposit elemental sulphur either inside their cell walls (e.g.
46☞ *Thiothrix*) or outside their cell walls (e.g. *Thiobacillus*). The photosynthetic bacteria are either green (e.g. *Chlorobium*) or purple (e.g. *Thiocystis*). Chapter 7 examines other kinds of sulphur bacteria.

The sulphate depth profiles are also a useful way to assess trophic status of a lake because they are determined by the amount of oxygen in the hypolimnion. Sulphate, being an oxidized form of sulphur, will be abundant in oligotrophic lakes and perhaps even show an increase at the mud-water interface (Fig. 5.13a). Eutrophic lakes have only reduced forms of sulphur in the hypolimnion, with a rotten egg smell of hydrogen sulphide present in shallow, anoxic lakes (Fig. 5.13c). Such lakes are often characterized

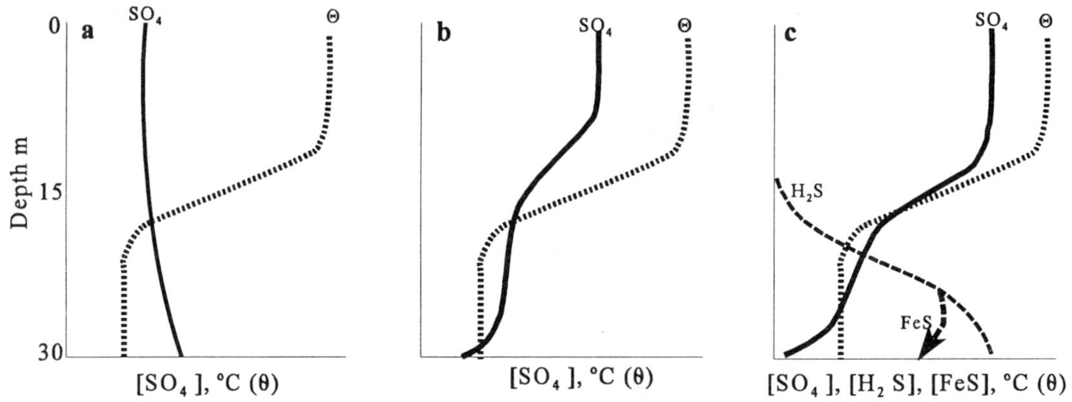

Figure 5.13. Summer depth profiles for sulphate and various forms of sulphur, when present, in oligotrophic (a), mesotrophic (b) and eutrophic (c) lakes.

by a black, oozy layer of sediment. Mesotrophic lakes have intermediate levels of sulphate in the hypolimnion (Fig. 5.13b).

Silica

The silica cycle (Fig. 5.14) is different from any of the cycles described above in that there is a one-way flow of silica from the edaphic supply to the lake sediments.
47☞ Only a few organisms require large amounts of silica; the rest need little or none. In freshwater, the most demanding of silica, mainly as reactive silicate (H_2SiO_4), are diatoms (a type of phytoplankton) which build their siliceous ***frustules*** (shells) with it, and sponges (a benthic, attached form), which build their protective needle-like ***spicules*** with it. The form of silicon used as a structural component in these organisms is ***opal***, a hydrated form of amorphous silica ($SiO_2 .nH_2O$).

Silica is released as soluble SiO_2 from rocks, particularly feldspars, when it is weathered by water containing carbonic acid in equilibrium with CO_2 and water:

$$2KAlSi_3O_8 + 2CO_2 + H_2O \rightleftharpoons 4\,SiO_2 + Al_2Si_2O_5\,(OH)_4 + 2K^+ + 2HCO_3^- \quad (5\text{-}17)$$

Potassium Feldspar	Rain, spring, stream water	soluble silica	Kaolinite (Clay)	Potassium ion	Bicarbonate ion

Some feldspars contain sodium instead of potassium. The weathering of feldspars is an important source of sodium and potassium ions in fresh waters. Kaolinite, a derivative of feldspars, has colloidal properties and, with some forms of silica, play an important role in sorption processes in the phosphorous cycle. Grazers of diatoms also release/excrete SiO_2 fragments, but the silica is unchanged chemically and is deposited on the lake floor.

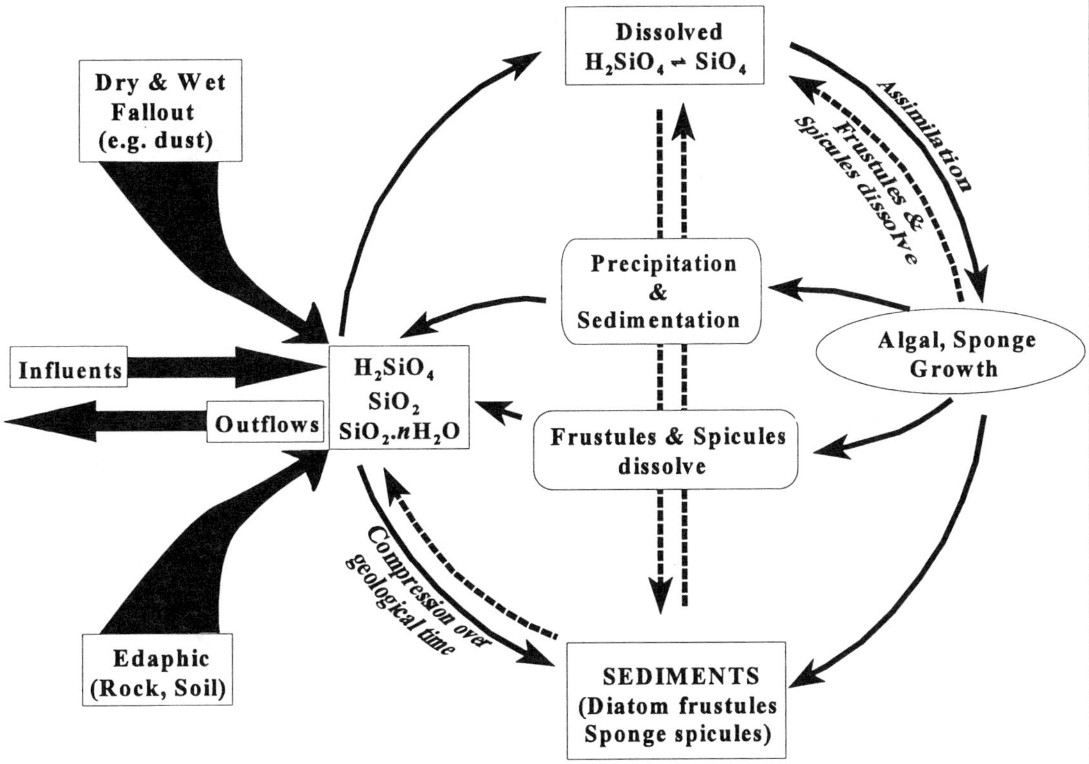

Figure 5.14. The silica cycle in deep lakes. Influents include groundwater, inflows, surface runoff, etc.

Diatom frustules are very resistant to decay and form huge deposits on the ocean floor where they have remained unchanged for millennia. For this reason, paleolimnologists use the types and numbers of diatoms for interpreting changes in trophic status over geological time. The diatom deposits are an excellent source of diatomaceous earth that has numerous uses, including filtration media for filters such as those used in swimming pools.

Iron

Most organisms require iron as a structural element in enzymatic and cellular processes, hemoglobin in blood, and enzymes. Iron is present in small amounts in
48☞ almost all clays, soils and granitic rock. Soluble iron ranges from about 0.1 to 3 mg/L in most freshwater, oligotrophic lakes having less than eutrophic lakes. Since most lakes have some oxygen, the oxidized form, ferric iron (Fe^{3+}), predominates over the reduced form, ferrous iron (Fe^{2+}). Because ferric iron is less soluble than ferrous iron, most iron precipitates out on the lake floor as ***ocher*** ($Fe(OH)_3$), a rust-coloured

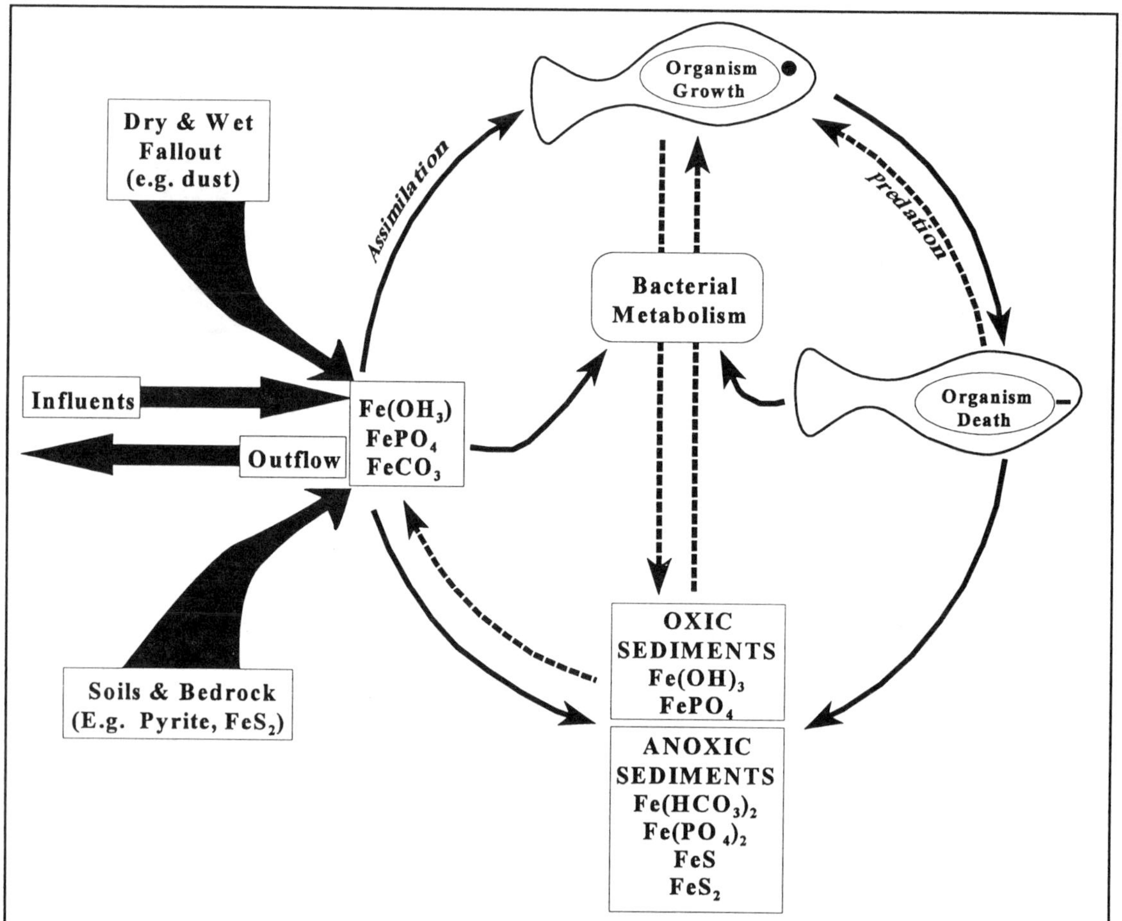

Figure 5.15. The iron cycle in deep lakes. Influents include groundwater, inflows, surface runoff, etc.

sediment layer.

The iron cycle is shown in Fig. 5.15. Most of the iron entering lakes is from inflows carrying water with crystalline forms of iron that have been weathered from soils and bedrock, such as pyrite (FeS_2), hematite (Fe_2O_3) and magnetite (Fe_3O_4), and is precipitated immediately on the lake floor. Even soluble ferrous forms that enter through groundwater are quickly oxidized to ferric hydroxide which precipitates to the lake floor. During the precipitation process, phosphate ions can be adsorbed onto the hydroxide and be carried out of solution. In fact, ferric hydroxide is often added to eutrophic lakes to remove phosphorous from solution and slow the eutrophication process. However, continued anoxia will result in the mobilization of phosphorous when ferric iron is reduced to more soluble ferrous iron. Sulphur also interacts under such anoxic conditions by precipitating iron as ferrous sulphide, leaving free phosphate in solution (Fig. 5.15). To prevent phosphorous mobilization, air is often bubbled into the lower layers of water using compressors, the oxygen that is introduced being used to oxidize ferrous iron to insoluble ferric iron.

CHAPTER 6

PLANTS - THE LAKE'S AERATORS AND FOOD BASE

Why Read This Chapter?

Aquatic (and terrestrial) ecosystems are powered by sunlight. Light energy drives many physical, chemical and biological processes, particularly photosynthesis in plants. Plants are at the bottom of the food chain. They, like bacteria and some protists, are ***autotrophs*** and can use elemental (e.g., carbon, nitrogen, phosphorous) materials to synthesize cellular tissues. ***Heterotrophs*** are unable to utilize elemental materials and must consume autotrophs or other heterotrophs in order to grow. This chapter describes who eats whom in an aquatic world, why aquatic plants are important, how plants use light energy and how plant diversity and photosynthetic rates can be used to determine the health of a lake.

Ecological pyramids are produced when organisms are lumped into feeding groups such as primary producers, herbivores, small carnivores, large carnivores, etc. Primary producers, or autotrophs (like plants), are at the bottom of the ecological pyramid in aquatic systems because they form the greatest biomass and provide the greatest amount of energy compared to any other trophic level. Large carnivores, such as mammals and birds of prey, are at the top of the pyramid because they produce the least amount of biomass and energy of all aquatic organisms. A pyramid results when biomass or energy produced is plotted according to the numbers, biomass or energy produced for each feeding group. Figure 6.1 depicts a typical ecological pyramid during the summer when production is greatest for all groups. During the winter, primary production almost ceases and plants form a much smaller biomass and provide less energy to consumers than in the summer. All other feeding levels also decrease in biomass during the winter, but much less than the autotrophs. This chapter examines different plants and their importance in the ecological pyramid.

Plants are divided into two fundamental groups, ***algae*** and ***macrophytes***. Most algae are microscopic, but some can be macroscopic, especially colonial and filamentous forms. They can be planktonic (freely floating in water column and are referred to as ***phytoplankton***), benthic (laying freely on the bottom) or attached to objects projecting from, or laying on, the bottom. The attached forms are referred to as ***periphyton***. Macrophytes (also called ***hydrophytes***) are the large plants; most have well developed root systems, flowering stages, fruiting bodies and simple to complex leaf structure. Both algae and macrophytes have one or

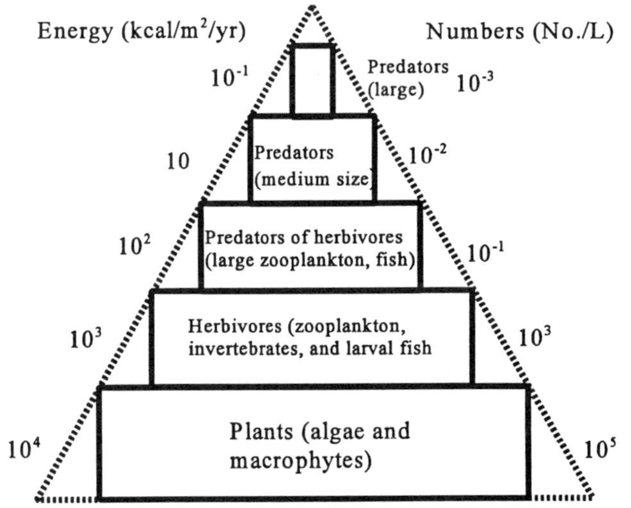

Figure 6.1. Ecological pyramid of a hypothetical temperate lake during summer.

more kinds of photosynthetic pigment called ***chlorophyll***, but all have chlorophyll *a*.

Macrophytes include the weeds, sedges, grasses, reeds, bullrushes, cattails, arrowheads, etc. The physical attributes and quality of water often determine the kinds and ***habits*** of species that make up the plant communities. A plant's habit is how it lives in the water. ***Emergent*** plants are erect, with most of the plant above water. ***Submergent*** plants remain submerged for most of the year, although the tips may penetrate the surface near the end of the summer. The leaves of many species float on the surface because the stem has no stiffening tissues to keep the emergent part upright. Most submerged species have well developed roots and are firmly anchored in the bottom. A few submerged species lack roots and sit on the bottom, leafy parts of the plant being anchored by mud. ***Floating*** plants live at the surface but some species, such as water lilies, have long stems and have roots that anchor the plant to the substrate. Other floating species, such as duckweed, have roots, but little or no stem and the roots float immediately below the leaves. In order for plants to live in streams, they must have an extensive root system that anchors the plant to the bottom to prevent them from being carried away by currents. This explains why plants without roots and floating species whose roots are not anchored in the bottom, are absent in streams with considerable flow.

The chapter begins with an examination of food webs. Then the different kinds of algae and macrophytes, and their value to organisms, including humans, are examined.

By the end of this chapter you will:

☞ 1. Recognize different types of feeding groups
☞ 2. Be able to draw a food web
☞ 3. Distinguish between bottom up and top down effects

☞ 4. Understand why population explosions occur
☞ 5. Understand the effects of an exotic species on the food web
☞ 6. Apply the indicator organism concept using algae to assess a lakes trophic status
☞ 7. Know the different kinds of chlorophylls and other photosynthetic pigments
☞ 8. Know the diagnostic features of the most common algal phyla, including:
☞ 8. Chlorophyta (green algae)
☞ 9. Cyanobacteria (blue-green algae)
☞ 10. Euglenophyta (euglenas)
☞ 11. Pyrrophyta (dinoflagellates)
☞ 12. Chrysophyta (yellow-green and golden-brown algae)
☞ 13. Bacillariophyta (diatoms)
☞ 14. Cryptophyta (cryptomonads)
☞ 15. Charophyta (stoneworts)
☞ 16. Rhodophyta (red algae)
☞ 17. Phaeophyta (brown algae)
☞ 18. Know the meaning and formula for primary production
☞ 19. Know how light and temperature affect primary production
☞ 20. Know how pH and alkalinity affect primary production
☞ 21. Know what spring pH depressions are and what causes them
☞ 22. Known the relationship between types of algal mats and clouds and the pH level
☞ 23. Know how buoyancy affects primary production
☞ 24. Know the importance of nutrient levels and availability
☞ 25. Know the seasonal and competitive interactions that occur among algal species
☞ 27. Be able to recognize periphyton
☞ 28. Know to measure periphyton biomass
☞ 29. Understand the factors that affect the kinds and abundance of algae in streams
☞ 30. Know that algae grow in and on sediments and what factors affect their growth
☞ 31. Know some qualitative methods for measuring primary production
☞ 32. Know some semi-quantitative methods for measuring primary production
☞ 33. Know how to use chlorophyll *a* levels to estimate trophic status
☞ 34. Know some quantitative methods for measuring primary production
☞ 35. Know the characteristics, adaptations and representative species of emergent plants
☞ 36. Know the characteristics, adaptations and representative species of surface floating plants
☞ 37. Know the characteristics, adaptations and representative species of rooted submergent plants
☞ 38. Know the characteristics, adaptations and representative species of unrooted submergent plants
☞ 39. Know the indicator value of macrophytes
☞ 40. Know the characteristics and life histories of mosses and lichens in the fringing plant community
☞ 41. Know the uses of *Sphagnum* moss

FOOD WEBS: Who Eats Whom?

Most, if not all, organisms are prey for other organisms. Plants, like algae and macrophytes, are at the bottom of the food chain and are consumed by ***herbivores***. If the plants are periphyton or phytoplankton, the consumers are called ***grazers***. Algae that are encrusted on surfaces of rocks, macrophytes, logs, etc. are eaten by ***scrapers***. Large plants (referred to in the previous chapter as CPOM) are reduced to small bits, or FPOM
1☞ (fine particulate organic matter) by ***shredders***. Some organisms have specialized mouth parts to pierce plant or animal cells and suck out the cell or body contents; these are known as ***plant piercers*** or ***animal piercers***, respectively. The FPOM ends up on the bottom or suspended in the water column and are consumed by ***collectors***. Collectors are divided into two groups, depending on what FPOM is eaten; FPOM suspended in the water column is consumed by ***filter feeders*** and FPOM laying on the bottom is consumed by ***gatherers*** (also called ***detritivores*** or ***deposit feeders***). These in turn are eaten by ***carnivores*** or ***predators***, which are often divided into several different levels, depending on their size. For example, some stonefly nymphs (level 1) are predators of other insects; stonefly nymphs are preyed upon by crayfish (level 2); crayfish are preyed upon by fish (level 3); fish are preyed upon by birds (level 4); and birds are preyed upon by other birds or mammals, including humans (level 5). Finally, all the consumer groups can be preyed upon by ***parasites***. Each of the consumer groups is referred to as a ***functional feeding group***.

The food web of most lake and stream ecosystems is represented by all functional feeding groups. However, certain functional feeding groups dominate in different parts of the lake or stream. For example, as described in the carbon cycle, shredders will dominate in the head waters of streams where there is plenty of leaf litter and other kinds of CPOM; filter feeders will dominate in the lower reaches after all the CPOM has been reduced to FPOM. Similarly, in lakes the profundal zone has mainly collectors (filter feeders and gatherers) and predators. Shredders and grazers are absent in the profundal zone because there is little or no CPOM, most of the organic material being FPOM, and no algae for grazers to feed on because of the lack of light. Nevertheless, most food
2☞ webs are very complex, especially when some of the organisms in a community are ***omnivorous*** (feed on a variety of feeding groups). Figure 6.2 is an example of a simple food web, the carp being omnivorous. A ***dynamic food web*** has energy values associated with each of the arrows.

Two concepts have been used to explain the transfer of energy from one level to the next. One focuses on the energy passed up from one trophic level to the next, stressing the importance of lower trophic levels to those higher up the pyramid. It is known as ***bottom up control***. Until recently, the control that upper trophic levels exerted
3☞ on lower trophic levels was largely ignored. However, we now know that ***top down control*** is also important. Top down control often has a ***cascade effect*** where each trophic level is affected sequentially down the pyramid.

Often feeding groups are described by the general class of organisms being fed upon. For example, herbivores feed on plants; ***planktivores*** feed on plankton;

benthivores feed on bottom organisms; and ***piscivores*** feed on fish. Occasionally, the groups are subdivided even further into more specific feeding groups. For example,

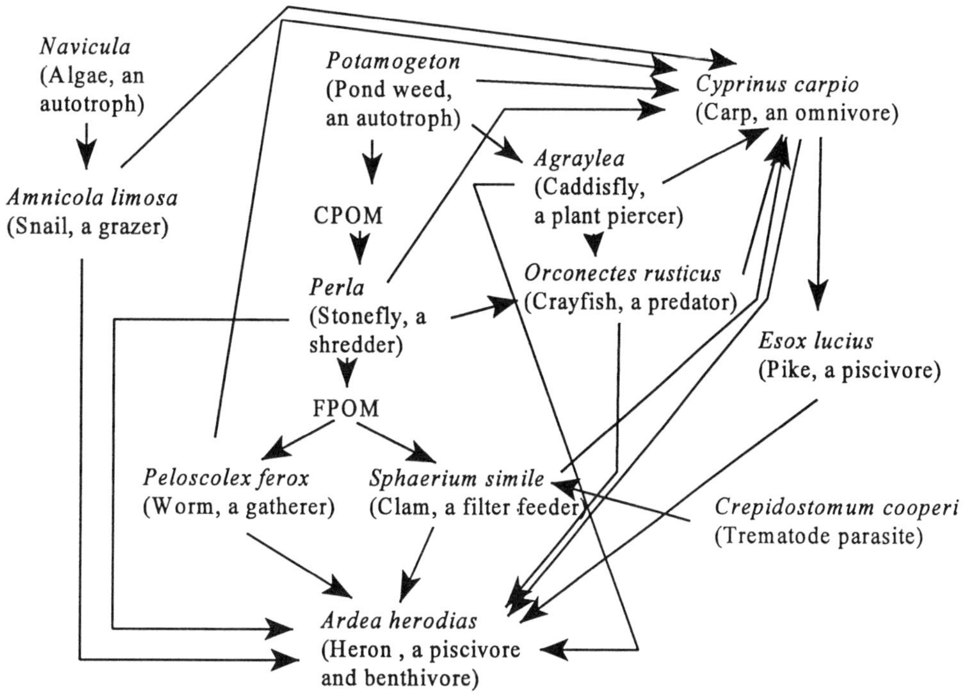

Figure 6.2. A simple food web showing different functional feeding groups. Notice that all arrows are directed toward the consumers.

benthivores can be subdivided into ***molluscivores*** if they feed on molluscs, namely clams and snails, or ***insectivores*** if they feed on insects.

The next three chapters describe feeding habits of many species of zooplankton, benthos and fish. Use the information in these chapters to learn the feeding habits of organisms in lakes and streams. Often it can help answer questions such as, "why is the rock bass population suddenly exploding in numbers?". It may be that population
4☞ numbers of predator species (e.g. pike, small mouth bass, pickerel) are declining. Perhaps an exotic, opportunistic species is replacing a native, generalist species. A decrease in diversity tends to weaken the food web because there are fewer members within a functional feeding group. If the exotic species is exceptionally opportunistic, it may replace all species within the same functional feeding group. In fact, this has happened with the zebra mussel. The zebra mussel populations became so large in Lake St. Clair that they out-competed the native clam species for food and all the native species populations were decimated. The zebra mussel is now the dominant filter feeder and when its population numbers decline (which occurs eventually with every new colonizer),

or if the species for some reason is eradicated, there will be few or no filter feeders to process the FPOM in Lake St. Clair. In the worst case scenario, there will be no filter feeders and FPOM will accumulate on the bottom and Lake St. Clair will become eutrophic with a thick layer of ooze on the bottom. Nevertheless, a food web with numerous species within a functional feeding group is much more stable than one with one or only a few species. A community with several species performing a similar function is less affected when one of it members is removed by predation, exploitation, or even pollution.

Figure 6.3 shows a "web" with an opportunistic, exotic species, such as the zebra mussel, invading the web. The question that we are unable to answer at present is, "At what point is the web weakened or tattered and is the native clam community suffering as
5☞ a result of the species invasion"? The zebra mussel is a filter feeder and has the potential to eliminate all other filter feeders in a lake. If this happens, the web will become weaker

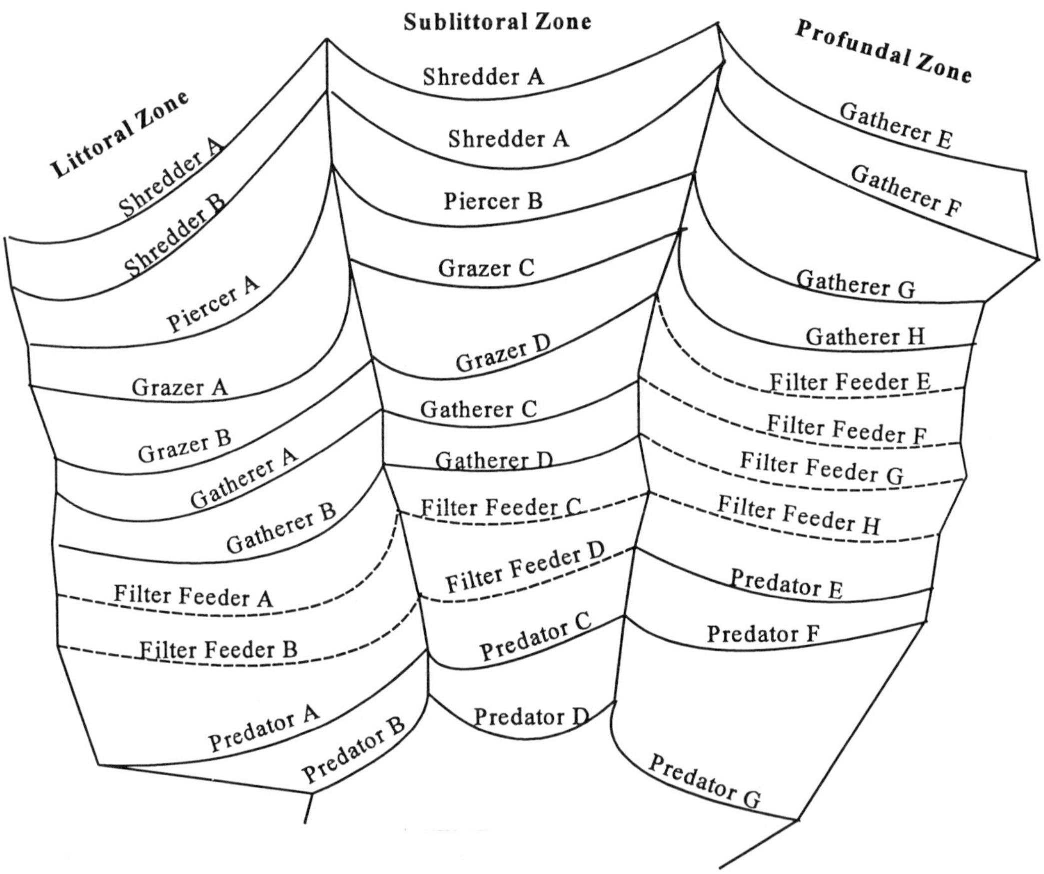

Figure 6.3. Part of a hypothetical web of benthic diversity in a deep lake. See text for explanation of broken line for filter feeding species.

because all the native species (filter feeders A to H) will disappear (hence the broken line) leaving big "holes" in the web in the littoral, sublittoral and profundal zones. In fact, this could ultimately affect other trophic levels. Some of the research at the University of Guelph has shown that the filter feeding activities of zebra mussels is affecting the transfer of energy from one trophic level to the next in the Great Lakes. The zebra mussels filter and remove phytoplankton from the water, but they cannot consume all that they filter. Instead, the mussels package up the undigested particles in mucous and deposit them on the bottom of the lake as ***pseudofaeces*** (undigested food) and the biomass of gatherers have increased as a result. The filtering activities of zebra mussels is also affecting the abundance of zooplankton that feed on the phytoplankton. It is anticipated that fish species, which feed on the zooplankton suspended in the water column, will decline in abundance in favour of those fish species which feed on bottom organisms that are feeding on the pseudofaeces. This is clearly a case of bottom up control.

If an exotic species enters your favourite lake, watch for changes in the food web, especially in numbers and kinds of its competitors (same functional feeding group). Chapter 14, "Biodiversity and Exotic Species", discusses how different traits allow "opportunistic" and "generalist" species to dominate communities.

ALGAE

The algae are represented by several phyla. Some are restricted to marine environments, others to freshwater environments. Still others have the ability to live in both freshwater and marine or estuarine environments. Most occur in all types of aquatic habitats. It is estimated that on average 99% of the solar radiation energy that enters the system is lost to the primary producers. Subsequently, as energy flows to the next trophic level, a loss of 90% occurs. This means that only 10% of the energy reaching a trophic level is successfully transferred to the next level where it is used to increase production.

Factors that affect the growth and production of algae and how to measure primary production are examined below. The information allows one to determine how physical and chemical variations in aquatic environments affect algal diversity and productivity, and how to integrate this information to assess the health of a lake. Like the following chapters, this one integrates physical and chemical concepts that relate to trophic status. For example, the Secchi disc (a physical concept relating to light transmission) is used as a simple, inexpensive tool to predict nutrient levels (chemical concepts), primary production and plant diversity (biological concepts). A species will dominate a particular habitat if its physical, chemical and biological needs are optimal. It follows that when the tolerances and requirements of dominant species are known, one should be able to predict the quality of the aquatic habitat. Particularly useful are those
6☞ species, called ***indicator organisms***, that always appear in greatest numbers or biomass under a specific set of conditions. For example, pollution indicators are those species which prevail in enriched (eutrophic) or organically polluted water; clean water

indicators are those which prevail in poorly enriched (oligotrophic) or pristine waters. Scientists have amassed a huge data base for several indicator species of algae, zooplankton, benthic invertebrates and fish. This chapter develops two lists of indicator species, one for algae, another for macrophytes. The application of indicator species for assessing the trophic status and fisheries potential of a lake is described in Chapter 13.

Table 6.1 lists some genera with contrasting species in clean and organically enriched water. Table 6.2 summarizes the indicator values of representative species in nine algae phyla. Two of the phyla, Phaeophyta and Rhodophyta, consist mostly of marine species; only a few freshwater representatives are present and most of these are attached forms in clean, running waters. The remaining seven phyla have numerous freshwater and marine representatives. Only freshwater species are examined here. The algae are represented by solitary (unicellular forms), colonial forms, or filamentous (either branched or unbranched) forms. Within each of these types are either planktonic forms (living freely in water column) or attached forms, discussed later under "periphyton".

Diagrams and taxonomic keys of different species of algae are beyond the scope of this text. If information on appearance and identification of algae is required, Prescott (1978)[1] is recommended. Only representative species within each of the phyla are illustrated here.

Table 6.1. Some genera and species with oligotrophic (clean) and eutrophic (organically enriched) indicator value. See Chapter 13 for additional species.

Phylum	Genus	Oligotrophic	Eutrophic
Euglenophyta	*Euglena*	*ehrenbergii*	*agilis, gracilis, oxyuris*
	Euglena	*spirogyra*	*polymorpha, viridis*
Cyanophyta	*Phormidium*	*inundatum*	*uncinatum*
Bacillariophyta	*Navicula*	*gracilis*	*cryptocephala*
	Nitzschia	*linearis*	*acicularis*
	Surirella	*splendida*	*ovata*

7 ☞ The phyla are characterized by several criteria, among the most important of which are the types of photosynthetic pigments present. Four kinds of pigments are recognized: ***chlorophylls***, ***carotenoids***, ***xanthophylls*** and ***biliproteins***.

---

[1]Prescott, G. W. 1978. How to know the freshwater algae. Wm. C. Brown Publ., New York, N.Y.

At least four types of chlorophylls are known. ***Chlorophyll*** **a** is the most common and is present in all plants, both algae and macrophytes. It is the primary photosynthetic pigment and receives light energy from most other pigments. Chlorophylls are identified on the basis of their ability to absorb certain wavelengths of light, measured in nanometers (nm). Chlorophyll *a* has two light absorption bands in the red part of the light spectrum, one at 660-665 nm and another near 430 nm wavelength. ***Chlorophyll*** **b** is present in higher plants as well, but within the algae it is found only in Chlorophyta, Euglenophyta and Charophyta. Maximum absorption bands occur at about 645 nm and 435 nm. ***Chlorophyll*** **c** appears to be an accessory pigment and has maximum absorption bands near 630 nm to 635 nm, 583 to 586 nm and 444 to 452 nm. ***Chlorophyll*** **d** has no known function and is present only in some species in the phylum Rhodophyta.

The ***carotenes***, of which α- and β-carotene are most common, are linear, unsaturated hydrocarbons. ***Xanthophylls*** (e.g. lutein, diatoxanthin, myxoxanthin, peridinin) are oxygenated derivatives of carotenes. Both occur throughout the algae kingdom. ***Biliproteins*** (e.g. phycocyanin and phycoerythrin) are a mixture of pigment and protein and are present only in blue-green algae (Cyanobacteria) and in some species within the Cryptophyta and Rhodophyta. As with chlorophylls *b* and *c*, light energy absorbed by carotenes, xanthophylls and biliproteins is transferred to chlorophyll *a* for its fluorescence and excitation.

ALGAE PHYLA

Phylum: Chlorophyta - the chlorophytes or green algae (Fig. 6.4)

The green algae are common in all freshwater systems, as well as in estuaries and oceans. They are single-celled, colonial or filamentous, the filaments being either
8☞ with or without branches. Many single-celled and colonial forms are flagellated. When present, there are usually two, sometimes four, or as many as eight flagella. All green algae have cellulose walls and chlorophylls *a* and *b*. The chlorophyll is contained within ***chloroplasts*** that often have a shape characteristic of a genus. For example, in *Spirogyra* (Fig. 6.4a) the chloroplasts are spiral in shape; in *Zygnema* (Fig. 6.4b) the chloroplasts are stellate (star-shaped); and in *Ulothrix* (Fig. 6.4c) the chloroplasts are ring-like, the two rings being separated by a narrow core giving it the appearance of an apple core with most of the apple eaten away.

Green algae have ***pyrenoid bodies*** for storing starch. This feature is used for positive identification of chlorophytes because the starch-storing body stains a dark blue when iodine is added to the water containing the algae. Only one other phylum, Cryptophyta, displays the ***iodine-positive test***.

Common genera including *Spirogyra*, *Ulothrix* and *Zygnema*, discussed above, are filamentous. *Mougeotia* (Fig. 6.4d) is another filamentous form, some species forming massive blooms in acidifying lakes. Many streams have branched filamentous forms, like *Cladophora* (Fig. 6.4e). *Cladophora glomerata* is a hard water indicator species that attaches itself to rocks in hard water streams where it forms long, massive,

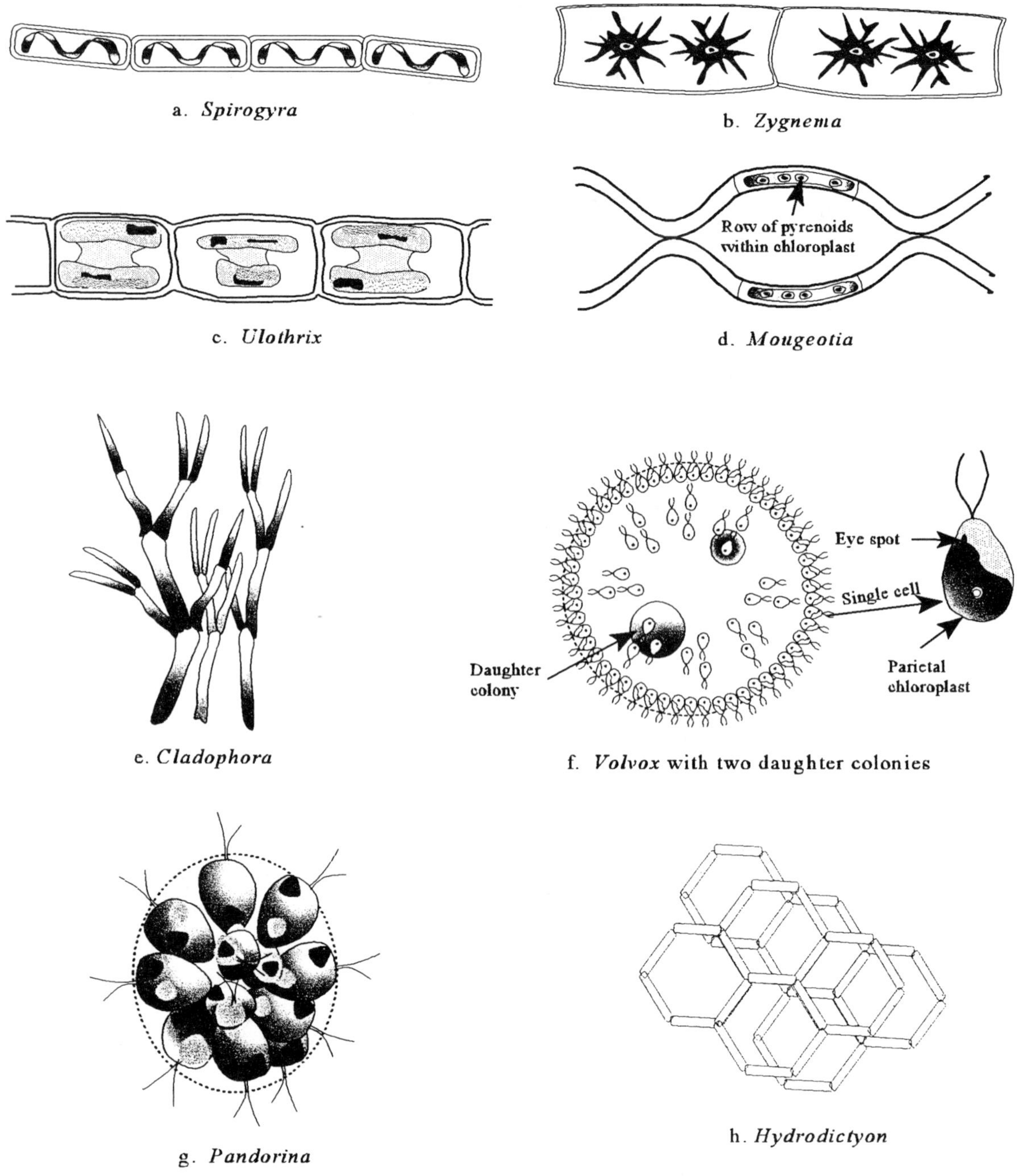

Figure 6.4. Some forms of green algae (Chlorophyta), including: Unbranched, filamentous forms (a-d) with characteristic chloroplast shapes, such as helical in *Spirogyra* (a), stellate in *Zygnema* (b), apple core-like in *Ulothrix* (c) and a plate-like axial shape in *Mougeotia* (d); branched filaments, as in *Cladophora* (e); colonial, as in *Volvox* (f) and *Pandorina* (g); net-like filaments in *Hydrodictyon* (h).

trailing strands. Usually the rocks are coated in travertine (calcium carbonate) because of the removal of carbon dioxide during photosynthesis by the algae.

Colonial forms include *Volvox*, a large form that grows to 1 mm in diameter. The colony consists of numerous biflagellate cells that are contained within several ***daughter colonies***. *Volvox* (6.4f) may be found in large numbers, usually producing blooms in waters rich in nitrogen. The blooms are of short duration during the warmer summer months but the water retains a fishy smell for a period long afterward. Often accompanying *Volvox* are other colonial forms, such as *Pandorina* (Fig. 6.4g) and *Eudorina*. Both genera have fewer cells in their colonies than does *Volvox*. *Pandorina morum* is considered an indicator of organically enriched waters. In all colonial forms the flagella extend outward, the flagella being used to move the cell about in the water column and maintain a planktonic life.

Many ponds have a green surface scum or dense mat that can cover the entire surface of the pond. *Hydrodictyon reticulatum* (Fig. 6.4h), the water net, is one such genus. The mat consists of numerous cells oriented into hexagons, each hexagon being attached to its neighbour to produce an expansive network of hexagonal cells. The mat not only prevents oxygen from diffusing from the atmosphere into the water but inhibits photosynthesis of planktonic forms by preventing light from entering the water. Most of the oxygen that is produced goes directly into the atmosphere because the mats are right at the air-water interface. As a result, the water quickly becomes anoxic and soon smells septic with the evolution of hydrogen sulphide.

Summary of key characteristics of Chlorophyta:
- Colour green to yellow-green
- Photosynthetic pigments are chlorophylls *a* and *b*, enclosed in chloroplasts
- Food reserve is iodine-positive starch
- Cell walls semi-rigid, smooth or with spines
- Nucleus present
- Flagella present in some
- No eye-spot

Cyanophyta or Cyanobacteria - the blue-green algae (Fig. 6.5)

The blue-green algae are represented by unicellular, colonial and simple-branched, filamentous forms. They are a primitive group of algae. Chloroplasts are
9☞ lacking and the chlorophyll, mostly chlorophyll *a*, is distributed throughout the protoplast. Accompanying the chlorophyll are three other pigments, carotene (three types), xanthophylls (up to fifteen types) and biliproteins. The cell wall is thin and somewhat membranous and usually has a gelatinous outer sheath. Often the protoplasm contains ***pseudovacuoles*** ("false sacs" filled with gas) which reflect light and obscure the true colour of the cells. The most common hues imparted to the water are red, gray, tan or brown, depending on the light waves being transmitted in the water. A nucleus is present but it is not surrounded by a nuclear membrane; instead the nuclear material is spread throughout the protoplasm. This is the only freshwater phylum of

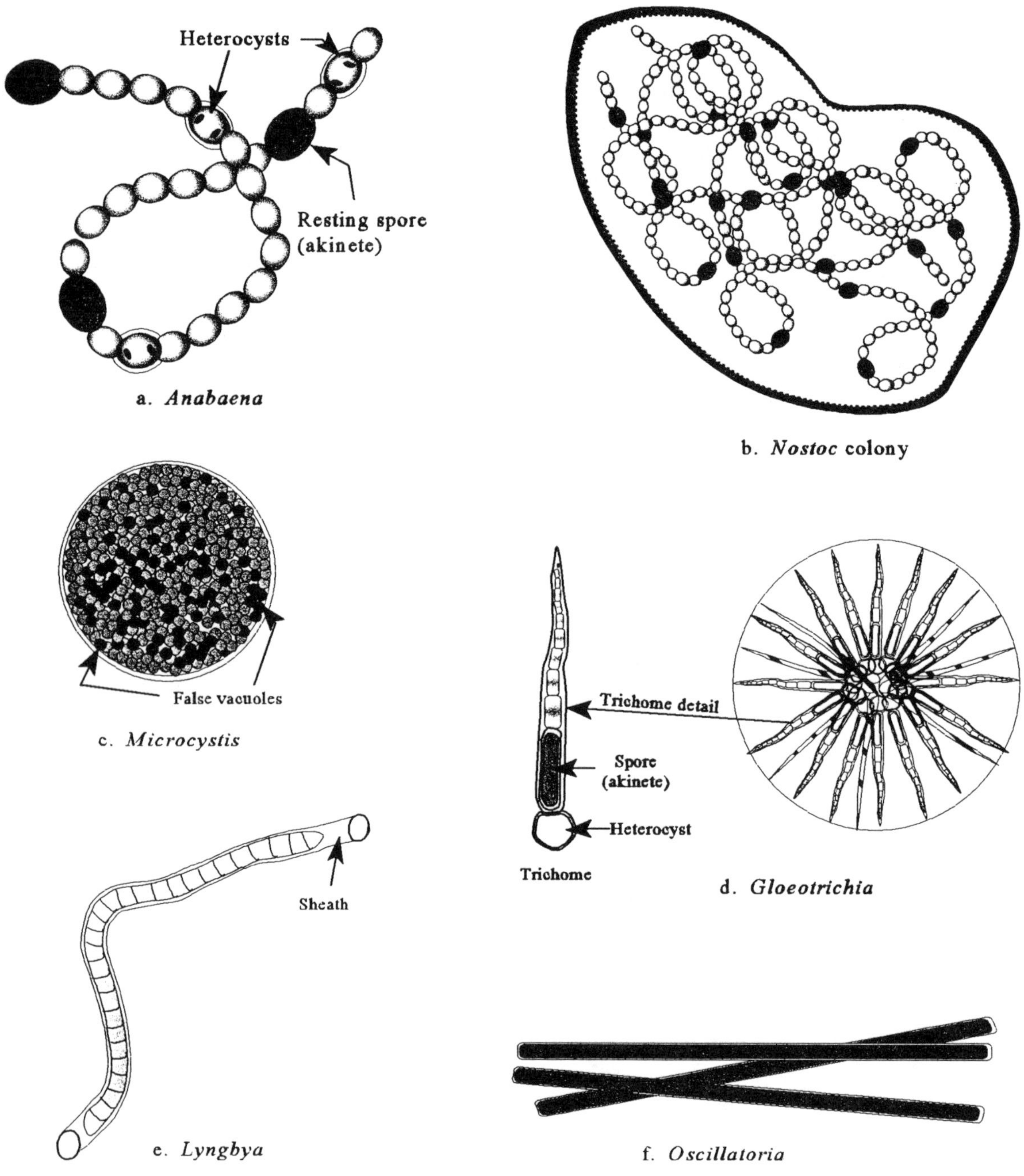

Figure 6.5. Some common blue-green algae (Cyanobacteria). Many species have heterocysts that play a role in nitrogen fixation, such as *Anabaena* (a), *Nostoc* (b) and *Gloeotrichia* (d). *Microcystis* (c), which forms a scum on ponds, has many false vacuoles that reflect light and give the scum a black, brown or purple colour. *Lyngbya* (e) is a sheathed form. While most species of cyanobacteria are blue-green in colour, some, like *Oscillatoria* (f), are yellow-green.

algae that lacks chloroplasts and a true nucleus.

Nearly all blue-green algae are immobile. Food storage is apparently glycogen, possibly a fluoridian type of starch, a food reserve typically produced by marine red algae. The iodine test for starch is negative. Some well known genera include the nitrogen fixers, *Anabaena* (Fig. 6.5a) and *Nostoc* (Fig. 6.5b), discussed in the last nitrogen cycle of the chapter. Their filaments consist of bead-like cells called ***trichomes***. Interspersed among every eight or so of these cells is a translucent organelle called a ***heterocyst*** which are involved in nitrogen fixation. Most species of *Anabaena* are colonial and surrounded in mucilage, but some are planktonic and form large blooms in circum-neutral (pH ˜ 7) northern lakes. The blooms are cloud-like and rarely form a surface scum like some of the chlorophytes. Other species of *Anabaena* produce a toxin that is potent enough to kill cattle that drink the water. In fact, most of the toxic algae are blue-green algae, such as *Anabaena circinalis, A. flos-aquae, Microcystis aeruginosa* (Fig. 6.5c), *M. flos-aquae, M. toxica, Gloeotrichia echulata* (Fig. 6.5d), and *Lyngbya contorta* (Fig. 6.5e). The toxic effects reported include hay fever, dermatitis and intestinal disorders in humans.

Nostoc is a colonial form with a tough outer tegument of mucilage within which are enclosed numerous tiny, coiled cells. The colony itself is large (˜ 8 - 40 mm diameter), mostly benthic and common everywhere, especially on the bottom of enriched ponds. A common planktonic form is *Oscillatoria* (Fig. 6.5f), so named because of their oscillating behaviour in water. The filaments are long and narrow and composed of short and small cells. The end of the filament is ***capitate*** (capped). The genus is especially common in eutrophic waters where they often form massive blooms that impart a blood red (as with *O. rubescens*) or purple (as with *O. prolifica*) colour to the water. Some species of *Oscillatoria* are associated with metalimnetic oxygen maxima, as described in Chapter 5; they are ***cold stenotherms,*** having a very narrow thermal tolerance and preferring cooler temperatures characteristic of the hypolimnion and metalimnion. Apparently the filaments become entrapped in the metalimnion because of the density difference between the epilimnion and the hypolimnion and are able to use light at the 1% incident light level. Many species of *Oscillatoria* are classed as indicators of enriched waters, for example *O. chalybea, O. chlorina, O. formosa, O. lauterbornii, O. limosa, O. princeps, O. putrida* and *O. tenuis*). Other enrichment indicators include *Anabaena constricta, Arthrospira jenneri, Lyngbya digueti, Phormidium autumnale* and *P. uncinatum.*

Another colonial form is *Microcystis*, the colony being somewhat globular in shape. The cells that make up the colony are closely compacted in mucilage. Each cell has pseudovacuoles that cause the plant to float fairly high in the water column, often producing a surface scum which lowers the diffusion rate of oxygen into the water from the atmosphere. Also, the CO_2 respired from the colony remains in the water so the ponds usually go anoxic fairly quickly resulting in a septic odour due to the evolution of hydrogen sulphide.

Gloeotrichia, is also a colonial form, but with tapering, radiating filaments enclosed in mucilage. Usually they are found fixed to branches of macrophytes, although the colonies can be floating freely in the water. The genus is considered a ***hard water indicator*** since it is found only in hard water. Colonies have the appearance of tapioca grains and give the water a buff colour. Some species become very abundant near bathing beaches and cause severe skin irritation, the symptoms of which are similar to ***swimmer's itch***, a disorder caused by larval ***cercariae*** of certain trematode parasites. The definitive host of the trematodes is waterfowl, but some cercariae accidentally find their way to humans and only partially enter their skin to cause the itchiness.

Summary of key characteristics of Cyanobacteria:

- Colour blue-green
- Photosynthetic pigment is chlorophyll *a*, lack chloroplasts
- Food reserve is non-starch and iodine negative
- Cell walls have an attached slimy coat
- Nucleus absent
- Flagella absent
- No eye-spot

Euglenophyta - the euglenas (Fig. 6.6)

All euglenoids are solitary cells, with one or two flagella arising from a gullet for motility. Some species (*Euglena orientalis, E. rubra, E, sanguinea*) form massive blooms in eutrophic waters and impart a reddish colour due to the presence of a ***red*** 10☞ ***eye-spot*** and pigments including chlorophylls *a* and *b*, four carotenes, as many as five xanthophylls, and ***haematochrome*** (a red carotenoid) in some. Food reserve is an insoluble starch called ***paramylum*** that is iodine negative. The starch is contained with ***pyrenoid bodies*** that are within or on the chloroplast.

As a group, euglenoids are often considered enrichment indicators because most prevail in enriched waters. However, some species, like *Euglena spirogyra*, are oligotrophic indicators (Table 6.1). Clearly, it is important to specify species when using *Euglena* as an indicator of water quality.

Summary of key characteristics of Euglenophyta:

- Colour green
- Photosynthetic pigments are chlorophylls *a* and *b*, enclosed in chloroplasts
- Food reserve is iodine-negative starch
- Cell walls semi-rigid
- Nucleus present
- Flagella present in all
- Eye-spot present

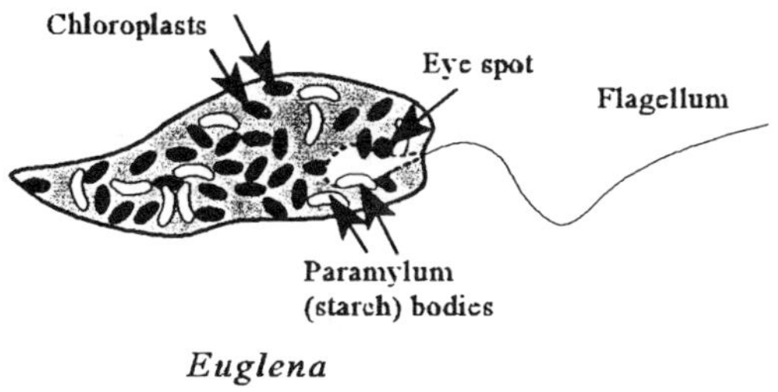

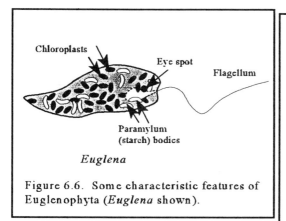

Figure 6.6. Some characteristic features of Euglenophyta (*Euglena* shown).

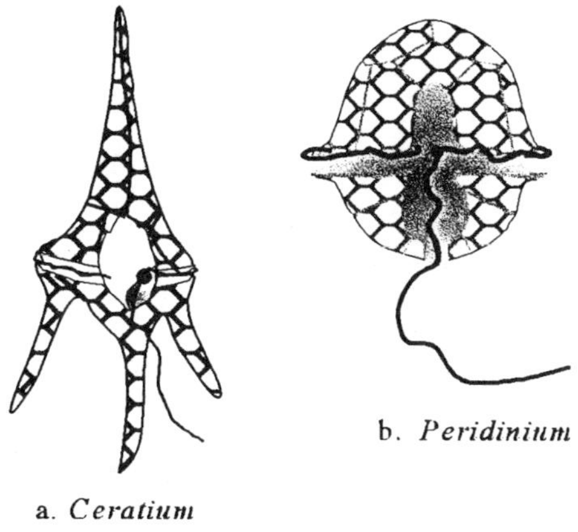

Figure 6.7. Pyrrophyta, with (a) *Ceratium* and (b) *Peridinium* shown.

Pyrrhophyta - the dinoflagellates (Fig. 6.7)

Cells are solitary, mostly with two flagella, one wound around the organism in a transverse furrow, the other extended posteriorly in a longitudinal groove. In freshwater
11☞ species, the cell wall, made of cellulose, is firm or consists of polygonal plates or ***theca***. Some species, like the ubiquitous *Ceratium* (Fig. 6.7a) and *Peridinium* (Fig. 6.7b), have thecal horns, usually one directed anteriorly and one to three of them directed posteriorly. ***Cyclomorphosis*** is exhibited by the thecal horns of some species whereby the horns increase in length through the summer months. The pigments are in chloroplasts and include chlorophylls *a* and *c*, two carotenes and four to six xanthophylls, phycopyrin often giving the cells a reddish colour. The food reserve is starch and oil. Massive blooms of *Ceratium* can lend a brown or grayish-brown colour and a fishy to septic taste to the water.

Summary of key characteristics of Pyrrhophyta:

- Colour grayish
- Photosynthetic pigment is chlorophyll *a* and *c*, enclosed in chloroplasts
- Food reserve is an iodine negative starch
- Cell walls naked or with cellulose plates and horns
- Nucleus present
- Two flagella
- Eye-spot in some

Chrysophyta - the yellow-green and golden-brown algae (Fig. 6.8)

12☞ Some taxonomists divide this phylum into two groups, the Xanthophyceae (yellow-green algae) and Chrysophyceae (golden-brown algae). Both groups have unicellular and colonial representatives, but golden-brown algae are rarely filamentous.

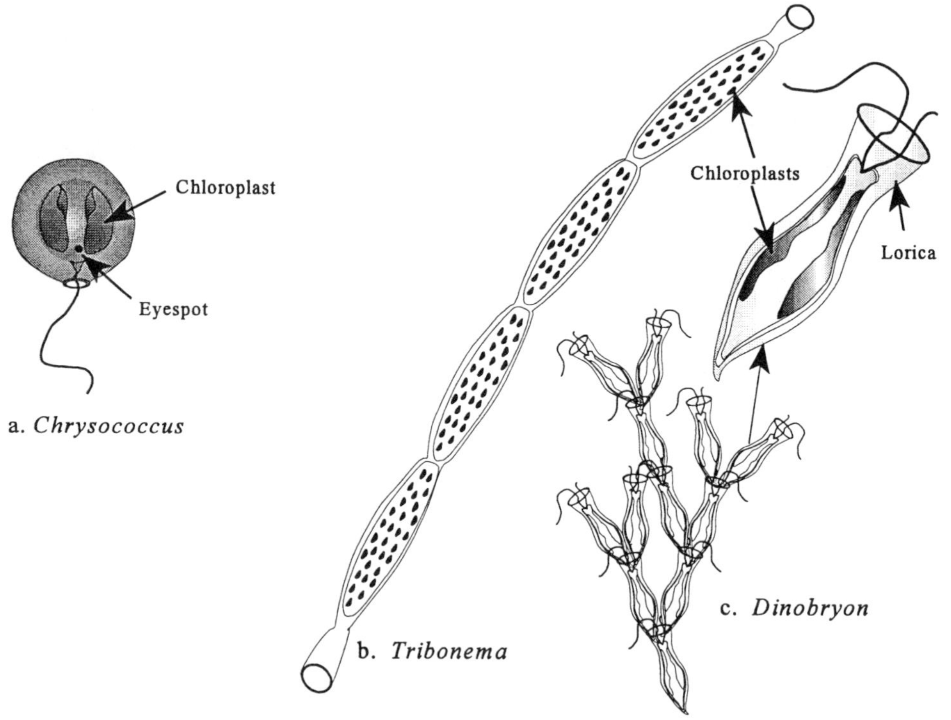

Figure 6.8. Some examples of yellow-green and yellow-brown algae (Chrysophyta), with *Chysococcus* (a) with 2 brown chloroplasts and an eyespot, *Tribonema* (b) , a filamentous form and *Dinobryon* (c), a colonial form, shown.

Both are characterized by significant inclusions of β-carotene and xanthophylls, along with chlorophylls *a* and *c* that result in their distinctive colouration. Nearly all unicellular forms have two flagella, one larger than the other, for mobility. The food reserve is in the form of oil or leucosin, the iodine test being negative. Some forms are ***epiphytic*** (attached to a substrate). Many are planktonic, for example *Chrysococcus rufescens* (Fig. 6.8a), *C. major* and *C. ovalis*, all considered clean water indicators. *Tribonema utriculosum* (Fig. 6.8b) is a filamentous form that produces a surface scum and a fishy odour when blooms occur. *Dinobryon sertularia* (Fig. 6.8c) is a flagellated colonial form and produces a fishy taste to the water when it blooms.

Summary of key characteristics of Chrysophyta:

- Colour yellow-green or golden-brown
- Photosynthetic pigments are chlorophylls *a* and *c*, enclosed in chloroplasts
- Food reserve is oils, iodine test is negative
- Cell walls thick
- Nucleus present
- Two flagella present in some
- No eye-spot

Bacillariophyta - the diatoms (Fig. 6.9)

The diatoms were once included with the Chrysophycaea because they have the same photosynthetic and accessory pigments but are considered by some to justify a phylum rank of its own. They are one of the major sources of silica in lakes and
13☞ oceans. The silica is contained within a shell, called a ***frustule***. The frustule is composed of a smaller "box", the ***hypotheca,*** and a larger "lid", the ***epitheca***. Both are adjoined by overlapping side pieces, the ***cingula*** (singular ***cingulum***). The broad, flat surface, called a ***valve***, has marginal flanges to which is attached the side pieces of the cingulum. Figure 6.9a illustrates the most common structures of a diatom frustule.

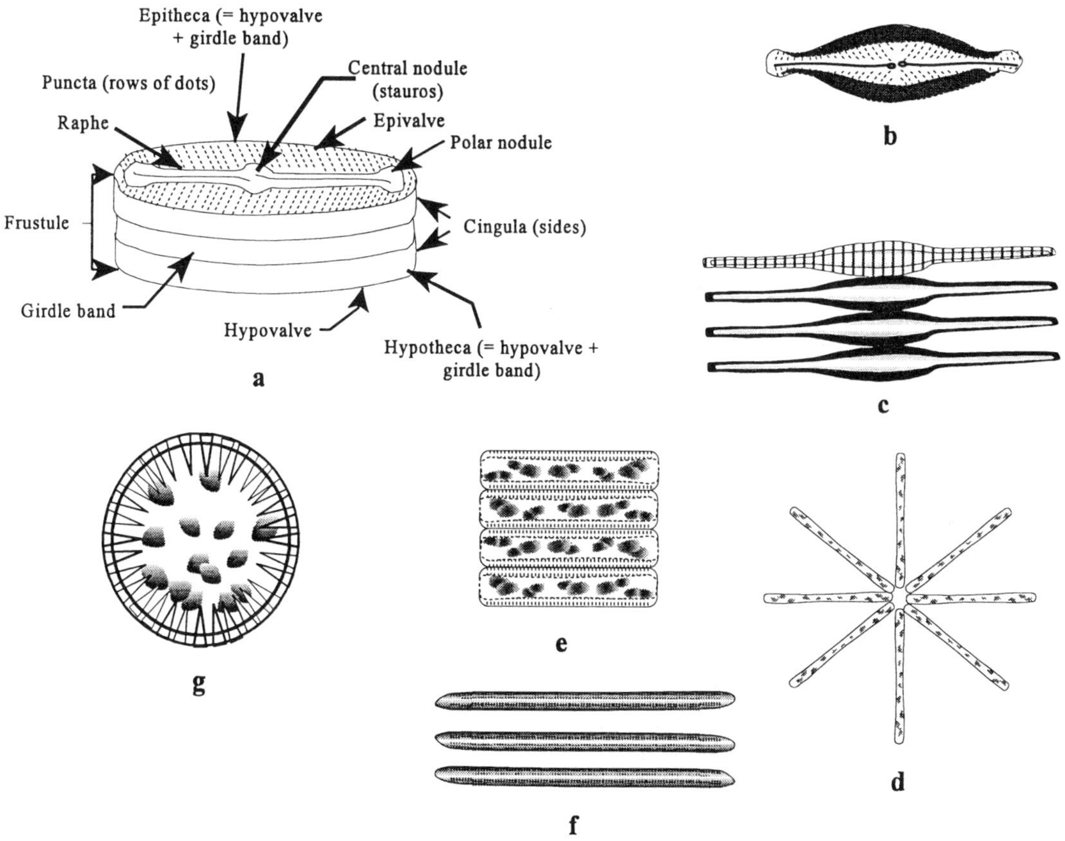

Figure 6.9. Bacillariophyta. (a) Structure of a pennate diatom with a raphe. (b) *Navicula*, with a raphe, and (c) *Fragillaria*, (d) *Asterionella*, (e) *Diatoma*, (f) *Tabellaria* , all with a pseudoraphe, and a centrales d iatom, (g) *Cyclotella*.

There are two groups of diatoms. The Pennales (often called ***pennate diatoms***) are elongate, have a "cigar-", "boat-" or "wedge"-shape, and have ornamentations arranged into patterns with bilateral symmetry; the Centrales (or ***centric diatoms***) are round or pill-box shaped and have radiate ornamentations. Some pennate species have a slit, or ***raphe***, running all or part of the cell wall; others have a depression instead of a distinct slit and is termed the ***pseudoraphe***. The frustule shape and its ornamentations are very diverse and diagnostic of species.

The diatoms are, arguably, the most used group in water quality surveys. Indices of trophic status have been developed solely on diatom diversity or the proportion of one diatom type to another. In general, centric diatoms are common in oligotrophic waters while pennate diatoms are more common in eutrophic waters. The relative diversity of diatoms within each group, has been used to indicate water quality. For example, the ratio of pennate forms with a pseudoraphe, collectively referred to as Araphidineae (e.g. *Fragilaria* (Fig. 6.9b), *Asterionella* (Fig. 6.9c), *Diatoma* Fig. 6.9d), *Meridion, Synedra* (Fig. 6.9e), *Tabellaria* (Fig. 6.9f)), to centric forms (e.g. *Melosira, Cyclotella* (Fig. 6.9g), *Stephanodiscus*), is known as the A/C index and is a method for assessing trophic status of lakes, but apparently only for lakes of low to moderate alkalinity (see Chapter 13 for details). With this caveat in mind, high A/C indices indicate eutrophy and low A/C indices indicate oligotrophy. Still, caution should be exercised in use of indicator species because other factors, such as pH, conductivity, ionic composition, grazing species of zooplankton, etc. all affect species abundance.

Summary of key characteristics of Bacillariophyta:

- Colour brownish to light green
- Photosynthetic pigments are chlorophylls *a* and *c*, enclosed in chloroplasts
- Food reserve is oils, iodine test is negative
- Cell walls silicified, frustules composed of two parts
- Nucleus present
- Flagella present in some
- No eye-spot

Cryptophyta - the cryptomonads (Fig. 6.10)

Cells are solitary, rarely colonial, with two flagella unequal in length in most species. For pigments they have chlorophylls *a* and *c*, two carotenes, four xanthophylls, and biliproteins (in some) that impart either a green, brown, blue or red
14☞ colour. Food reserve is a solid starch that is sometimes iodine positive. A few species are attached forms, such as *Rhodomonas lacustris* (Fig. 6.10a), which is characteristic of clean water lakes. Most cryptomonads are characteristic of oligotrophic waters, such as *Chroomonas nordstetii* (Fig. 6.10b) and *C. setoniensis*.

Summary of key characteristics of Cryptophyta:

- Colour green, brown, blue, red
- Photosynthetic pigments are chlorophylls *a* and *c*, enclosed in chloroplasts
- Food reserve is starch, iodine test is positive in some
- Cell walls thin to semi-rigid

- Nucleus present
- Two flagella usually present
- No eye-spot

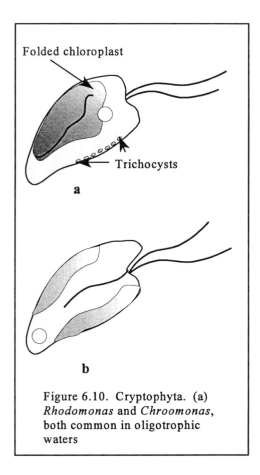

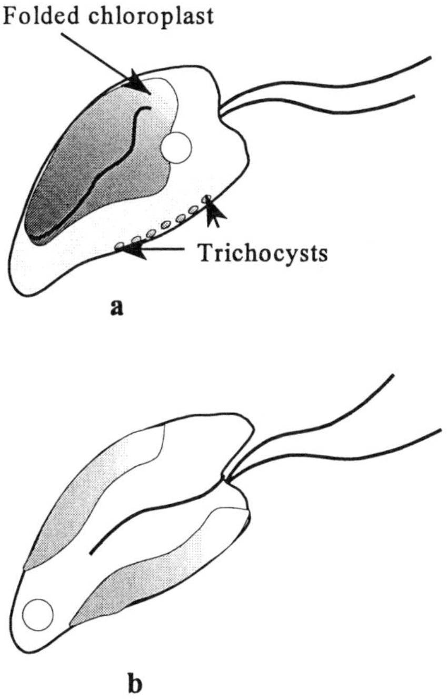

Figure 6.10. Cryptophyta. (a) *Rhodomonas* and *Chroomonas*, both common in oligotrophic waters

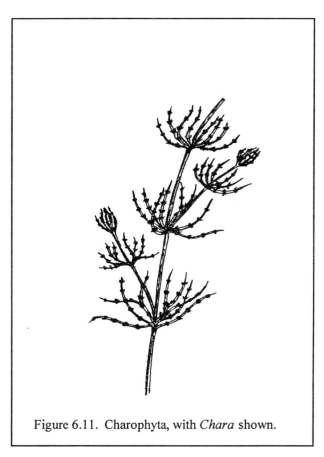

Figure 6.11. Charophyta, with *Chara* shown.

Charophyta - the stoneworts (Fig. 6.11)

Stoneworts were once included with the green algae because they share the same kinds of photosynthetic pigments, but most now consider it to be a phylum of its own. The charophytes are really macrophytes. They are large (up to 0.4 m in height),
15☞ submersed plants in the littoral and sublittoral zones. *Chara vulgaris* (Fig. 6.11) is characteristic of shallow, hard water ponds or stagnant areas of streams; it is absent in fast-flowing streams because it lacks roots (being an algae) and cannot anchor itself in the substrate. The plant usually has a thick coating of marl (lime), hence the name stonewort, which imparts a garlic or skunky odour to both the plant and water.

Nitella is the only other genus in the phylum but it tends to grow more deeply in circum-neutral or acidifying lakes. Some species occur in bog lakes stained brown with humic and tannic acids. The plant is greener than *Chara* because it does not form marl deposits.

Summary of key characteristics of Charophyta:

- Colour grayish green or green
- Photosynthetic pigments are chlorophylls *a* and *b*, enclosed in chloroplasts
- Food reserve is starch, iodine test is positive
- Cell walls semi-rigid
- Nucleus present
- Flagella absent
- No eye-spot

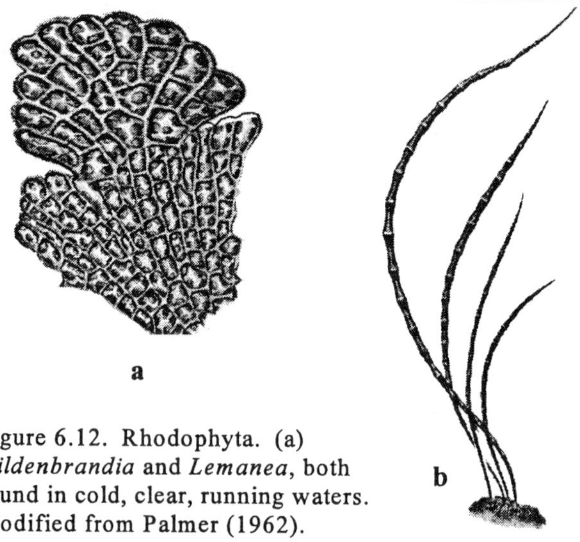

Figure 6.12. Rhodophyta. (a) *Hildenbrandia* and *Lemanea*, both found in cold, clear, running waters. Modified from Palmer (1962).

Rhodophyta - the red algae (Fig. 6.12)

Red algae are mainly a marine group, with only a few
16☞ freshwater genera. The plants have simple or branched filaments. Chlorophylls present are *a* and *d* and with at least two carotenes and up to fourteen xanthophylls give the algae a reddish colour. Food reserve is fluoridian starch which is iodine negative. Cell walls are relatively thick, often containing pores with intercellular connections. They are highly mucilaginous. The leaves are called *thalli* which can be huge in size, particularly in marine systems. Most freshwater forms are encrusting, such as *Hildenbrandia rivularis* (Fig. 6.12a)[2], *Lemanea annulata,* (Fig. 6.12b)[2] and *Bangia atropurpurea,* and all are characteristic of clean water streams.

Summary of key characteristics of Rhodophyta:

- Colour red
- Photosynthetic pigments are chlorophylls *a* and *d*, enclosed in chloroplasts
- Food reserve is fluoridian starch, iodine test is negative
- Cell walls semi-rigid
- Nucleus present
- Flagella absent
- No eye-spot

---

[2]Palmer, C. M. 1962. Algae in water supplies. U.S. Department of Health, Education, and Welfare, Public Health Service, Washington, D.C.

Phaeophyta - the brown algae (Fig. 6.13)

Brown algae are also mainly a marine group and include the brown seaweeds or kelps. They are essentially filamentous but mostly macroscopic and
17 very leathery in appearance. Pigments are chlorophylls *a* and c, with β-carotene and four to seven xanthophylls that impart a brownish colour to the algae. Food reserve is soluble carbohydrates, like mannitol. There are only a few freshwater genera. All are encrusting forms and most are in streams. *Lithoderma ambigua* (Fig. 6.13) is associated with fast-flowing, cold waters of head water or mountain streams.

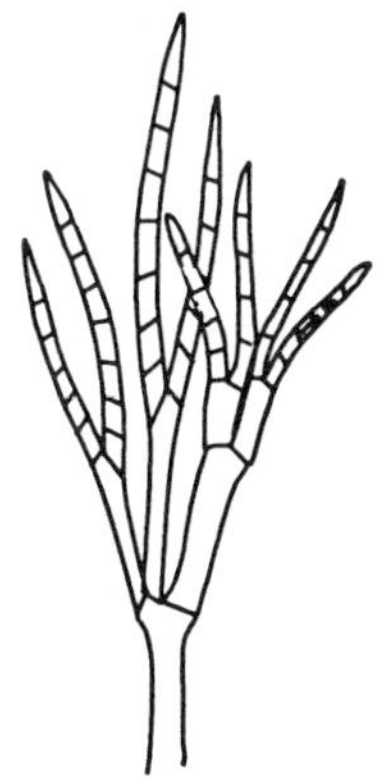

Figure 6.13. Phaeophyta, with *Lithoderma*, an encrusting form in swiftly flowing streams, shown.

Summary of key characteristics of Phaeophyta:

- Colour brown
- Photosynthetic pigments are chlorophylls *a* and *c*, enclosed in chloroplasts
- Food reserve is carbohydrates, iodine test is negative
- Cell walls semi-rigid
- Nucleus present
- Flagella absent
- No eye-spot

Primary Production

Production is the weight of new material formed over a period of time within a unit area or volume (= biomass) and includes any losses (e.g. due to injury, respiration,
18 excretion, death, grazing) that occurred during that period. Therefore, primary production is the amount of new material created by photosynthesis over a period of time. During photosynthesis, 6 molecules of carbon dioxide are combined with 12 molecules of water to produce 1 molecule of carbohydrate and 6 molecules of oxygen, or for every molecule of carbon dioxide used, 1 molecule of oxygen is produced:

$$6CO_2 + 12\ H_2O \rightarrow C_6H_{12}O_6\ \text{(glucose)} + 6H_2O + 6O_2 \qquad (6\text{-}1)$$

Productivity is the rate of formation of organic material per unit area averaged over a defined time period, usually a day or a year. ***Gross production*** refers to observed changes in weight within a unit area (= biomass), plus all losses (predatory and non-predatory), divided by the time interval. ***Net production*** is the difference between gross production and the losses, divided by the time interval. For algae, the terms ***gross photosynthesis*** and ***net photosynthesis*** are used for changes in amounts of photosynthetic products, such as carbon dioxide, dissolved oxygen, or chlorophyll *a*, but otherwise, the definitions are the same as for gross and net production.

Factors that Affect Primary Production

As described in Chapter 3, photosynthesis can occur only in the photic zone. Because of light absorption, the amount of primary production decreases with increasing depth. Each algal species has its own physiological tolerances and requirements, so that their growth rates may differ at different depths. Seasonal changes in temperature affect the supply of nutrients from the sediments, resulting in successional changes of species. The amount of primary production that occurs is a result of each algal specie's response to an array of ***abiotic*** (i.e. physical and chemical) and ***biotic*** (i.e. biological) factors. Among the most important factors controlling growth and succession are: (i) light and temperature; (ii) pH and alkalinity; (iii) algal buoyancy vs water currents; (iv) inorganic and organic nutrient availability; (v) competition among algal species; and (vi) grazing pressure.

Light and Temperature

Because temperature is intimately related to the amount of light, it is often difficult to separate the effects of the two variables. Hence, with increasing light intensity, one can expect an increase in photosynthesis due to both the increased level of
19☞ light and the resulting increase in temperature. However, each algal species has its own optima for light and temperature. Interactions of light and temperature frequently result in one of four vertical profiles of photosynthesis (Fig. 6.14 a-d). High light intensity, especially ultraviolet light, such as occurs at the surface, is detrimental to most algae, resulting in low levels of primary production near the surface. For this reason, the peak in primary production is generally below the surface (Fig. 6.14 a). However, if light intensity is not photo-inhibitory at the surface, a zone of maximum photosynthesis occurs, followed by an exponential decline in photosynthesis with increasing depth (Fig. 6.14 b). Highly coloured waters result in maximum photosynthetic rates at the surface, with rapid rates of decline immediately below (Fig. 6.14c). Maximum rates of photosynthesis within the metalimnion result in metalimnetic oxygen maxima (Fig. 6.14d), as discussed in Chapter 5.

The productivity of algae themselves affects the thickness of the ***trophogenic zone***. Oligotrophic lakes have low primary production but because of the resultant water clarity, photosynthesis can occur deeper than in eutrophic lakes (Fig. 6.15a,c). The thickness of the trophogenic zone of mesotrophic lakes is intermediate of these extremes (Fig. 6.15b). Highly coloured lakes, or ***hypereutrophic lakes***, have a very narrow trophogenic zone (Fig. 6-15d).

There is a great of variation in thermal tolerances and optima among the algae. The optimum temperature for most blue-green algae is 35 - 40 °C, for most green algae 30 - 35 °C, and for most diatoms 18 - 30 °C, Naturally, there are species in each group whose

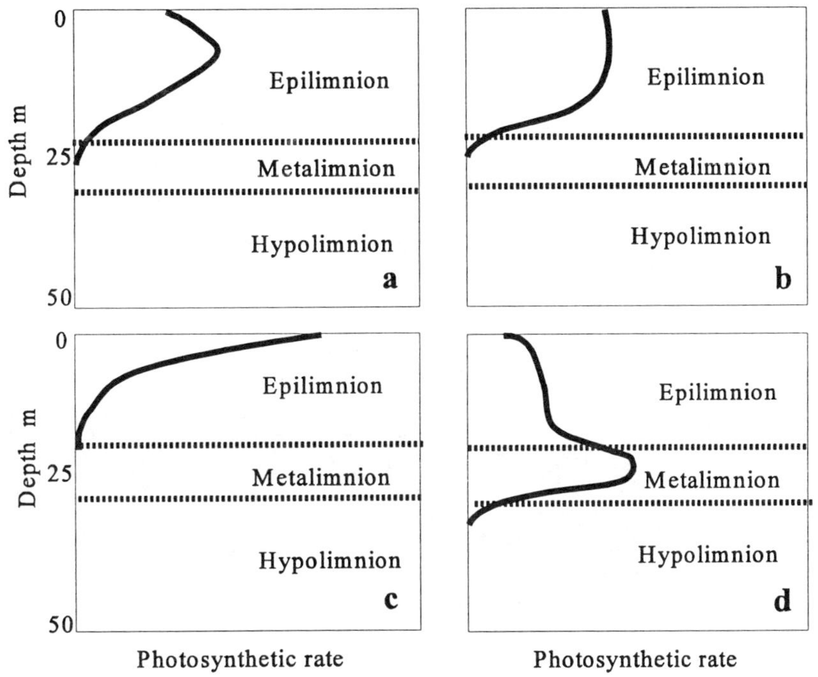

Figure 6.14. General patterns in vertical profiles of photosynthesis in lakes. See text for descriptions.

optimum temperature falls outside these ranges. An example of extremely high temperature tolerances are blue-green algae of hot springs, such as *Phormidium,* that occurs in waters up to 60 °C, followed by *Spirulina* and some species of *Oscillatoria* in waters up to 55 - 58 °C.

pH and Alkalinity

pH and alkalinity have a greater effect on the distribution of species than on the seasonal changes in productivity. Only in circum-neutral water is there a significant change in pH because of its poor buffering capacity, hard waters having high buffering
20☞ capacities to resist changes in pH. Owners of cottages on circum-neutral lakes can attest to large seasonal changes in algal biomass, with huge mats or clouds appearing in the summer. The mats or clouds increase in biomass as the lake acidifies.

As discussed in Chapter 5, pH and alkalinity are highly correlated. Most algae grow best at or near pH 7.0, but many, particularly the blue-green algae, grow best in water with a high pH. Certain species of *Microcystis* and *Coccochloris* exhibit optimal growth at pH 10.0 and do not grow below pH 8.0. *Cladophora glomerata* is found only in alkaline streams with pH greater than 8. On the other hand, many desmids, some green (e.g. *Oedogonium kurtzi*) and yellow-green algae (e.g. *Phaeosphaera perforata*) are found only in soft water streams with pH less than 7.

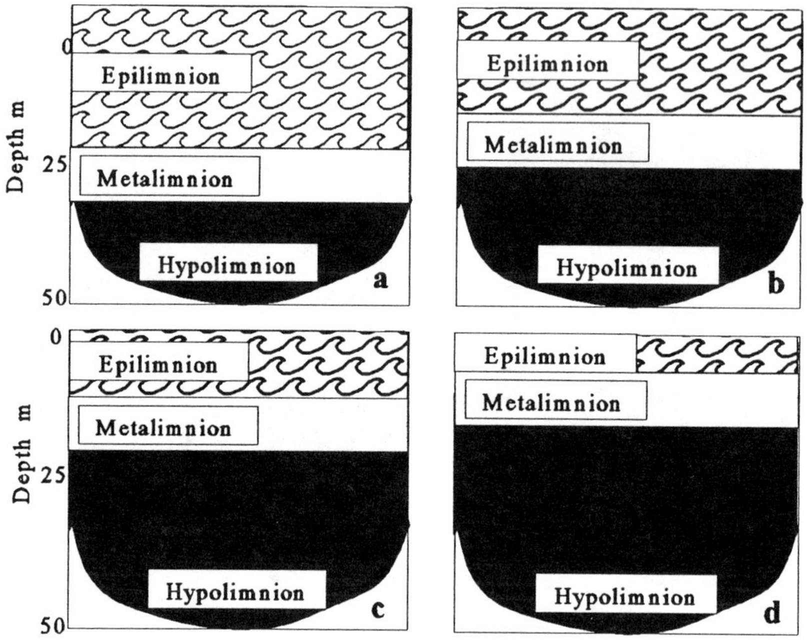

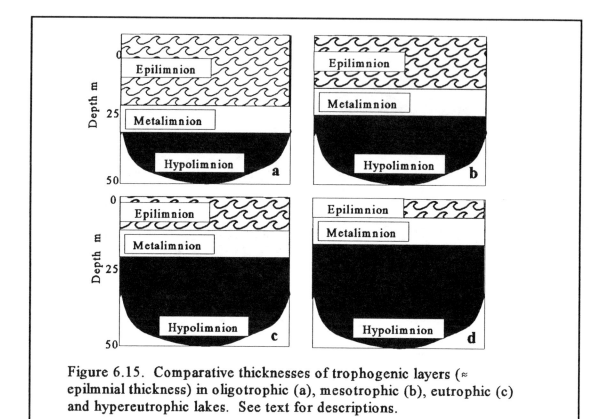

Figure 6.15. Comparative thicknesses of trophogenic layers (≈ epilmnial thickness) in oligotrophic (a), mesotrophic (b), eutrophic (c) and hypereutrophic lakes. See text for descriptions.

Lakes and streams with circum-neutral water experience annual ***"spring pH depressions"*** by as much as 2 or 3 pH units (e.g. pH drops from pH 6.5 to 4.5). The pH depression period lasts for only a few days, during the spring runoff period. As the ice
21 thaws, hydrogen ions migrate down to the base of the snow pack, resulting in an accumulation of hydrogen ions. In the spring, before the ground thaws, the layer of concentrated hydrogen ions is flushed into the lake or stream causing an instantaneous drop in pH. The softer the water (i.e. the lower the alkalinity), the lesser the buffering capacity, and hence the greater the pH depression. The pH of the remaining runoff is near 5.6, the pH of precipitation (≈ carbonic acid), but this is usually buffered somewhat by edaphic factors as the water infiltrates and percolates through the soil before it reaches the lake or stream. Concepts in Chapters 1 (hydrological cycle), 2 (spring discharge) and 5 (soil/bedrock chemistry) need to be integrated here.

Acidic lakes (pH < 7) have three types of algal communities: (i) ***blue-green mats***; (ii) ***green mats***; and (iii) ***epiphytic*** or ***periphytic green clouds***. The blue-green mats are
22 surface layers of felt-like, dark blue-green to blackish mats. Often they are orange coloured due to the presence of carotene-rich species. They usually occur in very clear waters 2 to 3 m deep with pH less than or equal to 5. Planktonic forms of algae are rare. The most common species associated with this type of mat are the blue-greens, *Lyngbya, Oscillatoria, Scytonema* and *Phormidium*, and the diatoms, *Tabellaria* and *Fragilaria*.

The green mats are bottom layers of loosely-packed mats of green algae, with colours ranging from green to reddish purple. They occur down to depths of 4 m in very clear water with pH ranging from 3.9 to 5.0. The most common species associated with this type of mat are *Pleurodiscus* (*Zygogonium*) and *Mougeotia.*

The epiphytic or periphytic green clouds are a euphemism for "elephant snot". The loosely attached clouds, or wefts of filamentous green algae, are often associated with submersed macrophytes in the littoral zone of lakes. Some become detached and remain submerged as free-floating clouds; during the winter they die (i.e. protoplasm in filaments die) and appear as whitish or greyish mats in the spring. The most common algae genera in lakes with pH < 5.0 are *Spirogyra*, *Zygnema*, *Pleurodiscus* and *Mougeotia* and in lakes with pH > 5.0, *Oedogonium* and *Bulbochaeta.* The macrophytes most commonly associated with the clouds are *Lobelia dortmanni, Isoetes echinospora, Utricularia* spp. and some species of *Potamogeton.*

The significant increases in sizes of algal mats or clouds with decreasing pH is a common phenomenon in many Ontario lakes, especially in the Sudbury, Parry Sound, Muskoka and Haliburton areas, Quebec lakes and New York lakes. So far there is no supportable scientific explanation for these mats and clouds, but the hypotheses are:

- the species have a preference for low pH
- invertebrate grazing is reduced at low pH
- microbial decomposition is reduced at low pH
- competition from less acid-tolerant species is reduced at low pH

Chapter 13 utilizes the different mats and clouds as indicators of lake acidification.

Algal Buoyancy vs Water Currents

Phytoplankton may be able to move but not against currents, even convection currents. All algae have some degree of buoyancy and can remain afloat, but most have a density greater than water (about 1.01 to 1.03 times greater) and will eventually settle out
23☞ of the photic zone. The sinking rate varies inversely with the viscosity of the fluid, directly with the square of the diameter or some linear measurement and directly with the excess density of the organism over that of the fluid. As the algal cell sinks, it creates its own minute currents that are large enough to disrupt nutrient gradients and increase its chances of obtaining nutrients. In other words, it maintains its availability of nutrients as it heads toward death. Upward water currents from the bottom not only restore nutrients but return algae to sufficient light levels to keep them alive.

Algae can alter their sinking rates using one of several mechanisms. (i) ***Form resistance*** describes the effect of cell morphology on the sinking rate. Cell extensions or protrusions, such as thecal horns or spines, decrease the settling rate. The increase in length of the thecal horns of *Ceratium* as the summer progresses is considered a mechanism for resisting sinking. As water temperature increases, water density and viscosity decrease. It is believed that the increase in temperature induces the changes in body form. Such seasonal changes in body form is called ***cyclomorphism***. It is present in some rotifer and crustacean zooplankton as well. (ii) ***Production of mucilage*** reduces the sinking rate. Several algae, especially blue-greens (e.g. *Gloeotrichia* and *Phormidium*) and some diatoms (e.g. *Achnanthes*) and green algae (e.g. *Chlamydomonas, Pandorina, Volvox, Eudorina*) have a mucilaginous sheath. While the sheath may reduce the sinking rate, it also reduces the efficiency of nutrient uptake. (iii) ***Inclusion of gas vacuoles*** in the protoplasm is unique to blue-green algae (and some bacteria). The gases contained within are in equilibrium with those dissolved in the water. The algal cell can control its rising or sinking rate by adjusting the volumes and numbers of its gas vacuoles. When light intensity is reduced and growth rate slows, the numbers of vacuoles increase and the cells rise. When light intensity increases and the photosynthetic rate increases, the oxygen produced creates turgor pressure on the vacuoles and causes the vacuoles to collapse and the cells to sink. (iv) ***Accumulation of fats*** reduces the density of the cell and causes the cell to rise. This mechanism is more common in species of phyla that have oils and lipids as food reserves, for example Chrysophyta and Bacillariophyta.

Inorganic and Organic Nutrient Availability

Inorganic nutrients include the elements, phosphorous (P) and nitrogen (N). Organic nutrients include carbonaceous (C) compounds. The ***Liebig's law of the***
24☞ ***minimum*** applies to inorganic nutrients. It states that the *nutrient in least supply directly determines the yield of organisms that require the nutrient.* Yield can be expressed in terms of abundance, biomass or growth rate, but growth rate is the most common measure of yield used. However, the use of growth rate requires detailed studies of growth on dominant algal species, which is beyond the scope of this chapter. The dynamics of phosphorous and nitrogen cycling has already been discussed in the previous chapter. The forms of nutrients measured are orthophosphate, nitrates and ammonia.

Organic nutrients have not been shown to be limiting to algae and include micronutrients that are essential (e.g. vitamins) and nonessential (e.g. hormones) organic compounds that are known to stimulate growth and cell development. Species that require vitamins for growth are called ***auxotrophic***. The vitamins required are vitamin B_{12}, thiamin and biotin. The blue-green algae, green algae, yellow-green algae, diatoms, dinoflagellates and red algae seem to be truly auxotrophic and require some vitamins (mostly B_{12} and thiamine). However, the vitamins required do not contribute greatly to the amount of carbon in the cell. The sources of these vitamins are moribund bacteria and wet and dry fallout, the latter of which carries pollen and living micro-organisms.

Most algae obtain carbon through reduction of carbon dioxide during photosynthesis. Such algae are called ***photoautotrophic***. Some algae are able to obtain carbon through metabolism of organic compounds in the dark. These are ***heterotrophic algae***, also called ***chemo-organotrophic*** algae. ***Mixotrophic*** algae utilize carbon both from carbon dioxide used in photosynthesis and from organic compounds metabolized in the dark.

Seasonal and Competitive Interactions Among Algal Species

The diversity of algal species in the plankton tends to decrease with increasing
25☞ fertility. Hence, competitive interactions tend to decrease as lakes eutrophy, with species

tolerant of eutrophy dominating the phytoplankton. It is believed that slower growth rates of algal species in oligotrophic lakes allow greater numbers of species with the same niche requirements to live together. Because currents in lakes are omnipresent, mixing nutrients to all levels in the photic zone and redistributing algal populations almost constantly, competitive interactions are probably avoided most of the time. Seasonal changes in light, temperature and availability of nutrients are accompanied by changes in algal species composition. In the fall and winter, algal species that are able to use lower light intensities at colder temperatures and a renewed supply of nutrients from the bottom as fall overturn occurs, replace species that prefer higher light intensities, warmer waters and decreasing supplies of nutrients in the summer as the lake stratifies. Phosphorous becomes most abundant during the winter and, as soon as the ice melts, algal species that are able to photosynthesize at low temperatures begin to utilize the renewed nutrient supply immediately after ice-out (i.e. spring turnover) and dominate the phytoplankton community. Biotic factors, such as parasitism and grazing, contribute to the seasonal variations in diversity of the phytoplankton community.

Grazing and Parasitism Pressures

Grazing may have a significant effect on phytoplankton biomass. Rotifers and microcrustaceans are able to almost decimate phytoplankton communities within a day.
26☞ However, in most cases, grazing maxima by zooplankton coincide with the decline of algal maxima. Yet in some instances, there appears to be no relationship at all between zooplankton abundance and algal abundance. Instead, zooplankton abundance correlates better with concentrations of detrital material. Some of the inconsistency in correlations between grazer numbers and algal biomass may be related to size-selective feeding behaviours. Grazers include some protozoans, especially ciliates, and several rotifers and crustaceans. Herbivorous ciliates graze upon very small algal species that belong to the ***picoplankton*** (< 2 μm) and ***nanoplankton*** (2 - 20 μm), the latter of which includes the ciliates. Rotifers, which themselves are part of the ***microplankton*** (20 - 200 μm) consume algae within their own size range. Crustaceans are part of the ***mesoplankton*** (200 - 2000 μm) and ***macroplankton*** (> 2000 μm), but the herbivorous forms consume algae of mesoplankton or microplankton size. Many grazers are consumer specific and graze only on certain species of algae.

Parasitism of algae is common in the plankton, with diatoms, desmids and blue-green algae being commonly infected by fungi, viruses or bacteria. Some parasites are host specific and attack only certain species of algae. Although the parasites appear to have little impact on seasonal patterns of the phytoplankton, they do affect interspecific competition by weakening dominant species of algae.

Periphyton

The ***periphyton*** are algae that grow on a variety of submerged substrates, such as rocks, plants or debris, in both lakes and streams. They are commonly referred to as attached algae but the term, periphyton, means "on plants", or the community of algae growing on plants. 27☞ ***Aufwuchs*** is a German term that refers to the community of organisms that grows on the surfaces of submerged rocks, plants or debris. This includes not only algae, but animals such as bryophytes (moss animals), sponges and cnidarians that also attach to substrate surfaces.

The periphyton are ubiquitous but especially common in streams. In fact, because of currents, only attached forms of algae can exist in streams. If planktonic forms are present they probably have been exported from pools or ponds upstream, including tributaries. The golden-brown colour of rocks in streams is due to films of diatoms, while the green or blue-green algae make up most of the long filamentous streamers, the most common being *Cladophora*, a branched form of green algae.

The study of periphyton is usually done with artificial substrates. Artificial substrates of uniform composition allows one to examine more carefully the colonization and growth of attached algae. Otherwise, habitat and community 28☞ structure are so variable that it is difficult to make any assessment of growth and population dynamics. The substrates used are typically glass slides, concrete blocks, rocks or ceramic plates. Smooth surfaces are easier to analyse than rough surfaces. However, each material is selective for certain species. The kinds and numbers of species that attach is also affected by the vertical or horizontal orientation of the plates. Because grazing can alter the biomass of the attached communities, most substrates are left in the water for only a very short period of time. The major shortcoming of artificial substrates is that they do not simulate exactly the community of organisms that occur naturally on the substrates. This is probably because of the effects of metabolites that one organism may have on another, and artificial substrates do not provide the array of organic and inorganic compounds found in natural substrates. However measuring the effects of heterogenous attached populations on variable substrates often exceeds the capacities of ones budget and time.

Distribution and Factors that Affect Primary Production in Streams

The interaction of light, temperature and current on periphyton succession is displayed prominently along the length of most streams. The water in the upper reaches (e.g. stream orders 1 to 3) of most streams is so turbulent, so well shaded and 29☞ so cool that periphyton are very poorly represented. Epilithic forms, such as the diatom *Gomphonema*, and the red algae, *Hildenbrandia* and *Batrachospermum*, predominate. Erosional processes are so dominant that if there is any photosynthesis occurring it is not enough to offset losses of oxygen through respiration of microorganisms. Hence, the ratio of the rate of oxygen produced (or carbon dioxide consumed) by photosynthesis (P) to the rate of oxygen consumed (or carbon dioxide produced) by respiratory processes (R), is usually less than 1 in stream orders 1 to 3. The P/R ratio changes seasonally due to the changing light conditions as the trees add more leaves to their limbs. The productivity pattern often shifts from one that is autotrophic in the spring to one that is essentially heterotrophic later in the season when the light is prevented from reaching the stream by the canopy of trees.

Periphyton do not begin to increase significantly in biomass until the streams are wide enough (e.g. stream orders 4 to 6) for sufficient light to penetrate and promote photosynthesis. Erosional processes are gradually replaced by depositional processes and the substrate consists mostly of small stones, gravel and some sand. The epilithic forms of upstream are replaced by episammic forms, such as the diatoms *Acanthes, Navicula* and *Cymbella*, the green algae *Ulothrix, Cladophora, Vaucheria* and some *Phormidium* and the red algae *Lemanea*. As water temperatures increase because of the increased light intensities, these periphyton develop to the point that primary production exceeds respiration and the P/R ratio is greater than 1.

In stream orders 7 and greater, the streams are wider and deeper, water velocity is slower, water becomes more turbid and depositional processes predominate with increased loadings of organic material on the sediments. As a result, light intensities at the bottom of the stream are not sufficient to promote photosynthesis and the P/R ratio falls below 1 again. Much of the photosynthesis that occurs is due to epipelic forms that can grow in organic sediments and to planktonic forms which increase in diversity as water velocity slows. The distribution and diversity of epipelic diatoms is determined in general by chemical features of the sediments and the overlying water, as well as by several biotic factors. Diurnal vertical migration rhythms have been demonstrated for some epipelic diatoms, flagellated green algae and blue-green algae in flowing water. Cell numbers on the sediment surface begin to increase before dawn and reach a maximum about mid morning. Surface cell numbers then decrease and reach a minimum before the onset of darkness. Photosynthetic rhythms persist at light intensities well below saturation. Minimum P values have been found in mid afternoon after which the rate increases again. This rise appears to precede the re-emergence of cells. Maximum P values occur when maximum cell numbers are present on the sediment surface.

Algae in and on Sediments

Just as in the interstitial water of the beach zone of lakes, are fairly large algal populations that exhibit a spring maximum in biomass. The percent water content of the sediments is an important determinant of the algal composition at the water's edge.
30☞ Within the substrate, light is attenuated fairly rapidly and is essentially absent within the first half centimeter. However, some algae have been found as deep as 20 cm and this is probably due to the mixing of the top 20 cm layer of sediment by turbulence or wave action. Light is attenuated much more rapidly by highly organic sediments than by inorganic sediments. Apparently some epipelic algae are capable of supplementing photosynthetic growth by heterotrophic utilization of organic materials. Such forms must be adapted to moving within the sediments in response to the availability of light.

In general, diatom masses predominate on sediments with a higher inorganic matter content, but the sediments must have an available form of silica. Moderate or low organic matter content of sediments appears to favour blue-green algal populations. Biomasses of attached algae tend to be fairly low in sediments with very low or very high organic matter content. The only attached algal forms that appear to prevail on organic rich sediments are some chlorophytes, especially the flagellated forms. Diatoms, blue-green algae and flagellated green algae populations appear to depend highly on the calcium content. Hence, algal growth of these forms tends to increase with increasing levels of calcium and phosphate.

Desmid flora (green algae) of the genera, *Pediastrum* and *Scenedesmus* are also common microflora on the sediments. Desmid flora are often fairly rich in species but not very high in abundance. Organic-rich sediments also appear to support a greater biomass of euglenoid forms. Epipelic blue-green algae appear to be equally well adapted to alkaline, neutral and slightly acidic waters, although different species appear to dominate at different pH levels and alkalinities. In lakes, algal population biomasses tend to be relatively constant on sediments between 1-6 m depth. Below this depth algal biomasses decrease quickly and are absent below 8-10 m.

Phototrophic diatoms are more sensitive to light than are blue-green algae which extend deeper in depth than do the diatoms. The seasonal variations in biomasses of epipelic algae tend to correspond closely to the curves for incident light and water temperature. The epipelic algae tend to increase in biomass fairly quickly and then decline equally as quickly. However, one species replaces the other with the result that species associations successively dominate during an annual cycle. Usually diatom growth begins in spring and reaches a maximum biomass in April to May in the temperate zone. Then a decline occurs in mid summer, prior to a smaller autumnal peak that disappears by the end of November. This pattern is similar to that of planktonic species, although the pattern for epipelic forms occurs more slowly than found in the planktonic community.

Just as in head water streams where attached algae biomass changes seasonally due to changing light conditions, the same phenomenon occurs in the littoral zone of lakes. The epipelic forms grow quickly when the submersed and emergent macrophytes are small and just beginning to grow. As the macrophytes begin to grow in height and biomass, the amount of light penetrating to the bottom decreases. This results in a corresponding decrease in the epipelic algal population biomass. In general, the biomass of epipelic algal forms in lakes increases as the organic carbon, nitrogen and phosphorus content of the sediments increase. As ammonia nitrogen (NH_4 -N) of the interstitial water of the sediments is quickly assimilated by the benthic algae, nitrogen fixation by epipelic and epilithic blue-green algae and bacteria predominates and constitutes a significant source of combined nitrogen to some lakes.

The chemistry of the overlying water is important, not only as a source of nutrients but for its effects on the penetration of light. High levels of dissolved organic matter (DOM) or particulate organic matter (POM) affect the quality and quantity of light penetrating to the bottom of the lake, and is followed by a decline in the biomass and diversity of the attached algal forms. The growth and productivity of the microphytes is also affected by grazing animals, such as gastropods, many insect larvae (e.g. riffle beetles, water pennies), crustaceans and herbivorous fishes. Much of the periphyton that is consumed by animals quickly enters the detrital pathways either as waste or spillage of pieces that are not ingested and fall directly to the detrital pool, or are digested with any excreta passed to the detrital pathway.

Methods For Measuring Primary Production

Why measure primary production? Researchers have found that fish productivity and diversity is linked to the amount of primary production. The hypolimnia of very productive lakes (eutrophic lakes) tend to go anoxic during the summer because the organic matter builds up on the bottom is constantly being degraded by aerobic bacteria which deplete the dissolved oxygen supply, and in the process produce copious amounts of CO_2. The lack of dissolved oxygen restricts fish to the upper, more highly oxygenated waters of the epilimnion during the summer. But fish, such as lake trout which prefer the colder water in the hypolimnion, cannot survive in the warmer epilimnion during the summer and they must move to colder, deeper water where most would eventually perish in eutrophic lakes.

There are several ways to measure primary production. Some methods are qualitative, some are quantitative and others are semi-quantitative. We will begin with a qualitative method using indicator species.

Qualitative methods: Use of biological indicators is a well known qualitative method for
31☞ assessing enrichment status of freshwater systems. It has greater value for enriched systems because only a few species are present and they are always present in enormous numbers. Remember, it is not the mere presence of indicator species that is important, it is their numerical dominance. Table 6.2 lists several species and their indicator value. Use the table when huge blooms of algae are present. A microscope is needed to identify the algae, along with the taxonomic key by Prescott, recommended above.

A good clue to trophic status is the diversity of algae present. If a lot of species are present the lake is probably mesotrophic to oligotrophic. Lakes that are eutrophic will be dominated by one or two species, especially blue-green algae. See Chapter 13 for a complete list of algal species common within clean and enriched waters.

Semi-quantitative methods: There are two semi-quantitative approaches to assessing trophic status. One is using biotic indices that are scores based on the presence and
32☞ abundance of algal species. Each species is given a rank based on its ability to tolerate enrichment. The ranking is done by qualified scientists, such as Dr. Vladimir Sladacek[1]. He ranked not only several algal species but species of zooplankton and invertebrates as well. Species with a high tolerance of organic enrichment were given a low ranking; species with no tolerance (i.e. oligotrophic species) were given a high ranking. Organically enriched waters have only a few species, each with a low ranking, yielding a low score. Oligotrophic lakes have a lot of species, each with a high ranking, yielding a

Table 6.2. Some algal species with indicator value and other significant contributions.

Genus Name	Common name	Phylum	Significance
Nostoc	blue-green alga	Cyanobacterium	Algal nitrogen fixer; enrichment indicator; septic odour
Anabaena constricta	blue-green alga	Cyanobacterium	Algal nitrogen fixer; enrichment indicator; septic odour
Oscillatoria putrida	blue-green alga	Cyanobacterium	Enrichment indicator; gives grassy to musty odour
Calothrix parietina	blue-green alga	Cyanobacterium	Attached form in clean streams
Ceratium	dinoflagellate	Pyrrophyta	Has cellulose plates; blooms give septic smell
Chlorella vulgaris	green alga	Chlorophyta	Enrichment indicator
Ulothrix zonata	filamentous green	Chlorophyta	Attached form in clean streams
Spirogyra communis	filamentous green	Chlorophyta	Enrichment indicator; bloom gives grassy odour
Cladophora glomerata	branching green	Chlorophyta	Indicator of clean, hard water stream
Volvox	colonial green	Chlorophyta	Common in eutrophic lakes; blooms give fishy odour
Euglena viridis	Euglena	Euglenophyta	Enrichment indicator; blooms give fishy odour
Euglena spirogyra	Euglena	Euglenophyta	Oligotrophic indicator
Tribonema	chysophyte	Chrysophyta	Causes surface scum; blooms give fishy odour
Hydrurus confusus	brown chysophyte	Chrysophyta	Encrusting stream form; gives off disagreeable odour
Rhodomonas lacustris	cryptophyte	Cryptophyta	Attached form in clean streams
Gomphonema parvulum	diatom	Bacillariophyta	Enrichment indicator; major source of silica
Navicula gracilis	diatom	Bacillariophyta	Attached form in clean streams; major source of silica
Nitschia palea	diatom	Bacillariophyta	Enrichment indicator; major source of silica
Cyclotella cryptica	Eurasian diatom	Bacillariophyta	Exotic species; major source of silica
Cocconeis placentula	diatom	Bacillariophyta	Attached form in clean streams; major source of silica
Chara vulgaris, lacustris	Chara	Charophyta	Indicators of enriched, hard water pool/lake
Hildenbrandia rivularis	red alga	Rhodophyta	Indicator of clean, hard water stream; attached form
Lemanea annulata	red alga	Rhodophyta	Attached form in cold, clean streams
Bangia atropurpurea	red alga	Rhodophyta	Exotic species in lakes

high score. Details of this approach are given in Chapter 13 as one of several methods for the biological assessment of water quality.

Another semi-quantitative method is to "guestimate" the chlorophyll levels using
32☞ Secchi disc relationships developed by scientists. The Secchi depth and amount of primary production (measured as μg Chlorophyll *a*/L) are inversely related because high primary production results in high phytoplankton biomass, which lowers the transparency of the water. Therefore, the Secchi depth should be useful to estimate the productivity, or the trophic status, of a lake. In fact, scientists have found good relationships between Secchi depth and average summer chlorophyll *a* level and spring overturn phosphorous levels (Fig. 6.17). Secchi depth varies considerably due to water colour, suspended sediments, season, and even among observers but the variation seems to be greatest between 1 and 6 m (Fig. 6.17). In eutrophic lakes (Secchi depth < 2 m)

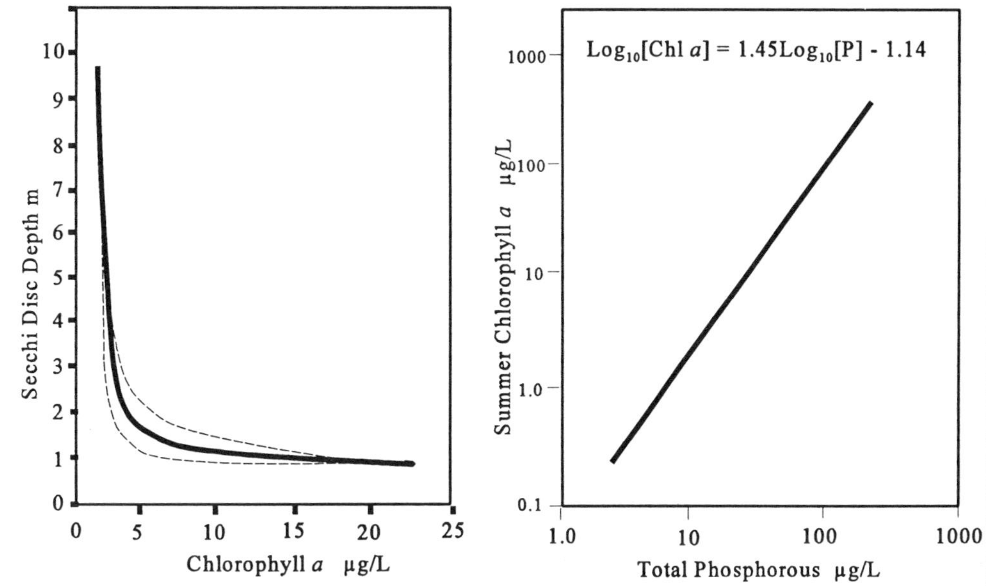

Figure 6.17. Relationships between average summer Chlorophyll *a* levels and Secchi depth (left) and spring total phosphorous levels.

chlorophyll a levels can change considerably with only a small change in Secchi depth. In oligotrophic lakes (Secchi depth > 5 m), the opposite occurs, with chlorophyll a levels changing very little as Secchi depth increases. The relationships are used to provide the data in Table 6.3 for "guestimating" the approximate summer chlorophyll *a* and total phosphorous concentrations based on the Secchi depth. The trophic status of a lake can then be determined from the chlorophyll *a* content; oligotrophic lakes have chlorophyll levels less than 2 μg/L, eutrophic lakes have more than 5 μg/L, and mesotrophic lakes have 2 - 5 μg/L.

Alternatively, the ***trophic state index*** (***TSI***) (Carlson 1977)[3] can be used to assess the trophic state of lakes. The TSI is based on either the Secchi depth, the average ***summer*** chlorophyll *a* level, or the average ***summer*** total phosphorous level. The relationship for summer chlorophyll *a* level is:

$$TSI = 10(6 - 2.04 - 0.68 \ln [Chl])$$

Oligotrophy is predicted if TSI is 0 - 35
or $[Chl\ a]_{ave.\ summer} < 1\ mg/m^3$
Mesotrophy is predicted if TSI is 35 - 50
or $[Chl\ a]_{ave.\ summer} = 1 - 6.4\ mg/m^3$
Eutrophy is predicted if TSI > 50
or $[Chl\ a]_{ave.\ summer} > 6.4\ mg/m^3$

(Note: 1 mg/m^3 = 1 μg/L)

Chapter 13 elaborates further on the use of TSI as a water quality assessment technique.

33 Table 6.3. Relationship between Secchi depth and spring overturn total phosphorous and average summer chlorophyll *a* levels in a lake. Data are estimated from relationships shown in Fig. 6.17.

Secchi Depth m	Spring Total Phosphorous μg/L	Chlorophyll *a* μg/L
< 1	>30	>10
1 - 2	30 - 18	10 - 5
2 - 3	18 - 13	5 - 3
3 - 4	13 - 11	2.5 - 3
4 - 5	11 - 10	2 - 2.5
5 - 8	10 - 8	1.5 - 2
> 8	< 8	< 1.5

Quantitative methods: The quantitative methods employ either measuring byproducts of photosynthesis (e.g. the amount of), or the end product of photosynthesis (i.e. t oxygen

[3]Carlson, R. E. 1977. A trophic state index for lakes. Limnology and Oceanography 22: 361 - 369.

evolved), gases used in photosynthesis (e.g. the amount of carbon dioxide utilized he amount of chlorophyll produced). An example of each is given.

1. Light and Dark Bottle Method:

34☞ This method measures the amount of oxygen produced by autotrophs. It is sensitive enough for measuring oxygen produced in mesotrophic and eutrophic waters, but not in most oligotrophic waters.

Three kinds of bottles are filled with water from a designated depth in the photic zone:

Light Bottle (L): Allows light to penetrate, therefore allows photosynthesis to occur, as well as respiration.

Dark Bottle (D): Wrapped in aluminum foil or black tape so that light cannot enter. Prevents photosynthesis, allows respiration only.

Initial Bottle (I): Measure dissolved oxygen content so that you know the initial level of oxygen present.

Both the light and dark bottles (usually done in triplicate for each) are placed in the water at a designated depth for 24 h . After 24 h the bottles are removed and the oxygen content in each is measured using either the Winkler method or a Hach kit.

Community Respiration = (I - D)/24 mg O_2/L/hr
Net Photosynthesis = (L - I)/24 mg O_2/L/hr
Gross Photosynthesis = Net Photosynthesis + Community Respiration
= (L - D)/24 mg O_2/L/hr

E.g. Calculate the Community respiration, net photosynthesis and Gross photosynthesis rates for the following example. The initial oxygen content = 8 mg/L; after 24 h, Light = 11 mg/L, Dark = 7.3 mg/L.

Community respiration= (Initial- Dark)/24 hours = (8-7.3)/24 = ***0.029 mg/l/hr***

Net photosynthesis = (Light- Initial)/24 = (11-8)/24 = ***0.125 mg/L/hr***

Gross photosynthesis = (Light- Dark)/24 = (11-7.3)/24 = ***0.154 mg/L/hr***

2. ^{14}C Method

This method measures the amount of carbon dioxide consumed by autotrophs. It is sensitive enough to use for oligotrophic waters, if left for sufficient time, and mesotrophic to eutrophic waters. The light (L) and dark (D) bottle method is employed but ^{14}C is injected into each bottle. The ^{14}C levels are determined on a Colter counter which measures the radiation emitted from the isotope. Autotrophs use both ^{14}C and ^{12}C, the normal isotope in water, and by knowing the amount of ^{14}C added and the total amount of ^{12}C present (from Table 6.4 relating $[CO_2]$ to pH and total alkalinity), we can estimate the amount of carbon dioxide consumed by autotrophs.

Primary Production (P) = $P_L - P_D$; $P_L = (r_L/R_L) \times C \times f$; $P_D = (r_D/R_D) \times C \times f$

where
r = uptake of ^{14}C in counts/min = counts/min x volume bottle/volume filtered

R = [^{14}C] injected into bottle in counts/min x microcurries added x efficiency of counter = 2.22×10^6 x microcurries added x efficiency

C = ^{12}C available in mg/m^3 = tot. alkalinity x conversion factor x 1000

f = correction factor of ^{14}C uptake relative to ^{12}C uptake = 1.06

E.g. A water sample has a pH of 7.5. Total alkalinity = 100 mg $CaCO_3$/L at 20°C. A 125 ml light bottle is filled with water and given 2 μcurries ^{14}C and placed in water for 24 h. 50 ml is filtered and put through Colter Counter which gave a reading of 4,500 counts/min with 85% efficiency in the light bottle and 3,400 counts/min with 90% efficiency in the dark bottle. Calculate gross primary production.

Conversion factor for C = 0.26 from Table 6.4

Primary Production in Light Bottle:

P_L = (4,500 x 125/50)(=r) / (2.22 x 10^6 x 2 x 0.85)(=R_L) (100 x 0.26 x 1000)(=C) x 1.06(=f)
= ***77.50 mg C/m^3/day***.

Primary Production in Dark Bottle:

P_D = (3,400 x 125/50)(=r) / (2.22 x 10^6 x 2 x 0.90)(=R_L) (100 x 0.26 x 1000)(=C) x 1.06(=f)
= ***55.31 mg C/m^3/day***.

Gross Primary Production rate = 77.50 - 55.31 = ***22.19 mg C/m^3/day***.

Table 6.4. Factors for conversion of total alkalinity to milligrams of carbon/litre.

pH	Temperature					
	0	5	10	15	20	25
6	1.15	1.03	0.93	0.87	0.82	0.78
6.2	0.82	0.74	0.68	0.64	0.6	0.58
6.4	0.6	0.56	0.52	0.49	0.47	0.45
6.6	0.47	0.44	0.41	0.4	0.38	0.37
6.8	0.38	0.37	0.35	0.34	0.33	0.32
7	0.33	0.32	0.31	0.3	0.3	0.29
7.2	0.3	0.29	0.28	0.28	0.28	0.27
7.4	0.28	0.27	0.27	0.26	0.26	0.26
7.6	0.27	0.26	0.26	0.25	0.25	0.25
7.8	0.25	0.25	0.25	0.25	0.25	0.25
8	0.25	0.25	0.25	0.25	0.24	0.24
8.2	0.24	0.24	0.24	0.24	0.24	0.24
8.4	0.24	0.24	0.24	0.24	0.24	0.24
8.6	0.24	0.24	0.24	0.24	0.24	0.24
8.8	0.24	0.24	0.24	0.24	0.24	0.23

3 . Chlorophyll a Method

This method measures the amount of chlorophyll *a* present in a water sample. It is sensitive enough to measure primary production in waters of all trophic levels.

A 1-L water sample is taken, preserved with 1 ml of 1% $MgCO_3$ solution and held at 4° C until measured for Chlorophyll *a*. The method employs a fluorometer (or spectrophotometer) and cuvettes with a 1-cm light path. First, the sample must be filtered through 0.8 µm maximum pore size; then the filter paper with the algae is folded and placed in a test tube to which 90% alkalized ($MgCO_3$) acetone solution (usually 20 ml) is added. This extracts the chlorophyll; the greater the chlorophyll content, the greener the solution. The sample is then transferred to a cuvette with a 1-cm light path which is placed in a fluorometer or spectrophotometer. The ***absorbance*** (amount of light absorbed) is read at wavelengths that are optimal for chlorophylls *a*, *b* and *c*. The greater the greeness, the more light is absorbed. For grass-green extracts, chlorophyll *a* dominates and absorbance need only be read at 663 nm:

[Chl *a*] in extract = (absorbance at 663 nm) x (13.4) = mg Chl *a*/L of extract.
[Chl *a*] mg/m^3 of lake water = (mg chlorophyll *a*/L of extract) x (extract volume (ml) x 1000 L/m^3)/(filtered volume (ml))

E.g. Calculate the chlorophyll a *level if 1 L of water is filtered, 20 ml of alkalized acetone is used to extract the chlorophyll, and the extract gives an absorbance of 0.10 at 663 nm*

[Chl a*] in extract = 0.10 x 13.4 = 1.34 mg/L of extract*
[Chl a*] in water mg/m^3 = 1.34 mg/L x 20 ml x 1000 L/m^3 / 1000 ml =* ***26.8 mg/m^3 = 26.8 μg/L***

If the sample is not grass green (i.e. has low chlorophyll levels), read the absorbances at 750, 665, 645 and 630 and subtract the absorbance at 750 nm from each of the other absorbances to correct for turbidity. Apply the following formulae for a spectrophotometer cell of 1-cm light path:

[Chl *a*] in sample = 116 (abs. @ 665 nm) - 13.1 (abs @ 645 nm) - 1.4 (abs @ 630 nm)
[Chl *b*] in sample = 207 (abs. @ 645 nm) - 43.3 (abs @ 665 nm) - 44.2 (abs @ 630 nm)
[Chl c] in sample = 550 (abs. @ 630 nm) - 46.4 (abs @ 665 nm) - 163 (abs @ 645 nm)
mg pigment / m^3 lake water = C x 2 / litres of lake water filtered

E.g. Calculate the levels of chlorophyll a, b and c collected by filtering a 4-L water sample and whose absorbance readings on 20 ml samples of extract are as follows: abs. at 665 nm = 0.05; at 645 nm = 0.02; at 630 nm = 0.015

[Chl a] in sample = 116 (0.05) - 13.1 (0.02) - 1.4 (0.015) = 5.80 - 0.26 - 0.02 = 5.52
[Chl b] in sample = 207 (0.02) - 43.3 (0.05) - 44.2 (0.015) = 4.14 - 2.17 - 0.66 = 1.31
[Chl c] in sample = 550 (0.015) - 46.4 (0.05) - 163 (0.02) = 8.25 - 2.32 - 3.26 = 2.67

[Chl a] in lake water = 5.52 x 2/4 = ***2.76 mg/m^3***
[Chl b] in lake water = 1.31 x 2/4 = ***0.66 mg/m^3***
[Chl c] in lake water = 2.67 x 2/4 = ***1.34 mg/m^3***

Knowing the chlorophyll *a* level not only allows you to determine the trophic status of a lake but its fisheries and recreational potential. Table 6.5 summarizes the recreational activity potential of lakes based on their chlorophyll *a* levels.

Table 6.5. Using chlorophyll *a* levels to assess a lakes recreational and fisheries potential. (See Chapter 13 for reference and use of the table).

[Chl a]	Recreational Potential	Fisheries Potential
2 mg m^{-3}	Very unproductive lakes; ideal for body contact water recreation; extremely clear with a mean Secchi disc visibility of 5 m (may be lower in brown water, dystrophic lakes)	Hypolimnetic concentrations of oxygen in excess of 5 mg $litre^{-1}$ will preserve a cold water fisheries (e.g. lake trout)
5 mg m^{-3}	Moderately productive; lakes can be used for body-contact water recreation; less clear, with a mean Secchi disc visibility of 2-5 m	Preservation of cold water fisheries difficult; ideal for walleye, pickerel, pike, bass fisheries
10 mg m^{-3}	body-contact recreation of little use; Secchi disc depths will be low (1-2 m); nuisance growths of submergent macrophytes common in some areas	Hypolimnetic oxygen depletion will be common but lake still good for bass, walleye, pickerel, pike, maskinonge, bluegill, yellow perch fisheries; danger of winterkill of fish in shallow lakes
25 mg m^{-3}	No body-contact recreational value; Secchi disc depth <1.5 m; nuisance growths of submergent and emergent macrophytes and organic bottom over most of lake	Suitable only for warmwater fisheries; hypolimnetic oxygen depletion beginning early summer, considerable danger of winterkill of fish except in deep lakes

AQUATIC MACROPHYTES

Keys for illustrating and identifying aquatic macrophytes are beyond the scope of this text. Prescott (1978)[4] is recommended for anyone wishing to pursue identification of macrophytes. Only a few more common species are illustrated below.

Macrophytes exhibit numerous adaptations for life in or on the water. The adaptations permit the plant to occupy different zones within the lake or stream. Three zones of plants are generally recognized: (i) the ***emergent*** macrophytes (Fig. 6.18) which occupy most of the littoral zone; (ii) the ***floating plant*** community (Figs. 6.19, 6.20) which shares the littoral zone with the emergent macrophytes; and (iii) the ***submerged*** (or ***submersed***) macrophytes (Figs. 6.21. 6.22, 6.23, 6.24) which occupy part of the littoral zone and all of the sublittoral zone.

---

[4]Prescott, G. W. 1978. How to know the freshwater plants. Wm. C. Brown Publ., New York, N.Y.

Macrophyte Communities

The Emergent Plant Community

Macrophytes within the emergent community tend to have stiffer leaves and stems than do plants within the submerged and floating communities. Submerged and
35☞ floating species rely on the buoyancy of the water to keep the leaves and stems erect or floating. However, none of the aquatic macrophytes have the true stiffening tissues, phloem or xylem, that characterize woody terrestrial plants. In woody plants the phloem is responsible for the transfer of food and the xylem for the transfer of water.

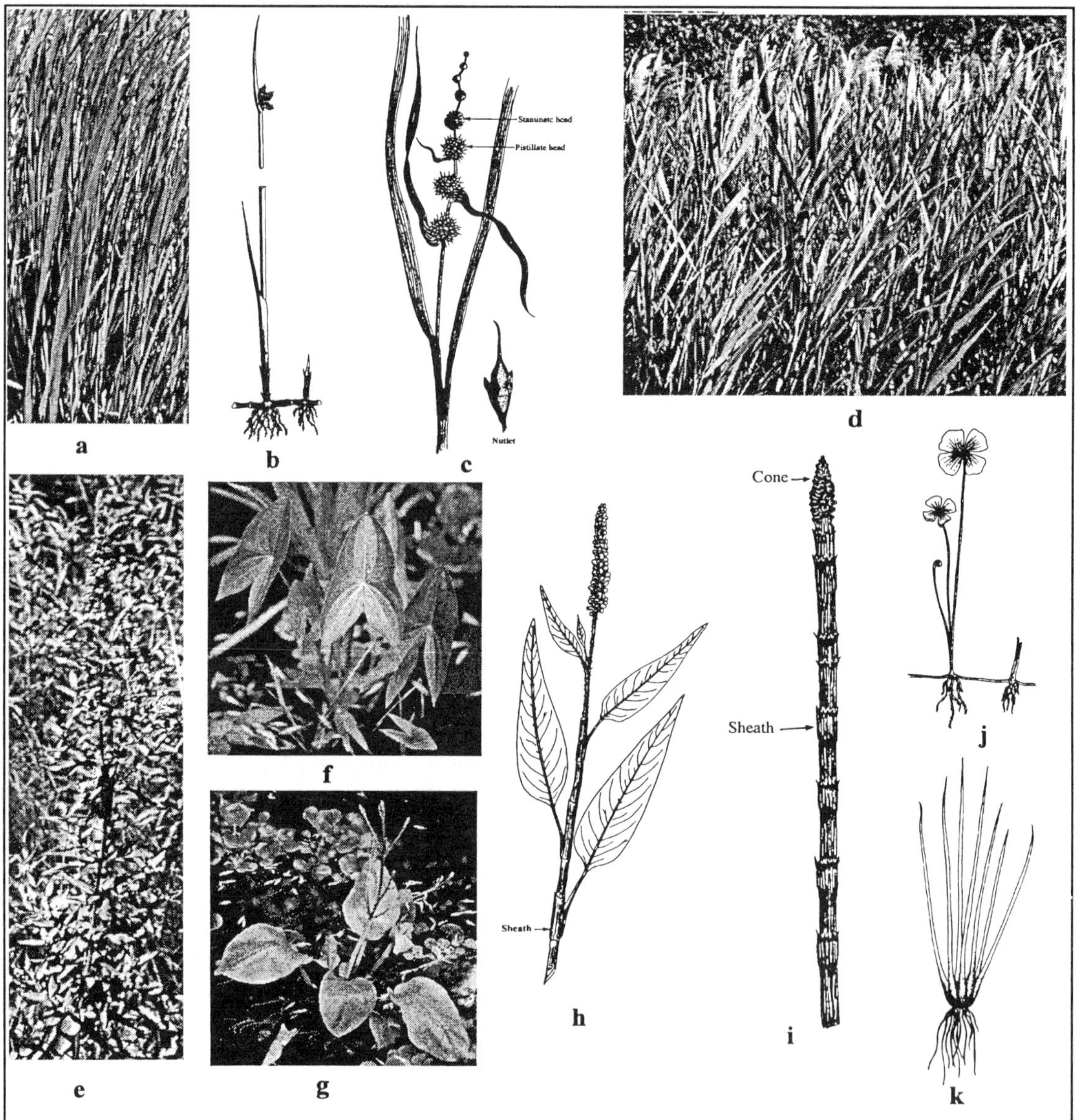

Figure 6.18. Some common genera of emergent macrophytes. (a) *Typha*, (b) *Scirpus*, (c) *Sparganium*, (d) *Phragmites* (see also Fig. 14.9), (e) *Lythrum* (see also Fig. 14.10), (f) *Sagitarria*, (g) *Pontederia*, (h) *Polygonum*, (i) *Equisetum*, (j) *Marsilea*, (k) *Isoetes*.

Stiffening tissues are especially evident in emergent plants such as *Typha* (the cattail, Fig. 6.18a), but they, like many other emergent plants, employ other adaptations to keep the leaves and stems rigid and erect. A common adaptation is the v-shaped cross-section of leaves. The v-shape gives the leaves rigidity, much like a fold or crease stiffens a page of paper. Without the crease the page simply collapses. V-shaped leaves are also present in *Scirpus* (bullrush, Fig. 6.18b), *Sparganium* (burr reed, Fig. 6.18c), and *Phragmites* (cane reed, Fig. 6.18d), plants that can grow up to 1 to 2 meters in height. All four genera usually occupy huge expanses in the shore zones of slow moving rivers and lakes.

The leaves and the roots of *Typha* provide a rich food supply to many shore birds and water birds. *Typha* is also well known for its ability to take up nutrients, particularly phosphorus, from the sediments. Of all the emergent plant species, *Typha* seems to be the most efficient at taking up nutrients from the sediments and temporarily preventing the nutrients from recycling back into the water. The Eurasian *Phragmites*, which can grow to 2 m in height, also is efficient at extracting phosphorous from the sediments. The plant can be somewhat remote from a waters edge because it sends out long (up to 10 m) ***stolons*** to supply the plant with water and nutrients. *Phragmites* was introduced to North America (see Chapter 14) and is out-competing *Typha* in many marsh systems. Although *Phragmites* offers excellent cover for birds and mammals, it is of less food value to them than is *Typha*.

Another emergent that has been introduced is purple loosestrife, *Lythrum salicaria* (Fig. 6.18e). The species is very prolific, reproducing both sexually by the numerous flowers and asexually by a spreading ***root stock***. Apparently, although evidence is mainly anecdotal, purple loosestrife is out competing native emergent plants in many marshes.

Another common emergent, and a favourite food of many aquatic birds and mammals, is the arrowhead, *Sagittaria* (Fig. 6.18f). The plant can grow over one meter in height. The leaves are distinctively shaped like an arrowhead, hence their common name. They adjust easily to changing water levels and are common in artificial impoundments. When water levels are lowered and the plants become stranded on shore, the narrow ribbon-like leaves of the submerged stem are replaced by the arrow-shaped leaves.

The stiffening tissues of some emergent plants are less apparent because as soon as they emerge from the surface of the water they collapse and float. Many of these species form extensive communities and clog waterways. In fact it is very difficult to walk in a bed of these plants because the plants get tangled around ones legs. The pickerel weed, *Pontederia* (Fig. 6.18g) is especially meddlesome. The leaves arise in clusters from a creeping root stock. It is because of the creeping root stock that they are almost impossible to remove from a lake where it proliferates. The smartweed or knotweed, *Polygonum* Fig. 6.18h), is similar in habit. It has jointed stems and is common in marshes, bogs, ponds and even in estuaries. The plants form tangled floating mats when the water levels drop due to evaporation over the summer.

Some emergent macrophytes have had, or still have, well-known domestic uses.

The horsetail or scouring rush, *Equisetum* (equi = horse, setum = hair) (Fig. 6.18i) has silica in cell walls of the stems, and many campers still use the horsetail for scouring their pots and pans. The horsetails are one of the few emergents that reproduce with spores, a very primitive method of reproduction. The spore-bearing branches that come up in the spring terminate in a cone which contains the spores. The horsetail appearance to these plants arises from the small leaves that encircle the stem at several "nodes" along the stem.

Two other spore-bearing emergents include the quillwort, *Isoetes* and the aquatic fern or pepperwort, *Marsilea* (Fig. 6.18j). *Isoetes* (Fig. 6.18k) produces cases, or ***sporangea,*** that bear the spores, in the swollen bases of their long awl-like leaves or quills, hence the common name, quillwort. Similar to terrestrial ferns, *Marsilea* produces two kinds of spores, female macrospores and male microspores, borne in clusters of four at the leaf axes. *Marsilea,* however, is noted more for its peculiar behaviour. Opposite pairs of its four clover-like leaves fold up at night. The plant is most commonly found in shallow, quiet water rooted in the muddy sediments.

Most emergent plants have the ability to reproduce, or spread, by vegetative means. The most common methods employed are spreading root stocks or stolons. When plants do reproduce sexually, they rely on gravity or wind to spread the male gametes. A common feature in many is the presence of the male flower, or a component of the flower, located above the female or the female component. In *Typha*, ***spikes*** carry the ***staminate*** or male portion above the lower portion which contains the ***pistillate*** or female flowers. In the bur reed, *Sparganium*, the small spheres or burs of staminate flowers are borne on the stem above the larger burs of pistillate flowers. *Sparganium* occasionally grows completely submerged. *Sagittaria* is unique in that its flowers have three petals arranged in three whorls, the lowermost of which is composed solely of pistillate flowers.

The Surface Floating Macrophyte Community

There are two groups of plants in this community; (i) floating plants with roots
36☞ but the roots are not anchored on the bottom; and (ii) floating plants with roots anchored on the bottom.

Two common plants in the group floating at the surface, with roots, are the duckweed, *Lemna* (Fig. 6.19a) and *Wolffia* (Fig. 6.19b). In many ponds and ditches *Lemna* forms an extensive green blanket. The plant consists of two or three flat fronds, each with one to five veins and a single trailing root. Sexual reproduction is rare. The normal method of reproduction is by fission or budding. Often, the blankets of *Lemna* and *Wolffia* are so extensive that they prevent light from penetrating to deeper water. This causes the water body to become anoxic, with no mechanism for oxygen to be restored to the lower water strata. The blanket quickly becomes a barrier that prevents oxygen in the atmosphere from passing at the air-water interface.

The water fern, or floating moss *Salvinia rotundifolia* (Fig. 6.19c), is a tropical fern that has established itself in ditches and ponds in more temperate climates. It is an

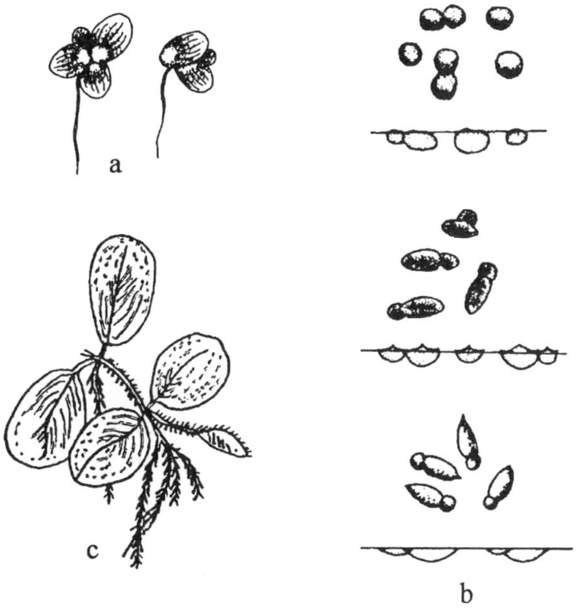

Figure 6.19. Floating plants with roots also floating, showing (a) *Lemna minor*, (b) three species of *Wolffia* (top = *W. columbiana*, middle = *W. papulifera*, bottom = *W. punctata*) with their profile on the surface shown below each species, and (c) *Salvinia*.

attractive plant and has become a common addition to home aquaria and artificial pools. *Salvinia* has paired leaves, often folded along the midrib to give it a heart-shape at the base. The plant has a brown, feathery, root-like appendage hanging down from the undersurface of the leaves. The root-like appendage is not a true root, but represents a third, vestigial leaf.

Other floating plants with roots include the introduced water hyacinth, *Eichhornia crassipes* (Fig. 14.6) and the water lettuce, *Pistia stratiotes* (Fig. 14.7) (these are discussed in more detail in Chapter 14). Both plants are tropical and are part of a lake or pond's ***pleuston*** (floating organisms). They propagate only in tropical areas, but are found in abundance in homes at temperate latitudes if they are brought indoors to be protected from freezing during the winter months. *Eichhornia* has broadly ovate leaves with short, bladder-like petioles. The thickened petioles give the plants buoyancy and effectively float on the surface of the water. It grows luxuriously in ditches, ponds and slow streams in the tropics, becoming a nuisance and covering the entire surface of the water body. The water lettuce has fleshy, pleated leaves that rise from a short vertical stem in the form of a rosette. The rosettes are usually connected to one another by root stocks.

The second group of floating plants are those that are rooted on the bottom but the leaves float on the surface. The two most common genera in this group are the white water lily, *Nymphaea* (Fig. 6.20a), and the yellow water lily, or spatterdock, *Nuphar* (Fig. 6.20b). Both genera live in stagnant pools and ponds, or in slow moving streams. The leaves of water lilies are large and disc-like, but they are much broader and more disc-like in *Nymphaea* than in *Nuphar*. Other distinguishing features between the two genera are the symmetry of the veins, the depth of the leaf notch and, of course, the colour of the flower. In *Nymphaea* the veins arise radially from the junction of the stem, the leaf notch is very deep and the flower is white; in *Nuphar* the veins arise from a midrib, the leaf notch is shallow and the flower is yellow. Other differences are more subtle. In *Nymphaea* the flowers are very fragrant and usually open from ~7:00 AM until shortly after midday; in *Nuphar* the flowers tend to be in bloom all day long, all summer long. In both genera, stems and flowers withdraw to the bottom during the winter so that the fruits may mature beneath the water surface.

The water lilies are ecologically important to many invertebrates. The undersides of the leaves serve as shelter and the attached algae are an important source of food for many species of snails and insects. Some aquatic caterpillars are so voracious that the leaves of both genera are thoroughly fragmented by the end of the summer. In addition, many species of snails and insects lay their cggs on the flower stems beneath the water level or on the undersurface of the leaves.

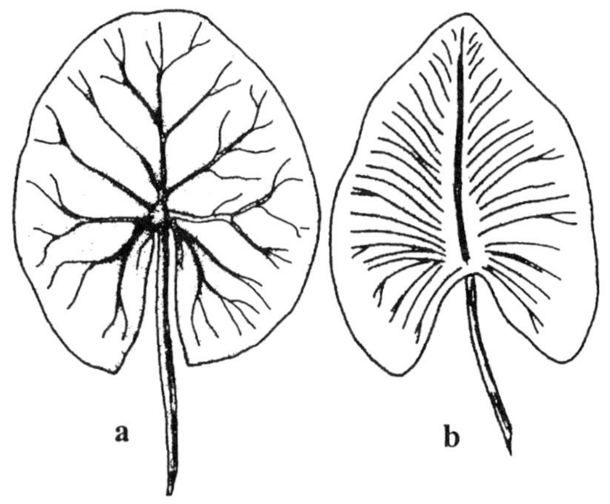

Figure 6.20. Two common examples of plants with floating leaves and roots on the bottom, showing (a) *Nymphaea* and (b) *Nuphar*.

The Submersed Macrophyte Community

There are two groups of plants that make up the submersed plant community: (i) plants rooted on the bottom; (ii) plants lacking roots but floating on the bottom
37☞ (sometimes the lower portions are buried in the mud) (Fig. 6.21).

The submerged plants exhibit several other adaptations for life in water. Although many of the adaptations are found on many plants in the emergent and floating communities, they are most prevalent within the submergent community. ***All aquatic plants lack root hairs on the roots***. In terrestrial plants root hairs are the main site of nutrient uptake, but most aquatic macrophytes obtain their nutrients through the leaves and stems, hence there is no need for root hairs. Terrestrial plants have stomata for exchange of respiratory gases and evapo-transpiration. ***Submersed plants lack stomata and emergent plants have reduced stomata because the plants would otherwise smother***. Submerged plants obtain CO_2 for photosynthesis through the tissues; the oxygen produced is released through the same tissues. ***The leaves of many submersed plants, and floating species, have fairly large voluminous cells that provide buoyancy***. In fact, if the plant is pulled from the substrate and released, it usually floats on the water. Also, ***the leaves lack a cuticle, or at most have a thin one***. The cuticle is important for terrestrial plants because it prevents evaporation of water from the plant tissues. But submerged plants are surrounded by water so there is no need for a cuticle.

The shapes of leaves often reflect the physical conditions in a habitat. Many species in the submersed communities of streams have long, linear leaves instead of broad flat ones. The long, linear leaves offer less resistance to flow than broad, flat leaves which float better in quiet waters. Many of the *Potamogeton* (Fig. 6.23a-d) species that occur in running water have long, linear leaves, such as *P. densus, P. pectinatus, P. pusillus,* and *P. robbinsii*. The eel grass, *Vallisneria americana* (Fig. 6.21e) is also common in rivers. It can grow up to three or four feet long, the limp, emergent leaves floating on the surface of the water. It has staminate flowers which grow in a cluster within a spathe at the base of the leaf cluster. When mature the spathe breaks loose and rises to float on the surface where the sperm will pollinate the pistillate flowers. The pistillate flower, which is born on a long coiled stem, then recoils to the bottom, pulling the developing fruit down into the water. In the spring

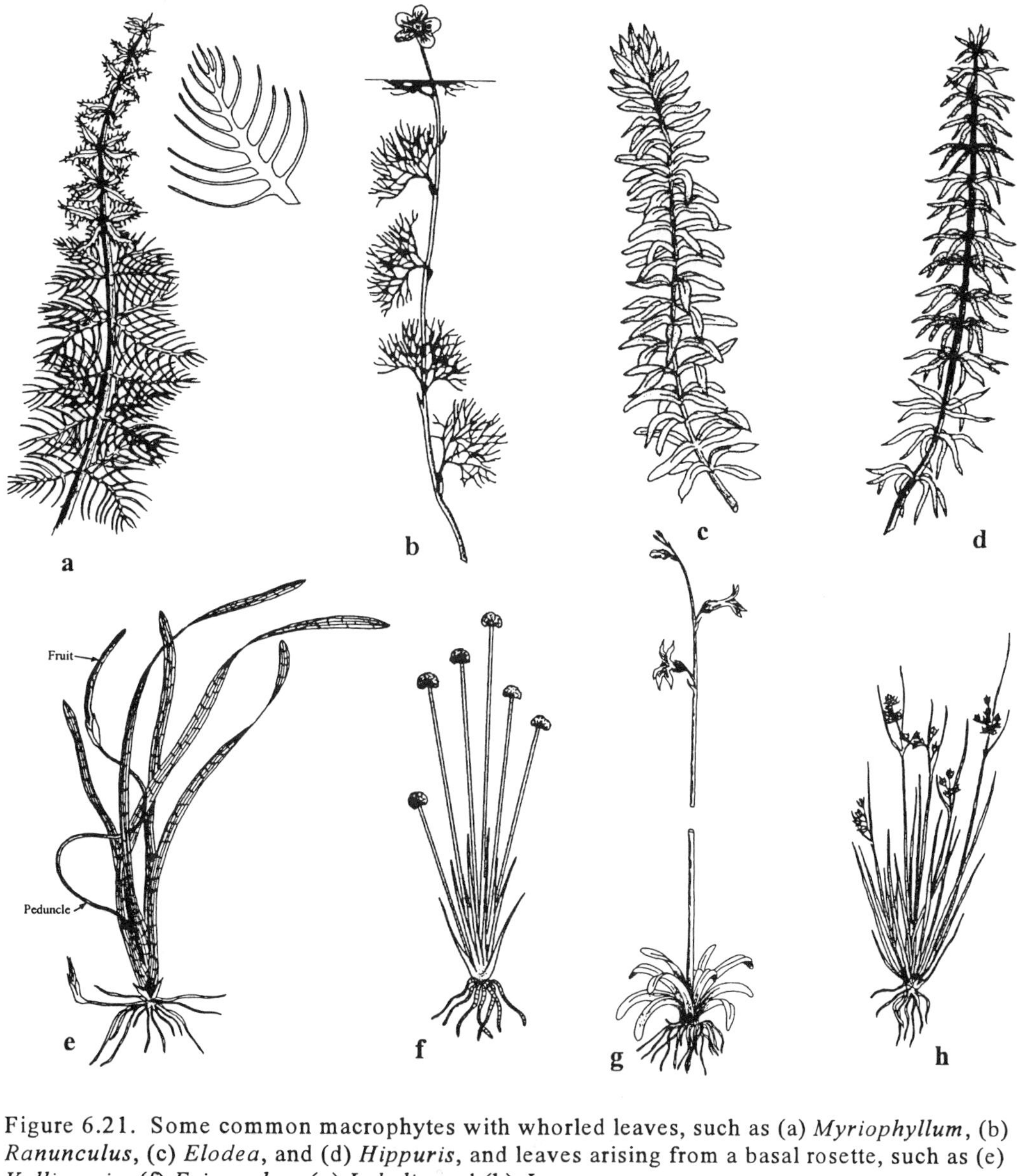

Figure 6.21. Some common macrophytes with whorled leaves, such as (a) *Myriophyllum*, (b) *Ranunculus*, (c) *Elodea*, and (d) *Hippuris*, and leaves arising from a basal rosette, such as (e) *Vallisneria*, (f) *Eriocaulon*, (g) *Lobelia* and (h) *Juncus*.

this fruit matures and takes root and germinates.

Plants with finely dissected leaves tend to occur in very quiet bodies of water. The water milfoil, *Myriophyllum spicatum* (Fig. 6.21a), an introduced species, has finely dissected leaves arranged in whorls around the stem. It is very prolific and a major nuisance, clogging waterways throughout North America. It was introduced from Asia in the early 1900's. There have been numerous attempts to control its growth and spread with chemical and biological methods, but with little success. When the plant dies in the fall, a small cylindrical roll of unexpanded shoots is produced and drops to the bottom where it overwinters. The following spring, new shoots arise from the cylindrical roll. Although the plant is a nuisance, it provides an abundance of food and shelter for a variety of aquatic organisms.

Ranunculus (Fig. 6.21b), commonly known as buttercup or crowfoot, also has dissected submersed leaves but the leaves are alternate with a basal stipule. The floating leaves are undissected and usually somewhat palmately lobed. *Ranunculus fluitans* is a common inhabitant of streams. Another macrophyte with small leaves (but not dissected) arranged in whorls, is the waterweed or frogbit, *Elodea* (Fig. 6.21c). It is a favourite plant for home aquaria and can be purchased in most pet stores that supply fish. The plant also forms large floating mats that often clog waterways. Segments of branches that break off often re-root themselves in other locations of the water body. *Hippuris* (Fig. 6.21d), which looks similar to *Elodea*, is also a common plant with small, whorled leaves.

Plants with tinsel-like, ribbon-like or awl-like leaves include the quillworts of the genus *Isoetes* (Fig. 6.18k), discussed as an emergent plant. However, it often is totally submerged. The buttonwort, *Eriocaulon* (Fig. 6.21f), has awl-shaped leaves that arise in a rosette from a very short stem. It is most common in fairly soft waters, or acidifying waters, and is considered a good indicator of soft water, like *Lobelia dortmanna* (Fig. 6.21g). *Lobelia* has clustered, de-curved, ribbon-like leaves that grow in a simple clump. The two flowers grow alternately on a long stem that rises from the water, the flowering stem being the only emergent part, the leaves always being submerged. Finally, *Juncus* (Fig. 6.21h), which looks more like a grass, is also a shallow water plant, although many species are found only in meadows.

Most of the submersed plants have simple leaves that are arranged opposite or alternate one another on the stem. Occasionally the leaves are compound. The arrangement of the leaves is important in classifying the plants. Among the more common plants with opposite leaves are the naiad, *Najas* (Fig. 6.22a), the water starwort, *Callitriche* and the false loosestrife, *Ludwigia* (Fig. 6.22b). *Najas* is commonly called the bushy pond weed because of its whorled leaves, but the whorls are on long petioles that are arranged opposite one another on the stem. The plant resembles the water weed (*Elodea*) in general appearance, but the linear leaves are widened at the base and are coarsely or finely toothed. *Callitriche*, which looks very much like *Hypericum* (Fig. 6.21c), has linear or spatulate leaves that are crowded toward the tip. It is an indicator of hard water and is common in streams. The stems are limp and need water to buoy the plant; otherwise it lies prostrate on the bottom.

Figure 6.22. Some submersed macrophytes with opposite leaves, showing (a) *Najas*, (b) *Ludvigia* and (c) *Hypericum*.

Ludwigia usually forms large mats and is widespread in ponds, streams and marshes.

There are several species with alternate leaves, but most are represented by one genus, *Potamogeton* (Figs. 6.23a-h). The water cress, *Nasturtium officinale* (Fig. 6.23i) has alternate leaves but the leaves are compound. Water cress was introduced from Europe but, unlike most other introduced species, has a beneficial use. Water cress is a favourite in salads for many people. In North America it has naturalized itself in cold spring-fed brooks and ponds, but it is also cultivated for sale in grocery stores. The plants often form tangled mats and break loose to form floating mats in sites downstream.

Potamogeton, or pondweed (Figs. 6.23a-h), is the largest genus with entirely aquatic species. It is ubiquitous, found in fresh to brackish water, roadside ditches, small and large lakes and rivers, and in acidic or alkaline waters. Pondweed is a favourite food of many species of waterfowl aquatic mammals (e.g. muskrat, beaver), deer and moose. *Potamogeton* has simple alternate leaves but exhibits great diversity in leaf form (Figs. 6.23a-h), so diverse that it is difficult to believe that some species belong to the same genus. However, there are three common features present in all *Potamogeton* species; (i) the leaves are alternate and simple, (ii) the leaves have either a sheathe or a stipule at the base, and (iii) the stems are jointed. While very few plants are confined to running waters (*Nasturtium* a possible exception), some of the *Potamogeton*s are usually found in running waters, especially *P. densus*. Beds of *Potamogeton* are usually extensive and dense and while they contribute oxygen to the water they also are major contributors to the CPOM pool. However, the leaf material

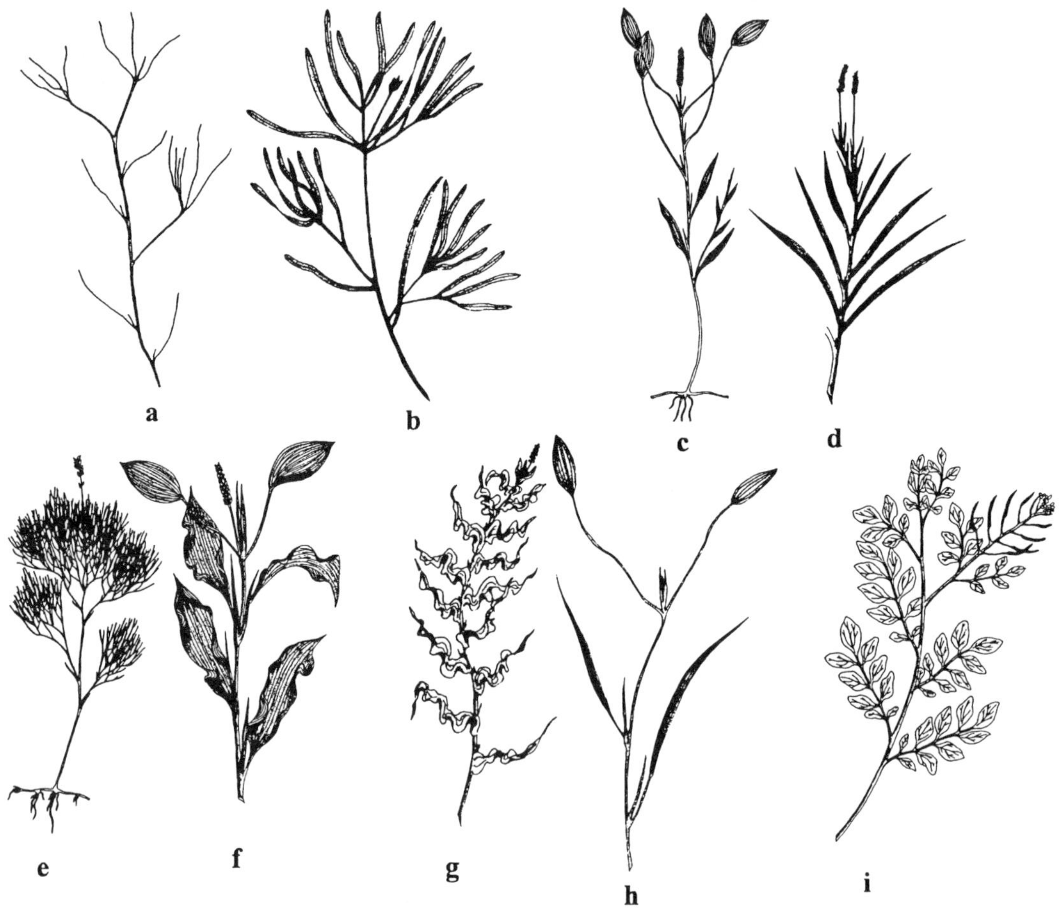

Figure 6.23. Some common submersed macrophytes with alternate leaves, showing (a) *Potamogeton gemmiparus*, (b) *Potamogeton pusullis*, (c) *Potamogeton gramineus* var. *graminifolius*, (d) *Potamogeton robbinsii*, (e) *Potamogeton pectinatus*, (f) *Potamogeton amplifolius*, (g) *Potamogeton richardsoni*, (h) *Potamogeton gramineus* var. *graminifolius* f. *maximus*, and (i) *Nasturtium officinale*.

is very labile and decays rapidly. Many species of caterpillars (e.g. *Paraponyx*) live in cases made of its leaves and ephydrid fly larvae live within the leaf tissues of *Potamogeton*.

The second group of submersed plants are those that lack roots and float beneath the surface, usually near or on the bottom. Often the stems or branches of the
38☞ plants are anchored in the mud. The group includes two algae, *Chara* and *Nitella* that belong to the phylum Charophyta. They are commonly called stoneworts but the name should be restricted to *Chara* which occurs only in hard waters and forms a crust of lime, called "***travertine***", on its leaves, hence the name stonewort. *Nitella* is found mostly in very soft or even acidic waters and never forms a layer of lime on its leaves. *Chara* has a skunky smell when it is freshly pulled from the water. Both genera reproduce sexually by producing female and male gametes in organelles called ***archegonia*** and ***antheridia***, respectively. They also reproduce asexually, mostly by fragmentation.

Two angiosperms also belong to the third group of floating and submersed. It has simple alternate leaves. They include the hornwort or coontail, *Ceratophyllum*

(Fig. 6.24a, b) and the bladderwort, *Utricularia* (Fig. 6.24c). *Ceratophyllum* may grow to nearly a meter in length. It is usually found in shallow waters of ponds or slow moving rivers and can be identified immediately by its forked, serrated leaves. The leaves grow in whorls around the floating branches. *Ceratophyllum* can be very prolific and become a pest in waterways. In late summer the ends of branches break off and fall to the bottom where they stay until the spring and then rise to the surface to be dispersed by surface currents. Several species of invertebrates inhabit the finely-divided leaves for shelter and probably a source of oxygen and food for grazers. The highly segmented and branched leaves provide greater leaf surface area for oxygenation of the water and for attachment of epiphytes. The plant is a favourite food of many species of aquatic waterfowl and mammals.

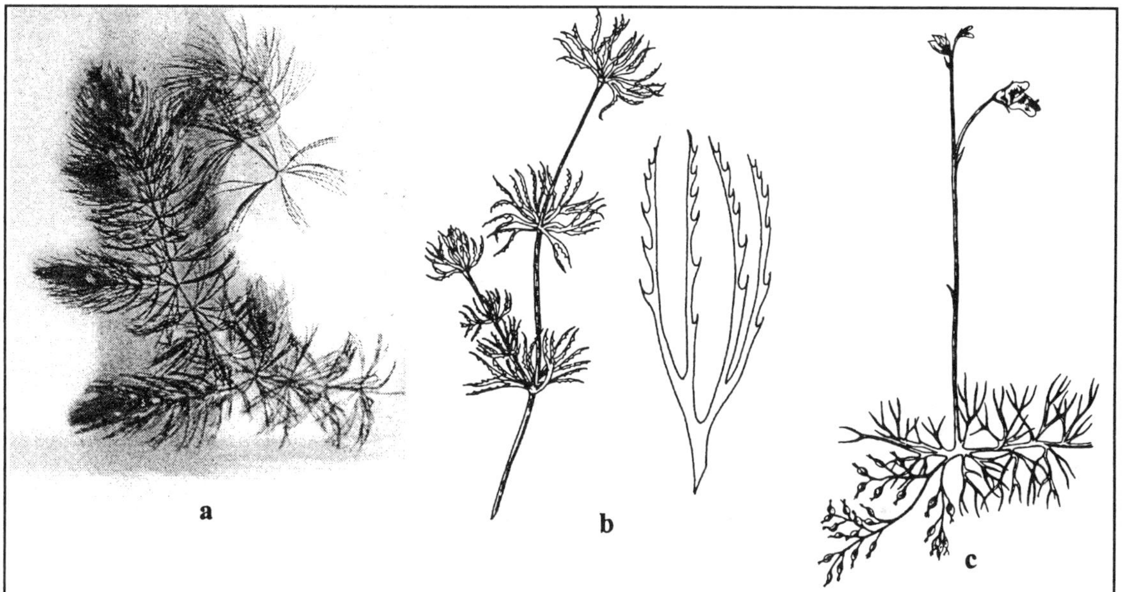

Figure 6.24. Two common submersed macrophytes without roots; (a) a photograph of *Ceratophyllum*, (b) an illustration of *Ceratophyllum* showing forked nature and serrated edges of a leaf and (c) *Utricularia*.

Bladderworts are smaller, more delicate plants than coontail. The leaves are somewhat filiform, (filamentous), finally dissected and arranged alternately or in whorls (Figure 6.24c). Attached to the leaves are tiny bladders which, when present, are the most diagnostic feature of the plant. The bladders not only aid in keeping the stems afloat, but they serve as traps for the capture of small organisms, usually insects, from which the plant gains nourishment. When organisms touch the bladders, a lid or ***operculum*** flips open; water quickly diffuses in carrying the organism(s) with it. The sap within the bladder contains numerous enzymes which digest the organism and the nutrients are then taken up by the plant.

Adaptations of Macrophytes to Aquatic Life

The different plant communities just discussed represent different plant forms or habits. Several adaptations of macrophytes to life in aquatic habitats are described but Table 6.6 summarizes the adaptations of four different plant habits (emergent; floating leaves and roots; leaves floating but roots anchored on bottom; and submerged) to life in aquatic habitats. Exceptions to each adaptation occur for each type of plant habit but, in general, the adaptations can be rationalized. Of the leaf features, simple, narrow, elongated leaves offer less resistance to flow than do compound, short and wide leaves. Finely divided leaves create greater resistance to flow than do ribbon-like leaves, but they provide for greater surface area for nutrient uptake and gas exchange (carbon dioxide uptake for photosynthesis and release for respiration; dissolved oxygen release from photosynthesis and uptake for respiration). Emergent leaves and stems can remain erect without greatly modifying their tissue strengths. As discussed earlier, stems and/or leaves with a v-shaped or circular cross section offer greater strength than does a flat plane. Leaves with a thin epidermis and cuticle permit more rapid exchange of dissolved nutrients and gases than do leaves with a thick epidermis and cuticle. Also, with the submersed leaves being constantly surrounded by water, the plants to not require a thick cuticle to conserve water, whereas emergent plants need to retain a thick cuticle to reduce evapo-transpiration. The same reasoning applies to non-functional stomata in submersed plants versus reduced numbers or no reduction in numbers of stomata in emergent plants, especially those that are often found far from the waters edge, like *Phragmites* and *Lythrum salicaria*.

Of the root and stem features, root hairs are reduced or absent in submersed forms because much of the nutrient uptake occurs through the leaves. Woody plants that live near water, for example, willows (*Salix*), alders (*Alnus*) and crowfoot (*Ranunculus*), have phloem and xylem for distributing water and nutrients throughout the plant and provide strength for supporting stems and branches. But truly aquatic macrophytes lack phloem and xylem, the rigidity being provided by thickening of stems as the plant grows in height. Otherwise, stems must be short to keep the plant erect.

In general, plants that have extensive root stocks, rhizoids or stolons rely less on sexual reproduction to disperse themselves within a lake. However, sexual reproduction must be used to disperse plants from one isolated lake to another, unless humans intervene and spread them accidentally or intentionally (see Chapter 14). Emergent plants often have staminate (male) portions located above pistillate (female) portions. This presumably enhances the chances of self-fertilization of eggs simply by gravity. This strategy is not commonly seen in submersed plants because water currents have a greater effect on sperm dispersal than does gravity.

Adaptations of Macrophytes to Running Waters

Only plants with both floating roots and leaves (Table 6.6, column 3) cannot perpetuate themselves in lotic habitats. In general, plants with other habits share the same adaptations in both lentic and lotic habitats, although plants in fast water tend to have:

- More extensive roots or rhizoid systems
- Smaller leaves and shorter petioles
- Tougher and more flexible stems
- Fewer floating leaf species
- Greater amounts of vegetative reproduction

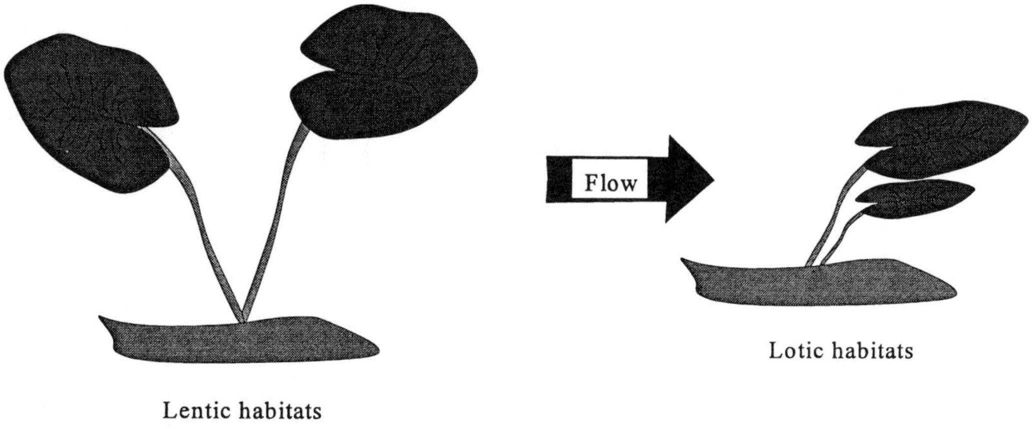

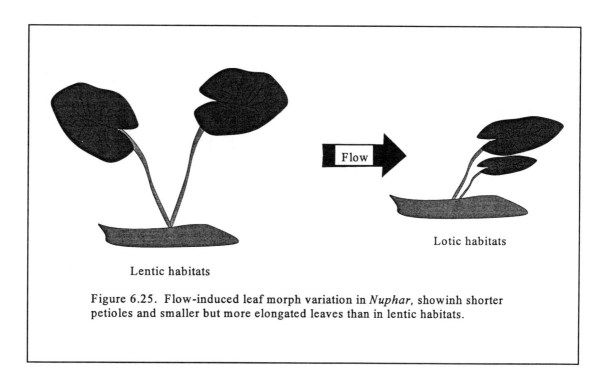

Figure 6.25. Flow-induced leaf morph variation in *Nuphar*, showinh shorter petioles and smaller but more elongated leaves than in lentic habitats.

than their counterparts in slower or stagnant waters. For example, *Nuphar*, the yellow water lily genus, often exhibits a flow-induced leaf morph that is smaller and more elongated than typically found in quiet waters (Fig. 6.25)

Factors that Affect Macrophytic Growth and Production in Streams

Fig. 6.26 summarizes most of the factors known to affect the growth of plants in lotic systems. The most significant factor is flow rate because it either directly or indirectly affects or modifies all other factors, as discussed below.

Very few plants are confined to running water habitats, the river weed, *Podostemum*, perhaps being an exception. River weed usually occurs in turbulent

Table 6.6. Characteristics of plants adapted to life in aquatic habitats. NA = not applicable; ✓ = feature typically applies; ✗ = opposite feature typically applies.

Adaptive Characteristics	Emergent	Floating leaves and roots	Floating leaves, plant rooted on bottom	Submersed
Leaf features				
Submersed leaves often ribbon-like or finely divided	NA	NA	NA	✓; ribbon-like (*Vallisneria*) and finely divided (e.g. *Myriophyllum, Ceratophyllum* common)
Emergent leaf/stem morphology modified	✓; leaves triangular (*Typha*) or stem rounded or triangular in cross section (*Scirpus*)	✗; leaf morphology little modified e.g. *Lemna, Wolffia, Salvinia*	✗; leaf size may be reduced in streams (*Nuphar luteum*) but morphology is similar.	NA
Epidermis and cuticle thin or modified	✗; emergent leaves usually have thick epidermis and cuticle e.g. *Typha*	✓, only on lower surface; upper surface has thicker epidermis and cuticle	✓; only on lower surface; upper surface has thicker epidermis and cuticle	✓; epidermis may have hairs or wax, but always thin to allow nutrient uptake and gas exchange
Stomata greatly reduced, non-functional or lacking	✗; stomata abundant but may be reduced in some	✓; reduced on upper surface, absent on lower	✓; reduced on upper surface, absent on lower	✓; absent to non-functional in all
Air-filled spaces often present	✗; air-filled spaces usually lacking	✓; petiole bulbous in *Eichhornia*; large spaces in *Lemna, Wolffia*	✓; large spaces in *Nymphaea, Nuphar, Brassenia, Nymphoides*	✓; spaces enlarged (e.g. *Potamogeton*) but not as large as floating species
Root and stem features				
Root hairs usually absent	✗; root hairs often present	✓; absent in most species	✓; absent in most species	✓; absent in most species
Conducting (e.g. phloem, xylem) and stiffening tissues reduced or absent	✗; stiffening tissues present but conducting tissues mostly absent	✓; conducting and stiffening tissues absent	✓; conducting and stiffening tissues absent	✓; conducting and stiffening tissues absent
Secondary growth in stem thickness absent	✗; stem thickness tends to increase with growth	NA, stems absent in *Lemna, Wolffia, Salvinia*	✓; buoyancy provided by growth in leaf , not stem	✓; stem thickness remains same
Reproductive features				
Sexual reproduction usually of minor significance	✗; Sexual reproduction prominent in some, e.g. purple loosestrife (*Lythrum salicaria*)	✓; binary fission of fronds (leaves) is typical	✗; Both sexual and vegetative (spreading root stalks or buried stems) are common	✓; many species spread by stolons, roots, and stem fragmentation
If sexual, staminate(♂) portion above pistillate (♀) portion	✓; e.g. spike of *Typha*; burs of *Sparganium*. But many exceptions occur as well (e.g. *Lythrum, Sagittaria*) where flower has stamens and pistils.	X; fruiting stages are very rare in plants with floating roots and leaves, *Eicchornia* and *Pistia* being exceptions.	X; white, yellow and purple flowers (water lilies) have stamens and pistils in same flower	X; even in *Vallisneria*, spathe (staminate portion) is at base of plant, pistillate is near surface of water.

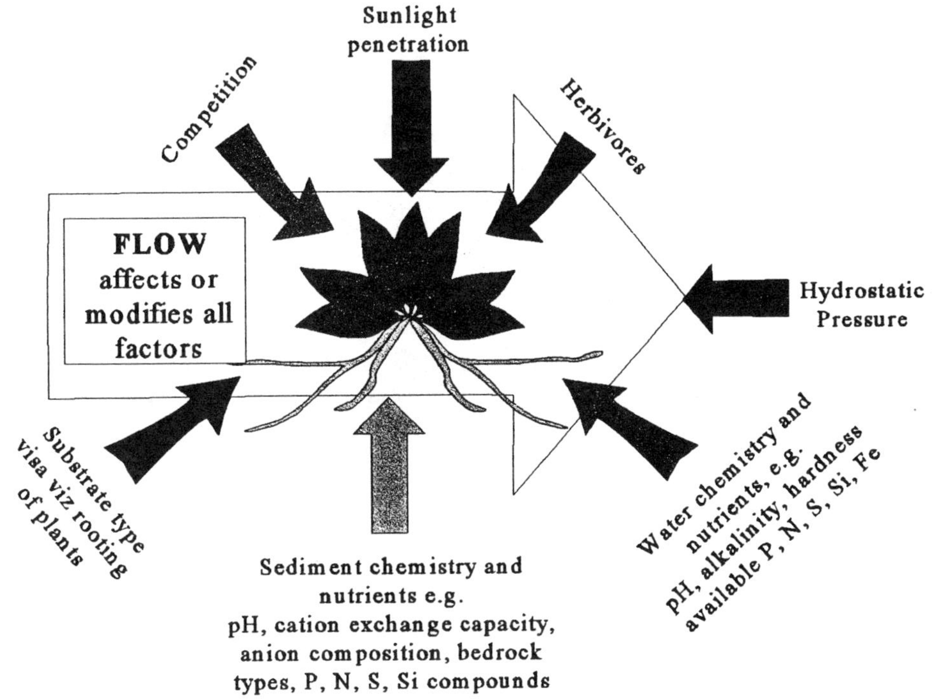

Figure 6.26. Factors that affect growth and production of macrophytes.

"white water" as a thick moss-like covering on rocks. *Podostemum ceratophyllum* attaches to rocks by disc-like processes. Many speedwells (genus *Veronica*) are also confined to streams, such as *V. glandifera,* while others are usually found in streams but may also be found in swamps, such as *V. americana.* The water cress, *Nasturtium officinale*, an exotic species, is found in clear, cold, springs in Europe, but is cultivated commercially in pond-like environments in North America and sold in stores as a salad delicacy. In general, macrophyte diversity and biomass tend to be inversely related to stream order size, larger, slow-moving streams having a greater diversity and biomass of plants than smaller, fast-flowing streams.

Sunlight penetration: Emergent and floating plants depend mainly on scattered and direct sunlight impinging on photosynthetic tissues extending above the surface of the water. However, after penetrating the water, the transmitted light is refracted (bent), absorbed and scattered. The amount of refraction, absorption and scattering depends on the quality (e.g. colour) and quantity (e.g. turbidity) of material suspended in the water column. Hence, water quality affects primary production of submerged plants much more than that of emergent or floating plants.

Water turbulence, especially if air bubbles are entrained, also greatly affects the quality and quantity of light reaching the leaf surfaces of submerged macrophytes. Generally, macrophytes receive light that is more greatly modified in turbulent streams than in quiet waters of lakes and ponds because air bubbles lessen the amount and kinds of light that is absorbed and, therefore, made available to submersed plants.

The depth of water greatly affects the amount and quality of light reaching the bottom where most submersed macrophytes occur. The deeper the water, the greater the extent of light absorption, the longer wave lengths (e.g. infra-red and orange) being absorbed more quickly than short wave lengths (e.g. blue and green). Hence, the effect of sunlight penetration increases with increasing size (order) of stream.

Herbivory: Grazing of submersed macrophytes occurs in all aquatic systems but is usually of minor significance, with less than 10% of the biomass consumed by herbivores, mostly waterfowl, mammals and fish. The common carp, *Cyprinus carpio*, feeds occasionally on macrophytes and algae (as well as invertebrates), but the amounts consumed pale compared to the grass carp, *Ctenopharyngodon idella*, which is used to control aquatic weeds in some African ponds and lakes. The grass carp is known to consume 50-100% of its body weight on a daily basis. Of the invertebrates that feed on living plant material, crayfish, aquatic lepidopterans and chrysomelid and curculionid beetles are among the most prominent.

The invertebrate herbivores of lotic systems often have morphological adaptations to prevent being carried away by currents. For example, many insect grazers and scrapers have long legs and stout claws (e.g. damselflies of the family Agrionidae) or lateral body projections or dorsal spines (e.g. the beetle *Berosus*) that hold them within beds of plants.

Invertebrates appear to be more important as shredders of moribund macrophyte material. The organic material of macrophytes is much more labile than that of leaves of most deciduous and coniferous trees.

Water Depth: Depth of water not only affects the amount and kinds of light reaching the plants, but it may also affect the hydrostatic pressure. Since it is difficult to separate the effects of depth and pressure, little is known about the influence of hydrostatic pressure on growth of plants. Nevertheless, some forms of *Potamogeton* are found only in deep water. *Potamogeton gramineus* var. *graminifolius* forma *maximus* is a deep-water form and *P. g. g.* forma *myriophyllus* is a shallow-water form. However, since most streams and rivers are relatively shallow, it is not likely that hydrostatic pressure is a factor in lotic environments.

Some plants grow only in very shallow water. For example, liverworts (e.g. *Fissidens*) grow mostly on damp surfaces of rocks and at the water's edge, being submerged only during spate events. The water moss, *Fontinalis*, is almost confined to flowing waters and is usually seen on the upstream edge of submerged rocks.

Water Chemistry: The distribution of many plants is limited by pH and associated "buffer variables" such as alkalinity and total hardness. Most species of *Sphagnum*, the bog moss, are found in soft waters with low pH (< 5), although some may form mats around the margins of hard-water, calcareous ponds and lakes creating a "quaking bog".

The moss, *Drepanocladus*, is often the only oxygenator in lakes and streams that cannot support other macrophytes due to low pH.

Many *Potamogeton* species are associated with hard-water environments, such as *P. pectinatus* in lakes and slow-flowing streams, *P. crispus* in enriched waters, and *P. filiformis, P. vaginatus, P. foliosus, P. friesii* and *P. strictifolius* in pools, streams and lakes. *Lobelia dortmanna*, the only strictly aquatic lobelia, is associated mostly with soft-water, or in acidifying rivers and lakes, but the brightly coloured, red lobelia, *L. cardinalis*, is usually found in swamps or on the banks of rivers and streams with medium to hard water. *Berula erecta*, often found associated with watercress, *Nasturtium officinale* is found in calcareous streams. Within the genus *Myriophyllum*, *M. alterniflorum* occurs in soft water, but *M. spicatum* occurs in hard water.

The foregoing demonstrates some influence of water hardness on the occurrence of at least some species. In general, macrophyte diversity and biomass tends to increase with increasing pH of the water. However, current velocity also determines the distribution of species within a range of water hardness. Table 6.7 illustrates the interaction of currents on the diversity of plants with increasing water hardness. The last column lists species that show some predilection for some current, but many species common in lakes and ponds could be assigned to the column, with *Sphagnum* in ponds with low pH (< 6.0), many *Juncus* species and *Nitella* in slightly acidifying lakes and ponds, and *Chara* in to moderately to highly calcareous lakes and ponds.

Substratum: The nature of the substratum is largely determined by current velocity, with coarse substrates in erosional habitats and fine substrates within depositional habitats. Hence, it is often difficult to separate the effects of substratum from water velocity. However, within rivers or lakes of similar water chemistry, the distribution of plants is primarily correlated with the nature of the substratum. It seems that substratum is more important in relatively slow waters. Sediments rich in nutrients will support a greater diversity of macrophytes than sediments poor in nutrients. Sediments with a good mix of organic (especially labile carbon) and inorganic (e.g. silt, mud, sand) materials support a greater biomass and diversity of plants than do purely inorganic substrates with gravel, rocks and/or bedrock. However, the inorganic substrates often reflect the basic chemical characteristics of the water, such as water hardness and conductivity, which in turn affect the kinds of plants that can occur in a body of water.

Competition: The importance of the role of competition affecting community plant diversity is not well understood, but competition for food and space probably occurs at all trophic levels. For example, if light conditions are not optimal, submerged species may be replaced by emergent species in running waters or floating species in stagnant pools. If floating species (e.g. *Lemna*, *Wolffia*, *Nuphar*, *Nymphaea*, *Brasenia*) begin to dominate the community, light is excluded from the water column, growth of submerged species declines, ultimately leading to the exclusion of all submerged species. Many emergent species also exhibit competition. Species with good dispersal mechanisms (e.g. purple loosestrife, *Lythrum salicaria*) are invading and occupying large areas of many marshes previously dominated by *Typha, Scirpus* and/or *Phragmites*.

Role of Macrophytes in Streams

Actually, the role of macrophytes is the same in all aquatic habitats (Fig. 6.27). However, flowing water modifies the importance of some of these roles. For example, all submerged plants contribute dissolved oxygen but since most of the oxygen enters turbulent water from the atmosphere, the importance of macrophytes as a source of oxygen is lessened considerably. Only in stagnant waters or pools will the importance of plants as oxygenators increase in value.

In reaches where plants are able to grow, an increase in the diversity of aquatic life occurs. There is a noticeable increase in the diversity of herbivorous species of invertebrates and fish. While one might expect an increase in the diversity of predators, only a change in the species assemblage may occur. Predators are found in

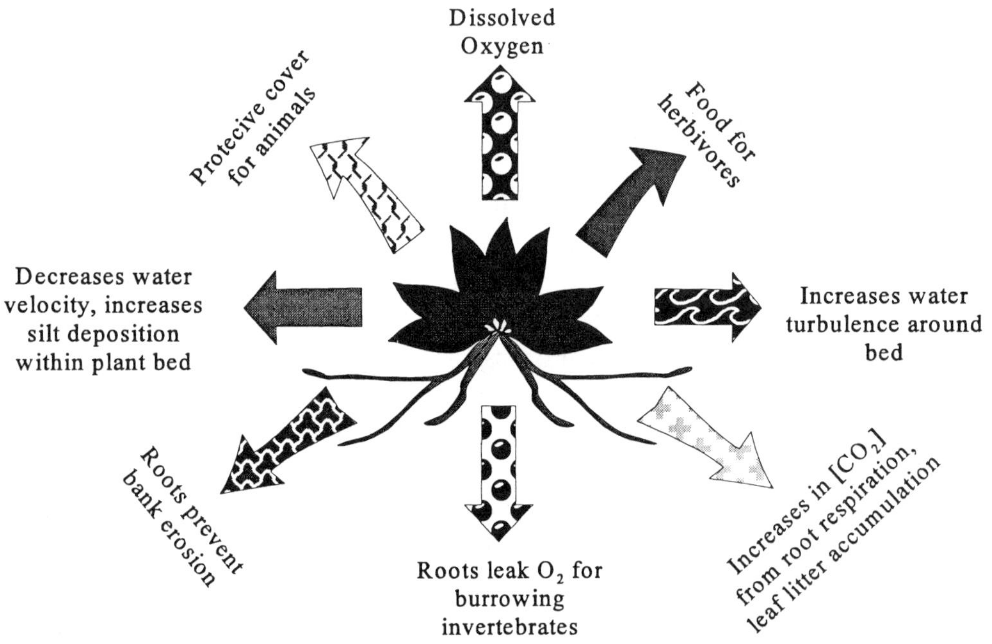

Figure 6.27. Effects of plants on streams.

all parts of a stream ecosystem, but among vegetation many species are adapted for life among the leaves and stems. For example, most damselfly nymphs are long and narrow, an adaptation for vertical movement on stems, and have long tarsal claws for clinging to leaves and stems. The mayfly, *Drunella grandis*, has several dorsal spines and hooks that catch and hold the animal when it is among finely divided leaves. The herbivores also show adaptations: lepidopterans have five pairs of abdominal prolegs with curved hooks, or ***crochets***, for attaching to stems of plants; the mouthparts of leaf-mining lepidopterans are directed horizontally (***prognathous***), an adaptation for slicing their way into plant tissues; the haliplid beetle, *Peltodytes*, has long lateral and dorsal filaments that keep the larvae entangled among stems and petioles; similarly, the hydrophilid beetle, *Berosus*, has lateral gill filaments that are mainly respiratory structures but otherwise, help prevent the larvae from being swept by currents from vegetation; and all chysomelid beetle larvae (e.g. *Donacia*, *Pyrrhalta*) use their pair of posterior spines to obtain air directly from vascular hydrophytes. While plants are a source of oxygen and food, they also offer protection for many species. *Berosus*, *Peltodytes*, and most lepidopterans and chrysomelid beetles are green and cryptically blend in with the macrophytes. Only when they move are they picked off by predaceous fish (e.g. sunfish, *Lepomis gibbosus*; smallmouth bass, *Micropterus dolomieu*) and macro-invertebrates (e.g. zygopterans) which themselves have greening cryptic colouration. Even the fish are adapted; many littoral species, especially centrarchids, are laterally flattened, an adaptation for swimming amongst weeds.

In producing oxygen, plants use carbon dioxide which affects the CO_2-bicarbonate equilibrium. Daily and seasonal variations in pH and alkalinity can occur, the magnitude of the variations depending on the pH, with greater variations in soft water than in hard water. In hard waters, travertine is deposited on rocks, plants and other solid substrates. The travertine roughens and thickens the surfaces, altering the nature and velocity of flow over the substrate. The chalk-like deposit on rock surfaces is also soft and less secure than the bare surface for invertebrates that attach with marginal spines and hooks. Hence, species with adaptations for attachment to rock surfaces are often found on the underside of rocks where travertine is absent.

Some plants "leak" oxygen from their roots into the soil and provide a source of oxygen for burrowing organisms. The oxygen also enhances oxidative metabolism of organic material by bacteria. However, roots also respire carbon dioxide. In addition, large plant "die-offs" contribute to accumulations of coarse and fine particulate organic matter which can also greatly increase carbon dioxide levels and the biochemical oxygen demand.

A common effect of macrophytes in running waters is their influence on stream hydrology. Plants with extensive root systems help to stabilize the sediments and prevent bank erosion. Midstream, plants can change the direction and rate of flow and contribute to the formation of shoals or islands (Fig. 6.28). Accompanying the plant-induced slowing of water velocity is the deposition of silt within and on the downstream edge of the plant bed. The plant community is extended downstream by extending their root systems or by producing adventitious roots form buried plant stems. At some

Table 6.7. Typical species of submerged macrophytes found in streams and rivers in relation to water hardness (and pH) and water velocity. Adapted from Hynes (1972)[5] which was based on British flora, but only North American species are given here.

River water hardness (total) and pH classification	**Type of habitat and approximate range in current velocity**				
	Highly Erosional > 60 cm/sec	**Erosional 70-25 cm/sec**	**Erosional/depositional 25-10 cm/sec**	**Depositional < 10 cm/sec**	**Littoral < 10 cm/sec**
Non-calcareous < 20 mg $CaCO_3$ /L pH < 6.5	None	*Myriophyllum alterniflorum* *Myriophyllum spicatum*	*Equisetum fluviatile* form *americanum* *Equisetum palustre* var. *natans* *Lobelia dortmanna* *Potamogeton tenuifolius*	*Lobelia dortmanna* *Potamogeton gramineus* *Eriocaulon septangulare*	*Eriocaulon septangulare* *Lobelia dortmanna*
Slightly calcareous 20-100 mg $CaCO_3$ /L pH 7.0-7.5	*Fontinalis antipyretica*	*Fontinalis antipyretica* *Myriophyllum spicatum* *Podostemum ceratophyllum*	*Potamogeton tenuifolius* *Sparganium simplex* *Podostemum ceratophyllum* *Nasturtium officinale* *Berula erecta*	*Elodea canadensis*	*Elodea canadensis*
Moderately calcareous 100-200 mg $CaCO_3$ /L pH 7.0-8.0	*Fontinalis antipyretica*	*Fontinalis antipyretica* *Myriophyllum spicatum* *Podostemum ceratophyllum*	*Berula erecta* *Nasturtium officinale* *Potamogeton amplifolius* *Potamogeton filiformis* *Potamogeton perfoliatus* *Ranunculus longirostris* *Sparganium simplex*	*Elodea canadensis* *Potamogeton crispus* *Potamogeton pectinatus* *Ranunculus longirostris*	*Elodea canadensis* *Ranunculus longirostris*
Highly calcareous > 200 mg $CaCO_3$ /L pH 7.5-8.0	*Fontinalis antipyretica*	*Fontinalis antipyretica*	*Berula erecta* *Nasturtium officinale* *Potamogeton amplifolius* *Potamogeton filiformis* *Ranunculus longirostris* *Sparganium simplex*	*Elodea canadensis* *Potamogeton pectinatus* *Potamogeton perfoliatus* *Ranunculus longirostris*	*Elodea canadensis* *Ranunculus longirostris*

[5]Hynes, H. B. N. 1972. The ecology of running waters. University of Toronto Press, Toronto.

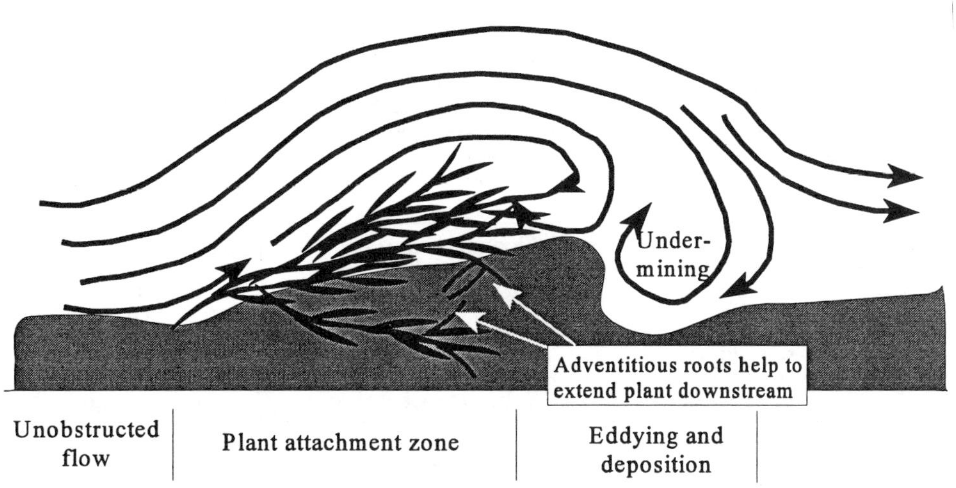

Figure 6.28. Role of macrophytes in altering current flow patterns and causing silt deposition

point, however, upward eddies are sufficient to undermine the streambed and create potholes. The entire shoal, however, can be destroyed or re-configured by a large spate event.

Indicator Value of Macrophytes

Macrophyte species respond to enrichment in different ways. Species that are submerged depend on light reaching their leaves, whereas floating species and emergent species depend on light at the surface of the lake. Hence, it is the community
39☞ of submersed macrophytes that can tell us more about the trophic status of a lake. As lakes become more eutrophic, the depth of light penetration decreases and submerged species of macrophytes and attached algae respond with a decrease in biomass and production. Some species are able to tolerate lower levels of light than others. The following discussion examines the indicator value of submersed macrophytes. Table 6.8 lists some of the more common macrophytes and their significance in a variety of freshwater habitats.

While production of emergent macrophytes may respond to organic enrichment, it is not because of light intensity in the water, it is because of increased nutrient levels the sediments. Some emergent species, such as *Typha*, may increase in productivity because of higher phosphorus levels. The roots of *Typha* are very efficient at extracting, or stripping, phosphorous from the sediments. For this reason, *Typha* is a valuable marsh species.

Table 6.8. Some common genera of macrophytes and their significance in freshwater habitats. The genera are listed alphabetically.

Genus name	Common name	Habit	Significance
Carex disticha	Eurasian sedge	Emergent/submergent	Exotic species
Ceratophyllum	coontail	Submerged, rootless	Enriched, stagnant pond genus; waterfowl food source
Eichhornia crassipes	water hyacinth	Floating with roots	Exotic nuisance & beauty plant
Elodea (Anacharis)	water weed	Submersed, whorled leaves	Common aquarium plant
Equisetum	scouring rush	Emergent	Eaten by moose; spore former
Eriocaulon	buttonwort	Submersed, basal rosette	Acidifying lake indicator
Isoetes	quillwort	Submersed, basal rosette	Spore former
Lemna	duckweed	Floating with roots	Enriched, stagnant pond genus
Lobelia dortmanna	lobelia	Submersed, basal rosette	Acidifying lake indicator
Lythrum salicaria	purple loosestrife	Emergent	Exotic nuisance
Marsilea quadrifolia	water clover	Submergent/emergent	Exotic species; spore former
Myriophyllum spicatum	European milfoil	Submersed, whorled leaves	Exotic nuisance species
Najas minor	European naiad	Submersed, opposite leaves	Exotic species
Nuphar	yellow water lily	Floating, rooted on bottom	Enriched, stagnant pond genus
Nymphaea	white water lily	Floating, rooted on bottom	Enriched, stagnant pond genus
Phragmites	reed grass	Emergent	Roots important in P retention
Pistia stratiotes	water lettuce	Floating with roots	Exotic nuisance & beauty plant
Polygonum	smartweed	Emergent/floating	Common in stagnant ponds
Potamogeton filiformis, amplifolius	pondweeds	Submersed, alternate leaves	Both are hard water indicators
Rorippa (Nasturtium) aquaticum	water cress	Submersed, alternate leaves	Exotic beneficial (salad) plant; cold, clear stream indicator
Sagittaria	arrowhead	Emergent	Tubers eaten by birds, mammals
Salvinia	water fern	Floating with roots	Enriched, stagnant pond genus
Scirpus	bullrush	Emergent	Roots important in P retention
Sparganium	bur reed	Emergent	Burs are effective dispersal agent
Typha	cattail	Emergent	Roots important in P retention
Utricularia	bladderwort	Submerged, rootless	Enriched, stagnant pond genus; bladders trap insects
Vallisneria	eel grass	Submersed, basal rosette	Waterfowl food
Zizania	wild rice	Emergent/submergent	Seeds sold as wild (Indian) rice

The diversity and biomass of floating species also offer clues to the trophic status of lakes. Floating species, like water lilies and duckweed, are mostly absent in oligotrophic lakes. The lack of floating species in unproductive lakes may be due either to their requirement for high nutrient levels in the water and sediments, or to their poor competitiveness with submersed species and other emergent species. The water lilies, such as the white (*Nymphea*) and the yellow (*Nuphar*) grow best in organic bottoms. Duckweed (e.g *Lemna*, *Wolffia*) is most common and prolific on small, wind-protected ponds and lakes that are enriched.

A British scientist, B. Seddon[6], examined aquatic macrophytes as limnological indicators and concluded that water chemistry, especially conductivity and water hardness, is the most important single factor influencing the general distribution of macrophytes. He used a hardness ratio, which is the total hardness (as mg $CaCO_3$/L) divided by the total weight of sodium and potassium ions (mg/L) in the water sample, as a measure for a "***trophic index***". The trophic index is the product of the hardness ratio and the conductivity. As a rule, high hardness ratios correlate well with high conductivities, yielding eutrophy; low hardness ratios generally correspond with low conductivities and yield oligotrophy. While water chemistry strongly influences the general distribution of macrophytes, the type of bottom and the physical nature of the body of water greatly influence the local distribution of a species within its range of chemical tolerance.

Table 6.9 lists several species of macrophytes according to their ionic and hardness ratio tolerances. Note the relationship between trophic status and conductivity and hardness ratio. While eutrophic waters generally have a high conductivity and hardness ratio, some oligotrophic waters may also have a high conductivity and hardness, but such waters are usually in northern latitudes so that primary production is kept low by cold temperatures and short summer seasons. Dystrophic waters are generally stained brown from humic materials and have low conductivities, but some dystrophic lakes can have higher conductivities and hardness ratios and border on mesotrophy to eutrophy. Seddon believes that, in general, restriction towards eutrophic conditions is considered to be an obligate relationship reflecting physiological demands; some dystrophic and oligotrophic species are known to have wide tolerances and are believed to be excluded from more enriched lakes by competition rather than by physiological limitation.

Fringing Plant Communities

40☞ Liverworts and mosses of phylum Bryophyta dominate the plant community at the water's edge, especially in marshes and bogs. The bryophytes are part of the

---

[6]Seddon, B. 1972. Aquatic macrophytes as limnological indicators. Freshwater Biology 2: 107-130.

Table 6.9. Aquatic macrophytes as limnological indicators. The hardness ratio is derived by dividing the total hardness by the weight of sodium (Na) plus potassium (K).

Species	Conductivity μS	Hardness Ratio
Widely Tolerant Species		
Nymphaea alba	< 400	< 12.0
Nuphar alba	< 400	< 12.0
Glyceria fluitans	< 400	< 12.0
Potamogeton natans	< 400	< 12.0
Myriophyllum alterniflorum	< 400	< 12.0
Eutrophic Species		
Potamogeton pectinatus	> 200	> 5.0
Potamogeton lucens	100 - > 400	> 3.0
Potamogeton crispus	> 150	> 3.0
Myriophyllum spicatum	> 200	> 5.0
Lemna trisulca	> 170	> 5.0
Ranunculus circinatus	> 100	> 2.5
Lemna minor	> 100	> 2.5
Mesotrophic Species		
Polygonum amphibium	~ 100	> 2.5 (usually > 3.0)
Ceratophyllum demersum	> 100	< 3.0
Potamogeton obtusifolius	100 - 220	2.5 - 4.0
Potamogeton perfoliatus	> 100	> 2.5
Oligotrophic Species		
Ranunculus aquatilis	> 60	> 2.0
Elodea canadensis	> 50	> 2.0
Potamogeton berchtoldii	~ 50	> 1.5
Nasturtium officinale	> 60	> 2.0
Dystrophic Species - Acid Tolerant		
Lobelia dortmanna	< 50	< 1.5
Sparganium angustifolium	< 50	< 1.5
Isoetes lacustris	< 50	< 1.5
Potamogeton polygonifolius	< 50	< 1.5
Isoetes echinospora	< 200	< 7.0 - 10.0
Callitriche intermedia	50-200	~1.5 - ~5.0

ontogeny (developmental history) of lakes as they pass from the aquatic to the terrestrial stage and progress through the intermediate stages of swamps, marshes and bogs.

The liverworts and mosses are the amphibians of the plant world and although they live on land, the land must be wet or moist at some stage in their development, particularly during the reproductive stages of the plants. Like the higher plants, bryophytes have multicellular sex organs and produce sex cells that are protected by a cellular layer. However, they do not have the conducting and stiffening tissues required of plants that grow above the water surface or must live on land. Nor do they have true roots. The bryophyte's life cycle is made up of two alternating stages, the ***gametophyte*** and the ***sporophyte***. The gametophyte is the greenish plant that is most require water to swim to the archegonia in order to fertilize the eggs, hence the requirement for an aquatic environment.

The embryo which develops upon fertilization forms the sporophyte stage. In liverworts, the sporophyte remains enclosed within the archegonium and is hidden by leafy appendages called ***bracts***. The gametophyte produces sex cells; the sporophyte produces spores. The gametophyte bears flat-shaped sex organs called ***archegonia,*** which produce eggs, and cylindrical ***antheridia***, which produce ciliated sperm cells. The ciliated sperm cells must swim to fertilize an egg. In mosses, the sporophyte shoots up from the fertilized egg, forming a long capsule-bearing stock or ***seta***. The spores develop within the capsule and are released through an opening that is covered by a lid, called the ***calyptra***. The spores germinate, under appropriate conditions, and form a juvenile gametophyte, called a ***protonema***. In mosses the protonema is small and branched but in liverworts it is greatly reduced and flattened. The protonema then grows into the familiar green plant , or gametophyte.

Most aquatic bryophytes also reproduce asexually by fragmentation. Liverworts are better adapted to terrestrial life than are mosses, although there are a few species of liverworts that live in slow, stagnant waters, float on the surface or lie flat on rocks. Most liverworts consist of a flat, scale-like or ribbon-like ***thallus*** of 1-6 cells thick. Root-like hairs, or ***rhizoids***, hang from the underside of many species (Fig. 6.29). Some have stems, but not true stems. The most common genera of liverworts are *Riccia* (Fig. 6.29a) and *Ricciocarpus* (Fig. 6.29b).

Unlike liverworts which are scale-like, mosses have their leaves evenly spaced around the stem or they occur in two opposite rows and although the sporophyte lasts several weeks, the gametophyte is the dominant stage throughout the year. As a rule, mosses occur in shallow water, but some species such as *Sphagnum fontinalis* and *S. antipyretica* have been found at depths down to 18-20 m. In hard water streams the mosses tend to form the calcareous deposits, called travertine, normally formed by algae. *Fissidens* (Fig. 6.29c) and *Fontinalis* (Fig. 6.29d) are common genera in streams, attached to rocks.

The most common and well-known genus of mosses is *Sphagnum* (Fig. 6.29e). It is a typical bog plant that grows widely in wet places as a thick, cushiony carpet but it lends very little support to the weight of humans and the unwary will sink to the waist, or even deeper.

Sphagnum has considerable economic and domestic importance. When dry,
41☞ moss is still used for packing fragile objects, such as glassware; it is commonly used as a soil conditioner; although it has little or no nutrients, it helps to condition the soil as an organic matrix for the inorganic soil particles; and it has antiseptic qualities that during the wartime was used for making dressings for wounds. The Scandinavians grow huge fields of *Sphagnum* for use as an excellent source of energy.

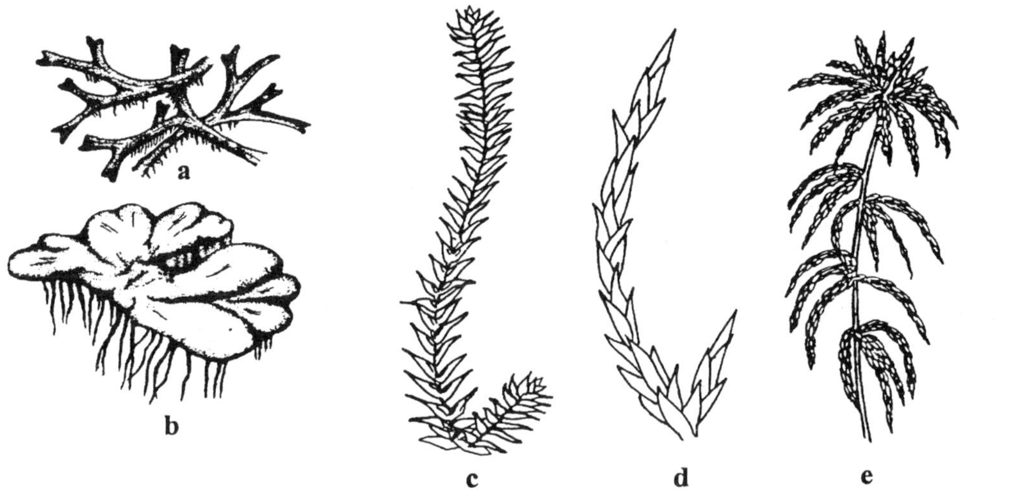

Figure 6.29. Some common bryophytes in aquatic environments, showing two liverworts, (a) *Riccia* and (b) *Ricciocarpus*, and three mosses, (c) *Fontinalis*, (d) *Fissidens* and (e) *Sphagnum*.

Swamps and marshes generally accumulate little peat because they have some external source of water. Bogs characteristically accumulate peat and are dominated by acidophilic (acid-loving) mosses and sedges. Peat bogs have little or no external water inflow and are dependent on a high water table. *Sphagnum* is essentially a water softener, exchanging ions as water is absorbed and processed through the plant tissues. Any calcium present in the water is adsorbed, and the moss yields hydrogen ions. As a result, acidic water is produced. Hence *Sphagnum* bogs tend to be highly acidic.

Sphagnum takes up cations differentially, any trivalent (e.g Fe^{+++}) and bivalent ions (e.g. Ca^{++}) being adsorbed much more readily than monovalent ions (e.g H^{+}). This is the same reaction used by resins in commercial de-ionizers. Although sulphuric acid is the main inorganic acid produced in bogs, there are also some organic acids produced, such as humic acid, fulvic acid and humins. Humic and fulvic acids are extractable with alkaline solutions; humins are not. Fulvic acids remain in solution when acidified to pH 2, but humic acids precipitate out at pH 2. Humic materials occur in most soils and reach lakes and streams by surface runoff. They are stronger than carbonic acid and will lower the pH down to about 4.0, as opposed to carbonic acid which will lower the pH down to 5.6 (i.e. pH of precipitation in equilibrium with carbon dioxide).

CHAPTER 7

PLANKTON - A KEY LINK BETWEEN PLANTS AND ANIMALS

Why Read This Chapter?

So often you hear the comments, "You mean, you really drink the lake water!" "Is it safe to drink?" "Do you know what's in the water?" This chapter examines the kinds of organisms in lake water. Anyone who drinks lake water is consuming pretty much what many fish eat, the plankton and, perhaps, much more. The more enriched the water is, the more plankton consumed, and the more likely that undesirable materials will be consumed.

The plankton are represented by mobile plants (***phytoplankton***) and animals (***zooplankton***), but their powers of locomotion are insufficient to overcome most water currents. Hence, the position of plankton in the water column is usually dictated by horizontal and vertical water currents. This chapter examines factors that affect the spatial (i.e. depth variations and across lake variations) and temporal (e.g. seasonal) variations in planktonic communities.

The inability of plankton to swim against even modest currents explains their absence in most streams, especially in stream orders 6 or less. Exceptions to this general observation are streams that receive water from lake outflows and discharges from artificial impoundments or any other body of standing water. Plankton begin to appear in streams that are wide enough and deep enough and the currents are slow enough that organisms can maintain their position without being carried downstream.

Plankton are a key link in the food chain between the algae and the ***benthos*** (bottom organisms) and fish. The fish that feed on plankton are called ***planktivores***, and the benthos that feed on plankton are called ***filter feeders***. The plankton and fish are part of the ***pelagic community***, also called the ***limnetic community***. Three levels of organization are usually recognized within the pelagic community. Besides the plankton is the ***microbial community*** that makes up most of the ***microbial loop***. The microbial loop is an integral part of the conventional autotrophic/heterotrophic food web. If there are bacteria and other kinds of microbial parasites in the water, they probably will be part of the microbial loop. The microbial loop does not include parasitic zooplankton. Fish, and other organisms that have strong powers of locomotion (strong enough to swim against currents), represent the ***nekton community***. Also part of the pelagic community is the ***neuston,*** or those organisms that are confined to the air-water interface. This chapter examines the importance of the microbial loop and the plankton to the ecology of a lake. The neuston and nekton are discussed in

Chapter 8.

The plankton are often classified according to their size. Table 7.1 lists the size classes of most members of the planktonic community.

Table 7.1. Size class designations of plankton and a summary of the major taxa in each class. Some members were described in Chapter 6, the rest are described in this chapter.

Size Class	Size Range μm	Members
Picoplankton	0.2 - 2	Bacteria, blue-green algae, some euglenoid algae
Nanoplankton	2 - 20	Some diatoms, euglenoid algae and ciliates
Microplankton	20 - 200	Dinoflagellates (= Pyrrophyta in Chapter 6), some rotifers, amoeboid forms, ciliates, and bivalve larvae
Mesoplankton	200 - 2000	Some rotifers, cladocerans and copepods
Macroplankton	> 2000	A few cladocerans and copepods

The chapter also cautions people not to drink lake water without boiling it first. Various pathogens and parasites that often occur in lake water are described. Even oligotrophic waters may have parasites. Parasitism is a normal, natural phenomenon in lakes. The requirements of the planktonic larval stages of parasites have evolved with the requirements of their hosts. For example, trout and salmonid fish, when present in large numbers, are diagnostic of oligotrophic waters, but their parasites have evolved with the same water quality requirements. As waters become more enriched, not only are trout replaced, but their parasites as well. Fish more tolerant of enrichment, and their associated parasites, begin to dominate. If enrichment is due to domestic and human waste, pathogens will be present. Parasites are found in hosts at all trophic levels. A broad and diverse community of parasites is just as indicative of unpolluted water as is a broad and diverse community of free-living organisms.

To help identify the various kinds of pathogens and parasites that occur in the plankton, look for the following, "don't drink the water" symbol:

By the end of this chapter you will:

☞ 1. Recognize the components and importance of the microbial loop
☞ 2. Be able to draw a pelagic food web with its microbial loop
☞ 3. Recognize the importance and roles of fungi in the microbial loop
☞ 4. Recognize the importance and roles of bacteria in the microbial loop

☞ 5. Know how to assess the trophic status of a lake using bacteria
☞ 6. Understand why the presence of coliform bacteria in lake water make it unsafe to drink
☞ 7. Know how to use bacteria to determine the kind and source of pollution
☞ 8. Know some of the pathogenic bacteria in lake water
☞ 9. Know how to reduce the risk of bacterial infections from lake water
☞ 10. Know how to recognize the five groups of protists in the microbial loop
☞ 11. Recognize the importance and roles of protists in the microbial loop
☞ 12. Know which protists make lake water unsafe to drink
☞ 13. Know why numbers of protists vary seasonally and spatially
☞ 14. Know how to use protists to assess water quality
☞ 15. Know how to recognize rotifers
☞ 16. Know the roles of rotifers in the pelagic food web
☞ 17. Know rotifers reproduce by cloning themselves
☞ 18. What factors cause seasonal and spatial variations in rotifer abundance
☞ 19. Know how to use rotifers to assess water quality
☞ 20. Know how to recognize Cladocera and Copepoda, two groups of crustaceans present in most plankton
☞ 21. Know the importance and roles of cladocerans (water fleas) in the plankton
☞ 22. Learn of some weird seasonal body changes that occur in Cladocera
☞ 23. Know how cladocerans clone themselves for most of the year
☞ 24. Know how to recognize three groups of copepods
☞ 25. Know the importance and roles of copepods in the plankton
☞ 26. Know how copepods reproduce and the names of their larval stages
☞ 27. Know how to estimate growth and production in crustacean zooplankton
☞ 28. Know the factors that contribute to temporal and spatial variations in abundance
☞ 29. Know how to assess water quality using crustacean zooplankton
☞ 30. Know which crustaceans are parasitic and make lake water unsafe for drinking
☞ 31. Know which crustaceans harbour larvae of freshwater tapeworm and fluke parasites
☞ 32. Know which larval stages of tapeworms are planktonic, making water unsafe for drinking.
☞ 33. Know some of the more dangerous freshwater tapeworms that can infect humans
☞ 34. Know which larval stages of flukes are planktonic, making water unsafe for drinking.
☞ 35. Know some of the more dangerous freshwater flukes that can infect humans
☞ 36. Know what crustaceans migrate from the bottom (benthos) into the plankton
☞ 37. Know what insects migrate from the benthos into the plankton
☞ 38. Know the planktonic life cycle stages of zebra and quagga mussels
☞ 39. Know what other exotic bivalves have planktonic larvae
☞ 40. Understand the factors that control settlement and the consequences of settlement by planktonic bivalves
☞ 41. Know the importance of zooplankton in nutrient cycling
☞ 42. Know how to sample zooplankton in freshwater

The Microbial Loop

Prior to the 1980s, energy budgets of food webs rarely balanced. But in the early 1980s scientists discovered that much of the imbalance was due to the omission of microorganisms of nanoplankton and picoplankton size, mostly because they passed through plankton nets and filters of previous investigations. Scientists have since discovered that microorganisms form a significant part of the food web. We now know that heterotrophic microbes, including bacteria, ciliates and flagellates, are key
1☞ organisms in processing ***dissolved organic materials*** (***DOM***) and that they form major components of the ***microbial loop***. The microbial loop is now considered a standard part of the pelagic food web, especially in eutrophic lakes where ***dissolved organic carbon*** (***DOC***) abounds.

Chapter 6 gives an example of a food web involving several communities of organisms. Figure 7.1 shows the classical autotrophic-heterotrophic pelagic food web and the flow of carbon, as dissolved organic carbon, through the microbial loop. ***Dynamic food webs*** are drawn to show the flow of energy from one trophic level to the next. The amount of energy that flows from one level to another will vary from one lake to the next, which explains the plethora of figures that appear in the literature describing food webs. Here, we will talk in general terms about the flow of energy from the conventional autotrophic/heterotrophic food chain through the microbial loop.

The planktonic bacteria and protistan ***detritivores*** (feeders of detritus) do not have immediate access to the ***detrital pool*** because much of the material is still living during the growing season. The detrital pool only becomes particulate in the autumn, at which time bacterial metabolism begins to slow because of decreasing temperatures. Not until the following spring and summer does the microbial loop begin to play an important role in converting particulate material to a soluble detrital form, or dissolved organic matter (DOM), that is ultimately is converted to DOC.

The main sources of soluble detritus are exudates, or metabolites, such as ***extracellular products of photosynthesis*** (***ECPP***). The DOC pool also consists of animal excreta, fragments lost (i.e. spillage) when organisms feed or prey upon other organisms and microbial decomposition itself. The soluble detritus is used by the heterotrophic microbes. As the bacteria, flagellates and ciliates utilize the materials, they themselves are available to the larger consumers and grazers, such as cladocerans, rotifers, copepods and larval fish, that make up the classical pelagic food web. In this manner, organic material is passed from one trophic level to the next. Eutrophication occurs when the supply of organic material exceeds the demands of the larger detritivores, grazers and herbivores and the organic material begins to accumulate on the bottom of the lake.

Although there has been an omission of the microbial loop DOC pool in energy budgets, it is still very difficult and time consuming to come up with a good quantitative estimate of energy derived from the microbial pool. Most methods require the use of ***radio-isotopes***, or ***tracers***, so that one can follow the incorporation of the detrital compounds into the microbial pool and hence the amount of energy cycled

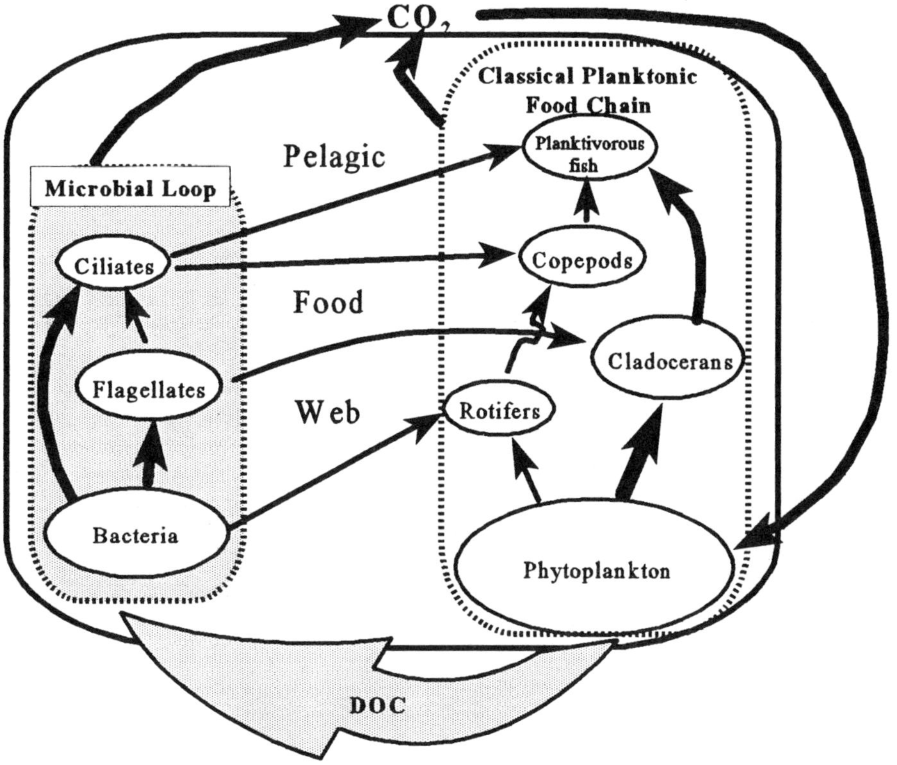

Figure 7.1. Carbon flow through the classical food chain and the microbial loop of the pelagic food web.

through the pool.

In most thermally stratified lakes, summer is characterized by the depletion of one or more of the limiting nutrients, carbon, nitrogen or phosphorus. Some zooplankton can be supported by organic material recycled through the microbial loop. Hence, if algae are not present or are rare, nutrient regeneration in the mixed layer may provide some inorganic material to the summer phytoplankton crop. ***Remineralization*** of organic
2☞ material by the microbial pool produces both new and recycled nutrients. ***New nutrients*** are those that have arrived from allochthonous sources, such as inflowing rivers or precipitation. ***Recycled nutrients*** are those that are recently excreted by zooplankton and fish.

Aquatic organisms excrete both inorganic and organic material which are then mineralized by the microbial pool. Dead material and body parts are decomposed by both bacteria and fungi. In streams and lakes, the fungi are important in conditioning both leaf litter of allochthonous (e.g. trees, shrubs, grasses) and autochthonous (e.g. macrophytes) origins before larger invertebrates tear or shred the coarse particulate organic matter (CPOM) to produce the fine particulate organic matter (FPOM). During the fungal and organic matter (CPOM) to produce the fine particulate organic matter (FPOM). During the fungal and bacterial conditioning of CPOM, DOC is leached from the leaf litter and contributes to the DOC pool for the microbial loop.

Table 7.2. Some important fungi in aquatic systems. Fungi are important microbes that follow the bacteria in conditioning leaf litter before consumption by shredding and collecting invertebrates. Only fungi (and a few bacteria) have appropriate enzymes for breaking down lignin in leaf litter.

Fungi		
Fungi genus	**Taxonomic division**	**Some typical aquatic habitats**
Actinoplanes	Actinomycetes	Damp leaves, streams, free living
Streptomyces	Actinomycetes	Benthic sediments, odour-producing forms, streams
Rhizophydium	Phycomycetes	Parasitic water mould, single cells
Allomyces	Phycomycetes	Saprophytic water mould, filamentous
Saprolegnia	Phycomycetes	Parasitic on fish
Cryptococcus	Ascomycetes	Aquatic yeast
Aurobasidium	Bascidiomycetes	Damp wood, sometimes in streams

The Role of Fungi

Fungi is a kingdom that includes molds and mushrooms. Most aquatic fungi
3☞ are benthic in nature but in water they indirectly provide DOC to the microbial loop. Table 7.2 lists some of the fungi that contribute to the DOC pool of lakes. Like plants, all fungi have rigid cell walls, but some are ***saprophytic*** (use organic substances for growth) and others are ***parasitic*** (feed on other organisms). All fungi spread and invade substrates or host tissues by root-like ***mycelia*** (singular ***mycelium***) that may be branched or unbranched

Fungi are represented by groups that have either a primitive level of cell organization or a more advanced level of cell organization. The Actinomycetes have a very primitive level of cell organization and have true branching mycelia. The group is represented by the odour-producing, *Streptomyces,* common in benthic sediments of lakes, and *Actinoplanes* that is found on damp leaves in headwaters streams. All other groups of fungi have a higher level of cell organization, but only some have true branching mycelia. Within this group are the Phycomycetes, Ascomycetes and Basidiomycetes. The phycomycetes are represented by both saprophytic (e.g.

Allomycetes) and parasitic (e.g. *Rhizophydium* and *Saprolegnia*) genera of water molds. *Saprolegnia* is a common parasite appearing as a fuzzy, white clump on dying or injured fish. It is especially common in home aquaria that are poorly maintained. The ascomycetes are represented by *Cryptococcus*, an aquatic yeast, and the basidiomycetes by *Aureobasidium*, common on wet wood in streams.

The Role of Bacteria

The bacteria and Cyanobacteria (recall the blue-green algae phylum discussed in Chapter 6) are included in the kingdom, Monera. They are either single-celled organisms or simple associations of cells.

Recycled nitrogen and phosphorus are usually critical in the development of the phytoplankton populations in the epilimnion of most temperate lakes. Nutrient recycling is particularly important where the autochthonous supply is at a minimum.
4☞ Table 7.3 is a partial list of the bacteria common in many aquatic habitats. The table lists bacteria according to their role in nutrient cycling. Some are photosynthetic and require light but most are non-photosynthetic. Photosynthetic forms are almost entirely devoted to the sulphate cycle and, occasionally, the carbon cycle. For example, the green and purple sulphur bacteria, such as *Chromatium*, convert hydrogen sulphide into elemental sulphur and the purple non-sulphur bacteria, such as *Rhodospirillum*, convert organic matter into carbon dioxide. Most, if not all of these bacteria, are associated with sediments in the hypolimnion of lakes or in surface muds of marshes and estuaries. They appear in the water column only if currents resuspend the bottom materials.

Within the non-photosynthetic group are aerobic (requiring oxygen) and anaerobic bacteria. The carbon, nitrogen, and sulphate cycles depend on both types. For example, under anaerobic conditions: *Methylomonas* and *Escherichia coli* convert methane and lactose sugar to carbon dioxide; *Methanobacterium* converts carbon dioxide into methane; *Beggiatoa* and *Thiodendron*, both convert hydrogen sulphide to elemental sulphur, and *Desulfovibrio* converts sulphate and sulphur to hydrogen sulphide; and *Azotobacter* and *Clostridium* fix nitrogen as organic nitrogen. Among the aerobic forms are *Nitrosomonas* and *Nitrobacter* which sequentially convert ammonia into nitrite and nitrate, and *Thiobacillus* converts hydrogen sulphide and sulphur into sulphates. While many bacteria involved in the carbon cycle directly utilize dissolved organic carbon, bacteria in the nitrogen and sulphur cycles utilize only nitrogenous and sulphurous compounds in dissolved organic carbon and are, therefore, only indirectly involved in processing DOC.

Sphaerotilus is known as "sewage fungus", but it is a sheathed bacterium attached to logs and rocks in waters carrying domestic or organic wastes. *Escherichia coli* is an indicator of faecal contamination because of its omnipresence in the colons of warm-blooded vertebrates. It is a member of the coliform group of bacteria. Water is unsafe for drinking if any *E. coli* are present. Beaches are closed if either *E. coli* is

present or if the total coliform count exceeds 200/100 ml.

Table 7.3. Some important bacteria in aquatic systems, all contributing in one form or another to the microbial loop. Cyanobacteria (blue-green algae) share features with both bacteria and algae, but are discussed in Chapter 6.

Carbon Cycle		
Bacteria genus	**Substrate**	**Some typical aquatic habitats**
Cytophaga	Cellulose, chitin	Detritus, sediments, bactivorous
Spirillum	Organics	Eutrophic fresh and saline water
Methylomonas	Organics, CH_4	Fresh and saline water, bactivorous
Escherichia coli	Parasitic, disease causing	In plants and animals (lactose $\rightarrow CO_2$)
Sphaerotilus	Organics	Sewage fungus attached to logs, rocks, etc.
Methanobacterium	CO_2	Anoxic hypolimnia, muds, marshes ($CO_2 \rightarrow CH_4$)
Nitrogen Cycle		
Nitrosomonas	Reduced nitrogen compounds	Oxic waters, sediments ($NH_4 \rightarrow NO_2$)
Nitrobacter	Reduced nitrogen compounds	Oxic waters, sediments ($NO_2 \rightarrow NO_3$)
Pseudomonas	Organic carbon	Anoxic waters, sediments, denitrifyers ($NO_3 \rightarrow N_2$)
Bacillus	Organic carbon	Anoxic waters, sediments, denitrifyers ($NO_3 \rightarrow N_2$)
Escherichia coli	Parasitic, disease causing	Anoxic waters, sediments, denitrifyers ($NO_3 \rightarrow N_2$)
Azotobacter	Organics, N_2	Anoxic waters, sediments (N_2 fixation)
Clostridium	Organics, N_2	Anoxic waters, sediments (N_2 fixation)
Sulphur Cycle		
Chromatium	H_2S, CO_2 (requires light)	Anoxic waters, green & purple sulphur bacteria ($H_2S \rightarrow S$)
Beggiatoa	Detritus, sulphur springs	Anoxic waters ($H_2S \rightarrow S$)
Thiodendron	H_2S	Anoxic waters ($H_2S \rightarrow S$)
Desulfovibrio	SO_4, S	Anoxic waters, sediments ($S \rightarrow H_2S$)
Thiobacillus	Reduced S compounds	Oxic waters, sediments (H_2S, $S \rightarrow SO_4$)
Thiothrix	Reduced S compounds	Oxic waters, sediments (H_2S, $S \rightarrow SO_4$)

In lakes with some hypolimnetic oxygen, the organic material is decomposed by aerobic heterotrophs. The numbers of aerobic heterotrophs decline greatly when anoxia develops, and anaerobic bacteria begin to increase in numbers. The ratio of the
5☞ numbers of aerobic heterotrophic bacteria to the total number of bacteria is a useful criterion for assessing the trophic status of lakes. A "***bacterial index ratio***", BIR, using plate counts of aerobic heterotrophic bacteria (#/ml) and epi-fluorescent microscopic counts of total bacterial densities (#/ml), is calculated as:

$$\text{BIR} = \frac{\text{Total number of aerobic heterotrophic bacteria}}{\text{Total number of bacteria}}$$

The criteria for determining trophic status from the BIR are:

BIR < 10% = Oligotrophy
BIR 10-25% = Mesotrophy
BIR > 25% = Eutrophy

Microbial Diseases in Water

The bacterial communities of most lakes are essential in nutrient cycling and are omnipresent and rather benign as far as affecting the drinking quality of water and body contact recreational activities. As eutrophication intensifies, the microbial communities intensify in abundance and the water becomes more unsuitable for drinking and body contact activities. If raw sewage contributes to eutrophication, then the water should not be used for drinking or swimming because the probability of a microbial disease
6☞ being present is greatly enhanced. *Escherichia coli* is present in the intestinal colon of all mammals and its presence indicates faecal contamination. Water is unsafe for drinking if *E. coli* is present in the water sample because it can cause urinary infections in humans. *Escherichia coli* is a member of the coliform group of bacteria, but not all coliforms are indigenous to the intestinal tract of humans and other warm-blooded animals. Some coliforms are indigenous to the carbon cycle and play a role in the microbial loop, as discussed above. Nevertheless, these aquatic coliforms can be consumed by humans either by eating vegetable matter (e.g. *Nasturtium*, or water cress) or drinking contaminated water. Non-faecal coliforms can then be discharged in
7☞ faeces with the faecal forms, such as *E. coli*. Therefore, it is important to measure the total coliform count in water as well as the *E. coli* count. In most communities, water is considered unsafe for swimming if the total coliform count exceeds 200/100 ml. Faecal coliforms are differentiated from other coliforms on the basis of growth at 44.5 °C, after incubation for 22 ± 2 h, which discourages growth of non-faecal coliforms. Also, specialized enrichment agar media are used for cultivation of *E. coli*.

Beaches are closed if the total coliform count exceeds the safe limit because the probability of disease forms being present is greatly increased. Faecal streptococci especially are important because they are indigenous to the intestinal tract of humans and other warm-blooded animals. One species, *Streptococcus faecalis* and its strains, are present in humans and warm-blooded animals, but only one, *Streptococcus salivarius* is diagnostic of human intestinal tract origin. Because *S. salivarius* dies off quickly in water, its presence is also a good indication of **recent** faecal contamination by humans. Dugan (1974)[1] suggests that the ratio of faecal coliforms to faecal streptococci is useful for determining the source of faecal contamination. He used the following criteria:

FC/FS < 0.7: probably from livestock or poultry waste
FC/FS 0.7 - 1.0: mixed pollution, predominantly of livestock origin
FC/FS 1.0 - 2.0: mixed pollution, but origin uncertain
FC/FS 2.0 - 4.0: mixed pollution, predominantly of human waste
FC/FS > 4.0: predominantly of human waste

The values are contingent upon following certain guidelines: (i) water samples must be taken within 24 h flow distance downstream of expected source; (ii) pH must be between 4.0 and 9.0; and (iii) faecal coliform counts, and not total coliform counts, must be used.

Some of the most dangerous bacteria are *Salmonella, Shigella, Leptospira,*
8☞ *Vibrio cholerae,* and *Mycobacterium tuberculosis*. The list is just a small sample of pathogens that may be present. The following is summary of their pathogenicity.

Salmonella - is present mostly in spoiled foods, such as chicken, and can be contracted by handling infected turtles. The bacterium causes ***salmonellosis***, an acute enteritis characterized by diarrhea, fever and vomiting. One form, *S. typhi*, causes typhoid fever. Cross infection can occur between humans and animals in polluted water. *Salmonella* can survive up to 120 km downstream of the source.

Shigella - belongs to the family of Enterobacteriaceae, a group of bacteria found in the intestines of humans. It is common in polluted water and can cause acute diarrhea in humans. Some species are fatal. As with *Salmonella*, only people with the disease, or are carriers of the disease-producing bacteria, would be expected to contribute significant numbers of pathogenic bacteria in their faeces.

Leptospira is a spiral-shaped, motile bacterium called a ***spirochaete***. It causes a disease

---

[1]Dugan, P. R. 1974. Bacterial, non-bacterial, and chemical indicators of environmental quality. *In*: Organisms and biological communities as indicators of environmental quality - A symposium. The Ohio State University, March 25, 1974. Ohio Biological Survey, Informative Circular No. 8.
pp. 34- 36.

known as ***leptospirosis***, an acute infection of the kidneys, liver and central nervous system. The primary hosts are rodents but humans contract the disease through cuts in the skin while wading through polluted water.

Vibrio cholerae is a short, rigid, motile bacterium with a single flagella and shaped like an "S" or a comma. The species causes cholera, an acute intestinal disease that may result in death within hours. It lives in the gut wall and produces toxins of two parts; one part binds onto cell surface receptors in the gut wall while the second acts as a hormonal mimic, causing the cell to excrete water, sodium and potassium, leading to severe dehydration.

Mycobacterium tuberculosis is rarely found in water, even polluted water, but when present will cause tuberculosis in mammals.

It is important to note that while the potential for infection by these disease forming microbes is extremely remote in most lakes, the potential increases with artificial eutrophication. Older cottages may still have outhouses too close to shore and the potential exists for some coliform bacteria to "leak" into the lake. It is always a 9☞ good idea to test the water at least twice a year for coliforms, especially *E. coli*, to reduce the risk of infections Many hardware stores now carry a do-it-yourself coliform testing kit as well as a do-it-yourself *E. coli* testing kit. Beware of cottage filters that claim to filter out microbes. Always ask if the filter removes picoplankton (0.2 - 2 μm), which includes bacteria and all protists. Even then, test the filtered water for coliforms. Do several tests in midsummer, when bathing and swimming activities are intense and can provide the greatest potential for infections. Above all, boil the water before drinking it.

There are some non-enteric (not of intestinal origin) bacteria that are of human origin as well. They enter the water from bathers and swimmers infected with bacteria in skin lesions or in the nose and throat. *Staphylococcus aureus* and *Pseudomonas aeruginosa* are the most hazardous non-enteric pathogens that can infect open wounds, the nose, the throat, upper respiratory tract, eyes, ears and the urogenital system. Their presence indicates contamination of human origin and inadequate disinfection procedures.

10☞ **The Role of Protists**

The Protista, also called Protozoa, are an important component of the microbial loop. The protists are represented by flagellates (both photosynthetic and non-photosynthetic forms), ciliates and, occasionally, sarcodines (amoeba-like organisms) and sporozoans (exclusively parasitic). Table 7.4 gives their common morphological features and aquatic habitats and lists several common genera in each of the five groups.

11☞ The protists have several types of feeding habits. The flagellates that have chlorophyll and a photosynthetic type of nutrition are called ***holophytic*** or ***autotrophic***.

Table 7.4. Protistan members of the microbial loop in most lakes.

Name	Common Features	Habitats	Common Examples
Phytoflagellates	1 or more flagella; 1 nucleus; chlorophyll *a, b, c*; cell walls of cellulose, peptides, or are a proteinaceous pellicle), some are loricate (shelled); asexual reproduction is by longitudinal fission	Lakes (*Synura*, a loricate form, occurs in acid waters) and ponds	Chlorophytes: *Volvox, Gonium, Chlamydomonas*, Dinoflagellates: *Ceratium, Peridinium, Glenodinium* Euglenoids: *Euglena, Phacus* Chrysomonads: *Dinobryon, Synura Ochromonas* Cryptomonads: *Rhodomonas, Cryptomona*
Zooflagellates	1 or more flagella; 1 nucleus; no chlorophyll; cell wall cellulose or pectin; asexual reproduction is by longitudinal fission; Choano-flagellates have collared cells.	Lakes, stagnant ponds (all genera listed are from stagnant ponds)	Cryptomonads: *Chilomonas, Cyathomonas* Chrysomonads: *Oikomonas* Collared flagellates: *Trepomonas, Urophagus* Parasites: *Giardia lamblia*. In faecal contaminated water, causes diarrhea,
Sarcodines	Amoeboid (pseudopodal) forms, some with shells (loricas); 1 or more similar nuclei; binary fission; classed according to pseudopod types - hairlike in Actinopoda; lobe-like or radiating in Rizopoda	Shelled forms are mostly benthic, occasionally planktonic; naked forms benthic or planktonic, feeding on or attached to algae	Actinopods: *Actinosphaerium, Lithocolla* Rhizopods with radiating pseudopods: *Protomonas, Pseudospora* Rhizopods with lobate pseudopods: *Amoeba* Rhizopods with loricas: *Difflugia, Pseudodifflugia, Difflugiella*
Sporozoans	All are endoparasites; move and feed using pseudopods	Aquatic animals	*Plasmodium*, causes malaria in man *Heneguya*, parasitizes fish *Cryptosporidium* spp. (see text)
Ciliates	Locomotion by cilia or ciliary organelles; two kinds of nuclei; cell wall a pellicle; asexual reproduction by transverse fission. Several types of sexual reproduction	Ubiquitous, in ponds (temporary and permanent) and oligotrophic to eutrophic lakes. Some are commensal in molluscs and crustaceans	Predaceous forms: *Dipilidium, Dileptus* Herbivorous form: *Frontonia* Filter feeders: *Paramecium, Vorticella* Omnivorous forms: *Spirostomum, Stentor, Blepharisma* Parasitic form: *Ophryoglena* Obligate anaerobes: *Bodo, Trepomonas, Saprodinium*

Such organisms synthesize carbohydrates from carbon dioxide during photosynthesis and with absorbed nitrogen, phosphorous, sulphur and silica synthesize cell components. Those protists that ingest solid particles of food, such as bacteria, algae, other protists and debris, are called ***holozoic***. ***Saprozoic*** nutrition occurs in all classes of protists; this type of nutrition involves absorbing dissolved salts and simple organic materials (e.g. DOC) from the water and synthesizing them into protoplasm. Some protists utilize two or more types of nutrition and are known as ***mixotrophs***. Finally, some utilize ***parasitic*** nutrition, much of which involves the uptake of dissolved materials, such as DOC, although many ingest particulate matter and are in essence holozoic in their feeding habits.

Most of the flagellates are phytoplankton (often called phytoflagellates); only a few are zooplankton (often referred to as zooflagellates). Protistan forms of algae (all flagellates) in the microbial loop include the phyla; Chlorophyta, Euglenophyta, Cryptophyta, Chrysophyta, and Pyrrophyta. The most diverse phylum is Chlorophyta (green algae). Euglenophyta (euglenoids) and Pyrrophyta, known as dinoflagellates to zoologists (botanists and zoologists often disagree on taxonomy!), can dominate in biomass, especially in eutrophic lakes, but neither phylum has the diversity of forms seen in Chlorophyta. The cryptomonads (Cryptophyta) and the yellow or brown-green algae (Chrysophyta, or the chrysomonads to zoologists) are the least common in plankton, but nevertheless contribute to plankton diversity. Refer to Chapter 6 for other characteristics of the phyla.

The only unflagellated algae phyla that are not represented, or at least poorly represented, in the microbial loop are Charophyta, which are mostly benthic, and
11☞ Phaeophyta and Rhodophyta, which are primarily marine or are attached forms in freshwater streams. The siliceous diatoms of the phylum Bacillariophyta (e.g. *Nitschia, Navicula, Tabellaria, Fragilaria*) are mobile, but not flagellated and form a diverse flora in the microbial loop in lakes of all trophic levels. None of the Cyanobacteria (blue-green algae) are flagellated. Most blue-green algae are filamentous but they too can be very common in plankton, especially in mesotrophic (e.g. *Oscillatoria*) and eutrophic lakes (e.g. *Anabaena*).

Non-photosynthetic flagellates (i.e. zooflagellates) occur in the plankton as well. Nearly all belong to the phyla Cryptophyta and Chrysophyta and many of them are amoeboid (move or feed by forming pseudopods - false feet). A third group is the ***Choanoflagellata***, or the collared flagellates. Their collared cells easily distinguish them from other flagellates. Like most flagellates, they reproduce by longitudinal cell divisions (compare to transverse fission of ciliates, discussed below). The zooflagellates have one type of nucleus and one or more flagella and contain all the cell components seen in other protists, including contractile vacuoles (a type of excretory organelle) and food vacuoles.

The phylum Sarcodina includes the amoebas and their shelled relatives. Sarcodines have two distinct layers, an outer ***ectoplasm*** and an inner ***endoplasm***, both of which are involved in the formation of pseudopods used for feeding and for their primary means of locomotion (some also have flagella). The endoplasm contains one or more nuclei, each with a nuclear membrane, and one or more contractile vacuoles and food vacuoles. Many forms have siliceous shells, such as those in the classes Rhizopoda and Actinopoda. In fresh waters, the shelled rhizopods are represented by such genera as *Difflugia* and the actinopods by *Lithocolla*, both forms contributing significantly to silica deposits in lakes. However, the siliceous armour forces many shelled forms to assume a benthic habit, with occasional movements into the plankton with the help of vertical water currents. An exception is *Difflugia limnetica* which regulates its buoyancy with fat inclusions in the protoplasm. The animal sinks when the fat globules are metabolized, usually by the autumn of each year. Most sarcodines are herbivorous, feeding on filamentous algae.

Some protists are parasitic, like *Actinosphaerium,* which causes white lesions or pustules on fish. Even more dangerous to humans are *Entamoeba hystolytica.* and
12☞ *Giardia lamblia. Entamoeba hystolytica* is rare in northern latitudes. It is a frequent inhabitant in polluted waters of tropical and subtropical climates where it causes amoebic dysentery. *Entamoeba* is an encysting form with four or eight nuclei in the cyst. *Giardia* is an amoeboid flagellate that inhabits the small intestine of humans and other animals. It is pear-shaped, has two nuclei and four pairs of flagella. They attach to the cells of the intestinal mucosa (lining) and absorb fats for nutrients. *Giardia lamblia* causes diarrhea in humans. It can encyst and cause recurring bouts of diarrhea.

The Sporozoa, closely related to Sarcodina, are a parasitic group that parasitize fish, waterfowl and mammals, including humans. Although the sporozoans are not an important part of the microbial loop, they are among the most dangerous to humans. *Plasmodium* causes malaria in humans, the parasite being transmitted by a mosquito. Ordinarily *Plasmodium* is not present in northern latitudes, although there have been some rare reports of infected mosquitoes being transported by humans from tropical climates to infect humans living in temperate climates.

Cryptosporidium is a more prevalent sporozoan parasite found in warm and cold climates in untreated and treated drinking water supplies, swimming pools, rivers, streams, lakes and impoundments. Its ubiquitousness is attributed in part to its broad range of host species, including mammals, birds, and fish. Infection in humans seems to occur mainly with *C. parvum.* Also, unlike most parasites, it is capable of completing its life cycle in a single host. The development cycle is complex and completed in 1 to 8 days (average = 3 days). Normally the life cycle involves asexual, followed by sexual multiplication, formation of a zygote, encystment of the zygote to produce an ***oocyst***, a spore formation outside of the host, release of numerous parasites into the water, and re-entry into another host to begin another cycle.

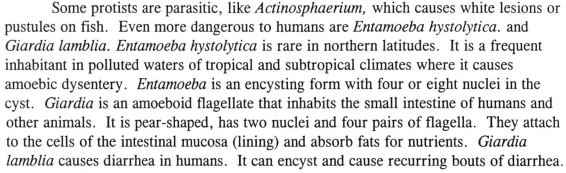

Cryptosporidium is extremely infective. Apparently a drop of water with as few as 10 oocysts is sufficient to cause ***cryptosporidiosis***, the disease causing gastroenteritis characterized by profuse, watery, greenish, ill-smelling diarrhea. Other symptoms are vomiting, nausea, headache, abdominal pain, low-grade fever and potentially significant weight loss. The parasite is among the most difficult to control because: (i) it is so ubiquitous in distribution; (ii) it is not host specific; (iii) it infects any tissue or organ; (iv) it has the capacity to complete an entire life cycle in one host; (v) it can be transmitted directly from person to person in spit or indirectly in water; (vi) it has a high infectivity; and (vii) oocysts are resistant to most disinfectants, including normal (1 ppm) chlorination. Oocyst infectivity can be destroyed by applying 70% bleach (domestic grades, like Chlorox and Javex, have ~ 10 ppm chlorine), 3% hydrogen peroxide, 10% formaldehyde, or 5-10% ammonia. However, boiling infected water for ten minutes will kill the oocysts.

The most complex protists are the Ciliophora, or the ciliates. The cilia, used for locomotion, have the same structure as flagella. Ciliates have two types of nuclei, a *macronucleus* which performs vegetative functions, and a *micronucleus,* which has the genetic material needed in reproduction. Many forms, like *Paramecium*, reproduce by ***conjugation*** where individuals pair up and exchange genetic material across a protoplasmic bridge formed during the mating process. Table 7.4 lists several genera according to their feeding habits. Most genera have some kind of gullet for swallowing particles or prey, which is usually free-living bacteria, fungi, yeasts, algae, and other protists, including ciliates.

Temporal and Spatial Variations in Protists

The population dynamics and productivity of protists is the least understood of all the planktonic groups. Protists generally constitute a relatively minor part of the
13☞ zooplankton, numerically and in biomass. Exceptions have been reported for some large lakes in Russia and Africa where biomasses of flagellates and ciliates equal or exceed those of phytoplankton during their decline in the midsummer. Bacterial densities increase as moribund algae become available for recycling carbon and nitrogen bacteria. Coincident with the increase in bacterial abundance is an increase in protistan biomass.

Population densities of *Difflugia limnetica* fluctuate seasonally in several German lakes. Sexual and asexual reproduction prompts an increase in their populations in mid summer, their densities peaking in late summer. In the fall, fat globules, which help to buoy the shelled animal in the plankton, are metabolized and the animals sink to the sediments. Many die, some encyst to carry the population over the winter, others remain active through the winter in littoral sediments. In the spring, feeding intensifies for the active species. In these species, fat globules are produced, along with gas bubbles, and the shelled forms rise. As they ascend, diatoms and sand grains become part of the animal's shell and helps to recycle silica into the epilimnial part of the lake. Temperature appears to play a major role in cueing the onset of reproduction and increasing the rate of reproduction.

Closely associated with temperature is light intensity. Some protistan species are ***photonegative*** (avoid light), others are photopositive (move toward light). Hence, differences in depth distribution among species can often be explained by their response to increasing light intensity.

Other than mechanisms employed for regulating their depth distribution, as with *Difflugia*, the spatial distribution of protists is pretty much dictated by water currents. Even ciliates, which are the fastest swimmers (e.g. 200 - 1000 μm/sec) of all the protists (flagellates move at 15 - 300 μm/sec; sarcodines at 0.5 - 3.0 μm/sec), have patchy distributions across the lake. Langmuir currents, especially, can create elongated patches of zooplankton.

Ecology and Indicator Value of Protists

The greatest number of species is found in ponds and pools. The greater the diversity of habitats in pools, the greater the diversity of species, particularly in oligotrophic to slightly mesotrophic bodies of water. As habitats become more 14☞ enriched, fewer species but greater numbers of each that do occur begin to accrue. Eutrophic (enrichment) indicator species occur in large numbers. Table 7.5 lists some genera of protists that have trophic indicator value. Note that many of the enrichment indicators are found as part of the microbial communities in trickling filters of water pollution control plants. Chapter 12 describes the use of protists in assessing water quality.

PLANKTON

Phytoplankton and Protists

Recall, organisms in the microbial loop are not isolated from other plankton, they are part of the planktonic food web. Hence, the plankton includes the phytoplankton (i.e. photosynthetic species), the zooplankton (i.e. the non-photosynthetic species) and protists in the microbial loop. The algae have been discussed in Chapter 6 and nothing else needs to be added here. The bacteria and protists were examined earlier in this chapter. All three groups (bacteria, phytoplankton and protists) are part of the plankton and play a major role in the microbial loop.

Zooplankton

The zooplankton are represented mainly by three groups, the protists discussed above, the rotifers and the crustaceans. Other members include larval stages of molluscs and insects.

Rotifers

Rotifers belong to the phylum ***Rotifera***. They are ubiquitous, occurring in damp soil, wet moss and vegetable debris, eaves troughs, temporary and permanent ponds, and lakes of all trophic states. Most rotifers are planktonic, a few are sessile, attaching themselves to filamentous algae and submerged macrophytes, logs, rocks, and even animals, like bivalves. Their most diagnostic feature is the ciliated disc at the anterior end (Fig. 7.2). Rotifers are often referred to as "wheel animals" because the cilia beat synchronously in either a clockwise or counter-clockwise direction, like a rotating

Table 7.5. A partial list of protists with trophic indicator value. Species are listed alphabetically within each phylum.

PROTISTAN ZOOPLANKTON			
Genus/species name	**Common name**	**Taxon group**	**Significance**
Amoeba proteus	amoeba	Kingdom: Protista P: Sarcomastigophora	Common in enriched waters
Arcella	shelled amoeba	P: Sarcomastigophora	Major source of silica; oligotrophic indicator
Bodo	flagellate	P: Sarcomastigophora	Enrichment indicator; sewage protist
Ceratium	flagellate	P: Sarcomastigophora	Enrichment indicator
Difflugia elegans	shelled amoeba	P: Sarcomastigophora	Major source of silica; oligotrophic indicator
Euglena gracilis	flagellate	P: Sarcomastigophora	Enrichment indicator
Monas	flagellate	P: Sarcomastigophora	Enrichment indicator; sewage protist
Pandorina morum	flagellate	P: Sarcomastigophora	Enrichment indicator
Trepomonas	flagellate	P: Sarcomastigophora	Enrichment indicator; sewage protist
Carchesium	ciliate	P: Ciliophora	Enrichment indicator; sewage protist
Colpidium	ciliate	P: Ciliophora	Enrichment indicator
Euplotes	ciliate	P: Ciliophora	Enrichment indicator
Litonotus	ciliate	P: Ciliophora	Enrichment indicator; sewage protist
Opercularia	ciliate	P: Ciliophora	Filter feeder of microorganisms; sewage protist
Paramecium	ciliate	P: Ciliophora	Enrichment indicator

wheel. The main morphological features are a ***corona*** on which the cilia and mouth are located, the main body in which the organs of digestion (complete, with a mouth, pharynx, ***mastax***, esophagus, stomach, a short intestine and an anus), reproduction (most are parthenogenetic with only an ovary or ovaries present), and excretion
15☞ (primitive flame cell structure) are located, and a foot with toes. The foot is prehensile in some species. The anus and the openings for the reproductive and excretory systems empty into a ***cloaca***, which opens on the posterior, ventral end of the animal. Many rotifers have a protective shield, or ***lorica***, around the body. The rest are naked, or ***illoricate***. The lorica is variously ornamented, from plain to highly sculptured, are of

various sizes that barely cover half the body to those that completely envelope the corona, body, and foot, and are of various thicknesses, from thin, elastic plate-like sections to thick, rigid structures.

The mastax is peculiar to rotifers. It lies between the pharynx and the esophagus and contains a complex set of teeth or jaws, collectively called ***trophi***. The trophi are used to seize, tear, grind or macerate food. Their structure is diagnostic of certain species and feeding habits. Some trophi have jaws shaped like pistons and are used for sucking body fluids from plants and animals; some are sieve-like for filtering phytoplankton; others are adapted for protrusion through the mouth to capture and then tear their prey; still others have protrusible trophi but they are pincer-like for grabbing zooplankton (other rotifers, ciliates, flagellates, small crustaceans). With the exception of ***raptorial*** (that is, those that swim after prey) species, most rotifers are filter feeders, using the cilia to create currents and draw food toward the mouth.

There are two classes of rotifers, ***Monogononta*** (one ovary, Fig. 7.2a) and ***Digononta*** (two ovaries, Fig. 7.2b). The Digononta are represented by two orders, although one, the Seisonidia, is commensal, strictly marine and represented by a single genus, *Seison*. The second order, Bdelloidea, is mostly a freshwater group. Only about 10% of freshwater rotifers belong to this group. All bdelloids lack a lorica (shell), are ***parthenogenetic*** there being no males present and eggs developing without

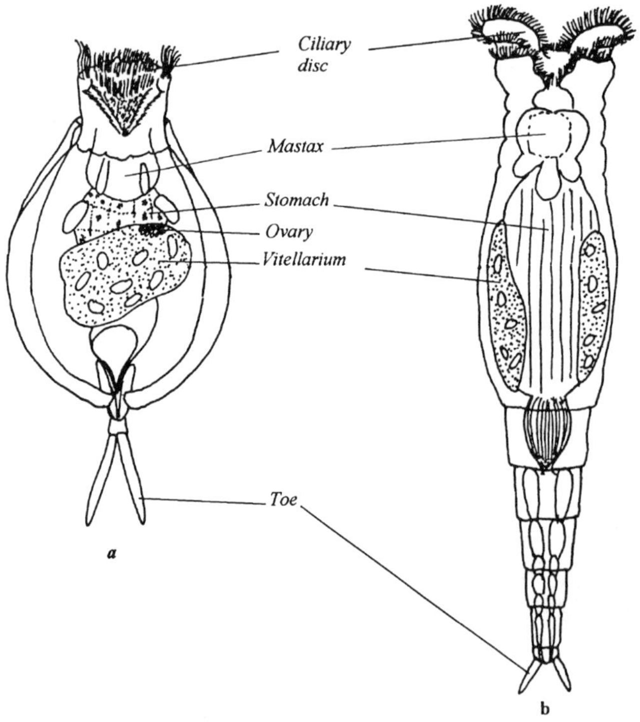

Figure 7.2. Common features of two classes of Rotifera, (a) Monogononta (*Epiphanes* shown) and (b) Digononta, (*Philodina* shown)

being fertilized, and the body is highly contractile. Bdelloidea, Greek for leech-like, alludes to the leech-like contractions of the body. The most common genera in the group are *Philodina* (Fig. 7.2a), with about 20 North American species, most occurring in sewage treatment processes, and *Rotaria*, an ***ovoviviparous*** (young are brooded by parents) genus with about 15 North American species, most being fairly large (about 1.5 mm long). *Rotaria* has eyes, usually two, on the end of its proboscis.

Only a few bdelloids are planktonic (e.g. *Philodina, Rotaria*), and nearly all are found in the littoral zone, as with rotifers in the class Monogononta. Most bdelloids occur in water or damp soils, sphagnum or mosses. With so little water present, it follows that most bdelloids tend to be sessile or are creeping in habit. However, nearly all types of trophi occur in the group and feeding habits range from filter feeding, that employs sucking and macerating type trophi, to carnivory, that employs pincer and tearing type trophi.

The class Monogonata is represented in both freshwater and marine environments. About 90% of the rotifers in fresh waters belong to this class. The class is divided into three orders, ***Collothecacea***, ***Flosculariacea***, and ***Ploima***, listed in order of increasing diversity. The orders are classified according to basic structure and modifications of the trophi present, and some morphological features. Members of
16☞ the Collothecaea and Flosculariacea are all filter feeders, but the mastax of the former is specialized for laceration and of the latter for grinding ingested plankton, periphyton and detritus. Both filter feeding and predatory behaviours are found among the Ploima, but the mastax of filter feeders is either adapted for sucking food into the mouth and then biting or nibbling the food prior to digestion, or for horizontal grinding of plankton and detritus. The mastax of predators are all protrusible and used for the capture and tearing of prey.

Growth and reproduction in some rotifers can be inhibited by some species of algae because of ***exudates*** (body secretions) that they produce. Some algae can be toxic to some species of rotifers. The feeding of rotifers on bacteria and detritus make them an important link to the microbial loop in the pelagic food web.

Of the Monogononta species, about 70% are littoral, 20% are pelagic and the rest are benthic. Both naked (illoricate) and loricate forms are common in the class. Nearly all are suspension feeders, with a few being raptorial predators. Among the suspension feeders are the illoricate genera, *Polyarthra* and *Synchaeta* (Fig. 7.3) and the loricate genera, *Brachionus, Keratella* and *Gastropus* (Fig. 7.4), all common in the plankton of most lakes.

The shape of the lorica is often diagnostic of a species (Fig. 7.4). In some species with spinous lorica, the spines and lorica exhibit ***cyclomorphosis***, with length of spine and width of lorica changing seasonally, especially in *Keratella* and *Brachionus*. In these genera, the production of spines is induced, in part at least, by low temperatures, starvation and chemicals or substances emitted by the predatory *Asplanchna*.

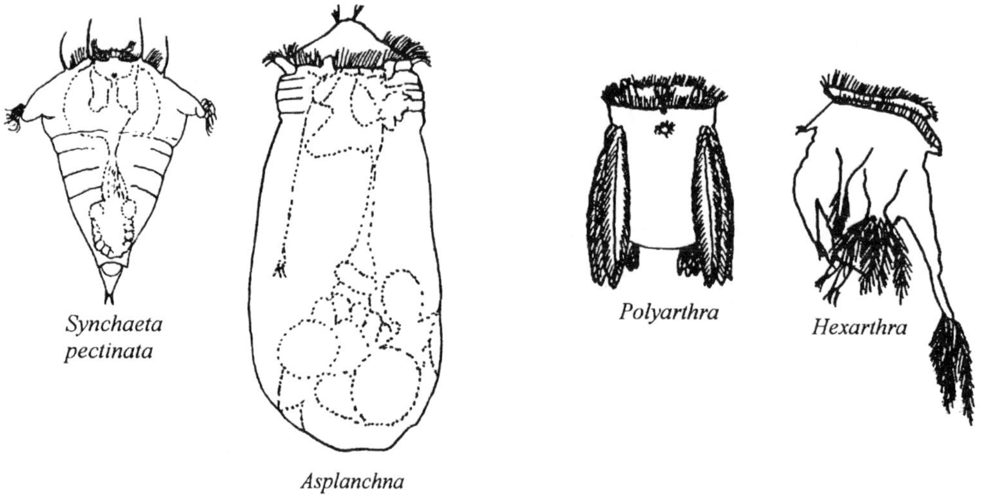

Figure 7.3. Four common illoricate rotifers.

Most species of monogonata are parthenogenetic (about 10% have males) and they undergo a unique type of annual life cycle (Fig. 7.5). In parthenogenesis, the
17 ☞ ovum develops without fertilization of the egg and only females are produced. It is a type of cloning where chromosome numbers are not halved, as in meiosis following sexual reproduction (i.e. $2N \rightarrow 2(N)$), but only meitotic cell divisions occur and the chromosome number is maintained (e.g. $2N \rightarrow 2N$). Parthenogenetic females, known as ***amictic females***, produce diploid (2N) eggs, called ***summer eggs***, that hatch into diploid females. In the fall, or when other adverse conditions arise, some (apparently about 1%) of the amictic females under go meiotic cell divisions to produce ***mictic* eggs** with a haploid (N) number of chromosomes. Some of these haploid eggs develop and grow into haploid males which undergo parthenogenesis to produce haploid sperm (Fig.7.5). A haploid sperm fertilizes the haploid egg produced by a mictic female to restore the diploid complement of chromosomes. The resulting zygote develops into a ***resting egg*** (sometimes called ***winter*** or ***fertilized egg***) that is heavy, thick-shelled and often sculptured. The resting eggs are very resistant to desiccation, temperature extremes, and pollution for long periods of time, even decades of time. In the spring, or when stressors are removed, the resting eggs hatch into amictic females and the life cycle repeats itself.

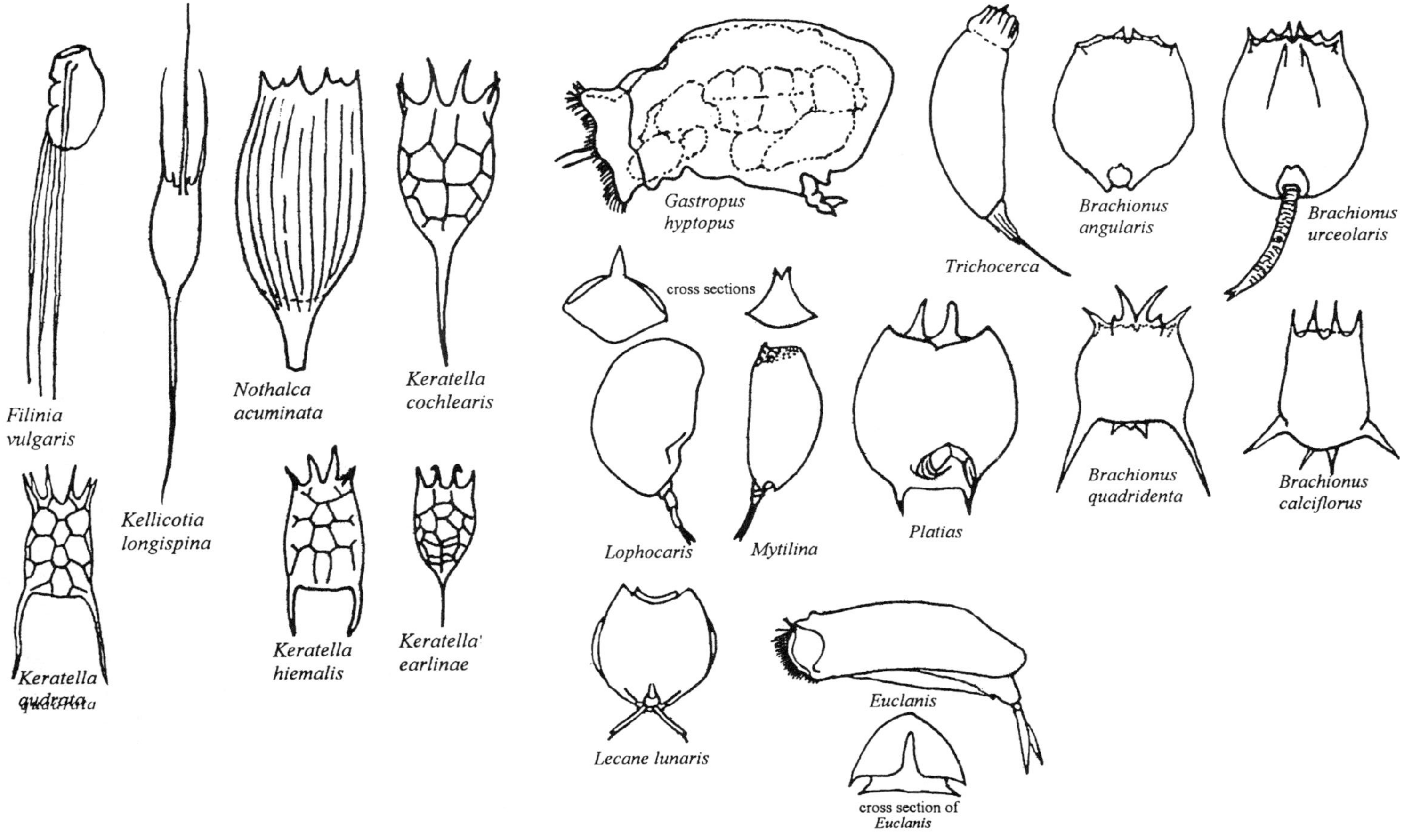

Figure 7.4. Some common loricate forms of rotifers, except *Filinia vulgaris* which has spines that make it appear loricate

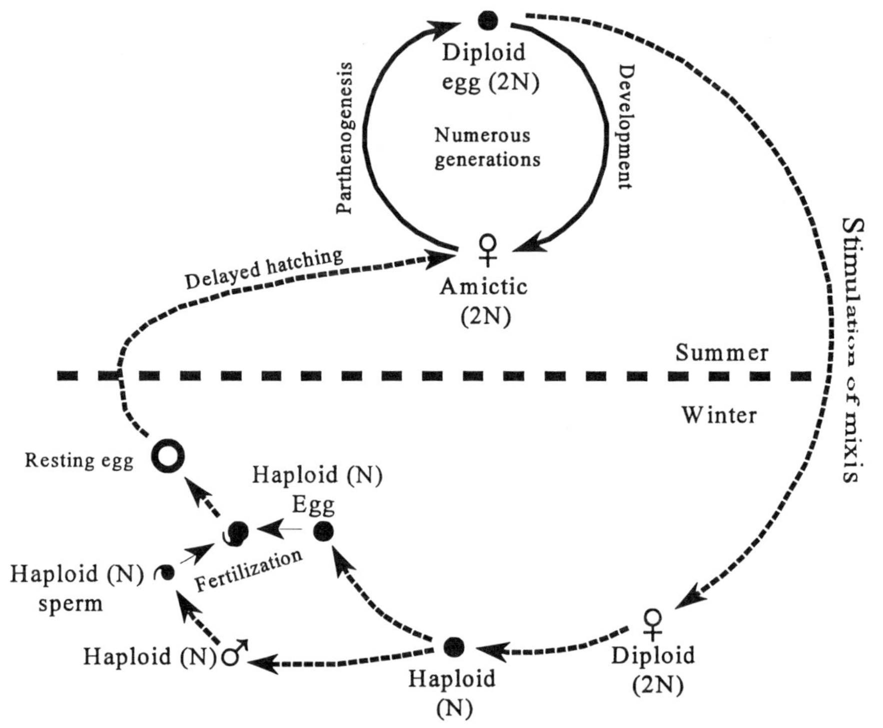

Figure 7.5. Typical annual life cycle of a Monogononta rotifer. Note that males are always haploid (N) and females are always diploid (2N).

Population Dynamics and Productivity of Rotifera

Up to 20 or 40 amictic generations may occur, whereas only one or two mictic generations occur in most Monogononta populations. Amictic females produce anywhere from about 4 eggs (e.g. *Brachionus*) to 45 eggs (e.g. *Epiphanes*) in their lifetime. Amictic eggs hatch within a day or two of ***partuition*** (release of young). Growth is rapid and life spans are short, usually less than 7-8 days (e.g. *Epiphanes, Brachionus, Lecane*). A few live for three weeks (e.g. *Keratella*), some for up to six weeks (e.g. many bdelloids).

Seasonal variations in densities vary from one to several peaks in abundance (some have even no peaks) per year. Some species of *Kellicotia* and *Chonochilus* have one peak, with a summer maximum; others, such as *Euchlanis*, have an autumn
18☞ maximum. Certain species of *Brachionus* and *Keratella* have two peaks, with spring and summer maxima. Most are variable in their cyclic patterns, being monocyclic one year and dicyclic or multicyclic in other years. Variations in seasonal patterns have been attributed to variations in temperatures, both mean summer temperature and rate of increase and decrease of temperatures in spring and autumn, light intensity, and competitive and predatory interactions. Light intensity directly affects temperature and algal growth, the primary food source of rotifers.

Ecology and Indicator Value of Rotifers

As stated earlier, bdelloids are most common in mosses and sphagnum; very few are limnetic (pelagic). *Philodina* and *Rotaria* are typical inhabitants of the littoral zone. However, about 75% of the monogonontoid species are limnetic. In fact, many species are found in either highly alkaline waters or very acidic waters. Species
19☞ confined to waters above pH 7.0 (and with high buffering capacities) are *Asplanchna, Brachionus, Filinia, Mytilina* and *Notholca*. Species confined to waters with pH less than 7.0 (usually less than 6.5, and low in buffering capacity) are *Cephalodella, Lecane, Monostyla* and *Trichocera*. Because many acidifying lakes are oligotrophic, the acid-loving species tend to be used as oligotrophic indicators as well (Table 7.6). Most species of rotifers occur in both alkaline and acidic water. Most of these are oligotrophic indicators (Table 7.6).

Crustaceans

Crustaceans, being ***arthropods***, have a chitinized exoskeleton, distinct body segments with a head, thorax and abdomen (the head and thorax are fused into a ***cephalothorax*** in most crustaceans), segmented appendages (usually more than five
20☞ pair in crustaceans) and antennae (two pair in crustaceans). The plankton is represented by two orders of crustaceans, ***Cladocera*** and ***Copepoda***. A third group of Crustacea, the ***Mysidacea***, or opossum shrimps, frequently migrate from the bottom sediments into the plankton. The mysids are discussed in Chapter 8.

Cladocerans

Cladocerans are a very diverse group characterized by a head and trunk, the abdomen being reduced to a postabdomen. The body, with 5 or 6 trunk appendages, is laterally compressed and covered with a folded shell, called a ***carapace***. The postabdomen, with its terminal claws, usually protrudes beyond the carapace when swimming but is withdrawn at rest. They range in size from about 0.2 mm (e.g. some *Bosmina*, *Chydorus*, *Alonella*) to 20 mm (e.g. *Bythotrephes*). Cladocerans are mainly a freshwater group, with only a few predaceous species in marine environments.

Most cladocerans are filter feeders, such as the omnipresent *Daphnia, Ceriodaphnia, Bosmina*, *Chydorus, Alona*, and *Simocephalus* (Fig. 7.6). The largest species, such as *Leptodora* (up to 18 mm long) and *Polyphemus* (up to 15 m long) (Fig. 7.7), and the European spiny water flea, *Bythotrephes* (up to 20 mm long, Figs. 7.7, 15.3), recently introduced to the Great Lakes, are all predaceous. The predaceous forms are rather bizzare and atypical of cladoceran morphology. All three lack the folded carapace and flattened appendages typical of most Cladocera, and all three have eyes that occupy about 90% of the head region. *Leptodora* has wing-like second antennae, a pair of posteriorly projecting, sac-like ***brood pouches*** and 6 pairs of subcylindrical legs modified for seizing prey that consists of rotifers and other planktonic crustaceans. *Polyphemus* also has subcylindrical legs (but only

Table 7.6. Trophic state indicator value of some common rotifers. The species are listed alphabetically.

ROTIFERAN ZOOPLANKTON		
Genus/species name	**Common name**	**Significance**
Asplanchna spp.	illoricate rotifer	Filter feeds on algae, microorganisms; most species are oligotrophic
Brachionus spp.	loricate rotifer	Nearly all species are most common in mesotrophic waters
Cephalodella sterea	loricate rotifer	Mesotrophic indicator
Cephalodella gibboides	loricate rotifer	Oligotrophic indicator
Conochilus spp.	colonial rotifer	Most species are most commo in oligotrophic waters
Epiphanes spp.	illoricate rotifer	Common in oligotrophic to slightly mesotrophic waters
Euchlanis spp.	loricate rotifer	Common in oligotrophic to slightly mesotrophic waters
Filinia spp.	loricate rotifer	Common in oligotrophic to slightly mesotrophic waters
Floscularia spp.	colonial rotifer	Common in oligotrophic to slightly mesotrophic waters
Hexarthra spp.	illoricate rotifer	Common in oligotrophic to slightly mesotrophic waters
Kellicotia longispina	loricate rotifer	Oligotrophic indicator
Keratella spp.	loricate rotifer	Filter feeds on algae, microorganisms; nearly all species are oligotrophic indicators
Lecane spp.	loricate rotifer	Nearly all species are most common in oligotrophic waters
Philodina spp.	bdelloid rotifer	Nearly all species are most common in oligotrophic waters
Polyarthra	illoricate rotifer	Nearly all species are most common in oligotrophic waters
Rotaria spp.	bdelloid rotifer	Most species are found in mesotrophic to eutrophic waters
Synchaeta	illoricate rotifer	Common in oligotrophic to slightly mesotrophic waters

4 pairs are present) modified for seizing prey (also rotifers and crustaceans), but its
21☞ second antennae are not as wing-like and the female broods her eggs in the upper part
of her abdomen, as in most cladocerans. *Bythotrephes* has sac-like brood pouches
directed anteriorly and stunted subcylindrical legs, and the two pair of antennae are

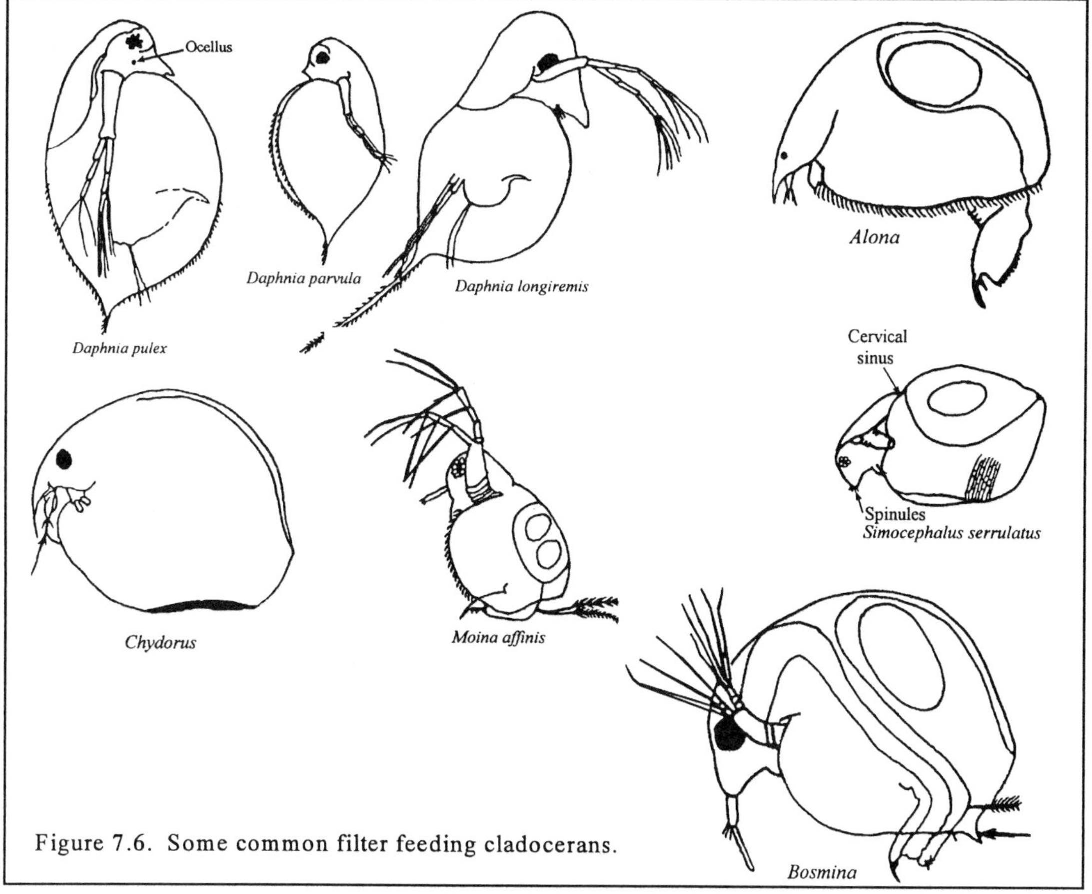

Figure 7.6. Some common filter feeding cladocerans.

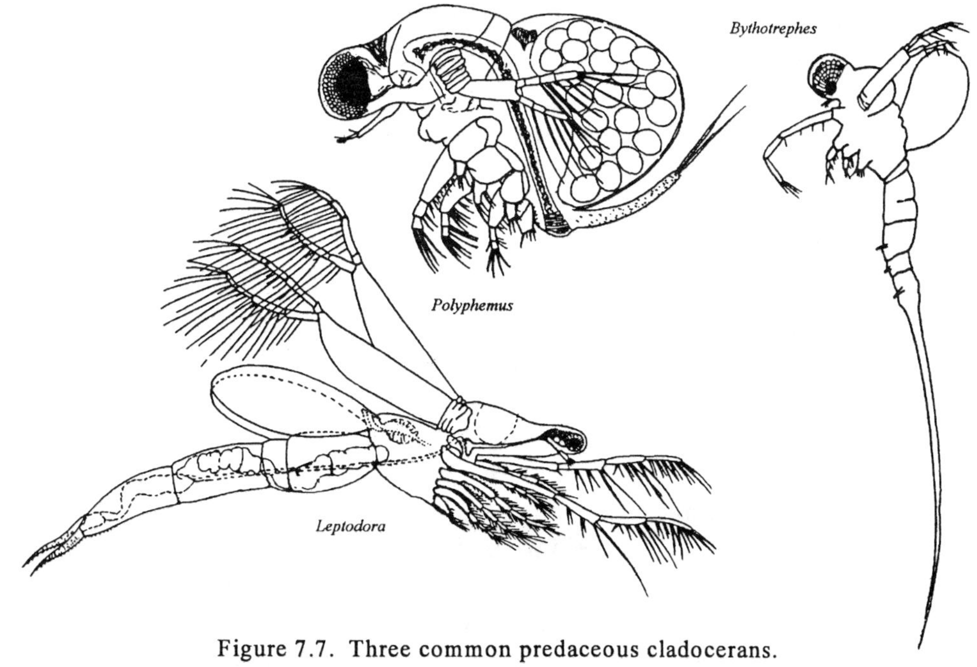

Figure 7.7. Three common predaceous cladocerans.

greatly reduced. Its most distinguishing feature is the elongated spinous tail that is at least as long as the cepahlothorax and abdomen together. *Polyphemus* is more ubiquitous than *Leptodora*, the former being found in northern lakes, ponds and marshes, the latter mostly in northern lakes; *Bythotrephes* was first observed in Lake Huron in 1984 and has since spread throughout the Great Lakes and is now present in many inland lakes. The species will have a much more cosmopolitan distribution than its two predaceous competitors. *Leptodora* seems to exhibit greater vertical diurnal (daily) migration than does *Polyphemus* or *Bythotrephes*. *Leptodora* is found in the lower water strata during the day but at night rises to prey on other zooplankton. It is also parthenogenic during the summer but males appear in the late fall and the fertilized eggs serve as the overwintering stage.

The remaining cladocerans are filter feeders, taking algae and suspended particles out of the water with their highly branched antennae. They filter feed by creating, with their antennae, a current of water towards the inside of the carapace. The food that is entrapped is then filtered out by the mouth appendages for ingestion.

Some species have unique features. *Holopedium* is readily identified by its huge carapace enclosed in a large gelatinous mantle (often transparent and almost invisible in life) and its fairly large size (1-2 mm across the gelatinous mantle), easily visible to the naked eye (Fig. 7.8). *Holopedium gibberum* is characteristic of acidifying lakes and waters low in calcium content.

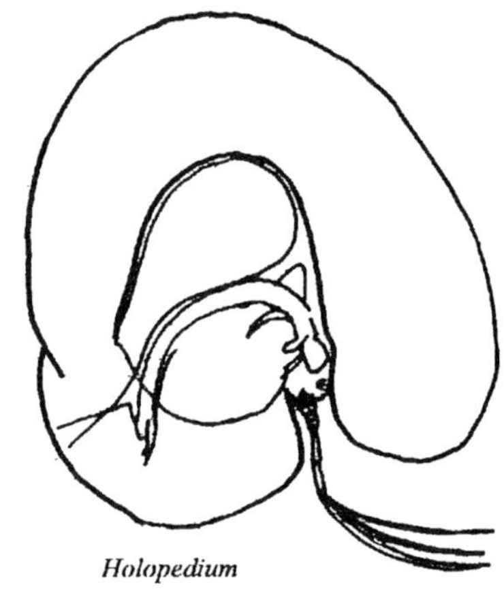

Figure 7.8. *Holopedium gibberum*, showing a large carapace enclosed in a gelatinous mantle.

Many cladocerans, especially females of *Daphnia pulex*, *D. retrocurva* and *D. galeata*, undergo ***cyclomorphosis***
22☞ (seasonal change in morphology) where the carapace on the head takes on a pointed helmet shape (Fig. 7.9). ***Cyclomorphism*** is generally believed to be triggered by temperature, although some scientists believe that accumulation of waste products, nutrition (e.g. availability of food), internal cycles and genetics also play a role in determining the onset of cyclomorphosis and the size of helmet formed. Regardless, the helmet grows longer with the summer so that the longest helmets are found during the warmest months, July and August.

Cladocerans are parthenogenetic and clone themselves during most of the year. When stressed, particularly when crowding occurs, parthenogenesis is interrupted by sexual reproduction (Fig. 7.10). During most of the year, females are diploid with a 2N complement of chromosomes. The parthogenetic females produce diploid eggs that
23☞ develop into more parthenogenetic females. The eggs are deposited into a brood

chamber, or pouch, a cavity located in the dorsal part of the carapace. Usually one clutch of eggs is released into the brood pouch during each adult ***instar***. The numbers of eggs produced per clutch varies according to the species and environmental conditions; usually between 10 and 20 laid, but the numbers vary between 2 and 40.

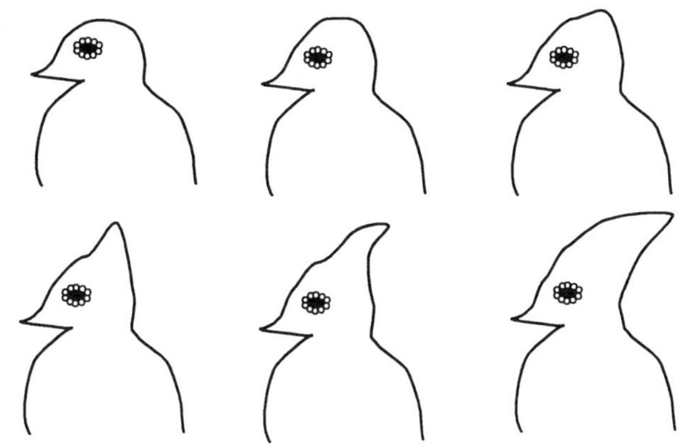

Figure 7.9. Cyclomorphosis in a cladoceran, showing variations in head form.

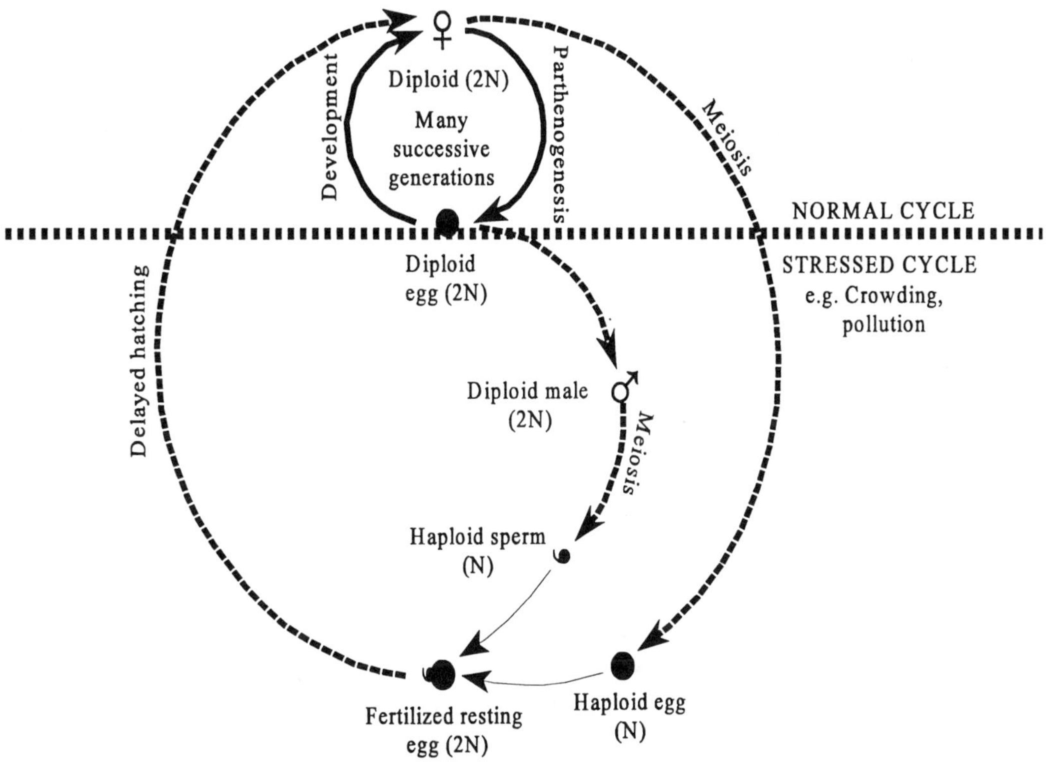

Figure 7.10. Typical annual life cycle of parthenogenetic cladocerans. Note: Both males and females are diploid (2N)

Development takes about two days within the pouch and is direct, with no larval stages. The eggs are released into the brood pouch as juveniles. The young are released from the brood pouch as first-instar ***juveniles*** by moving the postabdomen downward. The juveniles, which look like adults, have few moults but growth is greatest during this period. Between the last juvenile instar and the first adult instar is an ***adolescent*** stage. The animal sexually matures during the adolescent stage so that eggs are released into the brood pouch during the first adult instar.

The number of adult instars is highly variable, from 18-25 for *Daphnia pulex*, 10-19 for *D. rosea*, and 6-22 for *D. magna*. The average time between moults is about two days, but varies greatly according to temperature, food availability and environmental conditions. The duration of an instar varies inversely with the magnitude of most of these variables. For example, the duration of instar for *D. magna* is 26, 42, and 108 days at 28, 18 and 8 °C, respectively. The same applies for food availability, except for blooms of some species of blue-green algae which release metabolites that have been shown to inhibit growth and reproduction in some cladocerans.

Parthenogenesis continues until unfavourable conditions, such as crowding, decrease in food availability, drought, temperature extremes and/or pollution, induce sexual reproduction. Two sexual events occur: (i) the diploid female produces eggs that undergo meiotic divisions resulting in haploid eggs; (ii) some of the diploid eggs develop into males which produce haploid sperms (Fig. 7.10). The haploid sperms fertilize the haploid eggs producing zygotes that develop into resting eggs. However, these eggs are contained within a saddle-like thickening, called an ***ephippium***, of the pouch region. The ephippium is released with the next moult. The wall of the ephippium is thick and resistant to desiccation and environmental extremes for long periods of time. It is an effective dispersal stage, capable of intercontinental (between continents) and intracontinental (within a continent) dispersal (see Chapter 15 re: dispersal mechanisms of exotic species). Winds may carry the ephippium to other lakes or ponds and water currents may disperse the species throughout the watershed. Ephippia are even capable of surviving passage in intestines and can be transported great distances by waterfowl and then be regurgitated or defecated. Sexual reproduction and the resultant desiccation- resistant ephippium is clearly a prerequisite for cladoceran survival in temporary ponds.

Some species have sexual forms, but males tend to be smaller than females. The male form is generally short lived and is adapted for reproduction. For example, the antennules and appendages are modified for grasping the female during mating and are of little use as filter feeding organs.

Copepoda

Copepods are generally much larger than cladocerans, averaging about 2 mm in
24☞ length and occurring as large as 3.2 mm in length, although some are as small as 0.3 mm in length. They are ubiquitous, occurring in the plankton, benthos and littoral regions of lakes, as well as in the headwaters of some streams.

Copepods are divided into three suborders, ***Cyclopoida***, ***Calanoida***, and ***Harpacticoida***. The main features used to distinguish among the three suborders are body shape, first antenna length, the numbers and positions of egg sacs in females, the degree of constriction between the 4th and 5th segments, structure of the 5th leg, and habitat type. The most diagnostic morphological features, whether male or female are:

Cyclopoids: body is club-shaped
antennae is of moderate length, distinctly shorter than the body itself

Calanoids: body is of two uniform widths, an anterior (trunk) wide portion and a posterior (abdomen) narrow portion; antennae are very long, almost as long as the body, or longer

Harpacticoids: body is of uniform width throughout
antennae are very short, usually shorter than the first segment.

Other characteristics, including habitat and feeding habits are given in Table 7.7.

Table 7.7. Morphological and ecological characteristics of three suborders of freshwater copepods.

Criteria	Cyclopoida	Calanoida	Harpacticoida
Body shape	Club-shaped	Anterior much broader than posterior	Uniform width throughout
Antennal length	As long as or shorter than first segment	Nearly as long as entire body	Distinctly shorter than first segment
Egg sac number and location	Two, carried laterally	One, carried medially	One, carried medially
Degree of constriction and body location	Marked, between 4th and 5th legs	Marked, between 5th leg and genital segment	No constrictions
Structure of 5th leg	Vestigial	Similar to other legs	Vestigial
Habitat	Mainly littoral, some planktonic	Mostly planktonic, rarely littoral	Exclusively littoral, on macrophytes and sediments
Feeding habits	Raptorial predator or raptorial grazer	Filter feeders of algae	Filter feeders of algae or grazers of epiphytes

The mouth parts of grazing copepods, especially harpacticoids, are adapted for raking and scraping algae off the bottom or attached to surfaces of macrophytes and other solid substrates. A few harpacticoid species are parasitic. The planktonic forms,

including most calanoids, filter their food from the water by first creating water currents toward the mouth with their first and second antennae and then sorting the
25☞ food with their mouth parts. Many particles are rejected in the process, leading to the ***"leaky sieve" concept***. Some of this "***spillage***" contributes to the DOC pool in the microbial loop. Common filter feeding calanoids include *Diaptomus* and *Limnocalanus*.

The mouth parts of cyclopoids are adapted for seizing and biting zooplankton (including their own young), unicellular phytoplankton and organic debris. *Cyclops* and *Megacyclops* are raptorial and predaceous on animals and *Eucyclops* is raptorial on phytoplankton. Some species (especially diaptomids) have a reddish hue due to the accumulation of oil droplets as food reserves.

Unlike the parthenogenetic cladocerans, sexual reproduction is the rule in copepods. Males clasp the female with their modified 5th legs and transfer sperm packets, called ***spermatophores***, to the female using the 5th legs. The female stores spermatophores in a sac-like ***seminal receptacle*** and then fertilizes the eggs as they leave her reproductive tract. The fertilized eggs (embryos) are carried below the abdomen in either one or two ovisacs, depending on the suborder (see Table 7.7). The ovisacs contain from 5 to 40 eggs, depending on the season. The incubation period of the eggs varies from a few hours (e.g. 12) to a few days (e.g. 5-6). Shortly after the first brood is hatched, a second one is incubated. As many as thirteen ovisacs (or pairs of ovisacs in the case of cyclopoids) may be brooded from one spermatophore. Some species of calanoids and harpacticoids produce resting eggs in the fall, the eggs having thicker walls than the summer eggs.

The copepod egg hatches into a minute ***nauplius*** larva with three pair of appendages destined to become the first and second antenna and the mandibles. The
26☞ naulpius larva moults into a nauplius II stage, adding another pair of appendages during the moult. There are six nauplius stages, the nauplius growing in size and numbers of appendages with each moult (Fig. 7.11). The nauplius VI stage moults into a ***copepodid*** I stage. There are five to seven copepodid stages, the last one moulting into the adult (Fig. 7.11). Hence, a typical life cycle involves one egg, six nauplius stages, five to seven copepodid stages and an adult. The entire cycle, from egg to egg, usually lasts for five to six months, although some cyclopoids can complete the cycle in about one week under ideal conditions.

Population Dynamics and Productivity of Crustacean Zooplankton

Population densities vary seasonally due to variations in ***natality rates*** (numbers of eggs that survive to hatching during the year) and ***mortality rates*** (numbers of eggs, larvae and adults that die during the year). The amount of variation over the year
27☞ depends upon the availability and quality of food, predation pressure and physical and chemical factors of the water. Using short time intervals (t), scientists usually measure the daily ***instantaneous rate of population growth*** (numbers, N, per m^3), r, as:

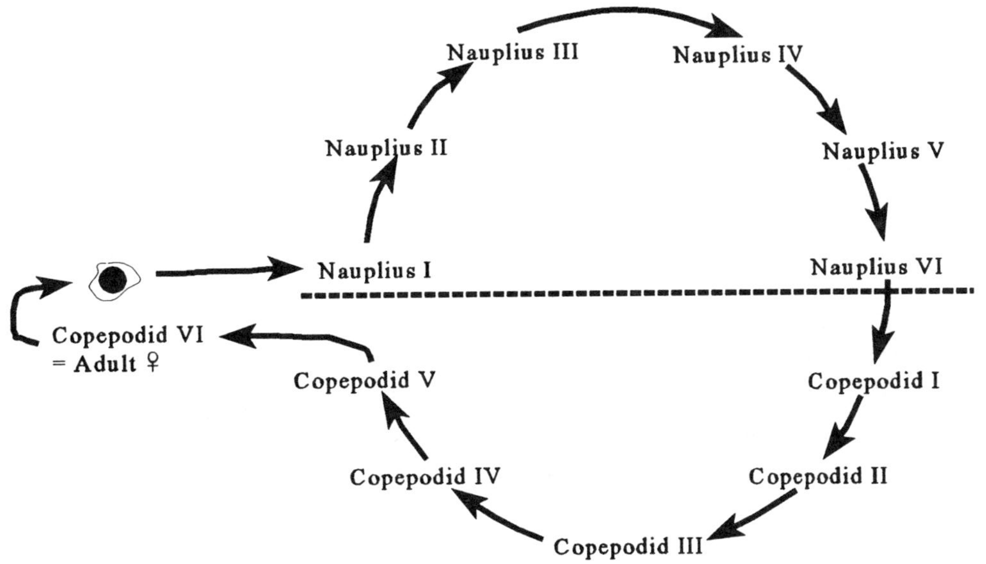

Figure 7.11. Typical life cycle of a copepod.

where the subscripts indicate numbers and times at time t_0 and time t_1, and ln is the natural logarithm of the N values. Use of the natural logarithm assumes that the population is growing exponentially ("ʃ"-shape growth), which is generally true, but not always. The value of *r* is estimated from weekly population estimates, *r* being expressed on a daily basis (i.e. divide by 7 if measurements are taken every week).

Birth rate, *b*, is estimated from:

$$b = \ln (E/D + 1)$$

where E = egg ratio = number of eggs divided by number of females, and D = egg development time in days. The value of E is easily determined by counting the numbers of eggs carried by each female. The value of D can only be determined experimentally by isolating 50 - 100 females with eggs and determining the time it takes for the eggs to hatch at a temperature similar to that at the time and depth of capture in the lake.

The ***death rate***, *d*, is estimated by subtraction of the instantaneous growth rate from the instantaneous birth rate, or:

$$d = b - r$$

An analysis of death rate provides information on predation pressures of higher trophic levels (e.g. fish) and the effects of environmental stressors on population densities. The same approach can be used for rotifers.

Seasonal population dynamics varies greatly among groups and species of crustaceans. Vertical migrations of some species, influence of water currents, especially Langmuir currents, and experimental error can introduce large variations into mean estimates of instantaneous growth, birth rates and death rates. After these are accounted
28☞ for, it is quite clear that monocyclic, dicyclic, tricyclic, and even multicyclic (i.e. single, double, triple and multiple population peaks, respectively) patterns are present in the crustacean zooplankton. Some perennial species exhibit peak densities only during colder periods in the spring, and in the metalimnion and hypolimnion during summer stratification. Monocyclic patterns may be seen in the spring, summer or autumn, depending on the species and the year's weather patterns. Most temporary pond populations of crustaceans exhibit monocyclic patterns. Dicyclic patterns typically have one population peak in the spring and another in the autumn (e.g. *Daphnia galeata mendotae*). In many cases, a species with a monocyclic pattern will co-exist with one that has a dicyclic pattern. For example, in some lakes *Daphnia retrocurva* is known to peak in numbers while *D. g. mendotae* declines in numbers during the summer, apparently to avoid competition. Population trends of predaceous species, such as *Leptodora kindtii*, usually correlate well to the population declines of its predators, mainly *Daphnia, Bosmina, Ceriodaphnia, Cyclops* and *Diaptomus.* Similar population trends have been reported for *Mesocyclops* which preys on nauplii of calanoids.

The population dynamics of many copepods is a bit more unusual. The annual cycle of many cyclopoid species is often interrupted by a ***diapause*** (rest period) that continues for one to several months, often for unknown reasons. Sometimes a period of ***aestivation*** is required. For example, nauplii may develop during winter and spring and copepodids during the early summer. Late copepodid stages aestivate in the sediments for a couple of months before emerging and metamorphosing into the adult in the autumn. Many species exhibit this type of diapause behaviour but the periods of growth and aestivation differ to avoid competition. At the other extreme are calanoids with long reproductive periods producing annual population density curves with numerous peaks that are difficult to separate into distinct generations, or ***cohorts***.

Ecology and Indicator Value of Crustacean Zooplankton

Cladocerans are ubiquitous and are most abundant in the limnetic zones of mesotrophic and oligotrophic ponds and lakes. They are absent or rare in fast running waters and in grossly enriched waters. The most cosmopolitan species are *Daphnia pulex*
29☞ and *D. magna*, occurring in temporary and permanent ponds, lakes and large rivers. The abundance and distribution of some species is limited by water hardness. As with most cosmopolitan species, they are among the most tolerant of enrichment and appear in large numbers in eutrophic lakes and large rivers. Other common forms in limnetic waters of most lakes are *D. galeata, Bosmina, Diaphanosoma* and *Ceriodaphnia*. Truly limnetic species, such as these, rarely occur in the littoral zone of lakes because many species of rooted macrophytes have a repellent effect. One exception is *Chydorus*, but it is a benthic

form, rarely occurring in the open water. It is uncommon to find more than one species of *Daphnia* in the limnetic zone; two or three species may be present, but only one usually dominates.

Many Cladocera species are characteristic of cold, oligotrophic waters, such as *Daphnia longiremis*, *Holopedium gibberum* and *Latona* spp (Table 7.8). All are ***cold stenotherms*** (prefer cold waters that vary little in temperature) and are more-or-less restricted to oligotrophic lakes in more northern latitudes. In fact, *D. longiremis* moves into the hypolimnion when the lake begins to stratify. *Holopedium* species are unique in that the animal is enveloped in a large gelatinous mantle. *Holopedium gibberum* is especially different in that it swims with its ventral side up and is limited in its distribution by pH, occurring only in poorly buffered waters or waters low in calcium levels.

Although copepods are very abundant in most lakes, there is usually a low diversity, with only one or two species of each of cyclopoids and calanoids. If more than one species of each is present, only one usually dominates. The most cosmopolitan cyclopoids are *Mesocyclops, Diacyclops* and *Acanthocyclops*, and of the calanoids, *Diaptomus, Lepidodiaptomus, Limnocalanus, Senecella* and *Epischura.* There are species within each genus that are characteristic of either oligotrophic lakes or mesotrophic to eutrophic waters. *Lepidodiaptomus minutus, Lepidodiaptomus siclis, Limnocalanus macrurus, Senecella calanoides* and *Epischura lacustris* are all confined to cold, oligotrophic waters (Table 7.8). *Acanthocyclops crassicaudis* is found only in warm, temporary ponds. Harpacticoids, such as *Harpacticus*, are exclusively littoral and are rarely found in the plankton.

Roles of Planktonic Crustaceans as Parasites and Hosts of Parasites

Some copepods are parasitic on fish and have no direct impact on humans. One
30☞ of the more common parasites in this class is the cyclopoid, *Ergasilus versicolor*. Only the adult females are parasitic, the immature females and all stages of males being free living and planktonic. The second antennae of the adult female are greatly enlarged with claws for clinging to the gills of fish. It obtains nourishment from the gill tissues and blood but usually does not kill the fish. Occasionally, however, they can cause mortality of fish in hatchery ponds when fish are overcrowded.

Other copepod parasites are more bizzare. *Lepeophtheirus salmonis*, the sea louse, is a marine species that attaches to the skin near the anus of migrating salmon.
31☞ It lives for only about a week in freshwater. The copepodid of species of *Lernaea* attaches to a temporary fish host until it becomes an adult and then seeks its final fish host. *Salmincola* is parasitic on Salmonidae (salmon and trout) and Coregonidae (whitefish). *Argulus* attaches itself to the branchial chamber or the surface of many species of fish.

Of more concern to humans are the copepods that serve as intermediate hosts to many species of flukes that parasitize fish, amphibians and waterfowl, tapeworms that parasitize fish, waterfowl and aquatic mammals, and a few nematodes of fish and birds. The flukes (known as digean trematodes) and tapeworms (known as cestodes) both have planktonic larval stages, but only the tapeworms use copepods as intermediate hosts.

Table 7.8. Some ecological characteristics and indicator value of planktonic crustaceans. The species are listed alphabetically within each of the suborders (SO).

CRUSTACEAN ZOOPLANKTON			
Genus/species name	**Common name**	**Taxon group**	**Significance**
Bythotrephes cederstroemi	spiny water flea	O: Cladocera	Nuisance exotic; predator of zooplankton; larval fish choke on spiny tail
Daphnia longiremis	water flea	O: Cladocera	Oligotrophic indicator
Daphnia pulex	water flea	O: Cladocera	Filter feeder of algae
Eubosmina (Bosmina) coregoni	water flea	O: Cladocera	Exotic species; filter feeder
Holopedium gibberum	gelatinous water flea	O: Cladocera	Acidifying lake and oligotrophic indicator
Latona setifera	water flea	O: Cladocera	Common in oligotrophic to mildly mesotrophic lakes
Leptodora kindtii	predaceous water flea	O: Cladocera	Predator of rotifers & crustacean plankton; common in oligotrophic lakes
Polyphemus	predaceous water flea	O: Cladocera	Predator of rotifers & crustacean plankton; common in oligotrophic lakes
Acanthocyclops robutus	cyclopoid copepod	O: Copepoda SO: Cyclopoida	Limnetic; acidifying, oligotrophic lake species.
Acanthocyclops vernalis	cyclopoid copepod	SO: Cyclopoida	Limnetic; hardwater species in mesotrophic waters
Eucyclops agilis	cyclopoid copepod	SO: Cyclopoida	Mostly littoral; grazers on attached algae or predators of crustacean plankton
Mesocyclops edax	cyclopoid copepod	SO: Cyclopoida	Limnetic; cosmopolitan in distribution
Eurytemora affinis	calanoid copepod	SO: Calanoida	Exotic species; all are pelagic filter feeders
Lepidodiaptomus minutus, L. siclis	calanoid copepod	SO: Calanoida	Cold water, oligotrophic indicator
Limnocalanus macrurus	calanoid copepod	SO: Calanoida	Cold water, oligotrophic indicator
Senecella calanoides	calanoid copepod	SO: Calanoida	Cold water, oligotrophic indicator
Mysis relicta	opossum shrimp	O: Mysidacea	A cold water, oligotrophic benthic species, migrates into surface water at dusk, down at dawn

The flukes have larval stages that are planktonic for a short time. The more hazardous species in each group are briefly discussed below. First, it is important to understand the life cycles of tapeworms and flukes.

The tapeworm has two larval stages, the ***procercoid*** and the ***plerocercoid***, but
32 ☞ usually only the former needs a copepod to develop. The life cycle begins with the release of eggs to the water with faeces. The eggs are fertilized and develop into embryos (called ***embryonated eggs***) that escape from there shells as ***coracidia*** (singular = ***coracidium***). The coracidium remains planktonic until eaten by a copepod, usually a cyclopoid. The coracidium sheds its cilia and transforms into a procercoid larva within the copepod. The procercoid develops into a plerocercoid when eaten by a second intermediate host. There are many variations in the cycle to this point, with the coracidium developing into an ***onchosphere*** (embryo with hooks to attach to body wall) in the copepod and then into the procercoid in the second intermediate host and the plerocercoid in the final host; in others the plerocercoid stage is omitted. The point is, there are two stages that are in the plankton, one as a ciliated coracidium and one as a larval stage within copepods.

Two of the more common tapeworms which use copepods as intermediate hosts are *Diphylobothrium latum* and *Proteocephalus amblopltitis*. The more dangerous of the two is *Diphylobothrium latum,* known as the broad fish tapeworm of humans. The
33 ☞ first intermediate hosts are the copepods, *Diaptomus* and *Cyclops*. Fish serve as second intermediate hosts. Humans are infected by eating poorly cooked fish that carry the larval plerocercoid larvae. Once consumed, the larval tapeworms can migrate from the stomach of humans to the muscle tissue where they can encyst. *Proteocephalus amblopltitis* (fish tapeworm) also uses copepods (*Cyclops* and *Eucyclops*) as first intermediate hosts and young fish as second intermediate hosts, the final hosts being smallmouth bass, rock bass, yellow perch, and bowfins. The worm invades gonadal tissue and completely eliminates reproduction by female fish.

The life cycle of flukes is very different from that of tapeworms. The ***definitive host***, or final host, is a fish, amphibian or bird. Infected hosts release embryonated (fertilized) eggs that release a ciliated ***miracidium***. This brief planktonic stage swims to an intermediate host, usually a snail, sometimes a bivalve, and enters the intermediate host through the skin to begin the asexual phase of reproduction. The miracidium metamorphoses into either a ***mother sporocyst*** or a ***redia***, depending on the species. The mother sporocyst is a sac-like structure that asexually produces ***daughter sporocysts*** or ***rediae***, but never both in the same species of fluke. There may be more than one generation of sporocyst or rediae. The daughter sporocyst or daughter rediae asexually produce ***cercariae*** which escape from the mollusc and swim until it finds either a second intermediate host, again usually a snail, or its final host and enters the host's body through the skin. The cercariae have a tail, usually forked and are good swimmers but remain planktonic only for a short time if they do not find a host. If they find a host, they penetrate the skin, lose their tail in the process, and develop into an adult fluke. If they do not find a host they can either encyst as ***metacercariae***, usually on macrophytes, or enter a second intermediate and encyst in the intermediate host.

The metacercariae enter the final host when it consumes either the infected vegetation or second intermediate host where it then develops into an adult. The adults are hermaphroditic and sexually produce eggs which are released with the faeces to begin the cycle again.

The planktonic stages are the miracidia and the cercariae. Table 7.9 lists
35☞ several trematodes that use aquatic animals as first or second intermediate hosts for the planktonic stages. *Trichobilharzia cameroni* and several other species within the genus causes dermatitis in humans. The cercariae cause "swimmers itch" to people who come in contact with them. The cercariae penetrate the human skin and lose their tails, but the cercariae can not enter the skin and develop into adults. *Echinostomum revolutum* is extremely common and cosmopolitan and the adult is known to infect humans. Very few flukes cause death of the final hosts. In fact, migrating birds which serve as definitive hosts can be effective dispersal agents of flukes.

> So, do you still want to drink unboiled lake water?

Table 7.9. Common flukes with planktonic larval stages in freshwater and their first (e.g. for miracidia, sporocysts) and second (e.g. cercariae) intermediate and final hosts. The fluke species are listed alphabetically.

Fluke Species	First intermediate hosts	Second intermediate hosts	Final hosts
Bucephalus papillosus	Bivalves	Metacercariae encyst on skin of sunfish	Intestine of walleye
Bucephalus elegans	Bivalves	Metacercariae encyst on skin of sunfish	Caecal pouch of rock bass
Clinostomum complanatum	Snails	Metacercariae encyst in muscles of perch, bass, sunfish	Mouth of herons
Cotylurus flabelliformes	Snails	Snails	Small intestine of ducks
Diplostomum baeri eucaliae	Snails	Blood vessels and brain cavity of sticklebacks	Small intestine of ducks
Echinostoma revolutum	Snails or bivalves	Tadpoles	Several species of birds and mammals, including humans
Spirorchis parvus	Snails	None	Brain, spinal cord, gut wall, spleen, lungs, heart, arteries of painted turtles
Trichobilharzia cameroni	Snails	None	Heart and lungs of ducks
Uvulifer amblopitis	Snails	Metacercarial cysts on skin of bass, rock bass, perch, sunfish	Intestines of belted kingfishers

Other Zooplankton Groups

36 ***Other Crustaceans - Mysidacea (Opossum shrimps)***

The mysids are mainly a benthic group, but they exhibit extensive daily vertical movements in lakes. *Mysis relicta* is usually confined to the benthic region, no more than about one meter above the sediments, during the day. They are adapted to the cold, 4 °C, well-oxygenated water of the hypolimnion of oligotrophic lakes and cannot survive indefinitely in temperatures exceeding 14 °C. At dusk, most of the older and larger individuals (> 12 mm) move into the surface waters only for a few hours to prey on other crustaceans, and then they return to their preferred temperatures of the profundal zone.

Other benthic forms that may appear in the plankton are *Hyalella azteca* and *Gammarus lacustris*. Both species are known to develop a planktonic (often referred to as ***nektoplankton***) mode in many arctic lakes that lack fish.

Insects

There are no truly planktonic insects in freshwater. Those that are found in the
37 water column, like mosquito larvae, are at the air-water interface and belong to the ***neuston*** community of organisms discussed in Chapter 9.

Some insects, however, frequently migrate from the bottom (benthos) into the water column on a diel (daily) cycle. The most common genus in this group is *Chaoborus*, the phantom midge. It contains two pairs of air sacs that are used to control the buoyancy of the animal. The animal typically has four larval instars. In some species the first three instars migrate into the plankton at night; the younger the instar the higher the animal rises in the water column. The last (fourth) instar is strictly benthic. The taxonomy of phantom midges and their role in the plankton is summarized in Table 7.10.

Bivalve Larvae

The only group of native clams that has a planktonic larval stage is the Unionidae (pearly mussels). However, the larvae, called ***glochidia***, are parasitic on fish and are planktonic only for a few minutes, or until the larvae attach to a fish host. The probability of finding a glochidium in a water sample is extremely remote.
38 However, with the introduction of the zebra mussel (*Dreissena polymorpha*) and the quagga mussel *(Dreissena bugensis*), many North American lakes now have a bivalve planktonic larval stage that can be found in the water throughout most of the summer and autumn months, and occasionally in the winter months. The larval life cycle of *Dreissena* usually takes about four weeks to complete but there may be several generations in the water. Three planktonic ***stages*** are recognized (Fig. 7.12, 7.13): a ***veliger stage***; a ***post-veliger stage***; a ***settling stage***. Each stage can be identified by the presence of certain structures. ***Veligers*** are identifiable by the presence of a tuft of cilia called the ***apical tuft.*** A "D"-shaped, bivalved shell is secreted shortly after the ***trochophore*** emerges from the developing egg.

Table 7.10. Other taxa commonly seen in the zooplankton - insects and bivalve molluscs.

INSECT AND MOLLUSCAN ZOOPLANKTON			
Genus/species name	**Common name**	**Taxon group**	**Significance**
Chaoborus punctipenis	phantom midge	P: Arthropoda, C: Insecta O: Diptera, F: Chaoboridae	Benthic, early instars migrate into plankton; predaceous on zooplankton
Corbicula fluminea	Asian clam	P: Mollusca, C: Bivalvia F: Corbiculidae	Larval stage is planktonic, for 3-5 d; only in water that does not freeze every year
Dreissena polymorpha, Dreissena bugensis	Zebra mussel Quagga mussel	F: Dreissenidae	Only larval stages are planktonic for 20-30 d; zebra mussel needs 12-15 °C to reproduce, quagga needs 7-10 °C

Development begins with the external fertilization of gametes. Fertilized eggs (zygotes) pass through an embryonic stage where the development is nourished by materials stored in the egg rather than by direct feeding. The embryo develops into the free-swimming trochophore larvae in six to twenty hours, depending on the ambient temperature. With the rapid development of the ***velum***, a ciliated swimming organelle, the trochophore larva becomes a veliger larva. The veliger stage includes any larva that possesses a velum. Several days after fertilization the veliger secretes the first larval shell (***Prodissoconch I***) from its shell gland. Prodissoconch I shells are unornamented and D-shaped in profile, hence the terms "***D-form***" or "***straight-hinged veliger***". The second larval shell (***Prodissoconch II***) is secreted later by the mantle tissue and can be ornamented in various ways. It is round or clam-like in profile with the typical ***umbonal*** (beak) region near the hinges. Because of the obvious umbone, the larval stage is often referred to as the "***umbonal veliger stage***". It is also known as a ***veliconcha***. The veliconcha (Fig. 7.13) is the last veliger stage that is free-swimming and is typically found in the plankton. As the veliconcha grows, the velum slowly develops into the siphons, the foot lengthens, and the blood and some organ systems begin development. The acquisition of a foot leads to a change in behaviour and the larva is now known as a ***pediveliger***. With the loss of the velum the larva enters the settling stage.

The pediveliger marks the beginning of the post-veliger stage. The foot can be used for swimming (by means of cilia on the base of the foot) near the bottom as well as for crawling on surfaces. This is typical behaviour of the post-veliger stage. Once a pediveliger encounters an appropriate surface, it secretes a ***byssal thread*** (a

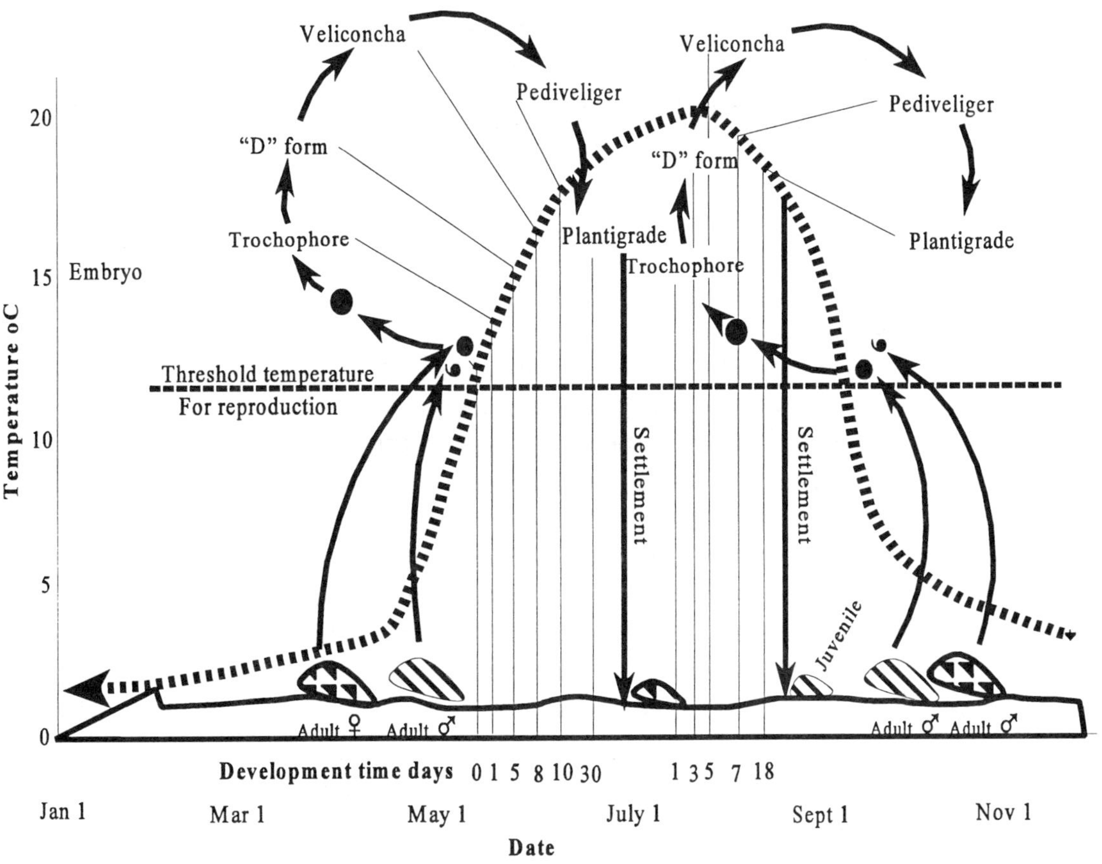

Figure 7.12. Bicyclic planktonic larval pattern of a dreissenid in relation to the annual temperature cycle of a temperate lake. Note that development times of different larval stages in the second cycle is much shorter than the first because of warmer temepratures.

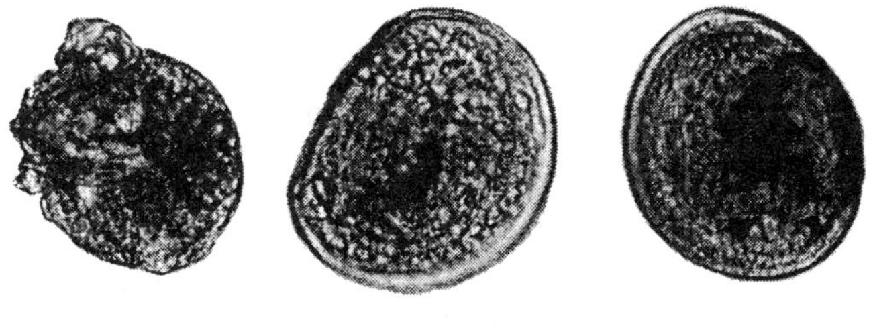

Figure 7.13. Planktonic larval forms of the zebra and quagga mussel (zebra mussel shown).

proteinaceous thread used to attach the bivalve to the substrate) and undergoes metamorphosis to become a ***plantigrade larva***, the primary settling stage. Without an appropriate surface to settle upon, such as filamentous algae, they, like marine bivalves, may delay byssal attachment and metamorphosis. Metamorphosis, which follows primary attachment, transforms the larval bivalve into an "adult" bivalve. The principal changes during metamorphosis are associated with the loss of the velum, the development of the gills, and the secretion of the dissoconch, or adult, shell. Growth of the foot and mouth are facilitated by the new growth axis of the dissoconch shell, which results in the familiar triangular mussel shape. The completion of these developments transforms the plantigrade larva into a ***juvenile***, which, with further growth and sexually maturity, becomes an adult. With the spawning of adults the life cycle begins again.

The life cycle of the Asian clam, *Corbicula fluminea* is similar to that of *D. polymorpha* and *D. bugensis* but the larval stages are planktonic for only a few days
39☞ (Table 7.10). Most of the development occurs in the gills of the parent and only pediveligers are released. After about 100 hrs the pediveligers lose the velum, the larvae begin to settle out as straight-hinged forms and take up a benthic existence. A single byssal thread is secreted during the umbonal stage but it is soon lost. The byssal gland becomes non-functional by the time the clam is a young adult.

Consequences of Settlement by Biofouling Bivalves

Since there is an inverse relationship between growth/development rate of veligers and temperature, the warmer waters of the southern states (relative to the Great
40☞ Lakes) greatly shortens the development time (i.e. increases the rate of development) required for individuals to reach maturity. For example, peak densities of veliger larvae in the Great Lakes occur when water temperatures are between 18 and 22 °C, usually late June to late August, and produces one or two settlement events, each one lasting two to three weeks. Such water temperatures occur in some of the southern states (e.g. Tennessee) from early May to mid-October. Hence, there is usually either a larger number of settlement events or the settlement events will be of much longer duration, if not continuous, in warmer waters. Moreover, recent studies at Ontario Hydro (now called Ontario Power Generation) have shown that there is significant ***translocation*** (detachment and relocation of mussels) movement of small adults (< 3-5 mm) during the winter and spring. It appears that not only plantigrade forms readily translocate but juveniles and adults readily move back into the water column after settling.

Cottagers, sports fishermen and swimmers will all experience biofouling problems with the zebra and quagga mussel. The final chapter (15) describes some of the consequences of biofouling reported to date by cottagers, sports fishermen, beach walkers and swimmers, and ways to remedy the problems. Briefly, the biofouling problems include plugging of foot valves that supply cottages with lake water, overheating of outboard motors with the cooling water intake plugged by zebra mussels, beach walkers cutting their feet on the sharp shells of dreissenids, swimmers cutting their hands on the shells attached to rocks and dock components, and degraded aesthetic value of boats and docks by numerous shells attached to them or by the byssal threads left behind.

Cnidaria

Jellyfish are primarily marine. *Craspedacusta sowerbyi* (Fig. 7.14) is the only freshwater jellyfish in North America. It may be considered as nekton but it is so small (2 cm diam) that its movements are usually determined by water currents. They are born from polyps, usually 2 to 4, sometimes up to 12. Some of the polyps produce bell-like, tentacle-fringed medusae, which beak off and swim away. Reproduction occurs in the summer, from July to September or October.

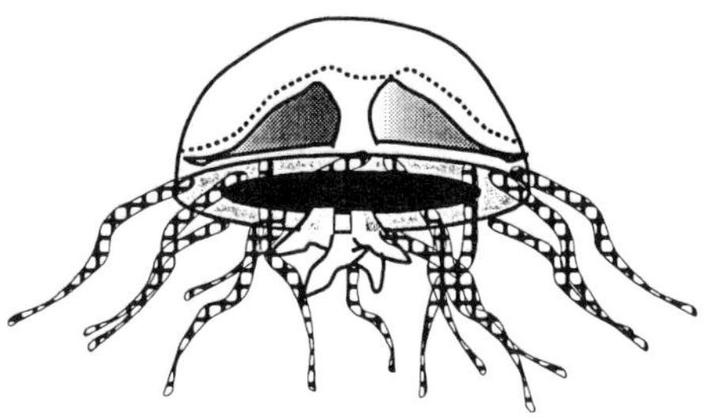

Figure 7.14. *Craspedacusta sowerbyi.*

Craspedacusta sowerbyi is an exotic species, introduced from Europe. They are most common in small lakes, large ponds, large rivers, reservoirs and even old quarries.

Effects of Zooplankton on Nutrient Recycling

Phytoplankton directly affect the levels of nutrients in lakes; levels of phosphorous decline as the biomass of algae increases in the epilimnion. However,
41☞ grazing intensity by zooplankton increases as phytoplankton biomass increases, resulting in lower uptake rates of nutrients by the phytoplankton. The end result is an increase in the levels of nutrients. In addition, excretion by zooplankton contributes to the nutrient base in the lakes. Algal flagellates, ciliates and bacteria recycle the dissolved organic carbon through the microbial loop.

The bottom up effect of zooplankton on primary production has been clearly demonstrated in several lakes. The magnitude of the bottom up effect is limited by the predation of zooplankton by fish and predaceous zooplankters. Planktivorous fish, such as perch *(Perca flavescens)* and alewife (*Alosa pseudoharengus*), have closely spaced "gill rakers" that effectively sieve zooplankton from the water. Other planktivores, like whitefish (*Coregonus clupeiformes*), gulp large amounts of water containing aggregations of zooplankton. Some planktivores like the rainbow smelt (*Osmerus mordax*) selectively feed on larger zooplankton.

How to Sample the Zooplankton

For picoplankton, nanoplankton and microplankton (includes bacteria and most single celled algae and protists), 100 ml to 1000 ml of water is collected by filling bottles
42☞ of appropriate size. Most regional health laboratories provide 100-ml bottles for bacteria samples and if they are to analyse the sample for total coliform and fecal coliform bacteria. The phytoplankton are analysed according to the methods described in Chapter 6. Protistan analyses require specialized equipment, such as a compound microscope

with at least 1,000 x magnification and scanning electron microscopy (abbreviated S. E. M.) as fundamental pieces of laboratory equipment.

The macroplankton are sampled with either a ***student plankton net*** or a ***Schindler-Patalas trap*** (Fig. 7.15). The plankton net has a mesh with openings that allow microplankton and smaller organisms to pass through the mesh but retain the macroplankton. Usually a mesh with openings of 150 μm is used. The diameter of the mouth of the net (Fig. 7.15) comes in a variety of sizes. The diameter used depends on the amount of plankton that one wishes to collect. In lakes with low numbers of plankton (e.g. oligotrophic lakes), a 0.5 m diam opening is often used because it collects more plankton in a given haul than does one with a smaller opening. Enriched lakes usually have a high biomass of plankton and a 0.2 m (20 cm) diameter openings is frequently used. The most popular plankton nets with 0.2 m mouths are the student plankton net and the Wisconsin plankton net (Fig. 7.15). The Wisconsin net has a removable bucket at the bottom of the net and makes the transfer of samples from the net to storage containers much easier than with the student net.

Either vertical, horizontal or oblique hauls are made with the plankton net. The volume of water sampled can be determined by attaching a meter to the opening of the net. The meter is calibrated to the diameter of the mouth and accurately measures the amount of water passing through the mouth. Alternatively, but less precisely, one can measure the distance that the net is pulled or hauled through the water and then multiply by the area of the mouth opening. That is:

$$\text{Volume of water sampled } (m^3) = \text{Area of mouth opening } (m^2) \text{ x distance hauled (m)}$$
$$= \pi r^2 \text{ x h } (m^3)$$

where r = radius of the mouth opening and h = distance net was hauled through the water

One can determine the amount of plankton in layers of water by sequentially hauling the net through each desired depth. For example, if one wanted to know the amount of plankton in the 0 - 2 m, 2 - 4 m, 4 - 6 m and 6 - 8 m depths, a haul is made first from the 2 m depth to the surface, then from the 4 m depth to the surface, then from the 6 m depth to the surface and finally from the 8 m depth to the surface. The amount of plankton in the 2 - 4 m depth equals the total plankton in the 0 - 4 m haul minus the total plankton in the 0 - 2 m haul; the amount of plankton in the 4 - 6 m depth is the total plankton in the 0 - 6 m haul minus the total plankton in the 0 - 4 m haul; the total plankton in the 6 - 8 m depth is the total plankton in the 0 - 8 m haul minus the total plankton in the 0 - 6 m haul.

The Schindler-Patalas trap is used to obtain grab samples from specific depths. It is available in two popular sizes, 12- L and 30-L. The 12-L is much lighter than the 30-L when filled with water and for that reason is probably more commonly used. A plankton sample is obtained by lowering the device to the desired depth. The trap has hinged doors that are forced up as the trap is lowered through the water column. As soon as the trap's descent is stopped and pulled upward, the doors close and trap a discrete sample (12 L or 30 L) that can be pulled to the surface and filtered through a detachable bucket (Fig.

7.15). The amount of plankton in each depth is much more easily estimated with the Schindler-Patalas trap than with the plankton net.

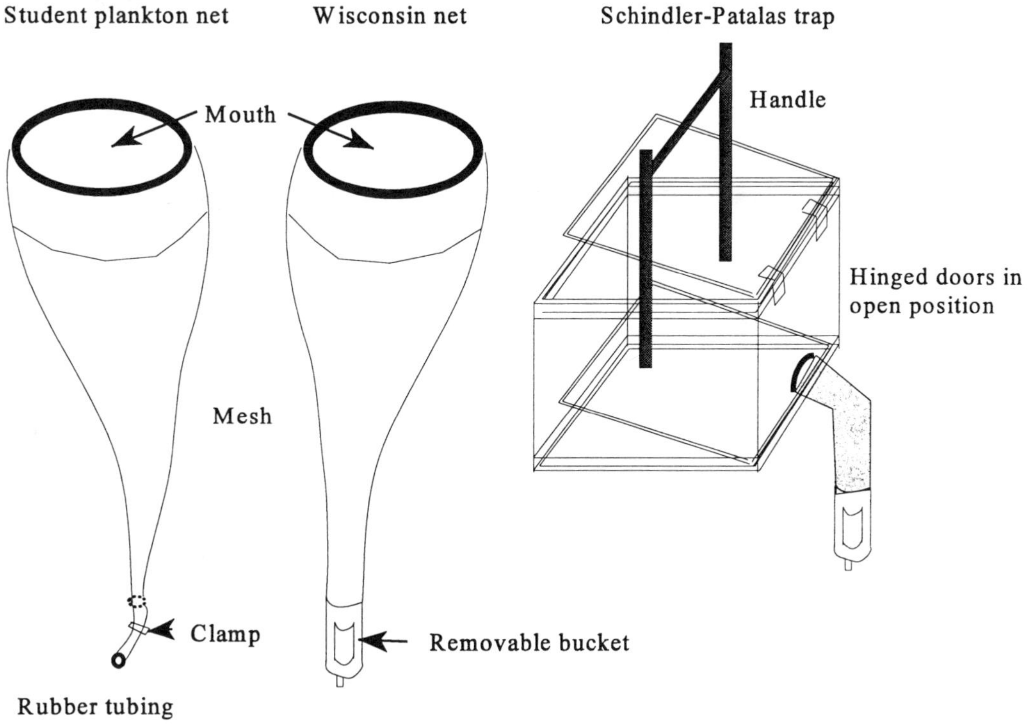

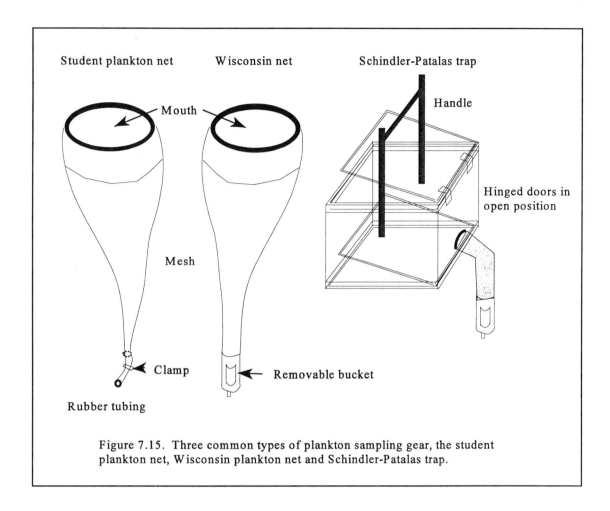

Figure 7.15. Three common types of plankton sampling gear, the student plankton net, Wisconsin plankton net and Schindler-Patalas trap.

CHAPTER 8

BENTHOS - A LINK BETWEEN WATER AND SEDIMENTS

Why Read This Chapter?

The benthos includes all organisms, plants and animals, that live on the bottom of lakes, oceans, rivers and other aquatic systems. This chapter examines the diversity of benthic organisms and their adaptations for life in various types of aquatic habitats. Particular emphasis is placed upon the use of the benthic community for assessing trophic status and environmental impacts. Only the benthic ***macroinvertebrates***, that is the invertebrates large enough to be seen by the naked eye, are examined here. The benthic macroinvertebrates are used more commonly than zooplankton and fish in water quality assessments because: (1) they are larger and more easily examined using low power standard microscopy than are most phytoplankton and zooplankton which require higher power and often specialized microscopy; (2) most of the species that make up the benthic community are more-or-less confined to a specific area and exhibit little movement out of the area, in contrast to zooplankton whose distribution is greatly affected by currents and wave action; (3) fish are able to swim away to avoid a stressor (e.g. a contaminant in an outfall), but macroinvertebrates are obliged to stay; (4) they are good integrators of water and sediment chemistry such that a level of a toxicant that is considered safe may, in fact, be sublethal enough to be detected by its effects on growth, reproduction, and/or physiology of sensitive species in the benthic community; and (5) the benthos cannot avoid even "slugs" or "spills" of effluent and will respond accordingly to the magnitude of the toxic event, which may be missed by chemists if they do not sample the water during the slug or spill event.

Many species of macroinvertebrates are diagnostic of certain kinds of habitats and their water quality. They are known as ***indicator organisms***, that is organisms that become numerically dominant only under a specific set of environmental conditions. For example, stream organisms that exhibit adaptations to life in flowing waters are indicators of stream environments. They are characterized by morphological adaptations that allow the animal to maintain its position in flowing waters. Adaptations such as a dorsal-ventral flattening of the body allows the animal to press itself close to the rock surface and "duck under" the water currents. There are a variety of mechanisms used to anchor the animal in place; spines on the body margin, large hooks or "grapples" for claws, adhesive discs or mucous secretions, or silk anchor threads all prevent the animal from being swept away by water currents. Even building heavy shells or cases to act as ballast, or using modified mouth appendages to filter food from flowing water, are all clues that the organisms are from erosional substrates in stream environments. In contrast, organisms that live in depositional substrates (e.g. pools of streams, sediments of lakes) have features characteristic of lentic environments. Many organisms have no mechanisms to resist flowing water and simply bury themselves in the substrate to avoid currents. These adaptations, and others, are described in this chapter.

Some benthic organisms are restricted to temporary ponds and each species has one or more adaptations to survive drought periods. The presence of large numbers of such species is a good indication that the pond will dry up sometime during the summer. The most common usage of benthic organisms is as indicators of water quality, especially a lake's trophic status, calcium hardness, alkalinity, pH and conductivity. This chapter describes only the more common species. If you are interested in pursuing the study of benthic organisms, two books are highly recommended. For aquatic insects, see Merritt and Cummins (1996)[1]. For all other invertebrates, refer to Pennak (1987)[2].

By the end of this chapter you will discover:

☞ 1. Different kinds of benthic communities
☞ 2. The characteristics and importance of freshwater sponges
☞ 3. The characteristics and importance of freshwater jellyfish and hydras
☞ 4. The characteristics and importance of flatworms
☞ 5. The characteristics and importance of roundworms
☞ 6. The characteristics and importance of horsehair worms
☞ 7. The characteristics and importance of moss animals
☞ 8. The characteristics and importance of freshwater tube worms
☞ 9. The characteristics and importance of aquatic earth worms
☞ 10. The characteristics and importance of leeches and blood suckers
☞ 11. The characteristics and importance of molluscs and how pearls are made
☞ 12. The characteristics and importance of snails and limpets
☞ 13. The characteristics and importance of clams and mussels
☞ 14. The characteristics, importance and impact of the Asian clam
☞ 15. The characteristics and importance of fingernail clams
☞ 16. The characteristics and importance of freshwater pearly mussels
☞ 17. The characteristics, importance and impact of zebra and quagga mussels
☞ 18. How to differentiate between spiders, mites, crustaceans and insects
☞ 19. The characteristics and importance of water mites
☞ 20. The characteristics and importance of crustaceans

---

[1] Merritt, R. W. and K. W. Cummins. 1996. An Introduction to Aquatic Insects of North America. Kendall-Hunt Publishers, Dubuque, Iowa.

[2] Pennak, R. W. (1987). Freshwater invertebrates of the United States. John Wiley and Sons, New York

☞ 21. The characteristics and importance of scuds and side swimmers
☞ 22. The characteristics and importance of freshwater sow bugs
☞ 23. The characteristics and importance of crayfish
☞ 24. The characteristics and importance of opossum shrimp
☞ 25. The characteristics and importance of insects
☞ 26. The characteristics and importance of dragonflies and damselflies
☞ 27. The characteristics and importance of mayflies
☞ 28. The characteristics and importance of stoneflies
☞ 29. The characteristics and importance of hellgrammites, dobsonflies and alderflies
☞ 30. The characteristics and importance of spongeflies
☞ 31. The characteristics and importance of aquatic butterflies
☞ 32. The characteristics and importance of caddisflies
☞ 33. The characteristics and importance of beetles
☞ 34. The characteristics and importance of true flies
☞ 35. Some physiological and ecological tolerances of common flies
☞ 36. Where most of the benthic diversity occurs in lakes
☞ 37. Thienemann's three principles governing benthic diversity
☞ 38. The potential impact of an exotic species on the benthic community
☞ 39. The adaptations of benthos to flowing waters
☞ 40. The different kinds of drift in streams
☞ 41. What organisms display behavioural drift in streams
☞ 42. The factors that affect drift rates of invertebrates
☞ 43. The adaptations of invertebrates for life in temporary ponds
☞ 44. The reasons for changes in benthic biomass in different regions of lakes in spring
☞ 45. The reasons for changes in benthic biomass in different regions of lakes in summer
☞ 46. The reasons for changes in benthic biomass in different regions of lakes in fall
☞ 47. The reasons for changes in benthic biomass in different regions of lakes in winter
☞ 48. The logic behind methods for measuring secondary production of invertebrates
☞ 49. The different kinds of life history patterns of invertebrates
☞ 50. How to measure benthic productivity using a formula method
☞ 51. How to measure benthic productivity using the Allen curve method
☞ 52. How to measure benthic productivity using a growth rate method
☞ 53. How to measure benthic productivity using a size-frequency method
☞ 54. Some quantitative methods used to sample benthos of streams
☞ 55. Some quantitative methods used to sample benthos of lakes
☞ 56. Some semi-quantitative methods used to sample benthos in any kind of habitat

TYPES OF BENTHIC COMMUNITIES

Benthic organisms may live either within the sediments or upon the sediments. The animals that live in the sediments are called ***infaunal***. They obtain their food and dissolved oxygen primarily from interstitial water held between the sediment particles. Some even engulf sediment, utilize the food that is taken in with it and then excrete or eliminate the indigestible sand particles.

Benthic organisms that live on top of the sediments, rocks, logs or plants are called ***epibenthos***. The prefix, "epi", simply means attached to the surface or living
1☞ freely upon the surface. The substrate to which the organisms are attached can be identified as a suffix in the term. For example, ***epifauna*** are those organisms attached to animals (e.g. zebra mussels often attach to clams, crustaceans, or snails). Organisms that live attached to rocks are called ***epilithic***; those upon plants are known as ***epiphytic***; and those living upon mud or sand are known as ***episammic*** organisms. The microscopic assemblage of organisms (mostly algae, bacteria, fungi and molds) that grow freely upon or attached to surfaces of submerged objects are called ***periphyton***. Since the periphyton include both planktonic and benthic forms, it is sometimes difficult to determine to which group they belong. They have been discussed in Chapter 6 as the periphyton are usually dominated by algae. Periphyton are common in both lakes and streams.

Lake benthos is often also classified according to the zone that they live in. ***Littoral benthos*** and ***sublittoral benthos*** are characterized by body appendages that allow the organisms to cling to plant stems and leaves. ***Profundal benthos*** are adapted for gathering or filter feeding the fine organic particles that typify the profundal zone. With the lack of coarse particulate organic matter, shredding invertebrates are absent. Likewise, since there is no light in the profundal zone, grazing (or scraping) animals are absent. In fact, the only functional feeding groups in the profundal zone are gatherers (e.g. worms), filter feeders (e.g. fingernail clams) and predators (e.g. midge flies). Occasionally a fourth group, the ***abyssal benthos***, are present. But the abyssal zone is present only in very deep lakes ($>$ 500 m) and many of the benthic species are blind.

In streams, the benthic assemblage in the ***riffle areas*** (rapids) is very different from that in the ***pools,*** or back eddies. The benthos of pools is often very similar in form, habits and species composition to that in ponds and lakes. The benthic communities of rivers, in general, are composites of assemblages from their tributaries, truly riverine species and cosmopolitan species that occur virtually every where.

BENTHIC MACROINVERTEBRATE GROUPS

Before describing other features of benthic communities, it is important to know the characteristics of taxa (groups) that make up the benthic community. We will begin by examining organisms with a simple level of body organization and end with the most complex, those with very highly developed nervous systems and other organ systems. The diagnostic features, mode(s) of reproduction, general ecology and role in the benthic community are described for each group.

Phylum: Porifera - The freshwater sponges

Sponges are primarily a marine group. There is only one family of freshwater sponges, the ***Spongillidae***, with about 30 species in North America. Sponges are mainly epibenthic, encrusting or enveloping logs, twigs, rocks and other firm substrates. They lack a distinct body form but can be recognized immediately by their "garlic" odour.

Sponges are characterized by the presence of ***spicules*** that provide skeletal
2☞ support for the flimsy tissues and a complex pore-canal system that is used to distribute food and water throughout the body (Fig. 8.1). The spicules of freshwater forms are made of silica and the kinds and shapes present are diagnostic of the genus and species. ***Megascleres*** are large, usually packaged into ***fascicles*** and form a network throughout and within the tissues and bind the tissue mass together. Some fascicles of megascleres penetrate the surface of the sponge making them gritty to the touch. ***Microscleres*** are small, flesh or dermal spicules that add to the gritty feel of sponges. Only a few genera have microscleres, *Spongilla* and *Eunapius* being the most common. Very few microscleres are found within the tissues. The spicules are secreted from small cells called ***silicoblasts***.

The tissues of sponges consist of several cell types that perform specific functions. For example, ***pinacocytes*** form the ***pinacoderm*** or epidermal layer and ***porocytes*** line pore openings. However, without the spicules the tissues would fall apart. The bulk of the body mass contains numerous canals and chambers lined with cells called ***choanocytes***, or ***collar cells***, that create a current of water through the sponge. Water enters the canal system through openings called **ostia** at the surface of the sponge. The water is filtered of its food as it passes through the sponge. Their food includes microorganisms, such as protists, algae and bacteria. The algae often lend a green colour to many species of sponge. The water is expelled through larger pores called ***oscula***, also at the surface of the sponge. All sponges are filter feeders.

The sponge colony grows throughout the entire summer and then collapses in the fall and winter leaving behind structures called ***gemmules.*** The gemmules are asexual bodies, ranging in diameter from about 200 to 1,000 μm, and contain within a thick protective coat all the cell types needed to regenerate a new sponge every spring. Gemmules may also form during unfavourable conditions, but usually ***reduction bodies*** carry the sponge over periods of adversity. Reduction bodies are formed from the shrinkage of the tissue mass leaving all the cell types and spicules needed to regenerate a new colony when appropriate conditions prevail. Sponges are also capable of sexual reproduction. Most species are hermaphroditic with eggs formed from specialized cells and sperm forming from choanocytes that lose their collars. It is believed that sexual reproduction occurs in the warmer months of the summer.

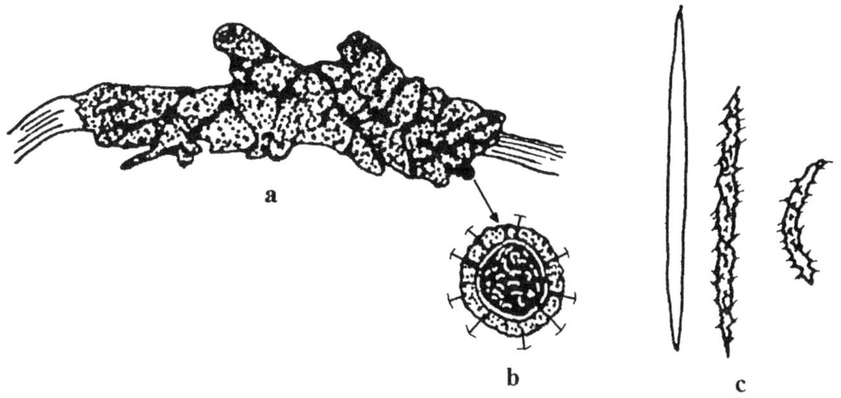

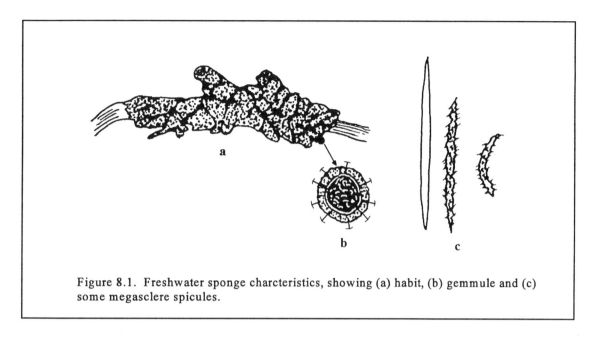

Figure 8.1. Freshwater sponge charcteristics, showing (a) habit, (b) gemmule and (c) some megasclere spicules.

Most sponges are very sensitive to enrichment and pollution and their presence in large biomasses usually indicates good water quality. Two of the most common species are *Spongilla lacustris* and *Ephydatia fluviatilis* but, unfortunately, both grow in all kinds of habitats and are among the more tolerant species. *Ephydatia muelleri* and *Eunapius fragilis* are alkaline species occurring in clean waters with pH greater than 7.5. *Heteromeyenia tubisperma* is restricted to clean, running waters. The main predators of sponges are spongilla flies of the family Spongillidae (order Neuroptera) and crayfish, especially *Orconectes*.

Phylum: Cnidaria - The Freshwater Jellyfish and Hydras

Another primitive group of freshwater macroinvertebrates is the cnidarians. They too are primarily a marine group. Cnidarians have two layers of tissues, called
3☞ ***ectoderm*** (outer layer) and ***endoderm*** (inner layer) with a jelly-like layer, called the ***mesogloea***, between them. *Craspedacusta* (Fig. 7.14) is the only genus of freshwater jellyfish in North America. It was introduced from Europe to North America in the early 1900s. The "jellyfish" form, or ***medusa***, is the adult and is mainly planktonic in life. The medusa arises as a bud on a short-lived benthic stage called a ***polyp*** that is similar in shape, but lacking tentacles, to the ubiquitous *Hydra*.

Craspedacusta sowerbyi is often collected in zooplankton hauls in mesotrophic and oligotrophic ponds, quarries, lakes and large rivers. The species exhibits vertical diurnal migration, rising in the water column at night and descending to near the bottom during the day where they occasionally occur as part of the benthos. It is predaceous, feeding on small (0.2 to 2.0 mm) zooplankton and fish and capturing its prey with ***nematocysts***, or stinging cells, on its numerous tentacles.

Hydra (Fig. 8.2) is epibenthic, living attached to the surface of the substrate or on plants, rocks, and animals, especially clams and mussels. Polyps exploit planktonic food in water currents created by the clams and mussels as they filter feed. The polyps are also filter feeders, using their tentacles to gather food. The tentacles have a variety of nematocysts, some that paralyse their prey, others with sticky secretions to which food particles adhere and still others that simply lasso their prey. Food consists mostly of zooplankton, particularly cladocerans. Once caught, the food is ingested by "licking" it from their tentacles. The medusa stage is absent in *Hydra*. Although predaceous on zooplankton, the polyps and medusae are often called filter feeders because of the sweeping action of the tentacles through the water and extracting food particles from the water column.

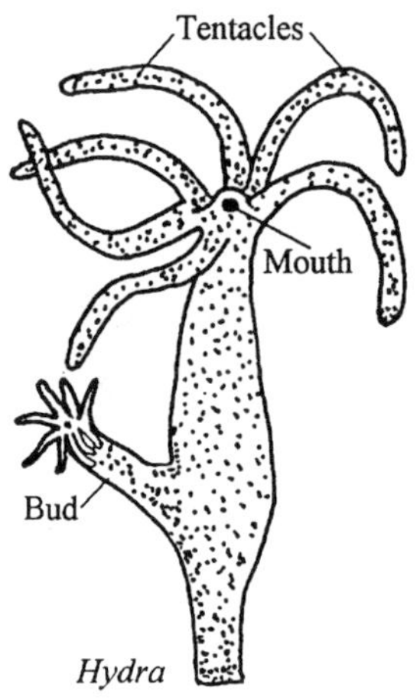

Figure 8.2. *Hydra* polyp with a bud.

Reproduction is either sexual or asexual. Medusoid forms have four gonads, either ovaries or testes, with one located on each of the four radial canals. Eggs and sperm are released to the surrounding water where fertilization occurs. The zygote develops into a branching, colonial polyp without tentacles. The polyp reproduces by either budding off new polyps from its body wall, budding off medusae larvae from its body, the larvae developing into sexual medusae, or budding off larvae that creep about on the bottom for a while and then develop into another polyp.

Polyps have three methods of reproduction, one sexual and two asexual. Sexual reproduction occurs by producing gonads from specialized cells on the body wall. Polyps may be dioecious or hermaphroditic, depending on the species. The testes are conical in shape and the ovaries are dome-like. A sperm from the same individual or from another individual fertilizes the egg. The zygote drops off, becomes enveloped in a protective covering, remains dormant for 1 to 3 months, and then the protective covering splits and releases the embryo which develops into another polyp. Asexual reproduction occurs by either budding of polyps off the body wall or by transverse or longitudinal fission.

Phylum: Platyhelminthes - The Flatworms

Flatworms have true tissue layers, with a cellular ***mesenchyme*** between the

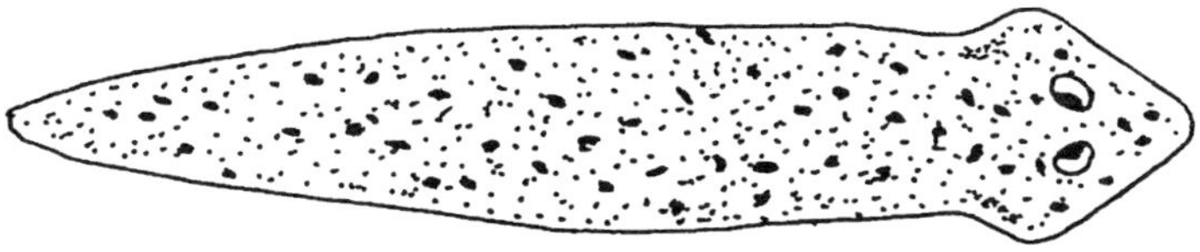

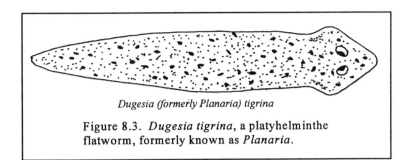

Dugesia (formerly Planaria) tigrina

Figure 8.3. *Dugesia tigrina*, a platyhelminthe flatworm, formerly known as *Planaria*.

outer epidermis and inner endodermis. However, they lack a body cavity or ***coelom***
4☞ (hence are termed ***acoelomate***) and a complete digestive system. The body cavity is filled with mesenchyme and the digestive tract ends blindly, with no anus. The phylum is represented by three classes: ***Turbellaria***, which includes the common *Planaria,* now called *Dugesia;* ***Trematoda***, or the parasitic flukes; and ***Cestoda***, or tapeworms. The flukes and tapeworms were discussed in Chapter 7 since the larval stages are planktonic. Although benthic invertebrates, especially molluscs, are intermediate hosts, all other life stages are parasitic. Only the Turbellaria are free living and truly benthic and are discussed below.

Flatworms are capable of both asexual and sexual reproduction. *Dugesia tigrina* (Fig. 8.3), *Procotyla fluviatilis* and *Phagocata vernalis* are capable of reproducing asexually by ***binary fission*** or ***fragmentation***. In fission, the animal constricts about midway and splits in half, each half regenerating the missing components. Fragmentation differs in that the body breaks into several parts and encysts in a slimy mass; after a short period of dormancy a small worm emerges from the cyst Apparently binary fission is used only when temperatures exceed 10°C. Sexual reproduction is the rule in most other turbellarians. In those species that also reproduce asexually, sexual reproduction typically occurs in the fall or winter months. Turbellarians are hermaphroditic and contain numerous spherical testes, two ovaries and numerous ***yolk glands***. The eggs are fertilized internally by sperm from either the same individual or sperm received from another individual and stored temporarily in a ***seminal receptacle***. The fertilized egg passes down an oviduct and accumulates yolk from the yolk glands. Zygotes (2 to 20) with their yolk cells are encased in a cocoon that is released through the genital pore and attached to the substrate. The cocoons of *Dugesia* have a long stalk.

Most turbellarians are ***detritivores***, feeding on dead particulate organic material, or ***zoophagous***, feeding on small living or moribund invertebrates (protists, rotifers, nematodes). Flatworms are photo-negative and most commonly occur on the under surfaces of rocks and leaves of macrophytes in streams and lakes, or occasionally in the mud. They tend to be associated more with mesotrophic and eutrophic bodies of water where detritus and decaying animal matter is abundant. Many species are diagnostic of peculiar types of habitats. *Pseudophaenocora sulfophila* is found only in sulphur springs where oxygen saturation rarely exceeds 5 to 40% whereas *Polycelis coronata* is found only in cold, well oxygenated streams. Most turbellarians require at least 70% oxygen saturation. *Hymanella retenuova* is found only in temporary seepages, ponds and ditches. Most turbellarians have eyes (one to numerous pairs) but cave species, such as most species of *Sphalloplana*, are blind.

The diversity of parasitic flatworms has some value in assessing environmental quality. Crites (1974)[3] compared the diversity and abundance of species in the genus *Bothriocephalus* between 1929 and 1970 and found that decreases in population sizes of its host fish species, mooneye, troutperch, sauger, blue pike, northern pike and walleye were accompanied by declining populations of *Bothriocephalus* species. When walleye population sizes increased between 1970 and 1974, the prevalence of *B. cuspidatus* also increased.

Phylum: Nematoda - The Roundworms

The nematodes have a greater degree of body organization than do the flatworms in that roundworms have a body cavity and a complete digestive system.
5☞ However, the mesenchyme lacks a ***peritoneum*** (inner body lining) typical of higher invertebrates. Also, their muscular development is poor, with only longitudinal muscle fibres; circular muscles, which help to extend or elongate worms, are absent. Hence, the undulations typical of true worms are absent and roundworms move by thrashing about aimlessly in the sediments.

The partially complete body cavity is referred to as a ***pseudocoelom*** and is present in many other phyla, including the ***horsehair worms***, or Nematomorpha, described next, and microinvertebrates such as rotifers, discussed in Chapter 7, and the relatively uncommon thorny worms of the Phylum ***Gastrotricha*** which are not discussed at all in this book. Some rotifers are benthic, such as numerous species of *Monostyla*, *Lecane* and *Euchlanis*, but their body features and ecologies are similar to the planktonic forms. Benthic rotifers, however, have a well developed and usually prehensile foot to attach itself to the substrate.

Nematode worms (Fig. 8.4) are mainly a parasitic group, with only a few free living forms. The ecology of free living nematodes is poorly understood because they are small and difficult to identify to species. Most are 0.5 - 1.0 cm long, less than 0.1 mm diam and pass through the 0.5 mm mesh openings of sieves used to separate them in the fine from coarser sediments. Roundworms are spindle-shaped, pointed at both ends and lack eyes and segmentation (Fig. 8.4). The mouth is at the anterior end and has three pairs of lips.

All nematodes have a thick, non-living proteinaceous cuticle that is usually ornamented with various kinds of sculpturing, including pits, scales, grooves, ridges and tubercles that lend rigidity to the cuticle. The type and combination of sculpturing on the cuticle is important in identification. The cuticle extends into the mouth cavity

---

[3]Crites, J. L. 1974. Parasitic helminths as indicators of environmental quality. *In:* King, C. C. and L. E. Elfner. 1974. Organisms and biological communities as indicators of environmental quality - a Symposium at The Ohio State University, March 25, 1974.

and lines the digestive tract, including the oesophagus and posterior part of the intestine. The anus opens ventrally in the posterior fifth of the animal.

The digestive system is ribbon-like in form. Depending on the species, nematodes feed upon dead or living plant and animal material. Those feeding on living plant material have a hollow stylet used to pierce the plant cells and to suck out the fluids. Predaceous and carnivorous forms have specialized lips for seizing and tearing the prey and a partially eversible pharynx armed with teeth for rasping or macerating the animal tissues. Animal prey includes protists, rotifers, oligochaetes or nematodes.

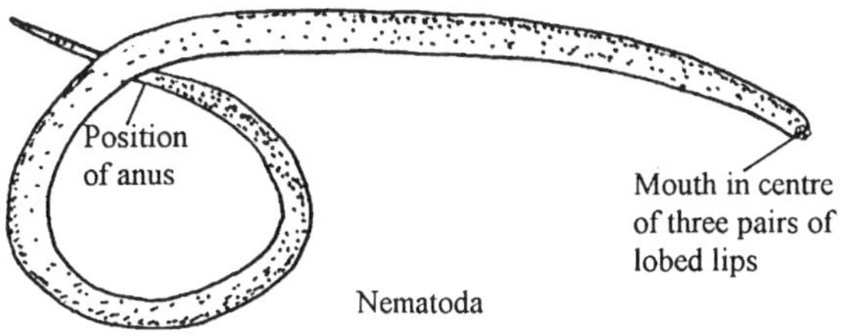

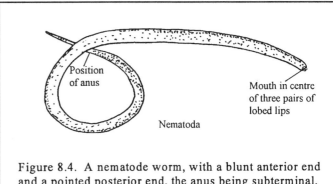

Figure 8.4. A nematode worm, with a blunt anterior end and a pointed posterior end, the anus being subterminal.

The sexes are separate and mating occurs to transfer sperm to the female. Wing-like flaps are present in the males of some species and are used for grasping the female in the mating process. Fertilization of the egg occurs within the female's uterus. Some species are ***ovoviviparous*** where the eggs are retained and hatched within the female, young worms being released upon hatching. Most are ***oviparous*** and the female releases fertilized eggs that develop and hatch in the sediments. Both oviparous and ovoviviparous forms appear to molt four times in their life span.

Little is known about the relationship between the abundance and diversity of free-living forms of roundworms and the trophic status or health of the aquatic environment. However, the diversity of parasitic forms can be an index of environmental quality. The parasitic forms are found in many host species at most trophic levels in all types of aquatic communities. Most parasitic species require benthic or planktonic invertebrates as intermediate hosts, including turbellarians, snails, bivalves, oligochaetes, crustaceans and insects. Predaceous species of fish and fish-eating birds serve as the definitive, or final, hosts. Therefore, a broad and diverse community of nematode parasites usually indicates a productive, unpolluted environment of good water quality. Many studies, particularly in the Great Lakes, have shown that acceleration of pollution, such as eutrophication by humans, decreases the diversity of intermediate and final hosts of parasites, as well as the diversity of the parasites themselves. The species of intermediate hosts that do survive increase in abundance, and the parasites respond with increases in both ***incidence*** (percentage of individuals that are parasitized) and ***intensity*** (numbers of parasites in each host). Some parasite species disappear altogether.

With these concepts in mind, Crites (1974, see reference previous page) suggested that two occurrence scenarios are possible: (1) parasites are absent; (2) parasites are present but they are either less abundant or more abundant than prior to environmental degradation. The absence of parasites usually indicates situations where lowered environmental quality has already occurred. Crites cites four examples from the Great Lakes; (i) exploitation - e.g. environmental changes and introductions of new species of fish in Lake Erie resulted in the removal or decrease in abundance of whitefish which was followed by a decrease in fish parasite diversity; (ii) waters polluted by acid mine wastes had a low diversity of benthic invertebrates and the low numbers of molluscs, crustaceans and insect larvae resulted in the loss of larval sustenance, transfer and host contact of nematode parasites; (iii) high pesticide levels in aquatic habitats eliminated many species of invertebrates and the parasites dependent upon them; and (iv) when the mayfly, *Hexagenia limbata*, was eliminated from Lake Erie, species of the nematode parasite, *Lanciomermis*, and digean trematodes dependent upon the mayfly as intermediate host, were also eliminated.

The relationship between numbers and varieties of parasitic species and habitat degradation has already been alluded to for the tapeworm, *Bothriocephalus*. An increase in abundance of parasites occurs when there are fewer host species remaining, but the survivors exhibit a huge increase in numbers. Pollution tolerant species of invertebrate and vertebrate host species have lower levels of competition and are allowed to increase in numbers. This increases the efficiency of transfer of parasites that utilize the same trophic relationship. For example, during the eutrophication process in Lake Erie, the increase in the cyclopoid copepod populations were followed by increases in planktivorous fishes, such as spottail and emerald shiners, as well as gizzard shad, alewife and rainbow trout. The latter three fish species were introduced during the eutrophication process. Two species of the nematode, *Philometra*, which uses cyclopoids as intermediate hosts, increased in prevalence during the same period. Other parasites also increased in prevalence, including the nematodes, *Camallanus oxycephalus, Hydromeris* and *Eustrongylides tubifex,* and the tapeworms, *Ligula intestinalis* and *Proteocephalus*. *Hydromeris* and *Eustrongylides* use the midgefly, *Chironomus* sp., and the tubificid worm, *Tubifex tubifex*, both enrichment indicators, as intermediate hosts, respectively.

Phylum: Nematomorpha - The Horsehair Worms or Gordian Worms

The adult horsehair worms are shaped much like the hair of humans, only thicker. Most are found twisted in a tangled knot, hence the common name, "*Gordian*
6☞ *worm*" (Fig. 8.5). The adults are long (1 to 7 cm) and thin (0.2 to 2.5 mm diam), usually brownish, yellowish or grayish in colour, and have a bluntly rounded or truncated anterior end and either a bluntly- rounded single, bilobed or trilobed posterior end. The numbers of lobes on the posterior end is a sexual characteristic, being single in the females of some common genera like *Gordius*, *Gordionus* and *Parachordodes,* and bilobed in the males of the same genera, or trilobed in the females of *Paragordius*. The anterior end may have a mouth opening or not, but the rest of the digestive tract is degenerate.

Like nematodes, horsehair worms have a thick cuticle and poor musculature in the body wall. The lack of circular muscles also limits their mobility and the worms show little translocation within sediments.

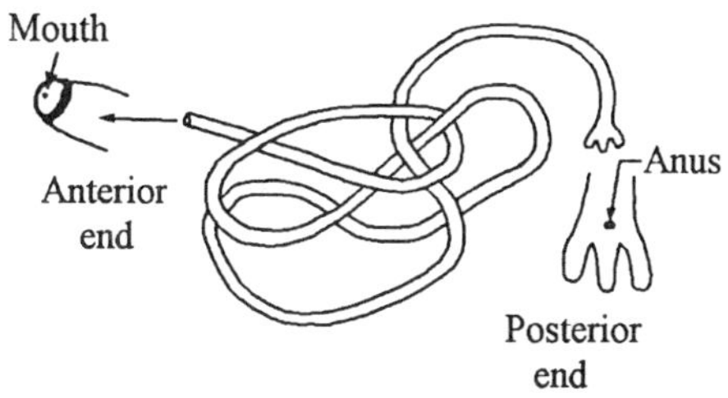

Figure 8.5. An adult horsehair, or Gordian, worm showing the blunt anterior end that lacks a mouth and the posterior end, which is a single lobe in females and single-, double-or triple-lobed in males.

The adult is mainly a reproductive stage and does not feed. The larval stages are parasitic upon insects which are required to complete development into the adult stage. Mating occurs, the eggs are fertilized internally and then released from the cloaca into the water. After 15 to 80 days, depending upon species and temperature, a larva emerges from each egg, swims for a short while, then encysts on vegetation. The cyst enters an insect when the vegetation is eaten. The larva, released from its cyst by the digestive juices of its host, penetrates the wall of the digestive tract and development into the adult stage occurs in the ***haemocoele*** (blood cavity) of the insect. The larva has an eversible ***proboscis*** (trunk) containing three stylets for piercing the host's intestinal wall. The adult is released and usually results in the death of the host.

Because adult horsehair worms do not feed, they play a minor role in the ecology of benthic communities and are not considered of great importance in assessing water quality. Adults occur in a variety of aquatic habitats in running and standing waters, usually in water less than 20 cm deep. The most common species are *Gordius robustus* and *Paragordius varius*.

Phylum: Bryozoa - The Moss Animals

While bryozoans are more highly evolved than any of the pseudocoelomate ("false" body cavity) forms, they are more primitive than others with a "true" body
7☞ cavity. A true body cavity has a peritoneum lining the mesenchyme. The body cavity itself contains a fluid called ***coelomic fluid***. The greatest diversity of benthic organisms occurs within the eucoelomate phyla, which includes Bryozoa, Annelida, Mollusca and Arthropoda.

The bryozoans are a colonial group of organisms that encrust or envelope twigs, logs, branches, rocks and other solid objects on or above the substrate (Fig. 8.6). The encrusting forms may be branching and thread-like or mat-like with a plant-like appearance. The enveloping forms are like balls of jelly that range in size from a few centimeters in diameter (e.g. grapefruit size) to a quarter meter in diameter (e.g. watermelon size). Each colony consists of thousands of individual ***zooids*** (Fig. 8.6)

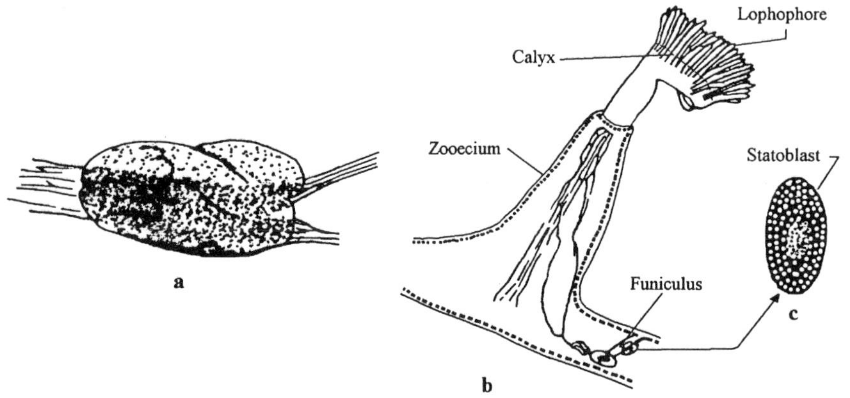

Figure 8.6. A bryozoan, showing (a) its habit, (b) a zooid and (c) a statoblast arising from a funiculus.

that live within tubes contained in a gelatin matrix. Each zooid bears a crown of ciliated tentacles that arise from a ***calyx***. In the centre of the tentacles is a mouth. The calyx with its mouth and tentacles are collectively called a ***lophophore***. Each zooid can retract within its tube when disturbed. The part of the zooid that can be extended out of the tube, that is the lophophore with its base, or stalk, is called a ***polypide***. The entire colony is protected by a layer called a ***zooecium*** and includes the tube and the part of the canal containing the reproductive elements and the retractor muscle used for pulling the zooid into its tube.

Zooids communicate with each other by a network of canals. Although there appears to be no nervous system connecting one zooid to another, the zooids display remarkable synchrony in their retraction and protrusion from their tubes. A touch of a single zooid will generate a wave of retraction over the entire colony within a few seconds. After a minute or two of quiet, the zooids protrude from their tubes and become active again.

Bryozoans are filter feeders, using their tentacles to extract food particles from the water that bathes the colony. The tentacles are almost motionless, but the cilia create currents that draw food to the tentacles and down to the mouth where they are ingested. Food consists of algae, protists, rotifers and detritus. Excessively large particles are swept away from the mouth by the tentacles. The digestive system is Y-shaped with a large caecum at the bottom of the Y and a mouth at one of the top parts of the Y and an anus at the other. The anus is located on the underside of the lophophore.

Bryozoans reproduce by both sexual and asexual means. Individual zooids are hermaphroditic in most species. The testes are located at the bottom of the stalk while the ovary is in the stalk itself. Eggs are fertilized internally in the coelomic fluid by sperm from either the same zooid or from other zooids in the colony via the canal system. The zygote develops in a marsupial sac near the ovary into a small, ciliated larvae that is 1 to 2 mm long. The larvae, each containing one or more polypides, escapes through a pore in the base of the lophophore or when the colony degenerates. The larva swims for about one day, settles on the substrate and then generates a new colony from the polypides.

Asexual reproduction by fission or proliferation is used for expanding the size of the colony throughout the summer. Asexual production of ***statoblasts*** and ***hibernacula***, are used to carry essential elements of a colony over periods of adversity and over the winter months. They are also the main dispersal element of bryozoans. Statoblasts are produced by most bryozoans and develop by ***internal budding*** from a ***funiculus*** (Fig. 8.7) attached to the body wall near the testes. Each statoblast contains a mass of

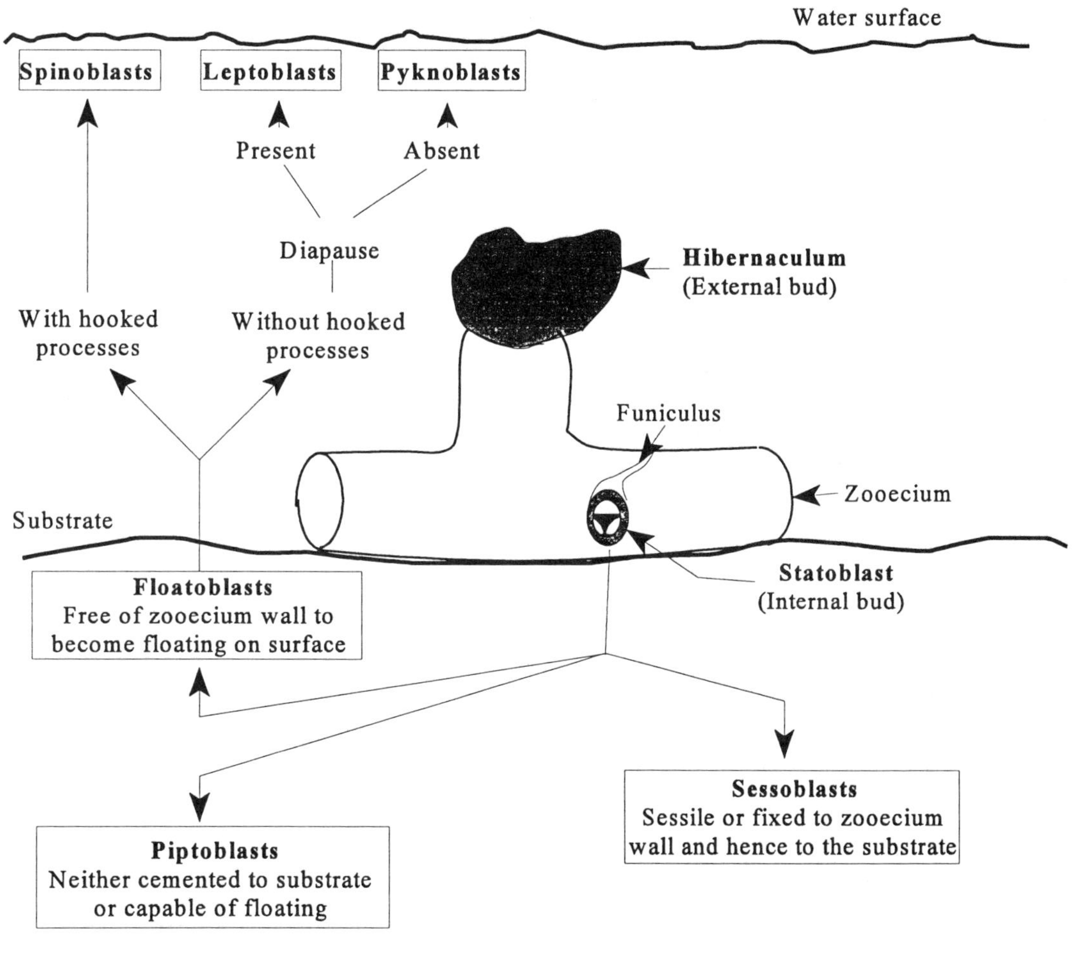

Figure 8.7. Two basic types of asexual structures (hibernacula and statoblasts) , and the three types of statoblasts and three types of floatoblasts produced by bryozoans to carry them over the winter months or periods of environmental stress.

undifferentiated germinative cells capable of generating a new colony when proper conditions arise. A statoblast is highly resistant to desiccation, temperature extremes and an array of environmental stressors, including most kinds of pollution. Hibernacula, produced by only a few bryozoan species (e.g. *Pottsiella* and *Paludicella*), are similar in function and cellular constituents to statoblasts but are ***external buds*** from the zooecium.

There are three types of statoblasts produced by most bryozoans (Fig. 8.7). ***Floatoblasts*** become free of the zooecium wall and float to the surface, after the zooecium disintegrates, to be dispersed by the wind and surface currents. ***Sessoblasts*** are sessile and attached or cemented to the zooecium wall. Since the bryozoan, with its thousands of zooecia, are fixed to the substrate, the statoblasts are like-wise fixed to the substrate. Sessoblasts can be produced during any part of the growing period. ***Piptoblasts*** are neither cemented to the substrate nor capable of floating but sink to the bottom when the zooecium disintegrates. Piptoblasts are more easily dispersed by water currents than are sessoblasts.

Three kinds of floatoblasts are recognized: ***spinoblasts*** have spines, hooks, barbs or other projections presumably for attaching to animals for dispersal; ***leptoblasts*** have little or no projections, do not need a diapause (rest period) and germinate immediately; ***pyknoblasts*** are thick-walled, without projections and require a diapause before germinating (Fig. 8.7).

Enveloping bryozoans are common in ponds, lakes and large rivers. Most occur in shallow water but many are found in the 5 to 10 m depth; only a few are found in depths greater than 10 m. *Lophopodella* and *Lophopus* are generally restricted to standing waters but *Cristatella, Paludicella, Pectinatella* and *Fredericella* are found in both standing and running waters. Encrusting types may be found in streams and other fast flowing aquatic habitats. *Plumatella* and *Urnatella* are representative of flowing waters and *Pottsiella* is restricted to rapid waters. No bryozoans are capable of tolerating pollution and their presence usually indicates good water quality with at least 50% oxygen saturation. Most are photonegative and develop best in shaded parts of the aquatic habitat. Some species (e.g. *Pectinatella magnifica*) are biofoulers and clog large water pipes and intakes but the water is slow flowing and clean. *Paludicella, Pectinatella* and *Fredericella* occasionally clog fishermen's nets with extensive growths. Bryozoans are fed upon by turbellarians, snails, oligocheates, water mites, crustaceans and insect larvae of caddisflies and midge flies.

Phylum: Annelida - The True Worms

Fresh waters have five classes of annelids. The most primitive of these are the tube worms of the class ***Polychaeta***. Polychaetes are mainly marine and only a few species are present in fresh waters. The class ***Oligochaeta***, or aquatic earthworms, is well represented in both marine and freshwater systems. The leeches and blood suckers of the class ***Hirudinea*** are entirely fresh water in habit.

The remaining two classes are specialized in their habitats. The ***Brachiobdellida*** are commensals (live in association with) on crayfish. They have features similar to leeches and represent an evolutionary link between oligochaetes and leeches and are only briefly discussed here. The ***Archiannelida*** are a recent addition to the annelid classes. They are microscopic worms that live in the ground water and interstitial waters of caverns and mountain streams. None have been reported from Canada yet but they are small and rare enough to be overlooked at this point in time. Because they are microscopic and rare, they are not discussed further.

All annelids have internal segmentation where each segment is isolated from the other. The first segment, called the ***prostomium***, may or may not bear eyes and tentacles. Annelids are entirely hermaphroditic and in some cases are capable of self-fertilization. Creeping and crawling locomotion is provided by setae of a variety of diagnostic shapes and fusion and by co-ordinated muscular movements. Some classes (e.g Hirudinea) lack setae and move by using suckers or by swimming.

Class: Polychaeta - The Freshwater Tube Worms

The polychaetes are represented in fresh waters mainly by a single species, *Manayunkia speciosa* (Fig. 8.8). Other species of *Manayunkia* are present but they are
8☞ very rare and mostly confined to brackish waters. Other genera of freshwater polychaetes are rare and confined to coastal areas, mountain streams or brackish water.

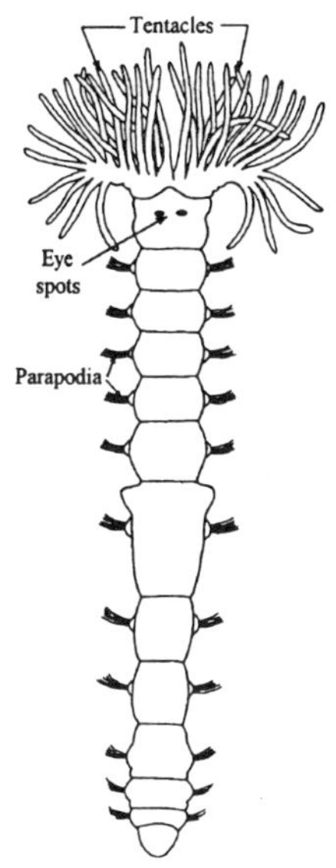

Figure 8.8. A freshwater polychaete, *Manayunkia speciosa.*

Manayunkia speciosa is found mainly in fine silty or sandy sediments in oligotrophic and mesotrophic lakes and large rivers. It is widely distributed in the Great Lakes, Ottawa River, St. Lawrence River and other numerous lakes and rivers in New York, Ohio, New Jersey, Pennsylvania, Vermont, Georgia, North and South Carolina, California, Oregon, Alaska and British Columbia. The adult grows to 5 - 6 mm in length and lives in a tube of sand and silt grains cemented together by mucous. The head bears two eyes and two groups of 18-20 ciliated tentacles used to filter food from the water at the mud-water interface. Behind the head are 11-12 segments, each with a ***parapodium*** containing 4-10 setae projecting from the base.

The animal reproduces by sexual and asexual reproduction. Asexual reproduction appears to be most common where the 6th segment, which is larger than the rest, divides by fission, each half regenerating its missing segments.

Little is known of its ecological tolerances and requirements, although Mackie and Quadri (1971)[4] found it only in water that was at least 60% saturated with oxygen and the sediments contained silt and sand with little or no organic material. Food trapped on the tentacles by mucous is directed to the mouth by cilia and is then swallowed.

Class: Oligochaeta - The Aquatic Earthworms

Of the freshwater annelids, the oligochaetes display the greatest diversity and have the greatest indicator value. There are ten families of oligochaetes, but two in
9☞ particular, the ***Naididae*** and ***Tubificidae***, form 80 to 100% of the annelid communities in the benthos of most streams and lakes at all trophic levels. These and the other eight families can also be found in wetlands, ditches, ponds, sewage lagoons, temporary pools and in semi-aquatic habitats. In lakes they occur from the water's edge to the greatest depths. Oligochaetes range in size from 0.1 cm in the Naididae to 3 or 4 cm in relaxed length in the Lumbricidae, the family that contains the earthworms.

The most diagnostic feature of oligochaetes are the four bundles of chitinous (protein made of chitin) ***setae*** (occasionally called ***chaeta***) present in most segments, two bundles dorso-laterally and two bundles ventro-laterally (Fig. 8.9). Setae are never present on the prostomium and they vary in shape, structure and function among the segments. Those on the reproductive segments are usually modified for transferring sperm, while those on the anterior segments are used for locomotion and burrowing, but the kinds and numbers present in each bundle vary, depending on the species.

The kinds of setae present on specific segments are often diagnostic for the species. The first segment (prostomium) bears tentacles, a proboscis and/or eyes in some species (especially in the family Naididae). Gills are present in one species, *Branchiura sowerbyi* (Fig. 8.9), a tubificid worm found in enriched rivers. The last segment bears the anus and is rounded, except in some naidids, like *Dero* species which have finger-like gills emerging from a ventral pad.

Oligochaetes are capable of both sexual and asexual reproduction, although some families (e.g. Naididae and ***Aelosomidae***) reproduce asexually most of the time. All annelids are hermaphroditic and cross-fertilization is the rule in most oligochaetes. The reproductive segments, usually indicated by a thickening of the epidermal tissue into a ***clitellum***, are in the mid to distal part of the anterior half of the worm, with the ovaries in the segment immediately behind that with the testes. Worms pair up head to tail and transfer sperm to each other's ***spermatheca*** (sac that stores sperm of mate) from its own ***seminal vesicle*** (sac that stores its own sperm). After separating, a "cocoon" is secreted by the clitellum around the worm and its own eggs and the sperm stored in the spermatheca are released into the cocoon. Fertilization occurs within the cocoon.

---

[4]Mackie, G. L. and S. U. Quadri. 1971. A polychaete, *Manayunkia speciosa*, from the Ottawa River, and its North American distribution. Canadian Journal of Zoology 49: 780-782.

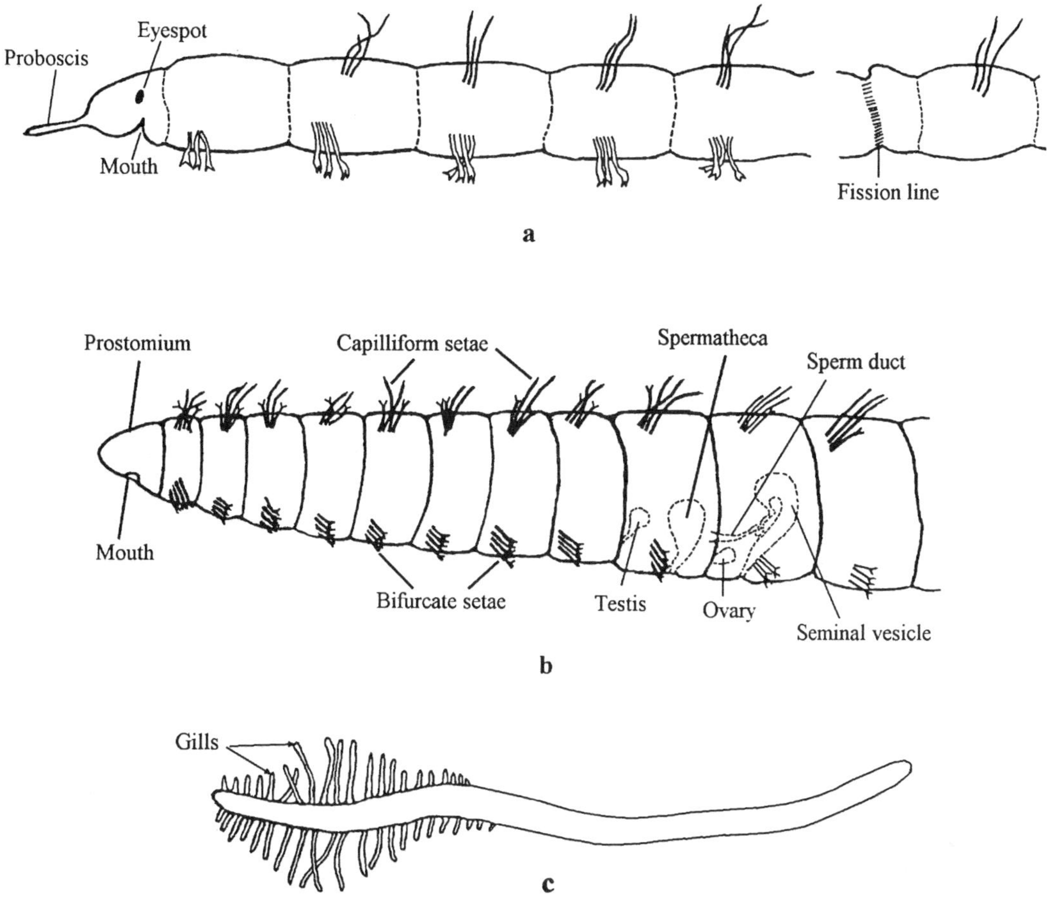

Figure 8.9. Some diagnostic features of (a) Naididae and (b) Tubificidae worms, with the exotic species, *Branchiura sowerbyi* shown in (c).

Sexual reproduction generally occurs in the summer months.

Asexual reproduction occurs by budding from a segment in the posterior region of the worm. The location of the budding zone is relatively constant within a species, but may vary according to environmental conditions within individuals and among habitats. Under favourable temperature, oxygen and food conditions, the budding zone moves anteriorly one segment or two; under unfavourable conditions, the budding zone appears to move posteriorly one segment or two. During fission, the "posterior worm" regenerates the missing anterior segments and the "anterior worm" regenerates the missing posterior segments. The whole process takes about 2 or 3 days. Some worms may bud "chains" of new individuals, especially in the spring and late summer.

Aelosomatids and naidids generally are found associated with submersed macrophytes and attached algae in the littoral and sublittoral zones. Some are algivores (e.g. *Amphichaeta americana*), detritivores (e.g. *Aelosoma* spp., *Specaria* sp.) and a few are carnivorous on insect larvae, protists and other oligochaetes (e.g. *Chaetogaster diaphanus*) are parasitic on snails and clams (e.g. *Chaetogaster limnaei*).

The tubificids are gatherers, feeding on detritus in the sediments. They are the only worms present in the deepest regions of lakes and are represented by several indicator species. The classical "pollution indicators" are *Tubifex tubifex* and *Limnodrilus hoffmeisteri*. Both species are able to survive periods of anoxia, such as occurs in the hypolimnia of eutrophic lakes during the summer and winter months. Most tubificids live in tubes with their tails extending out of the tube into the mud-water interface. Currents are created by wiggling the tail, the wiggling rate increasing with decreasing oxygen tensions. Most tubificids have ***erythrocruorin***, a red blood pigment, that effectively extracts low levels of oxygen dissolved in the water. The densities of *T. tubifex* and *L. hoffmeisteri* in sewage lagoons may be so high that the bottom appears pink.

Not all tubificids are pollution indicators. Some species, such as *Tubifex kessleri* and *Peloscolex variegatum*, require well oxygenated waters and reach their greatest densities in oligotrophic lakes.

*Class: **Hirudinea** - The Leeches and Bloodsuckers*

Although there is no general agreement on the usage of "leech" and "bloodsucker", this text uses leech for those forms that prey on invertebrates and
10☞ bloodsucker for those forms that prey on vertebrates and suck red blood of mammals, including humans. Many leeches (e.g. of the families ***Glossiphoniidae*** and ***Erpobdellidae***) attach to humans but they usually do not suck blood. In most instances, where leeches have been observed sucking blood from humans, the blood is taken from a cut or an abrasion, often between the toes.

Both leeches and bloodsuckers are characterized by an anterior, or ***oral sucker***, in the centre of which is a mouth, and a posterior, or ***caudal sucker***. The oral sucker is smaller in diameter than the caudal sucker. The body has 34 segments but each segment is subdivided by 3-14 false ***annuli***. The segments can be identified by internal ***septa*** that isolate one segment from another. Annuli lack internal septa. The first segment, with its sucker, usually bears two or more eyes. Some fish leeches have eyes on the caudal sucker. The mouth has either jaws (with or without teeth), a proboscis, both jaws and a proboscis, or no jaws at all. Figure 8.10 gives the main diagnostic features of four families of Hirudinea.

The leeches have a proboscis, with or without jaws, or no jaws at all. They are represented by three families (Fig. 8.11): ***Glossiphoniidae***, the broad leeches; ***Piscicolidae***, the fish leeches; and ***Erpobdellidae***, the linear leeches. The broad leeches can be recognized immediately by their broad width in the middle of the body, ventral oral sucker that is fused to the body, a proboscis, 1 to 4 pairs of eyes often arranged in a regular arch, and young that cling to the underside of the parent. They are

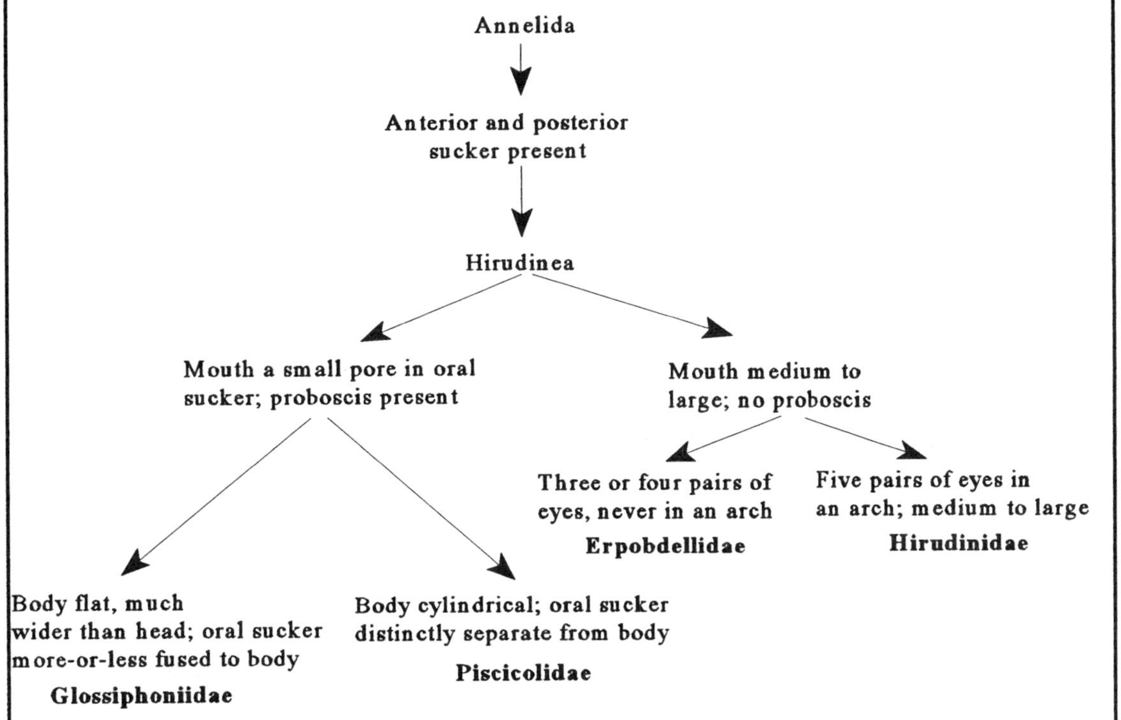

Figure 8.10. Characteristics of the four families of Hirudinea.

carnivorous, feeding mostly on insects, molluscs and oligochaetes, or scavengers, feeding on dead animal matter. The fish leeches have a cylindrical body, an oral sucker distinctly separated from the body and they are almost invariably found as parasites on fishes. The linear leeches are flattened and elongate, the mouth lacks jaws and a proboscis, and the head has 3 or 4 pair of eyes never arranged in a regular arch. They

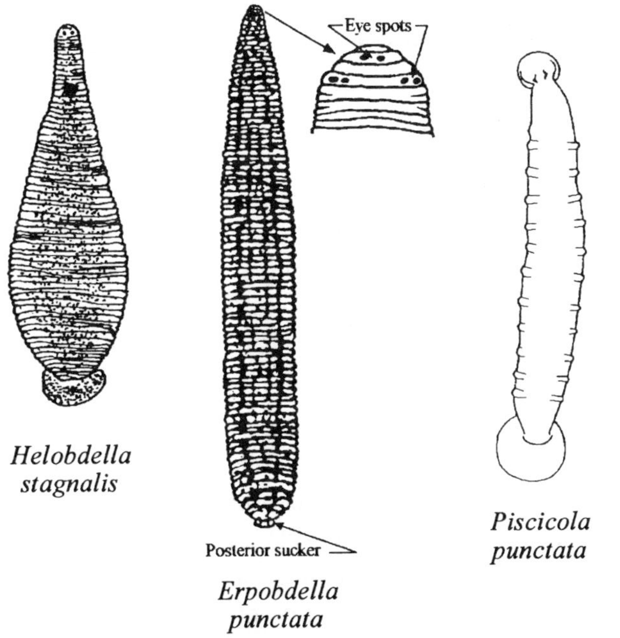

Figure 8.11. General appearance of three families of leeches showing Glossiponiidae (*Helobdella stagnalis*), Erpobdellidae (*Erpobdella punctata*) and Piscicolidae (*Piscicola punctata).*

usually eat their prey whole. If they do suck blood, it is from a wound because they have no jaws to make an incision in the skin.

Blood suckers are represented by the family Hirudinidae. They have five pairs of eyes arranged in an arch. Jaws with teeth are typically present, although a few are toothless or even jawless. Many species have elaborate colouration, such as *Macrobdella decora*, the American medicinal leech, which has a red or orange ventral surface and green dorsal surface. The true bloodsuckers have three sharp teeth that make a "Y" incision in the skin of the host. *Macrobdella* and *Philobdella* regularly feed on human blood. After making the incision, a salivary anticoagulant, ***hirudin***, is secreted to prevent the blood from clotting. The medicinal leech is able to take up blood to twenty times its body weight by storing the blood in the numerous gastric caeca. Some bloodsuckers have been starved for two years after a blood meal and survive. A few bloodsuckers, such as *Haemopsis*, have vestigial teeth and feed on small invertebrates or dead animal material. Some occasionally feed on mammals, including man. The horse leech, *Haemopsis marmorata*, obtained its common name by living in ponds and troughs and attacking cattle and horses that frequent them.

Hirudinea reproduce entirely by sexual means. They are hermaphroditic with two pairs of genital pores, the two male pores being on the ventral median line of the 11th segment and the two female pores being on the ventral median line of the 12th segment. Like oligochaetes, a clitellum is present in nearly all Hirudinea. Worms pair up head to tail during mating so that male and female genital pores are aligned. Sperm is packaged into ***spermatophores*** and then passed to the female through a penis. The individuals separate, secrete a cocoon and then pass the spermatophores and eggs into the cocoon as it is slid off the worm. Fertilization occurs in the cocoon. The cocoons of Erpobdellidae and Hirudinidae are attached to the substrate in the summer months and fully developed young emerge from the cocoon in the fall. Many bloodsuckers take 2 to 3 years to mature and some live for 10 to 15 years. The Glossiphoniidae make a membranous capsule instead of a cocoon and attach the capsule to the underside of the body. When the capsule breaks, the young can be seen attaching to the parent, feeding on mucous until they are developed enough to feed on small invertebrates. Most broad leeches live for only 1 to 2 years. Some species, like the common *Helobdella stagnalis*, may have two generations per year.

Medicinal leeches are used in Europe for ***bloodletting*** or "***leeching***". *Hirudo medicinalis*, the European equivalent of *Macrobdella decora*, is raised commercially for this purpose. In the mid 1800s, as many as 25 million leeches were used in hospitals in France for leeching purposes. Leeching, with *M. decora*, is used in North America mainly for remedying boils, contusions and abscessed teeth, and for removing blood from tissue transplants. Bloodsuckers are also being used for biochemical research and as biomonitors of contaminants (see Chapter 12). Their use as bait by sport fishermen is well known. The most common species used is *Nephelopsis obscura* because it is a scavenger or feeds on invertebrates, and rarely bites humans.

Helobdella stagnalis and *Erpobdella punctata* are considered pollution indicators. Others are *Glossiphonia complanata, Helobdella elongata, Helobdella lineata, Illinobdella moorei, Mooreobdella microstoma* and *Dina parva*. Indeed, "no leeches can be classified as a clean water - sensitive species" (Sawyer 1974)[5]. However, like all leeches, they need appropriate habitat to attain large numbers typical of indicator organisms. The presence and relative abundance of food organisms and firm substrates is a prerequisite for all leeches. They attain their maximum numbers in littoral and sublittoral zones or in areas with rocks and other solid substrates. The sucker is important for locomotion, feeding and for reproduction; substrates such as silt and mud restrict these activities. For this reason, leeches are not found in the profundal zones of lakes. Indeed, there is a high correlation between the kinds of food organisms present and the indicator leeches present. *Helobdella stagnalis* and *Erpobdella punctata* feed on chironomids and tubificid worms which themselves contain many indicator species.

Most species are found in waters with pH > 7.0 and a total alkalinity > 60 mg $CaCO_3$/L. Only the highly tolerant indicator species, such as *H. stagnalis* and *C. complanata*, are found in waters with pH < 6.0. The only physical characteristics that leeches cannot tolerate is high silt and turbidity loading. Such waters usually have a silt bottom that leeches cannot inhabit. Indeed, the tolerance of leeches to many chemicals, including soap, makes it difficult for bathers to discourage their presence. Leeches are sensitive to salinity, hence adding salt to an attached leech is effective at removing them. But adding salt to the water is not very effective. High salinity also kills other invertebrates. Similarly, adding lime will discourage the presence of leeches, but lime has to be added almost daily to keep the leeches away. The leeches simply move away from the area until the salt or lime has dispersed and then move back in. Some people "bait" an area with fresh meat, wait for bloodsuckers to attach, then collect the bait and kill the bloodsuckers by burning or burying them in a box of salt. Releasing the bloodsuckers several meters from shore probably is not effective because most can tolerate long periods of desiccation and can crawl back to the water.

Phylum: Mollusca - The Freshwater Shells

Only two classes of molluscs are represented in freshwater, the ***Gastropoda*** (snails and limpets) and the ***Bivalvia*** (clams and mussels). All freshwater molluscs
11☞ have a shell made of calcium carbonate. The shell is secreted by a ***mantle***. The shell grows in size (length, width and height) by secretions from the mantle's edge (Fig. 8.12); the shell grows in thickness by secretions from the mantle's outer surface. Pearls are made by placing a crystal of shell carbonate or a sand grain between the mantle and the shell of mussels. The crystal is sensed as foreign material and the mussel reacts by secreting calcium carbonate around the material. If the calcified

---

[5]Sawyer, R. T. 1974. Chapter 4. Leeches (Annelida: Hirudinea). *In:* Pollution ecology of freshwater invertebrates. Academic Press, New York, NY. pp. 82-142.

material is not imbedded in the shell and is free to roll around in the space between the mantle and shell, a perfectly round pearl will result. Otherwise, the material becomes fixed and trapped into the shell.

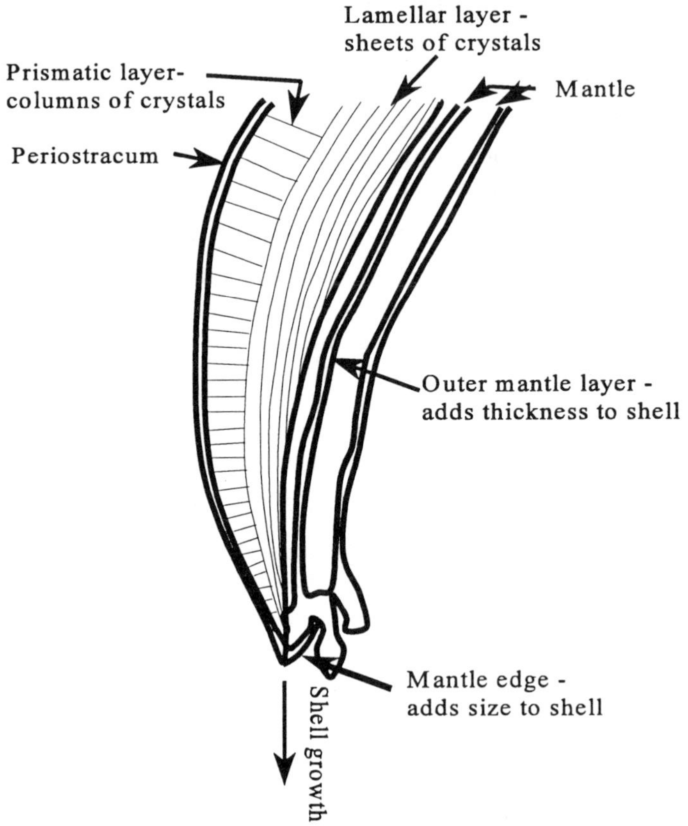

Figure 8.12. Structure of a bivalve's shell with columns and sheets of calcium carbonate in a conchiolin matrix.

The calcium carbonate crystals in the shell are bound together in columns or sheets, or sometimes it is randomly distributed, within a protein matrix. The protein, called ***conchiolin***, also makes up the outer organic layer of the shell, or the ***periostracum*** (Fig. 8.12). Pigments may be added to the periostracum to give the shell a characteristic colour pattern.

All freshwater molluscs move by a muscular foot. The foot extends out of the shell by blood rushing into the foot and filling the numerous spaces within. When blood is forced out of the foot and back into the body cavity, the foot can be retracted by contraction of retractor muscles. The foot is a creeping organ in gastropods but a burrowing organ in bivalves. In zebra and quagga mussels the foot is poorly developed because they spend most of their lives attached to the substrate. Zebra and quagga mussels attach to surfaces by means of ***byssal threads*** secreted by a ***byssal gland*** in the base of the foot. The mussel may secrete as many as 12 threads per day. As many as 500 to 600 threads are secreted by the time the mussel reaches 3 to 4 cm in length.

The snails and limpets are exclusively grazers, feeding on attached algae, or herbivores, feeding on leaf and stem tissues of macrophytes. The clams and mussels are exclusively ***filter feeders***. Some bivalves also ***deposit feed***, that is they use cilia on the foot to take up detritus, algae, bacteria and other food deposited on the sediments.

Fish and waterfowl, and occasionally aquatic mammals (muskrat, otters), are the main predators of freshwater molluscs. Many waterfowl seek invertebrates high in protein and calcium contents, like Mollusca, during egg production periods. Some turtles (e.g. *Graptemys pseudographica kohnii*) feed on Asian clams and small gastropods.

Class: Gastropoda - The Snails and Limpets

There are approximately 485 species of freshwater gastropods in North America. About 15 of these are introduced but only a few are of socioeconomic
12☞ concern or have quarantine significance, as discussed in Chapter 14. Freshwater gastropods are classified into two groups, the ***Prosobranchia*** and ***Pulmonata***. The prosobranchs are most easily recognized by the presence of an ***operculum*** in the aperture of the shell. The operculum is a lid, or cover, that closes the aperture and protects the animal within. It is attached to the hind part of the foot and is the last structure to be withdrawn when the animal retracts within the shell. Prosobranchs have gills for extracting oxygen dissolved in the water and the sexes are separate (in all families except Valvatidae). Pulmonates, as the name suggests, have lungs for respiring atmospheric oxygen. They are able to utilize oxygen dissolved in the water but will occasionally rise to the surface for atmospheric oxygen. Pulmonates can be recognized immediately by the lack of an operculum. In contrast to the prosobranchs, all pulmonates are hermaphroditic.

There are five families of prosobranchs. They can be distinguished from one another on the basis of their shell shape and the type of operculum present (Figs. 8.13, 8.14). The ***Valvatidae*** is the only family with a multispiral operculum. Two families have a semi-spiral, but ***Hydrobiidae*** species are small and globular in shape and the ***Pleuroceridae*** are tall with a pointed tip. Two families have a concentric operculum, but the operculum is withdrawn deeply into the rather large shell (> 1 cm high) of ***Viviparidae*** and is kept near the aperture of the smaller (< 1 cm high) ***Bythiniidae***.

The pulmonates are represented by four families. The limpets are easy to recognize because they lack a coiled shell (Figs. 8.13, 8.14) and the shell looks like one valve of a bivalve. Members of the family ***Planorbidae*** are also easy to recognize because the shell is "discoidal" (i.e. coiled in one plane). Snails in the remaining two families have a tall ***spire***, or vertical coiling (Figs. 8.13, 8.14), but the aperture opens on the left (if the shell is held with tip pointing up and aperture toward you) in ***Physidae*** and on the right in ***Lymnaeidae***.

Gastropods feed by scraping algae and detritus off the surfaces of hard substrates, such as rocks and macrophytes. Some, like *Physella* and *Lymnaea*, are scavengers and feed on dead animal material. The scraping is accomplished with a tongue-like ***radula*** unique to snails. The radula has several rows of sclerotized (hardened protein) teeth, like a rasp, the teeth being smaller in species that feed on algae than in those that feed on plant tissues. The radula moves back and forth, the anterior teeth gradually wearing away but being replaced bythe forward growth of the entire radula, like the human fingernail. Some common species, like the pleurocerids, *Elimia livescens* and *Pleurocera acuta*, and the viviparids, *Campeloma decisum* and *Viviparous georgianus*, take sand grains in with their detrital food and use the sand as a

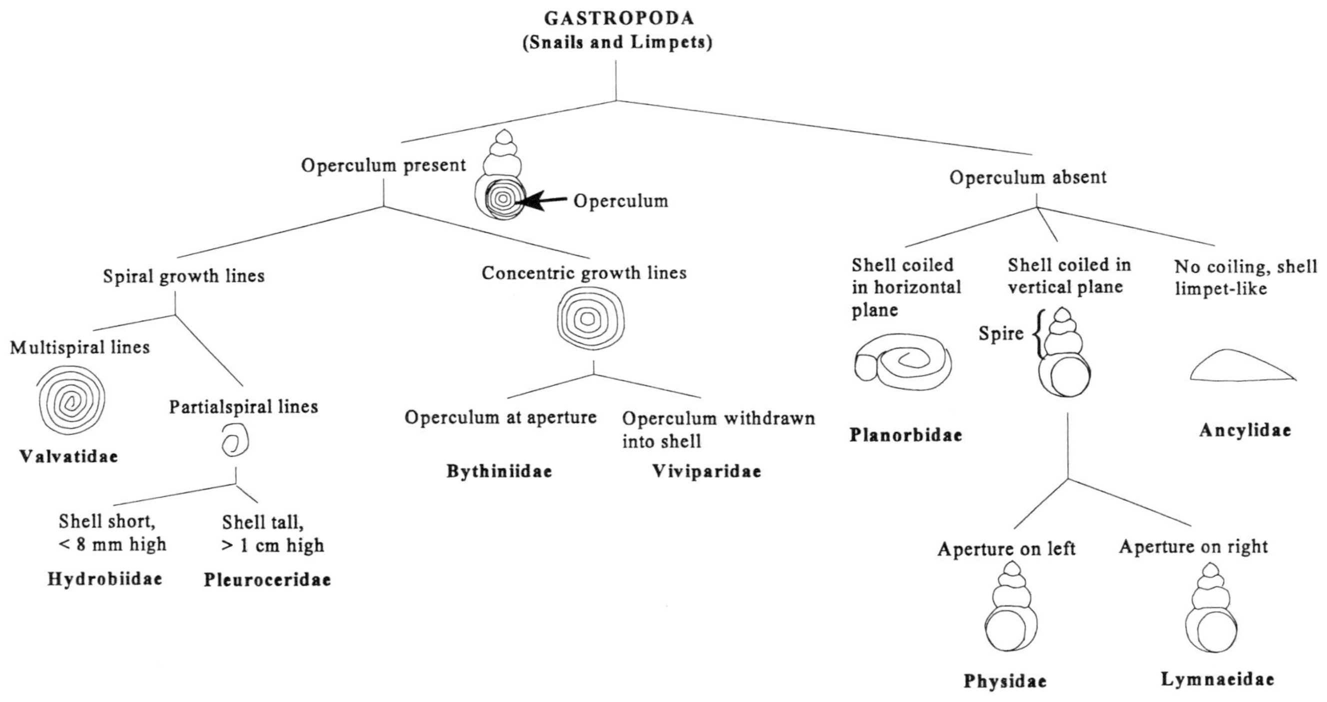

Figure 8.13. Families of freshwater snails

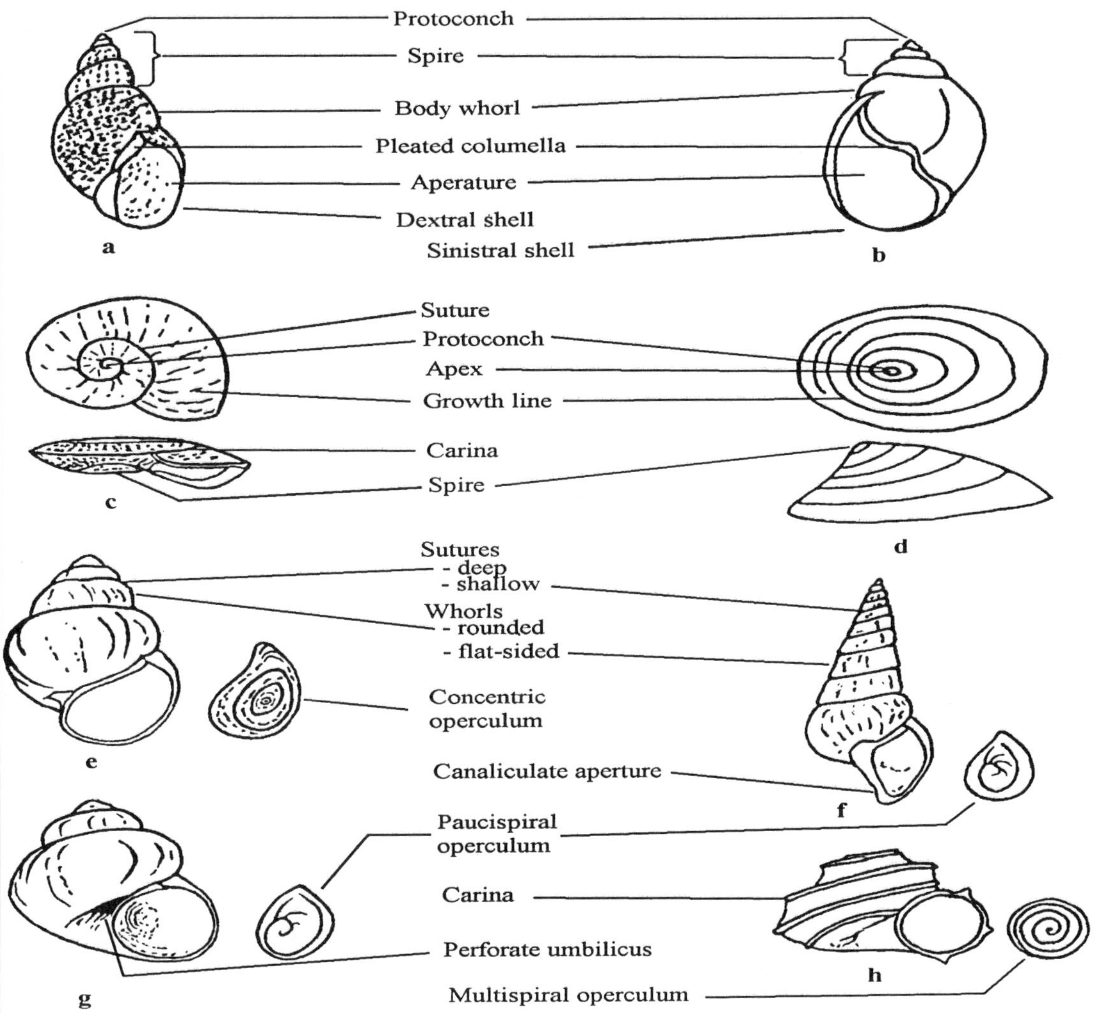

Figure 8.14. Some diagnostic features of eight families of freshwater gastropods showing (a) Lymnaeidae, (b) Physidae, (c) Planorbidae, (d) Ancylidae, (e) Viviparidae, (f) Pleuroceridae, (g) Hydrobiidae and (h) Valvatidae.

triturating (grinding or pulverizing) agent.

Most gastropods are ***oviparous*** (egg layers) laying eggs in masses on firm substrates such as macrophytes, rocks and logs. The numbers of eggs layed by oviparous forms varies considerably within and among species, but rarely are more than 100 produced by one female. Some species, like many lymnaeids, produce eggs all summer, with the young and the adults overwintering. In the following spring, the adults lay more eggs and then die. Most have one-year life cycles, with either one or two egg laying periods (spring and summer). Development is direct and tiny snails emerge from the eggs in the summer and fall.

Exceptions to this type of life cycle are the Viviparidae, which are ***ovoviviparous*** and brood their young, producing miniature adults in the late summer to early fall after breeding in the spring. The ovoviviparous forms are less fecund than oviparous forms and usually produce less than 10 young per parent, or a maximum of 50 per parent. Most viviparids live for 2 to 3 years. Some exotic species live for 3 to 4 years and do not need a mate to fertilize the eggs. This phenomenon, known as ***parthenogenesis***, is common in rotifers and other lesser developed taxa, but is rare in molluscs.

Considerable research has been done on the ecological and physiological tolerances and requirements of gastropods. Pulmonates tend to be more tolerant than prosobranchs of enrichment because pulmonates can rise to the surface to obtain oxygen when the dissolved oxygen supply is depleted. Most physids are known to tolerate anoxia for a short period of time but they, like all gastropods, need water well saturated with oxygen for proper development of eggs. Similarly, many prosobranchs, like some pleurocerids and viviparids, can tolerate near-anoxia, but only for short periods of time.

Because gastropods have a calcium carbonate shell, most require waters high in pH and alkalinity in order to grow and reproduce at optimal rates. Pulmonate species known to tolerate acidic waters (pH < 7.0, total alkalinity < 40 mg $CaCO_3$/L) are the physid, *Physella gyrina*; the lymnaeids, *Stagnicola humilis, Stagnicola columella* and *Radix auricularia*; and the planorbids, *Helisoma anceps, Planorbella trivolvis* and *Gyraulus parvus*. However, they also occur in waters with pH as high as 9.0 and an alkalinity exceeding 350 mg $CaCO_3$/L. Many prosobranchs, like most pleurocerids, hydrobiids, viviparids and valvatids, begin disappearing from waters when the pH falls below 7.0,.

Few, if any, gastropods are good trophic indicators because most prevail in the littoral and sublittoral zones of lakes. In these zones conditions can range from barely tolerable to optimal. For example, oxygen concentrations, nutrient levels, pH, alkalinity, light penetration, water currents and other chemical and physical factors vary hourly, daily, and seasonally, as well as with depth and distance from shore. The best trophic indicators tend to be those in the deeper waters, like the profundal zone, where conditions are somewhat predictable before and after thermal stratification occurs. *Bythinia tentaculata*, or the faucet snail, is an introduced species and in Europe it is commonly associated with enriched waters, clogging pipelines, some often coming through taps, hence its common name. But in North America the species seems to prefer clean, sandy sediments.

Gastropods are well known intermediate hosts of digean trematodes, as discussed in Chapter 7. Refer to Chapter 7 for the life cycle of trematodes. The larval trematodes, called rediae and sporocysts, usually invade gonadal and digestive gland (similar in function to our liver) tissue and either castrate and sterilize the snails or kill them directly. The adult trematodes are also capable of killing humans. This includes the tropical *Schistosoma mansoni*, that grows in and occludes the veins of humans, causing the disease known as ***schistosomiasis***.

Class: Bivalvia - The Clams and Mussels

About 270 species of freshwater bivalves are present in North America, 7 of which have been introduced, the most infamous of which are the Asian clam and the zebra and quagga mussels. There is no general agreement on the usage of “clams” and
13☞ “mussels”, but in the marine environment, “***mussel***” is used to describe any member of the family, ***Mytilidae***, such as *Mytilus edulis*, the common edible mussel. They, like the zebra mussel, attach to surfaces of rocks and other firm substrates by means of byssal threads. “***Clams***” is used to refer to burrowing forms and includes all the native freshwater bivalves in North America. Nevertheless, the term, “freshwater pearly mussel” is commonly used to refer to any member of the family ***Unionidae***.

The most characteristic feature of bivalves is the two shells, or valves, that articulate by a dorsal ***hinge*** (Fig. 8.15). The oldest part of a valve is the dorsal, central region, called the ***umbo***. Shell growth is concentric around the umbo, creating concentric grow lines. During the fall and winter, grow rate slows considerably and leaves concentric ridges, called ***annuli***. Since each annulus is layed down annually, the number of annuli represents the clam’s age. The surface of the valve is called the ***disc***.

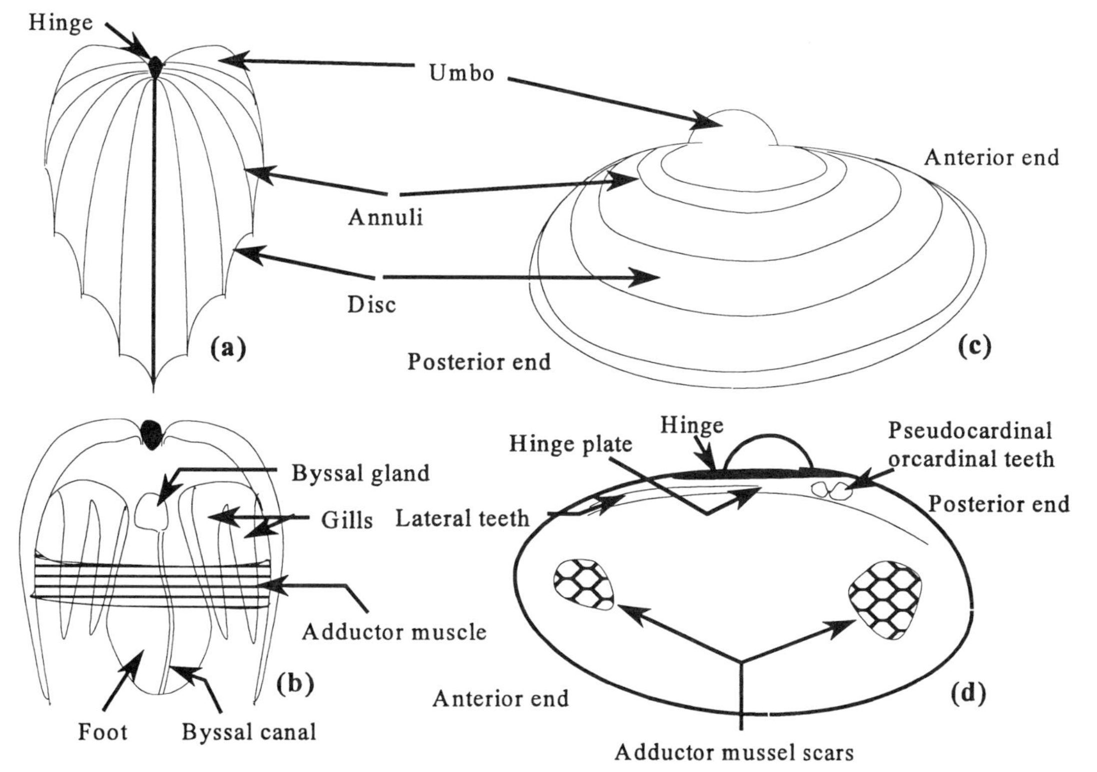

Figure 8.15. Features of a bivalve shell. (a) View from rear (posterior) end. (b) Cross section through a mussel showing soft parts discussed in text. (c) View from left side (left lateral view). (d) Inner features of left valve.

The disc may have various kinds of ornamentations, including ridges, tubercles and grooves that are often unique and diagnostic of a species. The periostracum on the disc may also be distinctly coloured or patterned and aid in the identification of bivalves.

The hinge is elastic and causes the valves to open, or gape. The valves close by means of two ***adductor muscles***, one anterior and the other posterior (Figs. 8.15 b, d). In clams the anterior and posterior adductor muscles are about the same size; in true mussels, the anterior adductor is considerably smaller than the posterior adductor muscle. When clams and mussels are found dead on the shore, the valves will be found gaping if the hinge is still intact (and the adductor muscle has disintegrated).

Clams move about by means of a muscular foot that extends between the gaping valves. Clams are able to burrow by first pushing their foot into the sediments. Blood then rushes to the tip causing the foot to swell. The swollen tip of the foot anchors the clam in the sediments and then retractor muscles pull the shell over the foot, thus pulling the clam deeper into the sediments.

Although zebra and quagga mussels do have some ability to crawl, the foot is poorly developed. Instead, the mussels are adapted to life on hard substrates. The adaptations include: (1) a flat ventral surface that allows the animal to be pulled tightly against the substrate by the byssal apparatus, making it difficult for predators to pry the

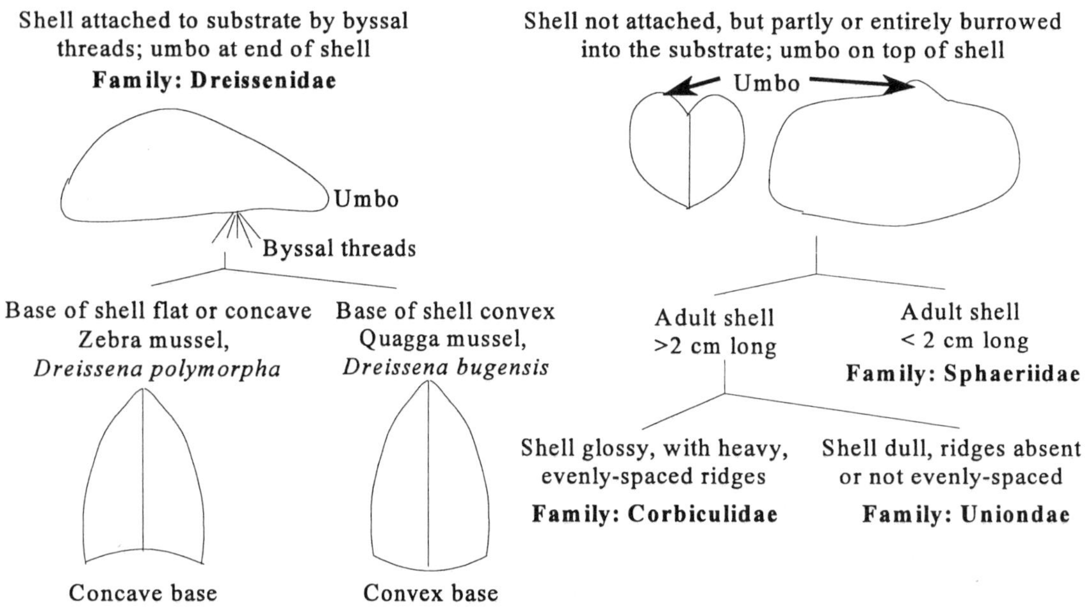

Figure 8.16. Diagnostic features of the four families of freshwater bivalves and of zebra and quagga mussels.

shell from the substrate (Fig. 8.16); (2) the umbo (Fig. 8.16) is lateral and adjacent to the substrate, giving the animal maximum upright stability at the surface of the substrate (Fig. 8.16); and (3) the shell is tapered dorsally (tent-shaped in cross-section, Fig. 8.16) making it difficult for predators to get a firm hold to pull the shell from the hard surface. In native clams, the umbo is rounded and dorsal in position (Fig. 8.16) and adapted for life within sediments. The ventral margin is also rounded and lacks a permanent opening because adults do not have a functional byssal gland.

There are two families of bivalves native to North America, the ***Sphaeriidae*** (fingernail clams) and the ***Unionidae*** (freshwater pearly mussels), and two families that were introduced from Europe, the ***Corbiculidae*** (Asian clams) and the ***Dreissenidae*** (zebra and quagga mussels) (Fig. 8.17). Corbiculidae is represented solely by the Asian clam, *Corbicula fluminea* and Dreissenidae by the zebra mussel, *Dreissena polymorpha*, and the quagga mussel, *Dreissena bugensis*. Figures 8.16 and 8.17 show the distinctive features of each family and of the exotic species.

The native freshwater bivalves all produce larvae within brood sacs or within pouches of the gill structures themselves. Zebra mussels and quagga mussels produce a free swimming larval stage. Each female mussel produces between 100,000 and 1,000,000 eggs. Hence, the numbers of adults may increase by orders of magnitude from one year to the next and increases their nuisance potential. The main feature that makes zebra and quagga mussels a nuisance is the byssus, used to firmly attach adults to the surface of the substrate. In addition, juveniles, as they settle out of the water column, settle upon adult mussels and grow so that the thickness of the colony also increases over time. The mussels are capable of reducing a three-foot diameter pipeline to one foot in diameter!

All bivalves are filter-feeding organisms. Water enters the mantle cavity through an incurrent siphon. Food suspended in the water is passed over the gills. Cilia on the gills sort the food into digestible and undigestible components. The former is passed to the mouth for digestion; the latter is enveloped in mucous and ejected out the lower siphon as ***pseudofaeces*** (false faeces). The material is called pseudofaeces because it has not been processed through the digestive tract. True faeces are passed out the excurrent (upper) siphon.

Because zebra mussels are so prolific in numbers and they are so efficient at filtering the water, there has been a noticeable increase in the clarity of water in the Great Lakes since their arrival in 1985. For example, the Secchi depth in Lake Erie increased from about 1.5 m to about 3.5 m in the first eight years that the mussels had been in the Great Lakes. The water clarity is suspected to have a profound impact on those organisms that feed on the plankton. This includes several zooplankton species, larval species of fish that feed upon the zooplankton, and adult planktivorous fish.

In general brooding species are less fecund than oviparous species because brooding space for the developing larvae within parents limits the numbers being reared. For example, the internal shell volume in sphaeriid clams remains relatively constant so that the reproductive output cannot be significantly altered and will always be relatively low. Although the numbers are highly variable, brooding forms are about an order of

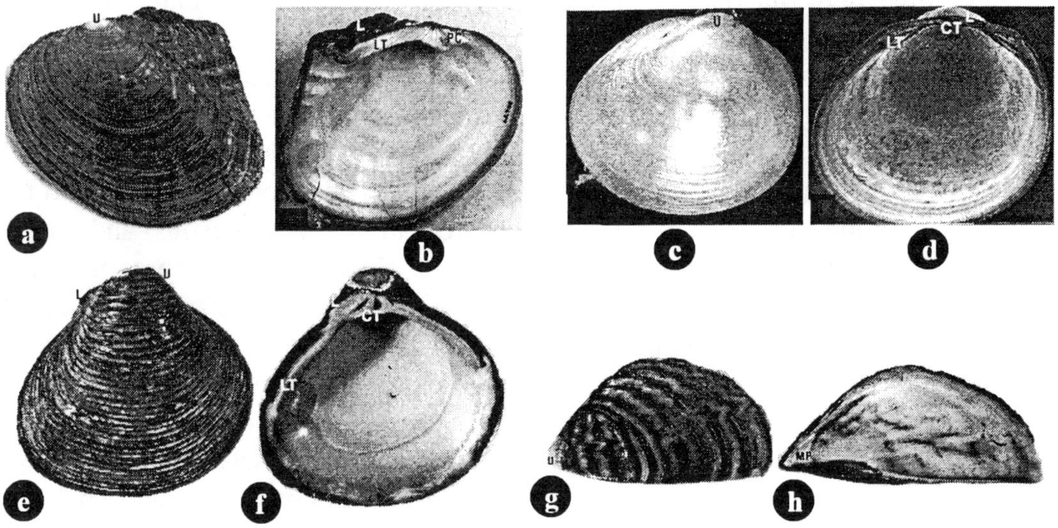

Figure 8.17. Four families of freshwater bivalves: Unionidae showing outer (a) and inner (b) views; Sphaeriidae showing outer (c) and inner (d) views; Corbiculidae showing outer (e) and inner (f) views; and Dreissenidae showing outer (g) and inner (h) views of valves. CT = cardinal teeth, L = ligament, LT = lateral teeth, MP = myophore plate, PC = pseudocardinal teeth, U = umbone.

magnitude (10 X) less fecund than oviparous forms (e.g. 10:100). Even though many brooding forms are hermaphroditic, reducing the risk of having to find a mate, the fecundities (egg numbers) are still relatively low. Even ovipositing, oviparous forms have low natalities (birth rates) relative to planktonic oviparous forms. Ovipositing oviparous forms are about three to four orders of magnitude less fecund than planktonic oviparous forms (e.g. 100:100,000 to 100:1,000,000).

The Asian Clams

Corbicula fluminea is medium in size, averaging 3 to 4 cm shell length (Fig.
14☞ 8.17 e, f). The shell is glossy with heavy, evenly-spaced ridges. The hinge plate has well developed hinge teeth, with serrated (saw-like) lateral teeth.

The Asian clam is a warm-water species and cannot survive waters that freeze. In the Great Lakes, the species occurs in areas that receive warm water plumes from hydro generating stations. The species is most common and abundant in rivers and lakes in mid and southern states of the United States. Asian clams are not likely to occur in inland lakes of Canada.

The sexes of Asian clams are separate, although some flowing water populations appear able to switch to hermaphroditism. Spawning typically occurs in the spring and early summer months. Sperm are released into the water and taken into the mantle cavity by the female. Eggs, fertilized internally, develop within marsupia of the female's gills. *Corbicula* seems to produce fewer eggs than *Dreissena*, with only 25,000 to 75,000 veligers produced in the lifetime of a single Asian clam. After about four days, the embryos develop into free-swimming ***veligers*** at which time they are released from the female. The pelagic or planktonic stage is common to all the exotic biofoulers (zebra mussels, quagga mussels and Asian clams), although the planktonic stage of *Corbicula* lasts only 2 to 4 days. Nevertheless, the planktonic stage allows the clams to enter pipelines and develop and mature into adults within pipes and holding tanks. Asian clams differ from zebra mussels in that the adult clams do not have a byssus to attach to the walls of structures; instead, the calms rattle around in the pipeline, growing, reproducing and eventually clogging the pipe. In lakes and rivers, the clams burrow into the sediments, like native clams. Asian clams appear to be capable of out-competing and displacing native species of clams from their habitats. Asian clams are common in enriched waters and can tolerate water with as little as 50% oxygen saturation, but not for prolonged periods.

The Fingernail Clams:

Fingernail clams are closely related to the Asian clams. They owe their common name to their size. There are four genera, but only three, *Sphaerium, Musculium* and *Pisidium*, are common. Most *Sphaerium* species are large, about 8 to
15☞ 20 mm; *Pisidium* species (Fig. 8.17 c, d) are the smallest, most ranging in shell length from about 2 to 6 mm; most *Musculium* species are intermediate in size, about 8 to 10 mm in shell length, and the shells are thin and fragile. Only *Musculium transversum* and *Sphaerium simile*, grow to 2 cm in shell length. Shell colours are rather monotonous, ranging from uniform cream to uniform brown, with no patterning.

Fingernail clams differ from freshwater pearly mussels not only in their smaller size but in their hinge plate that is finer and has ***cardinal teeth*** instead of ***pseudocardinal teeth***. Cardinal teeth are smooth and distinctly tooth-like; pseudocardinal teeth are rough and ridge-like or stump-like. The teeth serve only to align the two halves of the shell; the teeth have no feeding function.

Most species of *Musculium* can be found in temporary aquatic habitats. One species, *Musculium transversum*, is an enrichment indicator, reaching its greatest densities in organically enriched waters that may have as little as 25% oxygen saturation. *Musculium transversum*, is uncommon in temporary aquatic habitats.

Most other fingernail clams require clean water with high oxygen tensions. In fact, some fingernail clams are oligotrophic indicators. *Sphaerium nitidum* and *Pisidium conventus* reach their greatest densities in the profundal zones of oligotrophic lakes or in the shallow waters of oligotrophic lakes in high northern latitudes. While most fingernail clams are not assigned to any indicator group, they seem to be most abundant in sandy bottoms and waters with at least 75% oxygen saturation. Some, like *Sphaerium simile, Sphaerium striatinum*, *Pisidium casertanum, Pisidium compressum*, and *Pisidium adamsi* are abundant in organic sediments but the waters are usually well saturated with oxygen, as commonly occurs in most river and stream environments.

One species, *Sphaerium fabale*, is found only in running waters.

The fingernail clams are hermaphroditic with one part of the gonad used as a testis and the other as an ovary. They, like Asian clams, are brooders, but development proceeds through four larval stages within marsupial sacs attached to filaments of the inner gills (Fig. 8.18). As many as four sacs may be present on a gill. Each larva is attached to the gill by a single byssal thread. One to ten larvae may occur within each sac but usually only three to ten develop into the ***extra-marsupial stage***, the stage that breaks free from its byssal thread and the marsupial sac and is released from the parent through its excurrent siphon. The newborn closely resemble the parent but are one quarter to one fifth the size of the parent. The size of the newborn is one reason that few young are produced by fingernail clams.

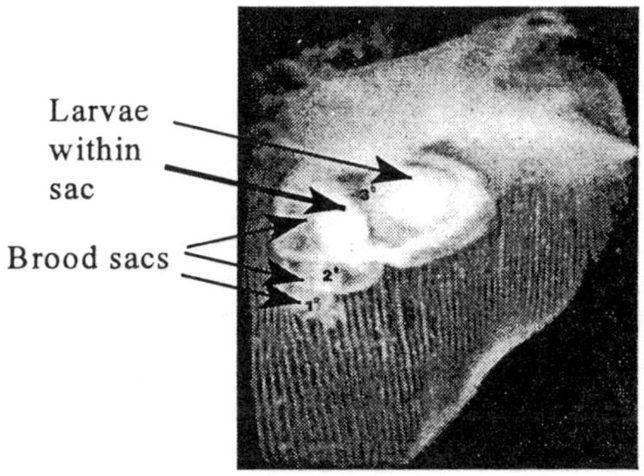

Figure 8.18. Primary, secondary and tertiary brood sacs attached to inner gil of a fingernail clam.

The Freshwater Pearly Mussels:

The most familiar bivalves are the pearly mussels (Fig. 8.17 a, b). Most species are large (30-150 mm), but some may grow to nearly 250 mm (~ 10 in) in shell
16☞ length. The shell surface varies from dull to glossy with either monotonous colouration or highly colourful patterns. The discs of some species are ornamented with wavy ridges, tubercles or grooves.

The Unionidae have separate sexes. All pearly mussels produce a parasitic larval stage, called a ***glochidium*** (Fig. 8.19). Anywhere from 1,000,000 to 2,000,000 glochidia are brooded in marsupia created within either the inner gills or both inner and outer gills, depending upon the species. Some species are ***bradytichtic***, in which the glochidia are brooded for a short term (e.g. April to August); others are ***tachytictic*** and brood the glochidia for a long term (e.g. mid-summer to the following spring or summer). The glochidia are released either through the lower siphon or through minute pores in the gills, depending on the species. Clouds of glochidia are released when an appropriate fish species passes by. The edges of the siphons in some species of pearly mussels are elaborately adorned to attract or lure fish toward the mussel. When the fish bites at the "lure" the mussel spews out the glochidia. Opercular respiratory movements draw most of the glochidia into the fish's gill spaces. The glochidia enter

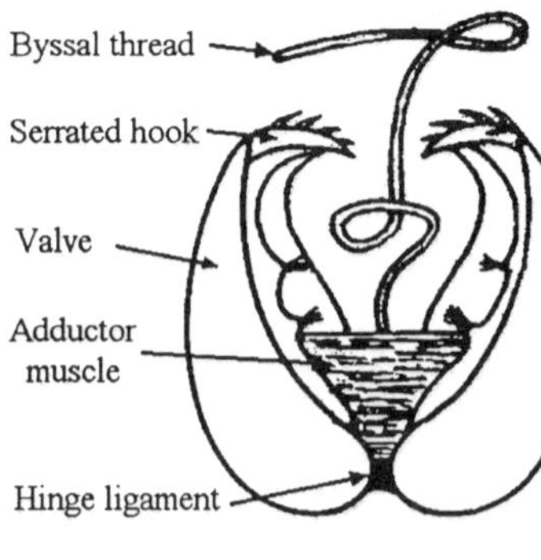

Figure 8.19. A unionid glochidium larva.

the gill tissue, encyst and live as parasites on fish blood. In some species, the glochidia attach to the tail or fins, but they are still parasitic on fish blood. During the parasitic stage (about two to four weeks), the glochidia develop most of the essential organ systems, break free of the gill filaments and then drop to the sediments to begin a free-living juvenile life that lasts for 1 to several (up to 100!) years, depending on the species. Many species secrete a single byssal thread to attach itself to the substrate for part of its juvenile life. The reproductive system is the last organ system to develop at which time it is called an adult mussel. The new juveniles lack fully developed gills and obtain food by "***pedal feeding***" whereby cilia on the foot direct food to the mouth. After about 30-60 days of juvenile life the gills and siphons are fully developed and ***filter feeding*** replaces most of the pedal feeding activities. Most pearly mussel species depend on fish as the dispersal agent for the glochidia. A few species parasitize amphibians, like mudpuppies.

The mortality of glochidia is high. Of the millions of glochidia brooded, it is estimated that less than 0.0007% develop to adulthood. Mortality can be attributed to: (i) inability of glochidia to find the proper fish host; (ii) the fish, perhaps host to thousands of glochidia, is preyed upon or caught by anglers; (iii) the fish disperses the juveniles to an inappropriate habitat; (iv) the glochidia and juveniles die due to a variety of natural biotic mortality factors, including parasitism and disease.

Zebra Mussels and Quagga Mussels:

Zebra mussels were first discovered in 1988 in Lake St. Clair but probably first arrived in 1985. Quagga mussels were first discovered in 1990 in Lake Ontario but probably first arrived in 1988 or 1999. The diagnostic features of zebra and quagga
17☞ mussels are shown in Figs. 8.16 and 8.17 g, h but they also differ in a few ecological and physiological traits, as shown in Table 8.1. Because of the quagga mussel's ability to reproduce in cooler waters and survive in soft substrates, they will be found in deeper, colder waters of deep lakes and occur further north than zebra mussels. Conversely, zebra mussels will probably prevail on hard substrates in the shallow waters of lakes and will be the main species in the southern United States where water temperatures are warmer than found at higher latitudes. The species cause similar kinds of impacts, as described in Chapter 14.

There are three important characteristics that separate zebra and quagga mussels from our native species. (i) The adults can produce byssal threads that allow the species to attach firmly to hard substrates, including rocks and walls of pipelines. The ability to attach has allowed the mussels to exploit the surfaces of hard substrates. No other native species has this ability, and zebra and quagga mussels arrived largely unchallenged for an epifaunal existence. (ii) Dreissenids are the only freshwater bivalves to produce a planktonic larval stage. All native species are brooders. Being planktonic means that larvae can be pumped into any pipeline, and once inside can develop and grow into adults and stay there by means of their byssal threads. Unless pumps take in large amounts of sediment with the water, or fish that are carrying glochidia, native species of clams cannot enter pipelines. (iii) All exotic species, including the Asian clam, produce enormous numbers of free-living planktonic larvae. Pearly mussels produce millions of parasitic glochidia but only a small percentage ($<$ 0.0007%) reach adulthood.

Table 8.1. Differences between zebra and quagga mussels.

Features	Zebra Mussel	Quagga Mussel
Shell size and shape	Up to 4 cm long, flat or concave ventral surface	Up to 3-4 cm long, ventral surface convex
Habitat	Mainly on rocks and hard substrates; chunks of colonies, or "druses", may break off and become a substrate in soft sand and mud. Otherwise, adults cannot grow and reproduce in soft substrates.	Adults can grow and reproduce on either hard or soft substrates.
Depth	Do better than quagga mussels in shallow (1-5 m), wave swept zones.	Do better than zebra mussels in deeper water.
Threshold temperatures for growth and reproduction	Need > 10 °C to grow Need 12-15 °C to release eggs and sperms	Can grow at 4-6 °C and reproduce at 7-9 °C.

Zebra and quagga mussels cannot survive in streams unless they are constantly replaced by populations that live within lakes or impoundments upstream. Here's why: Any organism that produces planktonic larvae (veligers in this case) cannot survive longer than one adult generation because the current would carry the larvae downstream. The distance that it is carried downstream depends on the duration of larval development and water velocity. Veligers are planktonic for 20 to 28 days (let's say 20 for ease of calculation here); with an average velocity of, say 0.01 m/sec (a slow stream!), the larvae would be carried 0.01 m/sec x 60 sec/min x 60 min/hr x 24 hr/day x 20 days = 17,280 m or 17.28 km downstream before they settle out! Without a population upstream to replenish the loss of adults downstream, zebra mussels cannot possibly survive longer than one life span in streams. The only reason zebra mussels are doing well in the Mississippi River is because of the numerous impoundments that allow the mussels to establish "seed" populations at regular intervals along the length of the river. Without the dams the zebra mussels would not be able to persist!

Sexes are separate in both species of mussels. At temperate latitudes, spawning may last three to five months. Longer spawning periods are being recorded at southerly latitudes and in thermally enriched waters. Depending on where the cottage is located, one can expect larvae in a water supply when waters reach 10-12°C (50-55°F). The warmer waters of many of the southern states may allow the zebra mussel to be even more prolific than in the northern latitudes. The chemistry (e.g. pH and calcium levels), physics (e.g. water temperature) and biology (e.g. trophic status), will determine how prolific the mussels will be. The roles of the chemical, physical, and biological variables are discussed in Chapter 14.

Gametes (eggs and sperms) begin to form when mussels reach about 5 mm shell length. Spawning occurs when the mussel attains about 10 mm shell length and temperatures are appropriate. Most mussels spawn at the same time (late May to early June) but a few spawn later (e.g July and August) and a few others spawn earlier (early to mid May). Hence, spawning can occur throughout the spring and summer months, with a peak in June. Once the ovaries and testes are ***spent***, the gonads begin ripening immediately and gonadogenesis may continue throughout the winter months. By early spring the ovaries and testes are ripe again for release of gametes a second time. Most adults die after 1.5 to 2 years of life. Spawning begins when water temperatures reach 10 to 12° C for zebra mussels and 7 to 9° C for quagga mussels. However, it is not unusual to find veligers in lake water that is well below these temperatures. Shallow bays may reach threshold reproductive temperatures sooner than the main body of water. Currents can transport veligers from the warmer bays into the main body of colder water. The colder water does not kill the veligers but it does slow their growth and development rates. The larval life cycle of *Dreissena* usually takes about four weeks to complete, and is described in detail in Chapters 7 and 14. The relationship between the various developmental stages and the annual temperature cycle is well known and described by Claudi and Mackie (1994)[6].

The adult secretes numerous byssal threads throughout its life. As many as 500 to 600 threads may be secreted in its life time. Some mussels (juvenile or adult) may detach by releasing the entire byssal apparatus from the base of its foot. Such mussels are known as ***translocators*** because they move from one substrate to another. Some adults can be carried great distances by water currents.

Phylum: Arthropoda - The Arthropods

The phylum Arthropoda is represented by three classes of macroinvertebrates in fresh waters: ***Arachnoidea*** (spiders and mites), ***Crustacea*** and ***Insecta***. The insects are
18☞ the most diverse of the three classes. Only the most common taxa within each of the
classes found in fresh waters are described. The diagnostic features of the three classes are given in Fig. 8.20.

Class: Arachnoidea -The Water Mites

The arachnids are represented in freshwater only by water mites of the dubious taxon, ***Hydracarina*** (Fig. 8.21). Hydracarina is a term of convenience, not an official
19☞ taxonomic term. Water mites are found primarily in the littoral zone and are mostly
predaceous on larval insects. A few (e.g. *Unionicola*) are parasitic or symbiotic on

---

[6]Claudi, R. and G. L. Mackie. A practical manual on zebra mussel monitoring and control. Lewis Publishers, Boca Raton, Fl.

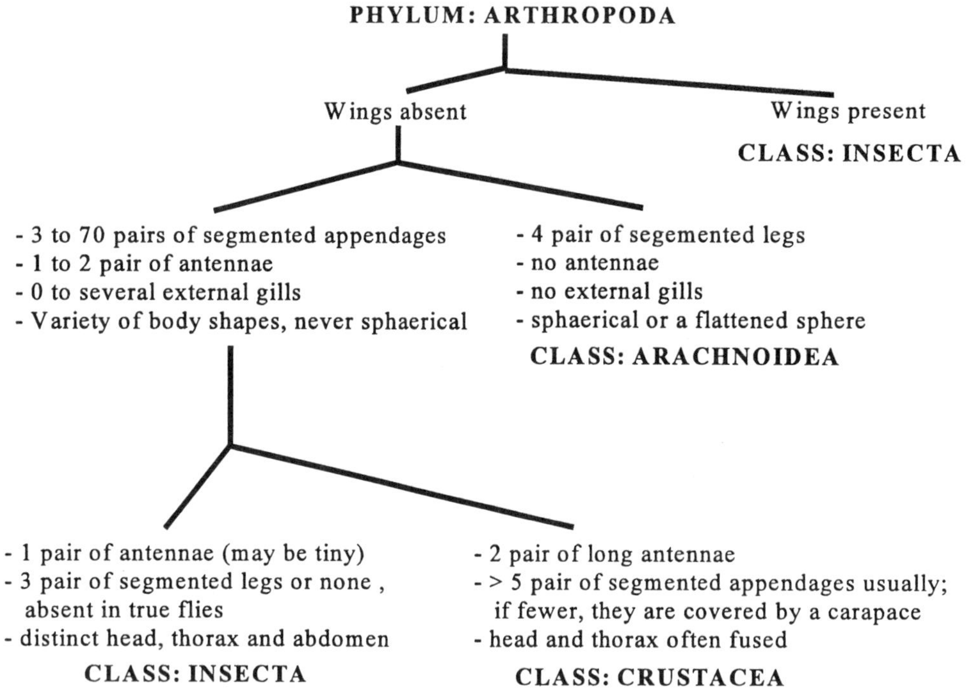

Figure 8.20. Diagnostic features of three classes of arthropods.

freshwater bivalves. Like all arachnids, water mites have four pairs of segmented legs and they lack antennae. They differ from spiders in that they have one main body part called the ***opisthosoma***. The anterior part of the opisthosoma bears one pair of sessile compound eyes and a mouth. The most prominent mouthpart is one pair of sensory organs known as ***pedipalps***. Other mouthparts are used for holding and chewing their prey. The legs are hairy and adapted for swimming from plant to plant or just above the surface of the sediments in the littoral zone.

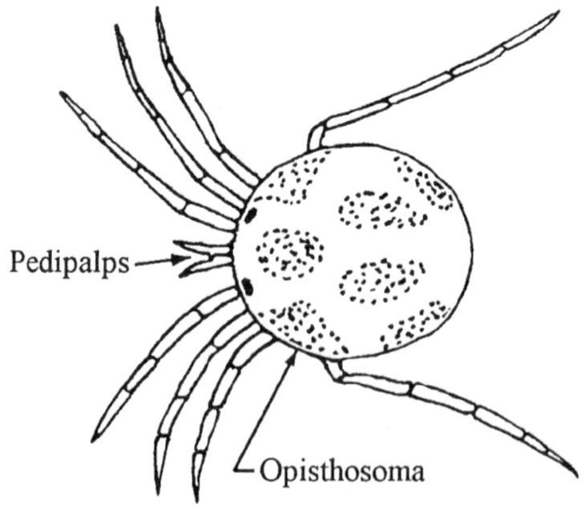

Figure 8.21. A typical water mite (Hydracarina).

The sexes are separate. The breeding season is usually from May through July but may extend throughout the year depending on species and temperature, the latter varying with latitude and season. Eggs, often reddish in colour, are deposited in clusters of 20 to 400 on rocks or vegetation. Eggs hatch in 1 to 6 weeks into larvae that bear 3 pairs of legs. After a short free-swimming period the larvae attach to and parasitize an insect and develop the fourth pair of legs and a reproductive system to become an adult. A few, such as *Piona* and *Limnesia*, species common in streams, do not require the parasitic stage to mature into an adult. Instead, they develop within the egg mass, obtaining nourishment from the egg.

A few genera, such as *Piona, Limnesia, Thyas*, and *Hydrophantes* are characteristic of temporary ponds. Some species (e.g. most species of *Hygrobates*) are adapted for life in running waters. These adaptations include dorso-ventral flattening, small size, strong claws and a reduction or absence of swimming hairs. Some genera, such as *Brachypoda, Torrenticola, Koenikea, Wettina*, and *Badakia*) are diagnostic of cold, clear, headwater streams. Most are found in standing waters, the most common being *Hydrachna, Limnochares, Eylais, Lebertia* and *Arrenurus*, the largest genus with about 120 species. Some mites are highly coloured, such as *Arrenurus* (red or green), *Neumania* (common in standing and running waters, has bright red or blue) and *Axonopsis* (of northern lakes and ponds, has a mix of red, green and yellow).

Class: Crustacea - The Freshwater Crustaceans

The class Crustacea is represented by 13 **orders** , but two are uncommon, found only in wells or subterranean waters of streams. Figures 8.22a and 8.22b give
20☞ diagnostic features of the 11 most common crustacean orders. The ***Cladocera*** and ***Copepoda*** are described in Chapter 7 and are not discussed further here. Of the remaining nine orders, the ***Anostraca*** (fairy shrimps, e.g. *Eubranchipus, Branchinecta*), **Notostraca** (tadpole shrimps, e.g. *Lepidurus*), and ***Conchostraca*** (clam shrimps, e.g. *Lynceus*) are found almost exclusively in temporary ponds. The Cladocera, Anostraca, Notostraca and Conchostraca all belong to the subclass, ***Branchiopoda***. All develop ***resting eggs*** that carry the populations over the dry period. Some produce ***summer eggs*** that hatch immediately and develop before the pond dries. Hatching and development of resting eggs begins shortly after the ice is gone. Life spans are more or less equal to the wet period of the pond.

The order, ***Branchiura*** (of the subclass Copepoda), is represented solely by fish parasites, hence the common name fish lice. They are not discussed further here. The subclass, ***Ostracoda*** (seed shrimps), is a diverse group, with over 300 species in North America. They have opaque shells, shaped like seeds or kidney beans. Seed shrimps are abundant and widely distributed and inhabit either temporary or permanent, standing or running waters. Some are parasitic on the gills of crayfish. Unfortunately, because of their small size and opaque shells, seed shrimps are usually ignored in benthic surveys. This probably explains why little is known about their physiological and ecological tolerances and requirements.

The remaining four orders of crustaceans are represented by species that are

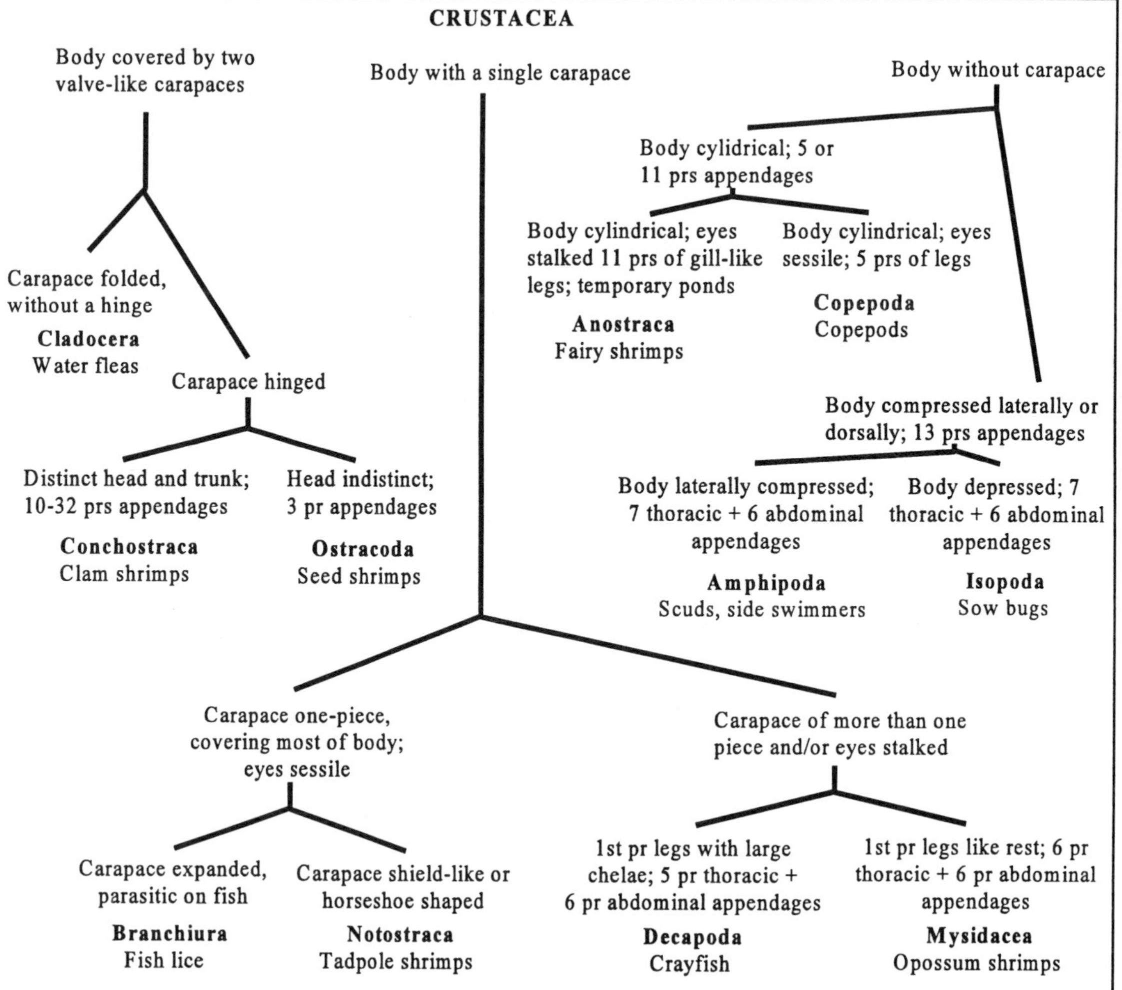

Figure 8.22a. Diagnostic features of eleven freshwater groups of Crustacea.

among the most common macroinvertebrates in lake and stream benthic communities. They all belong to the subclass, ***Malacostraca***, and include ***Amphipoda***, the scuds or side swimmers; ***Isopoda***, the aquatic sow bugs; ***Decapoda***, the crayfish; ***Mysidacea***, the opossum shrimps.

Crustaceans are characterized by a division of body segments into a **cephalothorax** which includes both the head (= prefix, cephalo) and the thorax, and a distinct **abdomen**. They also have several pairs of segmented appendages, the number of pairs being characteristic of each order. The most diagnostic features of the four macrobenthic crustacean orders are give in Table 8.2 and Fig. 8.22a. Refer to Fig. 8.24 for identification and location of the various body parts.

21☞ ***Order: Amphipoda - The Scuds or Side Swimmers*** (Fig. 8.23)

The order name refers to the double leg types; the prefix, "Amphi" is Greek for

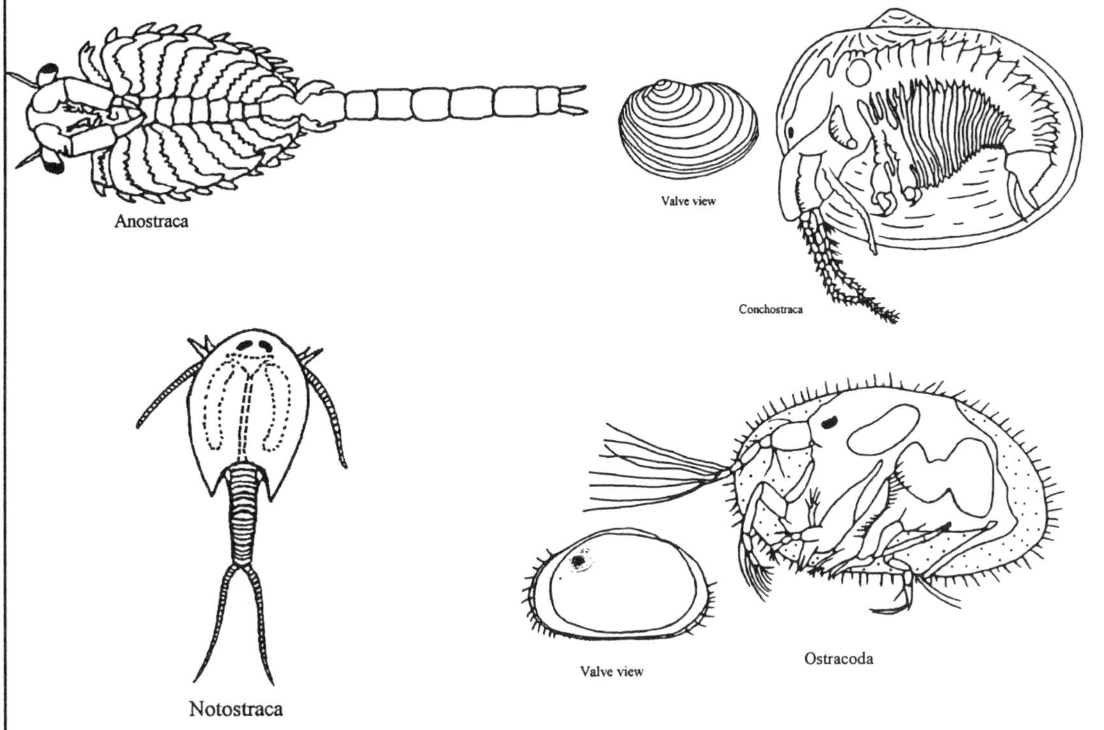

Figure 8.22b. Three orders of crustaceans common in temporary and permanent ponds: Anostraca (fairy shrimp); Conchostraca (clam shrimps); and Notostraca (tadpole shrimps), as well as the subclass Ostracoda (seed shrimps).

"both sides of" (that is both thorax and abdomen) or "double", and the suffix, "poda", is from the Greek word "podos", meaning "foot". They owe their common names to their habits and their swimming orientation. The term, scud, refers to their role as scavengers and detritivores of bottom sediments. Amphipods are somewhat laterally flattened, causing them to swim mostly on their sides. The vast majority of North American species are subterranean.

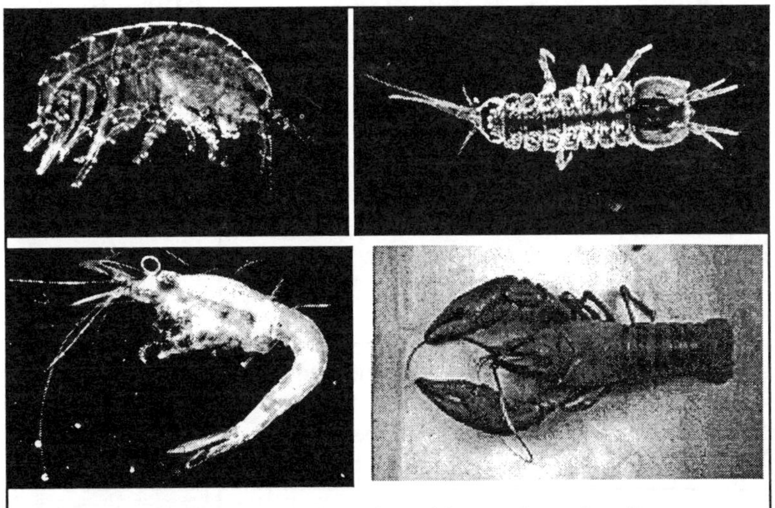

Figure 8.23. Representative species of four orders of malacostracan crustaceans common in lakes and rivers: Amphipoda (*Hyalella azteca*); Isopoda (*Caecidotea racovitzai*); Mysidacea (*Mysis relicta*); and Decapoda (*Orconectes propinquus*). First three photographs are courtesy of Jonathan D. S. Witt.

Amphipods have

separate sexes and the female broods her clutch below her abdomen. Mating occurs throughout spring and summer, when males reach their 9th instar and females their 8th instar. Pairing may continue for 1 to 7 days or until the female is ready to molt to the first adult instar. The female then passes the eggs into her abdominal marsupium. The male transfers sperm to the marsupium where the eggs are fertilized. Only after these exercises are complete does the pair separate. The eggs are incubated for about three weeks. Development is direct, with the eggs hatching into juveniles. The juveniles are retained for a few days and then released with the first molt following copulation.

Amphipods are represented by about 100 species in North America, but only a few are very common, like *Crangonyx gracilis, Gammarus fasciatus, Gammarus lacustris* and *Hyalella azteca* (currently under taxonomic revison). *Hyalella azteca* (in the strict sense) is so ubiquitous and abundant that their absence is considered a reliable indicator of lake acidification. They can tolerate pH's down to 6.5, at which point they begin to disappear. Hence, their absence is often considered a good indicator of the lake or stream acidifying below pH 6.5. *Diporeia hoyi* is found only in deep, cold, oligotrophic lakes. However, their preference for deep waters appears to depend upon their requirement for cold water because they have been found in profundal zones with less than 7% oxygen saturation. *Gammarus fasciatus* is an introduced species (prior to 1940), at least to the Great Lakes. The most recent introduction is *Echinogammarus ischnus*; it was first found in Lake Erie and Lake St. Clair in 1996. Additional details on these exotics are given in Chapter 14.

Table 8.2. Diagnostic features of four orders of benthic crustaceans.

Features	Amphipoda	Isopoda	Decapoda	Mysidacea
Body characteristics	Laterally compressed	Dorso-ventrally compressed	Large carapace covers entire thorax	Carapace partly covers thorax
Type of eyes	Sessile	Sessile	Stalked	Stalked
No. and type of thoracic appendages	7 pr, 1st 2 pr are chelae-like, last 5 pr have a claw on each leg	7 pr, 1st pr have chelae, next 6 pr are similar	5 pr, 1st pr are large and chelate, next 2 pr have two claws on each leg, last 2 pr have 1 claw	6 pr, all modified for swimming, legs getting increasingly longer with each segment
No. and type of abdominal appendages	6 pr, 1st 3 pr used for swimming, last 3 pr have uropods on each	6 pr but all are gill-like and placed one above the other	5 pr, 1st pr modified as sex organs, next 5 pr used for swimming	6 pr, smaller than thoracic legs; 4th leg much longer in male than female
Feeding type	Scavengers or detritus feeders, omnivorous	Scavengers of dead and dying organisms	Mostly omnivorous, will prey upon snails, insects, small fish, but mostly herbivorous	Mostly filter feed on zooplankton, sometimes are raptorial

Order: Isopoda - The Aquatic Sow Bugs (Fig. 8.23)

The Isopoda are dorso-ventrally flattened. When disturbed they roll up into a ball, much like the garden pill bug. The legs of isopods are similar, hence the order
22☞ name "Iso" from the Greek word "isos", meaning equal or similar, and "poda", from the Greek word "podos", as explained above for the amphipods. The similarity of leg types can be attributed to the fact that the legs are present only on the thoracic segments, which make up most of the animal's length. The abdominal appendages are hidden beneath the abdomen, the last segment on the animal. The tail-like appendages rising from the abdomen are called ***uropods***.

Sexes are separate in isopods. Sperm are transferred to the female ***seminal receptacle*** where they are held until the female releases her eggs into the oviducts. The eggs are fertilized as they pass through the seminal receptacle. The fertilized eggs are brooded in a marsupium between the bases of the thoracic legs of the female. The numbers of eggs brooded varies among species and populations, but averages about 25 to 30, with a maximum near 250. Eggs and newly hatched young are brooded for 3 to 4 weeks and then released as miniatures of the adult. Most isopods have a 1 year life span.

Isopods are scavengers and detritivores, feeding mostly on dead or dying animals. The common species, *Caecidotea* (formerly *Asellus*) *communis* and *C. racovitzai* can be found in large numbers in waters that are subpolluted with organic wastes. Although the order is often considered an indicator of moderate enrichment or subpollution, only certain species, such as *C. communis* and *C. racovitzai*, can be considered such. Other species are restricted to other kinds of habitats. For example, *Thermosphaeroma* is found only in hot springs, *Caecidotea kenki* is confined to cold streams and spring-fed creeks and *Lirceus garmani* is confined to temporary ponds and springs and are reliable indicators of those kinds of habitats.

Order: Decapoda - The Crayfish, or Crawdads, and Freshwater Prawns, or Glass Shrimps (Fig. 8.23, 8.24)

Arguably, the best known of the freshwater arthropods (including insects and mites), the decapods are immediately recognized by their 10 (= deca) legs (poda), or 5
23☞ pairs of legs, the first pair of which are enlarged into pincers, or ***chelae***, (like a lobster claw). The next two pair of legs are also chelate, but the pincers are much smaller. Only a single claw is present on the remaining pair of walking legs.

The head and thorax of crayfish are covered by a strongly sclerotized ***exoskeleton*** and their eyes are stalked. Crayfish also have an expanded tail (commonly called a tail fan) composed of a central ***telson*** and paired lateral ***uropods***. When disturbed, the tail fan is provided with powerful muscles that, when flexed, bends the telson down and under the abdomen for a lightning-fast, backward escape. The entire body, including the abdomen, is more or less dorso-ventrally compressed. There are more than 350 species of crayfish in North America, but the majority of freshwater species belong to two genera, *Orconectes* and *Cambarus*.

Included within the decapods are the ***freshwater prawns***, or ***glass shrimps***, that resemble the marine tiger shrimp. Being decapods, freshwater prawns have ten walking legs but only the first two pair are chelate and the animal is laterally compressed. The legs of freshwater prawns are adapted for swimming. Three genera are present in North America, but two are confined to brackish water or to the southern United States. Only two species are fairly common in Canada and northern U. S., *Palaemonetes paludosus* and *Palaemonetes kadiakensis*. Both are referred to as glass shrimp. Some attempts have been made to culture glass shrimp but with little success at the commercial scale.

Sexes are separate in decapods. Two forms of males occur during maturation. The forms are determined from the degree of sclerotization (hardening) of the first pair

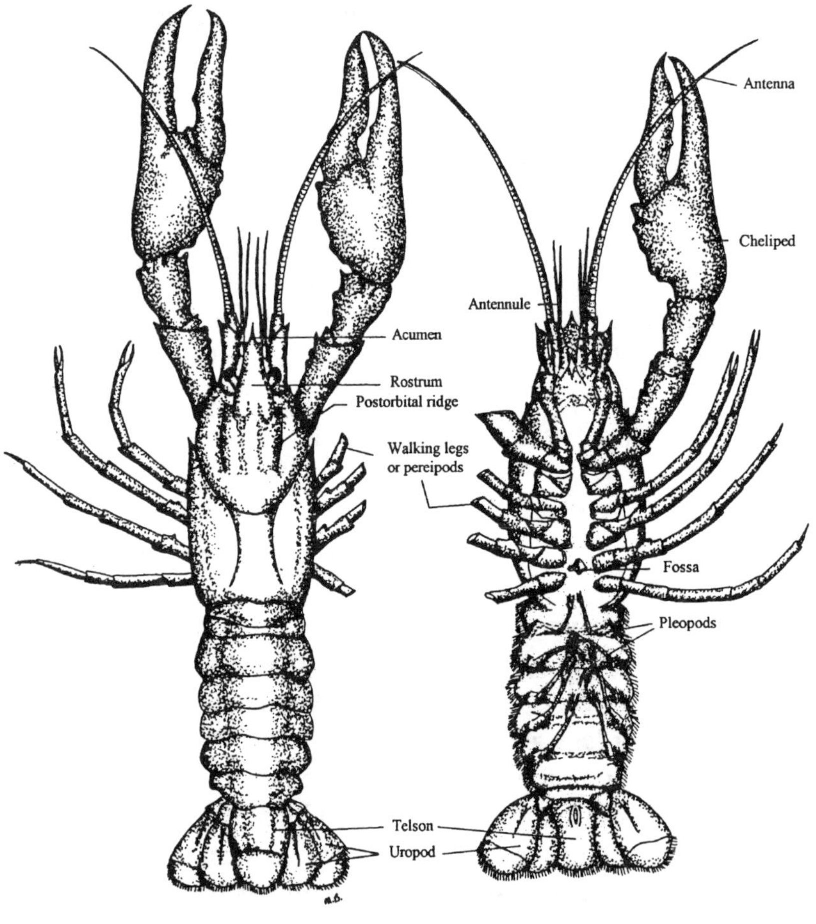

Figure 8.24. Dorsal and ventral views of a crayfish showing some major body parts.

of abdominal appendages (called ***pleopods***). In ***first form males***, the first pair of pleopods are firm and elaborately structured for copulating and transferring sperm to the female. In ***second form males*** the first pleopods are flimsy and poorly developed. The second form male develops after the first form male molts; the first pair of pleopods lose their rigidity and the male transforms into the ***second form***. A second form male can undergo a molt to become a first form male again. Copulation occurs between spring and autumn and may occur more than once, depending on the species. During copulation, the male transfers sperm to the seminal receptacle of the female and they separate. After several weeks the female cleans the ventral side of her abdomen, secretes a cement-like substance, called ***glair***, and then releases the sperm which stick to the glair. The female then lays on her back, curls her abdomen and releases the eggs. The eggs become dispersed among the glair and are fertilized by the sperm. The eggs are attached to the abdominal appendages and are bathed with well-aerated water by constant movements of the appendages. The female is said to be "***in berry***" while carrying the eggs. The larger the female, the greater the number of eggs, which usually number in the hundreds. Incubation may range from 1 to 20 weeks, depending on temperature. The young of crayfish may molt 6 to 10 times before reaching maturity; freshwater prawns may molt 5 to 8 times before sexual maturity. The life span of most crayfish is 2 years.

Crayfish live in burrows, the entrance of which can be identified by a "***chimney***" of pellets of mud. Chimneys normally have a height of about 15 cm (6 in) and are built only at night. Crayfish generally inhabit shallow waters, usually less than 1 m deep. Most crayfish are sensitive to low oxygen tensions and high silt loads. Stream-dwelling species appear to be more sensitive to low oxygen tensions than are pond or lake-dwelling species. On the other hand, some species are confined to turbid habitats. *Orconectes immunis* is called the "***mud crayfish***" because of its predilection for muddy substrates. Most crayfish inhabit both running and standing waters, like the ubiquitous *Orconectes propinquus*, which can be found on most rocky substrates in the Great Lakes drainage, and *Orconectes rusticus*, which was introduced to Canadian waters from states in the mid U. S. Others, like *Cambarus bartoni* and *Cambarus robustus*, are restricted to lotic environments, while *Cambarus diogenes* burrows only in wet fields and marshy areas, entering ponds and streams only to breed.

Crayfish have considerable economic importance in the U. S. As much as 40 million tons have been harvested from the Gulf Coast, Mississippi, Pacific, and Great Lakes drainages in a year. In addition to these "wild stocks", some states (e.g. Louisiana) culture crayfish commercially. Two species, *Procambarus clarkii* (the red crayfish) and *Procambarus blandingi* (the white crayfish) are cultured because they have rapid growth rates and mature sexually after 6 months. They have one year life spans and are harvested after the young are born.

Order: Mysidacea - The Opossum Shrimps (Fig. 8.23)

24☞ The mysids are mostly a marine group, with only three freshwater species, two of which are confined to the southern U. S. *Mysis relicta* is the only temperate species, found almost exclusively in cold, oligotrophic lakes. It is found mostly in the hypolimnion of deep lakes, occurring only in the epilimnion to feed on zooplankton at night. During the day, *Mysis* is found about a meter above the bottom, but at dusk they begin to ascend. They tolerate the warmer temperatures of the epilmnion just long enough to feed and descend again at dawn. When zooplankton is not available they filter feed on suspended organic material.

They resemble crayfish but the carapace is thin and does not completely cover the thorax; the 5 pair of thoracic walking legs are replaced by 6 pair of appendages modified for swimming by being biramous (each leg is split into two), setose (hairy) and long. As Table 8.2 shows, mysids have 6 pair of abdominal appendages, the 4th pair being modified for transferring sperm to the female; in females, the 4th abdominal appendage is reduced to make room for a marsupium, used to brood up to 40 eggs per clutch. The eggs and young are brooded for 1 to 3 months, the young emerging when they are 3 to 4 mm long. Adults grow to about 3 cm and live 2 to 4 years in cold northern waters and about 1 to 2 years in more temperate lakes.

Class: Insecta - The Aquatic Insects

The greatest diversity in form and habit is exhibited by the insects. They occupy every kind of freshwater habitat imaginable, including temporary streams and
25☞ ponds, the shallowest and deepest areas of lakes, the most pristine and most polluted rivers, roadside ditches, eaves troughs, moss, within and on macrophytes and all ranges of water chemistry, from acidified to alkaline bodies of water. They also represent all the functional feeding groups, including predators, shredders, grazers (or scrapers), filter feeders, gatherers, piercers and parasites.

Insects can be separated immediately from other arthropod classes by the presence of: (1) one pair of antennae; (2) 3 pairs of segmented legs in adults and most larvae (only the larvae of true flies lack segmented legs); (3) 1 to 2 pair of wings on the adults (Figs. 8.25). They are conveniently divided into three taxonomic groups based on the types of wings that develop from the larval or nymphal stages and on the type of life stages present. Those without wings are ***apterous*** (Greek: a = without; pterous = wings) and they have no change in body form after hatching from the egg. This type of development is called ***ametabolous*** (Greek: a = without; metabolous = change) metamorphosis. Only the springtails of the order, ***Collembola***, (Fig. 8.25, 8.26) have ametabolous development in aquatic environments. Springtails are easily distinguishable from other insects in that they live and are adapted for life on the surface of the water, they lack wings, and they have a ***furcula***, or "spring" that allows them to hop along the surface of the water

The remaining two groups have winged adults. The wings develop externally in the ***exopterous*** (Greek: exo = outside; pterous) forms, and internally in the

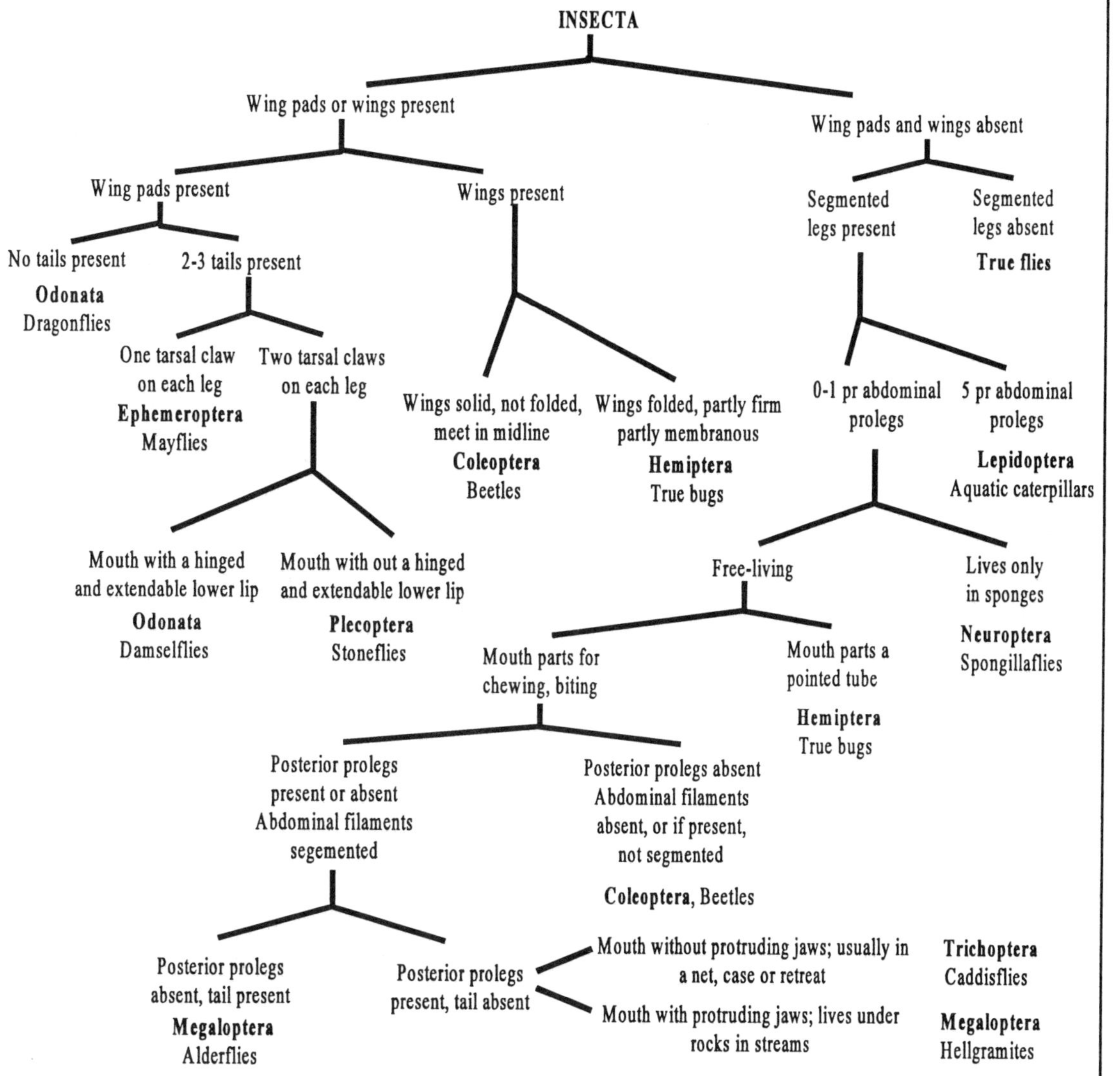

Figure 8.25. Diagnostic features of aquatic orders of insects.

endopterous (Greek: endo = inside; pterous = wings) forms. The development stages of the exopterous forms are called ***nymphs*** or ***naiads*** and they closely resemble the adults in appearance. The nymphs undergo several molts, or ***instars***, before transforming into an adult. The act of shedding or casting the old skin is called ***ecdysis*** and the cast skin is called an ***exuvium*** (plural = exuviae), commonly referred to as the "***shuck***" by fishermen. The adult is called an ***imago*** and it differs from the last instar in having fully developed wings and sexual organs. The adults and nymphs live in different habitats and feed differently. The entire process, from egg through several nymphal instars, is called ***hemimetabolous*** (Greek: hemi = incomplete or inseparable; metabolous = change) development. Only three ***orders*** of insects have hemimetabolous development, the ***Odonata*** (dragonflies and damselflies), ***Ephemeroptera*** (mayflies) and ***Plecoptera*** (stoneflies).

Very similar to the hemimetabolous forms are the ***paurometabolous*** (Greek: pauros = little, small; metabolous = change) forms in which wings develop externally on the nymphs but the nymphs and adults are often difficult to distinguish and both live in the same habitat and feed similarly. Development is gradual and the changes between instars are subtle and barely noticeable in most instances. Only one order, the ***Hemiptera*** (or true bugs) have paurometabolous development. Aquatic ***Orthoptera***, or aquatic crickets, also have this type of development but they are rare and not discussed here.

The endopterous forms have a **holometabolous** (Greek: holo = complete; metabolous = change) type of development in which the egg develops into a worm-like ***larva***. The larvae have no external evidence of wings. Segmented legs and antennae are usually present but may be missing in a few orders. Once the larva is fully grown it transforms into a resting stage, called the ***pupa***, during which it metamorphoses into an imago. The wings, legs, antennae and compound eyes develop in the pupal stage. Most species of aquatic insects have holometabolous development, including all the ***Megaloptera*** (dobsonflies, alderflies, hellgrammites), ***Neuroptera*** (spongeflies), ***Lepidoptera*** (aquatic butterflies), ***Trichoptera*** (caddisflies), ***Coleoptera*** (beetles) and ***Diptera*** (true flies). Note that "true flies" is two words; only flies of the order Diptera are spelled with two words, for example, crane flies, black flies, midge flies, etc.); all other "false flies" are spelled with one word (e.g. mayflies, caddisflies, etc.).

The characteristics of all aquatic orders of insects are given in Fig. 8.25. With development and life cycles already described above, the ecological characteristics of each order are described below. The information is taken from the authoritative reference on aquatic insects, Merritt and Cummins (1996), cited earlier.

Order: Collembola - The Springtails (Fig. 8.26)

Their size and water-repellant skin keep springtails afloat on the surface of the water. The few species (~ 50 in North America) live either in temporary habitats (ponds) or permanent habitats (lakes, rivers, etc.). Most springtails are less than 3 mm in length. All are detritivores feeding on a variety of dead plant materials, bacteria and algae trapped in the surface film. Spring runoff and floods are the main dispersal agents. Life on the surface film makes springtails of little indicator value of water quality.

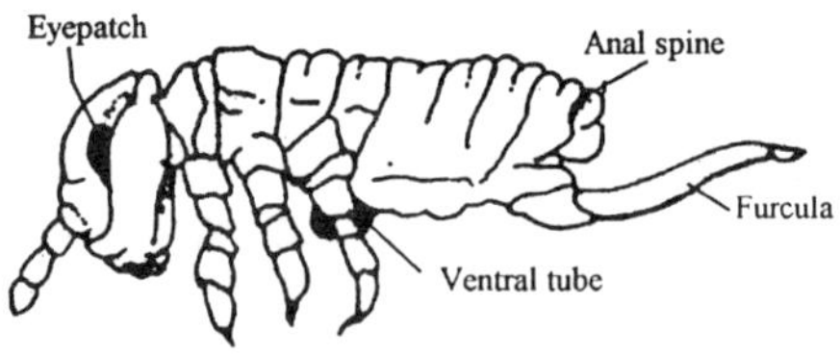

Figure 8.26. A springtail showing the furcula, or spring, used for hopping on the surface of the water.

Order: Odonata - The Dragonflies and Damselflies (Fig. 8.27, 8.28)

The dragonfly and damselfly nymphs share some
26☞ common features, such as a hinged, extendable lower lip, or labium, with powerful jaws used for grabbing and tearing prey (Fig. 8.27). All odonates are predaceous, feeding on small insects, crustaceans and annelids. They are found in both lentic and lotic environments, but only in the littoral and sublittoral zones of the former. Some damselfly and dragonfly nymphs are restricted to lotic waters, others to lentic waters. The damselflies belong to the suborder, Zygoptera, and the dragonflies to the suborder, Anisoptera.

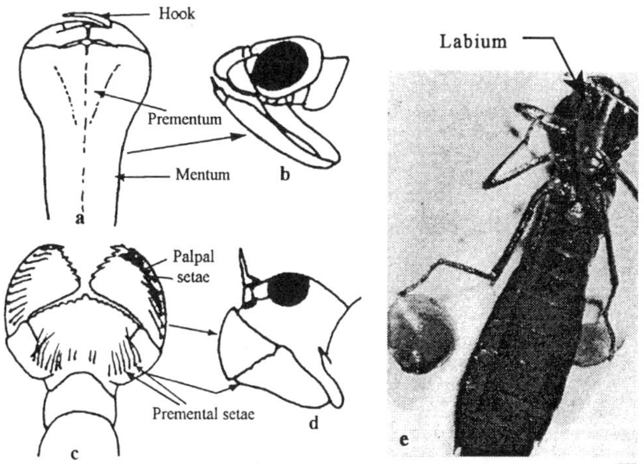

Figure 8.27. Labium of damselfly (a-b) and dragonfly (c-d) nymphs, with dorsal views in a and c and lateral views in b and d. (e) is a photograph of the ventral side of a gragonfly nymph; two *Pisidium* sp. are attached to its tarsal claws

Suborder: Zygoptera - The Damselflies (Fig. 8.28a)

Damselfly nymphs are sleek and slender with a long tapering body, the head being much wider than the abdomen, and have three feather-like or leaf-like tails, or gills at the posterior end. Most damselfly nymphs are climbers, some are clingers, confined to the shallow waters of lotic and lentic environments, like *Amphiagrion, Lestes, Coenagrion, Argia, Enallagma* and *Ischnura*. *Enallagma*, however, prefers more alkaline waters, even brackish waters. A few common genera are restricted to lotic environments, like *Calopteryx* (= *Agrion*) and *Heteraena*, others to lentic environments, like *Nehalennia*, common in bog mats.

Suborder: Ansioptera - The Dragonflies (Fig 8.28b)

Dragonfly nymphs are more robust than damselfly nymphs. The dragonfly's head is about a wide as its abdomen and instead of having feather-like tails it has spines on the posterior end. Most dragonfly nymphs are burrowers and sprawlers on the substrate. Only members of the family Aeshnidae are climbers on macrophytes. *Gomphus* and *Macromia* are very widespread burrowers in soft sediments of both lentic and lotic environments. *Ophiogomphus* is found only in cold, clear water and is generally considered an indicator of such environments. Many other common genera are found only in lentic waters, such as *Aeshna, Cordulia, Dorocordulia, Erythemis,*

Ladona and *Leucorrhinia*. *Aeshna* and *Leucorrhinia* are both climbers but the former stalks prey and the latter awaits its prey amongst the macrophytes.

Dragonflies and damselflies also differ in the adult stage. In damselflies the eyes are widely separated, the wings are equal in length and similar in form and the wings are held vertically at rest. In dragonflies, the eyes are close together, the hind wings are broader than the fore wings and the wings are held horizontally at rest.

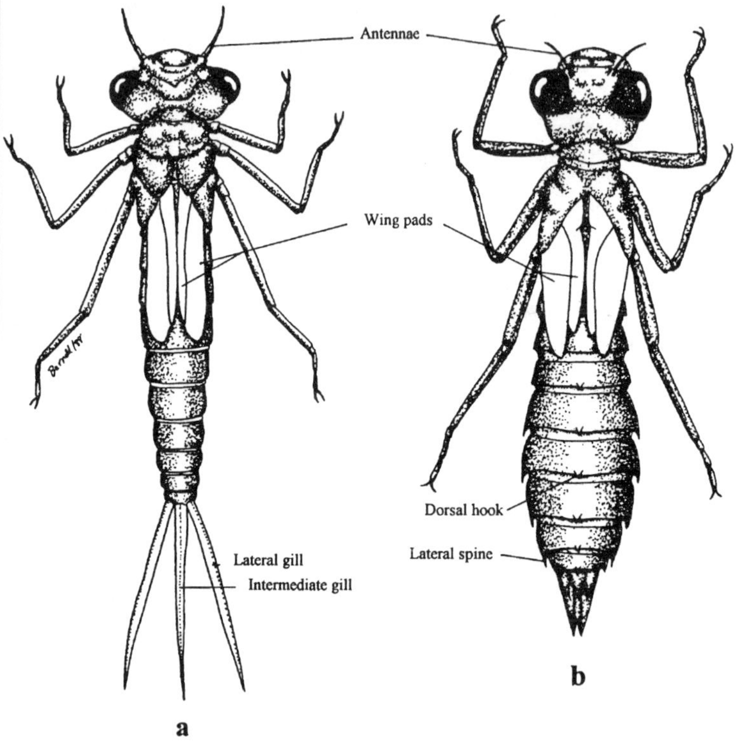

Figure 8.28. Some characteristics of the two suborders of Odoňata, (a) Zygoptera and (b) Anisoptera.

Order: Ephemeroptera - The Mayflies (Fig. 8.29)

The Ephemeroptera is one of four orders, the other three being Plecoptera, Trichoptera and Diptera, that are commonly, or perhaps always, used in environmental
27☞ impact assessments. For this reason, more emphasis is placed on these orders than on other orders of insects. Know the characteristics of at least these four orders and their ecological tolerances and requirements. Table 8.3 gives some ecological tolerances and requirements of common genera and species of mayflies.

The only order with which mayflies may be confused is Plecoptera, the stoneflies. Mayflies generally have three, multi-segmented tails, called ***cerci***, but some have two tails, like all the stoneflies. However, all mayflies have only a single tarsal claw on each leg and all stoneflies have two. Adult mayflies and stoneflies have two pair of wings but the hind wings are much smaller than the fore wings and the wings are held vertically at rest in mayflies. In stoneflies the hind wings are broader than the fore wings and the wings are directed rearward and held roof-like over their body at rest.

Most mayflies are lotic and display a great diversity in form, habitat requirements and functional feeding groups. Mayflies species exhibit a variety of morphological adaptations to life in flowing waters, such as: flattening of the body surface (e.g. *Stenonema*, *Epeorus, Rhithrogena*) to allow the insect to cling closely to the surface of the substrate; streamlining (e.g. *Baetis*) to reduce drag; reduction of projecting structures, such as the loss of gills and cerci in some *Baetis* species helps reduce drag; strong attachment claws to cling to rocks in fast waters (e.g. some *Ephemerella* species); hairy bodies to keep silt and sand off gills in depositional substrates (e.g. *Ephemera*); expanded femora (upper leg segment) to give greater marginal contact and clinging power with the substrate (e.g. *Rhithrogena*). Substrate types vary from boulders and rocks in erosional substrates to detritus, silt and mud in depositional substrates. Some species are indifferent and will occur in both lotic and lentic habitats on either erosional or depositional substrates. However, in lentic habitats nearly al mayflies are found in the littoral and sublittoral depths. One exception is *Hexagenia limbata* which is found in the profundal zones of oligotrophic lakes. Mayflies also represent five functional feeding groups; predators (e.g. *Ephemera*), shredders (e.g. some *Ephemerella* species), gatherers (e.g. *Dannella*), filter feeders (e.g. *Isonychia*) and scrapers (e.g. *Heterocloeon*). Some genera, like *Siphlonurus*, perform several functions (shredding, gathering, scraping and predating).

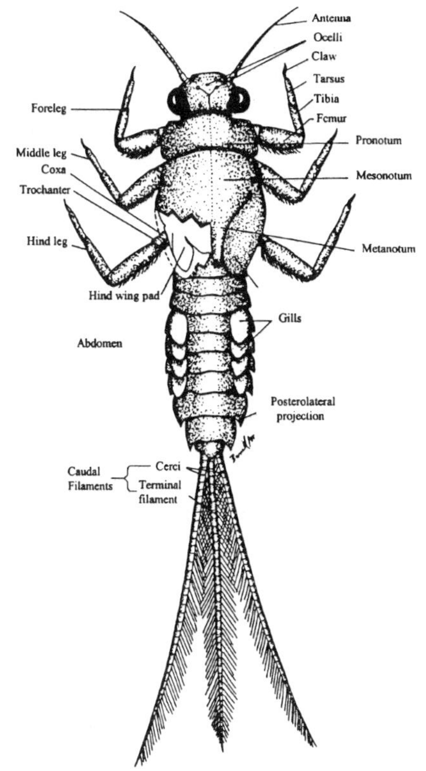

Figure 8.29. Some characteristic features of the order Ephemeroptera (mayflies).

As a group, mayflies are considered a sensitive group, intolerant of low oxygen tensions, low pH and enrichment (Table 8.3). None occur in acidic environments and only a few species (e.g. *Caenis* sp. and *Paraleptophlebia* sp.) Are found in water with oxygen levels less than 50% saturation. Cold, clean water indicators include *Ameletus* most *Baetis* species, *Caenis macrura*, *Epeorus, Rithrogena*, and *Hexagenia*. *Hexagenia limbata* is a well known indicator of oligotrophic lakes, reaching enormous densities in the profundal zone. See Table 8.3 for more information on these species and othe species of mayflies.

28☞ ***Order: Plecoptera - The Stoneflies*** (Fig. 8.30)

The stoneflies are primarily a stream-dwelling group, mostly in head water areas where the water is cool and water movement is fast enough to remove silt and

Table 8.3. Some physiological and ecological tolerances and requirements of common mayfly nymphs. General habitat describes where they occur and the type of substrate they prefer (L = lakes, P = ponds, S = streams); Feeding gives functional feeding behaviour (F = filter feeder, G = gatherer, P = predator, S = shredder, Sc = scraper); pH gives pH tolerance range; Oxygen is their requirement or tolerance of low oxygen (100% means species occurs in water near saturation, 50 means species can tolerate oxygen levels down to about 50% saturation, 25 means species can tolerate oxygen levels down to about 25% saturation); Trophic level is the indicator value of the species, if any (Oligo = oligotrophic, Dys = dystrophic, Eu = eutrophic). The species are arranged alphabetically.

Species	General habitat	Feeding	pH	Oxygen	Trophic level
Baetis vagans	gravel, S	Sc	≥ 7	100	Oligo
Epeorus vitreus	gravel, S	S	> 7	100	Oligo
Ephemera simulans	sand, gavel, L, S	P, G, S	≥ 7	50-100	Meso-Oligo
Ephemerella subvaria	gravel, S	Sc	~ 7	100	Oligo
Ephemerella cornuta	gravel, S	Sc	~ 7	100	Oligo
Heptagenia flavescens	wood, rock, S	S, G	?	50-100	Meso-Eu
Hexagenia limbata	mud, L	P	> 7	~100	Meso-Oligo
Hexagenia recurvata	mud, cold S	P	≥ 7	100	Oligo
Isonychia bicolor	swimmer, S	F	≥ 7	100	Oligo
Paraleptophlebia debilis	gravel, rocks, S	G, S	> 7	100	Dys-Oligo
Rhithrogena undulata	gravel, rocks, S	G	≥ 7	100	Oligo
Stenacron interpunctatum	rocks, L, S, P	G, Sc	<7->7	25-100	None, in all levels
Stenonema tripunctatum	rocks, S	G, Sc	≥7->7	50-100	None, in all levels
Stenonema femoratum	rocks, S	G, Sc	> 7	100	Oligo
Tricorythodes minutus	indifferent, S only	G	> 7	25-100	Meso, Dys

sand but leave gravel, rocks and boulders. They are found in stream segments that often dry up but only if water remains long enough for the nymphal life of the species. In general,
the more eutrophic the stream, the smaller the diversity and abundance of stoneflies. They thrive in waters that are well oxygenated and can tolerate only very short periods of low oxygen tensions. Most are cold stenotherms, requiring temperatures between 10 - 12 °C.

Stonefly nymphs are either predaceous or shredders of leaf litter. A few are scrapers. *Pteronarcys* is an exception and some species are known to be omnivorous, feeding by predating, shredding, scraping or gathering. Table 8.4 gives some physiological tolerances and requirements of the more common species, all of them being oligotrophic indicators.

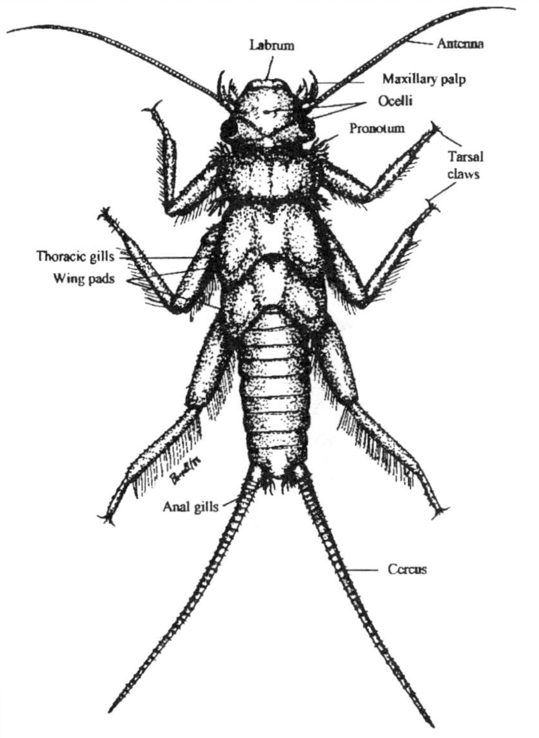

Figure 8.30. Some characteristic features of Plecoptera (stoneflies).

Order: Hemiptera - The True Bugs (Fig. 8.31)

True bugs are insects with two pairs of wings, the first pair being partly
29☞ chitinized (a hardened protein called chitin) and partly membranous and the second pair being entirely membranous. Both pairs are folded, like an accordion, over the back at rest. The nymphs closely resemble the adults, except they lack wings. Only two groups, called water striders (Families: ***Gerridae*** and ***Veliidae***), lack wings in the adult stage. For these groups, and in fact all true bugs, the most diagnostic feature that separates nymphs from adults is the number of segments in the **tarsi** (last leg segments terminating in claws); nymphs have one and adults have two tarsal segments. The hemipterans are represented by several families. Among the most common are: Gerridae, the water striders, e.g. *Gerris*; Veliidae, the broad shouldered water striders, e.g. *Rhagovelia*; ***Belostomatidae***, the giant water bugs, e.g. *Belostoma, Lethocerus*; ***Nepidae***, the water scorpions, e.g. *Nepa* and the water sticks e.g. *Ranatra*; ***Corixidae***, the water boatmen, e.g. *Corixa*; and ***Notonectidae***, the back swimmers, e.g. *Notonecta.*

Both nymphs and adults have piercing mouth parts within a ***beak*** that is usually held nearly horizontal to the underside of the head. All true bugs are carnivorous (predaceous), piercing tissues and cells and sucking out body fluids of insects (chironomids, mosquito larvae, mayfly nymphs, caddisflies, and some crustaceans (isopods and amphipods). Larger species, like *Belostoma* and *Lethocerus*, also feed on tadpoles and small fish.

Most hemipterans are powerful swimmers. The middle legs and/or the hind legs often are ***setose*** (hairy or spiny) or oar-like to stroke them quickly through the water. Some (e.g. water striders) are members of the ***neuston***, the assemblage of

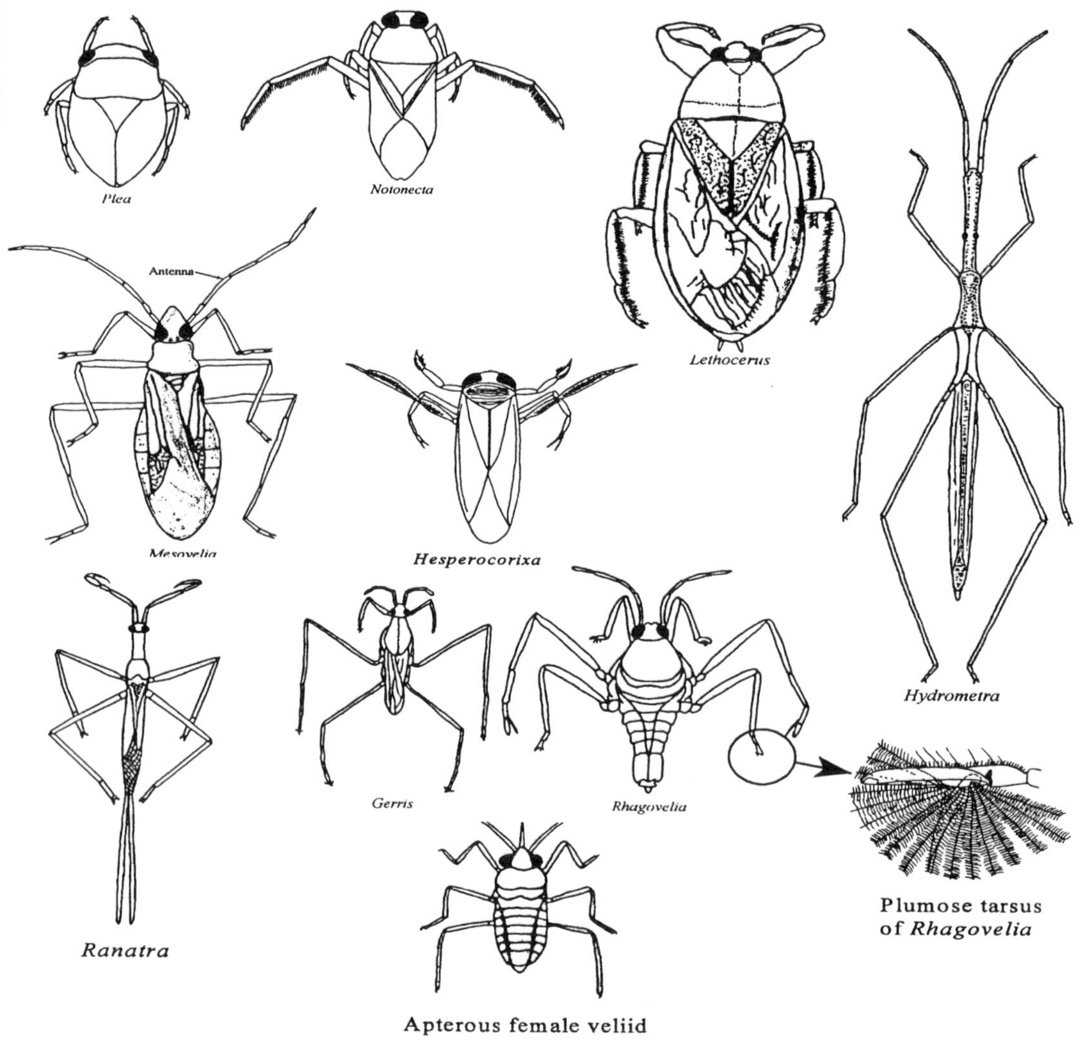

Figure 8.31. Most common families of Hemiptera, with Pleaidae (*Plea*), Notonectidae (*Notonecta*), Belostamatidae (*Lethocerus*), Hydrometridae (*Hydrometra*), Corixidae (*Hesperocorixa*), Mesoveliidae (*Mesovelia*), Nepidae (*Ranatra*), Gerridae (*Gerris*) and Veliidae (*Rhagovelia*). An apterous (wingless) form of a female and the plumose tarsus of *Rhagovelia* are also shown.

organisms that live at the surface of the water. The neuston is divided into ***epineuston***, those that live on the surface, and ***hyponeuston***, those that live just under the surface of the water. Water striders are members of the epineuston and are supported by the surface tension ("skin") of the water. To prevent rupture of the surface tension, the tarsi have a plume of unwettable hairs that offer low resistance to the surface "skin" of water. The claws are used to "dig" into the surface "skin" to provide traction for gliding over the water surface.

Most hemipterans are either lentic or slow-water lotic forms. All hemipterans

Table 8.4. Some physiological and ecological tolerances and requirements of common stonefly nymphs. See Table 8.3 for definitions of abbreviations. Species are arranged alphabetically.

Species	General habitat	Feeding	pH	Oxygen
Acroneuria lycorias	rocks, S	P of insects	< 7 - > 7	~ 100
Allocapnia spp.	rocks, S	S	> 7	~ 100
Amphinemura delosa	gravel, rocks, S	G, S	< 7 - > 7	100
Isoperla bilineata	plants, rocks, S	P of insects, G	> 7	100
Isoperla clio	plants, S	P of insects	> 7	100
Isoperla fulva	plants, rocks, S	P of insects, Sc, G	≥ 7	50-100
Nemoura trispinosa	plants, rocks, S	S	< 7 - > 7	100
Peltoperla maria	leaf litter, S	S	≥ 7	~100
Perlesta placida	rocks, leaves, S	P of insects, G	> 7	~100
Pteronarcys spp.	rocks, logs, leaves, S	P, Sc, S	≥ 7	~100
Taeniopteryx maura	rocks, logs, leaves, S	G, S	< 7 - > 7	~100

are air breathers and as such are more tolerant of environmental extremes than are most other insects. The water boatman, *Hesperocorixa*, and the water strider, *Gerris*, are among the few insects that can tolerate pH values less than 4.5 and are among the last to disappear when lakes and streams acidify. Of all aquatic organisms, the giant water bug, *Belostoma fluminea*, is considered by many to be among the most tolerant of extreme conditions (high chloride, high, B. O. D., low oxygen, low pH, etc.). However, it, like all hemipterans, has little or no indicator value because their life does not depend entirely on water quality.

Order: Megaloptera - The Hellgrammites, Dobsonflies, Alderflies and Fishflies (Fig. 8.32)

The megalopterans are unusual in that all eggs, pupae and adults are terrestrial. They are represented by two families, ***Sialidae*** (alderflies) and ***Cordulidae*** (larvae are called hellgrammites, adults are called dobsonflies or fishflies). The aquatic larvae are predaceous on small invertebrates, such as other insects, annelids, crustaceans and molluscs, and inhabit lotic and lentic habitats. Eggs are laid on branches and leaves overhanging streams or shorelines, the larvae dropping into the water when the eggs hatch. Larvae leave the water just prior to pupation. Adults are rarely seen because they are short lived and mostly nocturnal. Sialids have a 1 to 2 year life cycle, corydalids have a 2 to 5 year life cycle.

The most diagnostic features of megalopteran larvae are the 7 pairs of segmented lateral appendages, or gills, on the abdomen and large mandibles for grasping and engulfing prey. Sialids have a tail, or ***caudal filament***, whereas corydalids have a pair of ***anal prolegs*** with 2 claws on each. Of the two families, the corydalids have the larger species. *Corydalus, Nigronia* and *Chauloides* larvae attain 30-65 mm (more than 1 to about 3 in) in length . *Sialis*, the only genus of sialids, is typically 15 to 25 mm in larval length. *Corydalus* and *Nigronia* are lotic, under rocks, in fast flowing waters and *Chauloides* is lentic in habitat. All are intolerant of pollution. Although they do commonly occur in waters with pH levels near 5.5, circum-neutral or alkaline waters seem to have the largest populations. *Sialis* is considered to be more tolerant than the corydalids but cannot tolerate extreme conditions either. No species are recognized as good indicator organisms.

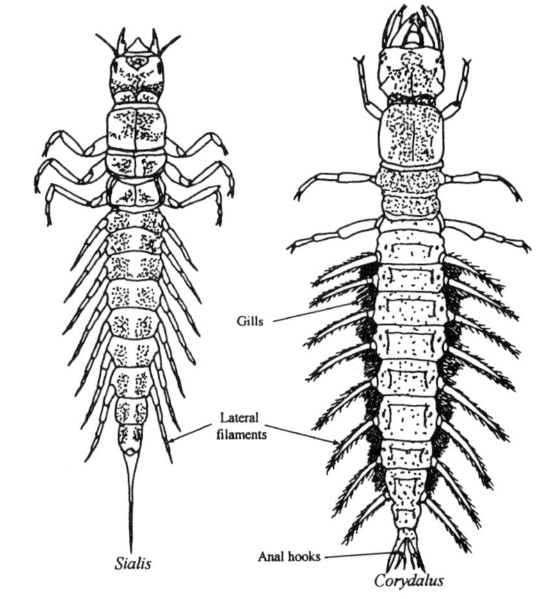

Figure 8.32. Two very common Megaloptera. *Corydalus* is found only in streams.

Order: Neuroptera - The Spongeflies (Fig. 8.33)

Most neuropterans are terrestrial. The only aquatic neuropterans are found parasitizing freshwater
30☞ sponges, mostly *Spongilla, Eunapius*, and *Ephydatia*, hence the common name, spongillaflies. The adults of terrestrial forms are commonly called lace wings, because of the lace-like venation in both pairs of wings. Aside from their presence in sponges, the most diagnostic feature of aquatic neuropterans is the stylet-like mandible used for puncturing sponge cells and sucking out the fluid contents. Only two genera occur, *Climacia* and *Sisyra*. Their life cycle is similar to the megalopterans. Neither have indicator value.

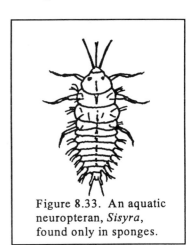

Figure 8.33. An aquatic neuropteran, *Sisyra*, found only in sponges.

Order: Lepidoptera - The Aquatic Butterflies (Fig. 8.34)

Another terrestrial group, the aquatic lepidopterans are represented by only a
31☞ few species. All are characterized by the presence of 5 pairs of abdominal prolegs, each with a circlet of small claws for attaching to plants. Most (e.g. *Nymphula,*

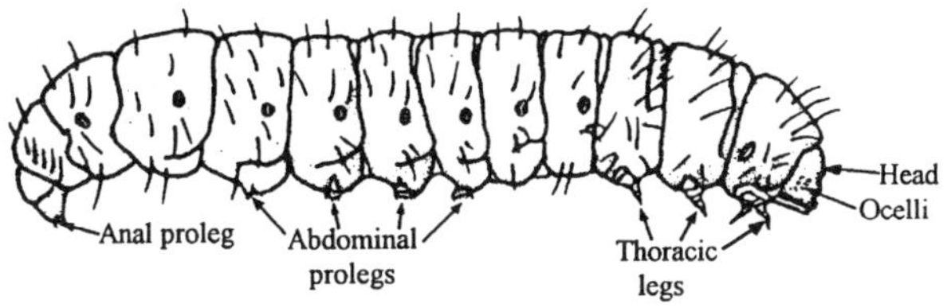

Figure 8.34. An aquatic Lepidoptera, showing some diagnostic features.

Parapoynx and *Synclita*) are herbivorous and are associated with and living in or on aquatic macrophytes. Larval habits include mining or boring into leaves, stems and roots, or feeding on foliage, flowers or seeds. A few (e.g. *Paragyractis*) occur in fast water streams living on rock surfaces and feeding on attached algae, mostly diatoms. One species, *Petrophila jaliscalis*, is considered an indicator of eutrophic conditions; it can tolerate low oxygen conditions, moderately high temperatures, reduced water flows and cultural organic enrichment.

32 ☞ ***Order: Trichoptera - The Caddisflies*** (Figs. 8.35, 8.36)

Caddisflies can be recognized immediately by their 2 prolegs, each with a prominent hook, or claw, on the last abdominal segment (Fig. 8.35) and their one-segmented antennae (which is very small). Some species have gills on their abdomen but the gills are numerous and never segmented and never arranged in a lateral row. Megalopterans also have the anal prolegs and abdominal gills but their antennae are more than one segment and the gills are segmented and arranged in a lateral row on each side of the abdomen. Except for some rare aquatic Lepidoptera and a few Diptera, Trichoptera is the only insect order that has cases or retreats (Fig. 8.36).

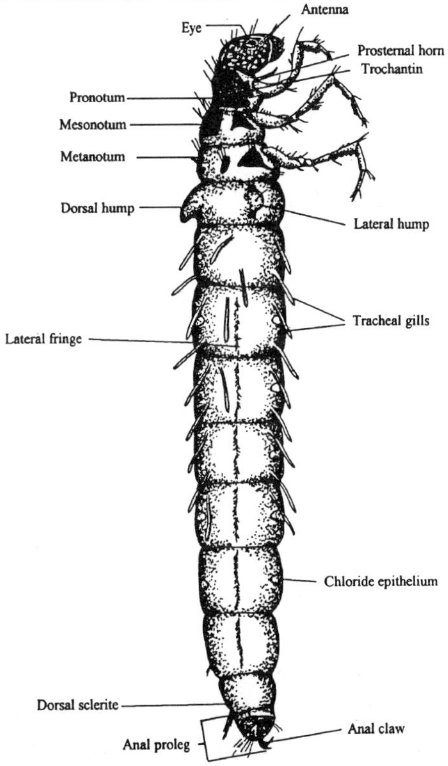

Figure 8.35. Some diagnostic characteristics of Trichoptera.

The caddisflies rival the mayflies in diversity of habitat and in functional feeding behaviour. All North American families are represented by species restricted to cool, lotic waters, while others are restricted to warmer, more lentic habitats. Still others are confined to springs, seepage areas, ponds, marshes or temporary pools. The functional feeding groups include predators, shredders, filter feeders, gatherers, scrapers and piercers. Many species are indicators of specific types of habitats and trophic status (Table 8.5).

Most caddisflies construct a case or

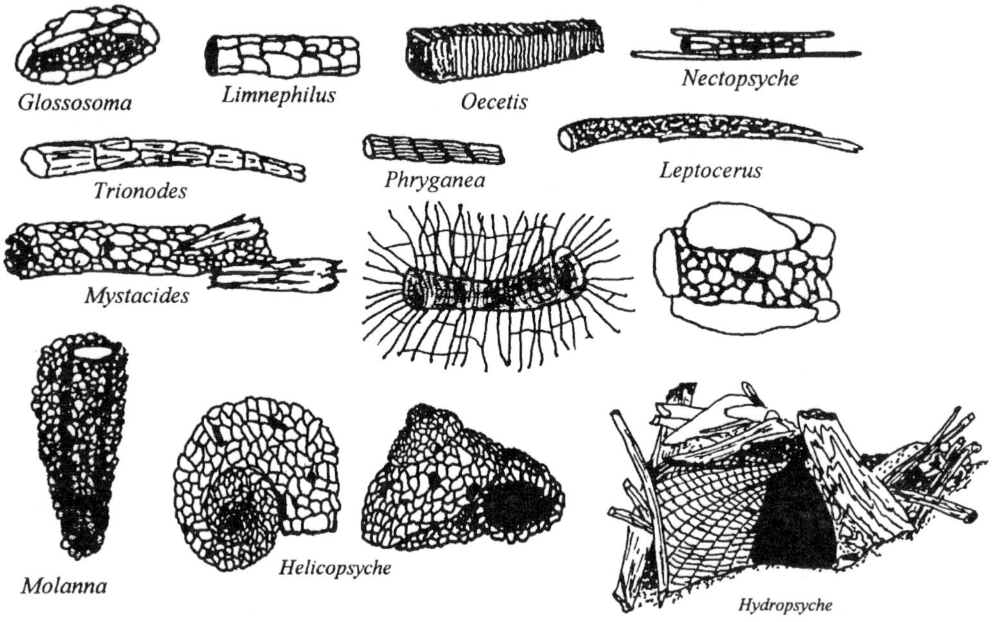

Figure 8.36. Some caddisfly cases and nets, all characteristic of the genera shown.

net-like retreat from silk secreted from a pore on the lower lip. The silk is used by itself or to fasten together rock fragments, pieces of plant material or mollusc shells. The retreats vary in shape, construction and portability (Fig. 8.36). Some are ***fixed retreats*** and include nets that are either finger-like (e.g. *Philoptomus*), trumpet-like (e.g. *Neureclipsis*) or box or fan-like (e.g. *Hydropsyche, Cheumatopsyche*), tubes of silk (e.g. *Psychomyia, Phylocentropus, Polycentropus*) and purse-like cases (e.g. many *Hydroptila* sp., *Agraylea*). ***Portable cases*** include **tube-like cases** that may be open at both ends or closed at the posterior end. Some are tapered and round in cross-section and made of leaf pieces (e.g. *Phryganea*) or sand grains (e.g. *Oecetis*); tapered and square in cross-section (e.g. *Brachycentrus*); snail-shaped cases (e.g. *Helicopsyche*); saddle or turtle shell cases of small stones (e.g. *Glossosoma*). In temperate regions, only one family, Rhyacophilidae, is free-living and all of its members inhabit streams.

Many species are restricted to streams, some in fast water in riffles on or under stones or wood (e.g.*Rhyacophila, Cheumatopsyche, Hydropsyche, Chimarra, Brachycentrus, Helicopsyche*) or attached to the upper surfaces of stones (e.g. *Leuctrichia*), others are in moderate flow in trailing plant masses (e.g. *Neureclipsis, Oecetis*), on rocks or twigs (e.g. *Pycnopsyche, Agapetus, Neophylax*), in tubes within sandy bottoms (e.g. *Phylocentropus*), or in cases on sandy bottoms (e.g. *Molanna, Athripsodes*). Other species are restricted to lentic habitats, like *Mystacides, Triaenodes, Limnephilus, Hydroptila, Ptilostomis*) that live on submerged vegetation or sticks. In lakes, all caddisflies are restricted to the shallow waters of the littoral and sublittoral zones.

As Table 8.5 indicates, most Trichoptera are clean-water insects. Among the most tolerant genera are *Limnephilus* and *Cheumatopsyche*. *Limnephilus* is represented by nearly 100 species, some of which can exploit temporary streams and pools. *Cheumatopsyche* and *Hydropsyche* frequently occur together in streams but the former is more tolerant than the latter. Chapter 12 discusses the importance and use of Trichoptera (and Ephemeroptera and Plecoptera) in environmental impact assessments.

Order: Coleoptera - The Beetles (Fig. 8.37, 8.38)

Of all the insects, the aquatic larval (Fig. 8.37) and adult beetles (Fig. 8.38)
33 ☞ display the greatest diversity of form. In fact, they are so diverse that it is difficult to identify them on single characters alone. They share features with Trichoptera and Megaloptera in that all larvae have 3 pairs of segmented legs on the thorax, but the lack of stump-like prolegs separates beetles from caddisflies and dobsonflies. Many species of larval beetles possess lateral gill filaments, but they are unsegmented, which

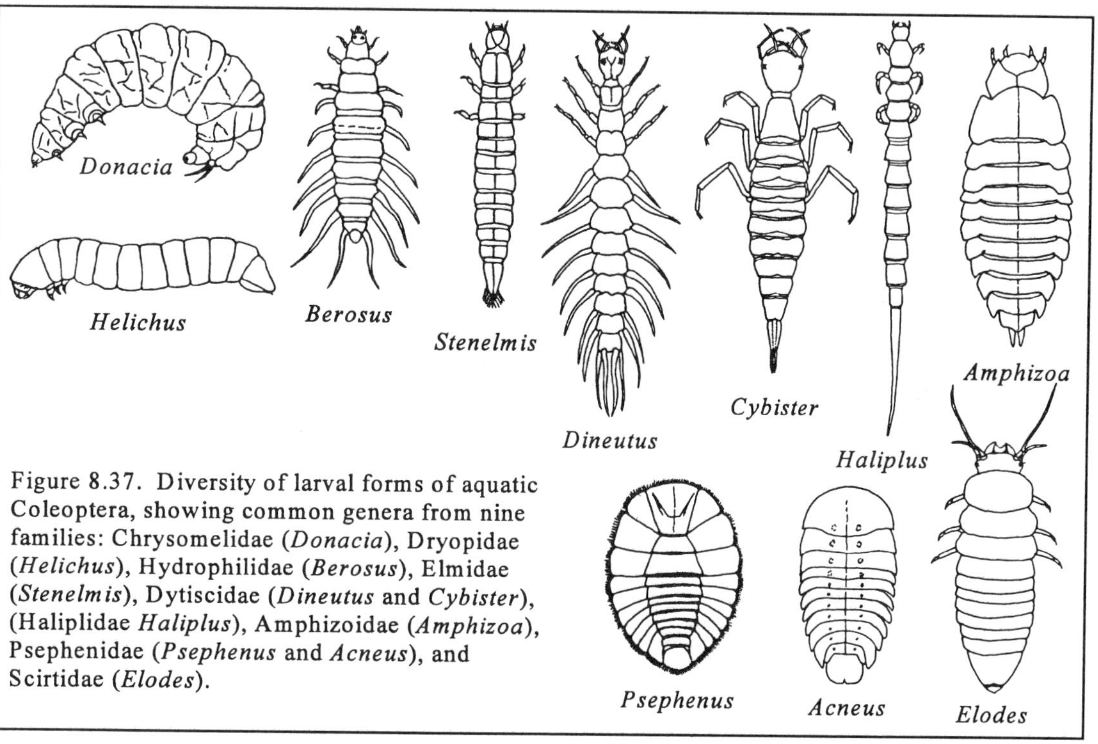

Figure 8.37. Diversity of larval forms of aquatic Coleoptera, showing common genera from nine families: Chrysomelidae (*Donacia*), Dryopidae (*Helichus*), Hydrophilidae (*Berosus*), Elmidae (*Stenelmis*), Dytiscidae (*Dineutus* and *Cybister*), (Haliplidae *Haliplus*), Amphizoidae (*Amphizoa*), Psephenidae (*Psephenus* and *Acneus*), and Scirtidae (*Elodes*).

separates them from larval alderflies with their segmented gills. Many species also have caudal filaments, but the filaments have only a few segments, while mayflies and stoneflies have numerous segments.

Only Megaloptera and Coleoptera have both aquatic immature and adult stages. The wings of adult beetles are entirely ***corneus*** (hardened protein) and meet in a mid-line down the back, while those of megalopterans are partly corneus and are folded over the back.

Table 8.5. Some physiological and ecological tolerances and requirements of common caddisfly larvae. Feeding: O = omnivorous; Pi = Piercer; see Table 8.3 for other definitions of abbreviations. Species are arranged alphabetically.

Species	General habitat	Feeding	pH	Oxygen	Trophic level
Agapetus spp.	turtle case, S	Sc	≥7	~100	Oligo
Agraylea spp.	silk purse, S, L	Pi, G	≤7 - >7	~100	upper Meso
Banksiola spp.	tapered cylinder of leaves in spiral; slow S, L	S, P (last 2 instars)	<7 - >7	> 50	upper Meso
Brachycentrus americanus	tapered square tube of plant mat'l; on logs & plants in S	F, Sc	≥7	> 50	upper Meso
Cheumatopsyche spp.	silk net, warmer S	F	<6 - >7	25 -100	lower Meso
Chimarra spp.	sac-like nets, warmer S	F	≥7	≥ 50	Meso
Frenesia spp.	tube of mineral, wood; cool springs	S	>7	~100	Oligo
Glossosoma nigrior	turtle case, S	Sc	≥7	~100	Oligo
Helicopsyche borealis	spiral case, S	Sc	<7 - >7	> 50	upper Meso
Hydropsyche spp.	silk net, S	F	≥7	> 50	upper Meso
Hydroptila spp.	silk purse, S	Pi, Sc	≤7 - >7	~100	Oligo
Lepidostoma spp.	tapered tube of sand, headwater S	S	~7	>50	upper Meso
Leptocerus americanus	silk tube, L	S	<7 - >7	>50	upper Meso
Limnephilus spp.	case variable, omnipresent, L, S	O	≤7 - >7	25 -100	None
Molanna blenda	tube case with lateral flanges, L, S	S, G, P	≤7	> 50	lower Oligo
Mystacides sepulchralis	tube of sand, plant mat'l, S, L	G, S	≤7 - >7	~100	Oligo
Neophylax spp.	tapered tube of sand, S	Sc	>7	~100	Oligo
Neureclipsis crespuscularis	trumpet-like net, S	F, S, P	≤7 - >7	~100	Oligo
Oecetis spp.	tapered, curved tube, S, L	P	≤7 - >7	> 50	lower Oligo
Phryganea cinerea	tapered cylinder of leaves in spiral, S, L	S, P	≤7 - >7	> 50	lower Meso
Phylocentropus placidus	silk tube, headwater S	F	~7	~100	Oligo
Polycentropus cinereus	silk tube, S	P, F, S	5 - >7	> 50	upper Meso
Psychomyia flavida	sac-like nets, S	G, Sc	<7 - >7	~100	Oligo
Rhyacophila spp.	free-living, S	P, G, S	<7 - >7	~100	Oligo

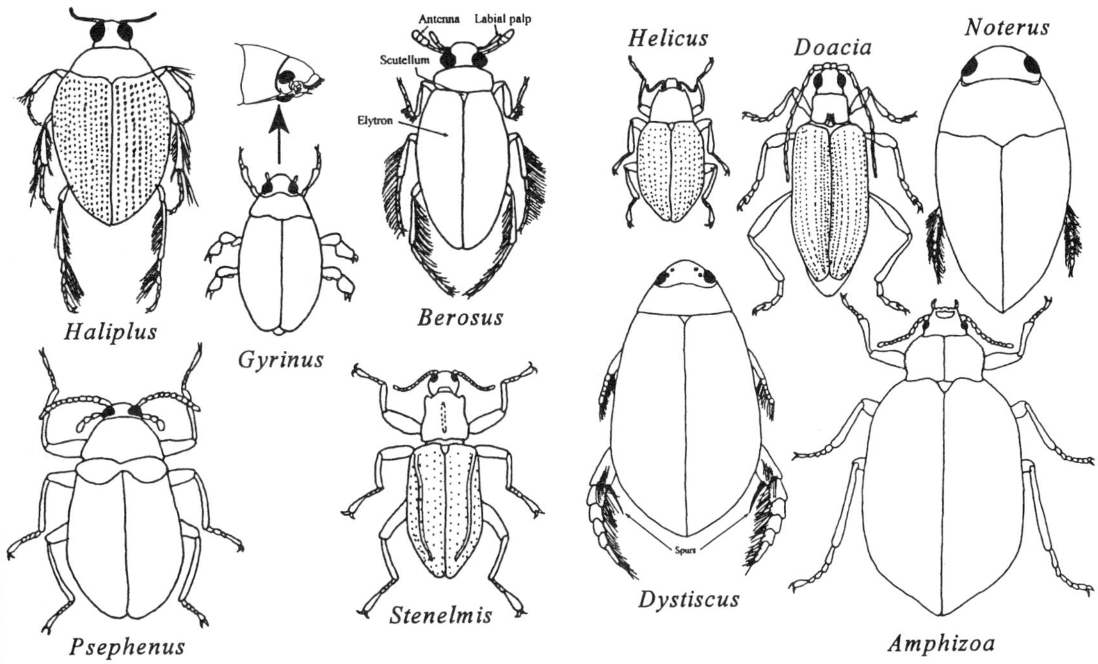

Figure 8.38. Diversity in form of adult aquatic beetles, showing representatives from eight families: Haliplidae (*Haliplus*), Gyrinidae (*Gyrinus*, with head enlarged to show double eye), Hydrophilidae (*Berosus*), Psephenidae (*Psephenus*), Elmidae (*Stenelmis*), Dryopidae (*Helichus*), Chrysomelidae (*Donacia*), Noteridae (*Noterus*), Dytiscidae (*Dytiscus*) and Amphizoidae (*Amphizoa*).

The great diversity in habit and habitats is accompanied by a great diversity in form of beetles. Larval beetles exhibit similar morphological adaptations to stream flow as do the mayflies. Water pennies (***Psephenidae)*** are dorso-ventrally flattened that helps to resist drag and have a fringe of spines which enhances marginal contact on the surface of stones to prevent being swept away by water currents. Species that live in macrophyte beds often have projections and spines to cling to stems and leaves and are greenish to provide cryptic colouration (e.g. *Berosus, Peltodytes*). Some larval riffle beetles (***Elmidae***) have three adaptations; they are dorso-ventrally flattened and streamlined, features that help to resist drag caused by water flowing over them and they have marginal spines to enhance marginal contact with rock surfaces (e.g. *Lara*). Adult whirligig beetles (***Gyrinidae***) are ideally adapted to life at the surface; each eye is split into an upper and lower half, the upper one for looking above the surface and the lower one for looking below the surface of water. The legs of adult diving beetles (***Dytisicidae***, e.g. *Dytiscus*) are oar-like and often have numerous spines that help to increase their power stroke through the water.

Most beetles are common in both lentic and lotic habitats. In lentic habitats they are found only in the littoral and sublittoral zones. Only two families are represented almost entirely by lotic species, ***Psephenidae***, the water pennies, and ***Elmidae***, the riffle beetles. If any elmid species do occur in lentic habitats, they are usually found in wave swept parts of the lake. The whirligig beetles (Family: ***Gyrinidae***) are confined to life at the surface where they scavenge or engulf food from the surface film. Likewise, two families are almost entirely lentic; the ***Chrysomelidae***, or leaf beetles and ***Curculionidae***, the weevils.

Aquatic beetles are either scrapers or predators. Predaceous forms can be recognized immediately by their large, powerful mandibles, as in larval and adult Dystiscidae, the predaceous diving beetles. Many of the predaceous species have piercing mouth parts for sucking out body fluids of their prey.

Most larval and adult beetles are tolerant of wide changes in pH and dissolved oxygen concentration. Many adults cannot use dissolved oxygen and must rise to the surface to respire atmospheric oxygen. Few beetles, if any, are recognized as indicator organisms of environmental health. Their main indicator value is in the physical type of habitat that they utilize.

Order: Diptera - The True Flies (Fig. 8.39)

Diptera is derived from Greek for two (Di) wings (ptera). Adults in all other orders of insects have two pairs of wings. The first pair of wings in dipterans is
34☞ reduced to a pair of knob-like structures called ***halteres*** that function as balancing organs or gyroscopes. The larvae are equally easy to recognize. Only dipteran larvae lack three pairs of segmented legs. Instead, they have stump-like ***prolegs***, either on the first segment only (e.g. black flies), last segment only (e.g. mosquitos), first and last segment (e.g. midge flies), or on several segments (e.g. some crane flies). Or they may lack legs entirely (e.g. sand flies).

The order Diptera is represented by nearly 3000 freshwater species, the greatest diversity of any freshwater invertebrate taxon. With the enormous diversity in species comes a plethora of forms (Fig. 8.39), habitat types and functional feeding behaviours. The habitats include brackish waters; littoral, sublittoral and profundal zones of lakes; fast and slow rivers; temporary and permanent ponds and streams; roadside ditches; clean and polluted waters; cold springs and seeps; thermal springs approaching 49 °C; acidic and alkaline waters; and deep wells and shallow sewage ponds. Virtually all the functional feeding behaviours are represented; engulfing predators, piercing predators, grazing herbivores, shredding herbivores and carnivores, filter feeding and gathering collectors, and commensal and parasitic symbionts.

Associated with confinement to certain types of habitats are morphological, physiological and/or ecological adaptations. The only way to survive life in sewage ponds where anoxia is the rule is to breath atmospheric oxygen. The rat-tailed maggot, *Eristalis* (formerly *Tubifera*) *tenax* of the family ***Syrphidae***, solves this problem with a telescopic respiratory siphon that can be extended to the surface for "fresh air" when required. At least three families are well adapted for life in erosional stretches of flowing waters. The black flies (Family: ***Simuliidae***) exhibit at least three adaptations.

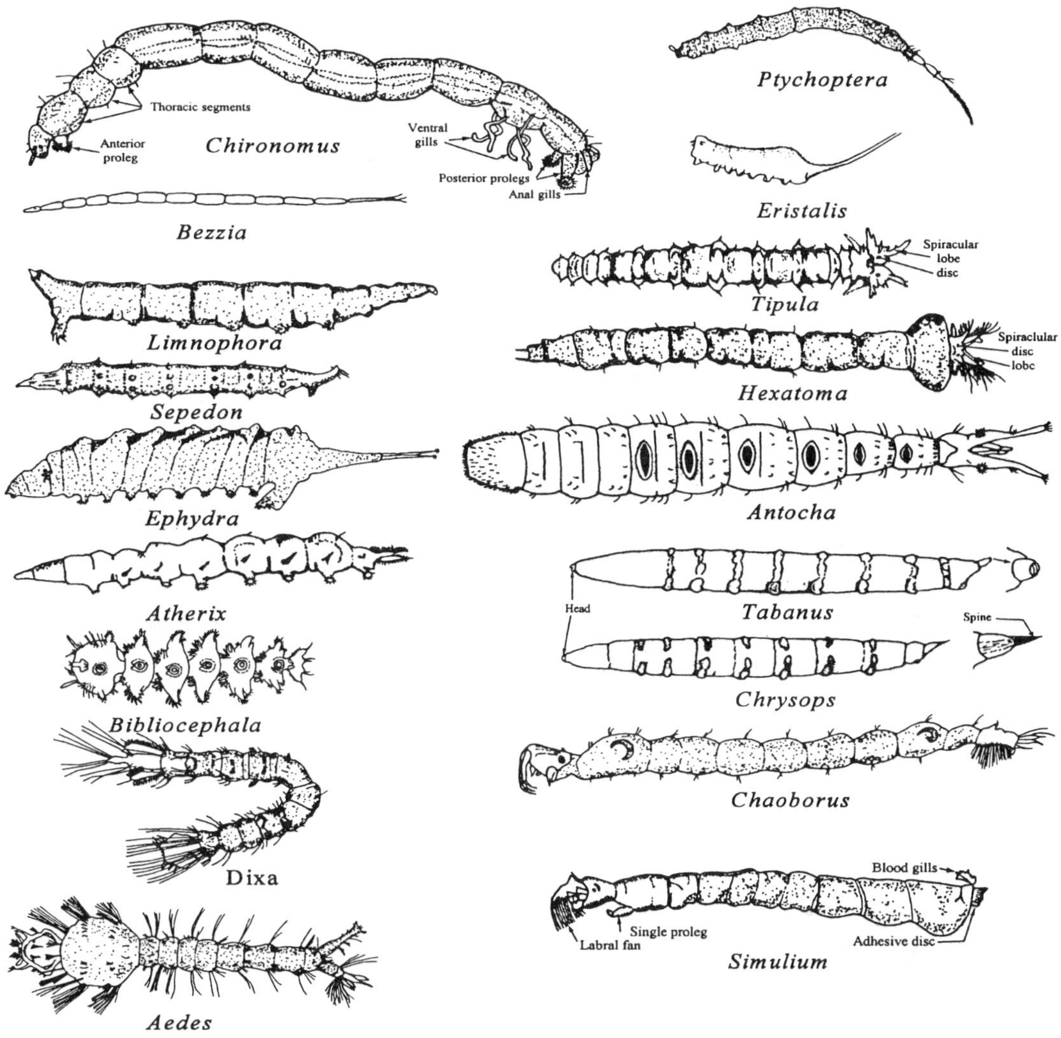

Figure 8.39. Diversity in form of Diptera larvae, showing common genera of fifteen families : Chironomidae (*Chironomus*), Ceratopogonidae (*Bezzia*), Muscidae (*Limnophora*), Sciomyzidae (*Sepedon*), Ephydridae (*Ephydra*), Athericidae (*Atherix*), Blephericeridae (*Bibliocepahala*), Dixidae (*Dixa*), Culicidae (*Aedes*), Ptychopteridae (*Ptychoptera*), Syrphidae (*Eristalis*), Tipulidae (*Tipula, Hexatoma, Antocha*), Tabanidae (*Tabanus, Chrysops*), Chaoboridae (*Chaoborus*) and Simuliidae (*Simulium*)

(i) The larvae attach themselves to rock surfaces using a sucker on the end of the last abdominal segment; (ii) they anchor themselves in place with a silk thread secreted from a silk gland located on the dorsal side of the last abdominal segment; (iii) to complement their sedentary life, the larvae have huge ***labral fans*** that strain food particles from water as it flows by. Mountain midges (Family: ***Deuterophlebiidae***) have 7 pairs of abdominal prolegs tipped with several rows of hooks for grasping crevices and ridges in rocks (e.g. *Deuterophlebia*). Larvae of the net-winged midges (Family: ***Blephariceridae***) have 6 main segments, each of which bears ventrally a

Table 8.6. Some physiological and ecological tolerances and requirements of common
35☞ dipteran larvae. See Table 8.3 for definitions of abbreviations. Species are arranged
alphabetically for Chironomidae.

Species	General Habitat	Feeding	pH	Oxygen	Trophic Level
F: Deuterophlebiidae *Deuterophlebia* spp.	Clingers on rocks in mountain S	Sc	<7 - >7	100	Oligo
F: Blephariceridae *Blepharicera* spp.	Clingers in S	Sc	<7 - >7	100	Oligo
F: Tipulidae *Antocha saxicola*	Clingers in silk tubes in S	Sc	>7	~100	Oligo
F: Psychodidae *Psychoda alternata*	Burrowers in L, S	G	5.5 - ~7	50-100	Meso
F: Athericidae *Atherix variegata*	Erosional S	P	6 - >7	~100	lower Oligo
F: Syrphidae *Eristalis tenax*	Burrow in organic bottoms of S, L	G	<7 - >7	<25	Eutr
F: Scathophagidae *Spaziphora* spp.	Sewage ponds	Sc	<7 - >7	<25	Eutr
F: Simuliidae *Simulium* spp.	Erosional S or wave swept shore of L	F	<7 - >7	~100	Oligo
Prosimulium spp.	Erosional S	F	<7 - >7	~100	Oligo
F: Chironomidae *Ablabesmyia* spp.	S, L	P	<7 - >7	25-100	Eutr
Chironomus plumosus attenuatus, riparius	Burrowers in tubes in S, L	G, S	<7 - >7	25-100	Eutr
Cricotopus exilis	On rocks in S	P	>8	25-100	Eutr
Cricotopus bicinctus	S, L	S	>7	25-100	Eutr
Cryptochironomus fulvus	Burrower, S, L	P	<7 - >7	25-100	Eutr
Dicrotendipes spp.	Burrowers, L, S	G, F, Sc	>7	25-100	Eutr
Polypedilum fallax	Clinger, S	S, G, P	≥7	25-100	Eutr
Procladius culiciformis	S, L	P	36011	25-100	Eutr
Rheocricotopus robacki	Erosional S	G, S, P	<7 - >7	25-100	Indifferent
Rheopelopia	Erosional S	P	<7 - >7	100	Oligo
Rheotanytarsus exiguus	In tube or net, fast S	F	<7 - >7	<50	Eutr
Tanypus punctipennis	L	P	>7	25-100	Eutr

median suctorial disc to firmly attach to rocks in erosional habitats (e.g. *Blephariceru*). Both mountain midges and net-winged midges are more or less confined to life on rock surfaces and have evolved mouth parts for this life style, that is scraping algae from the surfaces of rocks.

The mosquitos (Family: ***Culicidae***) are adapted for life just under the surface of the water (i.e. ***hyponeuston***). Most mosquitos deposit their eggs on still surfaces of ponds, small seeps or even lakes. Some deposit their eggs in moist ground depressions where they may lay dormant for years, or until flooded by rains or runoff. The eggs are capable of hatching within 24 hours in some species. Larvae complete their life cycle in 7-10 days, pupate for 3 or 4 days and then emerge as adults. The larvae are planktonic and must rise to the surface to obtain atmospheric oxygen through a dorsal siphon on the last abdominal segment. Most mosquitos are filter feeders, using their mouth brushes to collect algae, detritus and small invertebrates. Only the adult female takes blood meals, the blood protein needed for energy to produce eggs. The males feed on plant juices and nectar for their energy requirements. Adults live for about one week.

Another family adapted for a plankton existence in lentic habitats is ***Chaoboridae***, the phantom midges. Most species of *Chaoborus* exhibit diurnal vertical migrations in the water column, rising to the surface at night and returning to the bottom during the day. The larvae possess two pairs of air sacs used to regulate their rise and fall in the water column. Generally the smaller instars rise higher in the water column than do larger, older instars.

The dipteran groups with the greatest North American species diversities are the midge flies of the Family ***Chironomidae*** (about 2,000 species), the crane flies of the Family ***Tipulidae*** (about 1,500 species), the sand flies (also known as no-see-ums and biting midges) of the Family ***Ceratopogonidae*** (about 400 species), and the horse flies and deer flies of the Family ***Tabanidae*** (about 300 species). Of the three families, the chironomids are the most ecologically significant, partitioning ecological resources in an infinite number of ways under a variety of ecological conditions. They are shredders of leaf and wood, gatherers and scrapers of detrital bottom deposits, filter feeders of fine suspended detrital materials, gatherers of benthic and planktonic algae and fungi, and predators and parasites in cold and warm streams and in deep and shallow parts of lakes (Table 8.6). Their range of habitat types is the most extensive of any aquatic insect group. They not only occupy the same diverse habitats as the mayflies, caddisflies and beetles, but they do so over a much wider range of physical (temperature, depth, current) and chemical (pH, alkalinity, salinity, conductivity, oxygen concentration) gradients (Table 8.6).

Most crane fly larvae have on their last abdominal segment a ***spiracular disc*** through which atmospheric oxygen is taken in. Because the larvae must rise to the surface in order to respire, very few species are of any indicator value. However, one species, *Antocha saxicola*, lacks spiracular discs and has instead anal gills to extract oxygen dissolved in the water. It is sensitive to low oxygen tensions and is found only in waters saturated, or nearly saturated with oxygen (Table 8.6). Similarly, horse fly (e.g. *Tabanus* spp.) and deer fly (e.g. *Chrysops* spp.) larvae also utilize atmospheric oxygen and extend caudal spiracles out of the water to respire. The adult females of horse and deer flies are strong fliers and painful biters. Adult horse flies are larger and darker than the smaller, more patterned deer flies. Apparently, the horse flies are above-the-waste biters, are quiet and sneak up to get their blood meal. Deer flies are the "ankle-biters" and usually "buzz" before they bite.

The sand fly larvae are worm-like and most lack legs altogether, including prolegs. The mouth parts of adults are modified for piercing skin to obtain a blood meal. Only the females are blood feeders. Adults are tiny and gnat-like, able to penetrate most screen doors. They are extremely common, but unfortunately, the taxonomy is currently in disarray, making the ceratopogonids of little indicator value.

BENTHOS OF LAKES

Qualitative Aspects

In general ***species richness***, or the total number of species, increases from the water's edge to about the 2 to 4 m depth and then decreases again (Fig. 8.40). The
36☞ number of benthic species in the shallow shore water is low because waves constantly erode the substrate. Most of the macrophytes in this region are emergent species whose submerged parts are mostly stems that provide little protection from predators, are a poor source of food and contribute little oxygen from photosynthesis. However, oxygen concentrations are usually high because of water turbulence.

The greatest number of macroinvertebrate species usually occurs in the region where submersed macrophyte diversity is greatest. The sublittoral zone offers a great diversity of habitats and sediment types that benefit the benthos in many ways. Many submerged species of plants have finely divided leaves that offer protection from predators, provide an abundant supply of photosynthetic oxygen and offer a large surface area for epiphytic organisms that are consumed by grazing invertebrates. The litter provided by submerged macrophytes is generally much more palatable and more easily broken down (that is the plants are more ***labile***) by shredding invertebrates than is the tougher, fibrous tissues of stems of littoral emergent plants that are more resistant to decay (that is the plants are more ***refractory***). Figure 8.40 shows that in shallow waters maximum species richness is similar between eutrophic and oligotrophic lakes, but the depths at which the high diversity is maintained is greater in oligotrophic lakes than in eutrophic lakes. This correlates well with the greater width of the sublittoral zone in oligotrophic lakes than in eutrophic lakes.

Beyond the sublittoral zone species diversity tends to decrease rapidly (Fig. 8.40). The profundal zone offers less diversity in habitat types than does the sublittoral zone and the benthic community responds with decreasing species richness. However,

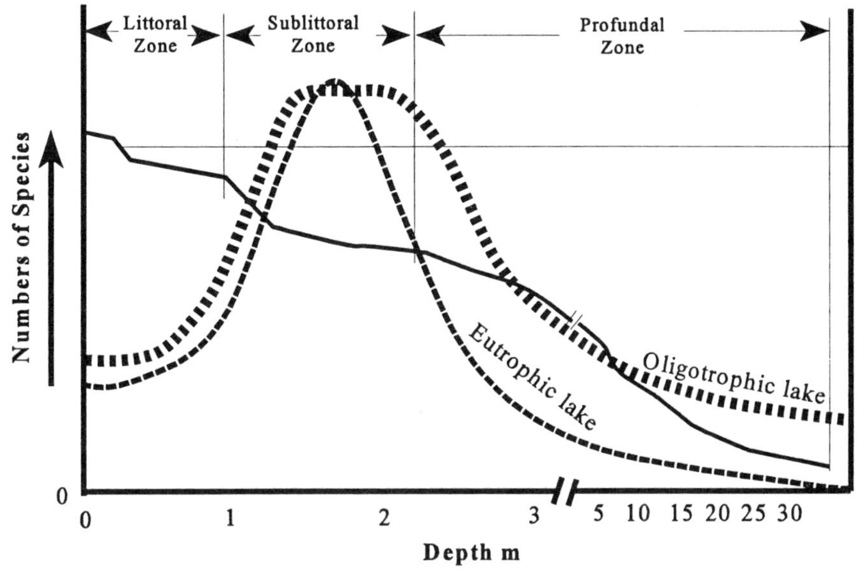

Figure 8.40. Variations in species richness of benthic macroinvertebrates relative to depth in eutrophic and oligotrophic lakes.

eutrophic lakes have a smaller diversity of benthos than do oligotrophic lakes because the low oxygen levels that prevail during the thermally stratified period can be tolerated only by eutrophic indicator species, such as *Chironomus plumosus* and *Tubifex tubifex*.

The seasonal and spatial variations in diversity of benthic species correspond closely to ***Thienemann's Principles***. There are three of them:

1. The greater the diversity of conditions in a locality, the larger the number of species that make up the community.

37☞

2. The more the conditions deviate from normal, and hence from the normal optima of most species, the smaller is the number of species which occur there and the greater the number of individuals of each species which do occur.

3. The longer a locality has been in the same condition, the richer is its biotic community and the more stable it is.

Thienemann's first principle helps explain why benthic diversity varies with depth. The second principle explains why the profundal zone of oligotrophic lakes has a greater diversity but lower numbers of invertebrates than do eutrophic lakes. Indeed, eutrophic indicator species are always present in enormous numbers, compared to oligotrophic indicator organisms. As lakes eutrophy, the diverse habitats that are characteristic of oligotrophic lakes become covered with organic detritus and soon the bottom becomes a homogeneous layer of ooze. The profundal zone soon becomes anoxic and only a few species can tolerate anoxia. But those organisms that can tolerate anoxia have the bottom pretty much to themselves, with very few competitors. The result is a homogeneous substrate containing a very low diversity of organisms (all of them being eutrophic indicator species) but a very high number of those species that are present.

The third principle is a little more difficult to demonstrate. It is easier to visualize if the benthic community is thought of as a web, as depicted in Fig. 8.41, with each of the functional feeding groups (shredders, grazers, filterers, gatherers, predators, piercers) represented by several species (A to P). Only a part of the web (benthic community) is shown, with only two habitats in each of the littoral, sublittoral and profundal zones. In a healthy lake, each zone would be represented by several habitats. Each of the benthic habitats are separated by longitudinal radii, or spokes. Each space between a pair of cross threads represents a niche occupied by a species with a specific functional feeding habit. Each habitat has its own assemblage of species. Hence, the shredding species in the littoral zone would be different than the

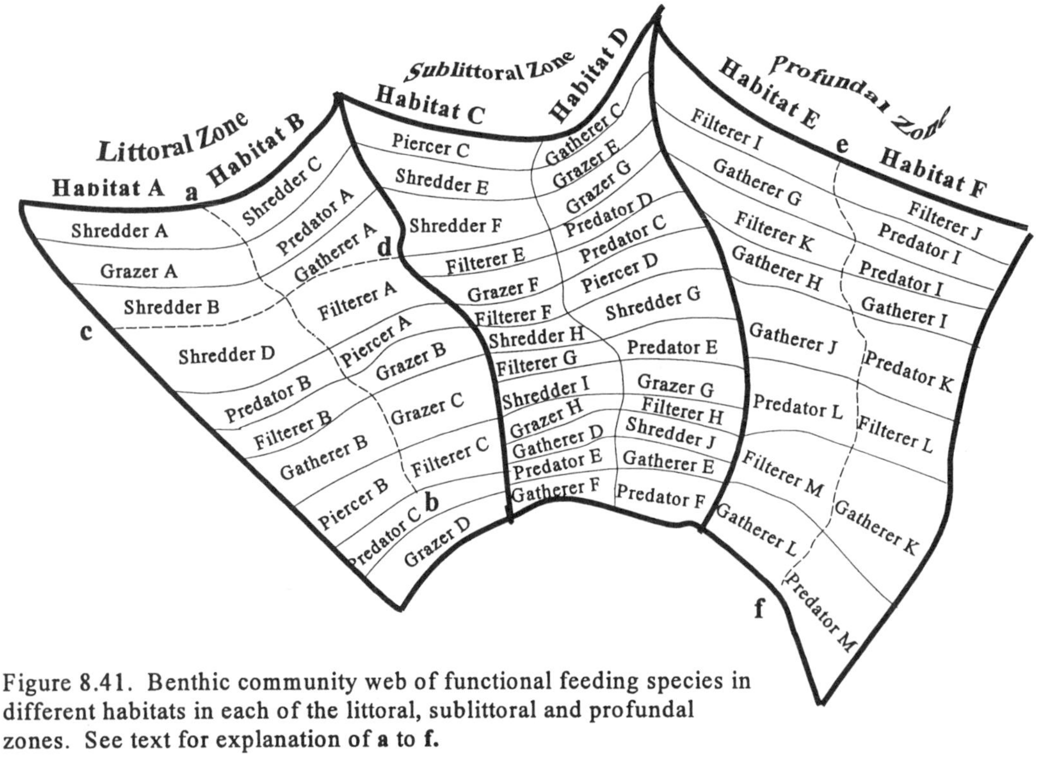

Figure 8.41. Benthic community web of functional feeding species in different habitats in each of the littoral, sublittoral and profundal zones. See text for explanation of **a** to **f**.

shredding species in the sublittoral zone because of different types of macrophytes (e.g. emergent species vs submergent species).

The web (benthic community) is very stable when all its radii (habitats) and all the cross threads (species with a specific niche) are present. The more radii and cross threads, the stronger and more stable the web. If a habitat is destroyed, or if we
38☞ remove one radius (e.g. line a to b in Fig. 8.41), several niches with their species would be lost and the web would become weaker, or the community less stable. If a cross thread (niche) is removed (e.g. line c to d), the community also becomes less stable, but it is more stable than removing a complete habitat. Notice that removal of niches has less affect on the community stability than does removal of habitats. In other words, we can remove one functional feeding species without greatly affecting stability because there are other species that can take over the functional feeding habit of the one removed. However, if too many cross threads (species) are removed, especially within one functional group, the web weakens, sags, and becomes less stable.

Removal of a habitat has a much greater devastating effect on stability than does removal of a niche. This can be demonstrated in the profundal zone of the lake shown in Fig. 8.40. The profundal benthic community has sixteen species in three functional feeding groups, that is filterers, gatherers and predators. If one habitat is removed (e.g. line e to f in Fig. 8.41 is removed), say by eutrophication, half the species would disappear and only the heartiest would remain. But not all functional feeding groups would necessarily be represented. For example, tubificid worms (described below), such as *Tubifex tubifex* and *Limnodrilus hoffmeisteri*, are gatherers and they replace other gatherers and filter feeders in the organic ooze of eutrophic lakes. Both species of worms are classical enrichment indicators (Table 8.7) and their presence in large numbers clearly indicates eutrophy. Hence, eutrophic lakes are less stable than oligotrophic lakes because the habitat diversity and niche diversity are reduced.

There are oligotrophic indicator species as well (Table 8.7). They are dominated by filter feeding species, with some gathering and predaceous species. The filter feeders are represented by fingernail clams, the gatherers by worms and the predators by insects. Characteristics of each group were described earlier in this chapter.

While species richness is a useful tool for assessing health of a benthic community, it relies entirely on the mere presence of a species. It gives equal weight to a species represented by one individual and a species represented by numerous individuals. Measures of diversity that account for numbers of individuals are more useful for assessing community health. A more useful measure of diversity is the Shannon-Weiner Index:

$$H = n_i/N \log_{10} n_i/N$$

where n_i is the number of individuals in species "i" and N is the total number of individuals present in a unit area of substrate. The index has been modified for use in

Table 8.7. Profundal macroinvertebrates with trophic status indicator value. Species are listed from primitive to most highly evolved forms in different phyla. P = phylum, C = class, O = order, F = family. Group characteristics were described earlier in this chapter.

Genus/species name	Common name	Taxon group	Function; Indicator Value
Manayunkia speciosa	freshwater polychaete	P: Annelida C: Polychaeta	Filterer, gatherer; Oligotrophic indicator
Limnodrilus hoffmeisteri	sludge worm	C: Oligochaeta	Gatherer; Eutrophic indicator
Tubifex tubifex	sludge worm	C: Oligochaeta	Gatherer; Eutrophic indicator
Peloscolex variegatum	oligochaete	C: Oligochaeta	Gatherer; Oligotrophic indicator
Tubifex kessleri	oligochaete	C: Oligochaeta	Gatherer; Oligotrophic indicator
Sphaerium corneum	European fingernail clam	C: Bivalvia	Filterer; Mesotrophic indicator
Sphaerium nitidum	Arctic-Alpine fingernail clam	C: Bivalvia	Filterer; Oligotrophic indicator
Pisidium conventus	Arctic-Alpine pea clam	C: Bivalvia	Filterer; Oligotrophic indicator
Caecidotea (Asellus) spp.	isopod	P: Arthropoda C: Crustacea O: Isopoda	Gatherer; Mesotrophic indicator
Diporeia hoyi	deep water amphipod	O: Amphipoda	Gatherer; Oligotrophic indicator
Mysis relicta	relict mysid	O: Mysidacea	Predator; Oligotrophic indicator
Chironomus plumosus See Table 8.6 also	blood worm	C: Insecta; O: Diptera F: Chironomidae	Gatherer; Eutrophic indicator
Hexagenia limbata	mayfly	O: Ephemeroptera	Predator; Oligotrophic indicator

assessing stream water quality and its calculation and application is left for discussion in Chapter 12, "Water Quality Assessment Techniques".

Quantitative Aspects

Benthologists have found that many lakes have a ***concentration zone***. This is the depth at which the peak abundance and biomass of benthos occurs. Usually the peak abundance occurs between the 2 and 4 m depths but can go as deep as 7 m. Benthic- feeding species of fish are known to congregate in this zone. Hence, if the location of the concentration zone is known, one will have a good idea of where to catch benthic feeding fish.

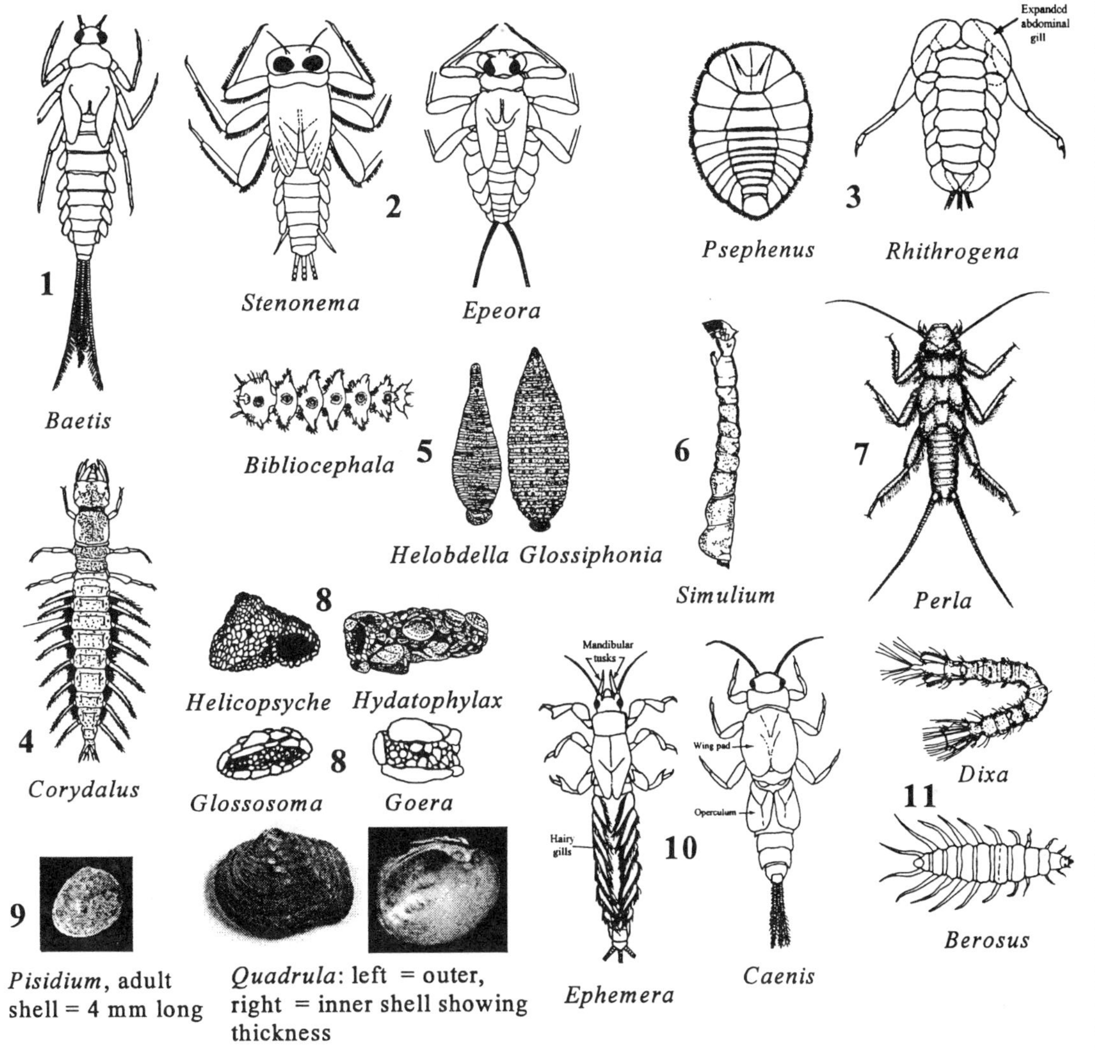

Figure 8.42. Illustrations of morphological adaptations listed in Table 8.8 (numbers correspond to adaptation number in column 1 of table).

Adaptations for Life in Flowing Waters

We have already described several morphological adaptations to life in flowing waters for several taxa, especially insects. Table 8.8 summarizes morphological adaptations and some physiological and behavioural ones as well. Figure 8.42
39☞ illustrates the adaptations, with the numbers in Table 8.8 (column 1) corresponding to the numbers in the figure. While many of the attributes can be argued to have some

Table 8.8. Morphological, physiological and behavioural adaptations for life in lotic environments.

Morphological Adaptations	Significance	Representative Taxa	Comments
1. Streamlining - most efficient is fusiform shape, with widest part ~ 1/3 from head end	Offers least resistance to flow	Best are *Baetis* and *Centroptilum* (mayflies)	Many *Baetis* spp. are lentic mayflies
2. Flattening of body surface	Allows animal to "duck under" turbulent flow to stay within laminar flow	*Psephenus* - water penny; *Rhithrogena* - mayfly; isopods, *Caecidotea*	Many species that live under rocks (e.g. *Stenonema* - mayfly) are also flattened.
3. Friction pads and marginal contact with substrate	Staying close to surface and clinging to crevices increases frictional resistance ; reduces risk of being dislodged	*Psephenus* - water penny; *Rhithrogena* - mayfly	Some water pennies (e.g. *Ectopria*) are lentic species
4. Hooks and grapples as provided by modified tarsal claws and posterior prolegs	Allows attachment to rough surfaces	Hellgramites, some stream caddisflies e.g. *Rhyacophila*	All caddisflies have posterior prolegs, including lentic species
5. Suckers, usually disc-shaped, adhere by hydraulic vacuum or mucous secretions	Provide attachment to smooth surfaces	Hydraulic vacuum - all Blephariceridae; mucous - black flies, *Simulium* and snails; vacuum and mucous - leeches	All leeches have oral and posterior suckers, even swimming forms; some snails are confined to temporary ponds
6. Silk and sticky secretions (from other than suckers)	Anchor animal to surfaces in fast water	*Simulium* secretes silk anchor thread	Lentic caddisflies use silk secretions also
7. Reduction of projecting structures from head, thorax and/or abdomen	Projections increase water resistance	Loss of central tail in some *Baetis* spp.; Most Plecoptera have smooth outlines	Many lotic species have lateral processes e.g. *Atherix* (Diptera) and hellgrammites
8. Ballast, as provided by heavy houses or shells	The heavier the body, the more force needed to move it	*Goera*, a caddisfly; Unionid mussel shells	Many of the unionids with thick shells live in lentic habitats
9. Small size	Permits life in crevices of rocks in fast water	*Pisidium* spp (Bivalvia); many lotic mites are smaller than lentic spp; most riffle beetles are small	Some of the smallest *Pisidium spp.* live in profundal zone of oligotrophic lakes
10. Hairy bodies and gill opercula	Keeps silt particles away from respiratory structures	The mayflies, *Hexagenia* and *Ephemera* with hairy gills; *Caenis* with gill opercula	Both mayfly genera found in lakes too
11. Stout claws & dorsal processes	Aid in attachment to plants	Stonefly, *Taeniopteryx* beetle, *Berosus* and the dipteran, *Dixa*	Not all plant dwellers have these
12. Reduction in powers of flight	Constant population size maintained for competition	Loss of hind wings in riffle beetles (Elmidae)	Reduces dispersal ability
Behavioural Adaptations			
13. Avoidance of current	Reduces risk of being carried away downstream	*Stenonema* (mayfly) lives under rocks; worms, many mayflies and true flies burrow in sediments	Many species e.g. *Stenonema* are flattened and adapted for flow, why avoid it?
14. Migrations to fast water for larvae, to slow water to emerge	Reduces risk of adult being caught and drowned	Many caddisfly larvae migrate to shore to slower currents to emerge	Many insects remain in fast water during emergence
Physiological Adaptations			
15. Dependent oxygen requirements	Lotic waters are usually saturated with oxygen ; no need for specialized oxygen uptake mechanisms.	Many species of stoneflies and lotic mayflies switch to dependent type of respiration under low flow conditions (Fig. 8.43)	Dependent respiration also exists in many lentic species. Some lotic species have independent respiration

16. Thermal adaptations	Cold stenotherms live in cold water; warm stenotherms & eutherms live in warm water	Stoneflies and many mayflies and caddisflies are cold stenotherms and emerge reproduce in spring	Similar thermal requirements occur for species.

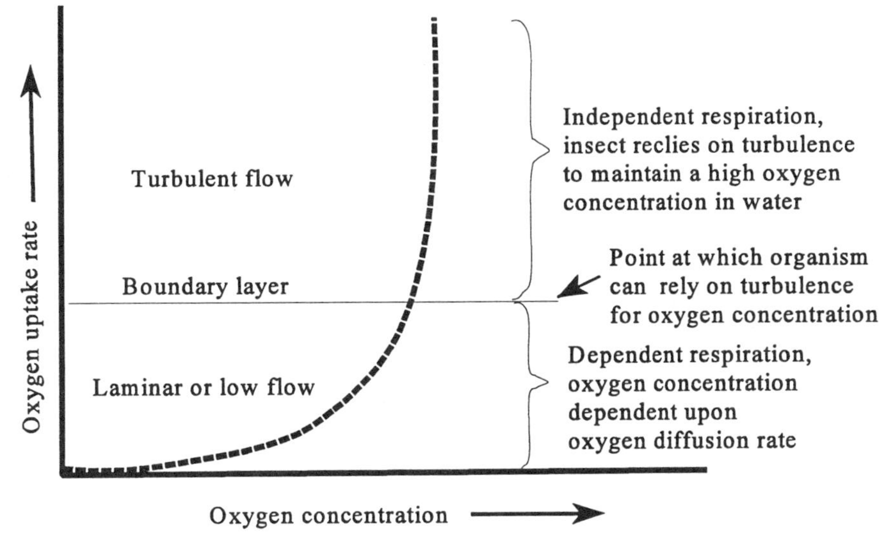

Figure 8.43. Dependent and independent types of respiration in relation to oxygen concentration in turbulent and laminar flow environments.

significance as an adaptation to flowing water environments, few if any are unique to lotic invertebrates. As indicated under comments in Table 8.8, many of the adaptations can be found in lentic species. However, the wave zones in lentic habitats often emulate lotic environments.

One of the more interesting behavioural adaptations is ***behavioural drift***, which is distinctly different from ***catastrophic drift*** and ***constant drift***. Catastrophic drift is the
40☞ result of flooding, spring runoff, ice or any other changes in physical conditions of the water. It is especially common in the winter and spring. In the winter, ***anchor ice*** forms in riffles during very cold nights, coating rocks in thin sheets of transparent ice. When the water temperature rises during the day, the ice breaks off and scours invertebrates from the rocks and substrate. In the spring, snow melt causes large spate events with torrential water flow churning up and eroding the substrate. Constant drift occurs at all times at low levels and, like catastrophic drift, is represented by most taxa in the stream. The reasons for constant drift are not known.

Behavioural drift is of two types, ***passive behavioural drift*** and ***active behavioural drift.*** Passive behavioural drift is an innate behaviour usually with some diel periodicity associated with it. Most invertebrates with this type of behaviour drift at night (e.g. *Baetis* and *Gammarus*), hiding during the day under rocks, logs or vegetation.

At sunset, the invertebrates become more active, searching for food and getting carried downstream by the currents. The greater the current, the more likely they are to drift and the further downstream they drift.

Active behavioural drift involves some specific movements that will move them downstream or upstream. There are five explanations for active behavioural drift. (i) First, it is probably density related. Competition for food will result in the exclusion of less competitive species or individuals which become part of the drift. (ii) If species are being carried downstream, there must be some upstream movement (especially by non-flying arthropods) to recolonize habitats vacated by drifters. Otherwise, upstream reaches would soon become depleted of invertebrates. Field experiments have shown that the rate of recolonization is directly proportional to the drift rate until the carrying capacity is reached, at which time no further increase occurs. (iii) Some life history stages are more prone to drift than others. Several species of mayflies *(Baetis rhodani, B. punilis, B. vernus, Rhithrogena hippopus, Hexagenia rigida)* and some black flies (*Simulium costatum*) become large components of the drift just prior to emergence. These species move to slower waters to emerge and in doing so, get caught in the currents. (iv) Avoidance of predators is bound to result in some organisms getting caught up in currents as they scurry from their hiding places across currents to find another hiding place. This probably explains some passive drift as well. (v) Seasonal variations result in less active behavioural drift in the winter than in the summer. The activity levels of most organisms is low in the winter when water temperatures are close to 0 °C. Invertebrate densities are also lowest in the winter, after summer and fall mortality factors (predation, parasitism, senescence, etc.) have taken their toll. The highest densities occur in the summer after recruitment has taken place.

The distances traveled by drifters is usually relatively small. Field measurements have shown that the mean distances traveled by *Baetis* and *Gammarus* are 21.6 and 28.5 m, respectively, with maxima between 1 and 5 km. Many die if scoured against rocks or substrata and many are eaten by fish and other invertebrates. Fish fry, in particular, depend on drifting invertebrates for a food source. This explains why evening is often the best time to catch fish.

The organisms most commonly involved in behavioural drift are:

The Crustacea:

Isopoda (e.g. *Caecidotea*)

41☞ Amphipoda (e.g. *Gammarus*)

and the Insecta:

Ephemeroptera (e.g. *Baetis, Stenonema, Leptophlebia, Rhithrogena*)

Plecoptera (e.g. *Nemoura, Isoperla, Leuctra*)

Trichoptera (e.g. *Hydropsyche*)

and occasionally:

Oligochaeta (e.g. Naididae)

Mollusca (e.g. *Ferrissia*, a limpet)

Most of the factors that affect the drift have already been discussed, but are summarized in Table 8.9 for the three kinds of drift. The question marks under constant drift indicate that each factor probably plays a role in affecting drift at low levels of magnitude. Suspended solids and pH can be both natural and anthropogenic factors. Suspended solids increase during spate events and the suspended particles act as abrasives in dislodging invertebrates. However, building of roads and pipelines across streams also increases sediment loads and rates of drift. Many streams have naturally low pH levels and low invertebrate densities and, hence, low drift rates. However, acid precipitation depresses pH levels, especially in spring snow melt, causing spring pH depressions that kill many invertebrates and even weaken the more resistant forms that end up as part of the drift. As snow melts, hydrogen ions migrate quickly to the bottom of the snow pack and are released into the runoff in a pulse event that lasts for a few days. During this short period, a stream can be denuded of its invertebrate fauna.

Table 8.9. Factors affecting drift of stream invertebrates

42☞

Factor	Catastrophic Drift	Constant Drift	Behavioural Drift
Water velocity	✓	?	
Water volume	✓	?	
Turbulence	✓	?	
Ice - anchor ice	✓	?	
Life cycle stage		?	✓
Light		?	+'ve or -'ve
Temperature		?	✓
Seasons	✓	?	✓
Substrate		?	✓
Competition		?	✓
Suspended solids	✓	?	
pH	✓	?	

Adaptations for Life in Temporary Ponds and Streams

Wiggins et al (1980)[7] give a thorough treatment of the subject and only a summary of the more common adaptations are given here. They divide the strategies

---

[7]Wiggins, G. B., R. J. Mackay and I. A. Smith. 1980. Evolutionary and ecological strategies of animals in annual temporary ponds. Arch. Hydrobiol./Suppl. 58: 97-206.

into four groups of animals. ***Group 1*** is overwintering residents capable of passive dispersal only. They aestivate and overwinter in the dry basin of the pool either as
43☞ drought-resistant cysts and eggs or as juveniles and adults. ***Group 2*** is overwintering spring recruits. They must reproduce in the pool in spring before surface water disappears. Residents aestivate and overwinter in the dry pool basin mainly as eggs or larvae, a few as adults. ***Group 3*** is overwintering summer recruits. Oviposition does not require water and the animals enter the pool basin after water disappears. They overwinter mainly in the egg stage or as larvae in an egg matrix. ***Group 4*** is non-wintering spring migrants. Oviposition requires water so animals enter the pool in spring. The new adults leave the pool before its dry phase and overwinter mainly in permanent water bodies.

While each group has its own adaptations, the following summarizes most of those used by all groups, with representative examples of taxa that employ each adaptation. Figure 8.44 illustrates some of the species adapted to life in temporary aquatic habitats.

- resistant eggs or cysts lying exposed on dry pool basin (flatworms, bryozoans, most crustaceans)
- embryonic development timed for hatching in spring (the fairy shrimp, *Eubranchipus*)
- secretion of protective coat of mucous (oligochaetes and leeches)
- secretion of a mucoid ***epiphram*** over aperture (the snails, *Aplexa hypnorum* and *Stagnicola elodes*)
- species with lungs rather than gills. All gastropods found in temporary ponds are pulmonates.
- rapid growth during a short period of activity (the fingernail clam, *Sphaerium occidentale* is restricted to temporary ponds)
- opportunistic species with wide physiological tolerances (chironomids)
- one-year life cycles (most taxa)
- oviposition independent of water (Group 3 trait - the damselflies, *Lestes* and *Sympetrum*)
- temperature-controlled egg diapause to prevent premature development if pond dries up too soon (the damselfly *Lestes dryas* and the dragonfly *Sympetrum obtrusum*)
- photoperiod-controlled egg diapause also prevents premature development if pond dries up too soon (the damselfly *Lestes dryas* and the dragonfly *Sympetrum obtrusum*)
- high thermal coefficient for growth. If appropriate threshold temperature is not attained, the eggs or larvae do not develop (*Lestes dryas* and *Sympetrum obtrusum*)
- modification of egg shape and chorion structure for terrestrial oviposition. Permanent pond species lay eggs in gelatinous masses, but temporary pond species lay spherical eggs in a bead-like string capable of rolling into damp

depressions of the pool basin (*Sympetrum obtrusum*)

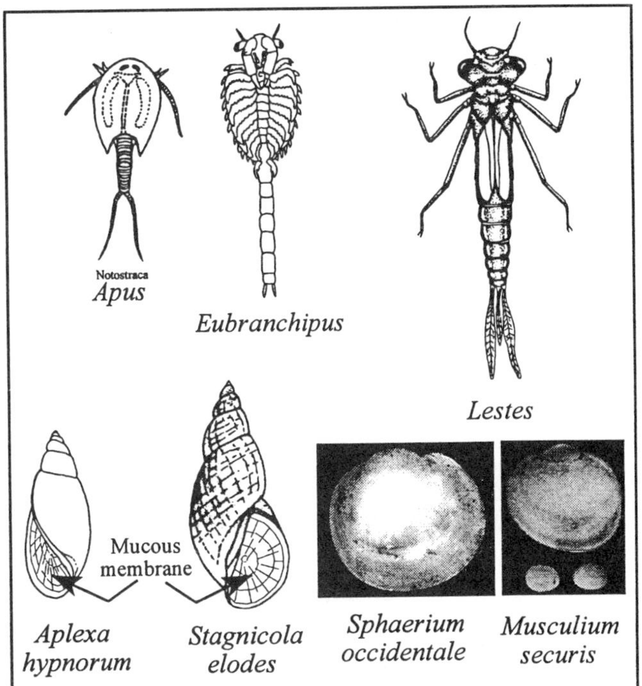

Figure 8.44. Some typical inabitants of temporary pools: Crustacea, *Apus* and *Eubranchipus*; Insecta, *Lestes;* Gastropoda, *Aplexa hypnorum* and *Stagnicola elodes*; and Bivalvia, *Sphaerium occidentale* and *Musculium securis*

Seasonal Variations in Benthic Biomass

Welch (1952)[8] described some of the seasonal changes in benthos in the littoral, sublittoral and profundal zones of lakes. These changes are summarized below, with some additional ones that have been described in more recent literature.

Spring Changes:

Littoral Zone:

- rapid growth of plants
- 44☞ periodic insect emergences
- annual shoreward migration of alderflies and snails from sublittoral zone
- rapid growth of sponges and bryozoans
- active growth, breeding and increase in population densities due to recruitment by numerous species
- Migration of adult unionids from sublittoral zone followed by brooding of glochidia by bradytichtic (short-term breeders) species and release of glochidia by tachytichtic species (long-term breeders) to parasitize fish

Sublittoral Zone:

- periodic insect emergences
- migration of alderflies and snails from upper profundal zone
- active growth, breeding and increase in population densities due to recruitment by numerous species
- build up of summer concentration zone

Profundal Zone:

- migration of some fingernail clams, oligochaetes and dipterans from sublittoral zone followed by marked insect emergence after spring overturn

---

[8]Welch, P. S. 1952. Limnology. McGraw-Hill Book Co., New York.

- beginning stages of summer concentration zone in sublittoral-profundal zone

Summer Changes:

Rapid growth of zebra mussels in littoral and sublittoral zones, and of quagga
45☞ mussels in profundal zone. Appearance of some new recruits due to early settlement in all depths.

Littoral Zone:

- continuation of growth of emergent plant species
- decreases in insect population densities due to large insect emergences
- increases in most species due to reproductive recruitment, and to final migrations of alderflies, snails and mayflies from sublittoral zone
- appearance of juveniles of tachytictic unionids and brooding by adults, release of glochidia by bradytictic species to parasitize fish

Sublittoral Zone:

- continuation of submersed plant growth
- fluctuations of insect population densities due to emergences followed by reproductive recruitment
- decreases in snail populations having littoral zone breeding habits but increases of others due to hatching of large egg masses
- continued growth and reproduction of most benthic invertebrates
- maximum development of sublittoral-profundal concentration zone

Profundal Zone:

- maximum development of sublittoral-profundal concentration zone
- gradual decreases in diversity but increases in densities of benthos in eutrophic lakes and increases in diversity and densities of benthos in oligotrophic lakes

Autumn Changes:

46☞ Appearance of new recruits of zebra and/or quagga mussels at all depths due to settlement of pediveligers and mortality due to senescence of adults.

Littoral Zone:

- cessation of growth and decline in biomass of emergent species of macrophytes
- gradual decline of benthos due to: (i) insect emergences; (ii) mortality factors, including senescence, parasitism, predation, wave action, etc.; (iii) migrations of alderflies, snails and mayflies into deeper water
- production of winter eggs by some crustaceans and of gemmules by sponges and statoblasts by bryozoans
- appearance of juveniles of tachytictic species of unionids

Sublittoral Zone:

- cessation of growth and decline in biomass of submerged species of macrophytes
- gradual decline of benthos due to: (i) insect emergences; (ii) mortality factors, including senescence, parasitism, predation, etc.; (iii) migrations of alderflies, snails and mayflies into deeper water
- production of winter eggs by some crustaceans and insects

Profundal Zone:

- increases in populations of tubificids and some pisidiids due to reproduction and migrations of some sublittoral species into profundal zone of oligotrophic lakes
- declines of population densities of some insects due to autumn emergences
- appearances of chironomid larvae due to hatching of eggs

Winter Changes:

47☞ Gametogenesis begins or continues in zebra and quagga mussels at all depths but growth rate slows.

Littoral Zone:

- minimum growth and dormancy of emergent plant species and hibernation of many invertebrates
- many plant dwellers become bottom dwellers
- disintegration of sponges, bryozoans and turbellarians
- migration of bivalves, snails, leeches, insects and others to, or below, depth limit of wave action or freezing
- burrowing of many species to depths to avoid freezing

Sublittoral Zone:

- population increases of insect species that laid eggs in autumn and of bivalves, snails and insect species migrating from littoral zone
- no insect emergences
- growth of winter insect larvae

Profundal Zone:

- midwinter maximum of tubificid and some phantom midge larvae
- no insect emergences
- rapid decline in population densities of benthos in eutrophic lakes due to winter anoxia but gradual decline in oligotrophic lakes

Productivity of Benthic Invertebrates

48☞ Measurement of benthic productivity is done using ***secondary production*** methods. As with primary production, secondary production is the unit change in biomass per unit area per unit of time. The most common units used for benthos is mg/m^2/yr. Biomass is simply the weight per unit area at any given moment in time and can be estimated as:

$$\overline{B} = \overline{N} \times \overline{w}$$

where N is the number of organisms in a unit area (usually /m^2) and w is the weight (usually in mg) of the organisms. Generally, several samples are taken for mean biomass estimates, hence the overbars.

Production is the *increment in population biomass plus the biomass that was eliminated* due to predation, excretion, respiration, etc. The increment in biomass can be estimated as:

$$B_t - B_0,$$

or the biomass at time t less the biomass at time 0. Hence, production can be calculated as:

$$P = B_E + (B_t - B_0)$$

The biomass eliminated (B_E) can be estimated by the product of number eliminated by the mean individual weight eliminated, or:

$$B_E = N_E \text{ x } w_E$$

The number eliminated is: $N_E = N_0 - N_t$

Therefore, the biomass eliminated is:

$$B_E = w\,(N_0 - N_t)$$

If the sampling time interval is kept short enough, one may estimate the mean weight as:

$$w = 1/2(w_0 + w_t)$$

Hence, production over a single interval of time is:

$$P = [(N_0 - N_t)(w_0 + w_t)/2][(B_t - B_0)]$$
$$= [(N_0 - N_t)(w_0 + w_t)/2][N_t w_t - N_0 w_0]$$

which simplifies down to: $\mathbf{P = [(N_0 + N_t)/2 \text{ x } (w_t - w_0)]}$

Production is usually measured for the most common species, with a production estimate for each one. Community secondary production estimates (i.e. estimates for each species population) are beyond the realm of consideration because of the time and money needed to make estimates for just a few species.

In order to measure production, it is first necessary to know whether the species has distinct ***cohorts***, or ***generations*** (Fig. 8.45). For many species, the generations overlap and it is difficult to determine when one generation begins and when one ends. Production estimates are easier for ***cohort production***, and a more arduous for ***non-cohort production***. We will examine three methods for cohort production and one for non-cohort production.

Cohort Production Methods

While use of cohort methods is popular because of their relative simplicity, care must be exercised in their application. One must be certain that the population is
49☞ composed of distinct cohorts. Populations may have one or more generations per year or in their life time. Cohorts in species with one generation per year, that is they are

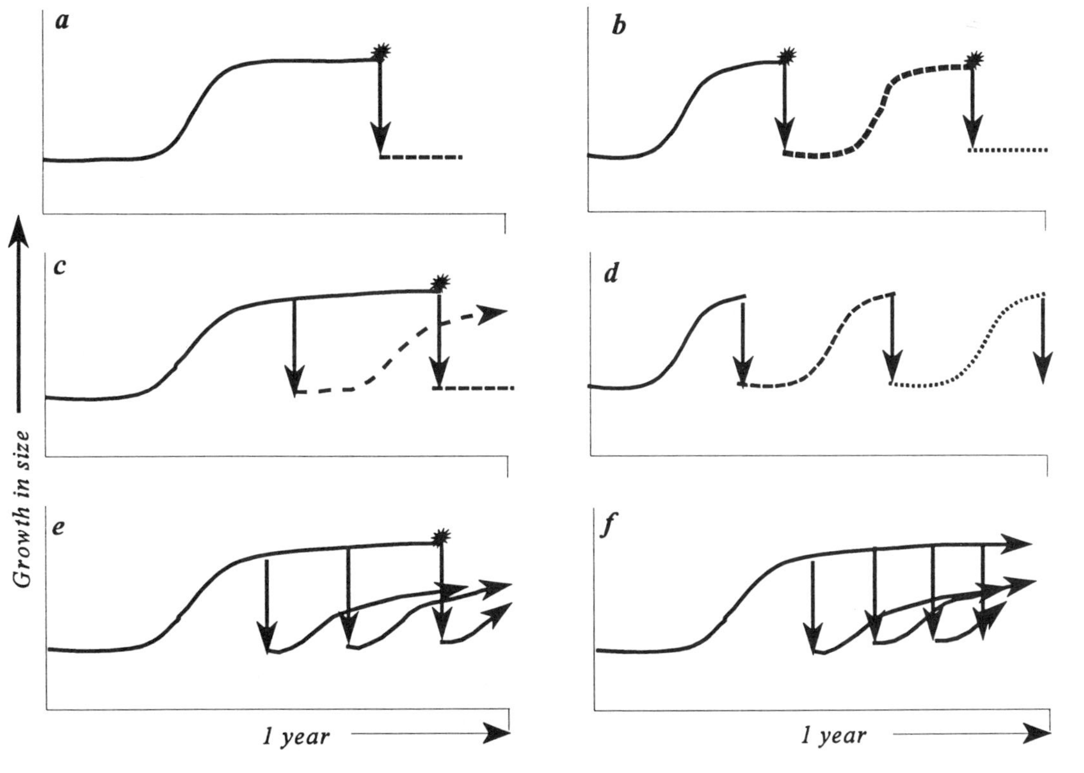

Figure 8.45. (a) Univoltine, semelparous; (b) Bivoltine, semelparous; (c) Bivoltine, iteroparous; (d) Trivoltine, semelparous; (e) Trivoltine, iteroparous; (f) Multivoltine, iteroparous.

univoltine, are easier to recognize than those in species with two generations per year (***bivoltine***), or several generations per year (***multivoltine***). In fact, some species may have several generations in their life time (called ***iteroparous*** species) and these usually have overlapping cohorts that require continuous cohort production methods for estimating production. Species that produce one generation in their life time (called ***semelparous species***) usually have distinct cohorts. Fig. 8.45 shows the different kinds of reproduction possible in benthic invertebrates. Cohort production estimates are possible with populations a, b and d in Fig. 8.45; non-cohort estimates are necessary for populations c, e and f (Fig. 8.45).

The Formula Method:

This method requires samples at frequent intervals to determine mean density per
50☞ unit area and the mean weight at each sampling event. For invertebrates, two week intervals is usually appropriate. For accurate production estimates, sample need to be taken over the life span of the species. Having obtained all the samples, one need only to sum all production estimates for each time interval, or:

$$\mathbf{P = \sum[(N_0 + N_t)/2 \times (w_t - w_0)]}$$

Table 8.10 gives an example calculation for a temporary pond population of *Musculium securis*, a small fingernail clam, that was sampled at two week intervals in the wet period and monthly intervals in the dry period. The species is hermaphroditic ovoviviparous, producing live newborn during the wet period. Adults die after producing their young, the newborn burying themselves in the mud before the pond dries and aestivate during the dry period. When water returns the following spring, the newborn break their dormancy and grow quickly into adults, reproduce as they grow and then release their young before the pond dries again. The data in Table 8.10 begin with aestivating newborn that exhibit little productivity. Most of the production occurs during the wet period (April to August). Note that during the t_{16} period, there was a loss in production. This represents the loss of tissues in senescent adults and of young from the and last surviving parents where all parents had died or were dying by August 15. The life span of the population was exactly one year. Also note that there was constant mortality of aestivating newborn, with slight growth in mean weight.

The ratio of production (P) to mean standing biomass (B) provides useful information regarding production dynamics of invertebrates. The P/B ratio corresponds to a turnover rate and reflects the part of the biomass that is replaced during one unit of time. The ***annual P/B*** ratio is the annual production divided by the mean annual biomass. The mean annual production in our example is 106,610.8 mg/m^2. The mean annual biomass is the product of the mean density (N) and mean individual weight (w). In Table 8.10, the mean density is the total density (135,345/m^2) divided by the number of sampling events (17), or 7,961.5/m^2. Similarly, the mean individual weight is the total

Table 8.10. Estimating production using the formula, $P = \sum [(N_0 + N_t)/2 \times (w_t - w_0)]$.

Time Period	Date	Mean No./m^2	Mean weight mg	Density change	Weight change	Production
t_0	Aug 15	8500	0.141	$(N_0+N_t)/2$	(w_t-w_0)	(col 5 x col 6)
t_1	Sept 15	8500	0.218	8500	0.077	654.5
t_2	Oct 15	8375	0.220	8437.5	0.002	16.9
t_3	Nov 15	8250	0.221	8312.5	0.001	8.3
t_4	Dec 15	8220	0.223	8235	0.002	16.5
t_5	Jan 15	8200	0.225	8210	0.002	16.4
t_6	Feb 15	8050	0.228	8125	0.003	24.4
t_7	Mar 15	8000	0.236	8025	0.008	64.2
t_8	Apr 15	7850	0.313	7925	0.077	610.2
t_9	Apr 30	7850	0.916	7850	0.603	4,733.6
t_{10}	May 15	7800	2.188	7825	1.272	9,953.4
t_{11}	May 30	7750	4.625	7775	2.437	18,947.7
t_{12}	Jun 15	7700	7.663	7725	3.038	23,468.6
t_{13}	Jun 30	7650	11.388	7675	3.725	28,589.4
t_{14}	July 15	7600	13.900	7625	2.512	19,154.0
t_{15}	July 30	7550	15.138	7575	1.238	9,377.9
t_{16}	Aug 15	7500	13.938	7525	-1.200	-9030.0
Totals		**135,345**	**71.781**	127,345	13.797	**106,610.8**
Means		**7,961.5**	**4.222**	7,959.1	0.862	

individual weight (71.78 mg) divided by the number of sampling events (17), or 4.22 mg. The mean biomass then is 7,961.5/m^2 x 4.22 mg = 33,598 mg/m^2. The annual P/B ratio, therefore, is 106,610.8 mg/m^2/yr /33,598 mg/m^2 = **3.17**.

Usually, however, the ***cohort P/B ratio*** is determined. This is the P/B ratio of a single cohort. In the above example, the cohort lived for exactly 1 year so the cohort P/B ratio would be the same as the annual P/B ratio. In other words, the ***cohort production***

interval, or ***CPI***, was 12 months. If, however, the cohort lived two months longer, one would have to add the additional production over this two month period to the 12 month production estimate. The new CPI is 14 months and must be used to correct for the annual P/B ratio. For example, suppose the total cohort production (from first appearance of newborn to death of the last adults) was 115,000 mg/m^2, and the mean biomass rose to 35,000 mg, the ***cohort P/B ratio*** would be 115,000/35,000 = ***3.29***. The ***annual P/B ratio*** would be adjusted to 12 months/14 months x 3.29 = ***2.81***.

The cohort P/B ratio can vary between 2 and 8, but usually lies between 3 and 5. The ratio tends to be constant for a species. Most univoltine species have cohort P/B ratios near 5. Annual P/B ratios are much more variable.

Having calculated the P/B ratio, it can then be used to estimate productivity. For example, by reversing the process using a cohort P/B = 3.17, and a mean biomass for the cohort of 33,598 mg/m^2, cohort P = 33,598 x 3.17 = 106,506 mg/m^2/cohort interval (total is not the same as table value because of decimal rounding). However, the method requires a good estimate of mean biomass and some knowledge on the time of year that mean biomass is attained. In our example, we would need to take samples near the end of May to be close to the mean biomass. Alternatively, we could strategically take several samples throughout the year, for example once in spring, three times in summer, and once in fall. For example, if we chose Oct 15, April 15, May 15, June 15 and July 15, we would average [(8375 x 0.22) + (7850 x 0.31) + (7800 x 2.18) + (7700 x 7.66) + (7600 x 13.9)]/5 = [1,843 + 2,453 + 17,063 + 59,001 + 105,640]/5 = 37,200. This would yield a production estimate of 117,925 mg/m^2/cohort interval, a value that is about 8% higher than the tabulated mean.

Allen Curve Method:

A fisheries scientist, Dr. K. R. Allen, developed this method for estimating fish production. The method described below uses the same data as in the formula method.
51 ☞ The number per square meter is plotted on the y-axis and the corresponding mean individual weight is plotted on the x-axis, as in Fig. 8.46. A smoothed curve is fit to the data points and production is estimated from the area under the curve. The area can be determined using a planimeter, by counting squares on squared graph paper, or by using the same paper weight method used for estimating lake areas in Chapter 3. The P/B ratio can also be determined from the Allen curve by dividing the area obtained for production by the area for mean biomass, obtained by plotting the mean number per square meter on the y-axis and the mean individual weight on the x-axis. Estimating the area of the mean biomass "box" is easy; it is the same as in the previous example, 7961.5 x 4.22 = 33,598 mg/m^2. Using the squared paper total production (cohort) is estimated at 106,000 mg/m^2/cohort year. This is slightly less than the formula method because of area error in using the squared paper method, but the error is acceptable (~ 0.1%), especially when rounding off to 106.0 g/m^2/cohort year.

52 ☞ *The Growth Rate Method*

To use this method one should know the seasonal size frequency distribution of

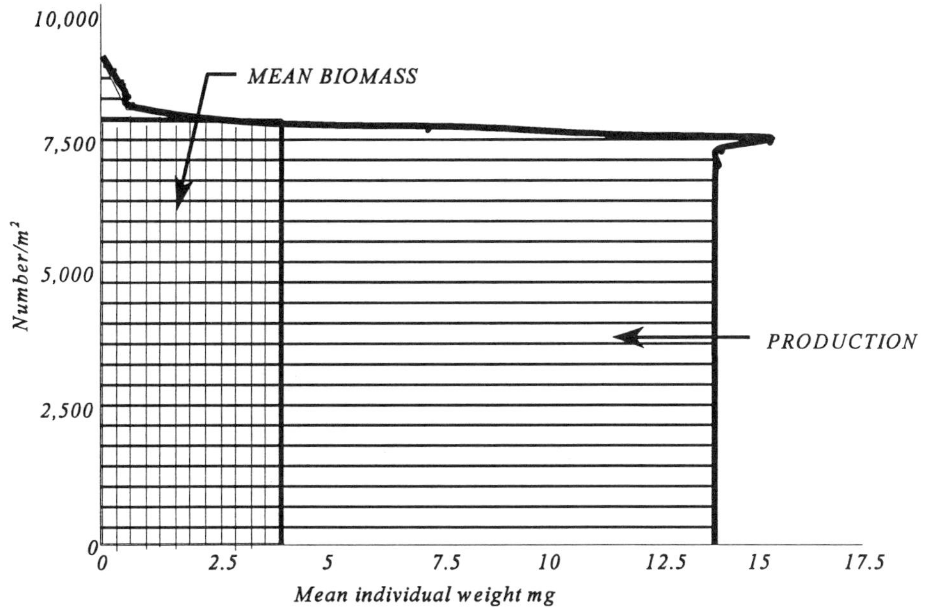

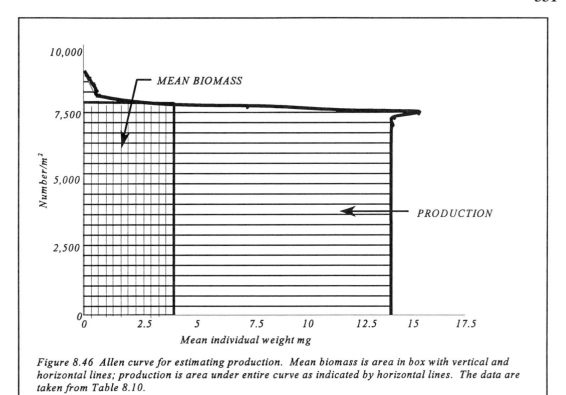

Figure 8.46 Allen curve for estimating production. Mean biomass is area in box with vertical and horizontal lines; production is area under entire curve as indicated by horizontal lines. The data are taken from Table 8.10.

the population. This will allow one to estimate changes in growth rate and the longevity of the population. Fig. 8.47 is an example for *Musculium securis* in a temporary pond near Carp, Ontario. It is the same population used for the previous two examples. Based on the growth curve, the population can be split into three classes; a very slow growing ***juvenile group*** from 1.2 to 1.5 mm; a fast growing ***adult group*** of 1.5 to 6.5 mm; and a ***senescent group*** of 6.5 to 7.0 mm.

Production per unit of time using the growth rate method is calculated as:

$$P = Bg, \text{ or } P = Nwg$$

where B is the mean biomass, calculated as Nw, as before, and g is the specific growth rate of each group. The specific growth rate is calculated as:

$$g=(1/t)(\ln w_2/\ln w_1)$$

where t is the time in each size group and w_2 and w_1 are the mean weights at the end and beginning of each size group. Table 8.11 gives growth rate data and the production of each size group of *Musculium securis* from Carp Pond using the growth rate method. The annual P/B is [(57.6 x 225) + (1133.5 x 92) + (432.0 x 48)]/(7819 x 6.60) = 12,964.5 + 104,282 + 20,736.0/ 51,605 = 137,982.5/51,605 = 2.67 using the growth rate method.

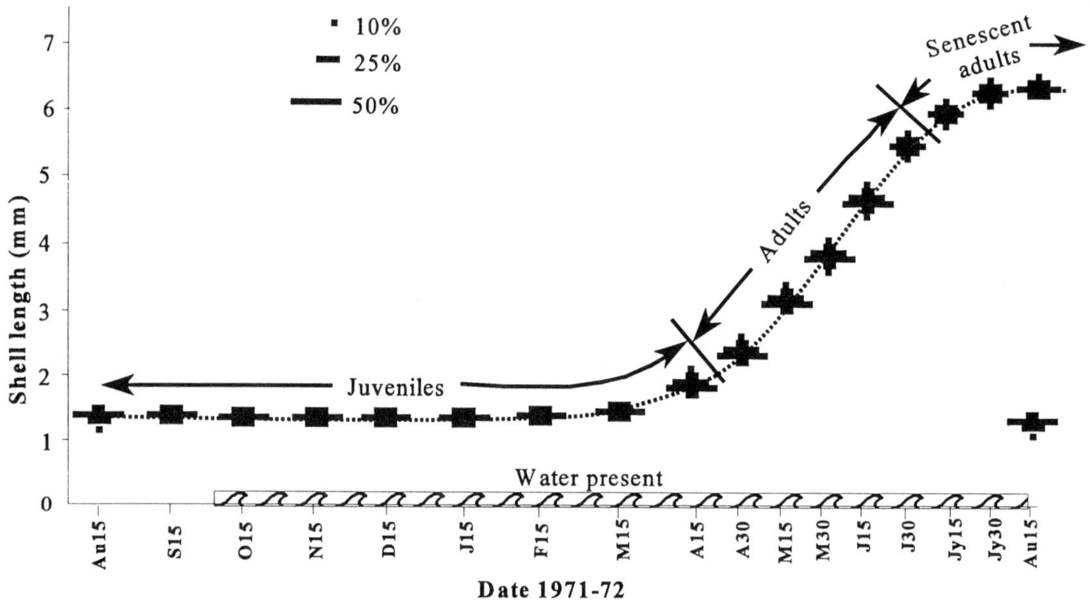

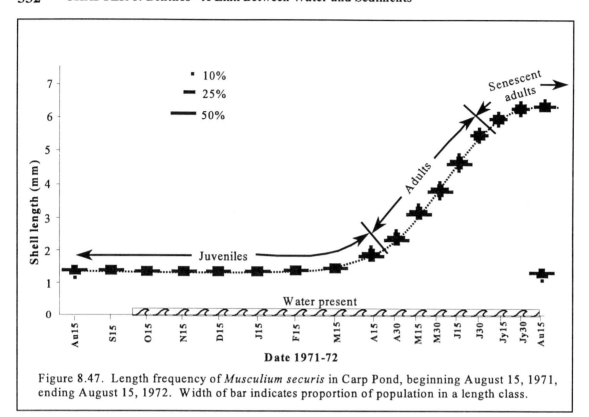

Figure 8.47. Length frequency of *Musculium securis* in Carp Pond, beginning August 15, 1971, ending August 15, 1972. Width of bar indicates proportion of population in a length class.

Table 8.11. Production of *Musculium securis* in Carp Pond using the growth rate method. The asterisks (*) indicate column means needed for the mean biomass estimate, not totals.

Size Group	Time days	w_1 mg	w_2 mg	$g = 1/t\ (\ln w_2/w_1)$	No. /m^2	Weight mg	P = Nwg mg/m^2/d
Juvenile	225	0.141	0.916	0.0083	8180	0.847	57.6
Adult	92	0.916	11.388	0.0274	7725	5.356	1,133.5
Senescent	48	11.388	13.938	0.0042	7550	13.591	432.0
TOTALS	365				7819*	6.598*	**1,623**

Note that the total annual production using the growth rate method is much higher than the estimates obtained using the formula and Allen curve methods. Most of the difference is in the estimate for adults where too much emphasis was placed on this size group. A more accurate estimate could be obtained by dividing the size groups into four instead of three. The estimate gets more precise with an increase in the number of size groups, but eventually one reaches the point that the formula method may as well be used.

Non-Cohort Method or Continuous Production Method

Unfortunately, most species do not display synchronous development and the population consists of several overlapping generations, none or few of which can be recognized in size frequency plots. However, survivorship of a population with overlapping generations (cohorts) can be approximated from its age (size) distribution at a single point in time. The following describes a size frequency method for estimating production from a population without recognizable cohorts.

Size Frequency Method

The basis for this method is that an average size frequency distribution plotted from samples taken over a period of one year will emulate the average survivorship of a
53☞ hypothetical ***average cohort***. The average cohort is not an ***actual cohort*** because it does not represent the actual number of individuals reaching each size class. To use the size frequency method one must make two assumptions. First, one must assume that the survivorship of a population with several overlapping generations can be approximated from its size distribution at a single point in time. This will allow one to approximate the number of individuals that reach each size in a one year period and to estimate production from changes in biomass based on changes between size groups (rather than on changes between sampling dates). One must also assume linear growth of a population that grows through N size classes over the one year period. This assumes that each size class lasts only 1/N part of the year. The mean number of individuals in each size class of the average cohort must then be multiplied by N to estimate the number of individuals that grew to that size class during the year. Even though one cannot recognize cohorts, it is important to know the life span of a cohort so that a cohort production interval (CPI) can be estimated.

Production for a non-cohort population is done below using another species of fingernail clam, *Sphaerium striatinum*. Most stream populations have overlapping generations without recognizable cohorts. However, plots of the size classes shows that recruitment occurs once a year in the spring or summer, with adults producing one cohort for each of two spring/summers and then dying after 18 months (i.e. the population is univoltine and iteroparous with a CPI of 18 months). Newborn are between 3 and 4 mm in shell length and grow to between 14 and 15 mm on average (Table 8.12).

Suppose a population is sampled over one year and the shell length is measured for each specimen in all samples. The samples can be pooled at the end of the year to determine the number of individuals in 1-mm length classes. All individuals in each length class are then weighed, the total weight is divided by the number of specimens in the length class to derive a mean weight. The size classes, number of individuals and mean individual weight in each size class are given for a creek population in Table 8.12. In the annual production estimate for this population, the size frequency method has two very different procedures from the cohort methods: (i) the changes in mean biomass values are based on changes between size groups rather than on changes between sampling dates; (ii) the time spent in each length class is 1/N = 1/12 so production for

Table 8.12. Estimating annual production for a population of fingernail clams, *Sphaerium striatinum* from Cox Creek, Ontario, using the size frequency method. The first three columns are observed observations. Column 4 is the product of columns 2 and 3; column 5 is the difference in N values between adjacent size classes; column 6 is the sum of average weights of adjacent length classes divided by 2; column 7 is the product of columns 5 and 6; column 8 is column 7 times the number of size classes (12).

Column 1	Col 2	Col 3	Col 4	Col 5	Col 6	Col 7	Col 8
Size Group Length, mm	**N No/m²**	**w mg**	**B mg/m²**	**ΔN No/m²**	**Weight at loss mg**	**Biomass loss wΔN**	**× 12 mg/m²**
3-4	450	2.12	954.0				
				-55	2.9	-161	-1,937
4-5	505	3.75	1,893.8				
				-247	4.7	-1161	-13,932
5-6	752	5.65	4,248.8				
				176	8.9	1,558.5	18,702
6-7	576	12.06	6,946.6				
				235	14.8	3,473.3	41,680
7-8	341	17.5	5,967.5				
				123	21.0	2,579.9	30,959
8-9	218	24.45	5,330.1				
				48	27.4	1,314.5	15,774
9-10	170	30.32	5,154.4				
				44	33.0	1,452.0	17,424
10-11	126	35.68	4,495.7				
				42	38.2	1,606.1	19,273
11-12	84	40.8	3,427.2				
				34	43.3	1,471.9	17,662
12-13	50	45.78	2,289.0				
				22	47.8	1,052.5	12,630
13-14	28	49.9	1,397.2				
				10	52.5	524.5	6,294
14-15	18	55	990.0	18	55.0	990.0	11,880
			Mean B = 3,591.2				**176,409**

each size class is multiplies by 12 (last column, Table 8.12). Also, the total production (176.4 g/m^2/18 mon) must be corrected for the CPI. This correction is 12/18 = 0.67, so the production value must be multiplied by 0.67, to yield an annual total production of 118.2 g/m^2.

Sampling Methods for Benthos

The sampling methods differ between lotic (Fig. 8.48) and lentic (Fig. 8.49) environments. In streams, one relies on currents to carry the organisms and debris into nets. In lakes, grab samples must be used. Some of the most common quantitative sampling devices are described for each habitat type. Quantitative samplers are needed for population density and biomass estimates on a unit area basis. A kick-and-sweep method is also described for use in semiquantitative sampling of lentic and lotic habitats.

Quantitative Sampling of Lotic Habitats (Fig. 8.48)

Surber Sampler - The classical stream bottom sampler is the Surber sampler. It has two square metal frames (usually made of brass). Each base is 30.5 cm (12 in) square and
54☞ hinged on one of the common frames. The hinged part merely allows collapse of the frame for storage. One of the squares has a net attached to it; the other is open and is placed on the stream bottom, with the net opening towards the stream so that water flows into the net.

The mesh openings in the net are 0.5 mm x 0.5 mm. The sampler is placed on the stream bottom, then the substrate with the organisms is kicked up or stirred up so that the current carries everything into the net. Particles and organisms smaller than 0.5 mm pass through the mesh and particles and organisms larger than 0.5 mm are retained. The net's contents are then emptied into a white tray and the organisms are separated by hand from the debris using fine forceps. The organisms are then preserved in 70% ethanol or 5% neutralized (with calcium carbonate) formalin. The formalin is acidic and should be neutralized so that mollusc shells do not dissolve during storage.

The disadvantage of the Surber sampler is that it takes so much substrate that several hours may be required to sort the invertebrates from the debris. Usually 3-5 replicates are taken so that sorting time is usually 10 to 15 times the sampling time. To express results on a per square meter basis, multiply the values per sample by 10.8 (i.e. there are 10.8 Surber samples per square meter).

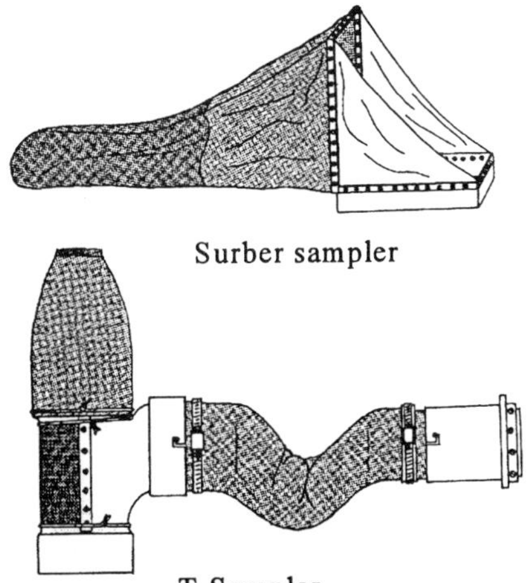

Figure 8.48. Two common quantitative stream sampling devices.

T-sampler - The T-sampler is made from 10 cm (4 in) diameter ABS plumbing pipe so the sampling area is about one ninth that of the Surber sampler and therefore requires considerably less time to sort invertebrates from the debris. The top part of the cross T has a window with a stainless steel screen containing 0.5 mm mesh openings. The vertical part of the T has a removable net (also with 0.5 mm mesh openings) attached to it. One end of the cross T has a sleeve for placing one arm of the sampling person; the other end is open and is placed on the bottom of the stream, with the stainless steel screen window facing the current. The bottom debris, with organisms, is stirred up and directed into the net. Once satisfied that all the organisms contained within the 4 inch diameter sampler opening are directed and retained within the net, the net is detached and its contents placed in a white enamel tray for sorting, as with the Surber sampler.

To express results on a per square meter basis, multiply the T-sampler values by 123.3 (i.e. there are 123.3 T-samples per square meter).

Quantitative Sampling of Lentic Habitats (Fig. 8.49)

Ekman Grab - The Ekman grab is light in weight and should be used on soft sediments only. It has a pair of spring-loaded jaws that are held open by wires attached to a
55☞ tripping mechanism. The top part of the sampler is covered by a screen (with 0.5 mm mesh openings) to prevent organisms from escaping through the top opening. A rope is attached to the grab at the tripping mechanism. The rope is fed through the centre of a heavy stainless steel weight. The grab is lowered to the bottom of the lake by means of the rope, while holding on to the stainless steel weight. The grab is allowed to fall freely and vertically so that it lands squarely and firmly on the bottom. Any slack in the rope is taken up and the stainless steel weight is released down the rope to trip and close the jaws of the grab. The jaws "grab" a sample of the bottom. The sampler with its trapped mud sample is raised to the surface and emptied into a box or pail with a sieve (0.5 mm mesh openings) attached to it. The mud is washed from the sample at the lake surface and the screen's contents are emptied into a jar for temporary

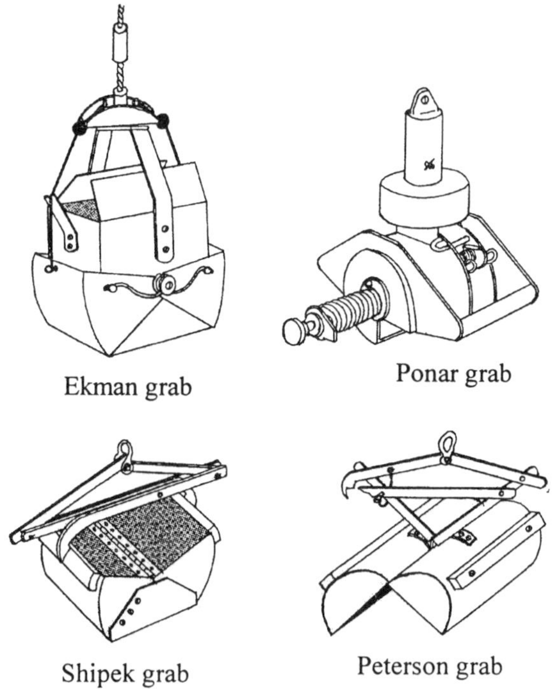

Figure 8.49. Four grab devices used for sampling benthos in lakes and large rivers. The Ekman grab is used in fine, loose substrates, the rest for coarser and more compact substrates.

storage. On shore, or in the lab, the screened sample is sorted in a white enamel tray.

The grab is available in two sizes, a standard 15 cm x 15 cm x 15 cm (6 in x 6 in x 6 in) sampler box and a larger 23 cm x 23 cm x 23 cm (9 in x 9 in x 9 in) version. To express densities on a per square meter basis, multiply the numbers of organisms in a grab sample by 44.4 for a standard grab and by 18.9 for the larger version.

Ponar Grab - The Ponar and Peterson grabs are used for sampling organisms in firm substrates, like gravel and packed mud or sand. The Ponar grab is more commonly used now than the Peterson grab. The Ponar grab has jaws with a rope, but it is tripped by its own weight. The tops of both jaws of the sampler are covered by a screen with 0.5 mm openings to prevent organisms from escaping. The sampler is "loaded" by pulling the articulating arms together (and the jaws apart) until the jaws of the sampler are held open by its own weight. The device is lowered to the bottom and when it strikes the bottom, the jaws are released to grab a sample of the substrate. The sampler is raised to the surface, emptied into a bucket or box with a screen and washed, stored and examined as for the Ekman grab sample.

The standard Ponar grab has an opening of 23 cm x 23 cm (9 in by 9 in) and is so heavy that a winch is needed to lift it. A "petit" model, with a 15 cm x 15 cm (6 in x 6 in) opening, is also available and, although heavier than the Ekman grab, can be used without a winch. Use the same areal conversions to square meters as for the Ekman grab.

Semi-quantitative Sampling of Either Lotic or Lentic Habitats

Semi-quantitative samples can be used for assessing environmental quality and for size frequency plots, as long as random samples are taken of the benthic communities
56☞ and all size classes are accurately represented in the sample. The most common method used is the kick-and-sweep method and is rapidly replacing quantitative samples for environmental impact assessments based on benthic community structure.

Kick-and Sweep Method - A D-frame net (Fig. 8.50) with a handle is used. The net has mesh openings of 0.5 mm. The D part of the frame is 30.5 cm (12 in) wide. Holding the net vertical, bottom debris in front of the net opening is kicked into the net. Usually about a 1 meter length of substrate is kicked into the net (i.e. the sample path is about 30.5 cm x 100 cm). The net's contents are emptied into a jar and the process is repeated about 5 times for each site. The jars with debris and organisms are stored on ice until they can be sorted and the animals preserved.

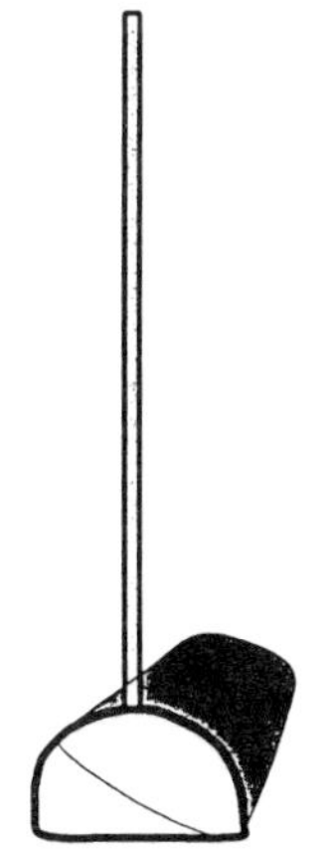

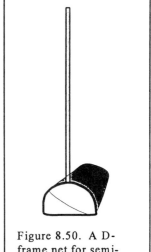

Figure 8.50. A D-frame net for semi-quantitative samples.

For environmental impact assessments, the trend now is to sort the first 100 or 200 organisms while in the field. It is important, however, that the selection be entirely random, with no conscious effort of selecting certain taxa, size groups or other features that might bias one's selection. Chapter 12 describes applications of the kick and sweep method for assessing water quality.

CHAPTER 9

FISHES - THE SEERS OF AQUATIC HEALTH AND VITALITY

Why Read This Chapter?

Fish are part of the open water community of strong swimmers called ***nekton***. Unlike the plankton, the nekton are capable of swimming against strong currents and their position in the water column is dictated by themselves, not by wind driven water currents. This chapter examines the different kinds of fish, how they partition resources to avoid competition, how they are distributed with respect to depth in lakes and distance along streams, factors that affect their distribution and abundance, how to sample them and estimate population sizes, and how to use them to tell us about the health and vitality of lakes and streams.

Other members of the nekton are large Crustacea, most of which (e.g. *Mysis relicta*) make daily migrations from the profundal sediments to the surface waters, insects with powerful swimming legs (e.g. *Dytiscus*), large blood suckers and other animals that frequently move from a bottom existence to the water column.

Many organisms are associated with just the surface film, taking advantage of the surface tension of water. Collectively, they are known as ***neuston*** and include several insects of the Orders Hemiptera (e.g. water striders) and Coleoptera (e.g. whirligig beetles). Because they live on the surface, they are more appropriately called ***epineuston***. In marine environments the term ***pleuston*** is used and includes the deadly Portuguese Man-of-War jellyfish, *Physalia*. The neustonic organisms that live just below the surface film are called ***hyponeuston***. Mosquito larvae are the best known members of the hyponeuston.

The chapter begins by examining the more common freshwater families of fish. Identification of fishes is beyond the scope of this text. Refer to Scott and Crossman (1998)[1]

---

[1]Scott, W. B. and E. J. Crossman. 1998. Freshwater fishes of Canada. Galt House Publications Ltd., Oakville, Ontario.

and Coad (1996)[2] for fish identification and details on the biology of each species.

By the end of this chapter you will know:

☞ 1. The three major taxonomic groups of fishes
☞ 2. The reproductive and feeding habits and value of lamprey
☞ 3. The reproductive and feeding habits and value of sturgeon
☞ 4. The reproductive and feeding habits and value of gars
☞ 5. The reproductive and feeding habits and value of bowfins
☞ 6. The reproductive and feeding habits and value of herrings and alewife
☞ 7. The reproductive and feeding habits and value of salmonids
☞ 8. The reproductive and feeding habits and value of whitefish
☞ 9. The reproductive and feeding habits and value of rainbow smelt
☞ 10. The reproductive and feeding habits and value of pike
☞ 11. The reproductive and feeding habits and value of minnows, dace and shiners
☞ 12. The reproductive and feeding habits and value of suckers
☞ 13. The reproductive and feeding habits and value of catfish and bullheads
☞ 14. The reproductive and feeding habits and value of eels
☞ 15. The reproductive and feeding habits and value of burbot
☞ 16. The reproductive and feeding habits and value of sticklebacks
☞ 17. The reproductive and feeding habits and value of sunfish and basses
☞ 18. The reproductive and feeding habits and value of perch, walleye and ruffe
☞ 19. The reproductive and feeding habits and value of exotic white perch
☞ 20. The reproductive and feeding habits and value of freshwater drum
☞ 21. The reproductive and feeding habits and value of sculpins
☞ 22. The reproductive and feeding habits and value of gobies
☞ 23. The meaning and different types of neuston
☞ 24. How insects walk on water
☞ 25. How some beetles can look below and above the water at the same time
☞ 26. Morphological adaptations of fish to life in flowing waters
☞ 27. Behavioural adaptations of fish to life in flowing water
☞ 28. The effect of current speed on distribution and occurrence of fishes
☞ 29. The effect of substrate and turbidity on distribution and occurrence of fishes
☞ 30. The effect of discharge fluctuations on distribution and occurrence of fishes
☞ 31. The effect of temperature on distribution and occurrence of fishes
☞ 32. How to use construct polygons to determine thermal tolerance in fishes
☞ 33. The effect of dissolved oxygen content on distribution and occurrence of fishes
☞ 34. The effect of dissolved materials on distribution and occurrence of fishes
☞ 35. Use of slope curves to determine fish zone
☞ 36. How to sample fishes in streams
☞ 37. How to sample fishes in lakes

---

[2]Coad, B. W. 1996. Encyclopedia of Canadian fishes. Canadian Museum of Nature and Canadian Sportfishing Productions Inc. Distributed by Key Porter Books Ltd., Ottawa, Ontario.

☞ 38. How to estimate population sizes using a catch per unit effort method
☞ 39. How to estimate population sizes using a mark and recapture technique
☞ 40. How to estimate secondary production in fish populations
☞ 41. Use of a morphoedaphic index to estimate fish yields in lakes

FISH CLASSIFICATION

Table 9.1 lists common fish species in freshwater environments. The list includes only about 10% of the total number of species in freshwater and is not meant to be comprehensive. It lists species that are useful for assessing the physical
1☞ characteristics of aquatic habitats and the quality of water. Some are listed because of their social (e.g. sport fishes) and economic importance (e.g. commercial fisheries). The fish are divided taxonomically into three groups, the ***Cyclostomes***, the ***Chondrichthyes***, and the ***Osteichthyes***. The Chondrichthyes include the sharks and their relatives but only a few Central American species live in freshwater and are not discussed further here.

Cyclostome Fishes

The Cyclostomes are primitive, jawless fish and include the lampreys of the **Family Petromyzontidae** (Fig.
2☞ 9.1, Table 9.1). The lampreys are among the most primitive of freshwater fishes. They lack a biting jaw and are adapted mainly to life as a parasite or scavenger on other fish species. The sea lamprey, *Petromyzon marinus*, was introduced into North America, initially the Great Lakes, in the 1930's and created havoc with their freshwater fisheries, especially salmonid (lake trout and salmon) and whitefish species. The introduction coincided with the opening of the St. Lawrence Waterway which provides a 3,000 km connection between the Great Lakes and the North Atlantic Ocean. Since then the lamprey has come under control by chemical treatment of streams where the larvae are produced. The sea lamprey is ***anadromous***, that is, it lives in the sea as an adult and returns to fresh water to spawn in the spring. Adult lamprey congregate in estuaries of rivers in late winter and move upstream in the spring when water temperatures exceed 4.4 °C. They are able to move against very strong currents by alternately swimming and attaching to rocks with their suctorial mouth. After spawning is complete, the female moves downstream and dies, the males dying a few days later. The eggs hatch in about two weeks into ***ammocoete larvae*** and immediately burrow

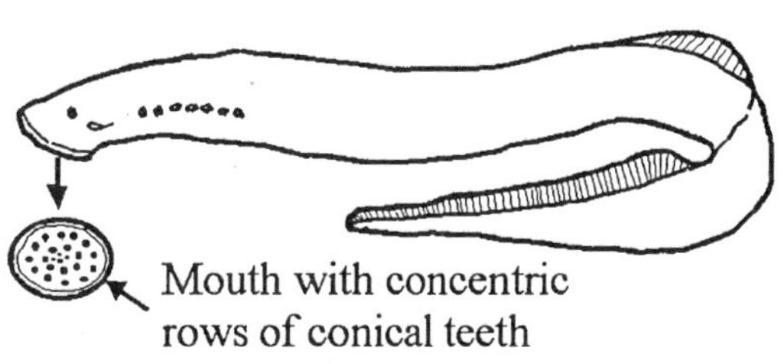

Figure 9.1. The family, Petromyzontidae (e.g. the sea lamprey, *Petromyzon marinus*), showing its diagnostic features.

Table 9.1. Egg-laying habits, feeding habits and habitat indicator value of common fishes in North America. Families are listed alphabetically. Feeding habits vary according to size and development stage and the habits listed are for adult fish.

Family Name	Egg-laying habits	Feeding habits	Habitat Indicator Value
Acipenseridae (Fig. 9.2)	On sand, gravel, rocks	Benthivores	Large, productive lakes, rivers
Amiidae (Fig. 9.4)	Nest a circular depression in bottom; adult guards nest	Piscivores	Swampy, vegetated bays of large lakes, rivers
Anguillidae (Fig 9.14)	Eggs spread on ocean floor	Benthivores, piscivores	Lakes that drain to sea
Catostomidae (Fig. 9.12)	Eggs spread randomly on stream bed	Benthivores	River redhorse in deep waters of large rivers
Centrarchidae (Fig. 9.17)	Adults guard nest in lake floor	Benthivores	Shallow waters of lakes, rivers
Clupeidae (Fig. 9.5)	Eggs broadcast at random, stick to objects on bottom	Planktivores	Alewife and herring in large, clean lakes, rivers
Coregonidae (Fig. 9.7)	Eggs spread randomly in shallows	Planktivore, benthivore	Cool, oligotrophic waters
Cottidae (Fig. 9.21)	Deposits adhesive eggs in a mass in a nest on stream bed	Benthivores	Slimy & mottled sculpins in deep or cold lakes and streams
Cyprinidae (Fig. 9.11)	Most scatter eggs at random; river chub makes nest of stones; carp, goldfish lay eggs on plants; creek chub guard eggs in circular nest	Herbivores (carp, creek chub); planktivores (emerald, golden shiner); benthivores (goldfish, river chub)	River and creek chub, pearl and long-nose dace and common shiner in cold, clear streams
Esocidae (Fig. 9.10)	Adhesive eggs spread randomly, attach to vegetation	Mostly piscivorous, also opportunistic omnivore	Warm, heavily vegetated lakes and rivers
Gadidae (Fig. 9.15)	Eggs scattered in shallows	Profundal benthivores	Deep, cold oligotrophic lakes
Gasterosteidae (Fig. 9.16)	Male builds and guards nests of plant material	Planktivores, benthivores	Brook stickleback in clear, cold streams
Gobiidae (Fig. 9.22)	Lay eggs under stones and empty bivalve shells	Planktivores	Shallow to deep lakes
Hiodontidae (Fig.9.8)	Scatter eggs on stream bed	Benthivores	Clean water lakes and streams
Ictaluridae (Fig. 9.13)	Guard nests of tunnels (channel cats) or depressions (bullheads)	Omnivores	Channel catfish in cool, clear large lakes and rivers
Lepisostidae (Fig. 9.3)	Scatters eggs on plants	Piscivores	Weedy areas of warm lakes and rivers
Osmeridae (Fig. 9.9)	Sticky eggs spread on bottom	Planktivores	Spawns in streams, lives in lakes
Percidae (Fig. 9.18)	Most scatter eggs randomly, perch lay strings of eggs, darters guard nests built under rocks	Young are planktivorous, adults benthivorous (perch) or mostly piscivorous (walleye, sauger)	Mostly turbid mesotrophic lakes rainbow darter in cool, clear streams
Perichthyidae (Fig. 9.19)	Adhesive eggs spread at random and stick to bottom objects	Benthivores	Warmer (~24 °C) large lakes and rivers
Petromyzontidae (Fig. 9.1)	Build nests in streams and desert their eggs	Parasitic on salmonids; American brook lamprey preys on cottids	Sea lamprey in large lakes, rivers that drain to sea; brook lamprey in cool, clear streams

Salmonidae (Fig. 9.6)	Build nests in lakes or streams and desert their eggs	Mostly benthivores, sometimes piscivores	Brook, rainbow, brown trout in cold, clear streams
Sciaenidae (Fig. 9.20)	Eggs spread randomly on bottom	Benthivores	Large, shallow lakes and rivers

into the sediments. The larvae burrow out of the nest after about 3 weeks of an infaunal existence. They quickly drift downstream and burrow again in U-shaped burrows, coming to the surface only to feed. Transformation into an adult begins after about 8 years in the burrow. Their parasitic life begins when they emerge from the burrow as an adult. The adults move downstream to the sea in late fall and remain in the sea until they are sexually mature.

The American brook lamprey, *Lampetra appendix*, is a small, secretive species, existing as an adult only long enough to spawn. It is a non-parasitic lamprey and has no effect on economically important species of fish.

Osteichthyes, The Bony Fishes

The remaining fish species in Table 9.1 belong to the Osteichthyes group. They are the bony fishes, most of which have gas bladders that enable them to regulate their buoyancy in the water. Sturgeon of the
3☞ family **Acipenseridae** (Fig. 9.2) are the most anadromous. The lake sturgeon, *Acipenser fulvescens,* is threatened in North America by pollution and by the creation of dams which prevent their migration from the sea into fresh waters. They are valued especially for their caviar. Like the lampreys, sturgeon are also primitive fishes. Lake sturgeon spawn from early May to late June in areas with swift rapids or at the foot of falls that prevent further migration. Most lake sturgeon mature sexually after 20 to 22 years of growth in the sea. Unlike lamprey which die after spawning, lake sturgeon return to the sea and repeat the migratory process until they reach 50 to 80 years of age. The lake sturgeon is among the largest freshwater fish species in the world, the record catch exceeding 230 pounds. See Table 9.1 for other characteristics.

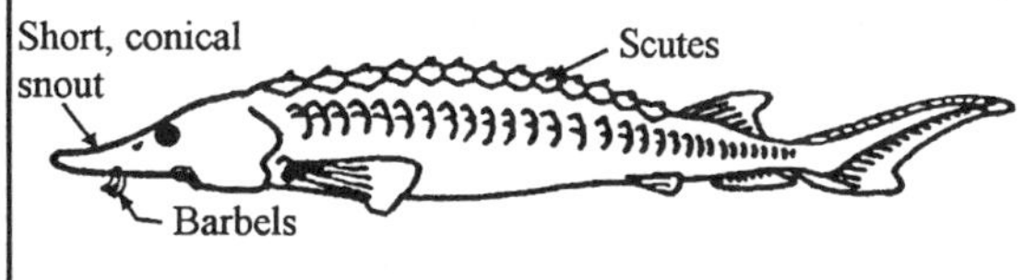

Figure 9.2. The family Acipenseridae (e.g. the sturgeon, *Acipenser fulvescens*) showing its diagnostic features.

The gars (family **Lepisosteidae**) (Fig. 9.3) are also a primitive fish, with two layers of
4☞ heavy, armour-like scales and spineless fins. The longnose gar, *Lepisosteus osseus*, is a voracious predator, feeding mostly on fish

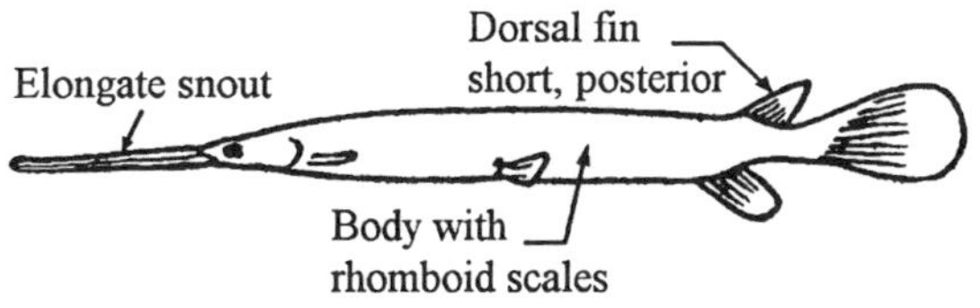

Figure 9.3. The family Lepisosteidae (e.g. long-nose gar, *Lepisosteus osseus*) showing its diagnostic features.

(sunfish, bass, perch, suckers, darters, pike, minnows) as adults and invertebrates as juveniles. Young gar have an adhesive pad on the tip of their snout by which they adhere to large leaves of macrophytes. See Table 9.1 for other characteristics.

Bowfins (family **Amiidae**) (Fig. 9.4) are also considered a
5 primitive fish, partly because they have spineless fins. *Amia calva* is the only freshwater species. It lives in lakes and rivers. The adults are piscivorous, like gars. Because of its voracious appetite for fish (Table 9.1), the bowfin is usually considered detrimental to sport fisheries. However, it is not a common species.

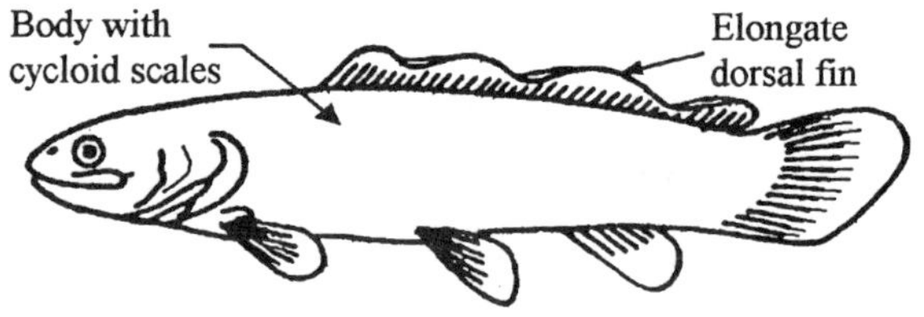

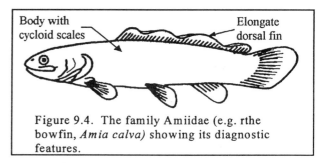

Figure 9.4. The family Amiidae (e.g. rthe bowfin, *Amia calva)* showing its diagnostic features.

The herrings (family **Clupeidae**) (Fig. 9.5) are silvery, pelagic fishes. Most herring species are anadromous, with
6 only a few, like the gizzard shad, *Dorosoma cepidianum*, living permanently in freshwater. The alewife, *Alosa pseudoharengus*, is an anadromous species. It was introduced to Lake Ontario in 1873, probably as an accidental introduction. Since then it has spread rapidly into other Great Lakes and has become landlocked in many other lakes. All herrings are planktivorous, feeding on cladocerans, copepods, mysids and insect larvae (Table 9.1).

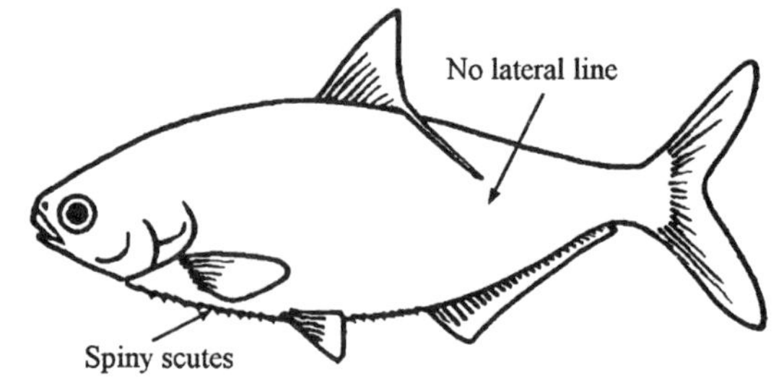

Figure 9.5. The family Clupeidae (e.g. Gizzard shad, *Dorosoma cepedianum)* showing its diagnostic features.

The **Salmonidae** (Fig. 9.6) are the dominant family in many temperate waters of North America. Many species are anadromous. Rainbow trout, also known as steelhead trout (*Oncorhynchus mykiss*), and brown trout (*Salmo trutta*) and brook trout
7 (*Salvelinus fontinalis*) are popular sport fish of streams. The lake trout, *Salvelinus namaycush*, are among the most popular deep water sport fish. Rainbow trout are native to drainages of the Peace and Athabasca rivers and were intentionally introduced into the Great Lakes in 1895. It has since spread throughout all Canadian provinces. The brown trout is native to Europe

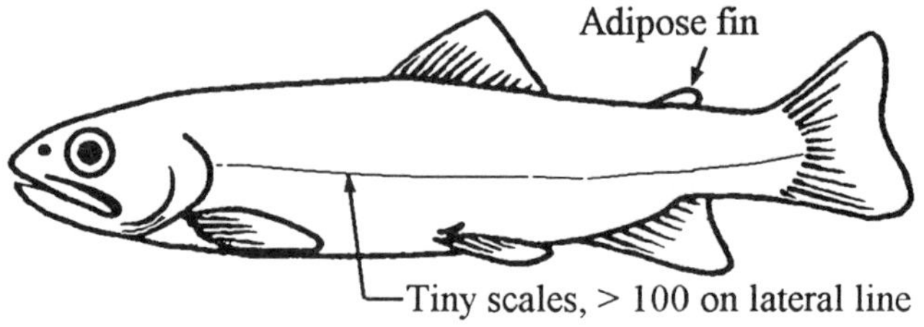

Figure 9.6. The family Salmonidae (e.g. rainbow trout, *Onchorhynchus mykiss)* showing its diagnostic features.

and was introduced into North America in the 1880s. Like the brook trout, which is native to North America, and the rainbow trout, brown trout spawn in cold, clear headwaters of streams. Spawning occurs during the spring for rainbow trout (but can extend from April to August), the fall (August to December) for brook trout and in the fall to winter (October to January) for brown trout. All three species feed mostly on benthic invertebrates, like insects and crustaceans, and occasionally on salamanders, frogs and a variety of fish. Lake trout spawn in the fall and feed mostly on bottom organisms, sometimes switching to small fish, especially smelt and sculpins (Table 9.1).

The whitefish and ciscos are often considered members of the family Salmonidae, although
8☞ some scientists assign them to their own family, **Coregonidae** (Fig. 9.7), as done here. The lake whitefish, *Coregonus clupeaformis*, is also a popular deep water sport fish. Like lake trout, it is a cool water species and is found in oligotrophic lakes only where oxygen tensions remain high enough in the hypolimnion during the summer for adult fish to feed on bottom organisms and on small fishes. The salmon fisheries in the Great Lakes is one of the largest in the world, and the serious and rapid declines in the lake trout and whitefish populations in the mid 1940's to mid 1950's was attributed to predation by the sea lamprey. The young are planktivorous and the adults are mostly benthivorous.

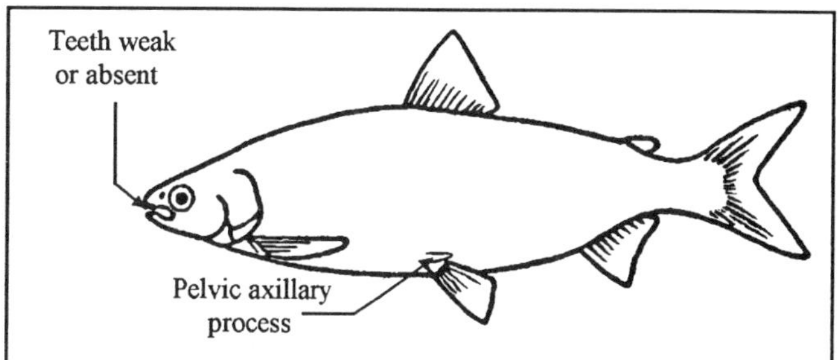

Figure 9.7. The family Coregonidae (e.g. whitefish, *Coregonus clupeaformis*) showing its diagnostic features.

Ciscos are mainly a northern fish, but the lake herring, *Coregonus artedii*, has the most extensive North American distribution of any cisco, since it is found throughout most of Canada and northern United States. It is a relatively small species, growing to 25 - 30 cm (8 - 12 in) total length. Like lake trout, spawning occurs in the fall (late November to early December). It is also a cold water species and moves to the hypolimnion during the summer to feed on plankton (mysids, amphipods, copepods). In the fall and spring they move into shallow water and feed on immature mayflies and caddisflies.

A relatively uncommon group, the mooneyes of the family, **Hiodontidae** (Fig. 9.8), are restricted to northern North America. They have a rounded snout with large, forward eyes and a dorsal fin located well over the anal fin. The mooneye, *Hiodon tergisus*, occurs in the Great Lakes and in Lake St. Clair in small numbers. It feeds mostly on benthic

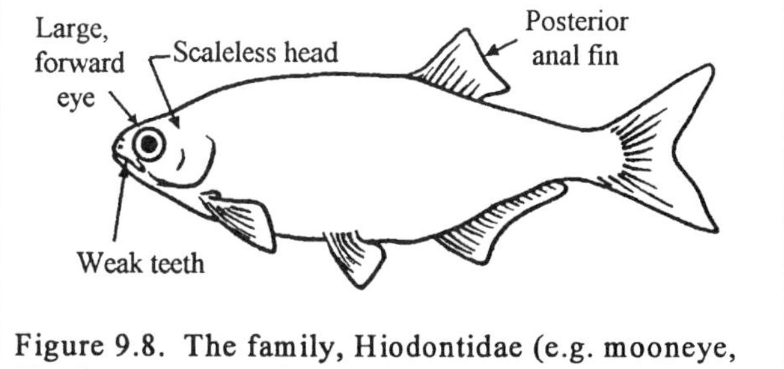

Figure 9.8. The family, Hiodontidae (e.g. mooneye, *Hiodon tergisus*), showing its diagnostic features.

invertebrates in swift waters (e.g. Detroit River, St. Clair River). In the southern parts of its range, mooneye migrate in large numbers up large, clear streams to spawn.

The rainbow smelt, *Osmerus mordax* (family **Osmeridae**) (Fig. 9.9), is native 9☞ to the Atlantic coast drainage but has been widely introduced into rivers and streams of the Great Lake's watershed and is now landlocked in many aquatic habitats. The rainbow smelt is an anadromous species, like the alewife, and migrates upstream in the spring to spawn, usually late April or early May, just after ice-out and before the water warms to 9 - 10 °C. The time of spawning is quite regular for each stream, lasting for about 3 weeks, or until water temperatures reach about 18 °C. The peak spawning period only lasts days, or at most, a week. Millions (about 13,000,000 across Canada in 1995) of smelt are netted by sport fishermen every year. Smelt are the only fish, besides baitfish, that can be caught legally after dark by methods other than angling; in Canada, the legal limit for a net is 0.557 m^2 (6 ft^2). Like the alewife, rainbow smelt are mostly planktivorous, but unlike the alewife, rainbow smelts will occasionally feed on small fish and benthic invertebrates (Table 9.1). Rainbow smelt are a favourite food of lake trout. They form about 8% of all the fish retained by sport fisherman in Canada.

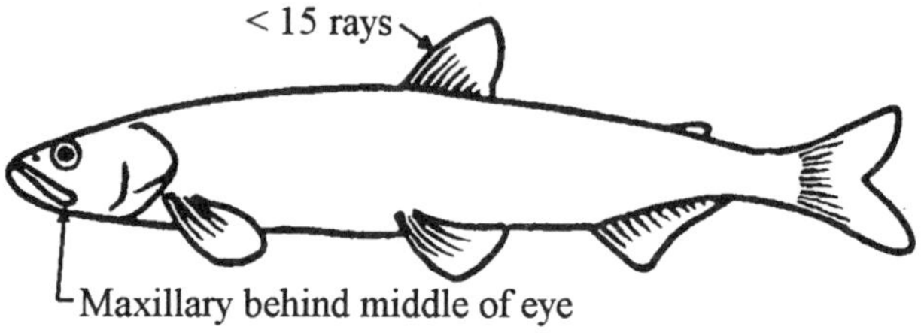

Figure 9.9. The family Osmeridae (e.g. rainbow smelt, *Osmerus mordax*) showing its diagnostic features.

The blueback herring, *Alosa aestivalis*, was first found in Lake Ontario in 1995. The species is closely related to the alewife. It is anadromous and dwells in the sea as adults and spawns in fresh waters. Blueback herring feed on small zooplankton, including veligers of the zebra and quagga mussels. Potential impacts include competition with, and declines in native planktivorous fish species.

The northern pike (*Esox lucius*) and muskellunge (*Esox* 10☞ *masquinongy*) of the family **Esocidae** (Fig. 9.10) are arguably the most voracious predators of fish. The northern pike has a circumpolar distribution, while the musky is restricted to fresh waters of eastern North America. The two species often hybridize in nature, producing fish called "tiger muskellunge, a miniature version of the gigantic musky which can grow to almost 2 m in total length. Both species are spring spawners, beginning immediately after the ice is out. Although both are basically piscivorous, the pike especially is opportunist, feeding on frogs, crayfish, and even mice, muskrat

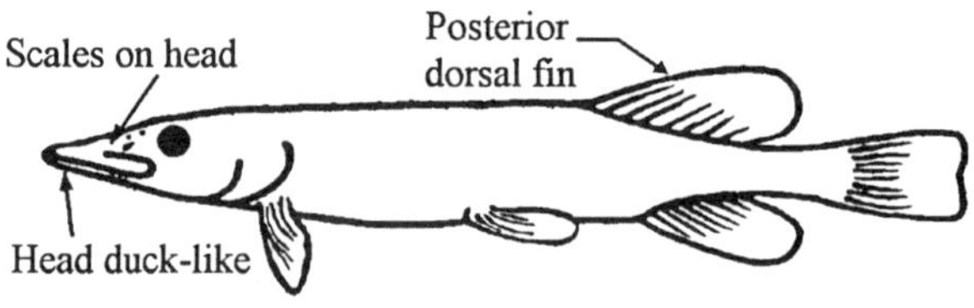

Figure 9.10. The family Esocidae (e.g. northern pike, *Esox lucius*) showing its diagnostic features.

and ducklings. Immediately after ice-out, pike move into shallow waters to spawn. Favourite spawning habitats are submerged vegetation on floodplains of rivers, marshes and bays of larger lakes. Spawning activity is exhausting and pike move into slightly deeper water to recover and wait for the water to warm. Only then do they begin to feed, the feeding period providing the best fishing for pike anglers. Good clues for depths to fish during cool weather are those with substrates that are dark and heat water quickly; when water is warm, pike move over lighter substrates that do not reflect as much heat as darker substrates. Otherwise, pike are most active in shallow water on warm, calm, sunny days and move into deeper water with prolonged hot weather. While in the shallows, pike seek shelter under logs, in sparse weed beds (dense vegetation restricts their movements), and/or by rocks. Best fishing times are late morning or early evening.

The minnow or carp family, **Cyprinidae** (Fig. 9.11),
11☞ is the largest of all fish families. About fifty species occur in Canadian waters. They are also among the most difficult to identify to species. The family includes the introduced carp (*Cyprinus carpio*) and goldfish (*Carassius auratus*), both native to Asia, and the numerous native species of dace (e.g. *Rhinichthys, Chrosomus*), chub (e.g. *Nocomis, Couesius, Semotilus*), shiners (e.g. *Notemigonus, Notropis*) and minnows (e.g. *Pimephales*). Most cyprinids spawn in the spring or early summer. Their feeding habits and habitats, however, vary considerably among species. Table 9.1 summarizes the features for many common species of cyprinids, including exotic species.

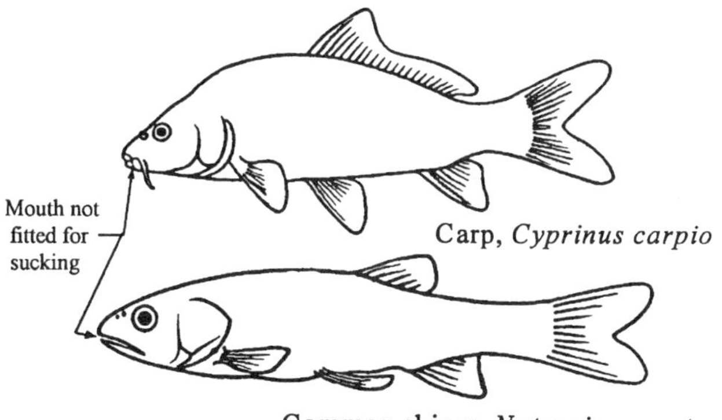

Figure 9.11. The family Cyprinidae (e.g. carp, *Cyprinus carpio*, chub, shiners, dace, etc.) showing diagnostic features of the family.

One of the most recognizable families is the
12☞ sucker family, **Catostomidae** (Fig. 9.12). The most common is the white sucker, *Catostomus commersoni*, being generally distributed throughout Canada and the northern half of the United States. The family's common name is derived from the sucker-like mouth, an adaptation for bottom feeding. They feed

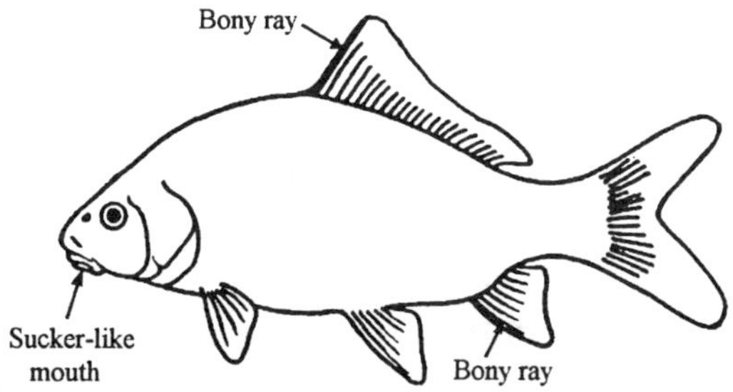

Figure 9.12. The family Catostomidae (e.g. white sucker, *Catastomus commersoni*) showing its diagnostic features.

heavily on benthic insects, crustaceans and molluscs, and occasionally on eggs of other fish species (Table 9.1), a habit for which the white sucker is often condemned. The species moves from lakes to streams in the spring to spawn. Some spawn in shallows of lakes, usually near the mouths of streams. The river redhorse, *Moxostoma carinatum*, however is restricted to lotic habitats, usually rivers and larger streams. Although the flesh of white suckers is edible, especially in soups and chowders, commercial catches are used primarily for making pet food. Some fresh sea food stores sell the fish as *mullet*.

The catfish family, **Ictaluridae** (Fig. 9.13), has a number of interesting features
13☞ associated with it: Their skin and "barbels" have numerous sensory cells that act as taste organs; venom is associated with their spines; they are scaleless; they have ***Weberian ossicles*** that play a role in reception and production of sound. The two most common species are the brown bullhead, *Ameiurus nebulosus* and the channel catfish, *Ictalurus punctatus*. The channel catfish is the largest of all catfishes, growing to weights of over 30 pounds. It is more demanding in habitat water quality than the bullhead, frequenting deep lakes and large rivers whose waters are cool and clear and the substrate is sand, gravel, or rubble. The brown bullhead prefers shallow, more turbid, vegetated areas. Both species are nocturnal in feeding habits and are omnivorous, feeding on a wide variety of plant (including tree seeds) and animal material (including invertebrates and fish) (Table 9.1). Brown bullheads are among North America's favourite pond-culturing species, especially in the United States.

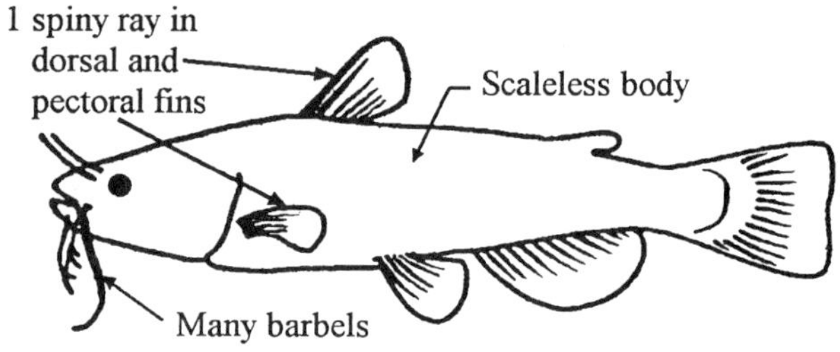

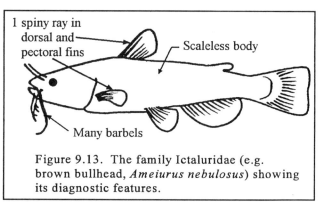

Figure 9.13. The family Ictaluridae (e.g. brown bullhead, *Ameiurus nebulosus*) showing its diagnostic features.

The eels of the family **Anguillidae** (Fig. 9.14) are
14☞ ***catadromous***, and migrate from freshwater to the sea to spawn. Adults of the American eel, *Anguilla rostrata*, live in muddy and silty bottoms of lakes, lying buried in the sediments during the day. They move into the water at night to feed on small fishes and invertebrates (Table 9.1). Eels are a delicacy to many North Americans and a staple food for many Europeans.

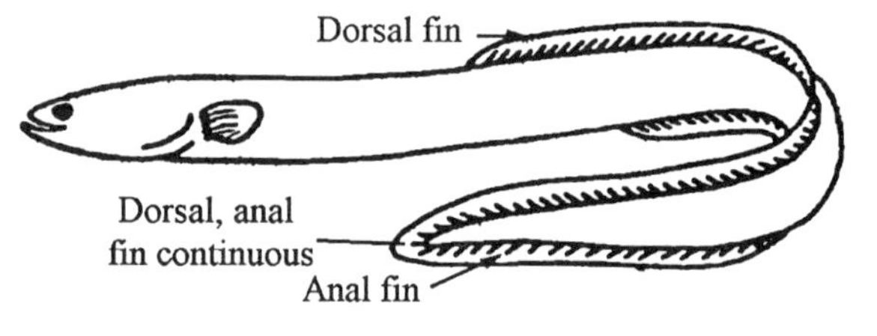

Figure 9.14. The family Anguillidae (e.g. American eel, *Anguilla rostrata)* showing its diagnostic features.

The burbot, *Lota lota* (family **Gadidae**) (Fig. 9.15), is one of the few species
15☞ that spawns in the winter, a trait that partly explains the species circumpolar distribution. The burbot is found only in deep, cold, well oxygenated waters of

oligotrophic lakes and large rivers. It is a voracious predator and night feeder, feeding on deep water insects, crustaceans and molluscs (Table 9.1). The Gadidae (common name, cod family) are primarily marine, bottom dwelling fish. The burbot is the only truly freshwater cod species.

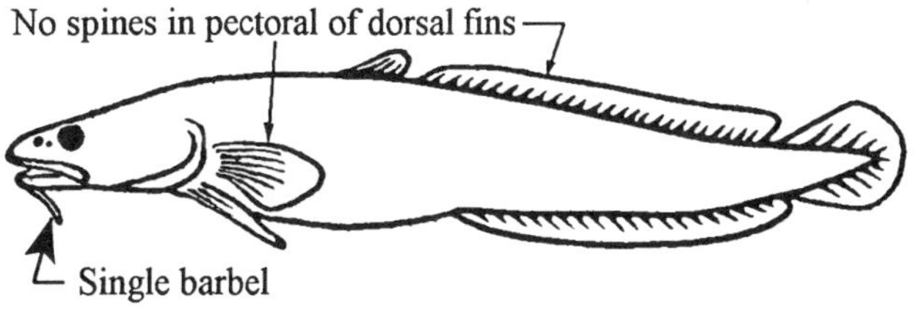

Figure 9.15. The family Gadidae (e.g. burbot, *Lota lota*) showing its diagnostic features.

The sticklebacks (family **Gasterosteidae**) (Fig. 9.16) are small, minnow-like fish occurring in shallow areas
16☞ of lakes and streams. They owe their common name to the stickles or spines in the dorsal fin, the numbers of spines varying among species. The brook stickleback, *Culea inconstans*, inhabits cool, clear, densely vegetated waters of small streams and spring-fed ponds. The preference for cool waters limits its distribution to Canada and northern United States. The species is carnivorous, feeding on insect larvae, crustaceans and eggs of fishes, including those of its own species (Table 9.1).

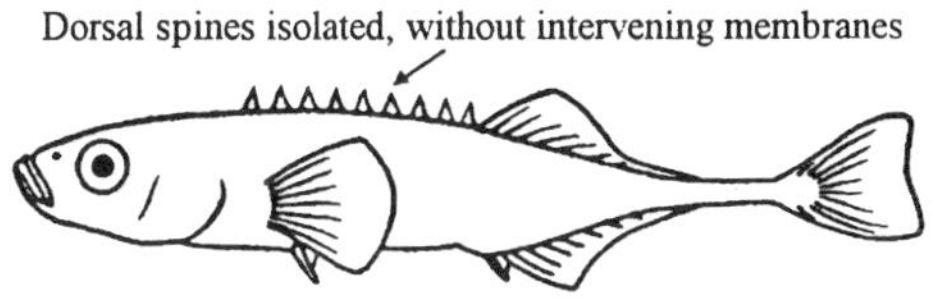

Figure 9.16. The family, Gasterosteridae (e.g. brook stickelbacks, *Culea inconstans*), showing its diagnostic features.

Among the most highly coloured and attractive freshwater fish are the sunfish and bass of the family, **Centrarchidae** (Fig. 9.17). They are
17☞ also popular to sport fisherman who refer to many of them as "pan fish" because of their laterally-compressed bodies and they are small and easy to pan fry . The most common pan fish sought after are pumpkinseed (*Lepomis gibbosus*), bluegill (*Lepomis macrochirus*) and rock bass (*Ambloplites rupestris*). Smallmouth bass (*Micropterus dolomieu*) and largemouth bass (*Micropterus salmoides*) are much larger species. Most centrarchids spawn in the spring and early summer. The rock bass, as its name implies, generally inhabits rocky areas in shallow lakes and rivers. The pumpkinseed and bluegill prefer clear water and cover from vegetation in ponds and small lakes and quiet waters of slow-moving streams. All three species are carnivorous, feeding mostly on larval insects, crustaceans, worms and molluscs. The habitats of smallmouth and largemouth basses vary with size and time of

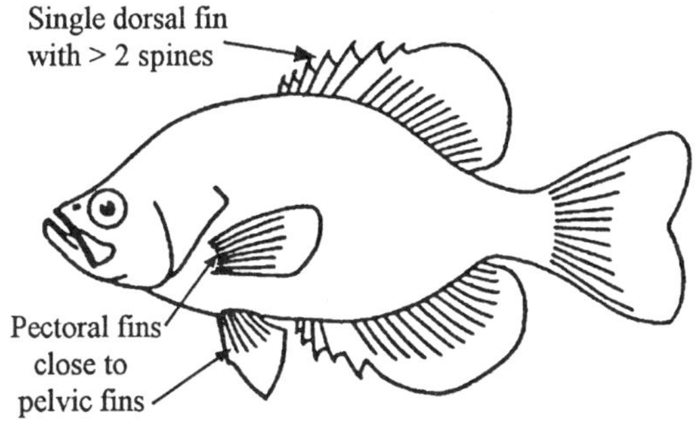

Figure 9.17. The family, Centrarchidae (e.g. pumpkinseed, *Lepomis gibbosus* and smallmouth bass, *Micropterus dolomieu*) showing its diagnostic features.

year. In the spring and early summer, adult fish congregate on spawning grounds in the shallows of lakes and rivers. When water temperatures rise, the fish move to greater depths or to shade and protection offered by sunken large logs and rocks of shoals. The food of small and largemouth basses consist of insects, crayfish and fishes, with food taken at the surface during the morning and evening and from the water column and bottom during the day. Both species can be caught still fishing with worms, minnows, crayfish or frogs, or by casting in weedy and stumpy areas with noisy surface plugs, jiggers, or poppers. In general, largemouth bass take surface lures more frequently and flies less frequently than do smallmouth bass. The attraction for anglers and their sporting quality are legendary and continue to attract anglers by the thousands every year. The reproductive and feeding habits of centrarchids are summarized in Table 9.1.

The perch family, **Percidae** (Fig. 9.18), is represented by several species, three of which are among the
18☞ favourite for sport and commercial fisherman. These three are the yellow perch (*Perca flavescens*), walleye (*Stizostedion vitreum*) and sauger (*Stizostedion canadense*). The darters (genera *Etheostoma* and *Percina*) are also members of the perch family. Most percids spawn in the early spring, usually before the end of May. The yellow perch is more adaptable than the walleye and sauger and is able to utilize a variety of warm to cooler habitats of ponds, lakes and rivers. They are opportunists feeding on a variety of invertebrates, especially crayfish, and small fishes and their eggs. Walleyes are more adaptable than saugers and reach their greatest abundance in large, shallow turbid lakes. The optimum transparency of walleye is 1-2 m Secchi depth, indicating that mesotrophic to eutrophic lakes are its preferred habitats. In clear lakes, walleye are usually found near the bottom, resting during the daytime. They rise to the top at night to feed. In these lakes, the oxygen tension of the deeper water must be high enough for them to respire for at least a few hours. Both walleye and sauger are extremely sensitive to bright light

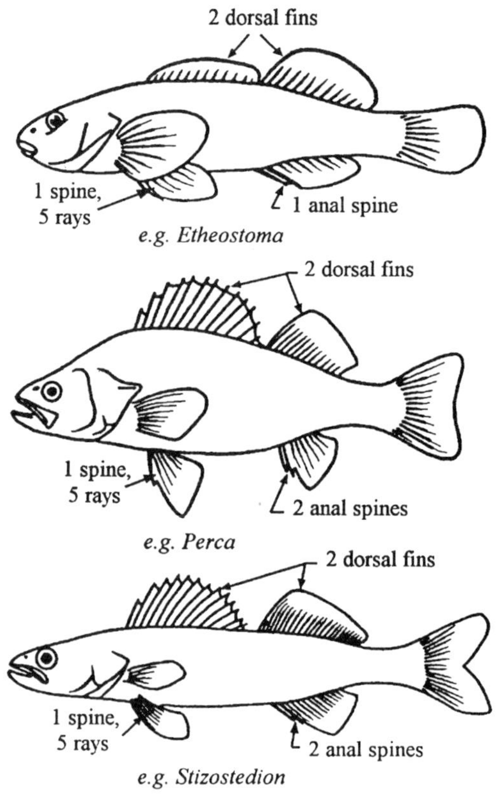

Figure 9.18. The family, Percidae (e.g. darters, (*Etheostoma*), perch (*Perca*), walleye (*Stizostedion*) and the European ruffe (*Gymnocephalus cernuus*)) showing its diagnostic features.

so feeding is restricted to twilight or dark periods. Sunken trees, boulders, weed beds, and to some extend turbidity, provide cover from bright light in the shallows of lakes and rivers.

Of the darters, the rainbow darter, *Etheostoma caeruleum*, is probably the most colourful of all fish, having brilliant blues and greens with inter-spaces of yellow and orange. It is a lotic species confined to erosional areas of streams where it feeds on invertebrates, especially larval insects, amphipods and molluscs.

The European ruffe, *Gymnocephalus cernuus,* first discovered in 1986 in plankton hauls from the St. Louis River, a tributary of Lake Superior, is also a percid. It is a voracious feeder of benthic invertebrates, zooplankton and fish eggs. Spawning occurs in spring and summer, on a hard bottom of clay, sand or gravel. Up to 200,000 eggs are laid. See Chapter 14 for other information.

The white perch, *Morone americana*, is not a member of the perch family; it belongs to the
19☞ family **Percichthyidae** (Fig. 9.19). The species is now common in Lake Ontario and Lake Erie but was originally introduced, apparently by its own movements through rivers and canals, from the brackish waters along the Atlantic coast southern Gulf of the St. Lawrence to Lake Ontario. Its reproductive and feeding habits are given in Table 9.1.

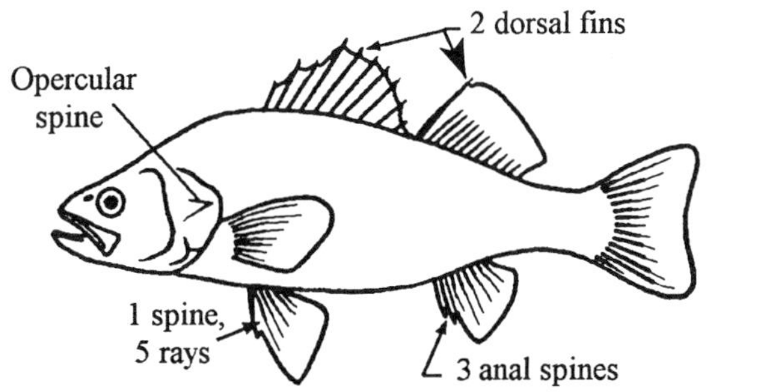

Figure 9.19. The family, Percichthyidae (e.g. white perch, *Morone americana*), showing its diagnostic features.

The drum family, **Sciaenidae** (Fig. 9.20), has only one freshwater species,
20☞ *Aplodinotes grunniens*, the freshwater drum. It resides in large, relatively shallow waters (e.g. Lake Erie, Lake Nippissing, Ottawa River). It is adapted to bottom feeding (Table 9.1), especially on molluscs because of the numerous small, round crushing teeth needed to break the heavy shells of molluscs. Although adult freshwater drum feed heavily on zebra and quagga mussels, the relatively small populations of drum in lakes and rivers are probably insufficient to make much of an impact on the enormously

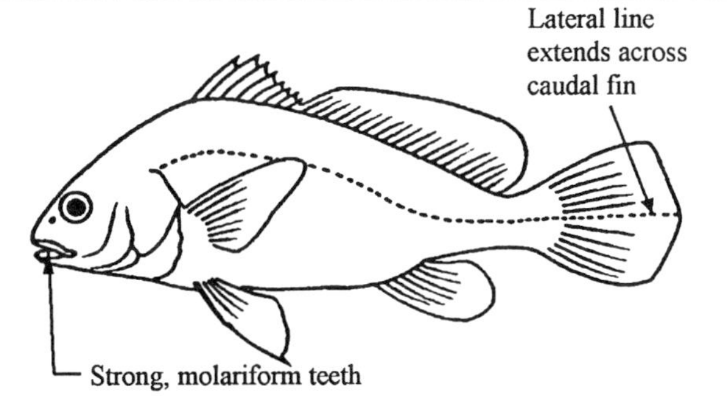

Figure 9.20. The family, Sciaenidae (e.g. freshwater drum, *Aplodinotes grunniens*), showing its diagnostic features.

large populations of zebra and quagga mussels.

The sculpins (family **Cottidae**) (Fig. 9.21) are small fishes (less than 18 cm, or 7 in),
21☞ dorsal-ventrally compressed and adapted for a bottom existence. The mottled sculpin (*Cottus bairdi*) and slimy sculpin (*Cottus cognatus*) are the most common sculpins. While the deepwater sculpin, *Myoxocephalus thompsoni* (formerly *quadricornis*), is distributed across Canada, it has a very spotty distribution and is confined to the profundal zones of oligotrophic lakes where it competes with salmonids and burbot for deep water mysids and amphipods (Table 9.1). The mottled sculpin is also restricted to cool waters but can be found in both headwater streams and deep lakes. However, it does not occur as far up headwater streams nor as deep in lakes as does the slimy sculpin. Sculpins are known to form a part of the diet of large brook trout in streams.

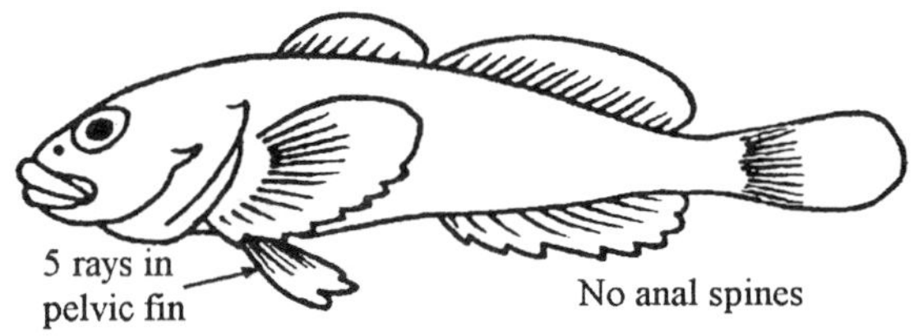

Figure 9.21. The family, Cottidae (e.g. the mottled sculpin, *Cottus bairdi*), showing its diagnostic features.

The round goby (*Neogobius melanostomus*) and
22☞ tubenose *goby (Proterorhinus marmoratus*) (family **Gobiidae**) (Fig. 9.22) are new exotic species in the Great Lakes (see Chapter 14). The round goby has robust molariform teeth used to crush the shells of molluscs, including zebra and quagga mussels. The tubenose gobi derives its name from tube-like nostrils that overhang the upper lip. Gobies are benthivores and compete for the same foods as sculpins and darters. They have several repeated spawnings, up to 6 times every 18 - 20 days, peaking near 15 °C. The eggs (up to 6,100) are laid under stones and in empty bivalve mollusc shells (Table 9.1). Round gobies are preyed upon by bass, walleye and trout.

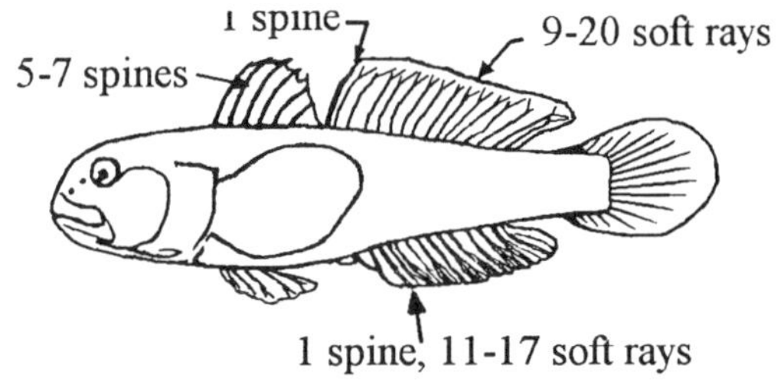

Figure 9.22. The family, Gobiidae (e.g. the round gobi, *Neogobius malanostomus*), showing its diagnostic features.

Other Nekton

Amphibians, seals and whales also constitute part of the nekton in freshwater and marine systems. Newts, salamanders and frogs inhabit the shores of ponds and lakes and their young spend all of their lives as nekton in small lakes and ponds. Seals and whales are important nekton in marine systems, but the seals in Lake Baikal are of a species unique to fresh waters and feed on fish. Other nekton vertebrates include otters, beavers and muskrats.

Invertebrate nekton are more common in salt water than in fresh water. Jellyfish (Phylum Cnidaria) are common nekton in the marine environment but *Craspedausta sowerbyi* is a rare freshwater jellyfish. However, it is small, about 1 cm in diameter, and is usually considered part of the plankton.

Within the phylum, Annelida, are many marine polychaetes (Class: Polychaeta) that are large enough to be called nekton. But in fresh waters, polychaetes are strictly benthic. Bloodsuckers (Class: Hirudinea) are strictly freshwater but many species are large, very strong swimmers of the nekton.

Within the phylum Mollusca are squids that are very powerful swimmers and feed on plankton in marine systems. Some naked gastropods, like nudibranchs (sea hares) are also effective swimmers in sea water. None of the freshwater molluscs belong to the nekton.

Within the phylum Arthropoda are numerous species of shrimp and krill that are large enough and have sufficient powers of locomotion to resist some of the strongest ocean currents. In fresh waters, only the mysid, *Mysis relicta*, is large enough to be part of the nekton, but it is more commonly associated with plankton. Many species of insects can become components of the nekton but most of them spend their life as benthos. Among the more common insects in the nekton are water bugs (e.g. *Notonecta, Belostoma, Corixa*) and beetles (*Agabus, Dytiscus*) that spend most of their time swimming or diving, some even against stream currents. Most of these insects are raptorial carnivores and must be effective swimmers to catch their prey.

Neuston

Neuston includes all the organisms that live at the surface of the water, those
living upon the surface being called ***epineuston*** and those living just under the surface
23 being called ***hyponeuston***. A few species are obliged to live at the surface and have
evolved unique adaptations that either allow them to move on the surface film or to
live just under the surface film without breaking the surface tension.

Marine scientists refer to epineuston as ***pleuston***, the best example being the
Portugese-Man-O-War, *Physalia*. It has an inflated bell that floats on the sea surface,
24 the long tentacles containing lethal toxins (capable of killing humans) dangling beneath
the surface searching for prey. In fresh waters, the water strider of the family
Veleidae (Order Hemiptera) is an excellent example of epineuston. They have a fan-
like structure of fine hairs that is stored in a groove in the tarsal segment of each leg
when swimming but is fanned out when they want to walk on the surface. The
25 numerous hairs spread the weight of the insect's body so that the surface tension of the
water is not broken. Whirligig beetles also live on the surface and have each eye
divided into two, a dorsal eye for looking at objects above the surface and a ventral eye

for looking at objects below the surface of the water.

Mosquitos of the family Culicidae (Order Diptera) lay upside down just under the surface film with a posterior respiratory tube penetrating the air-water interface and its head dangling down and filter feeding on plankton. Being able to breathe atmospheric oxygen enables mosquitos to live in enriched ponds, pools or ditches that frequently experience anoxia. Since mosquitos rely on surface tension to dangle, quiet ponds and pools are preferred habitats. They cannot risk being pulled under by currents, especially those of streams, and are obliged to live in quiet, standing waters. This is in stark contrast to black flies which are obliged to live in streams, relying on currents to bring food to them.

Many macrophytes are also epineuston. Examples include duckweed (*Lemna, Wolffia*), water fern (*Salvinia*), and the exotic water lettuce (*Pistia*) and water hyacinth (*Eicchornia*). The latter two species have inflated stems to float on the surface. All of four plants have roots that dangle below the surface and obtain their nutrients from the water. Interestingly, all four examples are very prolific and capable of coating the entire surface of the water. In most cases, ponds experience severe oxygen depletion as the plants prevent diffusion of oxygen from the air into the water; they also prevent light from penetrating the water so that photosynthesis by algae is inhibited.

Adaptations of Fishes to Running Waters

As with the benthic invertebrates, fishes exhibit morphological, anatomical, behavioural and/or physiological adaptations to life in rapidly flowing waters. Table 9.2 summarizes most of the adaptations and gives specific examples for each.

Generally, fish that are streamlined and round in cross-section are fast-swimming and similarly sized fish that are laterally flattened are slower. Also, round fishes tend to
26☞ be more elongate than flat fish, giving them a stronger and more powerful tail stroke; flat fishes tend to be shorter with less powerful tail strokes than round fish. Lateral flatness is a good adaptation for life in weed beds, allowing the fish to easily twist and turn around the stems of plants.

Fish that regularly swim in fast water are round in cross-section and streamlined. There is a good correlation between shape of the cross-section and the ability of fishes to resist current. The speed of flight from danger is faster in longer, more round fish. For example, in the sequence of yellow perch (*Perca flavescens*) > pikeperch (*Lucioperca lucioperca*) > northern pike (*Esox lucius*) are able to swim increasingly faster and each is more rounded and more elongate than the one before.

Not all stream fishes are rapid swimmers. Many are found closely associated with the bottom or take shelter under or behind stones. These include a great variety of cyprinids, catfishes, sculpins, darters and suckers. The morphological adaptations include dorso-ventral flattening (e.g. sculpins, catfishes, some cyprinids))or at least a flattened belly (e.g. suckers); laterally-placed paired fins (e.g. darters, sculpins); dorsal eyes (e.g. sculpins); laterally-placed gill openings (e.g. sculpins, darters); a reduction in amount of

Table 9.2. Adaptations of fishes to running waters.

Adaptation	Significance	Species examples
Streamlined and round	Best for speed and acceleration against currents	Rainbow trout, *Oncorhynchus mykiss*; brook trout, *Salmo trutta*; brown trout, *Salvelinus fontinalis*
Dorso-ventral flattening	Best for benthic fish, current presses animal down	Mottled sculpin, *Cottus bairdi* Slimy sculpin, *Cottus cognatus*
Flat belly	Best for benthic fish; allows contact of most of body, including head, with substrate	White sucker, *Catostomus commersoni*; River redhorse, *Moxostoma carinatum*; Northern hogsucker, *Hypentelium nigricans*
Dorsal eyes	Best for benthic fish; keeps eyes clear of substrate	Mottled and slimy sculpins; channel catfish (*Ictalurus punctatus*)
Laterally-placed paired fins	Best for benthic fish; act as hydrofoils to press fish down	Mottled and slimy sculpins; darters (*Etheostoma* spp.)
Laterally-placed gill openings	Best for benthic fish; keeps gills clear of sediments	Mottled and slimy sculpins; darters (*Etheostoma* spp.)
Scale reduction associated with poison spines	Best for benthic fish; less need for protection by scales but poison spines are added precaution	Mottled and slimy sculpins; channel catfish
Suckers or friction pads	Best for benthic fish; helps fish maintain position	Mottled and slimy sculpins
Swim bladder reduction	Best for benthic fish; less need for buoyancy in streams	Mottled and slimy sculpins; channel catfish; white sucker (*Catostomus commersoni*); darters
Colour change ability	Best for benthic fish; allows for cryptic colouration	Mottled and slimy sculpins; darters
Reaction to current	Behavioural response to minimize exertion; swim in bursts and rest	Rainbow trout; brook trout; brown trout
Guarding and territoriality	Results from the need for shelter from the current	Mottled and slimy sculpins; darters
Diurnal activities	Reduces risk of predation; night time feeding corresponds to night time drift of invertebrates	**Night time feeding**: Spotfin (*Notropis spilopterus*), sand (*N. stramineus*) and bigmouth (*N. dorsalis*) shiners; eels (*Anguilla rostrata*) **Day time feeding**: common shiner (*Notropis cornutus*)

scale covering correlated with the presence of poisonous spines (e.g. opercula of sculpins, pectoral and dorsal fins of catfishes); muscular pectoral fins (e.g. some cyprinids constantly move fins like hydrofoils to press and keep them on the bottom).

Other fishes from torrential waters use ventral suckers or friction pads to adhere to rocks to prevent them from being swept downstream. Mouth suckers are characteristic of lampreys and with their eel-like form enable the lamprey to migrate upstream. Lampreys are poor swimmers and the suckers must be used to adhere to rocks and maintain their position when they get tired of swimming against the current.

An anatomical adaptation to stream flow is a reduction in the size of the swim bladder. The swim bladder functions as a buoyancy organ, allowing the fish to rise or fall in the water column. In many bottom-dwelling species of cyprinids, catfishes, darters and suckers, the swim bladders are reduced and often divided and encased in bones.

Fishes living on sand tend to be paler in colour and more translucent than those living on stones. Fishes living on stony substrates (e.g. sculpins and darters) are usually dark-coloured or mottled so that they blend into the background. Many of these species can exert some control over their colour, becoming paler on pale backgrounds. The slimy sculpin is able to change from dark to light in 2 to 3 minutes.

Behavioural adaptations of lotic fish species include reactions to currents, including seeking shelter in quiet waters, guarding nests and territories and diurnal
27☞ rhythms. After continuous swimming, fishes will accumulate lactic acid and then tire quickly (explaining why anglers "play" a hooked fish to land it). Stream-dwelling trout and dace swim in short bursts and then rest. Stream fish always spend most of their lives facing into the current because it helps maintain their position and it makes respiration easier, having only to open their mouths to flush their gills with a fresh supply of oxygenated water.

Most sculpins and darters display ***nest guarding*** but trout exhibit ***territoriality*** and dominance, the latter two having little to do with breeding. Young trout tend to be territorial in shallow rapids and to form shoals in pools and back waters. Apparently, territoriality results from the need for shelter from the currents. Territorial behaviour includes snapping at and chasing off intruders. It is primarily intraspecific, affecting only individuals of the same species. Occasionally territoriality is interspecific, affecting individuals of other species, such as exhibited by rainbow, brown and brook trout. ***Shoaling*** is a gathering together of fishes and is the opposite of territoriality. The function of shoals is protection of the individual, the presence of a large number of fishes tending to confuse a predator. When one fish in a shoal is damaged, the rest disperse to safety.

Diurnal rhythms in activity are common among several species of shiners. Many species move into shallow water at night and back to deeper water during the day. These include the spotfin shiner, *Notropis spilopterus*, the sand shiner, *N. stamineus* and the bigmouth shiner, *N. dorsalis*. Migrating larval and adult eels and salmon smolt move only at night. Other species are more active in the daytime, such as the common shiner, *Notropis cornutus*. Some species show different periods for different activities. Young trout, for instance, swim actively at night, do most of their feeding early and late in the day, and show little activity of any kind about midday.

Factors Affecting Fish Ecology

Many physical, chemical and biological factors control the occurrence, interspecific relationships, abundance and growth rates of lotic fishes. Some abiotic factors interact, such as current and substratum or temperature, light and dissolved oxygen levels, but each is considered separately here.

Current Speed

For running water organisms, current speed is probably the most influential factor affecting an organism's ecology. Swimming fishes apparently maintain their positions by reference to fixed objects, which may explain why much of the downstream migration
28☞ occurs at night. Scientists have shown that blacknose dace, *Rhinichthys atratulus*, and common shiners, *Notropis cornutus*, maintain their positions in relation to stripes placed vertically up the sides and across the bottom of a running-water aquarium; when the stripes were moved, the fishes followed them. Benthic species, like the johnny darter, *Etheostoma nigrum*, on the other hand, maintain their position by physical contact with the substratum. In turbid rivers, fishes probably make greater use of their ***lateral-line system*** than their eyes. The lateral line functions as an echo-location device and is used as their primary distance receptor in turbid waters and at night to avoid obstacles.

Most, if not all, swimming fishes spend most of their time resting in shelter. But even closely related species adapted to fast-flowing waters differ in their optimal current speeds. For example, the following list gives the maximum swimming speeds of some North American fishes.

Lotic species	
Salmon, *Salmo salar*	800 cm/sec
Brown trout, *Salmo trutta* (introduced)	440
Creek chub, *Semotilus atromaculatus*	200
Lentic species	
Tench, *Tinca tinca* (introduced)	45-50
Pike, *Esox lucius*	45
Carp, *Cyprinus carpio* (introduced)	40

The distribution of fish species along a river system, as described in the river continuum concept first proposed by Vannote *et al* (1980)[3], is based largely on current speed and associated changes in substrate type, temperature, light, primary production and dissolved oxygen levels. Fig. 9.23 illustrates some of the changes that occur down the length of a stream and the characteristic fish species that are present.

---

[3]Vannote, R. L., G. W. Minshall, K. W. Cummins, J. R. Sedell, C. E. Cushing. 1980. The river continuum concept. Can. J. Fish. Aq. Sci. 37: 130137.

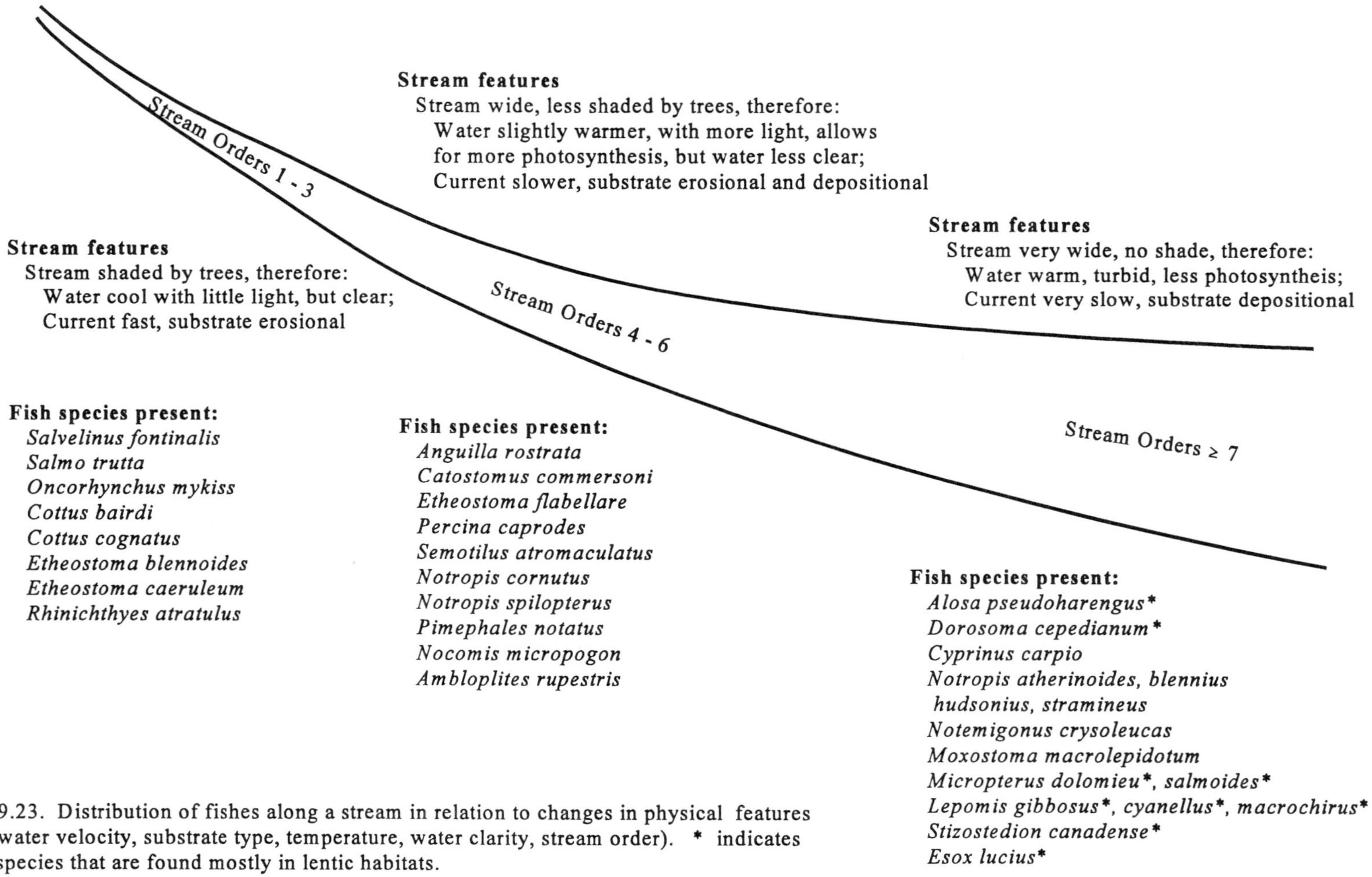
Stream Orders 1 - 3
Stream Orders 4 - 6
Stream Orders ≥ 7
Stream features
Stream shaded by trees, therefore:
Water cool with little light, but clear;
Current fast, substrate erosional
Stream features
Stream wide, less shaded by trees, therefore:
Water slightly warmer, with more light, allows for more photosynthesis, but water less clear;
Current slower, substrate erosional and depositional
Stream features
Stream very wide, no shade, therefore:
Water warm, turbid, less photosyntheis;
Current very slow, substrate depositional
Fish species present:
Salvelinus fontinalis
Salmo trutta
Oncorhynchus mykiss
Cottus bairdi
Cottus cognatus
Etheostoma blennoides
Etheostoma caeruleum
Rhinichthyes atratulus
Fish species present:
Anguilla rostrata
Catostomus commersoni
Etheostoma flabellare
Percina caprodes
Semotilus atromaculatus
Notropis cornutus
Notropis spilopterus
Pimephales notatus
Nocomis micropogon
Ambloplites rupestris
Fish species present:
Alosa pseudoharengus*
Dorosoma cepedianum*
Cyprinus carpio
Notropis atherinoides, blennius hudsonius, stramineus
Notemigonus crysoleucas
Moxostoma macrolepidotum
Micropterus dolomieu*, salmoides*
Lepomis gibbosus*, cyanellus*, macrochirus*
Stizostedion canadense*
Esox lucius*

Figure 9.23. Distribution of fishes along a stream in relation to changes in physical features (water velocity, substrate type, temperature, water clarity, stream order). * indicates species that are found mostly in lentic habitats.

Substratum and Turbidity

The current speed dictates in part the type of substrate present and the water
29☞ clarity. Fast water erodes the substrate leaving coarse gravel, rocks and boulders. The finer material is carried downstream resulting in clearer water upstream. Such habitats will support fishes that lay eggs in gravel, under rocks or attached to rocks, or fishes that normally seek shelter under stones (see Tables 9.1 and 9.2). Erosional substrates are characteristic of stream orders 1 to 3 (headwaters) and narrower reaches of stream orders 4 to 6 (Fig. 9.23). As a result, few macrophytes can become established and plankton populations are prevented from developing. Hence, the feeding guilds present are primarily benthivores and piscivores, with occasional visits from migrating lamprey that are parasitic on other fishes.

When water velocity slows, the materials eroded upstream begin to settle out downstream and the substrate becomes depositional in character with accumulations of silt, mud and detritus. Very large rivers take on features characteristic of lakes, except rivers do not thermally stratify. Currents are slow enough for littoral and sublittoral zones to become well established and for plankton to become abundant. Large streams and rivers (i.e. stream orders exceeding 6) have a greater diversity of fishes because of the additional food sources; benthivores that feed on burrowing forms of invertebrates; herbivores that feed on submersed macrophytes; planktivores that feed on plankton; parasitic fish and piscivores that feed on a wide variety of fish, and omnivores that have several kinds of feeding habits. The species present in each of the feeding guilds are different than those upstream because they must be able to tolerate murkier conditions, warmer temperatures and lower oxygen levels.

Fluctuations in Discharge

Spate (flood) events can be deleterious to stream fishes. Eggs and young of brook and brown trout can be destroyed by unusually high water in the winter. Yet, other species depend on high water for spawning. The bigmouth buffalo, *Ictiobus cyprinellus*,
30☞ spawns successfully in some streams only at times of high discharge, and fails to spawn in years when there is a lack of high water. Many species of minnows fall into three groups apparently to avoid high discharge levels: (i) spring spawners that breed in late spring to early summer; (ii) intermediate spawners that breed all through the summer; (iii) late spawners that breed in July and August. Spawning fails if water is not high enough, but if flash floods occur, much oviposition occurs on the flood plain and eggs and young fish become stranded when the water levels drop.

Some fish are adapted for occasional drying out of the stream bed, leaving only isolated pools or no water at all. Extended periods of drought eliminate most species of fish, at least in temperate climates. Some tropical species (e.g. lung fishes) burrow into the mud to aestivate. Fish can temporarily seek refuge in isolated pools if they exist, but the pools are small and will not support large populations. Only species that can tolerate the elevated temperatures and falling oxygen levels will survive in the pools.

Scientists have shown that when water returns to the stream in the fall, fish species tend to re-colonize in waves very quickly (about two weeks). Among the species in the first wave are the more tolerant bluntnose minnow (*Pimephales notatus*), white sucker (*Catostomus commersoni*) and creek chub (*Semotilus atromaculatus*). The redfin shiners (*Notropis umbratilis*), hog suckers (*Hypentelium nigricans*), common shiners (*Notropis cornutus*) and spotfin shiners (*N. spilopterus*) move in afterwards. These are followed by green sunfish (*Lepomis cyanellus*), rainbow darters (*Etheostoma caeruleum*), sand shiners (*Notropis stramineus*) and rock bass (*Ambloplites rupestris*), then smallmouth bass (*Micropterus dolomieu*).

Temperature

Each fish species has its own upper and lower thermal threshold values and its
31☞ own optimal temperature for growth and reproduction (spawning). Tropical species must
commonly endure temperatures in the low 30s, but the range of temperatures that tropical
fish experience is far less than that of most temperate species which normally endure
temperatures in the mid to high 20s. The rate of temperature change in nature is usually
not a factor but in the laboratory, care must be taken to slowly acclimate fish to the
desired temperature. A good rule of thumb is about 1 °C per day, but different species
have different rates of acclimation. Carp acclimate to changing temperatures much better
than do trout. ***Temperature polygons*** can be drawn, enclosing the lethal temperatures
which vary according to the previous exposure temperatures, as shown in Fig. 9.24. In
32☞ this example (data taken from Hellawell (1986)[4] for brown trout) an increase in
acclimation temperature from A to B raises the upper threshold value from C to D. A
further increase in acclimation temperature above B, however, does not raise the upper
threshold value (i.e. D = E = 24 °C). When acclimated to a high temperature (e.g. 22.5
°C, F), a sudden exposure to a low ambient temperature would put the trout outside its
temperature polygon and its thermal threshold (e.g. 1 °C, G) and die. Survival is just part
of a fishes activities; it must also feed, grow and reproduce, and each activity has its
temperature polygon. Fig. 9.24 shows the upper limits for feeding; feeding activities
cease at temperatures outside the temperature polygon and the fish would eventually die
through starvation. Hence, in nature species that normally experience wide changes in
temperature are more resistant to thermal shock than are species that rarely experience
wide fluctuations in temperature.

Trout are very definitely cold-water fish, spawning between 6 and 9 °C (brook trout) or 10 and 16 °C (rainbow trout, brook trout, lake trout). The upper lethal temperature of developing eggs is 12 -15 °C and of fingerlings is 24 - 25 °C. Brook trout tend to seek temperatures below 20 °C, while rainbow and brown trout prefer water near 21 °C. Mottled and slimy sculpins are even more demanding, preferring temperatures below 17 °C. Many darters that occur in headwater areas exhibit similar thermal preferences.

At the other extreme are fishes typically found in large rivers where temperatures may exceed 30 °C. The creek chub, *Semotilus atromaculatus*, the carp, *Cyprinus carpio*,

---

[4]Hellawell, J. M. 1986. Biological indicators of freshwater pollution and environmental management. Applied Science Publishers, London.

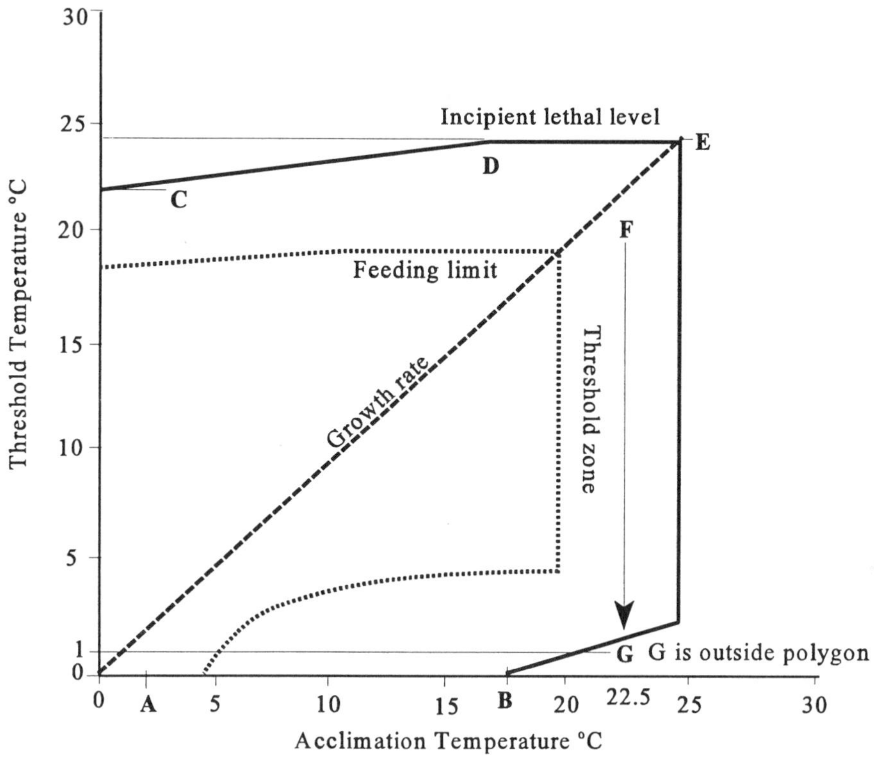

Figure 9.24. Temperature tolerance polygon for different activities of the brown trout, *Salmo trutta*. Data taken from Hellawell (1986). See text for explanation.

and the white sucker, *Catostomus commersoni* are known to survive temperatures up to 32 °C. The yellow perch, *Perca flavescens*, and the pike, *Esox lucius*, have upper thermal limits of tolerance near 30 °C.

The timing and extent of temperature changes are also important, particularly in the control of migration, breeding and spawning. Sea lampreys spawn in the spring and begin ascending streams when temperatures reach 4.4 °C. Spawning activities begin when temperatures reach 11.1 - 11.7 °C and peak at 14.4 - 16.6 °C, usually in June. The rainbow trout spawns in the spring between 10 and 15.5 °C, while brook and brown trout spawn in the fall at 6 to 9 °C and 5 to 10 °C, respectively. Brook trout eggs cannot survive in most reaches where rainbow trout spawn because the upper lethal temperature of developing eggs is about 11.7 °C.

Lake trout occur in deep lakes and rivers as long as there is cold, well oxygenated water or a source of cold, well oxygenated water. In some deep lakes, after they thermally stratify, oxygen may be depleted below 50% saturation and lake trout would perish. However, if the lake is fed by underground springs with cold, well-oxygenated water, the lake trout will congregate in the seepage areas. Most cottagers know such areas in their lake and successfully catch lake trout by trawling over the seepage zones.

Dissolved Oxygen

Active species (e.g. stream-dwelling trout) use more oxygen than inactive species (e.g. many chub species). Even within a species, like carp, stream-dwelling forms are more prodigal in their use of oxygen than are lake-dwelling forms. As we saw in
33 Chapter 5, the solubility of oxygen decreases with increasing temperatures. Hence, during the summer, species move to upper reaches in the stream where water temperatures are cooler and oxygen levels are higher than in lower reaches. Few species occur in the headwater origins where water seeps out of the ground because the water is devoid of oxygen. As water percolates upward, it picks up carbon dioxide from respiring micro-organisms and roots of plants, displacing any minute amounts of oxygen that may have been present. However, when the cold underground water reaches the surface, oxygen tensions are restored quickly to saturation levels by turbulent flow over and around rocks. This area of oxygen saturation in the headwaters can be readily identified by the first occurrence of trout, darters and/or sculpins.

Dissolved Salts

As discussed in Chapter 5, the chemical composition of stream water is determined largely by the soils and bedrock (edaphic factors) through which rivulets and tiny tributaries flow. Water flowing over hard, granitic bedrock (SiO_2) will have low
34 conductivites, pH, alkalinity and total hardness (some refer to these last three as "buffer variables" because they are tightly correlated with one another). Water flowing over softer, more soluble bedrock, like limestone ($CaCO_3$), gypsum ($CaSO_4$), or apatite ($CaPO_4$), will have a higher conductivity and higher levels of the buffer variables. Not surprisingly, there is generally a good correlation between growth rate of fish and the buffer variables. However, apparently there is no cause-effect relationship between the two. It is probable that the higher levels of salts in hard waters favour development of large populations of invertebrates for benthivores and plants (algae and macrophytes) for the herbivores. The productivity of the piscivores will depend on the productivity of the benthivores and herbivores on which they feed.

Most fishes can tolerate a wide range in pH. For example, brook trout occur throughout the pH range of 4.1 - 9.5, but the largest populations seem to occur in circum-neutral to alkaline waters. Chapter 13 examines the effect of spring pH depressions from snow melt on trout and the mechanisms that fish use to tolerate changes in pH.

Fishes and the River Continuum Concept

The river continuum concept (RCC) evolved from numerous stream studies on the physical, chemical and biological changes with distance downstream. Previous
35 chapters have examined the abiotic and biotic factors that affect the continuum of changes in diversity and biomass of plants and benthos from headwater tributaries to large rivers. Similar studies have been done for fish. Fish zonation concepts date back to

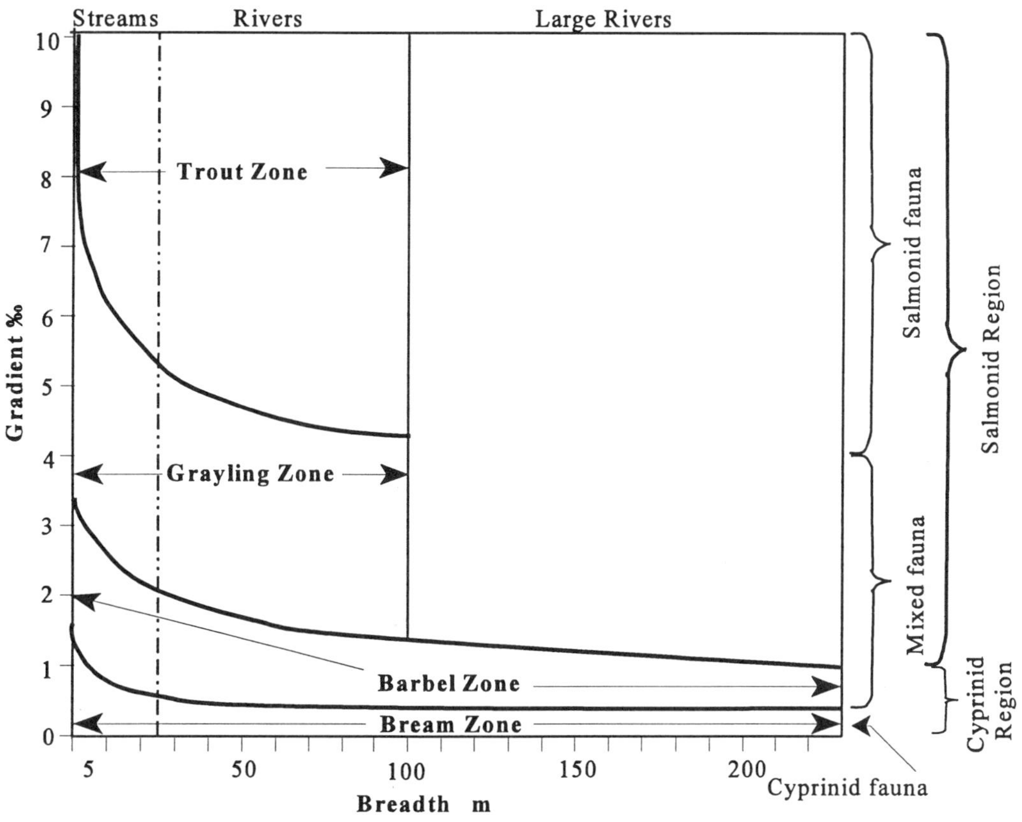

Figure 9.25. Slope graph showing relationship between gradient, breadth and fish faunal zones. Modified from Whitton (1975).

the early 1900s. Even a summary of the different schemes is beyond the scope of this text. However, most schemes relate to the effect of current speed, substrate type, temperature, dissolved oxygen content, nature and type of vegetation and composition of benthic communities. A French scientist, Huet, considered the gradient to be the primary feature characterizing the different zones, which to his point in time were "trout", "grayling", "barbel" and "bream" zones. The gradient affects directly or indirectly all the factors listed above (Whitton, 1975)[5]. Huet studied numerous European streams and noted that all rivers with similar size stretches with similar gradients had similar fish faunas. From this he formulated the ***slope rule***, "*In a given bio-geographical area, rivers or stretches of rivers of like breadth, depth and slope have nearly identical biological characteristics and very similar fish populations*". Huet constructed a ***slope graph*** (Fig. 9.25) showing the relationship between stream gradient, breadth and fish faunal zones. Hence, with a knowledge of the width and slope, one could predict the types of fishes that should be present.

While Fig. 9.25 is based on Huet's European data, the same concepts apply to the

[5]Whitton, B. A. 1975. River ecology. University of California Press, Berkeley, CA.

North American RCC. Headwaters (stream orders 1 to 3) are dominated by cold water species, such as darters, trout and sculpins that feed on benthic invertebrates; larger streams (stream orders 4 to 6) are dominated by warmer water cyprinids, centrarchids and suckers that are herbivores, benthivores and carnivores; and slow, warm, medium to large rivers (stream order > 7) are dominated by nearly all feeding guilds, including planktivores, as indicated in Fig. 9.23.

Methods For Sampling Fish

The methods differ for lakes and streams, but the key objective is to ensure that all species and all length classes are accurately represented in a sample. The methods in streams rely to some extent on current to carry the fish into nets or traps, with some help by the sampler. In lakes, a greater strategy is needed and the methods tend to be more labour intensive and costly than for streams.

Stream Sampling

In streams netting and electro-fishing (electro-shocking) are most commonly used. Poison (e.g. rotenone) was used years ago, and is still used in some instances, but
36☞ fortunately has been eliminated as a standard practice. The poisons used were non-selective and killed all species of fishes, and in many cases invertebrates.

Both netting and electro-fishing are harmless to fishes if properly employed. With netting, a stop net and seine are used. The stop-net is placed across the stream, ensuring that the lead line (bottom line of net with lead weights attached to it) is firmly set on the bottom. Then, beginning well upstream of the stop-net, two people work their way downstream toward the stop-net with a seine. The seine also has a lead line which often gets caught on logs or large rocks, so it is recommended that a third person follows and untangles the seine when caught. At the stop-net, a person at one end of the seine walks parallel to the stop-net towards the opposite shore, trapping all fish in the seine. The fish are identified, enumerated and released. Depending on the objectives of the study, fresh weights and fork lengths (tip of snout to notch in tail) or standard lengths (tip of snout to tip of tail) may also be taken before releasing the fish. If population size estimates are required, some fish may be marked (e.g. fin clips) or tagged.

Electro-fishing is much easier and probably more popular. A stop-net, seine or a dip net is needed to catch the electro-shocked fish. A light-weight generator capable of producing DC current with 200-300 volts between two electrodes is required. One of the electrodes is insulated (so is the operator, by wearing leak-proof waders!!). The other electrode lies in the stream or can be carried by the same person or another person. The direct current actually attracts the fish toward the positive pole (cathode, the insulated pole carried by the operator). Apparently, at increasing distances from the negative pole (anode) there are decreasing voltages, so the further away from the electrode, the less the voltage drop over the length of the fish. Therefore, at a certain distance away, a fish facing the electrode experiences only a small voltage drop (+'ve to -'ve) along its body.

This inhibits long nerve impulses running backwards from the brain to the tail muscles, and swimming ceases or is weakened. Closer to the electrode the shorter sensory nerve fibres running in series toward the brain are actuated by the voltage change along their length, forcing the brain to override the inhibition of the long motor fibres, and the fish swims vigorously forward. Near the electrode, the increasing voltage drop overrides the brain impulses and the fish becomes limp. Continued shocking forces the fish on its back and is carried downstream by the current into a net. Should the fish in the initial position be facing away from the positive electrode, the motor nerves are stimulated directly, but the short sensory nerve fibres are partly inhibited and the fish swims slowly away. Close to the positive electrode, complete inhibition results. Fish recover from the shock after a few minutes, and caught fish have to be placed in a trap or seine temporarily until they are identified, enumerated and processed for weight, length measurements and tagging, as required.

Fish species vary in their reaction to the electric field. Strong swimmers, like trout, are more susceptible to shocking than are secretive benthic species, like sculpins and some darters. Longer fish have a greater voltage drop than do shorter fish. The critical length at which efficiency falls steeply is about 15 cm. The conductivity of the water is also important; apparently conductivities exceeding 200 μS is recommended as the minimum for efficient electro-fishing. Most modern generators are capable of generating more than 300 volts to increase efficiencies in low conductivity waters. It is therefore necessary to match the machine to the water and to the species being caught. The only limitation of electro-fishing is that it can only be done in streams that are wadeable.

The data obtained from either netting or electro-fishing can be reported as "catch per unit effort" or as estimated population sizes if tagging or fin clipping was used to permit a "mark-recapture" estimate. Details of these methods are given in the next section.

Lake Sampling

The methods used for catching fish in lakes tend to be much more labour intensive than those used in streams. The methods vary from trapping and keeping fish
37☞ alive to netting and killing some or all the fish, as with gill netting. The latter was commonly used in the past, but trapping and seining seem to have come into more popular use. With gill netting, "gangs" of nets are stretched out from shore or parallel to shore, depending on the lake or species being sought after. A gang consists of several nets, each with a different mesh size and about 15 m (50 ft) in length. The mesh sizes vary from about 2.5 cm (1 in) to 10 cm (4 in). The meshes are randomly placed along the gang. Each net has a float line and a lead line so that the nets are kept in a vertical position. The depth of the gangs can be varied by tying anchors to the lead line, the length of the rope attaching them determining the height above bottom that the net will be located. A buoy is tied to the float line to mark its location in the lake and to help retrieve the net. Normally, gill nets are left for 24 h or less. Depending on when the fish are caught in the net and how they are handled when removing from the net, some fish can be

kept alive and released. Most, however, die because the mesh prevents opercular movements, or the investigator damages the fish when removing it from the net. Catfish and bullheads, in particular, are difficult to remove because their gills and pectoral and dorsal spines get entangled in the mesh.

Seines can be used to catch fish that come into shore. Because many fish are nocturnal and come to shore to feed at night, seining may have to be done in the evening as well. Seines differ in mesh size (6 mm to 12 mm =1/4 in to ½ in, usually) and length (3 m to 30 m = 10 ft to 100 ft), and in design. "Bag seines" are preferred because they have a bag-like extension in the middle where all fish are forced to congregate when the seine is brought to shore. The technique is fairly simple. Two people are needed, with one person on each end of the seine. Alternatively, one end can be tied to an object on shore. The seine has a float line and a lead line. The lead line keeps the net on the bottom so fish cannot escape below the seine. The float line keeps the top of the seine at the surface so fish cannot escape over the seine. One person, the lead, moves out, perpendicular to shore. The other remains on, or slightly off shore. Depending on the length of the seine, the lead person turns and walks parallel to shore for a few feet and then moves toward shore. Near shore the lead person swings toward the partner and both slowly and carefully drag the seine up on shore, leaving the bag in the water. The fish are identified, enumerated and measured, weighed and tagged if necessary. A few fish may be damaged and die in the process but most survive the ordeal and can be returned to the water.

A variety of hoop nets and trap nets are also available. Most are very expensive and are used for scientific purposes only. All, including gill nets and seines, require permits for use.

Population Size and Production Estimates

Regardless of what methods were used above for catching fish, abundances can be estimated from "catch per unit effort" or "mark-recapture" data.

Catch Per Unit Effort

These data are obtained by repeating a sampling technique over and over again in the same reach of stream or region of a lake. The catch per unit effort is then plotted for each species, or size class, against the number of trials (Fig. 9.26). This results in a
38☞ curve that tends towards zero. From this, one can estimate, by extrapolation, the total number of fish that would have been caught if collecting continued until no fish were left. The population size is the total number of fish caught in all the sampling efforts.

Mark-Recapture

39☞

The mark-recapture technique is based on the premise that marked fish that are released into the population will be re-caught in numbers proportional to their abundance in the population. The population size can then be estimated from the proportion marked to unmarked animals in random samples obtained from the entire population. The total number in the population is estimated from:

$$N_P = N_S N_M / M$$

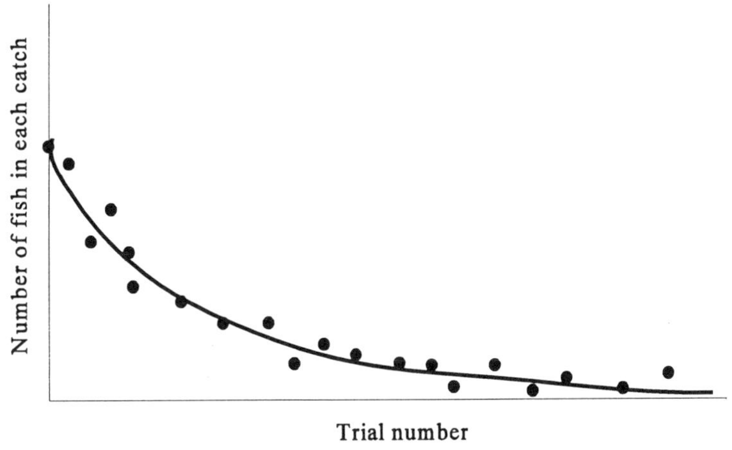

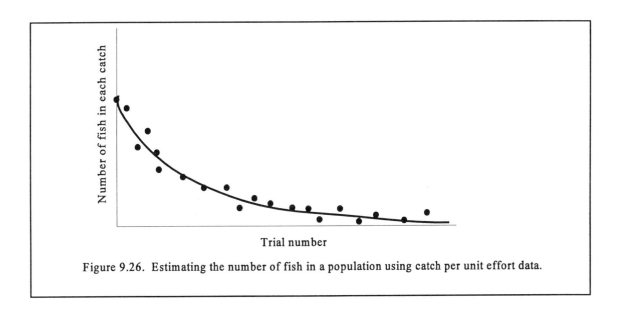

Figure 9.26. Estimating the number of fish in a population using catch per unit effort data.

where N_P = estimate of total number of fish in the population,
N_S = total number of fish in a sample from the population,
N_M = total number of marked fish in the population
M = number of marked fish in the sample.

However, some basic assumptions must be made for this estimate to be valid: (1) marked fish must be mixed randomly within the entire population; (2) there must be no bias in catchability between marked and unmarked fish; (3) marks (tags, fin clips) must be easily recognizable and not be lost; (4) there can be no difference in mortality or emigration between marked and unmarked fish; and (5) there can be no recruitment or immigration to the population.

An alternative method is to plot the proportion captured (M/N_S) as percent on the y-axis against the cumulative number tagged prior to each sample (Fig. 9.27). In this method, marked fish are returned to the population for recapture a second time, and more. The same data must be collected as for the mark-recapture estimate above, in addition to cumulative sums of tagged fish (plotted on x-axis). The total number in the population is the intersection of the x-axis by a perpendicular line drawn down from the 100% extrapolated value (Fig. 9.27).

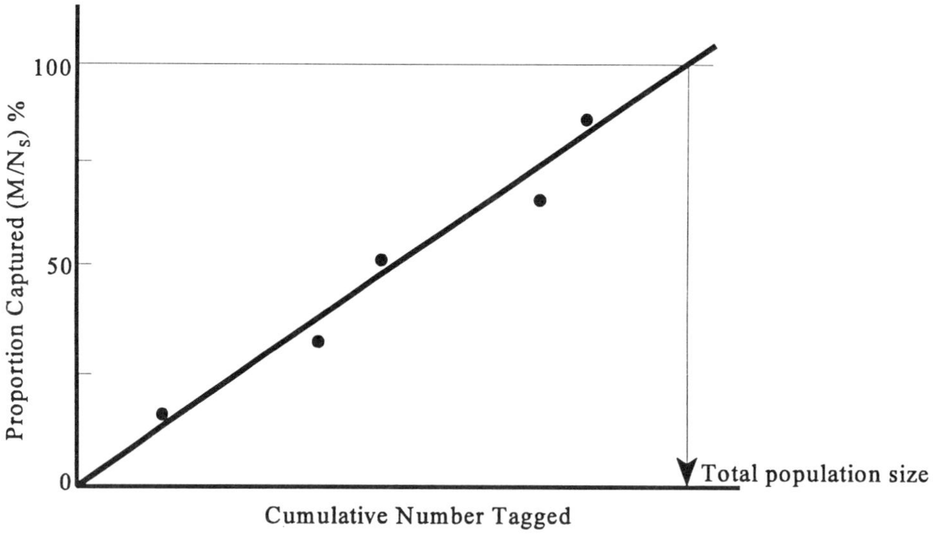

Figure 9.27. Estimating total population size from cumulative estimates of tagged fish and proportion captured each time.

Estimating Productivity

The methods used for estimating secondary production in benthic invertebrates
apply here as well. Of the methods described in Chapter 8, the Allen curve method
40 appears to be the most commonly used for estimating fish production. The methodology
need not be repeated here. It is important to remember, however, that the method is valid
only for populations that have recognizable cohorts so that production can be estimated
for each cohort. In order to use the Allen curve method (or any other production
estimate) mean weights of each catch per unit effort sample or mark-recapture sample is
needed. The mean individual weight is then plotted on the x-axis. The total number at
each sampling effort is derived from any of the population sizes estimated above.

The productive potential of a fisheries is related to the trophic state of the water
supporting them. In general, the harder the water the greater the fish productivity. The
morphoedaphic index (MEI; Ryder, 1965)[6] is a simple, useful way to approximate the
41 potential standing crop, or yield, from a lake or impoundment. The MEI is the quotient
of the concentration of total dissolved solids to the mean depth of water in meters:

---

[6] Ryder, R. A. 1965. A method for estimating the potential fish production of north-temperate lakes. Trans. Amer. Fish. Soc. 94: 214-218.

$$MEI = \frac{TDS}{\bar{z}}$$

The units are yield in kg per unit area. Since TDS is related to conductivity according to the equation,

TDS = 0.65 x conductivity (± 0.10) (mg/L)

the MEI estimate can also be estimated as:

$$MEI = 0.65x \frac{Conductivity}{\bar{z}}$$

Recall from Chapter 3 that mean depth is estimated from the lake's volume divided by its surface area.

In Canada the most productive lakes have indices lying between 10 and 30, the theoretically most favourable value being near 40. However, lake morphology and salinity may affect the theoretical optimum MEI value. A MEI of 40 could be found in a lake with a TDS of 4,000 mg/L and a mean depth of 100 m, or in another lake with a TDS of 400 mg/L and a mean depth of 10 m.

CHAPTER 10

SPECIALIZED STREAM HABITATS

Why Read This Chapter?

So far we have looked at "typical" habitats in streams, that is, habitats that would be predicted by the river continuum concept (RCC). The RCC predicts that habitats of a stream system, from source to mouth gradually change in a continuum from a cool, clear, well-oxygenated and erosional habitat to a warm, turbid, less oxygenated and depositional habitat. Some streams exhibit only a part of the continuum, do not follow the continuum at all or, in the case of impoundments, greatly accelerate the processes that occur in the RCC. Streams that deviate from the RCC are described in this chapter. Having learned the physical, chemical and biological concepts of aquatic environments, we can now integrate all the concepts in analyzing the specialized stream habitats.

The chapter begins by examining physical, chemical and biological characteristics of stream sources, or the "boil zones" of springs. Then we examine some of the specialized habitats that occur in the vertical dimension, from those that occur between the groundwater and sediment-water interface zones to those that live on the damp surfaces of rocks partially exposed to air. The habitat at the edge of streams is also quite different from "main stream" habitats.

While latitude and altitude also have a profound effect on the RRC, there is some similarity between the characteristics of headwater streams and high latitudes and altitudes, and between high-order streams and those at tropical latitudes. Yet, there are unique features of streams at high latitudes/altitudes and of streams at tropical latitudes. Hot springs in particular have a unique fauna.

The chapter ends with an examination of the physical, chemical and biological changes that occur in a stream when an impoundment is created. The changes that occur downstream of the dam are compared to those predicted by the RCC. We learn that, in most cases, impoundments accelerate processes predicted by the RCC.

By the end of this chapter you will:

☞ 1. Know the terminology and characteristics of springs, called the crenon.

☞ 2. Know the terminology and characteristics of the hyporheic zone, the zone between the groundwater and the sediment-water interface.

☞ 3. Know the terminology and characteristics of the madicolous habitat, the thin film of water on rock surfaces.
☞ 4. Know the terminology and characteristics of the psammon, the damp zone at the edge of streams.
☞ 5. Know the terminology and characteristics of arctic and high altitude streams.
☞ 6. Know the terminology and characteristics of glacier streams.
☞ 7. Know the terminology and characteristics of mountain streams.
☞ 8. Know the terminology and characteristics of spring streams.
☞ 9. Know the terminology and characteristics of tundra streams.
☞ 10. Know the terminology and characteristics of tropical streams.
☞ 11. Know the physical and chemical characteristics of tropical streams, like:
  ☞ a. Temperature.
  ☞ b. Dissolved oxygen.
  ☞ c. pH.
  ☞ d. Conductivity and salinity.
☞ 12. Know the adaptations of organisms to tropical stream environments.
☞ 13. Know the terminology of hot springs and thermal pollution.
☞ 14. Know characteristic of hot springs, including:
  ☞ a. Physical.
  ☞ b. Chemical
  ☞ c. Biological
☞ 15. Know the characteristics and uniqueness of intermittent streams.
☞ 16. Know the obvious effects of artificial impoundments on stream concepts.
☞ 17. Know the concepts and principles that govern the rates and kinds of changes that occur in impoundments and downstream, including:
  ☞ a. Thienemann's principles.
  ☞ b. The concept of succession.
  ☞ c. The concept of pulse stability.
  ☞ d. The concept of ecotone.
  ☞ e. The theory of island biogeography.
☞ 18. Know the inherent physical differences between lakes and streams.
☞ 19. Know the inherent chemical differences between lakes and streams.
☞ 20. Know the inherent biological differences between lakes and streams.
☞ 21. Know the role of the river continuum concept on changes downstream of the impoundment
☞ 22. Know the effects of the impoundment on:
  ☞ a. The physical continuum.
  ☞ b. The chemical continuum.
  ☞ c. The biological continuum.

SPRINGS (CRENON)

The ***crenon***, or springs and streams draining them, differ from normal streams in several ways: (i) they tend have a uniform temperature, usually close to the mean annual air temperature, except in volcanic areas where the water temperature may be higher than 1☞ air temperature; (ii) dissolved oxygen levels are usually very low; and (iii) deposits of iron are common place. If the spring is supplied with water from a relatively shallow underground source, the fauna are often phreatic (of groundwater origin) in composition.

Low oxygen levels result when waters rising from deep underground sources flush pore spaces rich in carbon dioxide that is constantly supplied by respiration of bacteria and other microorganisms and plant root systems. Under anoxic conditions there is usually no visible aquatic life, except for anaerobic bacteria such as sulphur-reducing bacteria and denitrifying bacteria that may occur in the boil zone. Macroinvertebrates, tolerant of low oxygen levels, including the mayflies, *Callibetes* and *Caenis*, begin to appear at 2.5 mg O_2/L, but the diversity increases rapidly in a short distance downstream. The fauna often consists of three elements: (i) ***phreatic*** animals, such as blind species of isopods (*Caecidotea*) and amphipods (*Crangonyx*) that persist in sheltered areas some distance downstream; (ii) wet soil fauna, such as the hydrophilid beetle, *Anacaena*, and the crane fly, *Pedicia*, that occur at the edge of streams in cool, shady places; and (iii) fauna unique to springs and cool streams, including the flatworm, *Phagocata gracilis*, the caddisfly *Agapetus*, the stonefly, *Leuctra nigra*, some beetles, e.g. *Helodes* and some hydrobiid snails e.g. most *Aphaostracon* spp., some *Cincinnatia* and *Marstonia* spp., *Bythinella hemphilli* and *Paludestrina bottimeri*.

Many species of springs are glacial relics, such as the flatworm *Crenobia alpina* and the snail *Melanopsis* of Europe, the caddisfly *Apatidea muliebris* of Denmark, and the isopod, *Caecidotea bivittatus*, of Kentucky. The faunas of springs in relatively recently glaciated times tend to be less diverse than those which have been in existence for a longer period. Part of the reason is that the number of dispersal mechanisms of fauna in isolated springs is rather limited. In Europe, many of the glacial faunal elements are hypothesized to have moved into groundwater when the springs dried up, and then to have re-appeared when the springs refilled with water. This explanation has been given for the flatworm, *C. alpina*m, and the flightless beetle, *Hydroporus ferrugineus*. Otherwise, upstream migration of purely aquatic forms and oviposition in the crenon by insects that have emerged and flown from downstream sites are probably the most common dispersal mechanisms.

THE HYPORHEIC ZONE

The hyporheic zone is the link between the rhithron and the groundwater or 2☞ ***phreatic zone***. ***Interstitial flow***, or flow of water through pore spaces in the substrate, is slow (e.g. cm/hr) but sufficient to replenish some of the oxygen that is used by the respiration of bacteria and other microorganisms. However, the oxygen content falls steeply with depth and the habitable layer is a thin one. The depth of the hyporheic zone is determined, in part, by current speed and water temperature, the depth increasing with faster water currents, cooler temperature and dissolved oxygen levels, the latter two parameters being closely related since oxygen is more soluble in cold water than in warm water.

The sizes and numbers of organisms that inhabit the hyporheic zone (collectively referred to as ***hyporhea***) depend largely on the soil particle size, which in turn determines the size of the pore spaces. The faunal assemblage is most abundant and diverse in hard crystalline rock, where pore spaces are large, and less abundant in softer rocks (e.g. limestone), where fine rock debris tends to create smaller pore spaces.

Typical hyporea are *Hydra*, Turbellaria, Nematoda, Oligochaeta, harpacticoid copepods, Ostracoda, chydorid cladocerans, small amphipods, and several small species of insects of the family Chironomidae and the orders Trichoptera, Ephemeroptera and Plecoptera. Some organisms, e.g. water bears (Tardigrada) and the water mite, *Porohalicarus*, occur nowhere else in running water environments.

There is a direct relationship between abundance of the hyporhea and the amount of organic detritus in the deposits. It is not surprising, therefore, that the main functional feeding group is collectors that feed on the detritus and microbes. Predators, such as water mites, are also present.

THE MADICOLOUS (HYGROPETRIC) HABITAT

The madicolous habitat, or the thin film of water flowing over rock surfaces, is most common at the edges of stony streams, at the sides of waterfalls or stony chutes and
3☞ in riffle areas. The habitat supports a specialized fauna (called ***madicoles***) that must be adapted to: (i) maintain their position; (ii) remain adequately submerged for essential life processes to occur; (iii) be able to move readily as conditions change. The madicoles consist of a variety of air-breathing Diptera, including the U-shaped *Dixa* (see Chapter 8), solitary midges of the genera *Thaumalea* and *Trichothaumalea*, and several types of crane fly larvae, Psychodidae and Stratiomyidae. Other components include lumbriculid worms, small species of hydrophilid beetles and some small caddisflies, e.g. *Tinodes*, which build fixed tunnels on submerged rocks. Nearly all are grazers, feeding on epiphytic algae and mosses, or gatherers.

THE PSAMMON

The psammon is the region of wet and damp soil at the margins of streams. It moves up and down with the water level. Water is drawn upwards by capillary forces,
4☞ the flow being created as water evapourates from the wet surface of soil particles at the upper edge of the psammon. Slow seepage through air-filled spaces maintains a supply of oxygen for organisms inhabiting the psammon zone.

As in the hyporheic zone, the abundance and diversity of life depends on the size of the interstitial spaces. Sands of grain diameters between 0.2 and 2.0 mm create pore spaces large enough to support large numbers and types of fauna. Small pore spaces with considerable amounts of organic material have insufficient oxygen to support a diverse faunal assemblage.

The highly variable conditions of flow in the psammon contrasts greatly with the more constant flow in the hyporheic zone. The psammon contains a specialized fauna, many of which are represented in the hyporheic zone: ciliates, Rotifera, Gastrotricha, Nematoda, Turbellaria, Oligochaeta, Tardigrada, Harpacticoida, Hydracarina, Chironomidae, Ephemeroptera and Plecoptera. The surface particles also contain a unique flora of algae that make the sediments green. Even though the psammon is a rather narrow zone, it can be continuous for long distances and play an important role in sheltering many young stages of truly aquatic species in the potamon.

ARCTIC AND HIGH ALTITUDE STREAMS

Arctic and high altitude streams have basically the same ecologies and differ only in that the former is latitudinal and the latter is altitudinal. For ease of discussion, they are both referred to here as arctic streams.

Much is still to be learned about arctic streams but Craig and McCart (1975)[1] and Harper (1981)[2], give excellent reviews and are primary sources of information here. Most stream ecosystem concepts apply, but are modified by the effects of (i) *extreme* 5☞ *photoperiods* and (ii) *a variety of temperature conditions*. The photoperiod changes predictably with latitude, with strong seasonality at mid-latitudes and polar-like beyond the polar-circle (66° 30'N) where continuous daylight occurs during the summer months. Unlike photoperiod, however, a variety of temperature and climatic conditions may occur within a latitude, and vary from a boreal climate with coniferous forests to an arctic climate with barren polar deserts. Streams of arctic areas and high latitudes have little snow-cover and in many cases are made snow-free by the wind. Hence, many arctic streams are not insolated, lose much or all of their heat and freeze to the bottom. An underlying permafrost will also cause heat loss from the stream bottom. During summer, most of the radiant energy is used in melting snow and ice. By the time the snow is melted, both the photoperiod and radiant energy are beginning to decline, resulting in a short summer with low stream temperatures, although extended periods of daylight still

---

[1]Craig, P. and P. J. McCart. 1981. Classification of stream types in Beaufort Sea drainages between Prudhoe Bay, Alaska and the Mackenzie Delta, N.W.T., Canada. Arctic and Alpine Research 7: 183-198.

[2]Harper, P. P. 1981. Ecology of high altitude streams. *In: Perspectives in running water ecology*. Lock, M. A. and D. D. Williams (Eds.). Plenum Press. Pp. 313-337.

exist.

Discharge is affected, indirectly, by photoperiod and radiant energy, especially in seasonal streams. Some mountain streams may exhibit two peaks in discharge, one during the spring thaw of snow and ice at low altitudes and another during the thaw of mountain glaciers. The spring spate events are highly erosional and capable of scouring several centimeters of substrate that is carried downstream. The scouring is aided by the shattering of rocks during freeze/thaw events, chunks of surface ice which gouge out channels in the stream bed and scouring effects of underwater ice, like ***frazil ice*** and ***anchor ice***. Frazil ice, or slush, results from the formation of ice crystals in supercooled water or entrainment of snow flakes. Anchor ice forms in shallow water in riffles on rock surfaces, first on the upstream faces of rocks, then on downstream faces. Anchor ice is a whitish deposit at first but becomes dense and transparent. It forms mostly during the evening and detaches and melts during the day. As the ice detaches, it lifts part of the stream bottom with it and carries the imbedded sediments downstream. Daytime flows may be twice the night-time flows due to anchor ice. Any or all of these scouring events contribute to high turbidities during periods of runoff and snow/ice melt in many arctic streams.

Arctic streams can vary considerably in their physical, chemical and biological characteristics, depending on their origin in three areas; mountains, foothills and coastal areas. Craig and McCart (1975) classify arctic streams into "***provinces***", specifically an ***Arctic Mountain Province***, an ***Arctic Foothills Province*** and an ***Arctic Coastal Province***. Four types of arctic and high altitude streams may occur within these provinces; ***glacier streams, mountain streams***, ***spring streams*** and/or ***tundra streams***. The chemical characteristics of the water may differ greatly between provinces, or even within provinces, depending on the geochemistry of bedrock and soils (called edaphic factors) in the drainage basin, patterns and chemistry of precipitation, forest types and biogenic processes. However, these are the same factors that affect water quality at all latitudes. Features that are more-or-less unique to the four arctic stream types are described below.

Glacier Streams

Glacier-fed streams are sometimes called ***kryon*** habitats. Because the water consists almost entirely of ice-melt, they are characterized by low temperatures, low conductance and low nutrient levels. The only source of ions are from erosion of
6☞ sediments from the bank and stream channel. Flow is reduced or even frozen during the winter. Anchor and frazil ice are common. Diurnal fluctuations in flow are the rule, with maximum flow during the day and minimum during the evening. The daily fluctuations in discharge are accompanied by daily fluctuations in temperature, and closely approximate air temperatures because of low flows.

Glacier streams have a very low diversity of life. Abundances of benthic invertebrates rarely exceed $100/m^2$. Mostly detritivores occur, with Diamasinae and Orthocladinae (both of Family Chironomidae) and Simuliidae (mostly *Prosimulium*) being the dominant benthic forms well downstream of the glacier. Life cycles are short, with all growth and productivity occurring in a very short summer period. Fish (grayling and arctic char) may frequent the streams when waters are running, but glacier streams are not used for spawning.

Mountain Streams

Mountain streams originate in the Arctic Mountain Province and extend into the Arctic Foothills Province. Most mountain streams are long and break up into
7☞ interconnecting channels, forming braided patterns as they enter the foothills. They can be perennial and/or seasonal, depending on the source of water. Perennial streams are fed by groundwater springs which provide perennial flows; most springs can be located by "***icings***" just downstream of the upwelling area during the freezing period. Seasonal streams are fed by melting snow and ice, providing flow only during spring and early summer. If water is present in the fall, the streams usually freeze to the bottom. Most streams above the tree line are of the seasonal type. Boreal mountain streams behave much like headwater streams of mid-latitudes.

Most mountain streams are very turbid owing to the extreme erosional effects of spring spates and floating and submerged ice. The removal of weathered material creates deeply entrenched channels, especially mid-channel in depositional areas. Stream banks are often undercut, with riparian vegetation being uprooted and carried downstream. "***Trash lines***" of riparian debris delimit the maximum height of spring floods. Formations of frazil and anchor ice is common and may cause as much as 60% reductions in discharge when they are formed and up to 60% increases in discharge when they melt.

Temperatures seldom exceed 10 °C in early July but may occasionally range up to 15 °C. By August, stream temperatures begin dropping and are close to zero. The stream is frozen by fall. Conductivity fluctuates widely, with lowest values during spring snow- and ice-melt. The mean and highest conductance values depend largely on the chemistry of the catchment basin, with limestone usually contributing to the highest mean and maximum values. Waters in riffle areas are usually 100% saturated with dissolved oxygen. Dissolved oxygen levels may drop in waters of low flow and deoxygenation may occur below ice in stagnant areas.

Submerged macrophytes are usually depauperate or absent in mountain streams, with perhaps some mosses (e.g. *Fontinalis*) present. Benthic invertebrate diversity is high but densities are low (average ~ 100/m^2). Dominant benthos include plecopterans (stoneflies), ephemeropterans (mayflies), trichopterans (caddisflies) and simuliids (black flies). The densities of benthos are inversely related to discharge, Craig and McCart (1975) finding the following relationship for many arctic streams in their study:

$$\text{Log Abundance (No./m}^2) = -0.69 \text{ Log (Discharge} + 2.96)$$

Fish are of the headwater types, with arctic char and grayling predominating. Arctic char spawn in late summer in perennial waters. Eggs cannot tolerate freezing but those that do survive hatch in early spring. Grayling spawn in spring and eggs hatch in the early summer, after arctic char. Hence, development of grayling eggs must be rapid in order to avoid freezing in late fall.

Spring Streams

The term, "spring" refers to the source of water, not to a vernal (spring-time) existence. In fact, most spring streams are permanent because they get their water from
8☞ underground seeps and springs. In areas with a permafrost, the supra-permafrost may give rise to seasonal seeps and springs which freeze over with the permafrost in the fall. Sub-permafrost water may occur in aquifers which provide a constant source of water to permanent springs at a fairly constant temperature. Hence, discharges are typically low, vary little over time and the water is clear and has a very low sediment load.

Many of the permanent springs arise in the floor of rivers and lakes and help to maintain flow and open water even in the winter. Downstream from the upwelling, the water may eventually freeze causing large buildups of ice, also referred to as "***icings***" (see mountain streams), "***Aufeis***" or "***wadeli***". The temperature of the water in the permanent springs is the same as the groundwater temperature (usually > 2 - 3 °C in winter, 4 - 11 °C in summer) and creates a "milder" microclimate in the upwelling area. The water in the upwell is poorly oxygenated due to the stripping of oxygen by roots and bacterial composition. But the water has a low biochemical oxygen demand and dissolved oxygen levels are quickly restored within a very short distance downstream. The conditions (e.g. low and constant flow, clear and well-oxygenated water, cool but not freezing water) are conducive to much higher diversities and densities of plants, benthic invertebrates and fish than are seen in other types of arctic streams. Seasonal successions of algae occur, with attached diatoms (e.g. *Cocconeis*), red algae (e.g. *Hildenbrandia, Lemanea, Paralemanea, Bangia*) and brown algae (e.g. *Hydrurus*) in June, and Chlorophyta (e.g. *Cladophora*) and Chryptophyta (e.g. *Rhodomonas*) in August. Attached blue-green algae (e.g. *Calothrix*) can dominate in biomass throughout the year in some spring streams.

Zoobenthos that occur in glacier and mountain streams are also found in spring streams, in addition to a more diverse assemblage of gatherers (e.g. oligochaetes, burrowing mayflies, caddisflies), filter feeders (e.g. black flies of genus *Simulium*, hydropsychid caddisflies), grazers (e.g. beetles of families Psephenidae and Elmidae, lepidopterans and limpets) and a variety of predators (e.g. Megaloptera, Plecoptera, Ephemeroptera).

Grayling and arctic char are also common in spring streams and are often accompanied by darters (*Etheostoma* spp.) and sculpins (*Cottus* spp.). Fish productivity is generally high owing to constant flow and cool temperatures with little risk of eggs freezing in the winter months.

Tundra Streams

Tundra streams are sometimes referred to as "***lowland streams***", "***foothill streams***" or "***floodplain streams***". Tundra streams is used here because most occur in
9☞ lowlands dominated by tundra. They have low slopes and therefore reduced flows which result in depositional substrates. Often many pools occur along the course of the channel to create a "***beaded stream***". Many tundra streams are also braided with numerous

parallel and intertwining channels.

Floods are less severe than in other types of arctic streams because of the absorptive capacity of the dense tundra vegetation. Summer water temperatures are higher than in other types of arctic streams because of the lack of shading by trees and lower altitudes. Tundra streams tend to freeze to the bottom as there is little or no replenishment from groundwater sources, the main source being from mountain streams. The pools and dense tundra vegetation act as filters to remove any turbidity that is present. The slow flows through tundra produce water that is dark brown and has a low pH, low ion and nutrient content, low alkalinity and calcium content and a high biochemical oxygen demand. Primary production is low because of the highly coloured waters that absorb most wavelengths of light needed for photosynthesis to occur.

Tundra streams have benthic diversities which are intermediate between those of mountain and spring streams, the regression given above (under mountain streams) applying to all three stream types. Chironomidae dominate in numbers, followed by Plecoptera, Ephemeroptera, and Oligochaeta. Other groups include mites, flatworms, caddisflies, crane flies, hemipterans and beetles, the latter two groups represented especially by surface-dwelling forms since flow is so slow. Drift from surrounding marshes often augments the invertebrate biomass and diversity.

Fish are represented mostly by cyprinids, with a few grayling and arctic char frequenting tundra streams during cool temperature periods.

ECOLOGY OF TROPICAL STREAMS

Much of the literature on tropical streams is from Africa, Australia and Hawaii. In many cases the concepts are unique to the island, but only concepts that apply in general to tropical areas are described here. Some excellent reviews are given by Maciolek (1978)[3], Awachie (1981)[4], Williams (1981)[5] and are relied upon here for describing tropical streams.

Climate and topography determine the stream type, the streams falling into one of
three hydrological regions. Streams in ***arheic*** regions are episodic and flow only during
10☞ wet seasons, but never reach the estuaries. Australia and Africa have immense

---

[3]Maciolek, J. A. 1978. Insular aquatic ecosystems: Hawaii. *In: Marmelstein, A. Classification, inventory, and habitat analysis of fish and wildlife habitat.* Proceedings of a National Symposium, Phoenix, Arizona, January 24-27, 1977. Pp. 103-120.

[4]Awachie, J. B. E. 1981. Running water ecology in Africa. *In: Lock, M. A. and D. D. Williams. Perspectives in running water ecology.* Plenum Press, New York, NY. Pp. 339-366.

[5]Williams, W. D. 1981. Running water ecology in Australia. *In: Lock, M. A. and D. D. Williams. Perspectives in running water ecology.* Plenum Press, New York, NY. Pp. 367-392.

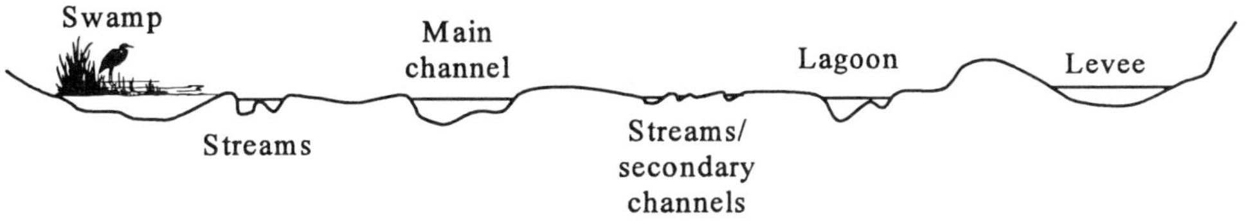

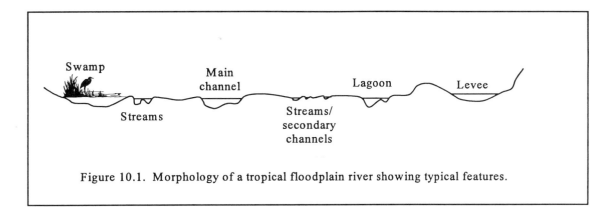

Figure 10.1. Morphology of a tropical floodplain river showing typical features.

endorheic basins in the centres of their continents and also have episodic flowing waters that occasionally may reach and drain into the ocean. Most of the coastal areas are ***exorheic*** and have fast-flowing streams that usually reach and drain into the ocean. Chapter 1 describes the hydrological cycle for aquatic environments in exorheic regions of al parts of the world.

Tropical streams fall into one of two morphological types, ***floodplain rivers*** and ***streams*** which are highly productive and have seasonally flooded banks, and ***non-floodplain streams***. Floodplain rivers consist of the main channel, floodplain pools, lagoons and swamps, tributary channels and creeks and levees (Fig. 10.1). The levees, which are raised banks bordering the main river channel, direct the flow of flood waters and control their rate of recession and, hence, the drying rate of the plain.

The main river channel can be either simple or braided, the latter being common along larger rivers, like the Niger, Zaire (Congo) and Zambezi Rivers. Flow through the floodplain varies from erosional during spate events to depositional and almost lacustrine during the height of the dry phase.

Smaller rivers typically exhibit one annual flow regime, but larger rivers expereience a biannual flow regime in its lower reaches. Earlier rains (January/February) upstream near source rivers create a siltless, "***black flood***"; later rains (August/September) on lower reaches create larger, turbid "***white floods***". The turbidity of white water is closely related to catchment erosion and discharge during flood events. Each phase of the annual water cycle lasts about six months for rivers near the equator, but the periods become increasingly unequal with increasing distance north and south of the equator.

There are three distinctive sources of flood water: (i) ***local rainfall*** from which water arises from the surrounding plain and forests and flows into the river channel; (ii) ***overspill*** from the river channel, that flows from the river to the lower fringing plain and occurs when a flood crest travelling down the river channel overflows the banks; (iii) ***tidal action***, that occurs only in coastal floodplains and estuaries. The varying gradients in tidal flow, current velocity and salinity tend to produce a longitudinal zonation of the biota.

11 ☞ As in temperate streams, current tends to even out physical and chemical variations in temperature, dissolved oxygen content, hydrogen ion concentration and

conductivity and salinity. However, the dynamics of the physical and chemical parameters follow rather closely on the flood cycle. In addition, differences in conditions between the river channel and floodplain arise during the dry season with their separation.

Temperature

11a☞ Surface water temperatures tend to follow air temperatures but, like temperate streams, are influenced by insolation, substrate composition, turbidity, vegetation cover, ground water and rain water inflows. As a rule, dry season temperatures are higher than those of the wet season for most tropical areas. Near the equator, running water temperatures are usually stable diurnally and seasonally, but rivers that traverse several latitudes exhibit strong north-south temperature gradients, from 28 - 33 °C near the equator to 17-19 °C in the headwater areas of north/south latitudes.

Dissolved Oxygen

Dissolved oxygen levels are usually near saturation in the running water sections of tropical streams but can be reduced or even absent in areas with organic loading from
11b☞ mats of vegetation like *Ceratophyllum*, *Pistia*, *Eichhornia*, *Cyperus papyrus* and *Echinochloa* and filamentous algal blooms in floodplains, levees and pools. Here, mass mortality of biota may occur during the anoxic periods of the dry season.

Hydrogen Ion Concentration

Levels of pH vary widely between different rivers and streams, depending on the carbon dioxide and bicarbonate-carbonate equilibrium processes. Forested rivers tend to
11c☞ have pHs between 4 and 7, while grassland rivers are slightly more alkaline. Floodplain lagoon systems in open savanna country have seasonally high pH levels due to evapouration of water and the resultant concentration of calcium salts. Phytoplankton metabolism may cause diurnal fluctuations in pH levels as well.

Conductivity and Salinity

Fluctuations in ionic contents of rivers occur mainly due to dilution by rainfall
11d☞ and discharge rates in the wet seasons, and to the geochemistry of the catchment area. Hence, conductivity tends to be highest during the dry season and lowest during the wet season.

Adaptations of Biota to Tropical Stream Environments

The seasonal pattern of rainfall and floods plays a major role in structuring the abundance and population dynamics of biota in tropical streams. Open rivers are
12☞ dominated by diatoms, while in backwaters, floodplains and reservoirs, flagellated green algae predominate. Phytoplankton and zooplankton production is directly correlated with conductivity and transparency and inversely related to water level and current velocity. Zooplankton communities are more highly developed in tropical streams than

temperate streams, largely because of the large numbers of pools, lagoons and levees associated with tropical systems. Crustaceans and rotifers dominate the zooplankton communities, with seasonal pulses due mostly to the latter group. Repopulation of rivers is facilitated by the reservoir populations of resting eggs of rotifers and ephippia of cladocerans in the pools and floodplains. Remarkably fast growth and reproduction of the zooplankters is another adaptation for a rapid repopulation process.

There is a high level of endemicity, notable absences and a diverse community which combine to produce a fluviatile fauna that is unique to each country. Dominant taxa of benthos include Chironomidae, Simuliidae, Ephemeroptera, Odonata, Trichoptera, Coleoptera, Gastropoda (Bulinidae, Planorbidae) and Bivalvia (Sphaeriidae and Unionidae), but the species are very different from temperate regions in North America. Notably absent in tropical steams are shredding Plecoptera. Altitudinal zonation of black fly, riffle beetle and triclads (*Dugesia*) species are prominent in high mountains. There is a close similarity in fauna between swift water streams and rivers and that on high mountains. In floodplain areas, many taxa survive periods of drought by employing one or more of numerous adaptations typical of temporary ponds and streams, as described in Chapter 8.

Two main groups of fish are often recognized, the freshwater and the brackish water groups. Within the freshwater group are two components: (i) a ***whitefish*** component that inhabits the savanna belt and avoids adverse conditions in rivers and floodplains by migrating into and within the main river channels; (ii) a ***black fish*** component that has adaptations to exist in hot and often dry and anoxic conditions during the dry period. The lung fish, *Protopterus*, lives in cocoons; *Polypterus*, *Heterotis* and *Gymnarchus* possess external gills in the relevant phase of their life cycles; *Clarias*, *Heterobranchus* and *Parophiliocephalus* possess accessory breeding structures; *Aphyosemion* and *Nothobranchus* produce dormant eggs.

The period of intensive feeding, reproduction and migration occurs during the flood phase. At least six phases in the migration and distribution of fish are recognized:

- longitudinal migrations within main channels
- lateral migration to the floodplain
- local movements within the floodplain
- lateral migration from the floodplain towards the main channel
- downstream migration within the main channel
- local movements within dry season habitats

As with invertebrates, flood-river fishes tend to have short life cycles. Many species (e.g. *Tilapia*) mature within one year and are ready to spawn by the next flood season. Faunal natalities are higher in floods that have a smooth and gradual rise in water level, a high amplitude and a long duration than in those with opposite features. Mortality is low during the wet season but rises sharply during the dry season as fish become stranded on the floodplain and in shallow depressions and as water temperatures rise and oxygen levels fall.

ECOLOGY OF HOT SPRINGS AND GEOTHERMAL STREAMS

The elevated temperatures characteristic of ***hot springs*** (the boil zone) and
geothermal streams (the hot stream below the boil zone) are due mostly to tectonic
13☞ processes (e.g. plate faulting). Elevated temperatures (typically 9 - 15 °C higher than
normal) due to the activities of humans is classified as ***thermal pollution***, or ***calefaction***,

14☞

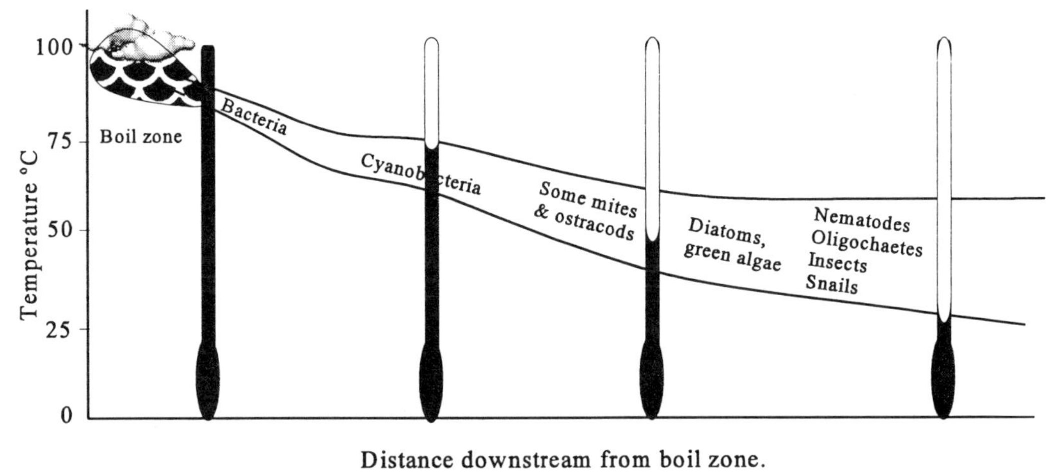

Figure 10.2. Longitudinal zonation of taxa in geothermal streams in relation to temperature. Examples of species within each taxon and temperatue range are given in text.

and is discussed in Chapter 13.

The temperature of geothermal streams is usually at least 35 °C higher than that of
normal temperate streams (Fig. 10.2). Hot springs are generated when water percolates
down through cracks and crevices in the earth's crust to depths greater than 2000 m,
becomes superheated and then rises, emerging as a ***boil zone*** or even a ***geyser*** (Fig. 10.3).
All springs have a boil zone, but only in hot springs do temperatures rise to boiling (100
°C). Typically, however, the temperatures of hot springs range from near 60 °C down to
14a☞ slightly higher than the average air temperature. Downstream of the boil, the rate of
water temperature decline depends on the rate and volume of the input, the grade of the
stream bed, the depth and whether the flow is turbulent or laminar. Atmospheric
conditions, such as air temperature, wind velocity and precipitation can also affect the
cooling rate of the stream. Most geothermal streams are shallow, ranging from a few
centimeters to less than one centimeter.

The chemistry of geothermal streams reflects the mineral composition of the rock
14b☞ that the water percolates and rises through. However, because the solubility of most
minerals increases with increasing temperatures, their leaching rate and concentrations

are generally higher than observed in most groundwater springs. The pH is usually much higher or much lower than the normal range of 6 to 8. Springs occurring in non-arid regions typically have high levels (50 to150 mg/L) of each of Ca^{++}, Na^{+}, Mg^{++}, HCO_3^-, SO_4^- and/or CL^-. Many hot springs have Na^+ as the dominant cation, with values ranging between 1000 and 2000 mg/L.

Figure 10.3. An Icelandic geyser. Photo provided by Christa Rigney.

Some springs have large deposits of materials that are diagnostic of the mineral content of the water. For example, when silicates are enriched and exceed 115 mg/L at pH levels greater than 5, it polymerizes to form ***opaline sinter***. In karst areas, formations of travertine are diagnostic of springs rich in limestone ($CaCO_3$) or gypsum ($CaSO_4$), the latter being readily identified by the presence of a strong odour of rotting eggs produced by hydrogen sulphide gasses that are released during deposition. The pH also differs between limestone karsts and gypsum karsts; the pH of limestone karsts exceeds 8, while that of gypsum karsts is generally lower than 5. When pH levels fall below 3, concentrations of heavy metals, especially Al, Mn, Fe, Cu and Ni, may become elevated.

With rare exceptions, such as the blue-green alga, *Synechococcus* (found in Japanese hot springs at 73 °C), only bacteria are present above 70 °C. Both
14c☞ chemoautotrophic and heterotrophic bacteria are able to flourish at 90 °C and grow up to 94 °C (Fig. 10.2). Species of sheathed bacteria, such as *Leptothrix,* are found at 75.5 °C in Tibet hot springs. Blue-green algae species of *Synechocyetis, Mastigocladus* and *Phormidium* and the photosynthetic bacteria of the genus *Chloroflexus*, occur up to 60 °C, and *Onconema*, *Spirulina* and *Oscillatoria* between 60 and 55 °C. Fauna do not begin to appear until about 50 °C, with the ostracod, *Cypris balnearea*, and the mite, *Thermacarus nevadensis*, tolerating up to 51.5 °C. In general, adult stages can tolerate higher temperatures than the young or larval stages. For example, larvae of *C. balnearea* are able to tolerate only 45 °C.

The diversity of flora and fauna rises sharply below 45 °C. Green algae begin to appear between 45 and 40 °C, with species of *Spirogyra* and *Mougeotia* and the diatoms, *Achnanthes* and *Pinnularia* especially common. Among the first faunal organisms to occur below 50 °C are the stratiomyid fly, *Hedriodiscus truqii* (47 °C) in western North America and the ephydrid fly, *Scatella thermarum*, and the chironomid, *Cricotopus,* in Iceland hot springs (47.7 °C). An increasing number of species within several taxa occur as temperatures fall to 30 °C and oxygen levels rise to near saturation. They include low oxygen-tolerant Nematoda, Oligochaeta, Ostracoda and Odonata and red blood worms

(*Chironomus*), beetles that utilize atmospheric oxygen, such as species of Dytiscidae and Hydrophilidae, dipterans with respiratory tubes, including Ephydridae and Stratiomyidae, and several pulmonate snails, like *Stagnicola*, *Melanopsis* and *Paludestrina*.

INTERMITTENT STREAMS

15☞ Intermittent streams may occur in high latitudes and altitudes, temperate latitudes and tropical latitudes. They are not common below impoundments because many reservoirs are created, in part, for low-flow augmentation purposes.

The physical and chemical characteristics of intermittent streams reflect those of the stream during the wet season. However, as the stream dries up, the water chemistry increasingly emulates that of the substrate under and adjacent to the stream. During the summer period, there is an increase in the magnitude of most parameters of the water, including: (i) temperature, to the point that it may be lethal to some organisms; (ii) conductivity, since evapouration of water increases the concentrations of ions; (iii) the concentrations of some dissolved gases, especially carbon dioxide and nitrogen, as low oxygen conditions develop. Conversely, dissolved oxygen levels gradually fall as the solubility of oxygen decreases with increasing temperature. Also, as temperatures increase, there is an increase in the metabolic activities of aerobic bacteria which consume oxygen and respire carbon dioxide.

There are four primary sources of biota for intermittent streams: (i) drift from upstream; (ii) upstream migration of animals; (iii) aerial sources (e.g. insects, ephippia of cladocerans); (iv) aestivating organisms (Fig. 10.4). The uniqueness of intermittent streams lies mostly in the kinds of aestivating organisms and their adaptations to surviving long periods of desiccation. The adaptations of organisms to life in intermittent aquatic habitats are described in Chapter 8 and need not be repeated here.

ARTIFICIAL IMPOUNDMENTS

In the construction of an artificial impoundment, a riverine sytsem is
superimposed upon a terrestrial one. The soils and geologic structure of the terrestrial
16☞ component contributes to the chemical character of the impoundment. The topography
affects the physical character, usually creating an impoundment with a dendritic pattern.
The flooded vegetation provides the first source of organic material in the sediments and
contributes to the fertility of the new ecosystem.

The rivers provide the initial materials for the new ecosystem, the water, the flora and the fauna. The result is an artificial lake ecosystem that develops its own limnological and physical characteristics and its own diverse populations of phytoplankton, zooplankton, macrophytes, benthos and fish.

Initially, the new ecosystem is an unsteady, undefinable state which, with time,

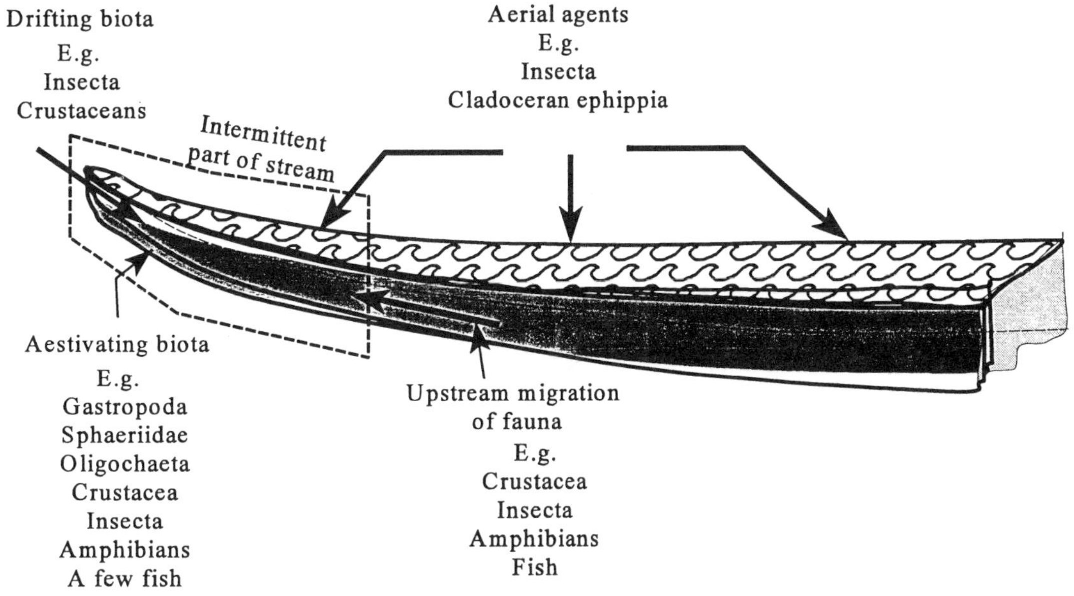

Figure 10.4. Sources and typical kinds of biota in intermittent streams.

becomes stabilized into a definable reservoir ecosystem that shares lotic and lentic characteristics (Fig. 10.5). The new water body has many predictable changes in water quality and biota, but the rate of change is dictated by four indices of the water balance regime: (i) the retention time of the lake, the internal flow regimes and degree of thermal resistance to mixing; (ii) the ratio of amount of precipitation on the surface to total inflow; (iii) the ratio of groundwater inflow to total inflow; (iv) the ratio of evapouration and evapo-transpiration from the lake water to the total discharge fluxes. Every impoundment has its own set of indices, hence every impoundment behaves differently.

The sequence of biological events that occur following the filling of the impoundment generally follows several principles (Baxter 1977, the quotes below are his)[6]. The principles are illustrated in Fig. 10.5.

Thienemann's Rules: There are three rules, all defined in Chapter 8. They predict that the numbers of species in a biological community vary directly with the diversity of
17a☞ habitats and with time, such that the longer a habitat has been in the same condition, the richer is the biological community.

---

[6]Baxter, R. M. 1977. Environmental effects of dams and impoundments. Annual Review of Ecology and Systematics 8: 255-283.

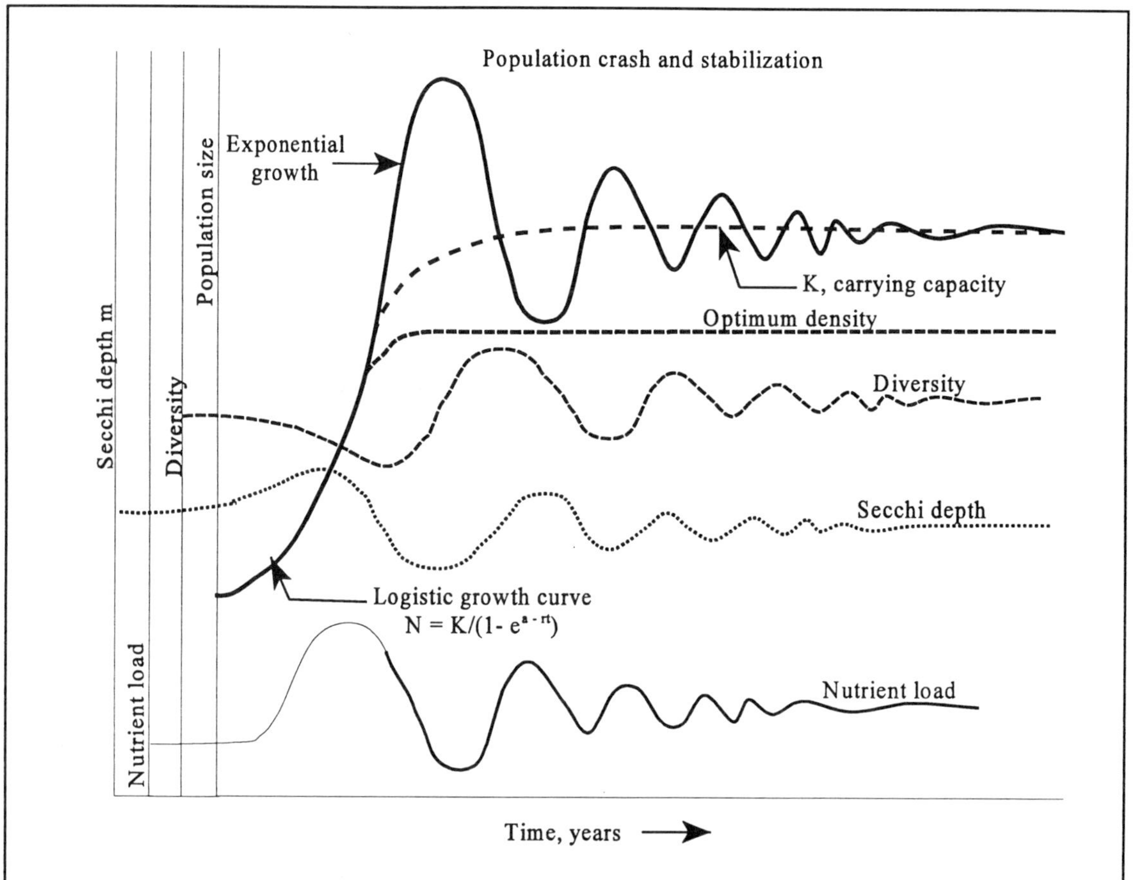

Figure 10.5. Predicted changes in production, community diversity, nutrient loading and Secchi depth in artificial impoundments over time.

The Concept of Succession: Originally described by Odum (1969)[7], the concept states that the development of any ecosystem, whether terrestrial or freshwater, follows a
17b☞ recognizable pattern. A balance is eventually approached between gross production and community respiration, with increases in species diversity and spatial heterogeneity, in the complexity of the food chains and in orderliness (i.e. a decrease in entropy, an increase in information content).

17c☞ **The Concept of Pulse Stability:** This concept, also Odum's, implies that the maintenance of an ecosystem (e.g. impoundment) in an immature state occurs by periodic, or quasi-periodic, impositions of physical perturbations (e.g. periodic flooding).

17d☞ **The Concept of Ecotone:** "An ecotone is the transition between two communities or environments. It is usually narrow and characterized by a greater diversity of species

---

[7]Odum, E. P. The strategy of ecosystem development. Science 164: 262-270.

than the communities or environments on either side". The littoral zone of the new impoundment may be regarded as an ecotone.

The Theory of Island Biogeography: The theory employs K-selection and r-selection traits, discussed in Chapter 14. K and r are parameters in the logistic equation,

17e☞ $$N = K/(1 + e^{a-rt})$$

where K represents the maximum population size possible (or the carrying capacity) within a habitat and r is the specific growth rate. N is the number of individuals in the population at time t and a = r/K. In the present context, the impoundment is considered the aquatic equivalent of an island. K-selected species tend to predominate when conditions stabilize; r-selected species are able to increase their numbers rapidly and predominate in the early stages of succession in the impoundment, exploiting the newly available habitats.

Inherent Physical Differences Between Lakes and Streams

There are several inherent physical differences:

18☞ 1. Streams are driven mostly by allochthonous inputs of energy and lakes mostly by autochthonous inputs. Artificial impoundments are a compromise, with allochthonous materials being provided at a much higher rate than in most lakes.
2. The greater contributions of allochthonous material is in part related to a high shoreline development inherent in a dendritic shape that characterizes impounded rivers. Lakes typically have many bays but are rarely dendritic in shape.
3. In streams, turbulent flow prevails; in lakes laminar flow prevails. Hence erosional processes predominate in streams and depositional process predominate in lakes. Artificial impoundments are "traps" for the sediment loads of parent streams, the greatest accumulation occurring near the dam. The ***trap efficiency*** (TE) of an impoundment is calculated as:

$$TE\ \% = ([\text{Total Suspended Solids}]_{outflow} / \sum[\text{Total Suspended Solids}]_{inflows}) \times 100$$

Inherent Chemical Differences Between Lakes and Streams

1. Turbulent flow enhances mixing processes, which ensures oxygen saturation at all depths, unless the water is grossly polluted. Laminar flow is conducive to the
19☞ thermal stratification of water, the hypolimnion becoming depleted of dissolved oxygen in eutrophic waters. However, even in eutrophic impoundments, the constant flow of turbulent water often replenishes dissolved oxygen supplies by limiting thermal stratification and lowering thermal resistance to mixing. The

amount of mixing in impoundments is affected greatly by the level of the outflow from the dam (Fig. 10.6).

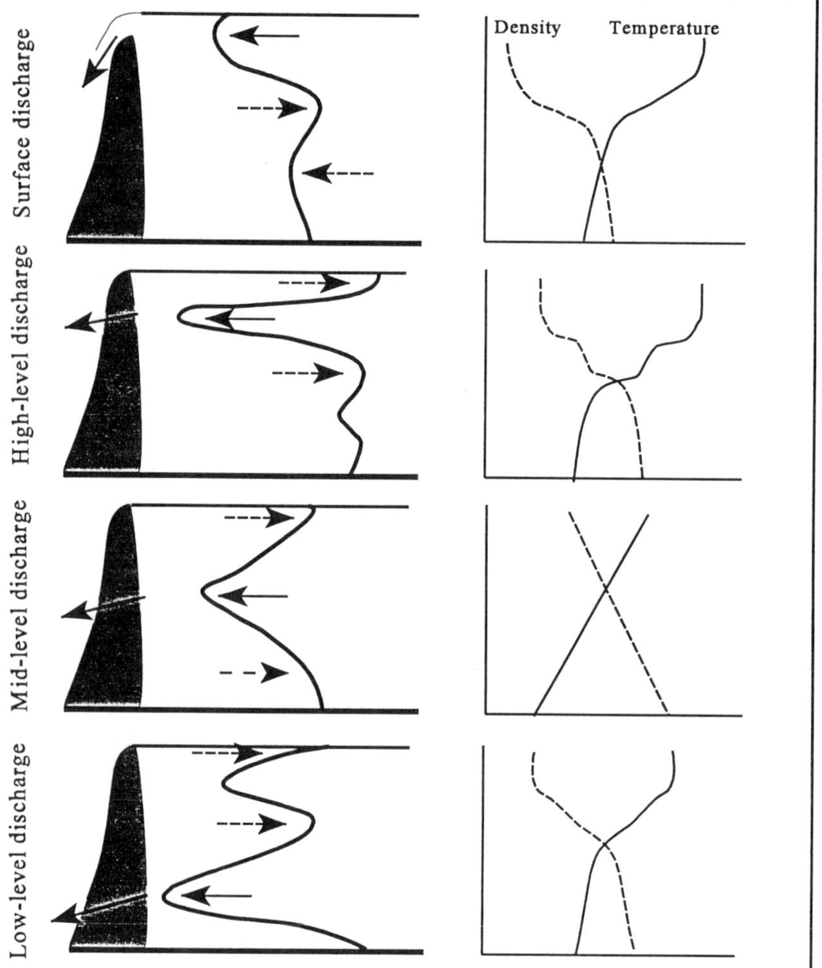

Figure 10.6. Effects of selective withrawal from different depths upon flow (left graphs; solid arrows indicate main flow through impoundment, dashed arrows indicate resultant currents), density and thermal profiles in stratified reservoirs.

2. Dissolved oxygen levels determine the availability of many nutrients, either directly, by chemical oxidation/reduction processes, or indirectly, by aerobic or anaerobic bacterial metabolism. Hence, the processes that occur in the different nutrient cycles differ between lakes and streams. Chapter 5 describes the different cycles in detail. Only a summary of the differences between lakes and streams is provided below.

(i) ***Nitrogen Cycle*:** Aerobic processes, such as nitrification, predominates in streams and either aerobic or anaerobic (e.g denitrification, nitrogen fixation) processes occur in the hypolimnion of lakes, depending on the amounts of dissolved oxygen.

(ii) ***Phosphorous Cycle*:** Phosphorous is mobilized from sediments under anoxic conditions. Hence, phosphorous mobilization rates increase as eutrophication proceeds. In the early stages of filling, phosphorous is released from the newly inundated sediments and provides the initial source of phosphorous for primary production in the impoundment. This allochthonous source is quickly replaced by autochthonous sources (e.g. moribund autotrophs, zooplankton, etc.).

(iii) ***Carbon Cycle*:** Most of the carbon in streams, especially headwater streams, is of allochthonous origin. Coarse particulate organic matter (CPOM) predominates in headwaters and fine particulate organic matter (FPOM) in lower reaches. In lakes, CPOM is most abundant in the littoral zone, FPOM in the profundal zone. Since water levels vary greatly in impoundments, littoral zones are poorly developed, hence FPOM predominates.

(iv) ***Sulphur Cycle***: The major source of sulphur is allochthonous in both lakes and streams, bedrock (e.g. gypsum) being the main source. However, under oxidized conditions, sulphur-oxidizing bacteria convert reduced forms of sulphur into sulphate; and under anoxic conditions, sulphur-reducing bacteria convert sulphates into sulphur dioxide, hydrogen sulphide and sulphur.

(v) ***Iron Cycle*:** Again, most of the iron is of geological origin (e.g. pyrite). The supply of ferrous (Fe^{++}) and ferric (Fe^{+++}) iron is also determined by the level of dissolved oxygen. Ferrous iron is more soluble than ferric iron, the latter precipitating and settling on the bottom as $Fe_2(CO_3)_3$, often with phosphate adsorbed to it. In the presence of hydrogen sulphide, highly insoluble ferrous sulphide (FeS) is formed. Such reduced forms of iron can occur only under anoxic conditions and will form only in eutrophic impoundments.

(vi) ***Silica Cycle***: Silica (SiO_2) is also of geochemical nature. It is required by diatomaceous algae, some protists (e.g. *Difflugia*), sponges (for spicule formation) and some plants (e.g. *Equisetum*). Most of these organisms are typical of lakes and occur in lower reaches of streams. Once the organism dies, the silica settles to the bottom. Since silica is very resistant to decay, very little is mobilized from the sediments and most accumulates to form large deposits, especially in oceans.

Inherent Biological Differences Between Lakes and Streams

If algae are to survive in lotic environments they must be attached forms or else they will be carried away in the flowing waters. Since most streams are rather shallow,
20☞ light is not a limiting factor. Most lacustrine algae are planktonic, where flagellated forms can swim toward their optimal light intensities. Periphytic forms are less prominent in lakes because light becomes a limiting factor with increasing depth.

Macrophytes also must be imbedded; only rooted, submersed and emergent forms can persist in streams. Rootless submersed forms and floating macrophytes (e.g. duckweed) can persist only in standing waters. Riverine forms tend to have elongate leaves, or leaves that offer little resistance to flow (see Chapter 6). If plants with highly dissected leaves (e.g. *Myriophyllum*) do occur, they are stunted and/or are in backwater areas. Most lakes are represented by plants with a great diversity of form and habit.

Just as planktonic algae cannot persist in streams, nor can zooplankton. Only pools and large rivers have plankton, but the greatest diversity and abundance occurs in lakes.

Stream organisms are adapted to life in flowing waters and lake organisms to life in standing waters (see Chapter 8 for details). In streams, most organisms can afford to sit and wait for their food and oxygen because they are continuously supplied in the flowing waters. Most lake organisms must actively go after their food or actively renew their oxygen supplies, for example by undulations of the body and/or gills or else be adapted to the lower oxygen levels.

Effects of Impoundments on the River Continuum Concept (RCC)

According to the RCC, the physical, chemical and biological attributes change with increasing distance downstream. The changes that would occur in the impoundment would depend on the point of insertion of the dam in the continuum. The retention time
21☞ of the new environment is almost infinitely greater than in a flowing one, the units being seconds or milliseconds in streams and days, weeks to months, or even years in larger impoundments. That is, organisms are continuously being bathed by a fresh supply of water in streams, but water movements are very slow to stagnant and organisms must create their own currents to renew their supplies of oxygen and nutrients and remove toxic waste products in lake ecosystems. The differences between the river and lake environments lessen with distance downstream. For example, if a dam is inserted at stream order 3, the differences in the physical, chemical and biological processes between the stream and the impoundment would be much greater than if a dam is inserted at stream order 6. In other words, the insertion of a dam in a stream greatly accelerates the rates of events and processes that occur in the stream continuum. Some of the more obvious effects of a dam on physics, chemistry and biology in the river continuum are described and illustrated below (Figs. 10.7-10.9). The illustrations assume a deep reservoir that thermally stratifies in the summer and the downstream effects are due to
22☞ bottom-level discharge. Effects on the RCC can be determined by comparing Fig. a with Fig. b for each attribute with increasing distance (stream order) downstream. Depth profiles for the various attributes are shown for the impoundment, water quality at the lower depths being discharged and reflected immediately downstream.

Effects on The Physical Continuum

Fig. 10.7 illustrates the changes that occur in temperature and turbidity, as
22a☞ predicted by the RCC. In the winter, warmer water (~4 °C) than the inflow (0-1 °C) is discharged; in the summer, colder water (~4 °C) than the inflow, which is close to air temperature, is discharged from the dam.

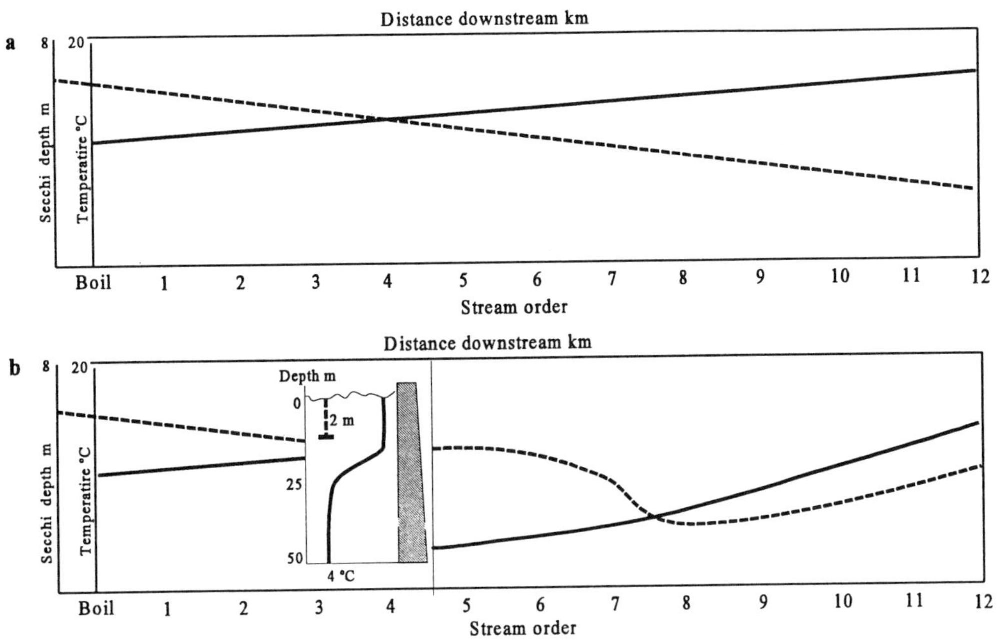

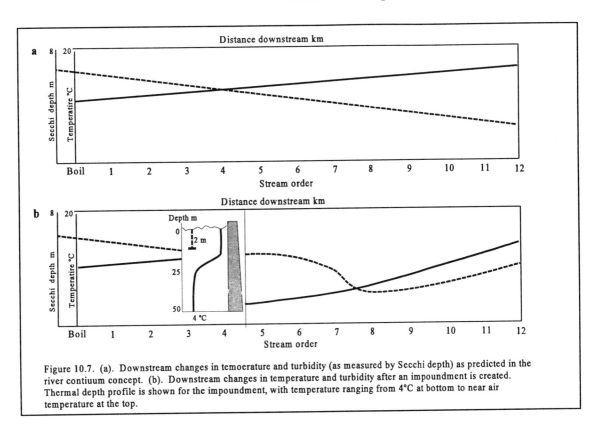

Figure 10.7. (a). Downstream changes in temoerature and turbidity (as measured by Secchi depth) as predicted in the river contiuum concept. (b). Downstream changes in temperature and turbidity after an impoundment is created. Thermal depth profile is shown for the impoundment, with temperature ranging from 4°C at bottom to near air temperature at the top.

The Effect on turbidity is illustrated as changes in Secchi depth. In practice, Secchi depths cannot be taken in streams but theoretically, the inflowing water would have a Secchi depth well above 5 m because most dams are not built on polluted rivers. Clearly, the impoundment is a sink for suspended solids, the water in the discharge being clarified by deposition of the solids on the bottom of the impoundment. However, the clarity of the water leaving the impoundment is rarely as good as the water entering. Plankton and other autochthonous materials are discharged before they settle out. Water clarity decreases downstream because turbulence erodes the sediments, but as water velocity decreases, some suspended solids settle out, resulting in an increase in water clarity again.

Effects on The Chemical Continuum

Impoundments dramatically alter the chemical conditions, mostly by affecting the concentration of dissolved oxygen, which directly or indirectly affect the concentration
22b☞ and availability of many of the nutrients. As water rises from seeps, the concentration of dissolved oxygen is low, but by atmospheric diffusion, the levels are brought to 100% saturation within a few meters downstream and are maintained until respiration rates exceed primary production rates in the higher stream orders (Fig. 10.8 a, b). Because most impoundments are mesotrophic to eutrophic, dissolved oxygen levels in the hypolimnion quickly become depleted or limited. Hence the water discharged is usually well below saturation, even in winter when the solubility of oxygen is near maximum.

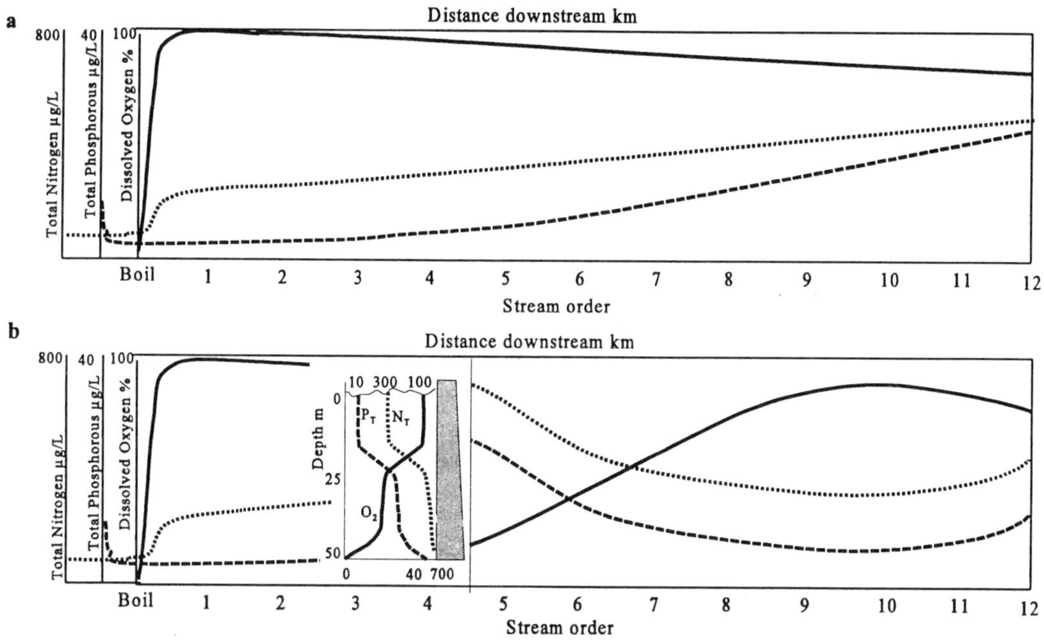

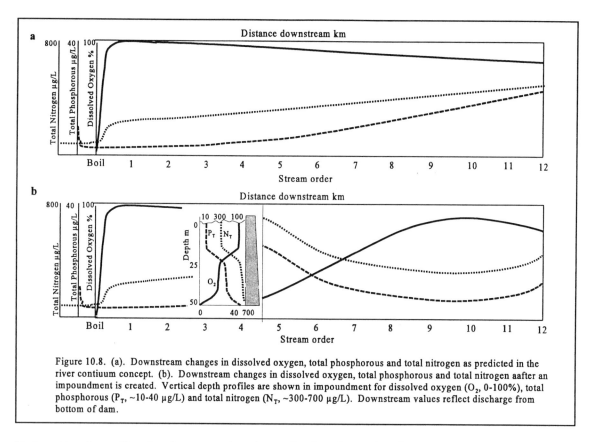

Figure 10.8. (a). Downstream changes in dissolved oxygen, total phosphorous and total nitrogen as predicted in the river contiuum concept. (b). Downstream changes in dissolved oxygen, total phosphorous and total nitrogen aafter an impoundment is created. Vertical depth profiles are shown in impoundment for dissolved oxygen (O_2, 0-100%), total phosphorous (P_T, ~10-40 µg/L) and total nitrogen (N_T, ~300-700 µg/L). Downstream values reflect discharge from bottom of dam.

Below the dam, dissolved oxygen levels again rise through primary production and air entrapment during turbulent flow.

The levels of nutrients, especially phosphorous, are greatly affected by the dissolved oxygen concentration. Under anoxia, phosphorous is mobilized from the sediments (Fig. 10.8 b) and discharged, only to be followed by immediate increases in primary production (Fig. 10.8 b). Autotrophic production in the impoundment also increases the total nitrogen levels being discharged. Nitrates and sulphates are microbially reduced to toxic ammonium and nitrite and hydrogen sulphide. Once dissolved oxygen levels are restored downstream, the reduced forms of nitrogen and sulphur are microbially oxidized to nitrates and sulphates, respectively.

Effects on The Biological Continuum

Many of the physical and chemical changes are due to the activities and metabolic processes of organisms. The sudden increase in retention time created by an
22c☞ impoundment increases the times for bacterial metabolism of organic materials, for chemical reactions and for exchange processes to occur between the water and sediments. The net result is an increase in the nutrient levels which is followed immediately by an increase in primary production. Hence, the P/R ratio changes quickly from < 1 to >> 1. This results in an immediate decrease in the ratio of allochthonous

inputs/autochthonous inputs. Upstream of the impoundment, shredders predominate but the sudden increase in fine particulate organic matter by autotrophs results in a decrease in the numbers of shredders and an increase in the numbers of collectors. Hence, the ratio of shredders/collectors decreases within and immediately downstream of the impoundment (Fig. 10.9). Collectors are found in large numbers upstream as well, but the riverine forms are replaced by lacustrine forms. As soon as the impoundment is filled, there is a shift from an invertebrate fauna adapted to flowing waters (e.g. flattened, streamlined forms, some with marginal contact, may with grapples and/or hooks, etc.) to one adapted to life in standing waters (e.g. adapted for life in vegetation, burrowing forms, etc.). Similar changes occur in the fish community. Upstream, the fish community is dominated by benthivores and a few piscivores, but within the impoundment the fish community could be represented by herbivores, planktivores, benthivores, piscivores and omnivores, provided that there is a source of fish species for each of the feeding guilds. The dam itself is a barrier to upstream fish migration, unless a fish ladder or spillway is provided.

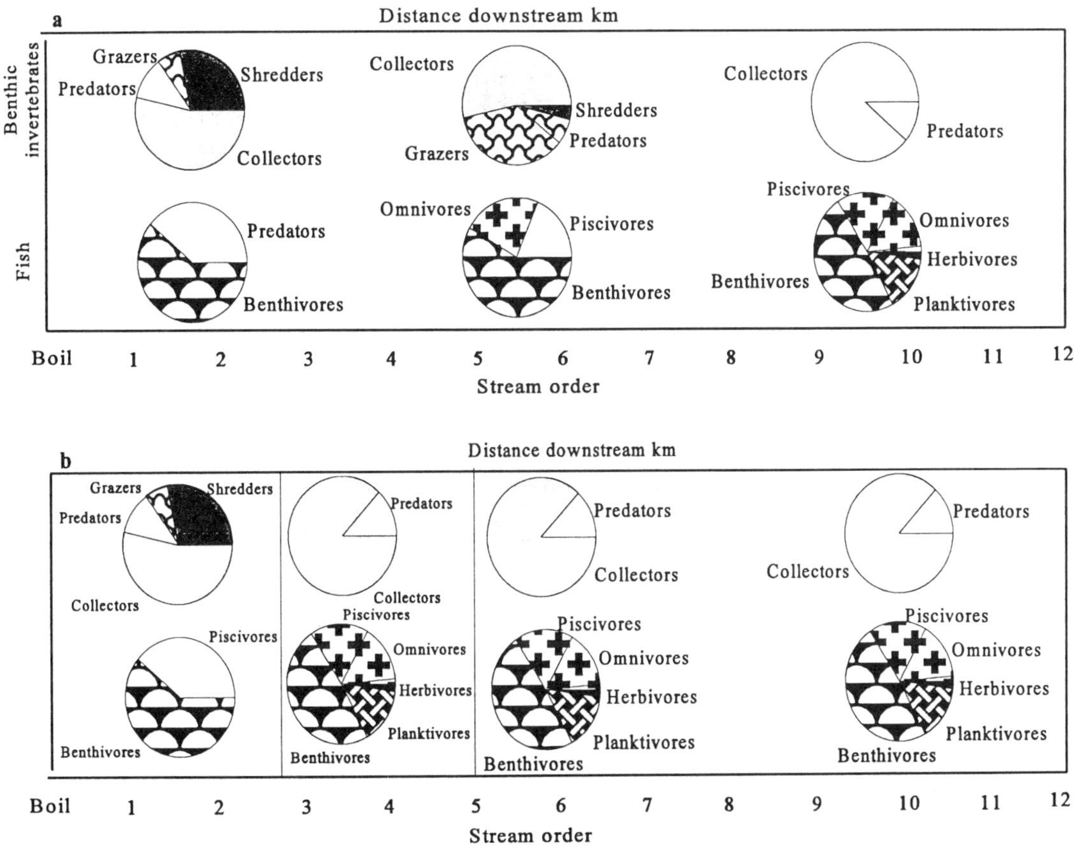

Figure 10.9. (a). Downstream changes in fproportions of functional feeding groups of benthic invertebrates and fishas predicted in the river contiuum concept. (b). Downstream changes in proportions of functional feeding groups of benthic invertebrates and fish after an impoundment is created.

CHAPTER 11

AQUATIC TOXICOLOGY

Why Read This Chapter?

During the period between 1960 and the early 1980s the chemical industries flourished and grew in proportion to the demand for pesticides and chemicals, including metals, for a plethora of uses. Indiscriminate use of the chemicals had resulted in the release of a variety of contaminants into aquatic environments, like the Great Lakes. The toxic effects were somewhat incipient at first, but as the concentrations increased and accumulated in organisms, the impacts became more noticeable. Dead fish began floating on the surface of the water, the water became discoloured and had a strong odour. Loss of species and changes in species composition, with only the most tolerant surviving (see Chapter 12), were direct evidence of the impacts. But what in the broth of chemicals was responsible? Was it just one chemical, or was it the mixture of chemicals that was toxic? The only certain way to identify the toxicant(s) was to perform laboratory toxicity bioassays on each of the chemicals and their mixtures. But, in order to accurately predict effects from one watershed to another and to determine the safe levels of toxicants to be released into a receiving water, the procedures for toxicity bioassays and the analytical methods had to be standardized. This enabled the toxicologists to provide criteria, objectives and standards (defined herein) that would protect the plants, invertebrates, fish, waterfowl, other aquatic life and humans.

While the objectives and standards may differ between states, provinces or even regions, the procedures and analytical methods are usually identical. Aquatic toxicology is now an exacting, applied science with strict rules and guidelines for procedures to be used. The terminology used in toxicity studies is mostly acronyms and terms unique to the discipline. This chapter begins with an examination of the terminology used and their definitions. The different kinds of toxic effects and the relationship between exposure and physiological effects are then reviewed. Methods for determining acute toxicity levels, times to 50% mortality and safe levels are then described. Finally, procedures for a full toxicity bioassay are described.

By the end of this chapter you will know:

☞ 1. The terminology used by aquatic toxicologists, including:
 - ☞ a. Acute toxicity
 - ☞ b. Application factor

☞ c. Bioaccumulation (= bioconcentration)
☞ d. Biomagnification
☞ e. Chronic toxicity
☞ f. Cumulative toxicity
☞ g. Effective concentration (EC)
☞ h. Half life
☞ i. Incipient lethal level (ILL)
☞ j. Lethal toxicity
☞ k. Lethal concentration (LC)
☞ l. Lethal dose (LD)
☞ m. Lethal time (LT)
☞ n. Lowest effect concentration (LOEC)
☞ o. Maximum acceptable toxicant concentration MATC)
☞ p. No effect concentration (NOEC)
☞ q. Resistance
☞ r. Safe level or safe concentration
☞ s. Sublethal level or sublethal toxicity
☞ t. Tolerance

☞ 2. The difference between direct and indirect toxic effects
☞ 3. Difference between local and systemic effects
☞ 4. The relationship between exposure and physiological effects
☞ 5. How to alter the tolerance of a contaminant to an organism
☞ 6. The difference between cause-effect and dose-response relationships
☞ 7. The uses and purposes of toxicity bioassays
☞ 8. How to perform an acute toxicity bioassay
☞ 9. How to determine lethal concentrations
☞ 10. How to determine the time to 50% mortality
☞ 11. How to determine the incipient lethal level of a toxicant
☞ 12. How to determine the safe concentration of a toxicant
☞ 13. What application factors are and how to use them
☞ 14. How to perform a sublethal test
☞ 15. The proper procedures for a full toxicity test, including use of proper:
  ☞ a. Dilution water
  ☞ b. Test organisms
  ☞ c. Test conditions
  ☞ d. Endpoints
☞ 16. The different kinds of toxic effects to expect from mixtures of toxicants

Definitions

The following terms appear frequently in this chapter and in the toxicological literature. They are listed alphabetically.

Acclimation: An organism's steady-state compensatory adjustment to changes in environmental conditions. It is a behavioural and/or physiological adaptation to experimental or artificial conditions that are only temporary in nature.

Acclimatization: An organism's normal seasonal or climatic compensatory adjustment.

1a☞ ***Acute toxicity:*** Coming quickly to a crisis, usually within 96 h.

Acute:Chronic Ratio (ACR): An indicator of toxicity, where ACR = LC_{50}/NOEL (both numerator and denominator are defined below). Ratio is commonly used to assess toxicity of an effluent. Alternatively, the NOEL can be estimated if the LC_{50} and ACR are known.

1b☞ ***Application factor (AF):*** A factor used to predict safe concentrations of a contaminant under field conditions. It is intended to provide an estimate of the relationship between a contaminant's chronic and acute toxicity and is calculated as the ratio of the MATC to LC_{50} values, or AF = MATC / LC_{50} (see below for definitions of these terms). Application factors vary from 0.0001 for persistent or accumulative compounds to 0.9 for compounds with short half lives and low toxicities. It is assumed that the AF is relatively constant over a range of test species. Hence, the AF derived from species A and the LC_{50} derived for species B could be used to estimate the MATC for species B, or $AF_{Sp.\,A} \times LC_{50\,Sp.\,B} = MATC_{Sp.\,B}$. It should be used only as an approximation of a safe level when an MATC value is unknown.

1c☞ ***Bioaccumulation (or Bioconcentration):*** The accumulation in body tissues of contaminants from the water.

1d☞ ***Biomagnification:*** The progressively increasing amounts of contaminants in body tissues of animals up the food chain. For example, carnivores will contain more than benthivores, which will contain more than planktivores, which will contain more than autotrophs.

1e☞ ***Chronic toxicity:*** Continuing for a long time, usually over the life span of the individual.

1f☞ ***Cumulative toxicity:*** Brought about, or increased in strength, by successive additions.

1g☞ ***Effective concentration (EC):*** Term used when an effect other than death is being used as the end point. For example, respiratory stress in fish, valve gaping in bivalves or byssal detachment of byssate mussels are end points for EC. Again, 50% mortality is usually used for the end point and an EC_{50}.is given for the result.

1h☞ ***Half life:*** The time required for a compound to degrade to half of its original level.

1i☞ ***Incipient lethal level (ILL):*** The concentration at which acute toxicity ceases. Organisms will eventually die above this value and live indefinitely below this value. It is usually taken as the concentration at which 50% of the population can survive for an indefinite time.

1j☞ ***Lethal toxicity:*** Causing death by direct action.

1k☞ ***Lethal concentration (LC):*** The concentration that causes death to a certain percentage of organisms. Usually 50% mortality is sought after and the lethal concentration is expressed as LC_{50}. The time of exposure is also important; for fish and benthic invertebrates, a 96 h exposure is used; for zooplankton, because of their short life spans, a 48 h exposure is used. Hence, the LC_{50} is usually preceded by the time exposure, e.g. a 96-h LC_{50}, a 48-h LC_{50}, etc.

1l☞ ***Lethal dose (LD):*** A dosage (e.g. applied orally or hypodermally) that causes death directly.

1m☞ ***Lethal time (LT):*** Time to mortality. Usually taken when 50% of the organisms are dead and is expressed as LT_{50}.

1n☞ ***Lowest Effect Concentration (LOEC):*** Concentration at which some significant deleterious effect is observed.

1o☞ ***Maximum acceptable toxicant concentration (MATC):*** The concentration of a toxic substance that does not harm the productivity and use of the receiving water. It is generally determined from chronic toxicity tests. The MATC is usually reported as being greater than the NOEC (see below) but less than the LOEC, or

$$NOEC < MATC < LOEC.$$

1p☞ ***No Effect Concentration or Level (NOEC or NOEL):*** Concentration of a contaminant below which no lethality is observed.

1q☞ ***Resistance:*** Time required for 50% of the population to respond to a specific concentration. The LT_{50} is a measure of an organism's resistance to a chemical.

1r☞ ***Safe level or concentration:*** The maximum concentration of a toxic compound that has no observable effect on a species after exposure over at least one generation.

1s ☞ ***Sublethal level or toxicity:*** Below the level that directly causes death.

1t ☞ ***Tolerance:*** The concentration required for response of 50% of the population. The LC_{50} is a measure of an organism's tolerance to a contaminant.

Toxicants

A toxicant is an agent that can produce an adverse effect in an organism sufficient to cause damage to its structure and functioning or to terminate in death. Toxicity can be
2 ☞ direct or indirect. ***Direct toxicity*** results from the toxic agent acting more or less directly at the site(s) of action in and/or on the organism. ***Indirect toxicity*** results from the influence of changes in the physical (e.g. decreased water clarity for plants), chemical (e.g. dissolved ion content) and/or biological (e.g. quality and quantity of food) environment. The toxic effects may be immediate or delayed until some time after exposure. This is determined by several factors, including the mode of toxic action, the chemical properties of the toxicant and the ability of the organism to metabolize and excrete the toxicant.

Some toxic effects are reversible and others are irreversible. All organisms have some ability to maintain their health through normal homeostatic mechanisms. If exposure to the toxicant continues and forces the organism to function outside this normal range, the organism will begin to experience some disturbed functioning but is often able to compensate by repairing any damage and reverse the effect(s). Failure results when exposure continues and the organism cannot reverse the effects. Under such circumstances the organism becomes weakened by disease(s) and ultimately dies. The sequence of physiological events that occur during exposure of an organism to a toxicant is shown in Fig. 11.1.

There are two kinds of effects, depending on the general site of action of action. ***Local effects*** occur at the primary site of contact, for example, skin or gill inflammation).
3 ☞ ***Systemic effects*** occur when the toxicant gains access to the internal medium of the organism, such as through the circulatory system. They require absorption and distribution of the toxicant to a remote site from the original contact or entry site. The toxicity of certain chemicals may also be ***selective*** or ***non-selective***. Selective chemicals affect only one type of tissue, cell or process. Organophosphate insecticides are selective chemicals, acting on the nervous system of the target organism. Molluscicides that act on the respiratory system are also selective chemicals. Non-selective chemicals affect the membranes of numerous cells and tissues of organisms. They induce ***narcosis*** (anaesthesia), a general disruption of cell membrane functioning. Many industrial solvents are non-selective chemicals. Chapter 12 describes several kinds of selective and non-selective chemicals.

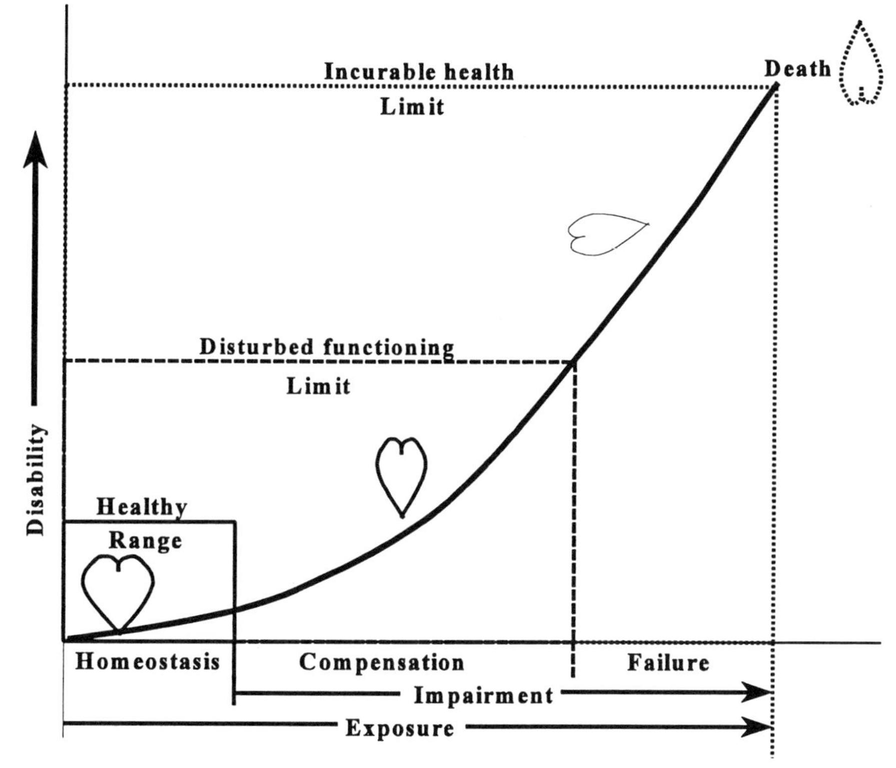

Figure 11.1. Relationship between exposure to a toxicant and physiological impairment.

Relative Toxicity

Often the toxicity of compounds is described as "extremely toxic", "slightly toxic", "harmless", etc. Table 11.1 gives the ranges of values for assessing the relative toxicity of compounds. The values should be used only as a guide for describing the relative toxicity of compounds and not be used as standards of toxicity.

Toxic Agents and Their Effects

Although toxic agents can be classified according to their use (e.g. pesticide), effects (e.g. cancer, mutation), physical state (e.g. liquid), or toxicity potential (e.g. extremely toxic, slightly toxic), they are classified below according to their general chemical features, mostly for convenience. They are divided into toxicants and pollutants, the former affecting the organism directly or indirectly and resulting in

Table 11.1. Values for the relative toxicity of a compound.

Relative Toxicity	EC_{50}/LC_{50} µg/L	LD_{50} mg/Kg
Super toxic	<0.01	<5
Extremely toxic	0.01 - 0.1	5 - 50
Highly toxic	0.1 - 1.0	50 - 500
Moderately toxic	1.0 - 10.0	500 - 5,000
Slightly toxic	10 - 100	5,000-10,000
Practically non-toxic	100-1000	10,000-15,000
Relatively harmless	>1000	> 15,000

impaired structure and function, or death of organisms, the latter affecting living resources, including commercial and sport fisheries, human health, and the quality of water for its intended use.

Tolerance to Toxicants

Prior exposure (i.e. acclimation) of an organism to a toxicant can greatly alter the toxicity. For example, an organism's tolerance to copper or zinc can be increased by exposing it to the metal in small increments. The metals induce the formation of
4☞ metallothioneines which bind up metals that are then stored in the liver and gills. However, metallothioneines can only be used once by an organism and any further increase in tolerance would be curtailed. In contrast, other metals, like nickel, may lower an organism's tolerance, for example, by causing gill damage. In addition, some metals, like nickel, do not stimulate the formation of metallothioneines.

Other mechanisms of tolerance are:

1. Increased metabolism and detoxification, for example by mixed function oxidase systems (MFOs).
2. By altering the site of action, for example by mutant acetylcholinesterase such as may occur in the presence of organophosphates that bind acetylcholinesterase to prevent breakdown at synapses.
3. By altering the penetration ability of compounds, such as changing the permeability of lipid membranes.
4. By enhancing excretion of the toxicants, as occurs with nicotine.
5. By altering an organism's behaviour.

Cross-tolerance occurs with some compounds. A decrease in the susceptibility to one pesticide or insecticide may alter the susceptibility to another. The compounds usually have a common tolerance mechanism.

Control of Tolerance

If an organism becomes tolerant of a certain concentration of a compound, its
5☞ tolerance can be altered by:

1. Decreasing the frequency of treatment.
2. Decreasing the extent of treatment, either by reducing the area of treatment or by reducing the percent of the population being treated.
3. Decreasing the persistence of a compound.
4. Avoiding the use of residual or slow-release formulations.
5. Treating the pest organism at only one life stage.
6. Using mixtures of compounds.
7. Replacing the chemical with new formulations.

Toxicity Testing

The most common goal of toxicity testing is to know the toxicity of a compound in relation to those of other compounds whose toxicities are known. The testing is often
6☞ done in response to observations that indicate either a ***cause-effect relationship*** or a ***dose-response relationship***. Cause-effect relationships can be direct or indirect and do not show a well defined relationship between amount of toxicant and the magnitude of the response (effect). Dose-response relationships exist when the magnitude of observed effects are related directly to the concentration of the toxic agent present in the ambient medium. Since the dose response relationship refers to doses applied orally or injected into the body, it is not well suited for aquatic toxicity testing. Instead, since exposure is usually via waterborne concentrations of the toxicant, aquatic toxicologists prefer the term, ***concentration-response***.

Bioassays are used to determine the toxicity of contaminants. The bioassay may be an acute, sublethal or chronic toxicity test. If nothing is known about the toxicity of a compound, an acute lethal bioassay is first performed in which the 96-h (usually) LC_{50} is determined. If the LC_{50} is known, a sublethal or chronic bioassay is performed to determine the MATC value. This allows one to estimate safe concentrations of the contaminant in a receiving water. Knowing both the LC_{50} and MATC will allow one to estimate application factors.

Uses and Purposes of Bioassays

7☞ The forgoing may imply that bioassays are used only for determining the toxicity of compounds. However, they have several other uses:

1 Regulate effluent discharges
2. Define the need for control or treatment of effluent
3. Monitor the effectiveness of controls
4. Compare the relative toxicity of effluents, processes and treatment
5. Compare the sensitivity of organisms
6. Examine modifying factors (discussed below)
7. To predict toxic environmental consequences of an existing or planned effluent discharge
8. Test for bioavailability

Bioassays are tests, not experiments. They are limited to a stated purpose. Some of these purposes are to:

1. Rank hazards
2. Set discharge limits
3. Regulate hazards
4. Protect important species
5. Protect ecosystem structure and function. This is probably an ideal purpose but is probably rarely realized because ecosystem structure and function are rarely fully understood.

Acute Toxicity Tests

Probit analysis is the standard method for determining a LC_{50} value. Probits are a
8☞ way of expressing standard deviations on a sigmoid curve with a normal distribution. This type of curve results when percent mortality is plotted against the concentration of a toxicant on ordinary arithmetic graph paper (Fig. 11.2a). The sigmoid shape can be "straightened by plotting the data on semi-log paper (Fig. 11.2b). The extremities of this ***concentration-response curve*** approach 0% at low concentrations and 100% at high concentrations, but theoretically never pass through 0% and 100%. The middle portion of the curve, between 16% and 84%, is linear and represents one standard deviation (SD) of the mean (50%). In normally distributed curves, ±1 SD = 68% of the concentration-response data, ±2 SD = 95.5% of the concentration-response data and ±3 SD = 99.7% of the concentration-response data. The units of deviation from the mean are called ***normally equivalence deviates (NED)***. The NED for a 50% response is 0, that for a 16 to 84% response is -1 to +1, that for a 4.5 to 95.5% response is -2 to +2, and that for 0.3 to 99.7% is -3 to +3 (Fig. 11.2c). But, if we add 5 to each of these deviates we do away

with negative values and end up with an axis that runs from 3 to 7 (Fig. 11.2d). These are the probits values and are used for estimating LC_{50} and LT_{50} values. Because probits are related to percentages, probit graph paper is usually available with probits on the left side and percentages on the right side of the y-axis. The x-axis is usually 3 log cycles. Probit/percentage-log paper can be purchased in almost any quality book store. Sometimes it is called "Probability scale × 3 cycle log paper".

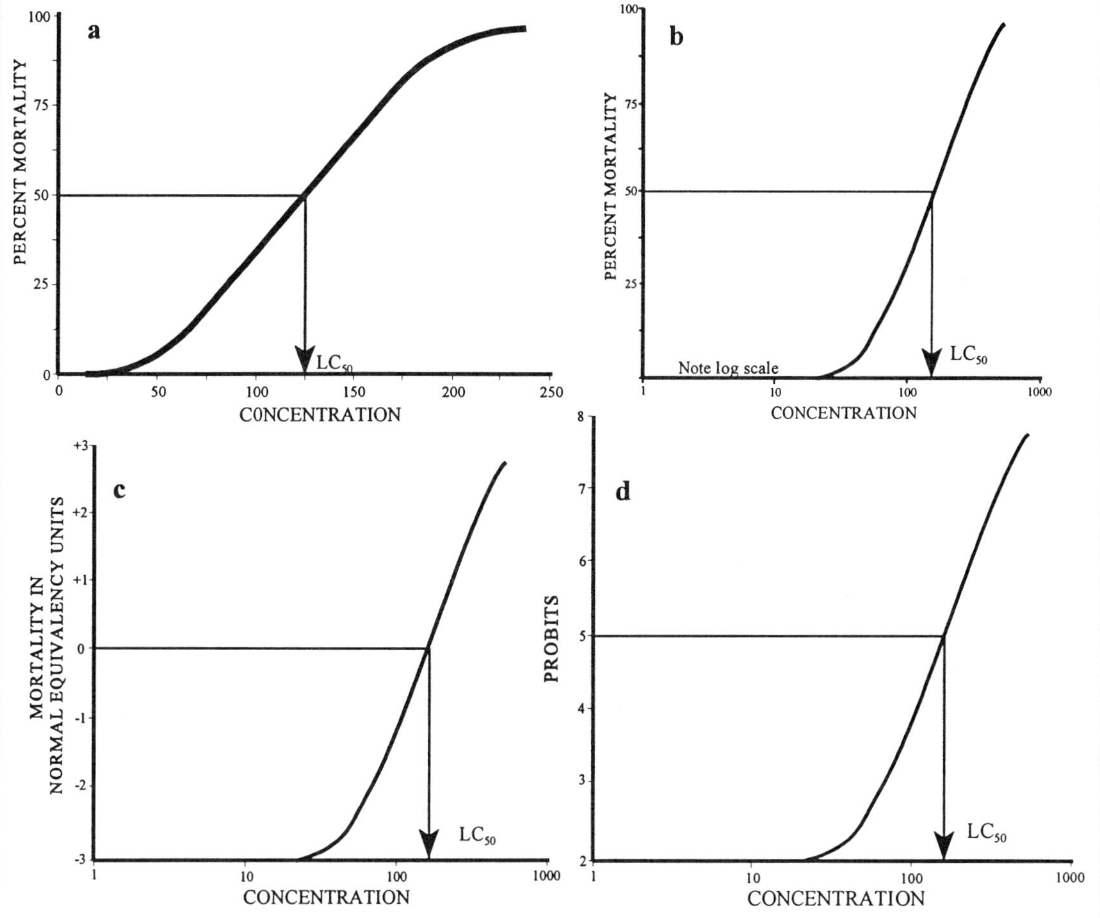

Figure 11.2. (a) Plot of percent mortality against concentration on arithmetic scales. (b) Same as (a) but x-axis is log transformed. (c) Same as (b) but percent mortality is converted to normal equivalency units (see text for explanation). (d) Same as (c) but normal equivalency units are converted to probits, as described in the text.

Here is an example of how to derive a LC_{50} value using probit analysis. Suppose a chemical manufacturer has developed a chemical that controls snails *(Biomphalaria glabrata*) that are intermediate hosts to a trematode that causes schistosomiasis but they now want to use it as a molluscicide to control zebra mussels. The molecular weight of the ***active ingredient*** (the component that is lethal to the molluscs) is 62.5. Only 50% of the compound is active ingredient so the total molecular weight of the compound is 125. The manufacturer wishes to know what concentration is needed to control (kill) zebra mussels and how long they would have to apply the chemical to achieve both 50% and 100% mortality. The 96 h LC_{50} of the chemical for *B. glabrata* is 0.5 mg/L.

First, perform a ***range-finding test*** to determine the minimum and maximum levels that should be tested in the acute lethal toxicity bioassay. Make five concentrations in a log series of, say 0.1, 1, 10, 100, and 1000 mg/L. A log series is preferred over a geometric series because organisms show clearer responses to concentrations that are ten times different than to concentrations that are only two or five times different. However, a geometric series does provide a more precise estimate once one has narrowed the lethal concentration to a level where a log series alone is no longer appropriate, as we shall see shortly. Prepare the solutions by making the 1000 mg/L solution first (e.g. 1000 mg of the compound is added to 1 L of water) and then diluting it 10 times to obtain the 100 mg/L solution; dilute this 10 times to obtain 10 mg/L; dilute the 10 mg/L to obtain 1 mg/L, and so on until all five concentrations are made.

Place 10 mussels in each bioassay vessel with 1 L of solution. Also run a control to ensure that the dilutant (e.g. well water with a pH of 8.0 and a hardness of 375 mg $CaCO_3$ /L) is not contributing to some mussel mortality. If all the mussels die in the control replace the dilutant with another solution and repeat the bioassay. If, however, only a few die (less than 10%), correct for control mortality using ***Abbott's formula***, discussed later. The usual end point for death in bivalves is gaping valves that do not close after gentle prodding.

Suppose after 96 h no mussels died in the control and 0.1 mg/L solutions, 1 died in the 1.0 mg/L solution and all mussels died in the 1000 mg/L solutions. In probit analyses it is desirable to have at least three or four solutions that have some mortality but not 100%; the more points on the graph, the more precise the LC estimate. Since some mussels were living in the 1 to 100 mg/L solutions, repeat the bioassay using toxicant levels between 2 and 100 mg/L. Make six solutions in a logarithmic and a geometric series, using 2, 5, 10, 20, 50, and 100 mg/L. Make three replicates of each solution and place 10 mussels in each replicate. Also run three replicates of a control with 10 mussels in each replicate. Record mortalities at 24 h, 48 h, 72 h, 96 h and every 24 h up to 7 days, or until all mussels are dead in the lowest concentration (2 mg/L). Normally, 96 h is sufficient, but since the manufacturer wants to know how long it will take to kill 100% of the mussels, let the bioassay go for 7 days, which is the usual period for a sublethal test, as discussed later. Table 11.2 gives the daily mortality of mussels in each concentration.

Determining the 96-h LC_{50}

The 96-h LC_{50} is determined by plotting the 96 h data in Table 11.2 on 3 cycle log-probit paper, as in Fig. 11.3. The line drawn through the data is of the form Probit Y
9☞ = log a + b log x, where a is the y-intercept, b is the slope of the line, and x is the concentration. To derive the LC_{50} simply substitute the probit value of 5 (= 50% mortality). For example, using the line drawn by eye, the LC_{50} value at probit 5 is approximately 11.1 mg/L. Using probit regression analysis (for those who have the

Table 11.2. Mortality of zebra mussels in 6 concentrations of a molluscicide. Numbers indicate total cumulative mortality out of 30 mussels (10 in each of 3 replicates).

Concentration mg/L	**Time h**						
	24	**48**	**72**	**96**	**120**	**144**	**168**
0 (control)	0	0	0	0	0	0	1
2	0	0	0	1	3	6	8
5	0	2	4	6	9	10	13
10	2	6	10	14	19	22	27
20	5	11	16	22	29	30	Ended
50	7	15	21	28	30	Ended	
100	9	18	25	30	Ended		

program or know how to perform regression analyses), the equation is, Probit Y = 2.468 + 2.423 log x. Since we wish to know the LC_{50} value (= x at probit 5, or 5 = 2.468 + 2.423 log LC_{50}), the equation to solve is, log LC_{50} = (5 - 2.468)/2.423, or log LC_{50} = 1.045. Since the log LC_{50} is its antilog, the LC_{50} is the antilog of 1.045, or 11.09, which is very close to the "eyeballed" LC_{50} estimate of 11.1 above.

Using regression analysis also allows one to provide a ***95% confidence interval* (CI)** for the LC_{50} estimate. The 95% CI indicates the range within which the LC_{50} estimate can be expected to fall 95% of the time. For the above probit analysis, the 95% CI of the LC_{50} estimate is 8.72 to 14.08. It can be estimated for all concentrations and be shown as ***confidence limits*** about the regression line. The confidence limits are shown as dashed lines on both sides of the concentration-response line in Fig. 11.3 and indicate the range of mortality to expect 95% of the time.

To derive the LC_{100} (actually $LC_{99.9}$ since, theoretically, there is no probit value for LC_{100}), substitute the probit value of 7.81 (= probit of 99.9%) into the equation. Using the same procedures above, but using a probit of 7.81, the LC_{100} is the antilog of 2.903, or 160.22 mg/L. Alternatively, one can "guestimate" the LC_{100} by drawing a straight line through the mortality plots to 98% mortality and obtain a value greater than 7.5 mg/L. This is considerably less than the calculated value because the graph paper has a maximum probit value of 7.1, a value much smaller than 7.81 used in the regression analysis. This illustrates the point that "eyeballing" LC values can lead to errors.

10☞ ***Determining the LT_{50} for Each Concentration***

Times to fifty percent mortality (LT_{50}) can be determined only for each

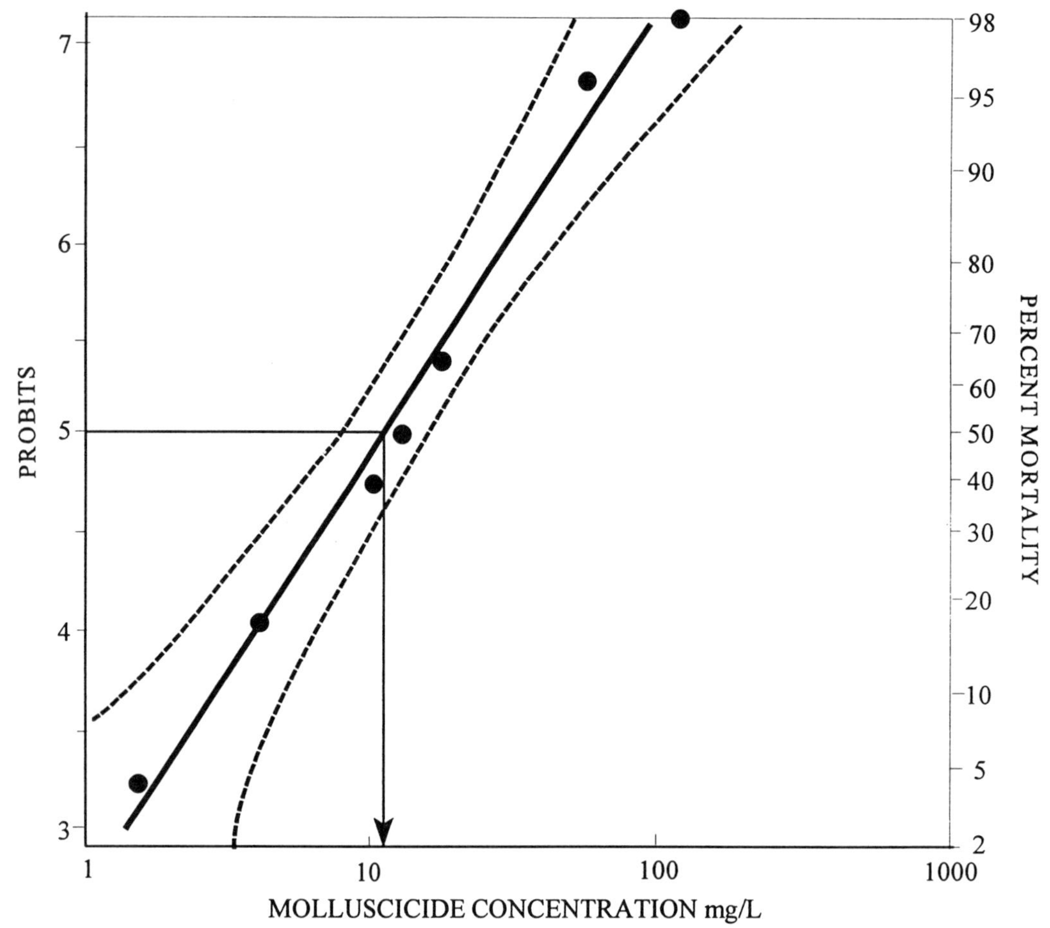

Figure 11.3. Estimating the LC_{50} of a molluscicide on zebra mussels using probit/log paper. The equation was determined from linear regression analysis of the probit-log data, as described in the text.

concentration in which there is close to 50% mortality for at least one of the time periods (24 h, 48 h, etc.). The best estimates for LT_{50} values are from tests in which there is more than one time period with mortality between 50% and 99%. The 95% CI for such LT_{50} values will be much smaller than for concentrations in which there is less than 50% mortality over the period of the bioassay. The LT_{50} is calculated the same way as described for LC_{50}, except time is on the x-axis instead of concentration (Fig. 11.4). Again, the LT_{50} can be estimated from a line drawn through the data points by eye, but more accurate estimates can be obtained by regression analysis which provides a line of best fit.

As one would expect, the LT_{50} decreases with increasing concentration. Table 11.3 shows the LT_{50} estimate for each concentration of molluscicide tested, and its 95% CI. Fig. 11.4 is a plot of the data in Table 11.3, yielding the time response line shown. The "eyeballed" fit yields a LT_{50} estimate of about 95 h. Probit-log regression analysis gives a value of 89.2 h with a 95% CI of 77.5 to 104.9 h (Table 11.2). Again, the

Table 11.3. Times to 50% and 100% mortality of zebra mussels and the confidence intervals (CI) in six levels of a molluscicide. Regressions are of the form, Probit Y = log a + b Log x.

Concentration mg/L	Probit Regression	LT_{50}	95% CI hours	$LT_{99.9}$	95% CI hours
2	Y = 4.89 log x - 6.46	221	176.8-850.8	855	424-17,420
5	Y = 2.44 log x - 0.63	203	156.7-419.5	3804	1290-58,414
10	Y = 3.01 log x - 0.87	89.2	77.5-104.9	962	608-1966
20	Y = 3.24 log x - 0.67	56.2	47.0-65.5	856	317-798
50	Y = 3.40 log x - 0.58	43.8	35.4-51.3	352	217-877
100	Y = 3.44 log x - 0.37	36.4	28.4-43.5	239	169-897

"eyeballed"estimate appears satisfactory because it is included within the 95% CI.

Note that the 95% CI is large for concentrations in which less than 50% mortality is observed (i.e. 2 and 5 mg/L). In fact, while the LT_{50} estimates for 2 and 5 mg/L look reasonable, the LT_{100} (actually $LT_{99.9}$) estimates look suspect because it takes about four times longer to kill all the mussels at 5 mg/L than it does at 2 mg/L! Relatively large 95% CIs would also occur for concentrations in which only more than 50% mortality is observed. However, if even one time period has less than 50% mortality, as for 100 mg/L at 48 h, Table 11.3, the 95% CI is fairly small. Clearly, one of the main causes of large confidence intervals in toxicity data is too few data points (i.e. small n values). Hence, the 95% CI increases with decreasing n values.

So far we have estimated 96 h LC_{50}s for bioassays in which there is no control mortality. If mortality exceeds 10% in the control, many investigators repeat the bioassay. However, one can account for control mortality by adjusting the mortality data with ***Abbott's formula***:

$$\text{Adjusted \% mortality} = \frac{\text{\% Test mortality - \% Control mortality}}{\text{100 - \% Control mortality}} \text{ X } 100$$

The formula is used more often for EC_{50} estimates than for LC_{50} estimates, in which case "response", such as a behavioural or physiological response, is substituted for "mortality" in the above equation. In estimating LC_{50}, the adjusted mortality values are used instead of the observed mortality values.

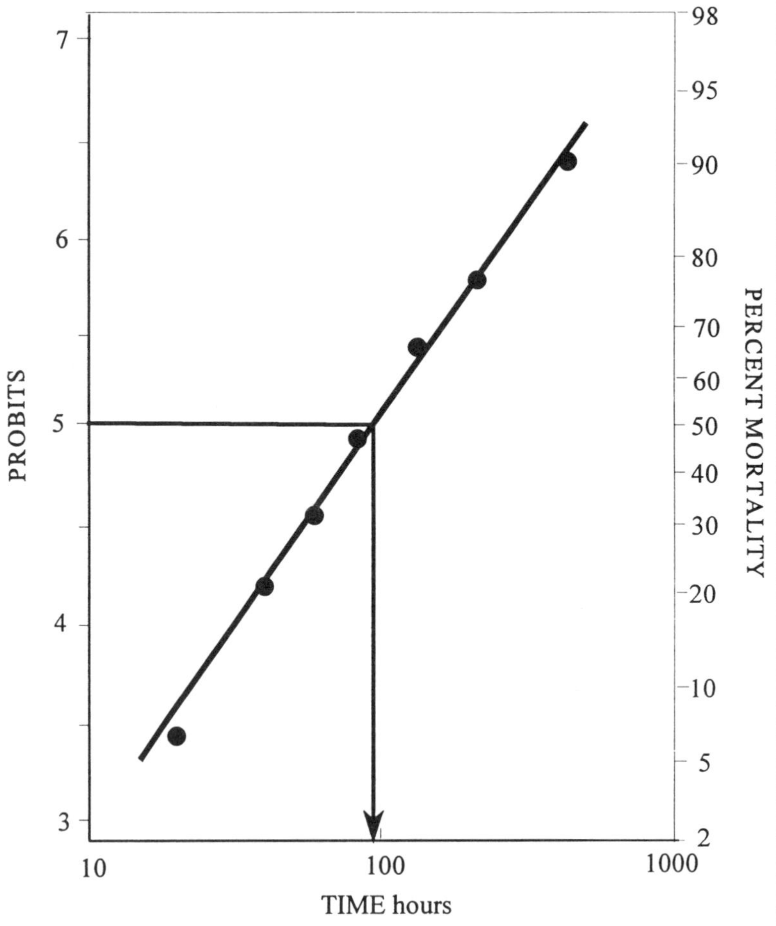

Figure 11.4. Estimating the LT_{50} of 1 mg/L of a molluscicide on zebra mussels using probit/log paper. The equation was determined from linear regression analysis of the probit-log data,as described in the text.

Toxicity Curves

When several bioassays have been performed using several exposure concentrations and periods, toxicity curves, such as those shown in Figs. 11.6 and 11.8,
11 can be drawn to determine the ***incipient lethal level (ILL)***. A minimum series of three or four logarithmic exposure concentrations spanning the expected toxicity range is recommended. The data for the two kinds of toxicity curves are derived either from exposure times or exposure concentrations. One is a plot of % mortality (probit transformed) against the logarithm of the exposure time for each concentration (Fig. 11.5). A toxicity curve (Fig. 11.6) can then be drawn using the median time to death (LT_{50}) for each exposure concentration, derived from the plots in Fig. 11.5. The other is a plot of % mortality (again probit transformed) against the logarithm of the exposure concentrations used for each time observation (Fig. 11.7). A toxicity curve (Fig. 11.8) can

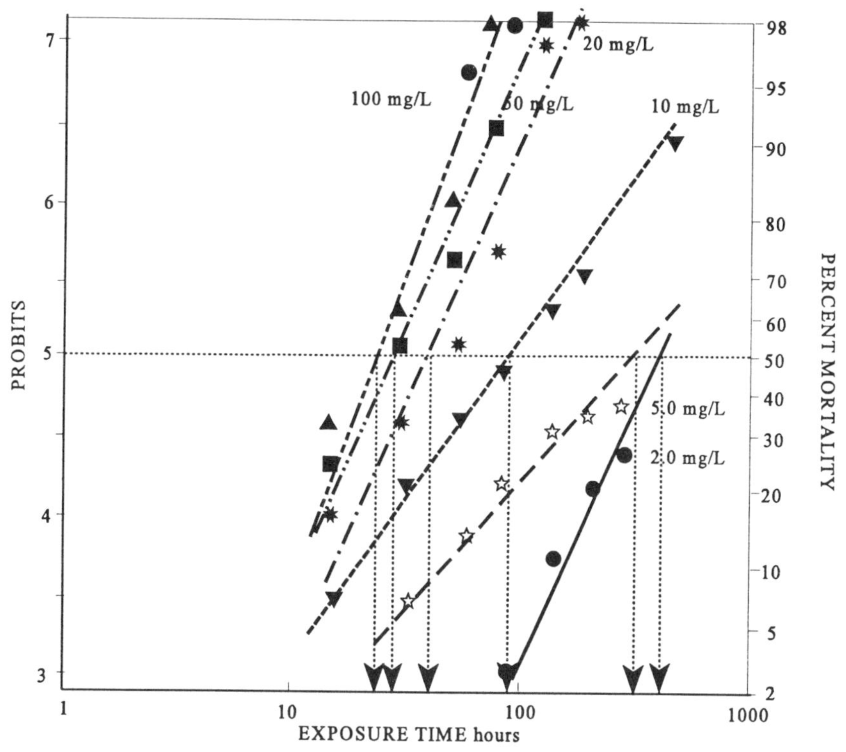

Figure 11.5. Plots of toxicity data for molluscicide in Table 10.7 showing plot of exposure time versus mortality at various exposure concentrations.

then be drawn using the inverse of the median effective concentration (i.e. $1/LC_{50}$) values and exposure periods, derived from the plots in Fig. 11.7. The ILL in each toxicity curve is the asymptotic value shown in Figs. 11.6 and 11.8. The ILL is sometimes referred to as the ***time-independent threshold value***. *If a sensitive life history stage of an organism is used, the ILL can be applied as an interim "safe concentration" until chronic bioassays confirm or modify the value.*

Chronic Toxicity Tests

A contaminant does not have to be acutely lethal to be toxic to aquatic organisms. Contaminants levels may be low enough not to be acutely toxic level but high enough to
12☞ impair an organism's physiology, behaviour, growth, development or reproductive ability. Chronic toxicity tests permit evaluations of the potential adverse effects of the contaminant under conditions of long-term exposure at sublethal levels. A ***full chronic toxicity test*** exposes the test organism to a chemical for an entire reproductive life cycle (e.g. egg to egg) to a minimum of five test concentrations of the chemical. Typically, the test exposure begins with a zygote or embryo, then proceeds through hatching of the embryo, growth and development of the newborn, and finally attainment of sexual maturity and reproduction to produce a second generation. Alternatively, the test may

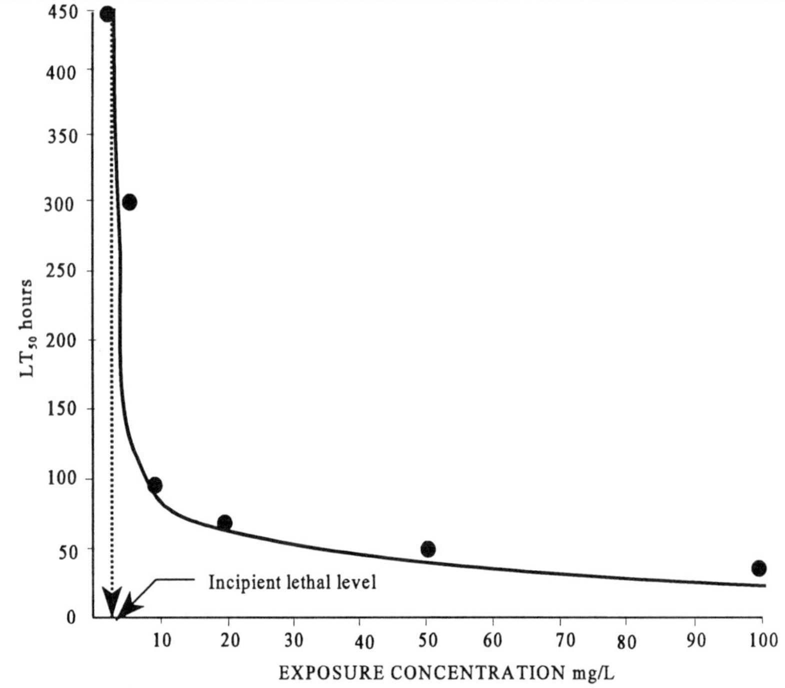

Figure 11.6 Toxicity curve showing plots of LT_{50} from Fig. 10-7.

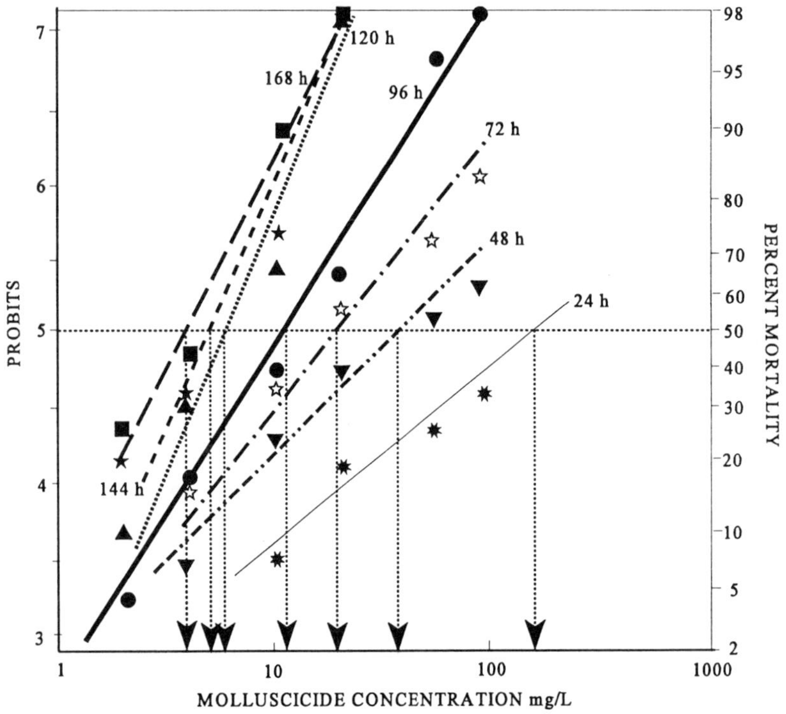

Figure 11.7. Plots of toxicity for the molluscicide data in Table 11.3 showing exposure concentration versus mortality at various exposure times.

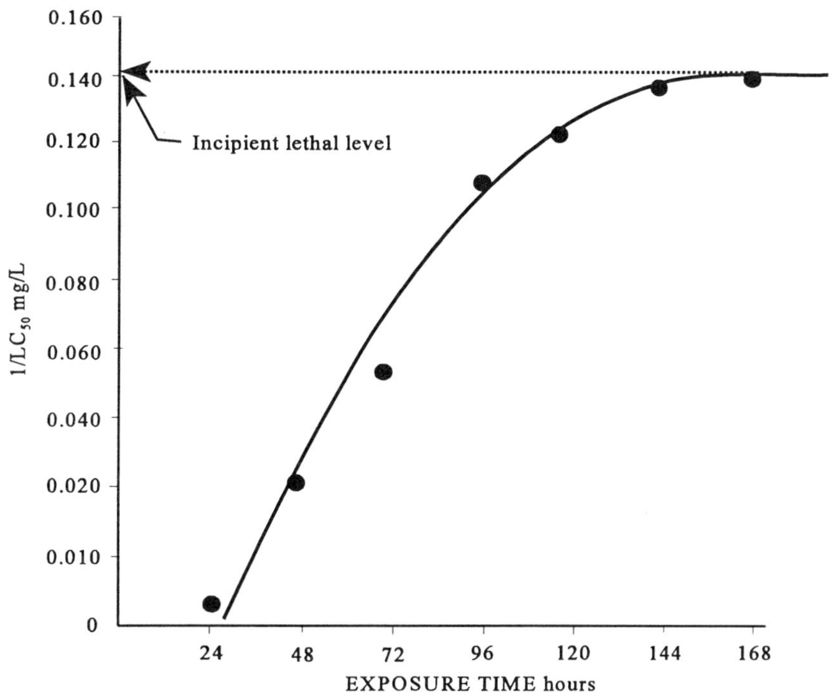

Figure. 11.8. Toxicity curve using data obtained from Fig. 11.7 with inverse of LC_{50} versus exposure time.

begin at any stage along the life cycle and end at the same stage. A ***partial chronic toxicity*** test involves only two or more sensitive life stages, at least growth and reproduction during the first year; early juvenile stages are usually omitted. The use of early life stages (e.g. egg, embryo, larva, fry) of aquatic organisms to various levels of a chemical for 1 to 2 mon has been shown to accurately predict the MATC value. Use of ***early life stage tests*** seems to be gaining increasing popularity and use because it generates MATC values much more quickly than full chronic toxicity tests.

Data obtained from either the full or partial chronic toxicity test are used to estimate the MATC, LOEC, NOEC, application factors (AF) and safe levels for aquatic
13☞ organisms. The AF for a given contaminant is presumed to be constant for a range of test species, but may range from about 0.90 to 0.00009, or by four orders of magnitude, depending on the chemical. Currently, water quality criteria for the protection of aquatic life are derived from MATCs.

Sublethal Effects Tests

Since acutely toxic levels of contaminants are not common in most aquatic environments, the effects on organisms are usually insidious and sublethal bioassays are needed to verify that the level of the chemical present will not affect the future survival of
14☞ the species. ***Short-term sublethal tests*** examine changes in growth, reproduction and survival over a 7-day period. Sublethal effects are generally divided into three classes: biochemical and physiological; behavioural; histological. Biochemical and physiological

effects include studies of respiration, enzyme inhibition, hematology and clinical chemistry. Since behaviour integrates complex biochemical and physiological functions and chemically-induced behavioural changes, it can be a sensitive indicator of sublethal effects. The most common behaviours studied are aggression and territoriality, locomotion and swimming, predator-prey relationships and learning. Changes in histological structure and occurrence of some pathologies usually indicate changes in tissue and/or organ functioning that ultimately affect the long-term survival of the species.

Toxicity Test Procedures

15☞ A detailed methodology is beyond the scope of this text and interested readers are encouraged to consult texts such as Rand (1995)[1] and references therein. Below is a "fast-track" description for performing a typical acute toxicity test.

Dilution Water

15a☞ The source of water for maintaining test organisms and for dilution of chemicals depends on the objectives of the study. For testing the inherent toxicity of a test chemical, a good freshwater supply from a well, spring or surface water is used. Dechlorinated tap water is used a last resort, dechlorination being achieved by aeration for 24 h, carbon filtration or use of sodium metabisulphite (1 mg/L reduces 1.5 mg chlorine/L). If the objective is to test the toxicity of an effluent, the dilution water should be representative of the receiving water. In this case, the dilution water should be collected from an uncontaminated source either upstream of the outfall or outside the zone of influence.

Test Organism

15b☞ Unless one is interested in determining the sensitivity of a certain species to a test chemical, the test species should be selected from Table 11.4. This will allow comparison of results and of the relative toxicity of a test chemical to that of other contaminants or toxicants. The majority of acute and chronic toxicity tests are performed with three species of invertebrates (*Daphnia magna, Daphnia pulex*, and *Ceriodaphnia dubia*) and four species of vertebrates (fathead minnows, bluegill, rainbow trout and brook trout). For acute effluent testing, *D. magna, D. pulex* and *C. dubia* are preferred. For chronic effluent tests, *C. dubia* and fathead minnows are recommended. For acute and chronic toxicity tests, *D. magna* and fathead minnows (*Pimephales promelas*) seem to be preferred. Fathead minnows are used more often than running water trout species because of the shorter life cycle that helps to accelerate obtaining results from chronic toxicity tests and they are easier to culture and maintain than trout that need a large, cold- and running-water holding facility.

---

[1]Rand, G. M., Editor. 1995. Fundamentals of aquatic toxicology. Effects, environmental fate, and risk assessment. 2nd Ed. Taylor and Francis Publ., Washington, D.C.

Table 11.4. Freshwater organisms recommended for use in toxicity testing. * indicate species most commonly used in each group.

Algae	
*Selenastrum capricornutum**	Green alga
*Scenedesmus subspicatum**	Green alga
Scenedesmus quadricaudata	Green alga
Chlorella vugaris	Green alga
Cyclotella spp.	Diatom
Navicula sp.	Diatom
Nitzschia sp.	Diatom
Synedra sp.	Diatom

Macrophytes	
*Lemna minor**	Duck weed
Vallisneria americana	Wild celery
Elodea canadensis	Waterweed
Ceratophyllum demersum	Coontail
Myriophyllum spicatum	Eurasian milfoil
Najas quadulepenis	Bushy pondweed
Potamogeton perfoliatus	Pondweed

Invertebrates - Plankton	
Brachionus calyciflorus	Rotifer
*Ceriodaphnia dubia**	Water flea
*Daphnia magna**	Giant water flea
*Daphnia pulex**	Water flea

Invertebrates - Benthos	
Dugesia tigrina	Planaria
Gammarus fasciatus	Amphipod
Gammarus limnaeus	Amphipod
Orconectes sp.	Crayfish
Cambarus sp.	Crayfish
Procambarus sp.	Crayfish

Invertebrates - Benthos (continued)	
Pteronarcys sp.	Stonefly
Baetis sp.	Mayfly
Ephemerella sp.	Mayfly
*Hexagenia limbata**	Mayfly
Hexagenia bilineata	Mayfly
Chironomus plumosus	Bloodworm
Chironomus riparius	Bloodworm
Chironomus tentans	Bloodworm
Amnicola limosa	Hydrobiid snail
Physella integra	Physid snail
Helisoma anceps	Planorbid snail

Vertebrates	
*Oncorhynchus mykiss**	Rainbow trout
*Salvelinus fontinalis**	Brook trout
*Salmo trutta**	Brown trout
Oncorhynchus tshawytscha	Coho salmon
Oncorhynchus kisutch	Chinook salmon
*Pimephales promelas**	Fathead minnow
Carassius auratus	Goldfish
Cyprinus carpio	Carp
Lepomis macrochirus	Bluegill
Lepomis cyanellus	Green sunfish
Ictalurus punctatus	Channel catfish
Catastomus commerconi	White sucker
Esox lucius	Pike
Gasterosteus aculeatus	Threespine stickleback
Brachydanio rerio	Zebra fish
Poecilia reticulata	Guppy

Immature organisms should be used whenever possible. For *Daphnia*, neonates < 24 h old are recommended for both acute and chronic toxicity tests. For fish, wet weights of 0.1 to 0.5 g are recommended, or the length of the smallest fish should not be more than twice as small as the largest fish.

Test Conditions

Most algal toxicity tests are chronic tests because effects are determined over several generations during the 96-h period. Under standard test conditions, algae (e.g.
15c☞ *Selenastrum capricornum*) can be expected to increase from about 1×10^{-4} cells/ml to nearly 4×10^{-6} cells/ml after 96-h. The test vessels for algae are usually 250 - 500 ml erlenmeyer flasks. The flasks need to be shaken by hand once or twice a day or placed on a variable-speed rotary or oscillatory shaker at 100 oscillations per minute.

Usually 1-L glass or plastic containers are used as bioassay vessels for invertebrates and fish. Loading of organisms in the bioassay vessels should be limited to ensure that dissolved oxygen levels are not fall below acceptable levels (see below). For fish, loading should not exceed 0.5 to 0.8 g/L of test solution in static bioassays and 0.5 to 1.0 g/L for flow-through tests. Usually 20 individuals of cladocerans or benthic invertebrates or 10 fish are added to each vessel.

At least five test concentrations and a control are recommended. Triplicates are made of all test solutions and the control for algal toxicity tests and duplicates for invertebrates and fish. Test concentrations should be either an appropriate logarithmic dilution series (e.g. 200, 20, 2, 0.2, 0.02) or a geometric series with a dilution factor of 0.5 or 0.6 (e.g. 100, 50, 25, 12.5, 6.25). Sometimes a mix of a geometric and a logarithmic dilution series is used (e.g. 25, 5, 2.5, 0.5, 0.25) .

Temperature and light need to be controlled for algal toxicity tests, and monitored for fish and invertebrate toxicity tests. The recommended temperatures for algal bioassays are between 20 and 25 °C. When possible, the test temperature should be selected from the series 7, 12, 17, 22, 27, and 32 °C when testing animals. For cold-water species, such as trout, 12 and 15 °C and for cladocerans and warm-water fish species, 20 and 25 °C are most commonly used. Test solutions usually are not aerated during the test. However, during the test dissolved oxygen should not fall below 40% saturation for warm-water species or 60% saturation for cold-water species. Gentle stirring is recommended to maintain oxygen levels above the minimum saturation values in the test solutions. Alkalinity, hardness and pH of the test solutions affect the toxicity of many chemicals. However, typically none of these buffer variables is adjusted unless they are one of the variables. If the objective of the chemical test is to determine the toxicity of a chemical, the pH need not be adjusted if it is between 6 and 9. Otherwise, if during the test the pH falls beyond this range, it may be adjusted to a neutral value (~ 7) or to its original pH value, which ever is less. For most static and static-renewal toxicity (acute and chronic) tests, total hardness, alkalinity, conductivity, pH, temperature and dissolved oxygen are measured at the beginning and at the end of the test.

Endpoints

The endpoints for algae include growth inhibition and stimulation and morphological and physiological changes. The most frequently reported is changes in
15d☞ biomass through growth inhibition. A coulter counter can be used to quickly count the numbers of individual cells but it cannot be used for colonial or filamentous forms. Since coulter counters count the numbers of particles in suspension, it cannot be used for counting algal cells in turbid water, or water (e.g. effluent) that has particles other than algae cells. Algal viability can be identified by use of mortal stains, such as Evans Blue. The most commonly used algae is green algae (Table 11.4).

A standard toxicity test for rooted macrophytes does not exist. Only *Lemna* has standard procedures for toxicity testing. It has several advantages over rooted forms, such as its small size, rapid growth and structural simplicity. The most commonly used endpoints for duckweed are changes in numbers of fronds and changes in biomass. Usually EC_{50}s and NOECs are estimated for duckweeds. The rooted macrophytes listed in Table 11.4 are used more commonly to determine toxicities of herbicides for weed control and for determining phytotoxic effects of contaminated sediments.

Of the invertebrates, planktonic species are used more commonly than benthic species. Of the planktonic forms, cladocerans, especially *Daphnia* spp. are most commonly used (Table 11.4). The endpoints of zooplankton is death as determined by their lack of gill or ciliary movement and settlement on the bottom of the bioassay vessel. Benthic invertebrates are most commonly used for testing the toxicity of sediments or of herbicides and piscicides, on non-target organisms. Mayflies of the genus *Hexagenia* are the most commonly used benthic invertebrate (Table 11.4). The endpoint is death as indicated by a lack of any gill movement or any other kind of movement after gentle prodding.

Mixtures of Toxicants

Among the most enigmatic problems is the development of a plethora of pesticides that have created in the environment a potpourri of chemicals whose interactions are
16☞ almost impossible to comprehend. So far, scientists have proposed relatively simple ***toxicological interactions*** in which exposure to two or three chemicals results in a biological response that is quantitatively different from that expected from the response of each of the chemicals alone. ***Biotransformation***, that is the interaction between chemicals or modification of potency by the ambient chemical milieu, can affect the toxicity of a chemical to an organism.

In the simplest of mixtures, exposure to two chemicals simultaneously may produce a response that is additive of the individual responses, or one that is greater than or less than expected from addition of their individual responses. Figure 11.8 illustrates the different responses that can be expected from mixtures of two chemicals. An ***additive effect*** occurs when the combined effect of two chemicals is equal to the sum of the effects of the individual chemicals applied alone (e.g. 1 + 1 = 2). Or, if 50% mortality occurs

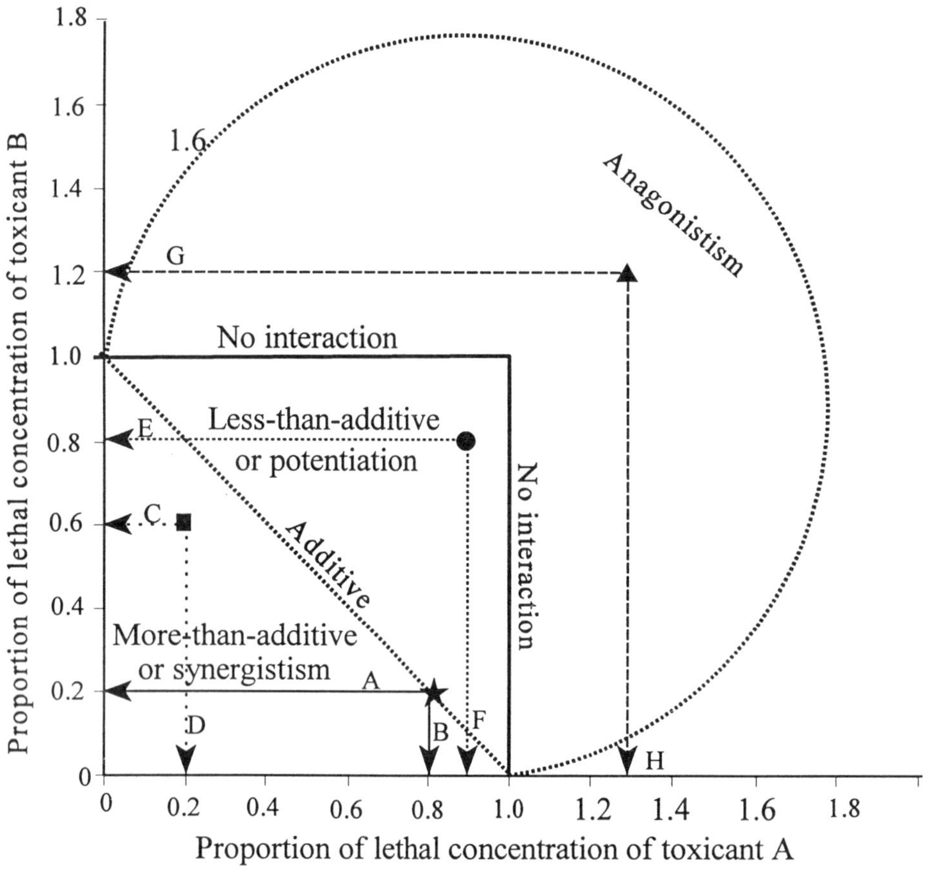

Figure 11.8. Effects of mixtures of chemicals.

with 0.2 parts of chemical A and 0.8 parts of chemical B, the interaction is additive, as shown by junction of A and B on the additive line in Fig.11.8. A ***more-than-additive***, or ***synergistic effect***, occurs when the combined effect of two chemicals is much greater than the sum of effects of the individual chemicals applied alone (e.g. 1 + 1 = 4). Or, if 50% mortality occurs with 0.6 parts of chemical C and 0.2 parts of chemical D, the interaction is synergism, as shown by C and D in Fig. 11.8. ***Less-than-additive effects***, or ***potentiation***, occurs when one chemical has a toxic effect only if applied with another chemical (e.g. 0 + 3 = 4, or 0.8 parts of chemical E and 0.9 parts of chemical F in Fig. 11.8). **Antagonism** occurs when two chemicals applied together interfere with each other's potency (e.g. 2 + 3 = 4) or one interferes with the other chemical (e.g. 3 + 0 = 2). This is depicted in Fig. 11.8 as chemicals G (1.2 parts) and H (1.3 parts).

CHAPTER 12

WATER QUALITY ASSESSMENT TECHNIQUES

Why Read This Chapter?

A plethora of methods for assessing the health of aquatic environments is available. Methods for assessing stream quality are different than those for measuring lake quality. This chapter examines methods for both lotic and lentic environments using not only water chemistry but indices developed from all trophic levels in the ecosystem, including bacteria, algae, macrophytes, invertebrates and fish. Benthic macroinvertebrates, in particular, have several advantages over other trophic levels in assessing water quality in lakes and streams. Many assessment techniques require experience with identification of organisms before they can be applied. Although such techniques provide detailed information on the source of pollution, they are expensive and labour intensive. Other methods can be applied much more cheaply, are much less labour intensive and many can be used by the layperson. Cottagers, in particular, will find some of the methods useful for assessing water quality and the fisheries and body contact recreation potential of their lake.

No single method can be relied upon to accurately assess water quality of a lake or stream. Several methods are recommended using a combination of physical, chemical and biological crteria. Previous chapters have already introduced some concepts for assessing the trophic status of lakes. This chapter summarizes the criteria developed so far and applies them as part of the methods for assessing water quality.

Emphasis is placed on the application of the different methods developed and used throughout the world. This is particularly true for the benthic invertebrates which is the community of choice for most bioassessments. Eighteen different metrics are examined. They include biotic indices developed in Canada, U.S.A., Britain, Australia and Germany. Each one is applied to a common data set from a benthic sample in order to compare and evaluate the different bioassessment methods.

By the end of this chapter you will know:

☞ 1. How to determine classes of water quality
☞ 2. How to assess chemical water quality for recreational use using a Walski and Parker index
☞ 3. How to assess the trophic state of a lake using morphological, physical, chemical and biological criteria
☞ 4. The importance of the Saprobien system as a foundation for the development of current water quality bioassessments in Candada, U.S.A., Australia, Britain and other European countries
☞ 5. The advantages and disadvantages of using bacteria, algae, invertebrates and fish to assess water quality
☞ 6. How to use bacteria to assess water quality of rivers
☞ 7. How to use algae to assess water quality of rivers
☞ 8. How to assess lake/stream acidification using kinds of algal “mats” and “clouds”
☞ 9. How to use macrophytes to assess water quality of rivers
☞ 10. How to use 18 macroinvertebrate metrics to assess water quality of rivers
☞ 11. How to use 13 fish metrics to assess water quality of rivers
☞ 12. How to use bacteria to assess water quality of lakes
☞ 13. How to use algae and phytoplankton to assess water quality of lakes
☞ 14. How to use algal productivity to assess water quality of lakes
☞ 15. How to use macrophytes to assess water quality of lakes
☞ 16. How to use macroinvertebrates to assess water quality of lakes
☞ 17. How to use fish to assess water quality of lakes
☞ 18. How to assess the recreational and fisheries potential of a lake

Water Quality Assessment

As discussed previously, "water quality" is used here to refer to the quality of water required for human use. Emphasis is placed in this section on biological criteria, but there are also some physical and chemical criteria that are useful, particularly for identifying the pollutants. Appendix 12.I lists chemical, provincial (Ontario) water quality objectives for the protection of aquatic life and recreation. Several physical/chemical indices have been developed for classifying water quality, but we will examine only two of them here.

Prati *et al* (1971)[1] used data from several countries, including England, Germany, Soviet Union, New Zealand, Poland and the United States, to develop a classification system with five different water quality classes, each of which corresponds to an ***implicit***
1☞ ***index of pollution (IIP)***. The term, pollution, is broadly used here and applies to a variety of uses for the water. The upper limits of the first indices are 1, 2, 4 and 8. The fifth exceeds an IIP of 8. The five classes correspond to conditions of "excellent", "acceptable", "slightly polluted", "polluted" and "heavily polluted". Prati *et al* (1971) developed explicit mathematical functions to determine the value of each IIP, the equations of which are given in Table 12.1, along with the ranges in values for each classification. To determine the index value, simply transpose the observed value for "X". For example, if the observed dissolved oxygen content is 88%, nitrates = 15 mg/L and B.O.D = 4.5 mg/L, the index values are 1.0, 2.3 and 3, respectively, classifying the water between excellent and slightly polluted since the indices given in Table 12.1 represent the upper limits.

Of particular interest here are indices that have been developed specifically for assessing the recreational use of water, such as swimming and fishing. For example, Walski and Parker (1974)[2] developed an index (called the ***Walski and Parker index***, or
2☞ ***WPI***) composed of several subindices employing variables that can be measured quickly. They selected variables that addressed four issues important to people for assessing a water body's worth for recreation: (1) those that affect aquatic life (e.g. dissolved oxygen, pH and temperature); (2) those which affect health (e.g. total coliforms); (3) those which affect taste and odour (e.g threshold odour number); (4) those which affect the appearance of the water (e.g. turbidity, grease, colour, Secchi depth). Mathematical models were then developed for each variable (Table 12.2) for assessing water quality for recreational purposes. For each variable, subindex values were calculated and then classified according to the criteria shown below Table 12.2.

---

[1]Prati, L. R., R. Pavanello and F. Pesarin. 1971. Assessment of surface water quality by a single index of pollution. Water Res. 5: 741-751.

[2]Walski, T. M. and F. L. Parker. 1974. Consumers water quality index. J. Environ. Eng. Div., Amer. Soc. Civil Eng.., pp. 593-611.

Table 12.1. Classification system of Prati et al (1971) for surface water quality. Only ten of the thirteen variables examined are listed here. B.O.D = biochemical oxygen demand. C.O.D = chemical oxygen demand. Some functions (e.g. dissolved oxygen, pH, chlorides, manganese)are segmented, non-linear and require more than one equation.

Condition	Equation	Functional Range	Excellent	Acceptable	Slightly Polluted	Polluted	Heavily Polluted
Index of Quality			**1**	**2**	**4**	**8**	**>8**
Dissolved oxygen (%)	$I = 0.00168X^2 - 0.249X + 12.25$ $I = -0.08X + 8$ $I = 0.08X - 8$	$0 \leq X < 50$ $50 \leq X < 100$ $100 \leq X$	6.5-8.0	6.0-8.4	5.0-9.0	3.9-10.1	<3.9, >10.1
pH	$I = -0.4X^2 + 14$ $I = -2X + 14$ $I = X^2 - 14X + 49$ $I = -0.4X^2 + 11.2X - 64.4$	$0 \leq X < 5$ $5 \leq X < 7$ $7 \leq X < 9$ $9 \leq X < 14$	88-112	75-125	50-150	20-200	<20, >200
5-day $B.O.D_{20\,C}$ (mg/L)	$I = 0.666667X$		1.5	3	6	12	>12.0
C.O.D (mg/L)	$I = 0.1X$		10	20	40	80	>80
Suspended solids (mg/L)	$I = 2^{[2.1 \log(0.1X - 1)]}$		20	40	100	278	>278
Ammonia (mg/L)	$I = 2^{[2.1 \log(10X)]}$		0.1	0.3	0.9	2.7	>2.7
Nitrates (mg/L)	$I = 2^{[2.1 \log(0.25X)]}$		4	12	36	108	>108
Chlorides (mg/L)	$I = 0.000228X^2 + 0.0314X$ $I = 0.000132X^2 + 0.0074X + 0.6$ $I = 3.75(0.02X - 5.2)^{0.5}$	$0 \leq X < 50$ $50 \leq X < 300$ $300 \leq X$	50	150	300	620	>620
Iron (mg/L)	$I = 2^{[2.1 \log(10X)]}$		0.1	0.3	0.9	2.7	>2.7
Manganese (mg/L)	$I = 2.5X + 3.9\sqrt{X}$ $I = 5.25X^2 + 2.75$	$0 \leq X < 0.5$ $0.5 \leq X$	0.05	0.17	0.5	1	>1.0

Table 12.2. Subindex functions for the Walski and Parker Index. A functional range is provided for those functions which have upper and lower limits. For example, for dissolved oxygen, I = 0 when X is greater than 8 mg/L. For temperature, Δ is the difference between the observed temperature and 20 °C. The value ranges are for intolerant (In), poor (Pr), good (Gd) and perfect (Pf) water quality conditions.

Pollutant Variable	Equation	Functional Range	Value Range			
			In	Pr	Gd	Pf
Dissolved Oxygen mg/L	$I = e^{[0.58\,(X - 8)]}$ $I = 0$	$0 < X \leq 8$ $8 < X$	0	4	7	8
pH	$I = 0$ $I = 0.04[25 - (X - 7)^2]$ $I = 0$	$X < 2$ $2 \leq X \leq 12$ $12 < X$	212	410	68	7
Total coliforms (#/100 ml)	$I = e^{-0.0002X}$		23,000	11,500	500	0
Average Temperature °C Actual: Deviation	$I = 0.0025[(400 - (X - 20)^2]$ $I = 0$ $I = 0.01(100 - \Delta X^2)$ $I = 0$	$0 \leq X \leq 40$ $\Delta X < -10$ $-10 \leq \Delta X \leq 10$ $10 < \Delta X$	4,010	1,400	14,263	200
Phosphates (mg/L)	$I = e^{-2.5X}$		1.8	0.92	0	0
Nitrates (mg/L)	$I = e^{-0.16X}$		29	14	0.7	0
Suspended Solids (mg/L)	$I = e^{-0.02X}$		230	115	5	0
Turbidity (JTU)	$I = e^{-0.001X}$		4,600	2,300	100	0
Colour C (units)	$I = e^{-0.002X}$		2,300	1,150	50	0
Grease Thickness (μ): Concentration (mg/L)	$I = e^{-0.35X}$ $I = e^{-0.016X}$		13,290	6.51	0.36	0
Odour (threshold odour #)	$I = e^{-0.01X}$		460	230	10	0
Secchi depth (m)	$I = \log(X + 1)$ $I = 1$	$X \leq 9$ $9 < X$	0	0.3	7	9

Walski and Parker Subindex Criteria

Subindex Value	Classification
0.01	Intolerable
0.1	Poor
0.9	Good
1.0	Perfect

For example, suppose one observes a dissolved oxygen level of 7.8 mg/L, a pH of 7.5, a total coliform count of 2/100 ml, a temperature of 18 °C and a Secchi depth of 5.8; transposing the values for X in each of the appropriate equations yields subindex values of 0.9, 1.0, 1.0, 1.0 and 1.0, indicating that the water is almost perfect for recreational use.

Chapters 3 to 9 have provided several clues (criteria) for assessing water quality (trophic status) in lakes and large rivers. All of them are re-examined here and others are introduced. This section basically integrates physical, chemical and biological data to determine the recreational and fisheries potential of a lake or large river.

First there is a synthesis of morphometric, physical and chemical criteria from earlier chapters. Then a variety of biological methods (i.e. bioassessments) using bacteria, algae, macrophytes, zooplankton, benthic macroinvertebrates and fish are examined. Most of the bioassessment methods are for streams and rivers, so we will begin with them and then examine some bioassessment methods for lakes. We will end with a description of how to determine the developmental capacity of a lake for body contact recreation and a cold- or warm-water fisheries.

Table 12.3 is a synthesis of information gleaned from previous chapters. The
3☞ discussion below follows the sequence given in the table. Criteria with numerical values are more useful in assessing trophic status than are those without numerical values, for a variety of reasons, as discussed in the following pages.

Morphometric Water Quality Criteria

The morphometric criteria listed in Table 12.3 are relative and are based upon the evolution of lakes as they naturally eutrophy over eons of time, from an oligotrophic state to a dystrophic one, with succession into land. Most lakes in the polar latitudes are oligotrophic or slightly mesotrophic and the morphometrics listed (basin shape, mean depth) do not apply. But in temperate latitudes, lakes that have a U-shape profile tend to be very deep and oligotrophic. Natural eutrophication of such lakes gradually fills the bottom with organic material making them less deep with less volume and lend a V-shape profile to the lake. However, the morphometric criteria are rather relative descriptions and there is no degree of "U-ness" or "V-ness", or specific depths or hypolimnial volumes that can be used to accurately assess the trophic status of lakes.

Physical Water Quality Criteria

The physical criteria can be divided into temperature, conductivity and Secchi depth, of which the latter is especially useful.

Temperature

Temperature is a relative criterion that has only a theoretical basis. Oligotrophic lakes, being generally deeper than eutrophic lakes, should have a shallower but colder

Table 12.3. Morphometric, physical, chemical and biological criteria for assessing the trophic status of lakes.

Features	Oligotrophic Lakes (Secchi depth > 5 m)	Mesotrophic Lakes (Secchi depth 2-5 m)	Eutrophic Lakes (Secchi depth < 2 m)
Lake basin shape	U-shaped	Deep V-shape	Shallow V-shape
Mean depth	Very deep	Moderately shallow	Usually shallow
Hypolimnion volume	Large	Moderately small	Small
Hypolimnion temperature	4° C	About 4° C	Usually >4° C
Thermocline depth	High	Moderately deep	Shallow
[Dissolved oxygen]	90-100% in hypolim.	25-90% in hypolim.	<25% in hypolim.
Dissolved oxygen profile	Orthograde	Clinograde	Clinograde
RAHOD mg O_2/m^2/day	< 0.017	0.017-0.033	> 0.033
[Total phosphorous]	< 10 µg/L	10 - 30 µg/L	> 30 µg/L
[Total nitrogen]	< 300 µg/L	300 - 600 µg/L	> 600 µg/L
[Calcium]	Low (< 10 mg/L)	Moderate (10-25 mg/L)	High (> 25 mg/L)
[Chlorophyll *a*]	< 2 µg/L	2 - 5 µg/L	> 5 µg/L
Macrophyte diversity	Many submerged spp. but low biomass, few/no floaters, emergents	Many submerged, floating & emergent spp. & high biomass	Few submerged sp; many floaters, emergents with high biomass
Dominant profundal benthic functional feeding groups	High oxygen requiring gatherers, filterers, predators	Some filterers, many gatherers, predators, few need high [oxygen]	No filterers, many low oxygen tolerant gatherers, predators
Profundal benthic diversity	High diversity (Log_2) > 3	Moderate diversity (Log_2) = 1-3	Low, mean diversity (Log_2) < 1
Profundal indicator species	*Pisidium conventus*, *Sphaerium nitidum*, *Tubifex kessleri*, *Peloscolex variegatum*, *Diporeia hoyi*, *Mysis relicta*, *Hexagenia limbata*,	Several tubificid spp., *Pisidium casertanum*, *Sphaerium corneum*, Caecidotea (*Asellus*) spp. several chironomid spp.	*Tubifex tubifex*, *Limnodrilus hoffmeisteri*, *Chironomus plumosus*
Profundal fish diversity	Most salmonids, white fish and cottids	A few salmonids and cottids, walleye, pike, bass; gizzard shad abundant	Coarse fish (perch, carp,suckers, minnows) or none

hypolimnion than a eutrophic lake. The reasons are two fold. First, because the water is more coloured and turbid, light (especially reds and infra-reds that heat the water) in eutrophic lakes cannot penetrate as deeply as light in oligotrophic lakes, so the epilimnion and the thermocline tend to be relatively shallow in eutrophic lakes. However, mixing by winds of warm, epilimnial water can occur at deeper depths in shallower (e.g. eutrophic) lakes than in deeper (e.g. oligotrophic) lakes. Hence, the hypolimnetic water of eutrophic lakes will receive some of the warmer water of the epilimnial region as the lake begins to stratify, resulting in a warmer hypolimnion than seen in most oligotrophic lakes. If mixing of epilimnial water with hypolimnial water is prevented, the temperature of the hypolimnion will always be near 4 °C.

Conductivity

By definition, oligotrophy means poor nutrient levels. Since conductivity is a measure of the total dissolved solids, oligotrophic lakes tend to have much lower conductivities than eutrophic lakes. However, like temperature and morphometric criteria, the conductivity level has only a theoretical basis with no specific levels that can be used to assess trophic status.

Secchi Depth

The Secchi depth is such an easy measurement to make and it correlates well with most other criteria shown in Table 12.3 as "column criterion" with trophic status. This is not meant to imply that Secchi depth is an independent variable. On the contrary, it is a variable whose value depends upon the nutrient (phosphorous and nitrogen) and resulting chlorophyll *a* levels. Chapters 4, 5 and 6 describe relationships between Secchi depth and levels of nutrients and chlorophyll *a*. It is also used as a key variable in determining the development capacity of a lake later in this chapter.

Chemical Water Quality Criteria

The chemical criteria can be divided into gaseous criteria and ionic or nutrient criteria. Of the gases, dissolved oxygen content is the most important because it is vital to the survival of most aquatic species. Of the nutrients and ions, levels of total phosphorous and total nitrogen are invaluable criteria, with calcium content generally related.

Dissolved Oxygen

Although some species (e.g. eutrophic indicators) are better adapted to low oxygen levels than others, all aquatic species need at least *some* dissolved or atmospheric oxygen in order to respire. In fact, the presence of several species capable of using atmospheric oxygen (e.g. pulmonate snails, most hemipterans, all adult beetles, mosquitoes, rat-tailed maggots and other dipterans) usually indicates that dissolved oxygen supplies are limited and the aquatic environment is stressed. The use of indicator species to assess trophic status is described later.

Hypolimnial Dissolved Oxygen Concentration

As stated in Chapter 5, it is the dissolved oxygen levels in the hypolimnion (not that in the epilimnion) during thermal stratification that is most useful for assessing the trophic status of a lake. As summer proceeds, dissolved oxygen levels drop in the hypolimnion because there is no mechanism for restoring any oxygen used by bacteria in metabolizing and breaking down any organic matter present. Since eutrophic lakes have greater accumulations of organic material than oligotrophic lakes, the loss of dissolved oxygen is more severe in the former than in the latter. Most cold-water species, like lake trout, require at least 60% dissolved oxygen saturation to survive and for that reason can be found in mesotrophic lakes whose oxygen levels do not fall below 60% saturation by the end of the summer. The 75-100% saturation value given as a criterion for oligotrophy is a fairly conservative range but assures that most cold-water species will persist in the lake. Many may argue that eutrophy has been attained only if the hypolimnion goes anoxic by the end of the summer. However, anoxia is considered good evidence of hypereutrophy and a rather conservative range is given for both eutrophy (< 25% saturation) and mesotrophy (25 - 75%saturation).

Oxygen Depth Profile

An orthograde oxygen profile (100% saturation at all depths) in a thermally stratified lake, especially at the end of summer, is positive evidence that the lake is oligotrophic. Clinograde oxygen curves occur in both mesotrophic and eutrophic lakes, except oxygen levels remain above 25% saturation and below 75% saturation in mesotrophic lakes. Although determining a dissolved oxygen profile is laborious and need not be determined for a lake in order to assess its trophic status, it is an additional criterion that can be used to support your assessment. The most important part of the profile is, clearly, the 1-m layer of water above the sediments. A special water sampler, such as a Van Dorn sampler or Kemmerer sampler, is needed. These are samplers that allow one to remotely (e.g. from a boat) obtain a water sample from any layer in the water column.

RAHOD

A remote water sampler is needed to determine the RAHOD ***(relative areal hypolimnial oxygen deficit***). The calculation of a RAHOD value is described in Chapter 5 and need not be repeated here. It does require measurements of dissolved oxygen from several depths near the end of spring overturn (but before thermal stratification begins) and at the end of thermal stratification (but before the lake begins to destratify). It is important to sample the same depths each time. The surface area of the hypolimnion also needs to be determined, using information in Chapter 3. Once the hypolimnetic surface area and dissolved oxygen concentrations for each layer are known, the RAHOD can be calculated.

Nutrient Levels

The values shown for total phosphorous and total nitrogen in Table 12.3 apply to levels during spring turnover. Because water is well mixed at all depths during this period, water samples can be taken from any depth. However, the measurements should be made on replicate samples, for example, one at 1 meter, another from mid-depth and a third from near the bottom. Spring turnover values are required because they represent the maximum amounts available to algae. Once stratification begins, algae begin utilizing and depleting the nutrient supplies.

Measurement of total phosphorous and total nitrogen levels requires specialized techniques and equipment. It is especially important to use glass-ware that has been acid washed (e.g. 5% sulphuric acid) to store the water samples. Ministries of environment or natural resources in most provinces (and states in the U. S.) have chemical and physical data for several thousand lakes. A telephone call to the nearest ministry will reveal if they have information on total phosphorous and total nitrogen for a specific lake.

Calcium Concentration

The level of calcium is generally related to the trophic status, eutrophic lakes typically having more calcium than oligotrophic lakes. However, there is no specific calcium level that distinguishes oligotrophic lakes from eutrophic lakes. For example, Lakes Superior, Huron, Ontario, Michigan, and Erie (listed in increasing order of eutrophy) have calcium levels of about 12, 23, 39, 32, and 37 mg/L, respectively. While there is a trend in increasing eutrophy with increasing calcium levels, calcium content itself cannot be used to assess trophic status.

Bioassessment of Water Quality

Bioassessment is the evaluation of water quality based on analysis of species assemblages of communities of aquatic organisms, or of their products (e.g. chlorophyll *a*). The underlying tenet is that biological communities reflect overall ecological integrity (i.e. chemical, physical, and biological integrity) and integrate effects of different stressors. A bioassessment can therefore provide an integrated measure of the cumulative impact of a variety of stressors applied either continually or in pulse doses. While physical and chemical analysis of water samples are useful, or even necessary, for identifying the nature of the stressor, the effects of the magnitude of the different physical and chemical parameters can only be determined from bioassessment techniques. Indeed, grab samples of effluent water may under estimate the severity of the stressor if a pulse event is missed, or over estimate the severity if the grab sample is taken during a pulse event.

Several bioassessment protocols are available. Most have been developed for streams and rivers, only a few for lakes. The most common taxonomic group used is the benthic macroinvertebrates, but protocols using bacteria, algae or fish have also been developed. The main objectives of bioassessments are: (1) to determine if an aquatic habitat is supporting or not supporting a designated use; (2) to characterize the existence and severity of impairment for its designated use; (3) to help identify sources and causes of use impairment; (4) to evaluate the effectiveness of control actions; (5) to characterize regional biotic components; and (6) to support studies on the attainability of a use designated for the body of water under investigation.

The basic approach in the bioassessment of water quality is to compare water quality at the test site(s) with water quality at ***control sites*** (also called ***reference sites***). The reference sites should be as similar as possible in physical characteristics to the test sites because both quality and quantity of available habitat affect resident community structure and composition. Usually the simplest approach is to select for reference sites areas immediately upstream of the outfall, if the outfall location or source of the stressor is known. However, often physically comparable, upstream reference sites are not available or the stressor is from a non-point source. In this case, one must select for reference sites, habitats in similar unimpacted streams within the same ***ecoregion***. Each ecoregion differs from the others because of regional differences in forest, agriculture, land use, soils and bedrock types. The biotic assemblages can be expected to differ naturally among ecoregions but be relatively similar within ecoregions. By selecting a relatively small number (3-5) of minimally-impacted regional reference sites, one has an ecologically and statistically valid means to establish baseline conditions. This assumes that the sites selected on each stream are similar in stream size (order), hydrological regime, riparian vegetation, etc. If site selection is tailored to accommodate these conditions, similar water quality standards, criteria and monitoring strategies are likely to be valid throughout a given ecoregion.

Historical Review of Bioassessment Methods

Historically, five taxonomic groups have been used to assess water quality; bacteria, algae, zooplankton, macroinvertebrates and fish. If listed in an order of
4☞ popularity, the current usage appears to be macroinvertebrates > algae > fish > bacteria > zooplankton. But in the evolution of bioassessment protocols, the ranking began as bacteria > zooplankton > algae ≥ macroinvertebrates > fish.

Two German scientists, Kolkwitz and Marsson (1908[3], 1909)[4], were the first to relate aquatic organisms to the purity and pollution of water using the ***Saprobien System***. For the next 70 years, European and North American scientists experimented with the system and many criticized it as being too arbitrary and over-complex. It was not until the late 1970s that the saprobic system began to gain popular usage, mainly because of

---

[3]Kolkwitz, R. and M. Marsson. 1908. Okologie der pflanzlichen Saprobien. Ber. Dt. Bot. Ges. 26A: 505-51.

[4]Kolkwitz, R. and M. Marsson. 1909. Okologie der tierischen Saprobien. Intern. Rev. Hydrobiol. 2: 126-152.

Sladacek's (1973)[5] review, synthesis and arguments supporting the use of indicator organisms (that he called ***saprobic organisms***) for assessing water quality. He considered the following principles as fundamental to the saprobic system: *"1. Dependence of the saprobic organisms on the environment; 2. Successions occurring in two directions. The primary succession goes on as a state of eutrophication" (either naturally or anthropogenically), "the secondary one as selfpurification".* He further added, "The main practical fundamentals are the continuous flow of the spontaneous selfpurification processes.....". "One of the main results is, that coliform bacteria develop in the anaerobic conditions of sewage and certain industrial wastes, that they must be considered one of the main constituents of the community and their decline follows as late as during the subsequent steps of decomposition". Sladacek divided the processes of eutrophication and self-purification into several saprobic zones, the main ones being:

Eutrophication

Katharobity (drinking water quality)
Limnosaprobity (clean to polluted surface and underground water)
Eusaprobity (contains sewage and industrial waste but not toxic)
Transsaprobity (toxic wastes, inaccessible to bacteria)

Self-purification

Antisaprobity (toxic wastes, slow or no natural recovery)
Metasaprobity (septic zone, highest coliform counts)
Polysaprobity (recovery has begun)
α-mesosaprobity
β-mesosaprobity
Oligosprobity
Xenosaprobity
Katharobity

Some bacterial and chemical characteristics of each of the self-purification zones are given in Table 12.4

Figure 12.1 is a synthesis of the progressive changes that occur as self-purification proceeds down stream. North American terminology is used to replace the Sladacek's saprobic zonation scheme. Katharobity is equated with conditions upstream of the outfall, while the zone of active decomposition ≈ eusaprobity, septic zone ≈ metasaprobity, zone of active decomposition ≈ α-mesosaprobity to oligosaprobity, and the clean zone ≈ xenosaprobity. It is presumed that with continued dumping of wastes that a river rarely attains the upstream katharobity status again. The key points of interest are that there is a succession of pollution tolerant forms, with the most tolerant near the outfall and the least tolerant further downstream. Immediately below the outfall, only

[5]Sladacek, V. 1973. System of water quality from the biological point of view. Arch. Hydrobiol. Beih. 7: 1-218.

Table 12.4. Relationship between saprobic classification of water quality and some bacteriological and chemical criteria. Total coliforms, dissolved oxygen and hydrogen sulfide values indicate levels less than those shown. Biochemical oxygen demand (B.O.D.) is a 5-day value at 20 °C.

Degree of Saprobity	Total Coliforms No./L	Dissolved Oxygen %	H_2S mg/L	B.O.D. Mg/L
Antisaprobity	0	various	0	0
Metasaprobity	10000000000	0	1	200-700
Polysaprobity	20000000	10	traces	80
α-mesosaprobity	1000000	20	0	9
β-mesosaprobity	100000	40	0	6
Oligosaprobity	50000	50	0	4
Xenosaprobity	10000	60	0	2
Katharobity	20	various	0	0

bacteria such as coliforms (e.g *Escherichia coli*) and sewage fungus (e.g. *Sphaerotilus natans*) increase in numbers, with the latter reaching peak densities in the zone of active decomposition. Sludge worms of the oligochaete family, Tubificidae (e.g. *Tubifex tubifex* and *Limnodrilus hoffmeisteri*), reach their peak densities in the septic zone, with the coliform bacteria. Blood worms of the insect genus *Chironomus* reach their peak densities immediately downstream of the sludge worms. Both blood worms and sludge worms have haemoglobin, needed for extracting minute amounts of oxygen is dissolved in the water. Sow bugs (Isopoda, e.g. *Caecidotea*) attain their greatest densities in the recovery zone at the end of which are other species that are characteristic of the clean-water communities upstream of the outfall. The clean-water species gradually increase in numbers, provided there is no additional discharge of effluent downstream. The downstream changes can also be viewed as changes over time as rivers (and lakes) proceed through oligotrophy, mesotrophy, eutrophy and hypereutrophy.

Sladacek (1973) developed a list of saprobic organisms that included protists, rotifers, cladocerans, copepods, macroinvertebrates and fish and assigned a ***valence*** or saprobic score that indicated their preference for one or more of the saprobic zones (see

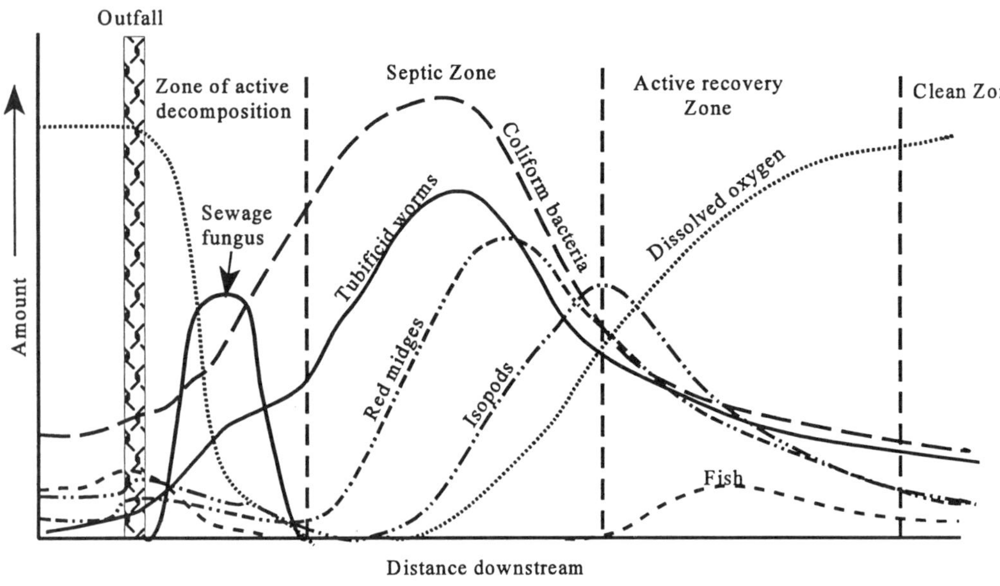

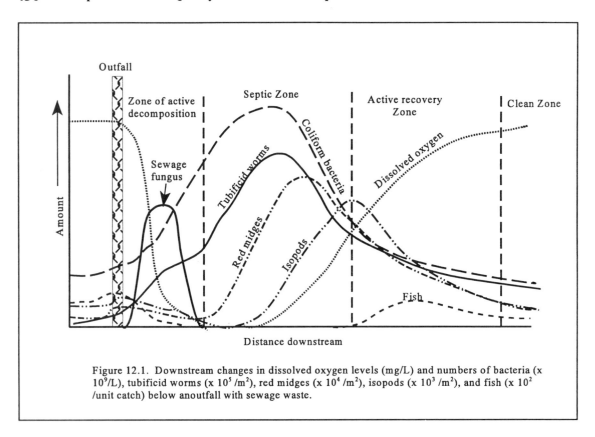

Figure 12.1. Downstream changes in dissolved oxygen levels (mg/L) and numbers of bacteria (x 10^9/L), tubificid worms (x 10^5 /m^2), red midges (x 10^4 /m^2), isopods (x 10^3 /m^2), and fish (x 10^2 /unit catch) below anoutfall with sewage waste.

Table 12.14 for a few examples). He then calculated a saprobic index (S) using the formula:

$$S = \sum (h \times s) / \sum h,$$

where, h = abundance according to the scale 1 for least abundant, 3 for moderately abundant and 5 for most abundant, s = the valence or saprobic score for each species.

Water quality was determined by matching the saprobic index value to a scale that varied between 0 and 4, 4 being the most polluted value. An example calculation is given later using benthic macroinvertebrates.

Bioassessment of Streams And Rivers

There is a plethora of methods for assessing water quality of lakes and rivers using indicator species or biotic indices based on either the structure of the entire community or on a portion of the community. The communities used most often are bacteria, algae (or phytoplankton), benthos and fish. Within the benthic community, the macroinvertebrates (those visible with the unaided eye) are the most commonly used. The macrophytes have some merit, as discussed below, but most of the discussion will centre around the macroinvertebrates.

The community selected often is dictated by the expertise of the investigator, but there are distinct advantages to using each group. Table 12.5 summarizes most of the
5☞ attributes of each type of community. However, it is highly recommended that any water quality assessment use at least three of the communities to make the water quality evaluation. One to three criteria should be evaluated for each community, with a minimum of five biological criteria AND three physical and/or chemical criteria. For example, if algae, benthos and fish are selected as the community types, one could use algal diversity, chlorophyll *a* level, profundal benthic community structure, profundal indicator species, and fish community structure for the biological criteria, and Secchi depth, dissolved oxygen depth profile and spring total phosphorous level for the physical and chemical criteria.

The analyses begins with an examination of some bacteriological criteria, then algae, macrophytes, benthic macroinvertebrate and fish communities, for example, using information in Table 12.3, Chapters 6 to 9, and other approaches.

Bacteriological Criteria

There are two approaches, use of indicator organisms and bacterial indices. Both require specialized techniques and expertise for identification and enumeration. However, most municipalities have a local health unit that will make the determinations for a nominal fee or even gratis. All have a specific protocol that needs to be followed if they are to analyze the samples. For example, five water samples must be submitted weekly during the summer period. The health unit provides a geometric mean number of total coliforms and faecal coliforms (i.e. *Escherichia coli*). Beaches are closed if the geometric mean exceeds provincial standards (see below) for two consecutive weeks.

Indicator Bacteria

There are several kinds of bacteria that can be used to diagnose faecal contamination or thermal enrichment. As described in Chapter 7, positive indications of human waste or faecal contamination are the presence of any one of the following genera:

Escherichia coli
6☞ *Pseudomonas aeruginosa*
Salmonella
Shigella
Leptospira
Pasturella
Vibrio

Table 12.5. Comparative advantages of different taxonomic groups for use in assessing water quality. *COSEWIC = Committee on Status of Endangered Wildlife in Canada.

Bacteria		Algae	Benthic invertebrates	Fish
Distribution - value for assessing local conditions in lakes and streams	Omnipresent. Good indicators of localized conditions, especially sediment quality.	Omnipresent, but most useful in lentic habitats. Only attached forms occur in streams. Not useful in headwater areas.	Limited migration patterns. Hence, good indicators of localized conditions; well suited for site-specific impacts in both lakes and streams. More diverse than fish in headwater streams.	Extensive migration abilities. More useful for assessing ecosystem impacts, except in headwater streams where fish fauna may be limited.
Response time - value for assessing short- and long-term effects. See Chapters 6-9 for overview of life cycles of the four groups.	Very rapid generation rates. Being at bottom of food pyramid, they are early integrators of short-term environmental effects.	Very rapid reproduction rates and life cycles. Most useful for short-term impacts.	Most species have 1-2 yr life cycles; integrate of effects of short-term environmental variations. Good for long-term analyses since historical data bases are common and some (e.g. unionid clams) live >20 yrs for good long-term biomonitors.	Relatively long-lived and mobile. Excellent integrators of long-term (several years) effects
Taxonomy - ease with which organisms can be identified. Stability of taxonomy	Requires specialists in most cases, but techniques are well established.	Requires a specialist in most cases, but often non-taxonomic methods are used (e.g. chlorophyll *a*, biomass)	Relatively easy to identify to family and genus. Heavy emphasis on insects for which taxonomy is relatively stable.	Easy to identify to species. Taxonomy relatively stable for most families.
Sampling ease - ease with which communities can be sampled and resultant impact on resident organisms.	Relatively easy, inexpensive to sample but use of sterile bottles essential. Few people required.	Easy to sample, inexpensive and requires few people.	Relatively easy to sample. Requires few people, inexpensive gear. Recolonization by species removed is rapid, with little impact on resident biota	Relatively easy to collect. If identified in field, fish can be returned unharmed.
Indicator value - extent of documentation for taxa within each group. See Chapters 6-9 for indicator value of several species.	Many bacteria have very specific functions (e.g. N_2 cycle bacteria) and can be used to identify cause-effect impacts	Many species useful for assessing organic pollution especially from sewage. Directly affected by physical and chemical factors.	Extensive documentation on indicator value for nearly all phyla.	Salmonids, cottids, some percids especially useful for lake and stream assessments. Many cyprinids also useful in headwater streams.
Trophic position - importance to humans. See Chapters 6-9 for details.	Base of food chain. Many are pathogenic to humans.	Autotrophs at bottom of food chain, vital to ecosystem functioning.	Primary food source for fish of sport and commercial value.	Top of food chain. Most important group since eaten by humans
Introductions - most exotic species are tolerant of a wide range of environmental variables. Chapter 14 has details.	Of those introduced, most have appeared as parasites of other exotic species.	Probably difficult or impossible to tell when or what alga species has been introduced and its potential impact on native forms.	Well documented for annelids, crustaceans and molluscs. Some species (e.g. zebra mussel) have altered benthic community structure.	Many species intentionally introduced for sport or forage value. Some (e.g. lamprey) have potential to alter native community structure.
Conservation status - Most endangered species are intolerant	Most pathogens are necessarily eradicated.	Poorly known, none on COSEWICs* list yet.	Molluscs have been placed on COSEWICs listing only since 1994.	Account for ~50% of COSEWIC endangered species list

The probability of pathogens being present increases with increasing numbers of faecal coliforms. For example, there is a 27.6% probability of *Salmonella* occurring in a water sample with < 200 faecal coliforms per 100 ml, 85.2% if 200-2000/100 ml and 98.1% if > 2000/100 ml. In most municipalities, ***water is unfit for human consumption if in three samples (taken over 3 weeks) the total coliform level has a geometric mean that exceeds 5/100 ml or an E. coli count > 0/100 ml; it is unfit for recreational use if the geometric mean count of E. coli is > 100/100 ml in two consecutive weeks, five samples per week.*** Health units also test for *Pseudomonas aeruginosa* in water from swimming pools, hot tubs and other pools used by several people. *Pseudomonas aeruginosa* is the main cause of eye, ear and skin infections.

Certain strains of *E. coli* can be lethal to humans if consumed in drinking water supplies. In May 2000, a heavy rainfall flushed cattle waste, containing the strain, into the drinking water supplies of Walkerton, Ontario. At least seven people died and scores were ill from drinking the infected water. This illustrates the importance of health inspectors maintaining a constant vigilance on drinking water standards and notifying communities of the presence of *E. coli* in drinking water supplies, a procedure apparently not followed by health officials in the community of Walkerton.

The ratio of faecal coliforms to faecal streptococci is a useful indicator of the source of contamination. The ratio is greater than 4 for human waste and < 0.7 for domestic animals (e.g. cattle). Because *E. coli* has a very short life span in water, its presence in surface waters indicates recent faecal contamination.

Sewage fungus, *Sphaerotilus natans*, is a sheathed bacterium that grows in chains or filaments. It grows prolifically in waters receiving organic wastes (i.e. the active decomposition zone (Fig. 12.1)), such as faecal, pulp and paper and sugar wastes.

Not all bacteria are pathogens and many species are required to mineralize substances needed to maintain good water quality. In fact, the activated sludge and trickling filter systems of sewage treatment plants rely on bacteria to decompose organic matter. As explained in Chapter 5, many nutrient cycles (e.g. nitrogen) depend on the presence of certain species of bacteria to oxidize or reduce nitrogenous compounds. Similarly, several species of sulphur bacteria are needed to process oxidized and reduced forms of sulphur. The presence of diverse bacterial communities, required for processing nitrogenous and sulphurous compounds and thus preventing the buildup of ammonia, nitrite, or hydrogen sulphide, indicates a healthy aquatic environment.

Some bacteria are unique to certain kinds of habitats. The bacterium, *Thermus aquaticus*, occurs in hot springs. However, its presence in surface waters indicates a nearby source of heated effluents.

Bacterial Indices

Europeans have used the following total bacterial densities (#/ml) as an index to
6☞ help determine trophic status of large rivers, although the index is not commonly used in North America:

Oligotrophy: 100,000 - 500,000
Mesotrophy: 500,000 - 2,000,000
Eutrophy: 2,000,000 - 10,000,000

Algae (and Phytoplankton) Criteria

Algae have been used for assessing water quality since the early 1900s. Bioassessments using algae can be divided into three basic categories. (1) Indicator species - there is a wealth of information on the indicator value of algae, not only for identifying degrees of organic enrichment in lakes and rivers but for diagnosing acidification, thermal pollution, taste and odour problems, and metal pollution. (2) Indices - there is a variety of algal indices that can be used to assess water quality of lakes and rivers, especially their trophic status. They include a modification of the saprobic index discussed earlier. (3) Primary productivity - several primary production criteria (e.g. chlorophyll *a* level, RAHOD) have been developed for assessing the trophic status of lakes and rivers.

Algal Indicators

Tables 12.6 to 10.9 list several species of algae according to their indicator value for a variety of stressors. The tables are based on information in Palmer (1962)[6].
7☞ Remember, it is not the mere presence or absence of a species that is important, it is its relative abundance. Species present in algal blooms are especially useful. Some species may be used to identify more than one problem, such as organic enrichment and taste and odour problems.

Table 12.6A lists genera of algae that cause taste and odour problems. The species of the genera listed in table 12.6A are given in Table 12.6B, organized alphabetically
7☞ within each phylum. It is organized according to the types of odours, with the genera listed alphabetically for each type of odour. Most of the algae that give a grassy odour are green algae (chlorophytes); those that give a septic smell are mostly blue-greens (Cyanobacteria); a geranium odour is almost entirely due to diatoms (bacillariophytes). All three phyla, with some yellow-green and yellow-brown algae (chrysophytes), contribute to a fishy smell.

Of greater interest to cottagers, swimmers and other users of fouled water is the
7☞ remedy needed to cure the taste and odour problems. The short-term (i.e. the "band-aid approach") is to simply scoop out the massive colonies if they are filamentous forms.

---

[6]Palmer, C. M. 1962. Algae in water supplies. An illustrated manual on the identification, significance, and control of algae in water supplies. U. S. Dept. Health, Education, and Welfare, Public Health Service Publication No. 657. 88 p.

Otherwise, the permanent cure is to eliminate the cause of the algal bloom(s). Most of the species that cause odour problems are those that form blooms under enriched conditions (cf. Table 12.7). Enrichment is usually caused by untreated sewage or domestic waste containing high levels of nutrients (e.g. phosphates and/or nitrates). Many cottages have old septic systems that leach organic wastes and bacteria into the lake. People bathing in lakes, washing their hair with shampoos and conditioners and using detergents to cleanse themselves, contribute to the nutrient loading. Some lakes are so eutrophic that phosphorous in the bottom sediments is mobilized annually during low-oxygen periods (winter, summer stratification). This "internal loading" of phosphorous is difficult to control but reducing the "external loading" from bathing activities is easy to control. The internal loading has been controlled in some lakes by artificially aerating the hypolimnial waters during the summer thermal stratification period.

Table 12.7 lists algae that are indicators of organically enriched waters, particularly those receiving untreated sewage wastes. The list represents species that occur in the self-purification process in streams and rivers but probably applies to ponds and outfall areas in bays of lakes. The same self-purification process occurs lakeward, from the shore line to the centre of the lake.

The algae utilize the byproducts of the natural purification process and are gradually replaced by species characteristic of clean waters (Table 12.8). The blue-greens, greens and euglenoids are especially well represented. Many of the species tolerant of organic pollution are also tolerant of other kinds of pollution, such as acidification, thermal enrichment, metal and oil pollution (compare Table 12.7 with Table 12.9).

There are several rather obvious differences between the assemblage of eutrophic indicator algae and the assemblage of oligotrophic indicator algae. First, there are few phyla (four) that represent organic pollution forms, while several phyla (eight) contribute to the list of pollution intolerant species (Table 12.8). Unique to the oligotrophic indicator phyla are the red (Rhodophyta) and brown (Phaeophyta) algae. Both phyla are consist mostly of marine algae, the brown algae being represented by the kelps, a major source of iodine and agar. *Hildenbrandia* is an encrusting form and *Lemanea* is an attached form, both requiring little light and cool waters, features characteristic of most headwater streams. The Chrysophyta and Cryptophyta consist of species mostly restricted to the head waters of streams. The red and brown algae are represented in fresh waters mostly by free-swimming forms found in the slower portions of rivers and in lakes. The diatom assemblage of oligotrophic lakes consists of some centrate forms (circular with radial symmetry, e.g. *Cyclotella, Cocconeis*). Pennate forms (elongate with bilateral symmetry, e.g. *Pinnularia nobilis, Nitzschia linearis, Navicula gracilis*) are also present but they are represented by different species than in eutrophic waters. The diatoms of eutrophic lakes are almost exclusively pennate forms. Indeed, as shown later, the ratio of centrate to pennate forms can be used as an index of water quality.

There are several genera of blue-green algae (e.g *Phormidium*), diatoms (e.g.

Table 12.6A. Algae causing taste and odour problems. Most taste sweet, a few taste bitter (e.g. *Ceratium, Nitella, Synura*). A few genera have more than one odour. Except for a spicy or
7 ☞ geranium odour, the odours apply when there are blooms present; the spicy and geranium odours apply when there is a moderate abundance of the species listed. The phyla are abbreviated, Bacillar = Bacillariophyta, Chloro = Chlorophyta, Chryso = Chrysophyta, Crypto = Cryptophyta, Cyano = Cyanobacterium, Pyrrho = Pyrrhophyta.

Genus	Phylum	Genus	Phylum	Genus	Phylum
Grassy odour		**Fishy odour**		**Septic odour**	
Actinastrum	Chloro	*Asterionella*	Bacillario	*Anabaena*	Cyano
Anabaenopsis	Cyano	*Chlamydomonas*	Chloro	*Anacystis*	Cyano
Closterium	Chloro	*Chrysosphaerella*	Chryso	*Ceratium*	Pyrrho
Cosmarium	Chloro	*Cyclotella*	Bacillar	*Cladophora*	Chloro
Gloeocystis	Chloro	*Dictyosphaerium*	Chloro	*Cylindrospermum*	Cyano
Gloeotrichia	Cyano	*Dinobryon*	Chryso	*Gloeocystis*	Chloro
Nitella	Charo	*Eudorina*	Chloro	*Gonium*	Chloro
Pediastrum	Chloro	*Euglena*	Eugleno	*Hydrodictyon*	Chloro
Scenedesmus	Chloro	*Glenodinium*	Pyrrho	*Nostoc*	Cyano
Spirogyra	Chloro	*Gonium*	Chloro	**Violet odour**	
Staurastrum	Chloro	*Malomonas*	Chryso	*Cryptomonas*	Crypto
Ulothrix	Chloro	*Pandorina*	Chloro	*Malomonas*	Chryso
Musty odour		*Peridinium*	Pyrrho	**Garlic/skunky odour**	
Chlorella	Chloro	*Pleurosigma*	Bacillar	*Chara*	Charo
Melosira	Bacillar	*Stephanodiscus*	Bacillar	**Geranium odour**	
Oscillatoria	Cyano	*Synura*	Chryso	*Cyclotella*	Bacillar
Rivularia	Cyano	*Tabellaria*	Bacillar	*Fragillaria*	Bacillar
Synedra	Bacillar	*Tribonema*	Chloro	*Melosira*	Bacillar
Spicy odour		*Uroglenopsis*	Chryso	*Staphanodiscus*	Bacillar
Asterionella	Bacillar	*Volvox*	Chloro	*Tabellaria*	Bacillar

Table 12.6B. Representative species of algae causing taste and odour problems.

7☞

Cyanobacterium (Blue-Green algae)	**Bacillariophyta (Diatoms)**
Anabaena circinalis	*Asterionella gracillima*
Anabaena planctonica	*Cyclotella compta*
Anacystis cyanea	*Diatoma vulgare*
Aphanizomenon flos-aquae	*Fragilaria construens*
Cylindrospermum musicola	*Stephanodiscus niagarae*
Gomphosphaeria lacustris	*Synedra ulna*
Oscillatoria curviceps	*Tabellaria fenestra*
Rivularia haematites	**Euglenophyta (Euglenas)**
Charophyta (Stoneworts)	*Euglena sanginea*
Chara vulgaris	**Pyrrhophyta (Dinoflagellates)**
Nitella gracilis	*Ceratium hirundinella*
Chlorophyta (Green algae)	*Peridinium cinctum*
Chlamydomonas globosa	*Glenodinium palustre*
Cladophora insignis	**Chrysophyta (yellow-green, yellow-brown algae)**
Cosmarium portianum	*Chrysosphaerella longispina*
Distyosphaerium ehrenbergianum	*Dinobryon divergens*
Gloeocystis planctonica	*Mallomonas caudata*
Hydrodictyon reticulatum	*Synura uvella*
Pandorina morum	*Uroglenopsis americana*
Pediastrum tetras	**Cryptophyta (Cryptomonads)**
Scenedesmus abundans	*Cryptomonas erosa*
Spirogyra majuscula	
Staurastrum paradoxum	
Volvox aureus	

Table 12.7. Pollution algae - algae common in organically enriched areas

7☞

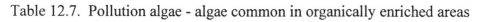

Cyanobacterium (Blue-green algae)	**Chlorophyta (Swimming green algae)**
Agmenellum quadriduplica - tumtenuissima type	*Carteria multifilis*
Anabaena consricts	*Chlamydomonas reinhardi*
Anacystis montana	*Chlorogonium euchlorum*
Arthrospira generi	*Cryptoglena pigra*
Lyngbya digueti	*Pandorina moum*
Oscillatoria chalybea	*Pyrobotrys gracilis*
Oscillatoria chlorina	*Pyrobotrys stellata*
Oscillatoria formosa	*Spondylomorum quaternarium*
Oscillatoria lauterbornii	**Bacillariophyta (Diatoms)**
Oscillatoria limosa	*Gomphonema parvulum*
Oscillatoria princeps	*Melosira varians*
Oscillatoria putrida	*Navicula cryptocephala*
Oscillatoria tenuis	*Nitzschia acicularis*
Phormidium autumnale	*Nitzschia palea*
Phormidium uncinatum	*Surirella ovata*
Chlorophyta (Non-swimming green algae)	**Euglenophyta (Euglenas)**
Chlorella pyrenoidosa	*Cryptoglena pigra*
Chlorella vulgaris	*Euglena agilis*
Chlorococcum humicola	*Euglena deses*
Scenedesmus quadricola	*Euglena oxyuris*
Spirogyra communis	*Euglena polymorpha*
Stichtococcus bacillaris	*Euglena viridis*
Stigeoclonium tenue	*Leptocinclis ovum*
Tetraedron muticum	*Leptocinclis texta*
	Phacus pyrum

Table 12.8 Algae characteristic of clean waters.

7☞

Cyanobacterium (Blue-green algae)	**Rhodophyta (Red algae)**
Agmenellum quadriduplicatum, glauca type	*Batrachospermum vagum*
Calothrix parietina	*Hildenbrandia rivularis*
Coccochloris stagnina	*Lemanea annulata*
Entophysalis lemaniae	**Bacillariophyta (Diatoms)**
Microcoleus subtorulosus	*Amphora ovalis*
Phormidium inundatum	*Cocconeis placentula*
Chlorophyta (non-swimming green algae)	*Cyclotella bodanica*
Ankistrodesmus falcatus	*Cymbella cesati*
Bulbochaeta mirabilis	*Meridion circulare*
Chaetopeltis megalocystis	*Navicula exigua var. capitata*
Cladophora glomerata	*Navicula gracilis*
Draparnaldia plumosa	*Nitzschia linearis*
Euastrum oblongum	*Pinnularia subcapitata*
Gleococcus schroeteri	*Surirella splendida*
Micrasterias truncata	*Synydra acus var. angustissima*
Rhizoclonium hieroglyphicum	**Chrysophyta (yellow-green, -brown algae)**
Staurastrum punctulatum	*Chromulina rosanoffi*
Ulothrix aequalis	*Chrysococcus major*
Vaucheria geminata	*Chrysococcus ovalis*
Chlorophyta (swimming green algae)	*Chrysococcus rufescens*
Phacotus lenticularis	*Mallomonas caudata*
Euglenophyta (Euglenas)	*Dinobryon stipitatum*
Euglena ehrenbergii	**Cryptophyta (Cryptomonads)**
Euglena spirogyra	*Chroomonas nordstetii*
Phaeophyta (Brown algae)	*Chroomonas setoniensis*
Lithoderma	*Rhodomonas lacustris*

Navicula, *Nitzschia*) and euglenoids (e.g. *Euglena*, *Phacus*) that are represented in organically enriched waters, but by different species, demonstrating the importance of identifying algae to species and not just to the genus level. Nevertheless, as discussed above, a high diversity of algal phyla (at least ten) and the presence of some centrate forms of diatoms are good clues that the water quality is good.

Algae may also be used to identify the probable source of water. Species restricted to hard waters are the charophyte, *Chara*, and the filamentous greens, *Spirogyra crassa* and *Spirogyra decimena*. Hard-water, landlocked lakes without an outlet and meromictic lakes have an equal abundance of green and blue-green algae with some euglenoids and yellow-greens (Chrysophyta), while soft-water, landlocked lakes have scant algal flora and filamentous forms are absent; a small diversity of blue-green algae is characteristic of such lakes. Bogs are characterized by a great diversity of desmids while fens are characterized by few algal species and numbers.

Table 12.9 lists algal indicators for different kinds of pollution, other than organic enrichment. Mine wastes with drastically low pH levels reduce the algal flora to a few acid-tolerant species, especially those listed in Table 12.9. Most algae grow best in waters with the pH near neutrality, although a few, such as *Coccocochloris* and *Microcystis*, appear to grow optimally at pH 10, with little or no growth below pH 8. The list is dominated by free-swimming forms of green algae, including desmids and diatoms.

Acidifying rivers (and lakes) are dominated by filamentous forms and exhibit
8☞ three types of benthic algal communities according to Stokes (1980)[7].

(1) *Cyanophycean mats, or blue-green mats*: Felt-like, dark blue-green to blackish mats, sometimes with an orange-coloured carotene-rich surface layer. Filaments of blue-green algae form the bulk of the mat, with some decomposed organic material. Usually in water 2 to 3 m deep. Typical species are those of the genera, *Lyngbya* and *Oscillatoria* (pH = 4.3 - 4.7), *Scytonema* and *Phormidium* (pH = 5.0) and *Phormidium tenue* with the diatoms *Tabellaria* and *Fragillaria* (pH = 4.8 - 5.1).

(2)*Chlorophycean mats, or green mats*: Coarser in appearance than (1) and less densely packed, colour varying from green to reddish purple and extend across the bottom of the lake to a depth of at least 4 m. The dominant (at least 90%) algae are *Pleurodiscus* (*Zygogonium*) and *Mougeotia*.

(3) *Chlorophycean epiphytic or periphytic algae, or green clouds*: Loosely attached clouds or wefts of filamentous green algae in the littoral zone of acidic or acidifying lakes, often associated with macrophytes. Many become detached and remain submerged

---

[7]Stokes, P. M. 1980. Benthic algal communities in acid lakes. *In:* Singer, R. (Ed.). *Effects of acidic precipitation on benthos*. Proceedings of a symposium at Colgate University, Hamilton, N.Y., Aug. 8-9, 1980. Published by the North American Benthological Association, Box 878, Illinois Environmental Protection Agency, Springfield, Illinois 62705.

Table 12.9. Algae tolerant of acidifying waters, thermal pollution, pulp mill wastes, oil wastes and metal wastes. See table 12.6A for definitions of abbreviated phyla.

Acid Tolerant Species	Phylum	Tolerant of Pulp Mill Waste	Phylum	Tolerant of Chromium	Phylum
Actinella	Bacillar	*Amphora ovalis*	Bacillar	*Closterium acerosum*	Chloro
Chlamydomonas	Chloro	*Caloneis amphisbaena*	Bacillar	*Euglena acus*	Eugleno
Chromulina ovalis	Chryso	*Cocconeis diminuta*	Bacillar	*Euglena oxyuris*	Eugleno
Cryptomonas erosa	Crypto	*Cocconeis pediculus*	Bacillar	*Euglena sociabilis*	Eugleno
Euglena adhaerens	Eugleno	*Cymatopleura solea*	Bacillar	*Euglena stellata*	Eugleno
Euglena hiemalis	Eugleno	*Cymbella ventricosa*	Bacillar	*Euglena viridis*	Eugleno
Euglena mutabilis	Eugleno	*Diatoma vulgare*	Bacillar	*Navicula atomus*	Bacillar
Euglena stellata	Eugleno	*Gomphonema herculaneum*	Bacillar	*Navicula cuspidata*	Bacillar
Euglena tatrica	Eugleno	*Navicula cryptocephala*	Bacillar	*Nitzschia linearis*	Bacillar
Euglena viridis	Eugleno	*Navicula radiosa*	Bacillar	*Nitzschia palea*	Bacillar
Eunotia exigua	Bacillar	*Surirella ovata*	Bacillar	*Stigeoclonium tenue*	Chloro
Eunotia lunaris	Bacillar	*Surirella o. var. salina*	Bacillar	**Tolerant of Iron Wastes**	
Eunotia trinacria	Bacillar	*Synedra pulchella*	Bacillar	*Chlorella varigata*	Chloro
Lepocinclis ovum	Eugleno	*Synedra ulna*	Bacillar	*Gomphonema acuminatum*	Bacillar
Navicula subtilissima	Bacillar	**Tolerant of Copper Wastes**		*Pinnularia microstauron*	Bacillar
Navicula viridis	Bacillar	*Achnanthes affinis*	Bacillar	*Stauroneis phoenicentron*	Bacillar
Penium cucurbitinum	Chloro	*Asterionella formosa*	Bacillar	*Stenopterobia intermedia*	Bacillar
Stauroneis anceps	Bacillar	*Calothrix braunii*	Cyano	*Surirella delicatissima*	Bacillar
Tabellaria flocculosa	Bacillar	*Chlorococcum botryoides*	Chloro	*Surirella linearis*	Bacillar
Ulothrix zonata	Chloro	*Cymbella naviculiformis*	Bacillar	*Trachelomonas hsipida*	Chryso
Xanthidium antipoaeum	Chloro	*Cymbella ventricosa*	Bacillar	**Tolerant of Oil Pollution**	
Tolerant of > 40 °C		*Navicula viridula*	Bacillar	*Amphora ovalis*	Bacillar
Mastigocladius laminosus	Cyano	*Neidium bisulcatum*	Bacillar	*Diatoma vulgare*	Bacillar
Oscillatoria filiformis	Cyano	*Nitzschia palea*	Bacillar	*Gomphonema herculaneum*	Bacillar
Phormidium bijahensis	Cyano	*Scenedesmus obliquus*	Chloro	*Melosira varians*	Bacillar
Phormidium geysericola	Cyano	*Stigeoclonium tenue*	Chloro	*Navicula radiosa*	Bacillar
Phormidium laminosum	Cyano	*Symploca erecta*	Cyano	*Surirella molleriana*	Bacillar
				Synedra acus	Bacillar
				Synedra ulna	Bacillar

as free-floating clouds which may overwinter and appear as whitish or greyish clouds in the spring. In lakes below pH 5.0, the dominant species are of the genera *Spirogyra* with *Zygnema*; *Pleurodiscus* (*Zygogonium*) and *Mougeotia*. In lakes with pH above 5.0, the dominant species are of the genera *Oedogonium* and *Bulbochaete*.

In all three types, the water is very clear and has characteristics of oligotrophic lakes. The increased water clarity with increasing acidity has lead to the use of the term, "***oligotrophication***". However, most species characteristic of truly oligotrophic lakes are absent in acidified lakes (cf Tables 12.8 and 12.9). The reasons for the sudden appearance of huge mats and clouds in acidified or acidifying lakes is still conjectural, but Stokes (1980) suggests that one or more of the following may apply:

1. The species have a preference for low pH
2. Invertebrate grazing is reduced at low pH
3. Microbial decomposition is decreased at low pH
4. Competition from other less acid tolerant algae is decreased at low pH

Many species (e.g. *Caloneis amphisbaena, Cymbella ventricosa, and Surirella ovata*) listed for pulp mill wastes in Table 12.9 are also tolerant of hydrogen sulphide, a common waste product of pulp mill effluent. Hydrogen sulphide at concentrations exceeding 3.9 mg/L is toxic to most diatoms. It is interesting to note that diatoms and a few species of euglenoids and green algae form the majority of the species that are tolerant to pulp mill wastes, oil and metal pollution. Species of *Stigeoclonium*, *Tetraspora, Closterium, Nitzschia, Navicula* and *Euglena* are remarkably resistant to the presence of chromium and iron.

Before examining some algal indices that are being used to assess water quality, we will look at an indicator concept that does not rely so heavily on species identifications. The concept relates bacterial and algal communities to saprobic zones of the saprobien system discussed earlier. Water quality is determined simply by relating the type of community present and matching it to a zone in Table 12.10. The letters "a", "b" and "c" are alternatives and the numbers "1", "2" and "3" are differences in degree of pollution. Zones I to IV correspond roughly to the zone of active decomposition, while Zones V and VI correspond to the septic zone, Zone VII to the recovery zone and Zones VIII and IX to the clean water zone in North American terminology. The main criticisms of the saprobien system are: (i) it assumes that all rivers have one of the communities listed; (ii) it does not have sufficient resolution to divide the continuum of downstream changes in self-purification into nine different zones. The coprozoic zone consists only of anaerobic bacteria or, *Boda edax*, a zooflagellate that occurs in stagnant water. The *Rhodo-Thio* bacteria and Chlorobacteria of the α-polysaprobic zone are purple and green sulphur bacteria that photosynthetically convert hydrogen sulphide into sulphur compounds. *Thiothrix* and *Beggiatoa* are chemosynthetic sulphur bacteria that also oxidize reduced forms of sulphur and represent the next step in self-purification of the water. The presence of sewage fungus (*Sphaerotilus natans*) is a good indication of septic conditions that characterize the γ-polysaprobic zone. The next step in

Table 12.10. Saprobic zones and corresponding bacterial and/or algal communities.

Zone	Community	Zone	Community
I. Coprozoic	a. Bacteria only b. *Bodo* only c. Both communities	VI. β-mesosaprobic	a. Mainly *Cladophora fracta* b. Mainly *Phormidium*
II. α-polysaprobic	1. Mainly *Euglena* 2. *Rhodo-Thio* bacteria only 3. Chlorobacteria only	VII. γ-mesosaprobic	a. *Batrachospermum or Lemanea* b. *Cladophora glomerata*
III. β-polysaprobic	1. Mainly *Beggiatoa* 2. Mainly *Thiothrix nivea* 3. Mainly *Euglena*	VIII. Oligposaprobic	a. Chlorophytes b. *Meridion circulare* only c. Any rhodophyte d. Mainly *Vaucheria sessilis* e. Mostly *Phormidium inundatum*
IV. γ-polysaprobic	1. Mainly *Oscillatoria chlorina* 2. *Sphaerotilus natans*		
V. α-mesosaprobic	a. Mainly *Ulothrix zonata* b. *Oscillatoria benthonicum (O. brevis, limnosa, splendida,* with *O. subtilissima, princeps, tenuis* present as associate spp) c. Mainly *Stigeoclonium tenue*	IX. Katharobic	a. *Chlorotylium cataractum* and *Draparnaldia plumosa* b. *Hildenbrandia rivularis* c. Lime-encrusting spp. (e.g. *Chamaesiphon* polonius, *Calothrix* spp.)

self-purification (zones α-mesosaprobic to γ-mesosaprobic) is the mineralization of organic compounds by highly tolerant autotrophs, or enrichment indicator algae (see Table 12.6). These are gradually replaced by low-oxygen sensitive species that make up the clean-water or oligotrophic indicator group listed in Table 12.8.

Algal Indices

The saprobic system has been variously modified and adopted as a preliminary standard procedure to assess water quality in rivers in many European countries but has not gained wide acceptance in North America. Nevertheless, it is given here because of its importance in the development of North American algal indices. Each species is assigned a number from 1 to 4 in each of several samples of water from a lake or river. The number is based on its ranking (s, not provided here) in the saprobity system and its relative abundance (n), 1 being rare, 3 being common, and 5 being very abundant. The abundance score can be determined by dividing the most abundant value by 5 to provide five abundance ranges and then assigning a value of 1 to 5 based on the abundance of the species in question. For example, suppose the maximum number observed for any species was 500. The abundance range for each ranking is 500/5 = 100. Therefore, the abundance ranking would be 1 for 1-100 organisms, 2 for 101-200 organisms, 3 for 201-300 organisms, 4 for 301-400 organisms and 5 for 401-500 organisms. The saprobity index, S, is calculated, as explained earlier for bacteria, from the formula:

$$S = \sum(s \times n)/\sum n$$

The water quality assessment is determined on the basis of the following criteria:

Saprobity Index	Degree of Pollution
1.0 - 1.5	Very slight
1.5 - 2.5	Moderate
2.5 - 3.5	Heavy
3.5 - 4.0	Very heavy

A ***biological index of pollution*** **(BIP)** has also been developed to assess water quality, again in rivers, based on the proportions of producers (algae = A) and consumers (non-chlorophyll-bearing animals = C) in a unit sample of water. The BIP is calculated as:

$$BIP = (C \times 100)/(A + C)$$

The assessment of water quality is then based on the following BIP criteria:

BIP Value	Degree of Pollution
0.6 - 12.0	Clean water
12.1 - 30.8	Moderate decomposition
30.9 - 55.1	Active decomposition
> 55.1	Septic conditions

In an effort to simplify assessments based on algal communities, a procedure was developed by Palmer (1969)[8] whereby an algal index was based on a composite of 20 common genera, each with a pollution index factor (Table 12.11). To use the index, all

---

[8]Palmer, C. M. 1969. A composite rating of algae tolerating organic pollution. J. Phycology 5: 78-82.

20 genera observed in a sample are recorded, providing more than 5 or more individuals (per slide sample) of a particular genus are present. The index factors of the genera present are then totaled for a pollution index. The water quality assessment (for rivers) is then based on ***composite pollution index (CPI)*** criteria, given below Table 12.11.

Table 12.11. Key algae genera for calculating a composite pollution index

Genus	Pollution Index	Genus	Pollution Index
Anacystis (Microcystis)	1	*Micractinium*	1
Ankistrodesmus	2	*Navicula*	3
Chlamydomonas	4	*Nitzschia*	3
Chlorella	3	*Oscillatoria*	5
Closterium	1	*Pandorina*	1
Cyclotella	1	*Phacus*	2
Euglena	5	*Phormidium*	1
Gomphonema	1	*Scenedesmus*	4
Leptocinclis	1	*Stigeoclonium*	2
Melosira	1	*Synedra*	2

CPI	Degree of Pollution
< 10	Clean water*
11 - 14	Moderate pollution
15 - 19	Probable organic pollution
≥ 20	High organic pollution

*Palmer cautioned that a low pollution index could also mean that some substance or factor interfering with growth of algae is present.

Toxicity of Algae

So far, only the toxicity of substances to algae have been considered, but algae can be toxic to organisms too, including humans. A toxin produced by the marine, armoured flagellate, *Gonyaulax*, has caused serious illness in people who have eaten clams that have fed on this algae. The toxin is reportedly ten times more toxic than strychnine to mice. Blooms of *Gonyaulax* and *Gymnodinium* create conditions known as "***red tide***", "***red water***" or "***yellow-green peril***" that cause serious illnesses to humans and are known to kill fish and other marine animals that feed on the armoured flagellates.

In fresh waters, nearly all of the toxic algae are blue-greens. *Anabaena flos-aquae* is known to produce toxins that are fatal to cattle and chickens and cause a contact type of dermatitis in humans. *Microcystis aeruginosa* and *Lyngbya* produce toxins that induce symptoms of hay fever. Other species that produce algal poisonings are *Aphanizomenon flos-aquae, Gloeotrichia echinulata, Coelosphaerium kutzingianum* and *Nodularia spumigena.* Apparently disintegration of large amounts of blue-green algae on sand filters of water treatment plants and the passage of toxic products into the distribution system can cause gastro-intestinal disturbances. The dinoflagellate, *Oodinium ocellatum,* parasitizes the gills of small fishes and, through interference with respiratory processes, is fatal to the fishes.

The factors that contribute to the development of concentrated blooms of blue-green algae are poorly understood, except they always occur in eutrophic waters during warm, sunny weather. Depending on the species, algal toxins are either excreted or released into the water during decomposition. Bacteria may also release toxins. The toxicity of water blooms depends on differences in algal and bacterial strains, species, age, accumulation, secretion and decomposition state, as well as differences in animal susceptibility and dosage.

Benthic Macrophytes

No indices have been developed to assess water quality based on the macrophyte community. Chapter 6 lists the probable trophic status of several species based on their
9☞ distributions with respect to hardness and conductivity. Stuckey (1975)[9] is among the few to have published a list of indicator macrophytes, a summary of which is provided here. The list is applicable to both rivers and lakes.

As rivers and lakes undergo natural eutrophication, the littoral community gradually transforms from one dominated by submersed species to one dominated by emergent species, including those with floating leaves, their roots either anchored on the bottom or floating with the leaves. The submersed community is also represented, but only the forms that can tolerate warm, turbid waters and organic sediments enriched with nutrients remain. While the remaining submersed species must be able to utilize the lower levels of light within the water column, the emergent forms rely only on light impinging above the surface of the water. Studies have shown that the submersed species which do survive the eutrophication process have wide ecological tolerances. Those that disappear in the eutrophication process tend to have very narrow ecological tolerances. Table 12.12 lists the two groups of submersed macrophytes, the sensitive species

---

[9]Stuckey, R. L. 1975. Submersed aquatic vascular plants as indicators of environmental quality. In: King, C. C. and L. E. Elfner (Eds.). Organisms of biological communities as indicators of environmental quality. Proceedings of a symposium at Ohio State University, Columbus, Ohio, March 25, 1974. Published by Ohio State University.

representing the probable indicators of "good" water quality and the tolerant species representing the probable indicators of "poor" water quality.

Table 12.12. Submersed species of macrophytes with probable indicator value. Species are listed alphabetically.

Probable indicators of "good" water quality	Probable indicators of "poor" water quality
Elodea canadensis	*Ceratophyllum demersum*
Megalodonta beckii	*Heteranthera dubia*
Najas flexilis	*Myriophyllum spicatum*
Potamogeton amplifolius	*Najas minor*
Potamogeton friesii	*Potamogeton crispus*
Potamogeton gramineus	*Potamogeton foliosus*
Potamogeton praelongus	*Potamogeton nodosus*
Potamogeton zosteriformis	*Potamogeton pectinatus*
Ranunculus longirostris	*Potamogeton pusillus*
Vallisneria americana	*Potamogeton richardsonii*
	Zannichellia palustris

Benthic Macroinvertebrates

The numbers of indices based on the benthic macroinvertebrate communities is
10☞ probably about five times that of any of the groups discussed so far, with about fifty indices currently in existence, and the number is still growing. Some of the benthic indices are based on specie's identification, the species assemblages being analyzed by a range of mathematical models, from a fairly straight forward species diversity index to more complex multivariate analyses, such as TWINSPAN. Many bioassessment methods require numerous quantitative samples (e.g. with Ekman grabs, T-samplers, Surber samplers, etc.) that need a great deal of time to sort and *separate* all the invertebrates, and more time and expertise (and money) to *identify* all the organisms. So recent trends have been towards the use of more rapid biological techniques, such as semi-quantitative collecting methods (e.g. kick-and-sweep) that require one to select (at random) and identify only the first 100 organisms in the sample. Some of these rapid bioassessment techniques have been standardized so that water quality comparisons can be made between streams. These standardized methods are in common use today and are termed

***RBPs* (*Rapid Biological Protocols*).** The Environmental Protection Agency (EPA)[10] developed five RBPs, the first three being based on benthic macroinvertebrates and the fourth and fifth on fish. The complexity of the protocol increased with the RBP number, RBP I being less complex than RBP II, which is less complex than RBP III. Similarly, RBP V is more complex than RBP IV. RBPs using benthic algae (periphyton) have also been developed by E. P. A. but are used in only three or four states but otherwise have not been widely used so are not discussed herein.

RBP I is used to discriminate obviously impacted and non-impacted areas from potentially affected areas requiring further investigation. It allows rapid screening of a large number of sites. Areas identified for further study can be rigorously evaluated using RBP II, III, and V (IV is a questionnaire survey). RBP III is examined here (RBP V is examined in the Chapter on fish), in relation to the new RBP protocol of Barbour et al (1999) . RBPs using benthic algae (periphyton) have also been developed, but they are used by only a few states and are not examined herein.

The RBP protocols were revised in 1999 (EPA)[11] but the basic principles developed by Plafkin et al in 1989 in developing the first RBP protocols (I to V) were retained. The revised RBP III uses two separate procedures that are oriented toward a "single, most productive" habitat and a multihabitat approach. The primary differences between the original RBP II and III are the decision on field versus lab sorting and level of taxonomy. In addition, a third protocol has been developed as a more standardized biological reconnaissance or screening and replaces RBP I of the original document. We will also examine the revised RBP III in this chapter.

The use of indicator organisms, species diversity, functional feeding groups, and some of the bioassessment methods currently being used by North Americans, Britains and Australians are also examined. The oldest concepts are examined first, beginning with the indicator organism concept and then the saprobic system, followed by species diversity. Then several other bioassessment methods that were once popular and instrumental in the development of current North American and European protocols are examined. The same data set (Table 12.21) is used to calculate the different indices and then the evaluation of water quality is made for each one.

---

[10]Plafkin, J. L., M. T. Barbour, K. D. Porter, S. K. Gross and R. M. Hughes. 1989. Rapid ioassessment protocols for use in streams and rivers: Benthic macroinvertebrates and fish. U. S. Environmental Protection Agency, Assessment and Watershed Protection Division, 401, M. Street, S. W., Washington, D.C. 20460.

[11]Barbour, M.T., J. Gerritsen, B.D. Snyder, and J.B. Stribling. 1999. Rapid Bioassessment Protocols for Use in Streams and Wadeable Rivers: Periphyton, Benthic Macroinvertebrates and Fish, Second Edition. EPA 841-B-99-002. U.S. Environmental Protection Agency; Office of Water; Washington, D.C.

Indicator Organisms

Chapter 8 describes the indicator value of several species of invertebrates and Tables 12.14 and 12.20 and Appendix 12.II give tolerance values of numerous species. 10☞ One can often "guestimate" the type of water quality by merely listing the indicator value (clean or oligotrophic, subpolluted or mesotrophic, polluted or eutrophic) of the ***dominant species***, or those that make up at least 5% of the benthic community and then assigning water quality based on the most abundant indicator type. For example, the benthic community shown in Table 12.13 would be assessed as clean to moderately polluted because it is dominated (i.e. ≥ 5%) by species that are either clean water indicators (~50%) or subpollution indicators (~50%). Use of indicator organisms gives about the same water quality evaluation as most other bioassessment methods (Table 12.13).

While the indicator organism metric is not advocated as a rigorous bioassessment method, it does provide a quick evaluation of water quality. However, it can be used only by experienced benthologists because species or genus identifications is required. It is also important to note the relative abundances of the indicator organisms because a different evaluation would be obtained based on mere presence or absence. For example, the mere presence of *Limnodrilus hoffmeisteri* would rate the sample from a polluted river, an assessment that clearly does not agree with most other assessments.

Saprobic Index

The saprobien system was described earlier. It relies heavily on bacterial 10☞ communities but also accommodates other communities, like the benthic community.

The benthic saprobic index was developed to elucidate conditions in slowly moving rivers with organic enrichment. The index tends to break down in streams where slow reaches are separated by riffle areas and in short turbulent stretches of streams. A major shortcoming is the valences assigned to the organisms are based solely on their sensitivity to biodegradable wastes, whereas more recent indices reflect sensitivities to other kinds of pollution. The index only works if the organisms are identified to species, or to genus in some cases. However, with these caveats in mind, the index gives comparable evaluations of organic enrichment to other indices, as indicated in Table 12.13.

The saprobic index is calculated in much the same way as for the algae, discussed earlier, using the formula:

$$S = \sum (s \times h) / \sum h,$$

where h = abundance according to the scale 1 for least abundant, 3 for moderately abundant and 5 for most abundant, and s = the valence or saprobic score for each species.

For the data in Table 12.13, the abundance values were assigned objectively, using h = 1 for 1-104 organisms, h = 2 for 105-208, h = 3 for 209-312, h = 313-416 and h

Table 12.13. Assessment of water quality based on the assemblage of macroinvertebrates in a hypothetical stream community. Ind = Indicator value, C = clean, S = Subpolluted for species that make up 5% or more of total; Sapro = Saprobic index; d = mean diversity with -p $\log_{10}$ p values given; Trent = Trent Index; Chan = Chandler Index; Beak = Beak Index; Sign = SIGNAL Index; BMWG = BMWG Index; ETP = EPT Index; HBI = Hilsenhoff's Biotic Index; PMA = Percent Model Affinity Index; Rich = Species Richness Index; SCI = Sequential Comparison Index. See text for descriptions of all indices.

Species	#/m²	Ind	Sapro	d	Trent	Chan	Beak	SIGN	ASTP	ETP	Hils. I	PMA	Rich	SCI
Pisidium fallax	522	C-S	6	-0.136	1	18	2.5 points				102		✓	see text
Pisidium compressum	245	C-S	4.5	-0.093	1	22		6	3		48		✓	
Sphaerium fabale	256	C	4.5	-0.095	1	22					54		✓	
Elimia flavescens	203	C-S	2	-0.083	1	25		5	4		42		✓	
Limnodrilus hoffmeisteri	52		3.6	-0.032	1	22	0 point	1	1		20	3%	✓	
Hyalella azteca	165	C-S	2	-0.072	1	40		5	2		48		✓	
Orconectes viridis	11		1	-0.009	1	84		6	8					
Taeniopteryx sp.	180	C	3	-0.077	1	98		8	10	✓	12	1%	✓	
Stenonema sp.	440	C	9	-0.126	1	94		7	10	✓	45		✓	
Rithrogena sp.	203	C	0.2	-0.083	1	80				✓	0	15%	✓	
Ephemera guttulata	82		3.8	-0.045	1	91	1.5 points	8	10	✓	21		✓	
Hydropsyche scalaris	142	C-S	4.8	-0.066	1	34		5	5	✓	10		✓	
Rhyacophila sp.	28		1.1	-0.020	1	65		9	7	✓	1	3%	✓	
Helicopsyche borealis	34		1.5	-0.023	1	75		8	7	✓	3		✓	
Optiocervus ovalis	143	C-S	3	-0.066	1	61		7	5		16	6%	✓	
Rheocricotopus sp.	18		2	-0.014	1	28		1	2			20%		
Simulium latipes	276	C	3.6	-0.099	1	73		5	5		36	1%	✓	
Totals	3000	5 Cs, 6 C-Ss	55.6	-1.139	18	932		81	79		458	94	16	
Index Value			1.5	3.783	9		4	5.8	5.6	7	4.58	53	18	11.2
Assessment		C-S	slight	C	c-sligh	moder	slight	Doubt	Doubt	Slight I	slight I	slight I	mod im	mod

= 417-522. The valence values were chosen from Sladacek (1970) (cited earlier). For species not listed by Sladacek (1970), valence values were subjectively assigned a tolerance score. The S values and water quality assessment shown in Table 12.13 were obtained from calculations and the criteria shown in Table 12.14. The $\sum (s \times h) = 55.6$ and $\sum h = 37$, yielding an S value of 1.5 and an evaluation of clean to slight pollution.

Table 12.14. Calculation of the saprobic index for evaluating water quality based on the benthic community shown in Table 12.13 and criteria for evaluating water quality. The usual levels of dissolved oxygen (% saturation) and 5 day, 20 °C biochemical oxygen demand (B.O.D. mg/L) associated with each evaluation are also provided.

Species	**#/m²**	**valence s**	**density h**
Pisidium fallax	522	1.2	5
Pisidium compressum	245	1.5	3
Sphaerium fabale	256	1.5	3
Elimia flavescens	203	2	2
Limnodrilus hoffmeisteri	52	3.6	1
Hyalella azteca	165	1	2
Orconectes viridis	11	1	1
Taeniopteryx sp.	180	1.5	2
Stenonema sp.	440	1.8	5
Rithrogena sp.	203	0.2	1
Ephemera guttulata	82	1.9	2
Hydropsyche scalaris	142	2.4	2
Rhyacophila sp.	28	1.1	1
Helicopsyche borealis	34	1.5	1
Optiocervus ovalis	143	1.5	2
Rheocricotopus sp.	18	2	1
Simulium latipes	276	1.2	3
Totals	3000	26.9	37

Saprobic Criteria			
S Value	**Evaluation**	**Usual values of:**	
		Diss. O_2 %	B.O.D mg/L
1.0-1.5	Clean, slightly polluted	> 50	< 2.5
1.5-2.5	Moderate pollution, lower recovery	40-50	2.5-5
2.5-3.5	Active decomposition, upper recovery	20-40	5-10
3.5-4.0	Heavy pollution, fouled water	0	10-50 or greater

Species Diversity Index

Wilhm and Doris (1968)[12] first advocated the use of information theory to assess
10☞ water quality. They calculated mean diversity (d) of benthic invertebrates in several streams that varied in water quality from clean to polluted, using the expression:

$$d = -\sum(n_i/N)\log_2(n_i/N)$$

where

n_i is the number of individuals in species "i"
N is the total number of individuals in the sample.

After correlating the mean diversity values to several physical and chemical criteria, Wilhm and Doris (1968) developed the following criteria to assess water quality:

Mean diversity	**Water Quality**
< 1	Polluted
1-3	Subpolluted
> 3	Clean

To calculate a "d" value for a species, simply divide the number of individuals for the species by the total number of individuals in the entire sample. Then multiply the n_i/N value by its logarithm (base 2). Because the value (n_i/N) is a fraction, the log value will be negative. Repeat the procedure for all other species and then total up the (n_i/N) log (n_i/N) values. Since logarithm tables to base 10 are much more common than logarithm tables to base 2, one can calculate the d value as above using $\log_{10}$ tables and then multiply by -3.322 to convert to log base 2. In fact, the procedure can be streamlined even more by using Table 12.15 for each n_i/N value, where $n_i/N = p$ in the table. For example, suppose one has four species (A, B, C, D) with the following abundances in the sample: A = 25, B = 93, C = 10, D = 30, for an N value of 148. The n_i/N values for each are: A = 0.168, B = 0.628, C = 0.067, D = 0.202. The $p \log_{10} p$ values (Table 12.15) for A to D are: A = 130, B = 127, C = 079, D = 140, which (see note above) are actually A = -0.130, -0.127, -0.079, -0.140. The sum = -0.476. To convert this $p \log_{10} p$ value to a $p \log_2 p$ value, multiply it by -3.322 to obtain 1.58. Using Wilhm and Doris's (1968) criteria, this particular benthic sample is predicted to have come from a subpolluted river because the d value is between 1 and 3.

The species diversity index has several advantages. It is a good, objective, numerical approach that is easily reported. The index has a reasonable theoretical basis

---

[12]Wilhm, J. L. and T. C. Doris. 1968. Biological parameters for water quality. Bioscience 18: 477-481.

Table 12.15. Table for converting n_i/N to p $\log_{10}$p when calculating diversity, where $n_i/N = p$. Top row is third decimal place, e.g. p $\log_{10}$p of 0.045 is -0.061.

	0	1	2	3	4	5	6	7	8	9	10
0.00	0.000	-0.003	-0.005	-0.008	-0.010	-0.012	-0.013	-0.015	-0.017	-0.018	-0.020
0.01	-0.020	-0.022	-0.023	-0.025	-0.026	-0.027	-0.029	-0.030	-0.031	-0.033	-0.034
0.02	-0.034	-0.035	-0.036	-0.038	-0.039	-0.040	-0.041	-0.042	-0.043	-0.045	-0.046
0.03	-0.046	-0.047	-0.048	-0.049	-0.050	-0.051	-0.052	-0.053	-0.054	-0.055	-0.056
0.04	-0.056	-0.057	-0.058	-0.059	-0.060	-0.061	-0.062	-0.062	-0.063	-0.064	-0.065
0.05	-0.065	-0.066	-0.067	-0.068	-0.068	-0.069	-0.070	-0.071	-0.072	-0.073	-0.073
0.06	-0.073	-0.074	-0.075	-0.076	-0.076	-0.077	-0.078	-0.079	-0.079	-0.080	-0.081
0.07	-0.081	-0.082	-0.082	-0.083	-0.084	-0.084	-0.085	-0.086	-0.086	-0.087	-0.088
0.08	-0.088	-0.088	-0.089	-0.090	-0.090	-0.091	-0.092	-0.092	-0.093	-0.094	-0.094
0.09	-0.094	-0.095	-0.095	-0.096	-0.097	-0.097	-0.098	-0.098	-0.099	-0.099	-0.100
0.10	-0.100	-0.101	-0.101	-0.102	-0.102	-0.103	-0.103	-0.104	-0.104	-0.105	-0.105
0.11	-0.105	-0.106	-0.106	-0.107	-0.108	-0.108	-0.109	-0.109	-0.110	-0.110	-0.111
0.12	-0.111	-0.111	-0.111	-0.112	-0.112	-0.113	-0.113	-0.114	-0.114	-0.115	-0.115
0.13	-0.115	-0.116	-0.116	-0.117	-0.117	-0.117	-0.118	-0.118	-0.119	-0.119	-0.120
0.14	-0.120	-0.120	-0.120	-0.121	-0.121	-0.122	-0.122	-0.122	-0.123	-0.123	-0.124
0.15	-0.124	-0.124	-0.124	-0.125	-0.125	-0.126	-0.126	-0.126	-0.127	-0.127	-0.127
0.16	-0.127	-0.128	-0.128	-0.128	-0.129	-0.129	-0.129	-0.130	-0.130	-0.130	-0.131
0.17	-0.131	-0.131	-0.131	-0.132	-0.132	-0.132	-0.133	-0.133	-0.133	-0.134	-0.134
0.18	-0.134	-0.134	-0.135	-0.135	-0.135	-0.136	-0.136	-0.136	-0.136	-0.137	-0.137
0.19	-0.137	-0.137	-0.138	-0.138	-0.138	-0.138	-0.139	-0.139	-0.139	-0.140	-0.140
0.20	-0.140	-0.140	-0.140	-0.141	-0.141	-0.141	-0.141	-0.142	-0.142	-0.142	-0.142
0.21	-0.142	-0.143	-0.143	-0.143	-0.143	-0.144	-0.144	-0.144	-0.144	-0.144	-0.145
0.22	-0.145	-0.145	-0.145	-0.145	-0.146	-0.146	-0.146	-0.146	-0.146	-0.147	-0.147
0.23	-0.147	-0.147	-0.147	-0.147	-0.148	-0.148	-0.148	-0.148	-0.148	-0.149	-0.149
0.24	-0.149	-0.149	-0.149	-0.149	-0.149	-0.150	-0.150	-0.150	-0.150	-0.150	-0.151

	0	1	2	3	4	5	6	7	8	9	10
0.25	-0.151	-0.151	-0.151	-0.151	-0.151	-0.151	-0.151	-0.152	-0.152	-0.152	-0.152
0.26	-0.152	-0.152	-0.152	-0.153	-0.153	-0.153	-0.153	-0.153	-0.153	-0.153	-0.154
0.27	-0.154	-0.154	-0.154	-0.154	-0.154	-0.154	-0.154	-0.154	-0.155	-0.155	-0.155
0.28	-0.155	-0.155	-0.155	-0.155	-0.155	-0.155	-0.155	-0.156	-0.156	-0.156	-0.156
0.29	-0.156	-0.156	-0.156	-0.156	-0.156	-0.156	-0.157	-0.157	-0.157	-0.157	-0.157
0.30	-0.157	-0.157	-0.157	-0.157	-0.157	-0.157	-0.157	-0.157	-0.158	-0.158	-0.158
0.31	-0.158	-0.158	-0.158	-0.158	-0.158	-0.158	-0.158	-0.158	-0.158	-0.158	-0.158
0.32	-0.158	-0.158	-0.158	-0.159	-0.159	-0.159	-0.159	-0.159	-0.159	-0.159	-0.159
0.33	-0.159	-0.159	-0.159	-0.159	-0.159	-0.159	-0.159	-0.159	-0.159	-0.159	-0.159
0.34	-0.159	-0.159	-0.159	-0.159	-0.159	-0.159	-0.159	-0.160	-0.160	-0.160	-0.160
0.35	-0.160	-0.160	-0.160	-0.160	-0.160	-0.160	-0.160	-0.160	-0.160	-0.160	-0.160
0.36	-0.160	-0.160	-0.160	-0.160	-0.160	-0.160	-0.160	-0.160	-0.160	-0.160	-0.160
0.37	-0.160	-0.160	-0.160	-0.160	-0.160	-0.160	-0.160	-0.160	-0.160	-0.160	-0.160
0.38	-0.160	-0.160	-0.160	-0.160	-0.160	-0.160	-0.160	-0.160	-0.160	-0.160	-0.159
0.39	-0.159	-0.159	-0.159	-0.159	-0.159	-0.159	-0.159	-0.159	-0.159	-0.159	-0.159
0.40	-0.159	-0.159	-0.159	-0.159	-0.159	-0.159	-0.159	-0.159	-0.159	-0.159	-0.159
0.41	-0.159	-0.159	-0.159	-0.159	-0.159	-0.159	-0.158	-0.158	-0.158	-0.158	-0.158
0.42	-0.158	-0.158	-0.158	-0.158	-0.158	-0.158	-0.158	-0.158	-0.158	-0.158	-0.158
0.43	-0.158	-0.158	-0.157	-0.157	-0.157	-0.157	-0.157	-0.157	-0.157	-0.157	-0.157
0.44	-0.157	-0.157	-0.157	-0.157	-0.157	-0.156	-0.156	-0.156	-0.156	-0.156	-0.156
0.45	-0.156	-0.156	-0.156	-0.156	-0.156	-0.156	-0.156	-0.155	-0.155	-0.155	-0.155
0.46	-0.155	-0.155	-0.155	-0.155	-0.155	-0.155	-0.155	-0.154	-0.154	-0.154	-0.154
0.47	-0.154	-0.154	-0.154	-0.154	-0.154	-0.154	-0.153	-0.153	-0.153	-0.153	-0.153
0.48	-0.153	-0.153	-0.153	-0.153	-0.153	-0.152	-0.152	-0.152	-0.152	-0.152	-0.152
0.49	-0.152	-0.152	-0.152	-0.151	-0.151	-0.151	-0.151	-0.151	-0.151	-0.151	-0.151
0.50	-0.151	-0.150	-0.150	-0.150	-0.150	-0.150	-0.150	-0.150	-0.149	-0.149	-0.149
0.51	-0.149	-0.149	-0.149	-0.149	-0.149	-0.148	-0.148	-0.148	-0.148	-0.148	-0.148

	0	1	2	3	4	5	6	7	8	9	10
0.52	-0.148	-0.148	-0.147	-0.147	-0.147	-0.147	-0.147	-0.147	-0.146	-0.146	-0.146
0.53	-0.146	-0.146	-0.146	-0.146	-0.145	-0.145	-0.145	-0.145	-0.145	-0.145	-0.145
0.54	-0.145	-0.144	-0.144	-0.144	-0.144	-0.144	-0.143	-0.143	-0.143	-0.143	-0.143
0.55	-0.143	-0.143	-0.142	-0.142	-0.142	-0.142	-0.142	-0.142	-0.141	-0.141	-0.141
0.56	-0.141	-0.141	-0.141	-0.140	-0.140	-0.140	-0.140	-0.140	-0.140	-0.139	-0.139
0.57	-0.139	-0.139	-0.139	-0.139	-0.138	-0.138	-0.138	-0.138	-0.138	-0.137	-0.137
0.58	-0.137	-0.137	-0.137	-0.137	-0.136	-0.136	-0.136	-0.136	-0.136	-0.135	-0.135
0.59	-0.135	-0.135	-0.135	-0.135	-0.134	-0.134	-0.134	-0.134	-0.134	-0.133	-0.133
0.60	-0.133	-0.133	-0.133	-0.132	-0.132	-0.132	-0.132	-0.132	-0.131	-0.131	-0.131
0.61	-0.131	-0.131	-0.131	-0.130	-0.130	-0.130	-0.130	-0.129	-0.129	-0.129	-0.129
0.62	-0.129	-0.128	-0.128	-0.128	-0.128	-0.128	-0.127	-0.127	-0.127	-0.127	-0.126
0.63	-0.126	-0.126	-0.126	-0.126	-0.125	-0.125	-0.125	-0.125	-0.125	-0.124	-0.124
0.64	-0.124	-0.124	-0.124	-0.123	-0.123	-0.123	-0.123	-0.122	-0.122	-0.122	-0.122
0.65	-0.122	-0.121	-0.121	-0.121	-0.121	-0.120	-0.120	-0.120	-0.120	-0.119	-0.119
0.66	-0.119	-0.119	-0.119	-0.118	-0.118	-0.118	-0.118	-0.117	-0.117	-0.117	-0.117
0.67	-0.117	-0.116	-0.116	-0.116	-0.115	-0.115	-0.115	-0.115	-0.114	-0.114	-0.114
0.68	-0.114	-0.114	-0.113	-0.113	-0.113	-0.113	-0.112	-0.112	-0.112	-0.111	-0.111
0.69	-0.111	-0.111	-0.111	-0.110	-0.110	-0.110	-0.110	-0.109	-0.109	-0.109	-0.108
0.70	-0.108	-0.108	-0.108	-0.108	-0.107	-0.107	-0.107	-0.106	-0.106	-0.106	-0.106
0.71	-0.106	-0.105	-0.105	-0.105	-0.104	-0.104	-0.104	-0.104	-0.103	-0.103	-0.103
0.72	-0.103	-0.102	-0.102	-0.102	-0.102	-0.101	-0.101	-0.101	-0.100	-0.100	-0.100
0.73	-0.100	-0.099	-0.099	-0.099	-0.099	-0.098	-0.098	-0.098	-0.097	-0.097	-0.097
0.74	-0.097	-0.096	-0.096	-0.096	-0.096	-0.095	-0.095	-0.095	-0.094	-0.094	-0.094
0.75	-0.094	-0.093	-0.093	-0.093	-0.092	-0.092	-0.092	-0.092	-0.091	-0.091	-0.091
0.76	-0.091	-0.090	-0.090	-0.090	-0.089	-0.089	-0.089	-0.088	-0.088	-0.088	-0.087
0.77	-0.087	-0.087	-0.087	-0.086	-0.086	-0.086	-0.085	-0.085	-0.085	-0.084	-0.084
0.78	-0.084	-0.084	-0.084	-0.083	-0.083	-0.083	-0.082	-0.082	-0.082	-0.081	-0.081

	0	1	2	3	4	5	6	7	8	9	10
0.79	-0.081	-0.081	-0.080	-0.080	-0.080	-0.079	-0.079	-0.079	-0.078	-0.078	-0.078
0.80	-0.078	-0.077	-0.077	-0.077	-0.076	-0.076	-0.075	-0.075	-0.075	-0.074	-0.074
0.81	-0.074	-0.074	-0.073	-0.073	-0.073	-0.072	-0.072	-0.072	-0.071	-0.071	-0.071
0.82	-0.071	-0.070	-0.070	-0.070	-0.069	-0.069	-0.069	-0.068	-0.068	-0.068	-0.067
0.83	-0.067	-0.067	-0.066	-0.066	-0.066	-0.065	-0.065	-0.065	-0.064	-0.064	-0.064
0.84	-0.064	-0.063	-0.063	-0.063	-0.062	-0.062	-0.061	-0.061	-0.061	-0.060	-0.060
0.85	-0.060	-0.060	-0.059	-0.059	-0.059	-0.058	-0.058	-0.057	-0.057	-0.057	-0.056
0.86	-0.056	-0.056	-0.056	-0.055	-0.055	-0.054	-0.054	-0.054	-0.053	-0.053	-0.053
0.87	-0.053	-0.052	-0.052	-0.051	-0.051	-0.051	-0.050	-0.050	-0.050	-0.049	-0.049
0.88	-0.049	-0.048	-0.048	-0.048	-0.047	-0.047	-0.047	-0.046	-0.046	-0.045	-0.045
0.89	-0.045	-0.045	-0.044	-0.044	-0.044	-0.043	-0.043	-0.042	-0.042	-0.042	-0.041
0.90	-0.041	-0.041	-0.040	-0.040	-0.040	-0.039	-0.039	-0.038	-0.038	-0.038	-0.037
0.91	-0.037	-0.037	-0.036	-0.036	-0.036	-0.035	-0.035	-0.035	-0.034	-0.034	-0.033
0.92	-0.033	-0.033	-0.033	-0.032	-0.032	-0.031	-0.031	-0.031	-0.030	-0.030	-0.029
0.93	-0.029	-0.029	-0.029	-0.028	-0.028	-0.027	-0.027	-0.026	-0.026	-0.026	-0.025
0.94	-0.025	-0.025	-0.024	-0.024	-0.024	-0.023	-0.023	-0.022	-0.022	-0.022	-0.021
0.95	-0.021	-0.021	-0.020	-0.020	-0.020	-0.019	-0.019	-0.018	-0.018	-0.017	-0.017
0.96	-0.017	-0.017	-0.016	-0.016	-0.015	-0.015	-0.015	-0.014	-0.014	-0.013	-0.013
0.97	-0.013	-0.012	-0.012	-0.012	-0.011	-0.011	-0.010	-0.010	-0.009	-0.009	-0.009
0.98	-0.009	-0.008	-0.008	-0.007	-0.007	-0.006	-0.006	-0.006	-0.005	-0.005	-0.004
0.99	-0.004	-0.004	-0.003	-0.003	-0.003	-0.002	-0.002	-0.001	-0.001	0.000	0.000
1.00	0.000										

To Convert $p_i \log_{10} p_i$ to $p \log_i p$, multiply the result by the following constant:

$$-\Sigma\, p_i \log_2 p_i = -3.322\, \Sigma\, p_i \log_{10} p_i$$

$$-\Sigma\, p_i \ln_2 p_i = -2.303\, \Sigma\, p_i \log_{10} p_i$$

that can be used to assess any kind of environmental stress. It takes into account all the species present and their relative abundance but it is not necessary to actually name the species. One needs only to distinguish species A from species, B, C, D and so on. In fact, identification to genus is sufficient because the diversity value changes only slightly at the generic level. For example, the mean diversity of the benthic community in Table 12.13 changes from 3.78 to 3.70 when the two *Pisidium* species are combined. However, the species diversity index becomes increasingly less reliable with higher taxonomic levels (e.g. family, order, class).

A major disadvantage of the diversity index is it ignores the "quality" of the species, for example, whether it is a tolerant or sensitive species. Also, the index is affected by factors other than pollution, such as habitat quality. Indeed, diversity values less than 3 are often obtained in the most pristine head water streams because few species re adapted to cold water and a shredding feeding behaviour.

The diversity index is somewhat sensitive to sample size as well. Studies have shown that at least 0.5 m^2 of bottom needs to be sampled, but above this value the species diversity index remains relatively constant. Either quantitative or semi-quantitative samples can be used. Theoretically, a diversity index can be calculated on the first 100 organisms selected from a sample (of at least 0.5 m^2). Some rare species may be missed but the diversity value is affected very little by rare species. For example, if we divide the densities of organisms in Table 12.13 by 30, to reduce the total number to 100, we obtain the numbers shown in Table 12.16. Two species, *Orconectes viridis* and *Rheocricotopus*, are assumed to be missed because fewer than 1 would be present. The mean species diversity is altered slightly (3.78 for 3000 organisms vs 3.54 for 100 organisms) and the sample would be deemed to have come from a river of clean water quality in either case.

Trent Index

The 1960s and 70s were active periods in the development of biotic indices,
10☞ especially in Europe. Woodiwiss (1964)[13], while working for the Trent River Authority, developed an index using benthic macroinvertebrates in riffle areas of Midland rivers in England. He devised a scheme in which the number of groups within defined taxa was related to the presence of six key organisms within the faunal assemblage. The key taxa were plecopteran and ephemeropteran nymphs, trichopteran larvae, *Gammarus, Asellus* (*Caecidotea*), tubificid worms and red blood worms (chironomids) (Table 12.17). The index was not only used by the Trent River Authority but was adapted by the Tennessee Stream Pollution Board. The Trent Index has since been modified and adapted by several countries. One modification is an "***Extended biotic Index***" which uses the Trent Index criteria to a maximum of 10 and then adds 5 more so the extended index can score a maximum of 15 points. However, because the original version and its modifications all have some major criticisms, as discussed below, the original index is applied to the data in Table 12.13. The index has been slightly modified in Table 12.17 to accommodate

---

[13]Woodiwiss, F. S. 1964. The biological system of stream classification used by the Trent River Board. Chemistry and Industry 11: 443-447.

Table 12.16. Recalculation of species diversity based on first 100 organisms selected, assuming that the probability of their occurrence is related to their abundance.

Species	#/m²	Ind	d
Pisidium fallax	522	17	-0.132
Pisidium compressum	245	8	-0.089
Sphaerium fabale	256	9	-0.091
Elimia flavescens	203	7	-0.079
Limnodrilus hoffmeisteri	52	2	-0.031
Hyalella azteca	165	6	-0.069
Orconectes viridis	11		
Taeniopteryx sp.	180	6	-0.073
Stenonema sp.	440	15	-0.122
Rithrogena sp.	203	7	-0.079
Ephemera guttulata	82	3	-0.043
Hydropsyche scalaris	142	5	-0.063
Rhyacophila sp.	28		
Helicopsyche borealis	34	1	-0.022
Optiocervus ovalis	143	5	-0.063
Rheocricotopus sp.	18	1	-0.013
Simulium latipes	276	9	-0.095
Totals	3000	100	-1.065
Index Value			3.54
Assessment			Clean

North American benthic fauna.

The index can be calculated on samples collected either quantitatively or qualitatively, and enumeration of individuals is not required. Kick-and-sweep collecting methods (in random fashion) greatly speeds up the collection process. Only riffle areas should be sampled; the index is not applicable to pool areas in rivers. Identification to species or genus is recommended for a rigorous assessment but not necessary, as long as one is able to recognize how many different species within the above taxa are present. The Trent index can be used to assess organic or mixed pollution.

The Trent index is a great improvement over the Saprobien system in that the index values are within a defined range (0 for extreme pollution to 10 for pristine conditions) and the sample sorting time is reduced because enumeration of individuals is not required. Some criticisms of the index are; (i) it is insensitive to determining improved water quality; (ii) it can only be used to assess streams with riffle areas, and grossly underestimates water quality in non-riffle areas; and (iii) it is not applicable to all geographical areas, which applies to nearly all indices, as we shall see. However, several elements of the Trent index have been used in the development of other indices, such as the Chandler index. Its main attractive features are the ease with which it can be applied and its simplicity. However, it does require some knowledge of invertebrate identification, as do most good indices.

To use the Trent index, count the total number of taxa present in the entire sample, according to the following rules:

Count 1 taxon for each known *SPECIES* in the grous below:

Plecoptera	Coleoptera	Mollusca	Odonata
Neuroptera	Crustacea	Hirudinea	
Hydracarina	Platyhelminthes	Diptera (except those in groups below)	

Count 1 taxon for each known *GROUP* below, regardless of the number of species in the group:

Stenonema nepotellum	Naididae
Each family of Trichoptera	Annelida (excluding Naididae)
Simuliidae	Chironomidae (except those below)
Each genus of Ephemeroptera (excluding *Stenonema nepotellum*)	Red chironomids and/or *Cricotopus bicinctus*

Once the total number of taxa is known, use Table 12.17 and follow these steps:

1. Determine which ***column*** to use
2. Determine which ***row*** to use
3. Using steps 1 and 2 provides a cell with a number that corresponds to the Trent index value. For example, using data in Table 12.13, there are 18 taxa based on group scores above so column 5 (15+) is used; the sample has only 1 plecopteran so row 2 (One species only) is used. The cell at column 5, row 1 reads and index value of “9" which is the index value.
4. Go to Table 12.18 to interpret the Trent index score. For example, an index value of 9 means the sample is probably from a river that is clean or slightly polluted, but pay heed to the comments following “Interpretation”.

Table 12.17. Trent index scoring system. See text for its use and explanation of high-lighted cell.

Types of organisms present		Total No. taxa present in entire sample				
		0	2-5	6-10	11-15	16+
Plecoptera nymphs	More than 1 species	–	7	8	9	10
	1 species only	–	6	7	8	9
Ephemeroptera nymphs	More than 1 species	–	6	7	8	9
	1 species only*	–	5	6	7	8
Trichoptera larvae*	More than 1 species*	–	5	6	7	8
	1 species only*	4	4	5	6	7
Gammaridae or Pelecypoda	All above organisms absent	3	4	5	6	7
Caecidotea, Lirceus or Gastropoda	All above organisms absent	2	3	4	5	6
Tubificidae, Lumbriculidae, red chironomid larvae or *Cricotopus bicinctus*	All above organisms absent	1	2	3	4	–
Rat-tailed maggot, other air-breathers may be present	All above organisms absent	0	1	2	–	–

**Stenonema nepotellum* is excluded from the Ephemeroptera and counted with the Trichoptera for the Trent index.

Table 12.18. Interpretation: The expected values for BOD and dissolved oxygen are ***average*** values found over 6 years in the Trent River system.

Index	Pollution Status	Expected BOD	Expected [D.O]	
			Average	Minimum
8 - 10	Clean or slightly polluted	< 3	10 - 11	7.0 - 8.5
7	Slight to moderate pollution	4	10	6
5 - 6	Moderate pollution	5 - 7	8.5 - 9.5	4.5 - 5.0
3 - 4	Moderate to heavy pollution	8 - 10	6.0 - 7.0	2 - 3
1 - 2	Heavy pollution	15 - 40	3.0 - 5.0	0 - 13
0	Severe pollution	> 60	1.5	0

The Trent index assesses water quality of the benthic community in Table 12.13 as clean or slightly polluted, the most common assessment of all the indices tested. Table

12.18 relates Trent index scores to expected B.O.D. values. However, they should be used with caution. They are merely approximations of what ***might*** be expected in similar rivers with similar kinds of pollution.

Chandler Index

Chandler (1970)[14], working on the River North Esk and other Lothian rivers in
10☞ Britain, used many elements of the Trent index, including the faunal groups and sampling protocol and some elements of the Saprobien system, like the ordering of organisms into tolerance levels (Table 12.19). The index can be used only for organic pollution. Chandler (1970) felt that measuring abundance accurately is a major technical problem and therefore absolute abundance is of little use. Instead, he recommended a subjective estimate of levels of abundance in a single, 5-min sample collected using a "stop-net" (his terminology), such as a D-frame net. Kick-and-sweep samples presumably qualify as a collection method. The levels of abundance (in a 5-min sample) used by Chandler were:

present	1 - 2/sample
few	3 - 10
common	11 - 50
abundant	51 - 100
very abundant	more than 100

To obtain the five abundance levels for the data in Table 12.13, the number in the most abundant organism was divided by five (i.e. present = 1-104; few = 105-108; common = 109-312; abundant = 313-416; very abundant = 417-522). Otherwise, Chandler's score system cannot be applied directly to the data set of 100 organisms in Table 12.13 without some kind of transformation.

As Table 12.19 shows, for a species characteristic of clean water, the score increases with increasing abundance, but tolerant species score fewer points with increasing abundance. *Gammarus*, a species found under a variety of conditions, occupies the centre of the table and scores 40 points under all levels of abundances.

To derive the index for a site, the benthos are counted and identified and each group is given a score according to its abundance (Table 12.19). The total score is the index for that site. Chandler (1970) did not provide criteria for assessing water quality but he did equate it to other biotic indices (Lothian and Chambers) and on this basis the criteria in the text box that follows Table 12.19 were derived:

---

[14]Chandler, J. R. 1970. A biological approach to water quality management. Water Pollution Control 69: 415-422.

Table 12.19. Biotic index for Chandler's score.

Groups Present in Sample	Points Scored with Increasing Abundance				
	Present	Few	Common	Abund.	Very abund.
Each species of Perlidae, Perlodidae, Chloroperlodidae, Taeniopterygidae	90	94	98	99	100
Each species of Nemouridae, Astacidae	84	89	94	97	98
Each species of Ephemeroptera (excl. *Baetis*)	79	84	90	94	97
Each species of cased caddis, Megaloptera, *Agrion* (Zygoptera)	75	80	86	91	94
Each species of *Ancylus, Ferrissia*, Unionidae	70	75	82	87	91
Each species of *Rhyacophila* (Trichoptera)	65	70	77	83	88
Genera of Muscidae, Empidae, Rhagionidae, Tipulidae, Stratiomyidae, Tabanidae	60	65	72	78	84
Genera of Simuliidae, *Pristina*, Coleoptera (excl. *Stenelmis*)	56	61	67	73	75
Genera of Nematoda, Nematomorpha	51	55	61	66	72
Genera of Ceratopogonidae	47	50	54	58	63
Baetis (Ephemeroptera), Anisoptera, *Stenelmis* (Coleoptera), Anisoptera, Sphaeriidae (Bivalvia)	44	46	48	50	52
Amphipoda, Hydrozoa Each species of uncased (exclud. Rhyacophila), Zygoptera	40	40	40	40	40
Each species of Tricladida	35	33	31	29	25
Genera of Hydracarina	32	30	28	25	21
Each species of Gastropoda (excl; *Ancylus, Ferrissia*)	30	28	25	22	18
Each species of Chironomidae	28	25	21	18	15
Each species of *Glossiphonia, Helobdella*	26	23	20	16	13
Each species of Isopoda	25	22	18	14	10
Each species of leech (excl. *Glossiphonia, Haemopsis*)	24	20	16	12	8
Each species of *Haemopsis*	23	19	15	10	7
Each species of *Tubifex, Limnodrilus*	22	18	13	12	9
Each species of *Nais, Dero*	20	16	10	6	2
Each species of air-breathing organisms	19	15	9	5	1
No animal life	1				

Chandler Index Value	Water Quality Assessment
0 - 100	Grossly polluted
101 - 200	Polluted
201 - 300	Moderate pollution to polluted
301 - 400	Slight to moderate pollution
> 400	Clean to slight pollution

Differences of diversity and abundance between clean and polluted parts of a river are immediately obvious. The index has a continuous gradation from polluted to clean conditions. The highest index values are obtained in a head water area with several species of stoneflies, mayflies and caddisflies.

Using the abundance criteria stated above, a Chandler index was calculated for the data in Table 12.13. An index value of 932 was obtained, rating the site as clean to sightly polluted. A different index value might be obtained by someone else, using different abundance criteria, but it is highly unlikely that a different assessment would be obtained, since values greater than 400 give the same assessment. Chandler (1970) obtained values near 900 in his clean water station.

Average Score Per Taxon (ASPT)

A team of scientists on the Biological Monitoring Working Party (BMWP) established a score system (Table 12.20) to assess the biological condition of rivers in the
10☞ United Kingdom. The new BMWP score system attempted to take the advantages of earlier biotic indices, including all those discussed to this point, and be able to apply it on a more nationwide basis to all river types and all types of pollution. Armitage *et al* (1983)[15] modified the BMWP score system slightly by dividing it by the number of scoring taxa to produce an average score per taxon (ASPT). They then assessed the performance of both systems in relation to physical and chemical features of the study sites and found that the ASTP score system explained a higher proportion of the variance in the environmental data.

Armitage *et al* (1983) sampled the benthos using a 3-minute kick-and-sweep method (mesh size used was 900 μm) at each site, three times a year (spring, summer, fall). All animals were sorted and identified to species where possible, but identification to family is all that is needed fore the BMWP score system and the ASTP. Site scores are obtained by summing the individual scores of all families present (if several species are present in a family, the family is scored only once). The score values (Table 12.20) for individual families reflect their pollution tolerance; pollution intolerant families have

---

[15]Armitage, P. D., D. Moss, J. F. Wright and M. T. Furse. 1983. The performance of a new biological water quality score system based on macroinvertebrates over a wide range of running water sites. Water Research 17: 333-347.

Table 12.20. Pollution sensitivity grades for common families of river macroinvertebrates used for calculating SIGNAL (S) and BMWP (B) scores. Families not occurring in North America have been omitted. N represents families found in North America and are graded according to the inverse of Bode *et al* (1991)[16] tolerance values to correspond to SIGNAL and BMWP scores (see text).

Family	**Grade**			**Family**	**Grade**			**Family**	**Grade**		
	S	B	N		S	B	N		S	B	N
Aeolosomatidae			2	Corixidae	5	5	5	Hydrophilidae	5	5	5
Aeshnidae	6	8	6	Corydalidae	4		6	Hydropsychidae	5	5	6
Agrionidae		8	4	Culicidae	2		1	Hydroptilidae	6	6	5
Ancylidae	6	6	4	Dixidae	8		10	Hygrobiidae	5	5	5
Anthomyiidae			4	Dolichopodidae			6	Idoteidae			5
Anthuridae			4	Dreissenidae			2	Isotomidae			5
Asellidae		3	2	Dryopidae		5	5	Lebertiidae			4
Arctiidae			5	Dytiscidae	5	5	5	Lepidostomatidae		10	10
Arrenuridae			4	Elmidae	7	5	5	Leptoceridae	7	10	6
Astacidae		8	4	Empididae	4		4	Leptophlebiidae	10	10	7
Athericidae	7		6	Enchytreidae		1	1	Lestidae	7		5
Atractideidae			4	Ephemerellidae		10	10	Leuctridae		10	10
Baetidae	5	4	5	Ephemeridae		10	8	Libellulidae	8	8	8
Baetiscidae			6	Ephydridae	2		4	Limnephilidae	8	7	7
Belostomatidae	5		5	Erpobdellidae	3	3	3	Limnesidae			4
Blephariceridae	10		10	Gammaridae	6	6	4	Limnocharidae			4
Branchiobdellidae			4	Gerridae	4	5	5	Lumbriculidae	1	1	2
Brachycentridae		10	9	Glossiphoniidae	3	3	3	Lymnaeidae		3	4
Caenidae		7	5	Glossosomatidae	8		10	Mesoveliidae	4	5	5
Calopterygidae			4	Gomphidae	7	8	6	Mideopsidae			4
Capniidae		10	8	Gordiidae	7	10	8	Molannidae		10	4
Ceratopogonidae	6		4	Gyrinidae	5	5	5	Muscidae	3		4
Chaoboridae			2	Haliplidae	5	5	5	Naididae	1	1	3
Chironomidae	1	2	1	Haplotaxidae	5	1	1	Nemouridae		7	8
Chloroperlidae		10	10	Helicopsychidae	10		7	Nepidae		5	5
Chrysomelidae		5	5	Helodidae		5	5	Nepticulidae			5
Coenagrionidae	7	6	2	Heptageniidae		10	7	Notonectidae	4	5	5
Corbiculidae	6		4	Hydridae	4		5	Odontoceridae	8	10	10
Corduliidae	7	8	7	Hydrobiidae	5	3	4	Oedicerotidae			4
Cordulegasteridae		8	7	Hydrometridae	5	5	5	Peltoperlidae			9

[16]Bode, R. W., M. A. Novak and L. E. Abele. 1991. Method for rapid biological assessment of streams. NYS Department of Environmental Conservation. 50 Wolf Road, Albany, NY. 12233-3503.

Perlidae	10	10		Pteronarcidae			10	Taeniopterygidae		10	8
Perlodidae		10	8	Ptychopteridae			1	Talitridae		2	
Philopotamidae	10	8	7	Pyralidae	6		5	Thiaridae	7	6	
Phryganeidae			7	Rhyacophilidae	7		9	Tipulidae	5	5	7
Physidae	3	3	2	Sabellidae			4	Tricorythidae			6
Piscicolidae		4	5	Scirtidae	8	5	5	Tubificidae	1	1	1
Planariidae	3	5	4	Sialidae	4	4	6	Tyrellidae			4
Planorbidae	3	3	3	Simuliidae	5		5	Unionidae		6	4
Pleidae		5	5	Siphlonuridae		10	8	Unionicolidae			4
Pleuroceridae			4	Sphaeriidae	6	3	4	Valvatidae		3	2
Polycentropodidae	8	7	4	Spurchonidae			4	Veliidae	4		5
Polymetarcyidae			8	Sisyridae			5	Viviparidae		6	4
Potamanthidae		10	6	Tabanidae	5		5				
Psephenidae	5		6								
Psychodidae	2	8	8								
Psychomyiidae		8	8								

high scores and pollution tolerant families have low scores. The ASTP is calculated by dividing the BMWP score by the number of scoring taxa. Criteria for water quality assessment were not provided, but are probably similar to the SIGNAL (stream invertebrate grade number - average level) biotic index, an Australian method (discussed next) that is based on the ASTP. These criteria are:

ASTP Value	**Water Quality Assessment**
> 6	Clean water
5 - 6	Doubtful quality
4 - 5	Probable moderate pollution
< 4	Probable severe pollution

The ASTP and BMWP score were calculated for the data in Table 12.13. A BMWP score of 79 was obtained, which when divided by the number of scoring taxa (14) yields an ASTP value of 5.6, or water quality that is doubtful. Although the evaluation agrees with evaluations from most other indices, the value is somewhat artificial because some families present in the sample are not represented in the BMWP score system. A family tolerance score was subjectively assigned to those not represented in the BMWP system based upon species tolerance scores of the rapid bioassessment protocols in North America (Appendix 12-II). Note that the North American species tolerance values are assigned in reverse order (high values for pollution tolerant taxa and low values for pollution intolerant taxa) relative to the BMWP score system and, therefore, are "inverted" in Table 12.20 to correspond to those used for the BMWP and SIGNAL indices. As with the BMWP score system (and the SIGNAL index), the scores assigned to the families are average species scores to help reduce possible bias associated with variations in stream size and sampling technique. In most cases, the inverted tolerance scores agree well with those of the BMWP and SIGNAL scores, some exceptions being Agrionidae, Astacidae (= Cambaridae), and Leptoceridae which score lower (more pollution tolerant) on the North American scale.

Stream Invertebrate Grade Number - Average Level (SIGNAL)

Signal is a modification of the British BMWP score system, discussed above, and
10☞ was adapted for use in Australian rivers. Since many British families of macroinvertebrates are absent in Australia, and many Australian taxa were not listed in the BMWP score system, Chessman (1995)[17] developed his own score system for the SIGNAL biotic index (Table 12.20). He awarded numerous families of macroinvertebrates from south eastern Australia sensitivity grades according to their tolerance and intolerance of pollutants.

Another modification was the use of only the first 100 organisms picked from a 10-m kick sample in riffles and a separate 10-m traveling sweep sample over the edges of pools. Chessman determined experimentally that riffles provide more consistent evaluations than pools and that the first 100 organisms, picked at random in the field, provide as good an evaluation as the first 150 organisms but better than the first 50 and opted for the first 100 because it was faster. All specimens were identified to family level only, which also contributed toward a more rapid bioassessment of water quality. As with the ASTP index, the SIGNAL index is calculated by summing the grades of all families present in a site sample and dividing by the number of families to provide an average grade. The assessment of water quality is based on the same criteria listed earlier for the ASTP index. The criteria are based on samples taken from unpolluted reference sites, and sites with mild to severe pollution by municipal effluent and urban water runoff.

When the SIGNAL index is applied to the data in Table 12.13, a water quality evaluation similar to ASTP is obtained. Note, however, that the value is again somewhat artificial, for the same reasons discussed for the ASTP index. A **weighted SIGNAL** score can also be calculated, where each sensitivity grade is multiplied by its relative abundance, using 1 for rare, 2 for scarce, 3 for common and 4 for abundant, then summing the products (= weighted scores) and then dividing by the sum of the abundance scores (as in for the Saprobic index). This has been done in Table 12.21, yielding a weighted SIGNAL score of 6, or a marginally clean water assessment.

Compared to the ASTP index, the SIGNAL index is much faster to use. Grown's

---

[17]Chessman, B. C. 1995. Rapid assessment of rivers using macroinvertebrates: a procedure based on habitat-specific sampling, family level identification and a biotic index. Australian J. Ecology 20: 122-129.

Table 12.21. Method for calculating a weighted SIGNAL biotic index. The abundance score is derived from the densities of species present, using 1 for 1-130/m², 2 for 131-260/m², 261-390/m² and 4 for > 491/m². See text for details.

Species	#/m²	Grade	Abundance score	Weighted score
Pisidium fallax	522			
Pisidium compressum	245	6	4	24
Sphaerium fabale	256			
Elimia flavescens	203	5	2	10
Limnodrilus hoffmeisteri	52	1	1	1
Hyalella azteca	165	5	2	10
Orconectes viridis	11	6	1	6
Taeniopteryx sp.	180	8	2	16
Stenonema sp.	440	7	4	28
Rithrogena sp.	203			
Ephemera guttulata	82	8	2	16
Hydropsyche scalaris	142	5	2	10
Rhyacophila sp.	28	9	1	9
Helicopsyche borealis	34	8	1	8
Optiocervus ovalis	143	7	2	14
Rheocricotopus sp.	18	1	1	2
Simulium latipes	276	5	3	15
Totals	3000	43	28	169
Index Value				6.0
Assessment				marg. clean

et al (1997)[18] recommended using the 100-animal sub-sample as the most cost-effective method for assessing sites affected by municipal sewage-treatment-plant discharges and urban stormwater runoff. The SIGNAL index gave an evaluation of marginally clean using sensitivity scores subjectively assigned to the various taxa based on North

[18]Growns, J. E., B. C. Chessman, J. E. Jackson and D. G. Ross. 1997. Rapid assessment of Australian rivers using macroinvertebrates: cost and efficiency of six methods of sample processing. J. North Amer. Benthol. Soc. 16: 682-693.

American fauna.

Beak Biotic Index

A plethora of biotic indices (e.g. Beak index, Beck index, Heister's modification of Beck's index, Brinkhurst's index, Hilsenhoff's index, to mention a few) have been developed to assess water quality of North American rivers. The Beak Index[19] is a
10☞ bioassessment technique that utilizes groups of taxa and species to assess not only water quality, like most other indices, but also the fisheries potential of North American rivers. The index ranges from 0 for severe pollution (usually toxic) to 6 for an unpolluted stream. It can be derived from samples taken by any method that permits a reasonably accurate estimate to be made of population densities. It is recommended that control samples from an unpolluted area be taken for comparison. Identification to species is essential for a rigorous assessment but an approximate result can be obtained by sorting and counting to families. To use the index to its fullest potential, the investigator needs to know the relative abundances of species with different functional feeding behaviours (i.e. grazers, filter feeders, predators, etc.) and their ecological tolerances and requirements. This information is more easily obtained today (e.g. see Appendix 12-II) than it was in the 1960s when the index was developed. Nevertheless, many of the tolerances and requirements are a matter of opinion and the index value assigned often "assumes" that the type of community listed in Table 12.22 is present. The Beak index is an advance over the diversity index in that species are also considered for their sensitivity to pollution.

To calculate the Beak index, "points" are assigned based on assemblage of the benthos present in a quantitative or semi-quantitative sample. For example, the community in Table 12.13 consists of several Trichoptera, Ephemeroptera and Plecoptera, but no megalopterans or odonates, so 2.5 points are assigned out of a possible 3. The sample also has a good complement of bivalves, a snail, an amphipod and a chironomid, but no isopods so 1.5 points are assigned out of a possible 2. Very few tubificids are present so no points are assigned (out of a possible 1). Hence, the total points for the sample is 4. According to Table 12.22, the community should have a good complement of filter feeders, grazers and gatherers, which it does; bivalves, *Hydropsyche* and *Simulium* are filter feeders, *Elimia, Stenonema, Rithrogena, Helicopsyche* and *Optiocervus* are grazers, and *Limnodrilus, Hyalella* and *Rheocricotopus* are gatherers. The numbers of shredders are reduced, with only *Taeniopteryx* present. *Orconectes* is a predator. *Helicopsyche* and *Optiocervus* are grazers, and *Limnodrilus, Hyalella* and *Rheocricotopus* are gatherers. The numbers of shredders are reduced, with only *Taeniopteryx* present. *Orconectes* is a predator. Beak further suggests that this section of stream would not support a good trout population, but it would support less sensitive

---

[19]Beak, T. W. 1965. A biotic index of polluted streams and its relationship to fisheries. In: Jaag, O. (Ed.). Advances in water pollution research. Vol. I. Pergamon Press, 373 p.

Table 12.21. Calculation of the Beak Index and its interpretation.

Sensitive groups: Odonata Trichoptera Megaloptera Ephemeroptera Plecoptera	A normal complement scores 3 points If only part of the group is found, score 1 or 2 points e.g. if only 1 order is present, score 1 point, if 2 orders, score 2 points
Facultative groups: **(in clean or polluted water)** Chironomidae Amphipoda Isopoda Gastropoda Bivalvia	A normal complement scores 2 points If most are missing, e.g. if only 1 or 2 groups are present, score 1 point
Pollution-tolerant groups Tubificidae Lumbriculidae *Procladius culiciformis* (Chironomidae)	A normal or supranormal complement scores 1 point

Interpretation of Beak Index:

Pollution Status	Biotic Index	Type of macroinvertebrate Community	Fisheries Potential
Unpolluted	6	Sensitive, facultative and tolerant predators, herbivores, filter and detritus feeders all represented, but no species in excessively large numbers	All normal fisheries for type of water well developed
Slight to moderate pollution	5 or 4	Sensitive predators and herbivores reduced in population density or absent. Facultative predators, herbivores and possibly filter and detritus feeders well developed and increasing in numbers as index decreases	Most sensitive fish species reduced in numbers or missing
Moderate pollution	3	All sensitive species absent and facultative predators (Hirudinea) absent or scarce. Predators of Pelopiinae and herbivores of Chironomidae present in fairly large numbers	Only coarse fisheries maintained
Moderate to heavy pollution	2	Facultative and tolerant species reduced in numbers if pollution toxic; if organic few species insensitive to low oxygen levels present in large numbers	If fish present, only those with high toleration of pollution
Heavy pollution	1	Only most toleranmt detritus feeders (Tubificidae) present in large numbers	Very little, if any, fisheries potential
Severe pollution, usually toxic	0	No macroinvertebrates present	No fish

species.

Rapid Bioassessment Protocols (RBPs)

Because of the costs and time required to complete a bioassessment that requires species 10☞ identifications of animals in large sample sizes, there has been greater focus toward the development of rapid bioassessment techniques in recent years. During the 1980s, several biotic indices have been developed and evaluated for their efficiency and cost effectiveness in accurately assessing water quality. Among the most cost effective and reliable biotic indices are those that utilize benthic macroinvertebrates and fish. The United States EPA (Environmental Protection Agency) (see Plafkin *et al* (1989) reference made earlier in this chapter) developed rapid bioassessment protocols (RBPs) in an effort to standardize the evaluation of water quality in all kinds of rivers and for all kinds of pollution. Because species assemblages differ naturally among different regions (ecoregions) in North America, and even between stream orders in the same ecoregion, the protocol requires a reference site for each evaluation. The reference site can be an unaffected reach in the same stream or in a neighbouring stream of the same order. Since the development of RBPs in the 1980s, several states and provinces in North America have adopted them, or variations of them, as their standard protocol for assessing water quality. However, many states and provinces (e.g. Ontario) have yet to establish a standard protocol, or they are in the process of developing them.

Many of the indices in the protocols use tolerance scores that were derived from large data bases of both published and unpublished studies of experts for all the major groups of taxa. Colonial taxa, like Porifera (sponges) and Bryozoa (moss animals), are not included in the scoring system. A complete list of the tolerance scores is provided in Appendix 12.II.

The techniques and level of assessment varies with the level of the rapid bioassessment protocol, as discussed earlier. RBP III, the highest level when using benthic macroinvertebrates, is examined here. It requires species or genus identification and provides a rigorous assessment of water quality (but see the revised protocol, discussed later). RBP II is faster and requires only family identification, but it uses a different set of criteria. RBP I is basically a site evaluation.

A less arduous alternative is the New York State protocol (NYSP) used by many states (Bode et al, 1991)[20]. It uses some of the same indices as RBP III.

For both the RBP III and NYSP, the standard sampling techniques are either: (i) a timed (usually 5 minutes), travelling (downstream) kick sample where rocks and sediments are dislodged by the foot upstream of a net so that the current carries the freed organisms into a net; or (ii) using a kick-and-sweep of about 1 square meter of bottom using a D-frame net, with 0.9 mm mesh openings, for each sample. Two riffle areas

---

[20]Bode, R. W., M. A. Novak and L. E. Abele. 1991. Methods of rapid biological asessment of streams. Bureau of Monitoring and Assessment, Division of Water, NYS Department of Environmental Conservation, 50 Wolf Road, Albany, NY 12233-3505.

should be sampled, one with fast flowing water and another with slower current. The two samples are combined (composited) for processing. If there is no riffle area, select runs with cobble or gravel substrates.

The first 100 organisms are randomly selected (without bias for size or other traits) from the composited sample, at the site, preserved in 70% ethanol and then identified later to species in the laboratory. To help ensure unbiased selection of organisms, it is recommended that a subsampling procedure be used. This entails evenly distributing the composite sample into a gridded pan with a light coloured bottom. Then all organisms are removed from a set of randomly selected grids until at least 100 animals are picked. Once identified, the functional feeding behaviour of each species is determined from Appendix 12.II, and recorded.

Data on width, depth and current speed are taken for each site. Notes on percentage of rock (> 25 cm), rubble (6.5 - 25 cm), gravel (0.02 - 6.5 cm), sand (0.06 - 2.5 mm), silt (0.004 - 0.006 mm) and clay (< 0.004 mm) should also be taken to help characterize each site. A series of sieves are needed to measure the proportions of finer materials. Data for dissolved oxygen content, pH, conductivity and temperature are also needed. All this information, with any notes on abundance of algae and macrophytes at each site, will: (i) help in the selection of other additional reference and test sites; (ii) allow one to make a quick assessment of habitat quality; (iii) help to determine if a rigorous assessment protocol is required; for example, one need not perform an RBP III if the sites are clearly grossly polluted or clearly pristine. RBP III is needed to distinguish between, for example, moderately impacted and slightly impacted, or non-impacted and slightly impacted.

A CPOM (coarse particulate organic matter, such as leaf litter) sample is also required from each site. The sample is collected with a D-frame net in the same manner as described above for a riffle sample, except it is taken from areas with a lot of twigs and leaf litter. The composited sample is field sorted in a pan, as described above. The sample is used for determining the numbers of shredders present.

The biotic indices (or **metrics**) most commonly calculated are: (i) taxa richness; (ii) a Hilsenhoff biotic index (also called a modified family biotic index); (iii) an Ephemeroptera, Plecoptera, Trichoptera (EPT) index; (iv) a percent model affinity (PMA) metric; (v) percent contribution of dominant taxon; (vi) a ratio of EPT and chironomid abundances; (vii) a community loss index; (viii) a ratio of numbers of scraping individuals to filter feeding individuals; (ix) a ratio of numbers of shredding individuals to total individuals. We will examine the rational for each and the method for their calculation. Then we will first examine the metrics used for the NYSP and then examine those for the RBP III.

Metrics for the New York State Rapid Biological Assessment Protocol

10☞ ***Species Richness***

The underlying principle here is the greater the diversity of habitats, the greater the diversity of species (recall Thienemann's Principles, Chapter 8). However, some pristine headwaters may have low diversity and some enrichment may result in higher diversity, especially of EPT families and species. To determine species richness, simply count the numbers of different species in the 100-organism sample. The criteria used to assess water quality are:

Species richness	Water Quality Assessment
>26	non-impacted
19-26	slightly impacted
11-18	moderately impacted
<11	severely impacted

Using the above criteria, the water quality of the benthic community in Table 12.13, with a species richness of 15, is moderately impacted. Note that the metric is based on a 100-organism sample and two species (*Orconectes viridis* and *Rheocricotopus*) would not likely be collected in such a small sample size. See Table 12.16 for numbers of organisms in each species for a sample size of 100.

Hilsenhoff Biotic Index (HBI)

Appendix 12-II is needed to calculate this metric. The metric is a modification of
10☞ the Hilsehoff biotic index and includes species in most phyla, not just arthropods. It is calculated the same way as the family biotic index for RBP III, but accounts for species only. The biotic index is calculated using the formula:

$$HBI = \sum (s_i\, t_i)/N$$

where
s_i = number of individuals in a species
t_i = tolerance value for a species (see Appendix 10-II)
N = total number of organisms in a sample

The criteria for assessing water quality based on the modified Hilsenhoff biotic index is:

Biotic Index Value	Water Quality Assessment
0 - 4.50	non-impacted
4.51- 6.50	slightly impacted
6.51- 8.50	moderately impacted
8.51 - 10.0	severely impacted

The biotic index for the benthic community in Table 12.13 was calculated as shown in Table 12.22, the values in Table 12.13 being the sum of the products of the last two columns of Table 12.22, each product being divided by 100. The sum of the products is 4.58. Based on the modified biotic index of Hilsenhoff, the water quality for the benthic community shown in Table 12.13 is slightly impacted, the most common assessment of the metrics examined so far.

Table 12.22. Method for calculating the Hilsenhoff biotic index using data from Table 12.13. The numbers of organisms per 100 animals assumes they would be picked in direct proportion to their abundance in the original sample of 3000.

Species	**#/m^2**	**#/100**	**Tolerance**	**HBI**
Pisidium fallax	522	17	6	1.02
Pisidium compressum	245	8	6	0.48
Sphaerium fabale	256	9	6	0.54
Elimia flavescens	203	7	6	0.42
Limnodrilus hoffmeisteri	52	2	10	0.20
Hyalella azteca	165	6	8	0.48
Orconectes viridis	11			0.00
Taeniopteryx sp.	180	6	2	0.12
Stenonema sp.	440	15	3	0.45
Rithrogena sp.	203	7	3	0.21
Ephemera guttulata	82	3	0	0.00
Hydropsyche scalaris	142	5	2	0.10
Rhyacophila sp.	28	1	1	0.01
Helicopsyche borealis	34	1	3	0.03
Optiocervus ovalis	143	4	4	0.16
Rheocricotopus sp.	18			0.00
Simulium latipes	276	9	4	0.36
Totals	3000	100	64	4.58

Ephemeroptera, Plecoptera, Trichoptera (EPT) Index

The EPT index usually increases with increasing water quality because it consists
mostly of species that are very sensitive to pollution, like most stoneflies, mayflies and
10☞ caddisflies. To calculate the EPT index, simply count the numbers of families present in
each order, Ephemeroptera, Plecoptera, Trichoptera.

Based on the EPT metric, the water quality for the benthic community (with 7 families in the three orders) in Table 12.13 is assessed as slightly impacted.

Percent Model Affinity (PMA)

10☞ PMA is a measure of similarity to a model non-impacted community based on percent abundance in seven major groups. It measures similarity to a non-impacted community consisting of:

Ephemeroptera	40%
Plecoptera	5%
Trichoptera	10%
Coleoptera	20%
Chironomidae	10%
Oligochaeta	5%
Others	10%

To calculate the PMA metric: (i) Determine the % contribution for each of the 7 major groups shown above; they must add up to 100. (ii) Determine the absolute difference between the model value and the sample value. (iii) Multiply the total absolute difference by 0.5 and subtract this from 100. Table 12.23 shows how the PMA values for the benthic community in Table 12.13 were calculated. The numbers are based on abundances of species in a 100 organism sample.

Table 12.23. Method for calculating PMA values for the data in Table 12.13.

Taxon	% in Sample	% in Model	Absolute Difference
Ephemeroptera	15+7+3=25	40	15
Plecoptera	6	5	1
Trichoptera	5+1+1=7	10	3
Coleoptera	4	10	6
Chironomidae	0	20	20
Oligochaeta	2	5	3
Others	56	10	46
Total	100	100	94
			94 x 0.5 = 47
			100 - 47 = 53

To assess water quality based on the PMA metric, use the following criteria:

PMA Value	Water Quality Assessment
> 64	non-impacted
50 - 64	slightly impacted
35 - 49	moderately impacted
< 35	severely impacted

Based on these criteria, water quality for the benthic community in Table 12.13 is still assessed as slightly impacted, with a PMA value of 53.

Criteria for Assessing Water Quality Using RBP III Metrics

Three of the four metrics described above are used in RBP III; (i) a modified (from that described above) Hilsenhoff biotic index, which is called a family biotic index here; (ii) the EPT metric; and (iii) the taxa richness metric, which again is modified from that described above. Since the EPT metric is calculated in the same way as described for the NYSP, it is not discussed again here.

To use RBP III, two kinds of sites are required, a reference site and a study site. The condition of the study site is determined by calculating ***biological condition scoring criteria***, using ratios of the different metrics, with the reference site forming either the numerator or denominator, as defined in Table 12.24a. The value obtained is a ***percent composition to reference*** score that is used to determine the water quality of the study site by matching it with one of the ranges in column 1 of Table 12.24b.

The benthic community in Table 12.13 is used for the reference site and another hypothetical community was created for the study site (Table 12.25). First, all the metrics used for RBP III will be examined and then applied to the data set in Table 12.25. Finally, the metrics will be interpreted using Tables 12.24a and 12.24b to determine the water quality of the study site.

Taxa Richness

Unlike the NYSP in which all the different species are enumerated, only the
10☞ different families are enumerated here. The rationale is the same as for NYSP.

Based upon the data in Table 12.25, the study site, with 12 species, is less rich than the reference site, with 15 species, implying that habitat diversity and water quality at the study site is more impaired than the reference site. However, when the ratio of the study/reference site × 100 (= 80%) is calculated for bioassessment, water quality at the study site would probably be assessed as slightly impaired, relative to the reference site (Table 12.24b).

Table 12.24a. Criteria for characterization of biological condition for RBP III. Superscripts indicate the following: a - score is a ratio of study site to reference site x 100; b - score is a ratio of reference site to study site x 100; c - determination of functional feeding group is independent of taxonomic grouping; d - scoring criteria evaluate actual percent contribution, not percent comparability to the reference site; e - is range of values obtained and already incorporates a comparison to the reference station.

Metric	Biological Condition Scoring Criteria			
	6	4	2	0
1. Species richness[a]	>80%	60-80%	40-60%	<40%
2. Hilsenhoff's biotic index[b]	>85%	70-85%	50-70%	<50%
3. EPT index[a]	>90%	80-90%	70-80%	<70%
4. Ratio of EPT and chironomid abundances[a]	>75%	50-75%	25-50%	<25%
5. % contribution of dominant taxa[d]	<20%	20-30%	30-40%	>40%
6. Ratio of scrapers to filter feeders[a,c]	>50%	35-50%	20-35%	<20%
7. Ratio of shredders to total[a,c]	>50%	35-50%	20-35%	<20%
8. Community loss index[e]	>0.5	0.5-1.5	1.5-4.0	>4.0

Table 12.24b. Interpretation of water quality metrics. Note that the percent composition to reference score (first column) is not a continuum. Percentage values obtained that are intermediate to the ranges shown will require subjective judgement as to the correct placement. Use of the habitat assessment and physiochemical data may be necessary to aid in the decision process.

% Comp. to Ref. Score	Biological Condition	Attributes
> 83%	Non-impaired	Comparable to the best situation to be expected within an ecoregion. Balanced trophic structure. Optimum community structure (composition and dominance) for stream size and habitat quality.
54 - 79%	Slightly impaired	Community structure less than expected. Composition (species richness) lower than expected due to loss of some intolerant forms. Percent contribution of tolerant forms increases.
21 - 50%	Moderately impaired	Fewer species due to loss of most intolerant forms. Reduction in EPT index.
<17%	Severely impaired	Few species present. If high densities of organisms, then dominated by one or two taxa.

Table 12.25. Hypothetical benthic macroinvertebrate communities at a reference site and a study site. Data for the reference site are from Table 12.13. The functional feeding behaviours (Feed) are from Appendix 12.II; Ff = filter feeder, Sc = scraper, Ga = gatherer, Pr = predator. Data for ETP (Ephemeroptera, Plecoptera, Trichoptera) and FBI (Family biotic index) are explained in text.

Reference Site Species	#/ 100	Feed	ETP	FBI	Study Site Species	#/ 100	Feed	ETP	FBI
P. fallax	17	Ff			*Pisidium compressum*	4	Ff		8x4
P. compressum	8	Ff		8x34	*Sphaerium striatinum*	6	Ff		8x6
S. fabale	9	Ff			*Limnodrilus hoffmeisteri*	31	Ga		10x31
E. flavescens	7	Sc			*Aulodrilus americanus*	10	Ga		8x10
L. hoffmeisteri	2	Ga		10x2	*Helobdella stagnalis*	4	Pr		10x4
H. azteca	6	Ga		8x6	*Caecidotea racovitzai*	10	Ga		8x10
O. viridis					*Stenonema femoratum*	3	Sc	✓	7x3
Taeniopteryx	6	Sh	✓	2x6	*Hydropsyche betteni*	2	Ff	✓	4x2
Stenonema sp.	15	Sc	✓	4x22	*Ablabesmyia sp.*	5	Pr		6x5
Rithrogena sp.	7	Sc	✓		*Chironomus riparius*	10	Ga		8x10
E. guttulata	3	Pr	✓	4x3	*Dicrotendipes sp.*	10	Ga		6x8
H. scalaris	5	Ff	✓	4x5	*Probesia*	5	Pr		6x5
Rhyacophila	1	Pr	✓	0x1					
H. borealis	1	Sc	✓	3x1					
O. ovalis	4	Sc		4x4					
Rheocricotopus									
S. latipes	9	Ff		6x9					
Totals	100		7	545		100		2	807

Family Biotic Index

Originally, the index was based mostly on insect families, but was modified by
10☞ EPA for their RBPs to include mites, crayfish, molluscs, oligochaetes, leeches and flatworms. The tolerance values range from 0 for the most sensitive species to 10 for the most tolerant species. The formula for calculating the family biotic index (FBI) is:

$$FBI = \sum (x_i t_i)/N$$

where,
x_i = number of individuals in a taxon (family or genus)

t_i = tolerance value for a taxon (see Appendix 12.II)
N = total number of organisms in a sample

The rationale for this metric is the same as for the modified Hilsenhoff biotic index discussed above. The procedure for calculating the index is also the same, except the tolerance values need to be taken from Appendix 12.II.

Table 12.25 shows the calculation for the numerator of the FBI index. Since the denominator is 100 for both sites, the FBI for the study site is 8.07, compared to 5.45 for the reference site. The scoring criterion is (5.45/8.07) × 100 = 67.5%, which assesses water quality at the study site as slightly impaired (Table 12.24).

Percent Contribution of Dominant Taxon (PCDT)

PCDT focuses on the evenness/redundancy of the benthic community, regardless of taxa composition. It is an indication of community balance at the lowest positive
10☞ taxonomic level (i.e. family). To calculate PCDT, simply divide the numbers of organisms in the numerically dominant family AT THE STUDY SITE by the total numbers of organisms in the sample and multiply by 100.

The dominant taxon in our example study site is *Limnodrilus hoffmeisteri*, which represents 31% of the benthic community. According to Table 12.24, water quality based on PCDT is moderately impaired.

EPT Index

The EPT index has already been calculated for the reference site (see NYSP
10☞ above) as 7. The value for the study site is 2 (Table 12.25). The condition score for the study site is (2/7) x 100 = 28.6%, a value that ranks its water quality as moderately impaired (Table 12.24).

Ratio of EPT and Chironomid Abundances

The ratio uses the relative abundances of four indicator groups, mayflies, stoneflies, caddisflies and midges. The numerator consists mostly of pollution sensitive
10☞ species, while the denominator consists mostly of pollution tolerant species. Hence, a high value indicates non-impacted waters while low values indicate stressed habitats. The metric is especially useful in areas receiving increasing enrichment of heavy metals because many chironomids (e.g. some species of *Cricotopus*) become increasingly dominant in terms of relative numbers and proportions to the other taxa in the metric.

The metric has a value of 0.20 at the study site and about 38 at the reference site, if 1 individual is assigned to *Rheocricotopus* (Table 12.25). Whether 0 or 1 individual, the biological condition score (0.25/38 x 100 = 0.53%) for the EPT/chironomid metric assigns the site as severely impaired (Table 12.24).

Community Loss Indices

10☞ Two community similarity indices are examined, the ***community loss index*** and the ***Jaccard coefficient of community similarity***. The community loss metric is an index of compositional dissimilarity and measures the loss of benthic taxa between a reference station and the study station. Values range from 0 to infinity. It is calculated using the formula,

$$\text{Community Loss} = \frac{\text{total no. taxa present in Sample A - total no. taxa common in both samples}}{\text{total no. taxa present in Sample B}}$$

The community loss for our example study site is (15 - 2)/12 = 1.08, which, according to Table 12.24, represents a benthic community from slightly impaired water quality, relative to the reference site.

The Jaccard coefficient measures the degree of similarity in taxonomic composition between two or more stations in terms of taxon presence or absence. An example is given for each of a comparison between two sites and comparisons among several sites. The coefficient is calculated using the formula:

$$\text{Jaccard coefficient} = \frac{a}{a + b + c}$$

where
a = number of taxa common in both samples
b = number of taxa present in sample A but not in sample B
c = number of taxa present in sample A but not B

The Jaccard coefficient for our example study site is (2)/(2 + 10 + 14) = 0.077. Water quality criteria have not been provided for this metric.

Ratio of Scrapers to Filter Feeders

The ratio reflects the dominance of one functional feeding group over another.
Since some families represent several functional feeding groups, the metric is
10☞ independent of taxon. Scrapers reflect an abundance of plant material, especially algae,
and dominate areas that receive large nutrient inputs. Filter feeders dominate areas with
large inputs of fine particulate organic matter (FPOM) and availability of attachment
sites (e.g. for simuliids and some trichopterans) or soft sediments for burrowing
bivalves. An overabundance of filamentous algae can also provide FPOM for the
filterers.

Filter feeders are sensitive to toxicants bound to fine particles and should be the first group to decrease in abundance with prolonged exposures to the toxicants. The metric is especially useful when point source discharges contain toxicants that adsorb readily to dissolved organic material and form a flocculant that becomes available to filter feeders as FPOM. The ratio may not be a good indicator of organic enrichment if adsorbing toxicants are present, but other metrics such as EPT and FBI may provide additional insight.

Appendix 12.II is needed to calculate the scraper/filter feeder ratio. As inferred from the ratio, it is the number of individuals with scraping feeding behaviour divided by the number of individuals with filter feeding behaviour.

Table 12.25 lists the feeding behaviours for the different species in each of the example reference and study sites. The ratio is 34 scrapers to 39 filter feeders at the reference site (= 0.87) and 3 scrapers to 12 filter feeders at the study site (= 0.25), yielding a biological condition score of (0.25/0.87) x 100 = 28.7%. Table 12.24 attributes the scraper/filter feeder metric from a habitat with moderately impaired water quality.

Ratio of Numbers of Shredders to Total Numbers

This ratio is estimated from the CPOM (coarse particulate organic matter) sample. Shredders are sensitive to riparian zone impacts and are especially useful indicators of
10☞ toxic effects when the toxicants involved are readily absorbed to the CPOM. The toxicants may directly affect the shredders through toxicity or indirectly through effects on the microbial communities that condition the CPOM for the shredders. The metric may be omitted from the assessment if the site(s) selected do(es) not normally have shredders.

Table 12.25 does not permit an evaluation of water quality for the example study site using this metric because the estimate must be based on a CPOM sample. However, for the sake of demonstration, assume that the study site had 1 shredding individual, the shredder numbers/total number ratio would be 0.01 for the study site and 0.06 for the reference site, yielding a condition score of (0.01/0.06) x 100 = 16.7 %.

Summary Bioassessment Evaluation of Hypothetical Benthic Sample

Of the seven RBP III metrics used (shredder numbers/total numbers ratio not included), three (scraper number / filter feeder number ratio, percent contribution of dominant taxon, and Ephemeroptera, Plecoptera, Trichoptera index) classified the study site as moderately impaired, three (community loss index, family biotic index and taxa richness) classified it as slightly impaired and one (EPT/chironomid ratio) classified it as severely impaired. The low value of the last metric cited is probably due to the selection of species in the hypothetical sample, where chironomids normally make up a large proportion of species under almost any conditions. To evaluate water quality of the study site, the percentages are totaled and an average is taken. Not including the community loss index (done for study site only) or the shredder/total ratio (requires a CPOM sample), the remaining six metrics provide an average of 39.4%, rating the study site overall as moderately polluted.

When all metrics are considered, including the Saprobic index, Trent index, Beak index, Chandler score, SIGNAL, ASPT, SCI, and indices of the NYSP, it is clear the reference site is slightly impacted. But the study site is even more degraded and would have to be classified as moderately impaired on both an absolute scale (i.e. compared to a pristine site) and, according to RBP III, even on a relative (to the reference site) scale.

Revised RBP III Protocol

Now that we know the protocol, the different metrics available and the criteria used for assessing water quality in the old RBP III, let's examine the revised protocol and its water quality evaluation method.

One of two procedures may be used, a "single, most productive habitat" survey or a "multihabitat survey". The primary differences between the original RBP II and III are the decision on field versus lab sorting and level of taxonomy. A third protocol has been developed as a more standardized biological reconnaissance or screening and replaces RBP I of the original (1989) document. The following is a summary of the revised RBP protocol described by Barbour et al (1999, see reference cited earlier). Any quotations are from that document. The document may be seen on the web site,

http://www.epa.gov/owow/monitoring/rbp/index.html

Single Habitat Approach: 1 Meter Kick Net

"The original RBPs (Plafkin et al. 1989) emphasized the sampling of a single habitat, in particular riffles or runs, as a means to standardize assessments among streams having those habitats. This approach is still valid, because macroinvertebrate diversity and abundance are usually highest in cobble substrate (riffle/run) habitats. Where cobble substrate is the predominant habitat, this sampling approach provides a representative sample of the stream reach. However, some streams naturally lack the cobble substrate. In cases where the cobble substrate represents less than 30% of the sampling reach in reference streams (i.e., those streams that are representative of the region), alternate habitat(s) will need to be sampled (See Section 7.2 in Barbour 1999). The appropriate sampling method should be selected based on the habitat availability of the reference condition and not of potentially impaired streams. For example, methods would not be altered for situations where the extent of cobble substrate in streams influenced by heavy sediment deposition may be substantially reduced from the amount of cobble substrate expected for the region.

Field Sampling Procedures for Single Habitat Approach

1. Select a 100-m reach representative of the characteristics of the stream and at least 100 meters upstream from any road or bridge crossing to minimize its effect on stream velocity, depth, and overall habitat quality. There should be no major tributaries discharging to the stream in the study area.
2. Document site description, weather conditions, and land use. After sampling, review this information for accuracy and completeness.
3. Draw a map of the sampling reach. Include in-stream attributes (e.g., riffles, falls, fallen trees, pools, bends, etc.) and important structures, plants, and attributes of the bank and near stream areas. Show direction of flow and areas that were sampled for macroinvertebrates on the map. If available, use hand-held Global Positioning System (GPS) for latitude and longitude determination taken at the furthest downstream point of the sampling reach.

4. A composite sample is taken from individual sampling spots in the riffles and runs representing different velocities. Generally, a minimum of 2-m^2 composited area is sampled.
5. "Sampling begins at the downstream end of the reach and proceeds upstream. Using a 1-m kick net, 2 or 3 kicks are sampled at various velocities in the riffle or series of riffles. A *kick* is a stationary sampling accomplished by positioning the net and disturbing one square meter upstream of the net. Using the toe or heel of the boot, dislodge the upper layer of cobble or gravel and scrape the underlying bed. Larger substrate particles should be picked up and rubbed by hand to remove attached organisms. If different gear is used (e.g., a D-frame or rectangular net), a composite is obtained from numerous kicks".
6. Composite the jabs or kicks collected from different locations in the cobble substrate into a single homogeneous sample. "After every kick, wash the collected material by running clean stream water through the net 2 to 3 times. If clogging does occur, discard the material in the net and redo that portion of the sample in a different location. Remove large debris after rinsing and inspecting it for organisms; place any organisms found into the sample container. Do not spend time inspecting small debris in the field".
7. Transfer the sample from the net to sample container(s) and preserve in enough 95 percent ethanol to cover the sample. Forceps may be needed to remove organisms from the dip net. Label the container(s) with site identification, date, etc.
8. "Record the percentage of each habitat type in the reach. Note the sampling gear used, and comment on conditions of the sampling, e.g., high flows, treacherous rocks, difficult access to stream, or anything that would indicate adverse sampling conditions".
9. Document observations of aquatic flora and fauna. Make qualitative estimates of macroinvertebrate composition and relative abundance as a cursory estimate of ecosystem health and to check adequacy of sampling.
10. Perform a visual habitat assessment by walking the reach after sampling has been completed. Conduct the habitat assessment with another team member, if possible.
11. Return samples to laboratory and process as described below.

Field Procedures for Multihabitat Approach: D-frame Dip Net

A method suitable to sampling a variety of habitat types is desired in this approach. Benthic macroinvertebrates are collected systematically from all available instream habitats by kicking the substrate or jabbing with a D-frame dip net. A total of 20 jabs (or kicks) are taken from all major habitat types in the reach resulting in sampling of approximately 3.1 m^2 of habitat. For example, if the habitat in the sampling reach is 50% snags, then 50% or 10 jabs should be taken in that habitat. An organism-based subsample (usually 100, 200, 300, or 500 organisms) is sorted in the laboratory and identified to the lowest practical taxon, generally genus or species.

The major habitat types to look for are ***cobble*** (hard substrate in riffles), ***snags*** (woody debris that have been submerged for a relatively long period but not recent deadfall), ***vegetated banks*** (emergent vegetation), ***submerged macrophytes*** (sample rooted plants by drawing the net through the vegetation from the bottom to the surface of the water, maximum of 0.5 m each jab), ***sand*** (and other fine sediment, by bumping the net along the surface of the substrate rather than dragging the net through soft substrates)

Proceed as follows:

1. Select a 100-m reach representative of the characteristics of the stream and at least 100 meters

upstream from any road or bridge crossing to minimize its effect on stream velocity, depth, and overall habitat quality. There should be no major tributaries discharging to the stream in the study area.

2. Document site description, weather conditions, and land use. After sampling, review this information for accuracy and completeness.
3. Draw a map of the sampling reach. Include in-stream attributes (e.g., riffles, falls, fallen trees, pools, bends, etc.) and important structures, plants, and attributes of the bank and near stream areas. Show direction of flow and areas that were sampled for macroinvertebrates on the map. If available, use hand-held Global Positioning System (GPS) for latitude and longitude determination taken at the furthest downstream point of the sampling reach.
4. "Different types of habitat are to be sampled in approximate proportion to their representation of surface area of the total macroinvertebrate habitat in the reach. For example, if snags comprise 50% of the habitat in a reach and riffles comprise 20%, then 10 jabs should be taken in snag material and 4 jabs should be take in riffle areas. The remainder of the jabs (6) would be taken in any remaining habitat type. Habitat types contributing less than 5% of the stable habitat in the stream reach should not be sampled. In this case, allocate the remaining jabs proportionately among the predominant substrates. The number of jabs taken in each habitat type should be recorded on the field data sheet".
5. "Sampling begins at the downstream end of the reach and proceeds upstream. A total of 20 jabs or kicks will be taken over the length of the reach; a single *jab* consists of forcefully thrusting the net into a productive habitat for a linear distance of 0.5 m. A *kick* is a stationary sampling accomplished by positioning the net and disturbing the substrate for a distance of 0.5 m upstream of the net".
6. Composite the jabs or kicks collected from the multiple habitats to obtain a single homogeneous sample. Follow same procedure as # 6 for "Single habitat Approach".
7. Follow # 7-10 for "Single habitat Approach".

Laboratory Processing of Macroinvertebrate Samples

Keep accurate records of all jars and vials, with labels that identify contents and date of sampling on all of them. "The RBPs use a fixed-count approach to subsampling and sorting the organisms from the sample matrix of detritus, sand and mud. *The following protocol is based on a 200-organism subsample, but it could be used for any subsample size (100, 300, 500, etc.).* The subsample is sorted and preserved separately from the remaining sample for quality control checks.

1. Thoroughly rinse sample in a 500 μm-mesh sieve to remove preservative and fine sediment. Visually inspect large debris for organisms and discard after preserving any organisms found. Soak contents preserved in 95% ethanol in water for about 15 minutes to hydrate the benthic organisms. If more than one container was used for a composited sample, mix together in one tray.
2. "After washing, spread the sample evenly across a pan marked with grids approximately 6 cm x 6 cm. Note the presence of large or obviously abundant organisms; *do not remove them from the pan*".
3. "Use a random numbers table to select 4 numbers corresponding to squares (grids) within the

gridded pan. Remove all material (organisms and debris) from the four grid squares, and place the material into a shallow white pan and add a small amount of water to facilitate sorting. If there appear (through a cursory count or observation) to be 200 organisms ± 20% (cumulative of 4 grids), then subsampling is complete". Any organism that is lying over a line separating two grids is considered to be on the grid containing its head. In those instances where it may not be possible to determine the location of the head (worms for instance), the organism is considered to be in the grid containing most of its body. If the density of organisms is high enough that many more than 200 organisms are contained in the 4 grids, transfer the contents of the 4 grids to a second gridded pan. Randomly select grids for this second level of sorting as was done for the first, sorting grids one at a time until 200 organisms ± 20% are found. If picking through the entire next grid is likely to result in a subsample of greater than 240 organisms, then that grid may be subsampled in the same manner as before to decrease the likelihood of exceeding 240 organisms. That is, spread the contents of the last grid into another gridded pan. Pick grids one at a time until the desired number is reached. The total number of grids for each subsorting level should be noted on the laboratory bench sheet.

4. Save the sorted debris residue in a separate container. Add a label that includes the words "sorted residue" in addition to all prior sample label information and preserve in 95% ethanol. Save the remaining unsorted sample debris residue in a separate container labeled "sample residue"; this container should include the original sample label. Length of storage and archival is determined by the laboratory or benthic section supervisor.
5. Place the sorted 200-organism (± 20%) subsample into fully labeled glass vials and preserve in 70% ethanol.
6. Midge (Chironomidae) larvae and pupae should be mounted on slides in an appropriate medium (e.g., Euperal, CMC-9); slides should be labeled with the site identifier, date collected, and the first initial and last name of the collector. As with midges, worms (Oligochaeta) must also be mounted on slides and should be appropriately labeled.
7. Identify all invertebrates to lowest taxon (e.g. species) possible using peer-reviewed keys.

Metrics for Revised RBP III

The metrics described for RBP III by Plafkin et al (1989) are all ecologically sound, but may require testing on a regional basis. An attempt should be made to include metrics that assess four ecological components: Taxa richness measures; composition measures; tolerance/intolerance measures; feeding measures or trophic dynamics. The rationales for these four measures are given in Barbour et al (1999) but can be summarized as follow:

Taxa richness measures: A measure of the number of distinct taxa. It represents the diversity within a sample. "Taxa richness usually consists of species level identifications but can also be evaluated as designated groupings of taxa, often as higher taxonomic groups (i.e., genera, families, orders, etc.) in assessment of invertebrate assemblages. Richness measures reflect the diversity of the aquatic assemblage. Increasing diversity correlates with increasing health of the assemblage and suggests that niche space, habitat, and food source are adequate to support survival and propagation of many species.

Number of taxa measures the overall variety of the macroinvertebrate assemblage. No identities of major taxonomic groups are derived from the total taxa metric, but the elimination of taxa from a naturally diverse system can be readily detected. Subsets of "total" taxa richness are also used to accentuate key indicator groupings of organisms. Diversity or variety of taxa within these groups are good indications of the ability of the ecosystem to support varied taxa. Certain indices that focus on a pair-wise site comparison are also included in this richness category". Taxa richness measures include: Species richness; Total No. taxa; No. EPT taxa; No. Ephemeroptera Taxa; No. Plecoptera Taxa; No. Trichoptera Taxa.

Composition measures: This measure is characterized by several classes of information, i.e., the identity, key taxa and relative abundance. "Identity is the knowledge of individual taxa and associated ecological patterns and environmental requirements. Key taxa (i.e., those that are of special interest or ecologically important) provide information that is important to the condition of the targeted assemblage. The presence of exotic or nuisance species may be an important aspect of biotic interactions that relate to both identity and sensitivity. Measures of composition (or relative abundance) provide information on the make-up of the assemblage and the relative contribution of the populations to the total fauna". Composition measures include: % EPT; % Ephemeroptera.

Tolerance/Intolerance measures should be representative of relative sensitivity to perturbation and may include numbers of pollution tolerant and intolerant taxa or percent composition . Tolerance is generally non-specific to the type of stressor. However, some metrics such as the Hilsenhoff Biotic Index are oriented toward detection of organic pollution. The Biotic Condition Index is useful for evaluating sedimentation. "The tolerance/intolerance measures can be independent of taxonomy or can be specifically tailored to taxa that are associated with pollution tolerances. For example, both the percent of Hydropsychidae to total Trichoptera and percent Baetidae to total Ephemeroptera are estimates of evenness within these insect orders that generally are considered to be sensitive to pollution. As these families (i.e., Hydropsychidae and Baetidae) increase in relative abundance, effects of pollution (usually organic) also increase. Density (number of individuals per some unit of area) is a universal measure used in all kinds of biological assessments". Tolerance/intolerance measures include: No. of Intolerant Taxa; % Tolerant organisms; % Dominant Taxon.

Feeding measures or trophic dynamics: This measure includes "functional feeding groups and provides information on the balance of feeding strategies (food acquisition and morphology) in the benthic assemblage. Examples involve the feeding orientation of scrapers, shredders, gatherers, filterers, and predators. Trophic dynamics (food types) are also included here and include the relative abundance of herbivores, carnivores, omnivores, and detritivores. Without relatively stable food dynamis, an imbalance in functional feeding groups will result, reflecting stressed conditions. Trophic metrics are surrogates of complex processes such as trophic interaction, production, and food source availability". "Specialized feeders, such as scrapers, piercers, and shredders, are the more sensitive organisms and are thought to be well represented in healthy streams. Generalists, such as collectors and filterers, have a broader range of acceptable food materials than specialists .... and thus are more tolerant to pollution that might alter availability of certain food. However, filter feeders are also thought to be sensitive in low-gradient streams".

Bioassessment Methods for Revised RBP III

The revised RBP III does not use criteria to evaluate the level of impact. Instead, either an ***a priori*** or ***a posteriori*** approach to classifying sites is used. "To provide a broad comparison of the 2 approaches, it is assumed that candidate reference sites are available from a wide distribution of streams. In the first stage, data collection is conducted at a range of reference sites (and non-reference or test sites) regardless of the approach. The differentiation of site classes into more homogeneous groups or classes may be based initially on *a priori* physicochemical or biogeographical attributes, or solely on *a posteriori* analysis of biology (Stage 2 as illustrated in Figure 9-1). Analysts who use multimetric indices tend to use *a priori* classification; analysts who use one of the multivariate approaches tend to use *a posteriori*, multivariate classification". An *a priori* classification can be used with multivariate assessments, and vice-versa.

Clearly, the emphasis in the revised RBP III is on statistical analyses, mostly using a multimetric approach that incorporates biological concepts to test significant differences between a reference site and a test site. Description of the multimetric approach is beyond the scope of this text and interested readers are encouraged to read Chapter 9 of the new U.S.E.P.A. RBP protocols, provided on the web site listed earlier.

The multivariate approach is commonly used in Canada and Australia. For example, in Canada, Reynoldson et al (1995)[21] use BEAST (BEnthic Assessment of SedimenT) for assessing quality of a habitat. The BEAST technique uses a probability model based on taxa ordination space and the "best fit" of the test site(s) to the probability ellipses constructed around the reference site. Another approach is the use of a standarized normal curve to illustrate the mean and plus-or-minus one, two and three standard deviations that enclose 68%, 95% and 99.9% of the area under the curve. This approach is used by David et al (1995)[22] for assessing water quality in Ontario streams

---

[21]Reynoldson, T.B., R.C. Bailey, K.E. Day, and R.H. Norris. 1995. Biological guidelines for freshwater sediment based on BEnthic Assessment of SedimenT (the BEAST) using a multivariate approach for predicting biological state. Australian Journal of Ecology (1995) 20:198-219.

[22] David, S. M., K. M. Somers, R. A. Reid, R. J. Hall and R. E. Girard. Sampling protocols for the rapid bioassessment of streams and lakes using benthic macroinvertebrates. Queen's Printer for Ontario, Toronto, Ontario, Ref. No. PIBS 3519E01,

and lakes; they employ many of the metrics described above for the old RBP III and any test sites that fall between the 95th and 99.9th percentile are considered atypical (e.g. impacted) relative to the reference sites. Additional details are provided below under, "Selection of a Bioassessment Method".

BioMAP (Griffiths, 1998)[23] uses stream sampling and satellite imagery of vegetation cover as sources of information to estimate water quality in Ontario streams. Griffiths (1998) suggests that as benthic invertebrates encounter the effects of pollution in streams they die off and are replaced by more pollution-tolerant forms. Relying on concepts of many of the biotic indices described earlier (e.g. Trent Index, Beak Index, Hilsenhoff Index), BioMAP detects differences in tolerant/intolerant benthic communities from a water quality index (WQI), calculated as:

$$WQI = [\sum e^{sv_i} * \ln (x_i + 1))] \div [\sum \ln (x_i + 1)]$$

where

sv_i = the sensitivity value of the ith taxon
x_i = the density of the ith taxon
n = number of taxa in sample
ln = natural logarithm
e = 2.718

The BioMAP water quality index is calculated as the average sensitivity value of the top 25% quartile of taxa (a minimum of 4):

$$WQI = 1/k [\sum \ln (sv_i)], \text{ with } k = \text{integer } (n/4),\ k \geq 4$$

where

sv_i = sensitivity value of the ith ranked taxon (ranked from highest to lowest)
n = number of taxa at the site.

The sensitivity values for benthic taxa are provided in an appendix. Many sensitivity values differ from those recommended by U.S.E.P.A. for RBP protocols.

Stream water quality is evaluated as impaired or unimpaired, an unimpaired stream having a BioMAP water quality value greater than 12 and an impaired stream having a value of less than 10. If the water quality value lies between 10 and 12, then no conclusion can be drawn based solely on the BioMAP water quality value.

Sequential Comparison Index (SCI)

10☞ This is the last biotic index examined for assessing water quality in streams and

---

[23]Griffiths, R. W. 1998. BioMAP. How to manual. Ministry of Municipal Affairs and Housing, Planning and Policy Branch, 777 Bay St., 13th Floor, Toronto, Ontario.

rivers using macroinvertebrates. It was developed for engineers and other people who have no experience at identifying organisms. In fact, the accuracy of the index decreases with increasing benthological experience of the user. The index is based on the "sign test" and the "theory of runs". That is, it relies on the innate ability of the user to recognize differences in size, shape and colour (i.e. signs) of organisms. It is an expression of community structure since it depends upon both the species richness of the community and on the distribution of individuals among the species. Only two individuals are compared at a time. The current individual need only be compared to the previous one. If it looks similar it is part of the same "run"; if not, it is part of a new run. The greater the number of runs, the greater the diversity. Organisms of the same appearance are assigned to the same taxon. The number of different looking "signs" represents the number of different taxa. The formula was developed by Cairns and Dickson (1971)[24]:

$$SCI = \frac{\text{Number of runs} \times \text{Number of taxa}}{\text{Total number of individuals}}$$

where,

a run is a set of organisms that looks similar
a taxon is a different looking organism.

The criteria for assessing water quality with the SCI is as follows:

SCI Value	Water Quality
≤ 8.0	Polluted
8.1 - 12	Moderately polluted
> 12	Clean

The index is used by first randomly selecting 100 organisms and placing them side by side in another random order of selection. The first organism is taxon 1. The second (adjacent) organism is compared to the first; if it is the same, it is assigned taxon 1 in run 1; if different, it is assigned taxon 2 in run 2. Next, compare the 3rd to the 2nd organism; if it is the same, it is assigned taxon 2 of run 2; if it is different, it is assigned taxon 3 of run 3. The process is continued until all organisms are processed. When finished, decide which taxa are the same and which are different, because taxon "x" may

---

[24]Cairns, J. and K. L. Dickson. 1971. A simple methods for the biological assessment of the effects of waste discharges on aquatic bottom-dwelling organisms. J. Water Pollution Contr. Fed. 43: 755-772.

be the same as taxon "i" seen in an earlier run. For example, suppose the following set of 15 organisms are drawn at random:

Organism #	1	2	3	4	5	6	7	8	9	10	11	12	13	14	15
	☺	☻	☻	◘	◙	◙	◙	◙	☺	☹	☹	☹	◘	☻	◙
Run	1	2	2	3	4	4	4	4	5	6	6	6	7	8	9
Taxon	1		2	3			4		1		5		3	2	4

The SCI for this particular sample is: (9 x 4)/15 = 2.4, a value that ranks it from polluted water, according to the criteria above.

A SCI was calculated for the data in Table 12.13 after randomly picking numbers from a hat. Numbers 1 to 15 were written on pieces of paper, with #1 being for species 1, 2 for species 2, 3 for species 3, etc. Of the 15 species, 17 pieces of paper had # 1 (*Pisidium fallax*), 8 had #2 (*Pisidium compressum*), 9 had #3 (*Sphaerium fabale*), and so on until 100 pieces of paper were made. All 100 pieces of paper were placed in a hat and then drawn at random in the order shown below:

1,12,7,1,11,6,6,8,11,15,15,15,7,7,4,14,15,3,2,1,15,15,8,1,1,14,1,15,8,1,1,1,14,7,7,3,9,10,3
9,6,3,3,10,14,4,11,2,8,8,7,3,8,8,1,1,2,2,1,1,1,5,9,5,8,1,4,8,15,15,4,3,8,3,6,6,13,11,9,8,8,11
6,2,9,1,8,10,3,2,4,4,8,1,2,2,4,8,9,9

The SCI for the sample is (79 x 15)/100 = 11.9, or a water quality assessment of moderately polluted, but more likely slightly impaired since it is very close to the "clean" category.

The theory behind the SCI is good. However, it works best at the elementary level of identification. It does not work for experienced benthologists because they would probably classify different larval stages of a single species as one taxon, whereas an inexperienced person would probably call them different taxa, or two organisms of different sizes may be classified as the same by the benthologist but differently by the novice. It is essential that the sorting be done randomly. For example, if all the large animals are sorted first and then all the small ones, the sample would probably be biased in the ordering of taxa, resulting in fewer runs and a smaller index value. The index would probably work well for inexperienced cottagers, sport fishermen, etc., with the above caveats in mind.

Selection of a Bioassessment Method

Having reviewed numerous metrics and methods using the benthic community to assess water quality, which are the best to use? Most provinces and states have developed their own protocol and in most cases a RBP is used but modified or adapted to suit the needs for the region. Nearly all employ comparisons of metrics between a test site and reference sites. A single index is rarely used because different stressors may impact upon components of the benthic community in different ways. A single index may indicate an impact, but the type of impact cannot be determined.

In some cases, protocols are still being developed and evaluated. Barbour et al (1992)[25] examined the EPA's metrics and recommended the use of EPT, taxonomic richness, Hilsenhoff and percent shredder metrics; the remaining metrics were too variable to be useful. Barton and Metcalfe-Smith (1992)[26] evaluated eight biological indices and recommended taxonomic richness, a modified Hilsenhoff index and percent oligochaetes because they best discriminated sites with known degradation. David et al (1998, cited earlier) have adapted rapid bioassessment protocols for Ontario streams. They provide the following list of ten indices in four summary categories (with some modifications):

Simple summaries

(1) Counts (richness measures)

(A) Taxonomic richness (total number of different taxonomic groups)

(B) Number of insect groups (number of taxonomic groups that are insects)

(2) Enumerations (abundance measures)

(A) Total number of individuals (total number of invertebrates in a sample)

(B) Total number of EPTs (total number of mayflies, stoneflies and caddisflies)

(3) Compositional indices

(A) Percent amphipods (ratio of number of amphipods to total number of individuals)

(B) Percent Diptera (ratio of number of flies to total number of individuals)

(C) Percent oligochaetes (ratio of number of worms to total number of individuals)

(4) Diversity indices

(A) Shannon-Wiener diversity (evenness of counts among groups)

(B) Percent dominants (ratio of most abundant taxon to total number of individuals)

Weighted summaries

(5) Pollution-tolerant indices

(A) Trent, Beak, Chandler, etc. Biotic Indices (pollution-tolerance weighted richness)

(B) Hilsenhoff biotic index (pollution-tolerance weighted abundance

Multivariate summaries

(6) Pairwise similarity indices

(A) Jaccard's coefficient of community (degree of taxonomic similarity between 2

---

[25]Barbour, M. T., J. L. Plafkin, B. P. Bradley, C. G. Graves and R. W. Wisseman. 1992. Evaluation of EPA's rapid bioassessment benthic metrics: metric redundancy and variability among reference stream sites. Environ. Toxicol. Chem. 11: 437-449.

[26]Barton, D. R. and J. L. Metcalfe. 1992. A comparison of sampling techniques and summary indices for assessment of water quality in the Yamaska River, Quebec, based on benthic macroinvertebrates. Environ. Monit. Assess. 21: 225-244.

samples)

(B) Percent similarity (degree of compositional similarity between 2 samples)

(7) Pairwise distance indices

(A) Euclidian distance (absolute difference between 2 samples)

(B) Bray-Curtis distance (distance complement of percent similarity (1-PS)

(8) Comparison to a standard

(A) Index of biotic integrity (tolerance-weighted sum across a selection of indices)

(B) Percent model affinity (relative difference between a sample and a target)

Other summaries

(9) Trophic or functional feeding groups

(A) Number of predators (a feeding habit classification)

(B) Percent shredders (ratio of shredders to total number of individuals)

(10) Groups of reproductive guilds

(A) Number of psammophils (number of taxa reproducing on sandy substrates)

(B) Percent phytophils (ratio of individuals that reproduce on plants to total)

The list employs metrics discussed above and others, for example from the BioMAP protocol of Griffiths (1998, cited earlier) which employs 10 indices. David et al's protocol assesses impacts by comparing six of the indices (from the above list) in test sites to reference sites. The comparisons are based on critical percentiles derived from the reference site. For example, if the reference site had critical values for the percent EPT (in their example) of 9% (in the 5th percentile) and 7% (in the 0.1 percentile), test sites with more than 9% EPTs would be judged similar to the reference site, with less than 7% would be judged as extreme values (potentially impacted) for the catchment and with between 7% and 9% would be judged as atypical (beyond the normal range and require further assessment and monitoring). The protocol uses revised RBP III methodology in general, with some modifications. Three riffle sites, all less than 1 m deep, are sampled in streams. If no riffles are present, snags and debris dams are sampled. The most downstream site is sampled first. Kick-and-sweep methods are used in riffles and grab samplers or corers in depositional areas. For lakes, because they generally have a greater variety of habitats than streams, five sites are selected randomly from a series of candidate sites in proportion to the different types of near-shore habitats. For example, if 60% of the near-shore areas are macrophytic beds, then 3 of the five sites should have macrophytes.

Fish Indices

Again, there are indicator species, already discussed in Chapter 9, that can be
11☞ used to assess the trophic status of lakes. However, they are of little value in assessing other kinds of stresses. The main disadvantage with using fish is they can often avoid the stress by swimming away. However, the absence of fish, if properly applied, can be

used to assess water quality. The main advantage with using fish is that they are at the top of the food chain and almost everyone relates to deteriorating health and/or disappearance of fish as a reliable indicator that something is wrong in a lake.

The most rigorous test of water quality using fish is EPA's rapid bioassessment protocol (RBP) V (see Plafkin *et al*, 1989 cited earlier). It is based on the ***Index of Biotic Integrity (IBI)*** which analyses fish communities for twelve metrics (Tables 12.26 and 12.27). The IBI incorporates zoogeographic, ecosystem, community, population and individual descriptors and accommodates natural differences in distribution and abundance of species that result from differences in water body type, size and region.

As with all RBPs, the fish metrics of the study site are compared with those of a reference site, the habitats (stream order, water velocity, depth, substrate type, etc.) of which must be comparable. At least one (preferably two) of each of riffle(s), run(s) and pool(s) should be sampled. Usually the fish are sampled by electro-fishing. Typical station lengths are 100-200 m for small streams to 500-1000 m in rivers. Plafkin *et al* (1989) suggests that the size of the reference station should be sufficient to produce 100-1000 individuals and 80-90 percent of the species expected from a 50 percent increase in sampling distance. Sampling can be done day or night, whatever yields the best results for a particular stream. Use of block nets is recommended for capturing large species. All fish, not just game species must be collected.

Species may be separated into juveniles and adults by size and colouration, with total numbers and incidence of any anomalies recorded for each group. Reference specimens from each group should be kept and preserved in 10% formaldehyde. Otherwise, any captured fish are kept alive in live-wells or any other effective means. For abundant species, subsampling is recommended to keep mortality to a minimum. Fish are released immediately after the necessary data have been collected.

The metrics used to calculate an IBI vary from state to state and province to province. The total list of metrics is given in Table 12.26, with those used in Ontario highlighted. Thirteen metrics are given but only twelve are used in the assessment. The metrics used depend on the ecoregion. Those highlighted for Ontario may not necessarily apply throughout the province and other metrics may be substituted, depending on the professional judgment of the assessor. The following discussion gives the rationale for each of the thirteen metrics. Water body size affects the first five metrics and it is important to examine the species/water-body size relationship in each zoogeographic region. It is recommended that plots be made of species richness (y-axis) vs log watershed area in each region. A line is fit by eye through the data points to include 95% of the points. Then the area under the line is trisected and scored as 5, 3 and 1, as shown in Fig. 12.2. The scores are used to help determine the ecoregion effects on the reference site (see Table 12.27). Details are provided by Plafkin *et al* (1989) and Barbour et al (1999) the revised RBP protocols. The only difference between the old and revised RBP protocols is in the evaluation of water quality; in the old system and IBI is calculated and compared to values in column 1 of Table 12.28; in the revised protocol, multimetric statistics are used to test for differences differences between reference sites and the test site.

Table 12.26. Different metrics available for use in calculating IBIs from different states and provinces in North America. Those used in Ontario are in bold face and italics.

1. Total number of species	**4. Number of sucker species**	**8. % Insectivorous cyprinids**
a. # native fish species	a. # adult trout species	a. % insectivores
b. # salmonid age classes	b. # minnow species	b. % specialized insectivores
2. Number of darter species	***c. # sucker and catfish species***	c. # juvenile trout
a. # sculpin species	**5. Number of tolerant species**	***d. % insectivorous species***
b. # benthic insectivore species	a. sensitive species	**9. % top carnivores**
c. # Darter and sculpin species	b. amphibian species	a. % catchable salmonids
d. # salmonid yearling numbers	***c. presence of brook trout***	b. % catchable wild trout
e. % round-bodied suckers	**6. % green sunfish**	c. % pioneering species
f. # sculpin individuals	a. % common carp	d. density catchable wild trout
3. Number of sunfish species	b. % white sucker	**10. Number of individuals**
a. # cyprinid species	c. % tolerant species	***a. density of individuals***
b. # water column species	d. % creek chub	**11. % Hybrids**
c. # sunfish and trout species	***e. % dace species***	a. % introduced species
d. # salmonid species	**7. % Omnivores**	b. % simple lithophils
e. # headwater species	***a. % yearling salmonids***	c. # simple lithophil species
		d. % native species
		e. % native wild indiviuals
		12. % Diseased individuals
		13. Total fish biomass

Metric 1. Total Number of Fish Species

11☞ Species diversity decreases with increasing degradation. The metric does not include hybrids or exotic species. Metric 1 should be used in cold, headwater streams where fish diversity is low; the age classes of the species found reflect the suitability of the stream for spawning and rearing.

Metric 2. Number and Identity of Darter Species

11☞ Darters and sculpins are benthivores and thus are sensitive to silt loading and oxygen depletion. Many smaller species inhabit the spaces among rubble. Since most darters and sculpins remain within small areas (100 - 400 m^2) of the stream, the metric is useful for assessing localized impacts.

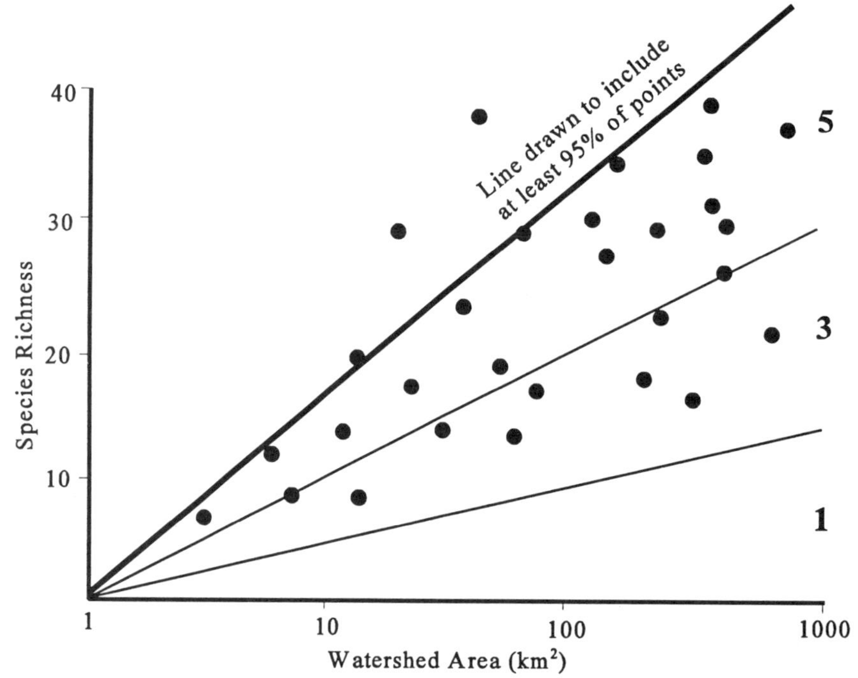

Figure 12.2. Total number of fish species in relation to watershed area for a hypothetical ecoregion.

Metric 3. Number and Identity of Sunfish Species

Metric 3 is a pool species health indicator. The pool species diversity decreases
11 with increasing pool degradation. Most pool species feed on drifting and surface invertebrates. Both sunfish and salmonids are excellent sport fishes. Sunfish are used mainly in southern streams for metric 3.

Metric 4. Number and Identity of Sucker Species

Fish species in this metric (suckers, trout, cyprinids, catfish) make up most of the fish biomass in streams and all are sensitive to physical and chemical habitat degradation.
11 A metric with suckers and/or catfish would be used for evaluating medium and large streams, cyprinids and trout for small streams, with trout used especially for cold-water streams. All these species (except cyprinids) are long-lived and provide a long-term integration of physico-chemical conditions.

Metric 5. Number and Identity of Intolerant Species

Metric 5 discriminates between high and moderate quality streams, with
11 intolerant species being the first to disappear after physical or chemical perturbations. The intolerant or sensitive species selected should form no more than 5-10 percent of the most susceptible species, otherwise the metric is less discriminating. Some states use

Table 12.27. Index of biological integrity scoring criteria. See text for description and rationales for each metric. Metrics 1 to 5 are scored relative to the maximum species richness line in Fig. 12.2. Metric 10 is drawn from reference site data.

Metric	Scoring Criteria		
	5	3	1
1. Number of native fish species	>67%	33-67%	<33%
2. Number of darter or benthic fish species	>67%	33-67%	<33%
3. Number of sunfish or pool species	>67%	33-67%	<33%
4. Number of sucker or long-lived species	>67%	33-67%	<33%
5. Number of intolerant species	>67%	33-67%	<33%
6. Proportion of green sunfish or tolerant individuals	<10%	10-25%	>25%
7. Proportion of omnivorous individuals	<20%	20-45%	>45%
8. Proportion of insectivores	>45%	20-45%	<20%
9. Proportion of top carnivores	>5%	1-5%	<1%
10. Total number of individuals	>67%	33-67%	<33%
11. Proportion of hybrids or exotics	0%	0-1%	>1%
12. Proportion with disease/anomalies	<1%	1-5%	>5%

amphibians as the sensitive taxon. Trout, especially brook trout, are used for Ontario streams. Table 12.29 gives the relative tolerances of several common species of fish.

Metric 6. Proportion of Individuals as Green Sunfish

In contrast to metric 5, this metric distinguishes low from moderate quality waters
11☞ because green sunfish species increase in biomass and abundance with increasing enrichment of the habitat. Dace are used for Ontario streams but creek chub could be substituted in regions where they dominate.

Metric 7. Proportion of Individuals as Omnivores

This is the first of three metrics that considers the trophic composition of the fish fauna for assessing water quality. The proportion of omnivores in a community increases
11☞ with increasing habitat deterioration. Omnivores are defined as fish that consistently feed on plant and animal material. If a stream lacks omnivores, the proportion of generalized feeders may be substituted.

Metric 8. Proportion of Individuals as Insectivorous Cyprinids

11☞ This metric evaluates the mid-range of biotic integrity. Benthivores are the dominant trophic guild of North American streams. Since insects tend to dominate the benthic community, most fish are insectivores. As the benthic macroinvertebrate food

Table 12.28. Interpretation of index score from Table 12.27.

IBI	Integrity Class	Characteristics
58-60	Excellent	Comparable to pristine conditions, exceptional assemblage of species
48- 52	Fair	Decreased species richness, especially intolerant species; sensitive species present
40-44	Fair	Intolerant and sensitive species absent, skewed trophic structure
28-34	Poor	Top carnivores and many expected species absent or rare; omnivores and tolerant species dominant
< 25	Very poor	Few tolerant species and individuals present; tolerant species dominant; diseased fish frequent

resource is decreased, the benthivorous fish are replaced by omnivorous forms.

Metric 9. Proportion of Individuals as Top Carnivores

Metric 9 discriminates between systems with high and moderate integrity. Top carnivores are piscivores (as adults), feeding on fish, amphibians and crayfish (although,
11☞ in reality, crawdads are invertebrates and part of the benthos). The percent of large salmonids may be substituted for percent piscivores in trout streams.

Metric 10. Number of Individuals in Sample

This is the first of three metrics that evaluates population recruitment, mortality, condition and abundance. Metric 10 is an abundance evaluator and is expressed as
11☞ catch-per-unit-effort either by area, distance or time sampled. Unusually low numbers generally indicate toxicity. However, some ecoregions have characteristically low fish abundances and some artificial enrichment may increase fish abundances. The metric is considered a good discrimination of water quality at the low end of the biological integrity scale.

Metric 11. Proportion of Individuals as Hybrids

This metric evaluates the suitability of a habitat for reproduction. It is an
11☞ estimate of reproductive isolation. As a rule, the percent of hybrids and introduced species increases but the proportion of lithophils (fish that lay eggs in interstices of sand, gravel, cobble) decreases as a habitat is degraded. Hybridization is absent in highly impacted sites. The metric is used mostly by western and midwestern states to measure the loss of species segregation that existed before the introduction of midwestern species to western rivers. Ontario does not use this metric in their IBI analyses.

Metric 12. Proportion of Individuals with Disease, Tumours, Fin Damage and
11☞ ***Skeletal anomalies***

Metric 12 evaluates the health and condition of individual fish. The conditions

Table 12.29. Feeding guilds, tolerance and habitat of common species of fish in temperate latitudes of North America. Trophic habits include: Ben = benthivore, Fil = filterer, Ins = insectivore, Pis = piscivore, omn = ominvore, her = herbivore, par = parasite, gen = generalist, pla = planktivore. Tolerance (Toler) includes: Tol = tolerant and int = intolerant, mod = moderately intolerant. Habitat (Habit) includes: Len = lentic, Lot = lotic, Both = lakes or rivers.

Family	Trophic	Toler	Habit	Family	Trophic	Toler	Habit
Petromyzontidae				Catastomidae			
Northern brook lamprey	Fil	int	lot	Shorthead redhorse	Ben	mod	both
Sea lamprey	Par	mod	both	River redhorse	Ben	int	lot
Acipenseridae				Northern hogsucker	Ben	int	lot
Lake sturgeon	Ben	mod	both	White sucker	Ben	tol	both
Lepisostidae				Longnose sucker	Ben	mod	len
Longnose gar	Pis	mod	both	Ictaluridae			
Amiidae				Channel catfish	Gen	mod	both
Bowfin	Pis	mod	both	Brown bullhead	Ben	tol	both
Hiodontidae				Tadpole madtom	Ben	mod	both
Mooneye	Ins	int	both	Anguillidae			
Clupeidae				American eel	Pis	mod	both
Alewife	Pla	mod	both	Cyprinodontidae			
Gizzard shad	Pla	mod	both	Eastern banded killifish	Ins	tol	both
Salmonidae				Gadidae			
Brown trout	Ins	int	lot	Burbot	Pis	int	len
Rainbow trout	Ins	int	lot	Percopsidae			
Brook	Ins	int	lot	Trout-perch	Ins	mod	both
Lake trout	Ben, Pis	int	len	Percichthyidae			
Osmeridae				White perch	Pis	mod	len
Rainbow smelt	Pla	mod	both	Centrarchidae			
Umbridae				Black crappie	Ben	mod	both

Central mudminnow	Ins	tol	both	Rock bass	Pis	mod	both
Esocidae				Smallmouth bass	Pis/ben	mod	both
Grass pickerel	Pis	mod	both	Largemouth bass	Pis/ben	mod	both
Northern pike	Pis	mod	both	Green sunfish	Ben	tol	both
Muskellunge	Pis	mod	both	Bluegill	Ben	mod	both
Cyprinidae				Pumpkinseed	Ben	mod	both
Common carp	Omn	tol	both	Percidae			
Goldfish	Omn	tol	both	Sauger	Pis	mod	both
Golden shiner	Omn	tol	both	Walleye	Pis	mod	both
Hornyhead chub	her/ben	int	lot	Yellow perch	Pis	mod	both
Lake chub	Ins	mod	len	Logperch	Ins	mod	both
River chub	Ins	int	lot	Johnny darter	Ins	int	both
Creek chub	Gen	tol	lot	Iowa darter	Ins	mod	both
Northern redbelly dace	Omn	mod	len	Rainbow darter	Ins	int	lot
Longnose dace	Ben	int	len	Least darter	Ins	mod	both
Blacknose dace	Ins	int	lot	Sciaenidae			
Pearl dace	Ben	int	lot	Freshwater drum	Ben	mod	len
Rosyface shiner	Ins	int	lot	Cottidae			
Emerald shiner	Gen	mod	len	Mottled sculpin	Ben	int	lot
Common shiner	Ins	mod	len	Slimy sculpin	Ben	int	len
River shiner	Ins	mod	lot	Deepwater sculpin	Ben	int	len
Spottail shiner	Ins	mod	both	Gasterostidae			
Sand shiner	Ins	mod	lot	Brook stickleback	Ins	mod	lot
Bluntnose minnow	Omn	tol	len	Ninespine stickleback	Ben	mod	both
Fathead minnow	Omn	tol	both	Atherinidae			
Grass carp	Her	mod	both	Brook silverside	Ins	int	both

listed are rare in healthy populations but the incidences increase significantly below point sources and in streams receiving toxic chemicals. The metric is an excellent measure of subacute effects of chemical pollution and the aesthetic value of sport fish.

Metric 13. Total Fish Biomass (optional)

11☞ This metric may be used as a substitute in larger rivers where fish size may vary by orders of magnitude.

Evaluating Biotic Integrity

Any 12 metrics can be used, the choice being dictated by the geographic locality and the fish species composition. First, it is recommended that plots be made of species richness (y-axis) vs log watershed area in each region. Fit a line by eye through the data points to include 95% of the points. Then trisect the area under the line and score the three areas as 5, 3 and 1, as shown in Fig. 12.2. The scores are used to help determine the ecoregion effects on the reference site (see Table 12.27). Then determine the IBI scoring criteria for each of the 12 metrics following instructions in the caption of Table 12.27. An evaluation of the stream is made by summing up the scoring criteria and comparing to the values in column 1 of Table 12.28.

In the revised RBP IV protocol, the same metrics are calculated but the scores of each are statistically compared to those from reference sites, rather than summing up the criteria, using similar procedures described above for macroinvertebrates.

BIOASSESSMENT OF LAKES

Unfortunately, the techniques for the biological assessment of lakes are not as diverse as for running water systems. Those that have been developed are for assessing trophic status, with few for assessing the impacts of lake acidification. We have already reviewed morphological, physical and chemical criteria (see Table 12.3) for assessing trophic status of lakes and they far outnumber the biological criteria. Appendix 12.I provides chemical PWQO (provincial water quality objectives) for Ontario. The biological criteria are examined within different groups of organisms, beginning with bacteria.

Bacterial Indices

12☞ Although not commonly used, a bacterial index ratio (BIR) can be used to assess water quality (trophic status) in lakes. The BIR was developed from water samples taken from the Great Lakes but probably has application in most lakes. The BIR was developed by Rao and Jurkovic (1977)[27] specifically to assess trophic status of lakes. The BIR is the ratio of aerobic heterotroph densities (H) to total bacterial densities (TB), expressed as a percent, in 1 ml of water, or:

$$\text{BIR} = (\text{H}/\text{TB}) \times 100$$

The total bacterial densities (#/ml) are based on the acridine orange epi-fluorescence technique. The following BIR criteria (means and ranges) were derived from correlating more than 30 bacteriological samples, taken near the end of summer in each of the Great Lakes, with trophic status:

Oligotrophy: BIR < 7 (0.1-11.0)
Mesotrophy: BIR = 7 - 25 (3.0-26.0)
Eutrophy: BIR > 25 (14.0-41.0)

The criteria have not been tested rigorously and have rarely been used to supplement other trophic status criteria, probably because the technique requires some bacteriological expertise that is not part of most limnologist's repertoire of trophic assessment skills. Nevertheless, the theory behind the BIR is good. Bacteria are the basic mineralizers of organic materials and the relative numbers of heterotrophic aerobes should be a good indicator of the oxidation state of the water.

Algae and Phytoplankton

Most algal indices, as discussed earlier, have been derived from stream and river
13☞ data. Some running water indices, particularly those useable in large rivers, can be applied to lakes also. This includes lake acidification indicators.

Few algal indices have been developed specifically for assessing the trophic status of lakes. Of those that have been developed, most are based on species assemblages, especially of desmids and diatoms. The desmids and centric diatoms represent primarily oligotrophic groups, while the pennate diatoms represent the entire range of trophic states. Several quotients have been developed to determine the trophic status of lakes. These include:

Myxophycean quotient
= No. myxophycean species/ No. desmidacean species

---

[27]Rao, S. S. and A. A. Jurkovic. 1977. Differentiation of the trophic status of the Great Lakes by means of bacterial index ratio. J. Great Lakes Res. 3: 323-326.

Chlorophycean quotient
= No. chlorococcales species/ No. desmidacean species

Diatom quotient
= No. centric diatom species/ No. pennate diatoms species

Euglenoid quotient
= No. euglenoid species/(Nos. myxophyceans + chlorococcales species)

Compound quotient
= (Nos. myxophyceans + chlorococcales + centric diatoms + euglenoids species)/ No. desmid species

Apparently the quotient that is most reliable in assessing trophic status is the compound quotient, its criteria being:

Compound Quotient	Trophic Status
0.0 - 0.3	Dystrophy
0.4 - 1.0	Oligotrophy
1.0 - 5.0	Eutrophy
5.1 - 20.0	Hypereutrophy to organic pollution

Algal Productivity

14☞ The following chlorophyll *a* levels are used as criteria for assessing the trophic status of lakes.

Chlorophyll *a* μg/L	Trophic Status
< 2	Oligotrophic
2 - 5	Mesotrophic
> 5	Eutrophic

The method for measuring chlorophyll *a* and the development of the criterion and its relationship to phosphorous levels and Secchi depth are discussed in Chapter 6. The three variables are also used for calculating Carlson's trophic state index (TSI), also discussed in Chapter 6. Carlson developed three equations, each one, except total phosphorous, being based on the average summer concentration, not the spring turnover level as used in Table 12.3. The predictive equations for each variable are:

Secchi depth (SD): $TSI = 10(6 - \ln SD/\ln 2)$

Chlorophyll *a* (Chl): $TSI = 10(6 - [(2.04 - 0.68 \ln Chl)/\ln 2])$

Total phosphorous (TP): $TSI = 10(6 - [\ln(48/TP)/\ln 2)]$

The TSI trophic criteria are:

TSI	**Secchi depth m**	**Chl a μg/L**	**Total P μg/L**	**Trophic Status**
0 - 35	> 6	< 1	< 10	Oligotrophy
36 - 50	2 - 6	1 - 6.4	10 - 24	Mesotrophy
51 - 100	< 2	> 6.4	> 24	Eutrophy

Secchi depth and chlorophyll *a* and total phosphorous levels are also used later for determining the development capacity of a lake for body contact recreation and fisheries.

Macrophyte Indices

15☞ Chapter 6 provides lists (Tables 6.7 and 6.8) of macrophytes with indicator value with respect to trophic status and lake acidification. However, there have been no rigorous tests that relate species abundance to phosphorous levels, chlorophyll *a* levels or Secchi depth, at least in North American waters. Table 6.8 lists macrophyte species according to their abundance (biomass) in relation to the conductivity and hardness ratio (total hardness / concentrations of sodium and potassium) of British lakes. While many of the relationships appear to apply to North American waters, none have been verified. At best, we know that the diversity of submerged species is much greater in oligotrophic waters than in eutrophic waters, mainly because light intensity decreases with increasing eutrophication. Also, the biomass of submerged macrophytes tends to be low. Emergent and floating macrophytes become more abundant as sediments become enriched and waters become highly coloured, the leaves no longer depending on light energy in the water column.

Benthic Macroinvertebrates

16☞ The benthic community structure in deeper parts of lakes, such as the profundal zone, reveals more about the trophic status of a lake than does that from either the littoral or sublittoral zones. The epilmnial region of both eutrophic and oligotrophic lakes is usually saturated, if not supersaturated with dissolved oxygen, and benthic invertebrates are rarely limited by oxygen supplies or food availability. However, once a lake thermally stratifies, the benthic community structure in the deeper parts of the lake, especially in the profundal zone, is dependent upon the amount of dissolved oxygen, the type and availability of food and the nature of the substratum. Eutrophic lakes have an abundance of organic detritus that bacteria reduce to carbon dioxide and dissolved organ carbon, reducing the oxygen levels in the process. Only low-oxygen tolerant invertebrate species are present in large numbers. The species present must also be able to utilize the bottom materials, so collector gatherers predominate. The only other functional feeding group present is predators but the predaceous species must also be able to tolerate low dissolved oxygen levels. Filter feeders, such as fingernail clams, could also utilize the materials but all species require at least 50% dissolved oxygen saturation and are absent in the deeper parts of eutrophic lakes. Hence, the deeper and profundal zones of eutrophic lakes are dominated by tubificid worms, such as *Limnodrilus hoffmeisteri* and *Tubifex tubifex*, both of which are gatherers, and chironomid larvae, including several species of *Chironomus*, all of which are gatherers, and predaceous species of *Procladius* and *Tanytarsus*. Some *Tanytarsus* species are able to utilize a filter feeding habit.

The profundal and deeper zones of oligotrophic lakes, on the other hand, have a great diversity of gatherers, filter feeders and predators. But the gatherers are represented by amphipods (e.g. *Diporeia hoyi*) and many species of low-oxygen sensitive tubificid worms (e.g. *Tubifex kessleri, Rhyacodrilus* and *Peloscolex variegatum*) and chironomids (e.g. *Heterotrissocladius subpilosus, Protanypus caudatus, Paracladopelma obscura*); the filter feeders are represented by several species of fingernail clams, particularly *Pisidium conventus* and *Sphaerium nitidum*, the most sensitive of the group, and *Pisidium lilljeborgi* and *Pisidium ferrugineum*; and the predators, *Hexagenia limbata* (mayfly) and *Mysis relicta* (crustacean). While low-oxygen tolerant forms (e.g. *L. hoffmeisteri, Chironomus*) may also be present, they are far outnumbered by the low-oxygen sensitive species.

With these trends in mind, it should be possible to develop indices based on the proportions of functional feeding groups or on taxa at the family level. The three functional feeding groups are gatherers, filter feeders and predators; the primary family taxa involved are Tubificidae, Chironomidae, Sphaeriidae, Mysidae, Gammaridae, Ephemeridae, Chaoboridae and, occasionally, Sialidae in water just below the metalimnion. The tubificids and gammarids are gathering groups; chironomids are represented by both gatherers and predators and some filter feeders; sphaeriid clams obtain food by filter feeding; the remaining four families, Mysidae, Ephemeridae, Chaoboridae and Sialidae, are mostly predaceous.

For example, if a ratio is developed for an index, the denominator should have all three feeding groups. Since bivalve filter feeders and crustacean gatherers and predators
16☞ are found in oligotrophic and mesotrophic lakes, and are absent in eutrophic lakes, a numerator incorporating them will yield an index close to “0" for highly eutrophic lakes and a value much greater than 0 in mesotrophic and oligotrophic lakes. Hypothetically,

such a ***Profundal Functional Feeding Index (PFFI)***, could be expressed as:

$$\text{PFFI} = \frac{(\text{Sphaeriidae} + \text{Mysidae} + \text{Gammaridae} + \text{non-chironomid insects})}{(\text{Tolerant Tubificidae} + \text{red Chironomidae})}$$

where the number of species is substituted for each group. For oligotrophic lakes a maximum of near 10 for the numerator (e.g. 4-5 sphaeriid species, 1 mysid species, 1 amphipod species, 1 -2 mayfly species) could be found in the profundal zone and 2 - 4 for the numerator (e.g. 1-2 tubificid species, 1-2 red chironomid species). In eutrophic lakes, the numerator should be close to zero.

Fish Indices

Chapter 9 discusses the use of sport fish yield as an index of trophic status
17☞ because it is closely correlated to primary productivity, as determined by summer chlorophyll *a* levels. The relationship is described by the following regression:

$$\text{Annual sport fish yield (kg/ha)} = -1.8 + 2.7\ [\text{chl } a]_{\text{summer ave.}}\ (\mu\text{g/L})$$

Chlorophyll *a* levels in this regression explain 91% of the variation in angler sportfish yields. Hence, if angler sport fish yield is closely monitored for a particular lake, the following criteria may be used to assess its trophic status:

Angler Sport Fish Yield kg/ha	Trophic status
< 4	Oligotrophy
4 - 11.7	Mesotrophy
> 11.7	Eutrophy

The profundal fish community structure is also a useful criterion for assessing trophic status of lakes. The profundal zone consists largely of two feeding guilds, benthivores and carnivores (predators), at least in oligotrophic and mesotrophic lakes. The benthivores are the dominant feeding guild in deep waters. Some planktivores, especially whitefish and ciscos, also occur in deep water but they feed on mysids and phantom midges that move from the benthos into the water column at night. The truly deep-water forms, like the deep water cisco (*Coregonus johannae*), lake whitefish (*Coregonus clupeaformis*) and deep-water sculpins *(Myoxocephalus thompsoni*), feed mostly on profundal benthos typical of oligotrophic lakes, like fingernail clams, mysids, and amphipods. Burbot (*Lota lota*) is a vicious predator, feeding on young fish, crayfish, and benthos. Table 12.29 lists some tolerance levels and feeding guilds of the more common species in northern latitudes (above ~ 40° latitude) of North America.

The profundal zone of eutrophic lakes is either anoxic, or close to anoxic, and only a few, if any fish are present. Those that are present are very tolerant forms (e.g. some Cyprinidae like carp, some Ictaluridae like brown bullheads, and some Catastomidae like the white sucker, Table 12.29) and cannot remain in the profundal zone for extended periods of time after the lake thermally stratifies. The dominant families of fish present in profundal zones of oligotrophic lakes are Salmonidae (e.g. *Salvelinus namaycush*), Cottidae (e.g *Cottus cognatus, Myoxocephalus thompsoni*), Coregonidae (e.g. *Coregonus artedii, Coregonus johannae, Coregonus clupeaformis*) and, occasionally, Gadidae (e.g. the burbot, *Lota lota*). Families that spend most of their time at the sediment-water interface or in very deep water (e.g. Cottidae and Coregonidae) disappear first as lakes approach mesotrophy. Other families of fish, like some Percidae, Esocidae, Osmeridae, Sciaenidae) may frequent the deep waters but they are not characteristic of the oligotrophic assemblage of fish. Hence, an index that reflects the differences in profundal fish community structure between oligotrophic and mesotrophic lakes (profundal fish absent in eutrophic lakes) should be useful for assessing trophic status.

For example, using the same arguments as applied to the benthic invertebrates, a
Profundal Fish Index (PFI) that relies on the presence of profundal fish species could
17☞ also be developed to assess trophic status. Like the PFFI, it could be a ratio composed of the most sensitive fish families in the numerator to any other of the remaining less sensitive fish families in the denominator, for example:

$$\text{PFI} = \frac{\text{Cottidae} + \text{Coregonidae} + \text{Gadidae} + \text{Salmonidae}}{\text{Any other family}}$$

where each species in a family would provide a score. Again, eutrophic lakes should have a score close to 0 because no fish could survive the summer in the hypolimnion.

So far methods for assessing the trophic status of lakes using the assemblages of fish communities have been examined. But once the trophic status is known, one is often interested in knowing the fisheries and body contact recreational potential of the lake. Dillon and Rigler (1975)[28] developed a stepwise procedure to determine the development capacity of a lake. It was formulated from data on lakes in southern

---

[28]Dillon, P. J. and F. H. Rigler. 1975. A simple method for predicting the capacity of a lake for development based on lake trophic status. J. Fish. Res. Bd. Canada 32: 1519-1531.

Ontario but is applicable to lakes elsewhere, provided data for precipitation, runoff, phosphorous loading, etc. for the appropriate ecoregion are used. The following is a summary of Dillon and Rigler's (1975) procedures. The procedure is currently being updated and revised (Dillon, pers. comm.).

Stepwise Procedure for Calculating the Development Capacity of a Lake

Step 1. Based on long-range plans for the lake, decide what the maximum permissible
18☞ average summer chlorophyll *a* concentration will be:

Level 1: 2 mg m^{-3}; for lakes to be used primarily for body contact water recreation, and where it is desirable to maintain hypolimnetic concentrations of oxygen in excess of 5 mg $liter^{-1}$ to preserve cold water fisheries. The lake will be extremely clear with a mean Secchi disc visibility of 5 m and will be very unproductive. (Note - the Secchi disc visibility may be lower in brown water, or dystrophic, lakes).

Level 2: 5 mg m^{-3}; for lakes to be used for water recreation but where the preservation of cold water fisheries is not imperative. The lake will be moderately productive and correspondingly less clear, with a mean Secchi disc visibility of 2-5 m.

Level 3: 10 mg m^{-3}; for lakes where body-contact recreation is of little importance, but emphasis is placed on fisheries (bass, walleye, pickerel, pike, muskellunge, bluegill, yellow perch). Hypolimnetic oxygen depletion will be common. Secchi disc depths will be low (1-2 m), and there is a danger of winter kill of fish in shallow lakes.

Level 4: 25 mg m^{-3}; suitable only for warmwater fisheries. Secchi disc depth <1.5 m, hypolimnetic oxygen depletion beginning early in summer, considerable danger of winter kill of fish except in deep lakes. The planning agency may pick any intermediate level should it so desire.

Step 2: From the chosen summer average chlorophyll *a* concentration, calculate the permissible spring phosphorus concentration, [P] from:

$$\text{Log}_{10}\,[\text{chl } a] = 1.45 \log_{10}\,[\text{P}] - 1.14$$

$$\text{i.e. } \log_{10}\,[\text{P}] = (\log_{10}\,[\text{chl } a] + 1.14)/1.45$$

e.g.

[chl *a*]	[P]
2 mg m^{-3}	9.9 mg m^{-3}
5	18.5
10	29.9
25	56.3

Step 3: Determine the lake surface area (A_0 in m^2), mean depth ($\bar{z}$ in m), and volume (V in m^3) from available information if possible. If such data are not available, the lake must be sounded and a contour map drawn. The area (A_0) is obtained by planimetry from an aerial photograph of known scale.

Step 4: Outline the lake's drainage area on a 1:50,000 scale topographic map or on aerial photographs and calculate the area (A_d in m^2) by planimetry.

Step 5: Using Figure 10.3 (from Pentland, 1968)[29] determine the total annual unit runoff (r) in cfs mi^{-2} and convert to $m^3\ yr^{-1}\ m^{-2}$ or $m\ yr^{-1}$ by multiplying by 0.345. If lake is not in southern Ontario, use data for ecoregion containing your lake.

Step 6: If $A_d > 10\ A_0$ calculate Q, the total outflow volume as:

$$A_d \cdot r\ (m^3\ yr^{-1})$$

and calculate the flushing rate (ρ) as Q/V or

$$(A_d \cdot r)/V\ (yr^{-1})$$

If A_d is $< 10\ A_0$, determine the mean annual precipitation (Pr) from Fig. 12.4 (from Brown *et al.*, 1968)[30] (Note: if your lake is not from southern Ontario, use data from the ecoregion that contains your lake), and the mean annual lake evaporation (Ev) from Fig. 12.5 (from Bruce and Weisman, 1966)[31], convert to $m\ yr^{-1}$ by multiplying by 0.0254 and calculate Q:

$$Q = A_d \cdot r + A\ (Pr-Ev)$$

$$\therefore\ \rho = (A_d \cdot r + A_0\ (Pr-Ev))/V \qquad (yr^{-1})$$

Step 7: Calculate the areal water load as (q_s) as Q/A_0 $\qquad (m\ yr^{-1})$

Step 8: Calculate retention coefficient (R) as

---

[29]Pentland, R. L. 1968. Runoff characteristics in the Great Lakes basin. Proc. 11th Conf. Great Lakes Res. 1968: 326-359.

[30]Brown, D. M., G. A. McKay and L. J. Chapman. 1968. The climate of southern Ontario. Environment Canada, Atmospheric Environment, Climatological Studies, Toronto, Ontario. Number 5, 2nd Ed. 50 pp.

[31]Bruce, J. P. and B. Weisman, 1966. Provisional evaporation maps of Canada. Canada Dept. Transport, Meteorological Br. Toronto, Ontario. 21 pp.

$$R = 0.426 \exp(-0.271\, q_s) + 0.574 \exp(-0.00949\, q_s)$$

Step 9: Calculate the response time of the lake to a change in phosphorus loading:

$$\text{Response time} = (3 \rightarrow 5)\, t_{1/2}$$
$$= (3 \rightarrow 5)\, 0.69/(\rho + 10/\bar{z}) \qquad \text{(yr)}$$

This will provide an indication of the time required for a lake to "respond" to development and will give an idea of when follow-up studies (if any) should be carried out. Conversely, the response time can assist in the interpretation of present water quality. For example, a lake with 250 new cottages may appear to be in good condition, but if one can calculate that it has a response time of 6-10 yr, then caution is necessary before additional development is allowed.

Step 10: Calculate the permissible phosphorus load (L_{perm}) to the lake:

$$L_{perm} = [P]\cdot\bar{z}\cdot\rho/(1 - R) \quad (\text{mg m}^{-2}\ \text{yr}^{-1})$$

and the permissible supply (J_{perm})

$$J_{perm} = L_{perm}\cdot A_0 / 10^6 \quad (\text{kg yr}^{-1})$$

Step 11: Divide the watershed of the lake (on the topographical map) into subunits for all inflows and determine the area for each one (A_{di}). From the most recent aerial photographs available determine if >15% of the area of each watershed is either cleared land or marsh. Determine whether the watershed (not just the lake) is on Precambrian igneous rock or on sedimentary rock, and estimate the export (E mg m^{-2} yr^{-1}) for each subwatershed from Table 12.30. Calculate the total supply of phosphorus from the land to the lake:

$$J_E = (\textstyle\sum_i A_{di} \cdot E)$$

and the load

$$L_E = (\textstyle\sum_i A_{di} \cdot E)/A_0 \quad (\text{mg m}^{-2}\ \text{yr}^{-1})$$

Take into consideration an upstream lake by reducing the supply to the downstream lake from the watershed containing the upstream lake by multiplying by $(1 - R^1)$ where R^1 is the phosphorus retention coefficient of the upstream lake. R^1 is calculated as in Step 8, using q_s for the upstream lake.

Step 12: The phosphorus load from precipitation, L_{PR}, is 75 mg yr^{-1}. Calculate the supply from precipitation as:

$$J_{PR} = 75 \cdot A_0 / 10^6 \ (m^2 \ kg \ yr^{-1})$$

Step 13: The natural supply and natural loading are:

$$J_N = J_E + J_{PR}$$
$$L_N = L_E + L_{PR}$$

Table 12.30. Ranges and mean values for export of total phosphorous (E) from 43 watersheds (mg/m²/yr).

Land Use	Geological Classification	
	Igneous	Sedimentary
Forest		
Range	0.7-8.8	6.7-18.3
Mean	4.7	11.7
Forest + Pasture		
Range	5.9-16.0	11.1-37.0
Mean	10.2	23.3

If $J_N \geq J_{perm}$, i.e. if the natural supply is greater than the permissible supply, allow no (further) development.

Step 14: Determine the present number of cottages (N_C) and permanent dwellings (N_D) within 300 m (1000 ft) of the lake or any of the inflowing streams or rivers from recent aerial photographs or field surveys. For cottages, assume 253 capita-days per year (0.69 capita-years yr^{-1}). Assume 4.3 people per dwelling, and calculate N_{CY}, the number of capita-years yr^{-1} spent at the lake:

$$N_{CY} = 0.69 \times N_C + 4.3 \times N_D$$

i.e. one permanent unit = 6.2 cottage units.

Step 15: Calculate the phosphorus supplied to the lake from the cottage units (artificial supply) as:

$$J_A = 0.8 \times N_{CY} (1 - R_s) \qquad (kg \ yr^{-1})$$

where R_s = 0 for conventional septic tank-tile field systems on the Precambrian Shield. If there is firm evidence that holding tanks are used for all household wastes and the systems are pumped and removed to a treatment plant outside of the watershed, neglect such cottage(s) in the calculations. If the septic tile filter beds are situated off of the Shield on soils that correspond to those of Table 12.31, use the appropriate values for R_s.

Table 12.31. Retention coefficients of total phosphorous for septic tile filter beds of different characteristics.

Filter Bed	R_s
22 in. sand (D_{10} = 0.24 mm) or 8 in. mixture 4% red mud, 96% sand	0.76
30 in. sand (D_{10} = 0.24 mm)	0.48
30 in. sand (D_{10} = 0.30 mm)	0.34
30 in. sand (D_{10} = 0.60 mm)	0.22
30 in. sand (D_{10} = 1.00 mm)	0.01
30 in. sand (D_{10} = 2.50 mm)	0.04
15 in. sand (D_{10} = 0.24 mm) or 15 in. mixture 10% red mud, 90% sand	0.88
15 in. sand (D_{10} = 0.24 mm) or 15 in. mixture 50% limestone, 50% sand	0.73
15 in. sand (D_{10} = 0.24 mm) or 15 in. mixture 50% clay-silt, 50% sand	0.74
30 in. silty sand	0.63

Step 16: Calculate the present total supply of phosphorus to the lake:

$$J_T = J_N + J_A \qquad (\text{kg yr}^{-1})$$

If $J_T \geq J_{perm}$, allow no further development.

Step 17: If $J_T < J_{perm}$, calculate the total permissible number of cottage units:

$$N_{perm} = (J_{perm} - J_N)/0.69 \text{ x } 0.8\ (1 - R_s)$$

Step 18: The additional number of cottage units permitted is:

$$N_{add} = N_{perm} - N_{CY}$$

Fig. 12.3. Annual runoff (cfs) in Great Lakes drainage area.

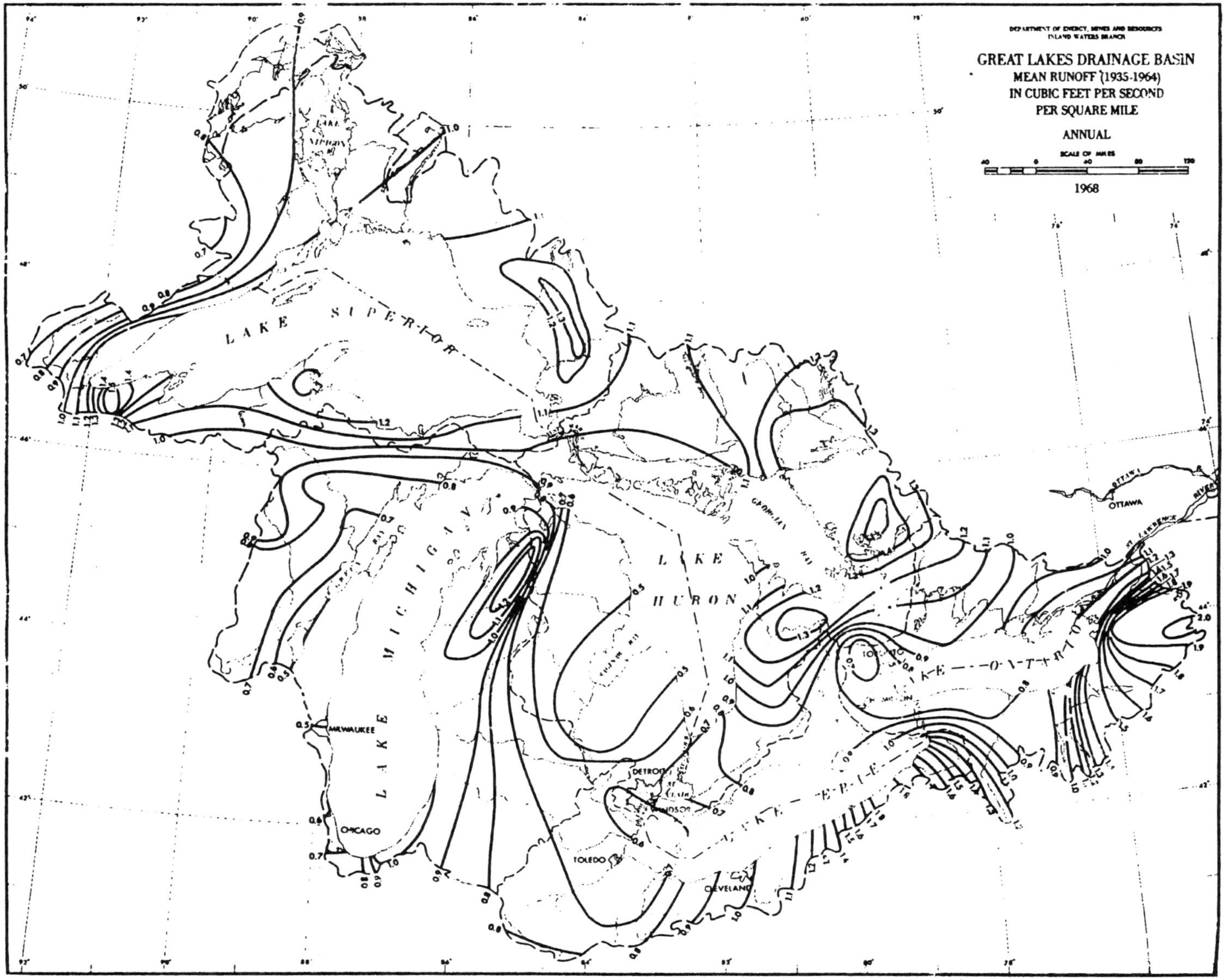

Fig. 12.4. Mean annual precipitation (inches) in southern Ontario.

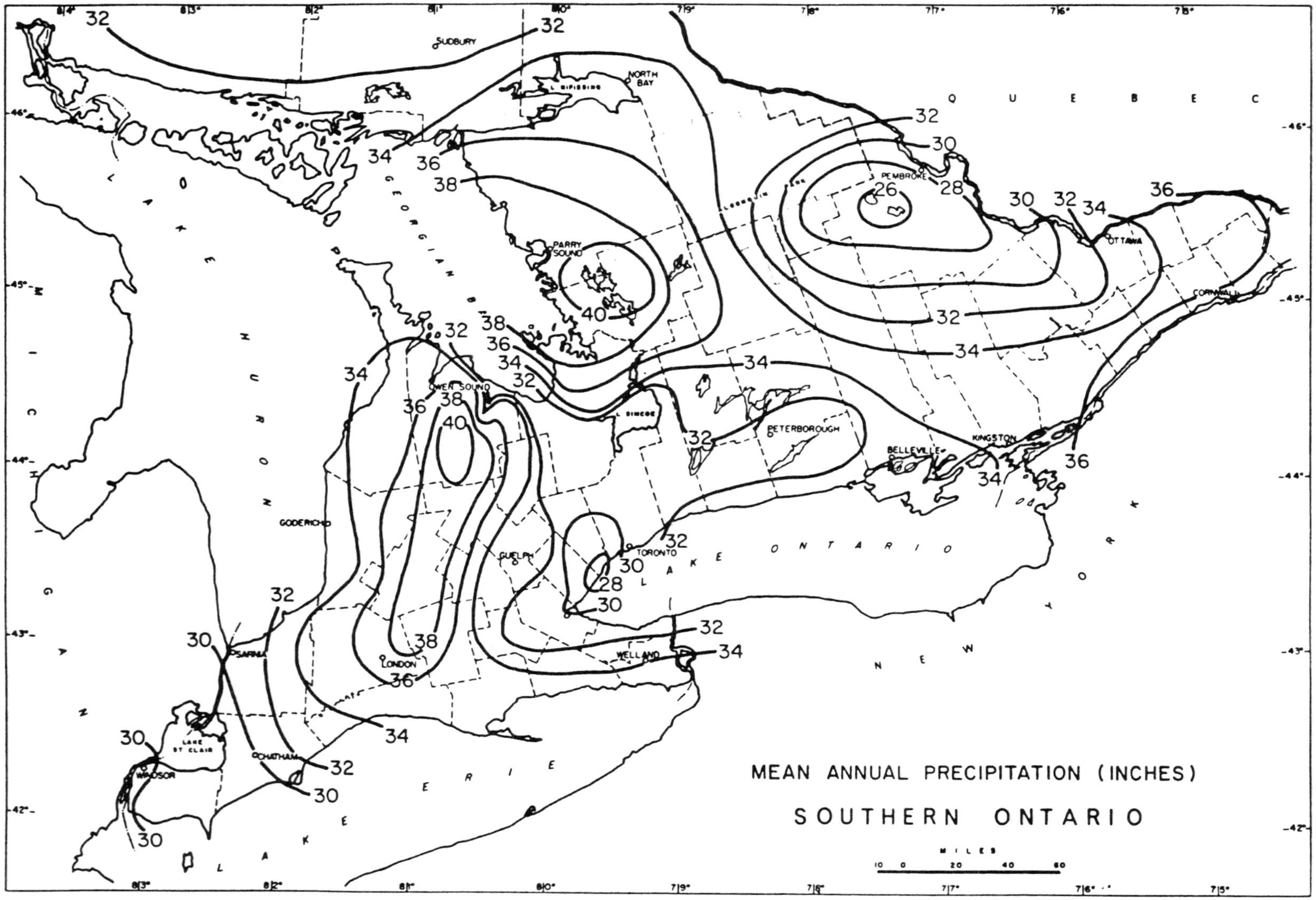

Figure 12.5. Mean annual evapouration (inches) in Canada.

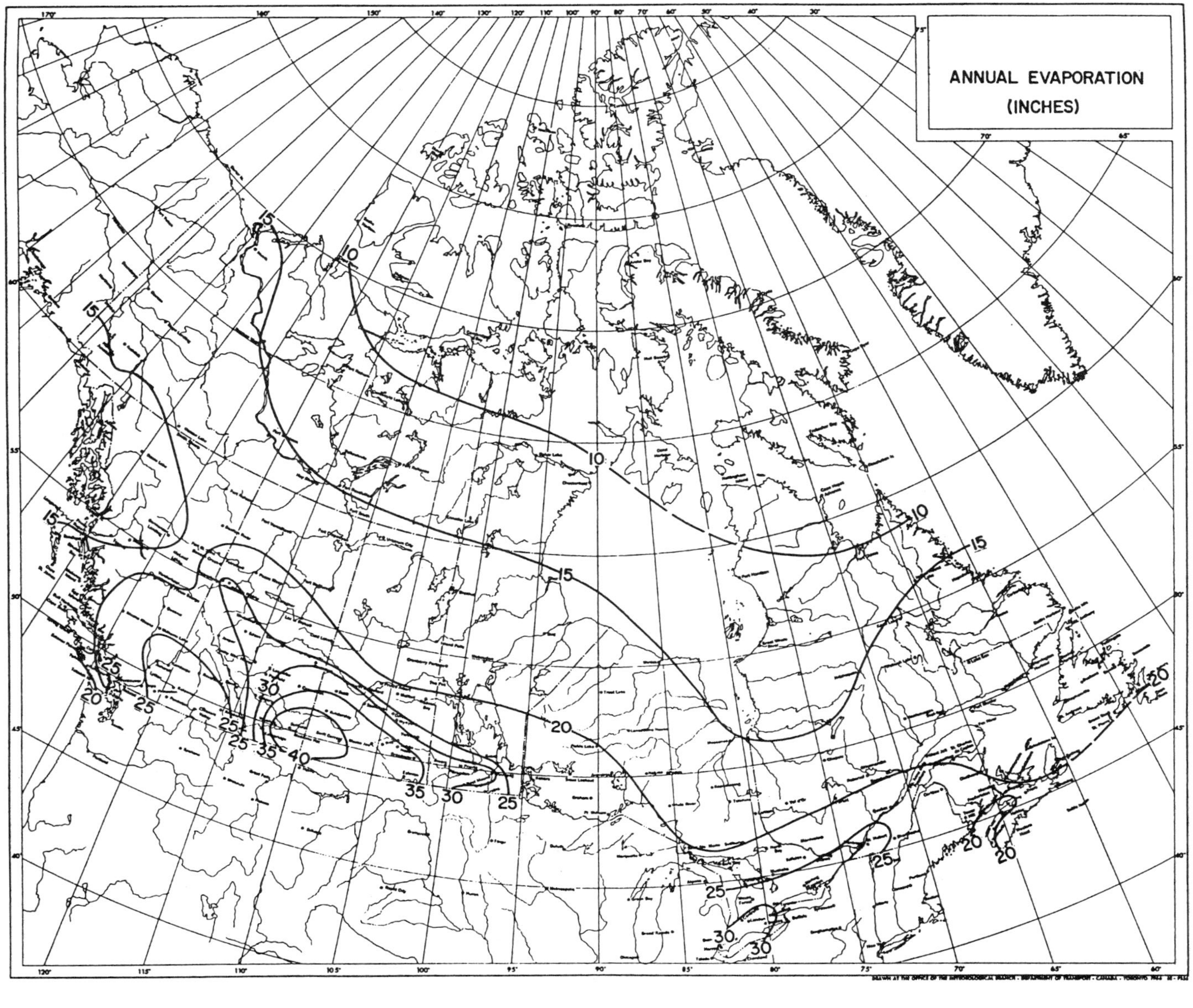

APPENDIX 12.I

Provincial Water Quality Objectives (PWQOs) for protection of aquatic life and recreation in Ontario. Superscripts are as follows: 1-11, PWQO value is taken from documents cited at end of this appendix; a, PWQO value is currently being revised, use with caution; b, PWQO value is based on most recent information and is currently being evaluated.

Aesthetics	PWQO[1]: Water used for swimming, bathing and other recreational activities should be aesthetically pleasing. The water should be devoid of debris, oil, scum and any substance which would produce an objectionable deposit, colour, odour, taste, or turbidity.
Aldrin/Dieldrin	0.001 μg/L PWQO[1] PWQO is for the sum of the concentrations of aldrin and dieldrin in water.
Alkalinity	PWQO[1]: Alkalinity should not be decreased by more than 25% of the natural concentration.
Aluminum	Interim PWQO[4]: · At pH 4.5 to 5.5 the Interim PWQO is 15 μg/L based on inorganic monomeric aluminum measured in clay-free samples. · At pH >5.5 to 6.5, no condition should be permitted which would increase the acid soluble inorganic aluminum concentration in clay-free samples to more than 10% above natural background concentrations for waters representative of that geological area of the Province that are unaffected by human-make inputs. · At pH >6.5 to 9.0, the Interim PWQO is 75μg/L based o total aluminum measured in clay-free samples. · If natural background aluminum concentrations in water bodies unaffected by human-made inputs are greater than the numerical Interim PWQO (above), no condition is permitted that would increased the aluminum concentration in cay-free samples by more than 10% of the natural background level. Note: pH values of 6.5 and 8.5 are outside the range considered acceptable by the PWQO for pH.
Ammonia (un-ionized)	20 μg/L (PWQO)[1] The percentages of un-ionized ammonia (NH_3) in aqueous ammonia solution for different temperature and pH conditions are listed in the Table 12.IA. For example, at 20 C and pH of 8.0, a total ammonia concentration of 500μg/L would give an un-ionized ammonia concentration of 500 x 3.8/100 = 19μg/L which is less than the un-ionized ammonia Objective of 20 μg/L.
Antimony	20 μg/L (Interim PWQO)[b]
Arsenic	100 μg/L (PWQO)[1]
Arsenic (revised)	5 μg/L (Interim PWQO)[b]
Bacteria	See *Escherichia coli*
Benzene	100 μg/L (Interim PWQO)[b]
Beryllium	(PWQO)[1]: Hardness as $CaCO_3$ (mg/L) — PWQO μg/L <75 — 11 >75 — 1100
Biphenyl	0.2 μg/L (Interim PWQO)[a]
Boron	200 μg/L (Interim PWQO)[a]
Cadmium	0.2 μg/L (PWQO)

Cadmium (revised)	Interim PWQO[b]: Hardness as $CaCO_3$ (mg/L) — Interim PWQO (μg/L) 0 - 100 — 0.1 > 100 — 0.5
Chlorine	2 μg/L (PWQO)[1] · Total residual chlorine, as measured by the amperometric (or equivalent) method.
Chromium	100 μg/L (PWQO)[1]
Cobalt	0.6 μg/L (Interim PWQO)[b]
Copper	5 μg/L (PWQO)[1]
Copper (revised)	Interim PWQO[b]: Hardness as $CaCO_3$ (mg/L) — Interim PWQO (μg/L) 0 - 20 — 1 > 20 — 5
Cyanide	5 μg/L (PWQO)[1] · PWQO is for free cyanide in an unfiltered water sample.
2,4-D (BEE)	4 μg/L (PWQO)[1] · chemical name 2,4-dichlorophenoxyacetic acid - (2 - butoxyethyl) ester
DDT& Metabolites	0.003 μg/L (PWQO)[1] ·· PWQO is for thesum of DDT, DDD and DDE. The sum of the concentrations of DDT and its metabolites in whole-fish should not exceed 1 ug/g (wet weight basis) for the protection of fish-consuming birds.
Diazinon	0.08 μg/L (PWQO)[1]
Diquat	0.5 μg/L (PWQO)[1]
Dissolved gases	(PWQO)[1]: To protect aquatic organisms, the total dissolved gas concentrations in water should not exceed 110 percent of the saturation value for gases at the existing atmospheric and hydrostatic pressure.
Dissolved Oxygen	(PWQO)[1]: · Dissolved oxygen concentrations should not be less than the values specified below in Table 12.IB for cold water biota (e.g. salmonid fish communities) and warm water biota (e.g. centrarchid fish communities): see Table 12.29 · In waters inhabited by sensitive biological communities, or in situations where additional physical or chemical stressors are operating, more stringent criteria may be required. For example, a sensitive species such as lake trout may require more specific water quality objectives. · In some hypolimnetic waters, dissolved oxygen is naturally lower than the concentrations specified in Table12.IB. Such a condition should not be altered by adding oxygen-demanding materials causing a depletion of oxygen.
Diuron	1.6 μg/L (PWQO)[1]
Endrin	0.002 μg/L (PWQO)[1]
Endosulphan	0.003 μg/L (PWQO)[1]

Escherichia coli	100 E. *coli* per 100 ml (based on a geometric mean of a t least 5 samples) ·Based on a recreational water quality guideline published by the Ontario Ministry of Health in 1992. This Ministry of Health guideline was specifically intended for application by the local Medical Officer of Health to swimming and bathing beaches. It is based upon a geometric mean of levels of E. *coli* determined form a minimum of 5 samples per site taken within a given swimming area and collected within a one month period. If the geometric mean E. *coli* level for the sample series at a given site exceeds 100 per 100mL, the site should be considered unsuitable for swimming and bathing. E. *coli* was selected for the guideline because studies have determined that, among bacteria of the coliform group, E. *coli* is the most suitable and specific indicator of faecal contamination. · An analytical test with a high degree of specificity for E. *coli* regardless of water sample source, requiring no confirmation procedures, and which produces results in 21 hours has been developed and adopted by both the Ministry of the Environment and Energy laboratories. · Where testing indicates sewage or faecal contamination, a site-specific judgement must be made as to the severity of the problem and the appropriate course of action. · As of May 1, 1994, MOEE staff have been advised to base all new compliance, enforcement and monitoring activities on the E. *coli* test. Some water managers may find it necessary to continue testing for faecal coliforms or total coliforms. For example, where testing at a long term water quality monitoring station requires a continuous record of results using either the faecal or total coliform test to monitor trends in water quality. As a benchmark of the long term monitoring results, the former objectives for faecal coliforms and total coliforms are referenced for your information. For faecal coliforms the objective was a 100 counts per 100ml (based on a geometric mean density for a series of water samples). For total coliforms the objective was 1000 counts per 100 ml (based on a geometric mean density of a series of water samples).
Hydrogen sulphide	2 μg/L (PWQO)[1]
Iodine	100 μg/L (Interim PWQO)[a]
Iron	300 μg/L (PWQO)[1]
Lead	(PWQO)[1]: Alkalinity as $CaCO_3$ (mg/L) — PWQO (μg/L) < 20 — 5 20 to 40 — 10 40 to 80 — 20 > 80 — 25
Lead (revised)	Interim PWQO[b]: Alkalinity as $CaCO_3$ (mg/L) — PWQO (μg/L) < 30 — 1 30 to 80 — 3 > 80 — 5
Lindane	0.01 μg/L (PWQO)[1] · chemical name: gamma - 1,2,3,4,5,6-hexachlorocyclohexane
Malathion	0.1 μg/L (PWQO)[1]
Mercury	0.2 μg/L (PWQO)[1] · in a filtered water sample
Methoxychlor	0.04 μg/L (PWQO)[1]
Mirex (Dechlorane)	0.001 μg/L (PWQO)[1]
Molybdenum	10 μg/L (Interim PWQO)[b]
Nickel	25 μg/L (PWQO)[1]

Oil & Grease	PWQO[1]: Oil or petrochemicals should not be present in concentrations that: · can be detected as a visible film, sheen, or discolouration on the surface; · can be detected by odour; · can cause tainting of edible aquatic organisms; · can form deposits on shorelines and bottom sediments that are detectable by sight or odour, or are deleterious to resident aquatic organisms.
Parathion	0.008 μg/L (PWQO)[1]
pH	PWQO[1]: The pH should be maintained within the range of 6.5 - 8.5 · to protect aquatic life; and · both alkaline and acid waters may cause irritation to anyone using the water for recreational purposes.
Phenol	5 μg/L (Interim PWQO)[b] · common synonym - monohydroxybenzene
Phenols	1 μg/L (PWQO)[1] · Determined by the total reactive phenols test - the 4-AAP (4-amino-antipyrine) test. · This objective should be used primarily as a screening tool. · The isomer specific PWQOs for various phenolics should be employed where possible.
Phosphorus, total	Interim PWQO[a]: · Current scientific evidence is insufficient to develop a firm Objective at this time. Accordingly, the following phosphorus concentration should be considered as general guidelines which should be supplemented by site-specific studies: · To avoid nuisance concentrations of algae in lakes, average total phosphorus concentrations fro the ice-free period should not exceed 20 μg/L; · A high level of protection against aesthetic deterioration will be provided by a total phosphorus concentration for the ice-free period of 10 μg/L or less. This should apply to all lakes naturally below this value; ·Excessive plant growth in rivers and streams should be eliminated at a total phosphorus concentration below 30 μg/L.
Radionuclides	PWQO[8]: · Radiation exposure should be kept as low as reasonably achievable, economic and social factors being taken into account. ·The Provincial Water Quality Objectives for radionuclides are based on drinking water requirements, which are derived from dose-response relationships as recommended by the International Commission on Radiological Protection (ICRP) in Publication 26. The Objectives* are as follows: Radionuclide** — PWQO*** (Becquerels/Litre) ^{137}Cesium — 50 ^{131}Iodine — 10 ^{226}Radium — 1 ^{90}Strontium — 10 Tritium — 40 000 * The radionuclide objectives are based on the total concentration in an unfiltered water sample. ** If two or more radionuclides affecting the same organ or tissue are found to be present, the following relationship based on ICRP Publication 26 should be satisfied: $\frac{c_1}{C_1} + \frac{c_2}{C_2} + \cdots \frac{c_I}{C_i} \leq 1$ where c_1, c_2 and c_I are the observed concentrations, and C_1, C_2, and C_i are the maximum acceptable concentrations for each contributing radionuclide. *** Radionuclide concentrations that exceed the maximum acceptable concentrations may be tolerated for a short duration, provided that the annual average concentrations remain below this level and meet the restriction for multiple radionuclides. Note: for further information on the radionuclide objectives as related to potable water supplies, consult the publication *Ontario Drinking Water Objectives* (Revised, 1994).

Selenium	100 μg/L (PWQO)[1]
Silver	0.1 μg/L (PWQO)[1]
Simazine	10 μg/L (PWQO)[1]
Temperature	PWQO[1]: 1) General The natural thermal regime of any body of water shall not be altered so as to impair the quality of the natural environment. In particular, the diversity, distribution and abundance of plant and animal life shall not be significantly changed. 2) Waste Heat Discharge (a) Ambient Temperature Changes The temperature at the edge of a mixing zone shall not exceed the natural ambient water temperature at a representative control location by more than 10 C (18 F). However, in special circumstances, local conditions may require a significantly lower temperature difference than 10 C (18 F). Potential dischargers are to apply to the MOEE for guidance as to the allowable temperature rise for each thermal discharge. This ministry will also specify the nature of the mixing zone and the procedure for the establishment of a representative control location for temperature recording on a case-by-case basis. (b) Discharge Temperature Permitted The maximum temperature of the receiving body of water, at any point in the thermal plume outside a mixing zone, shall not exceed 30 C (86 F) or the temperature of a representative control location plus 10 C (18 C) or the allowed temperature difference, which ever is the lesser temperature. These maximum temperatures are to be measured on a mean daily basis from continuous records. (c) Taking and Discharging of Cooling Water Users of cooling water shall meet both the Objectives for temperature outlined above and the "Procedures for the Taking and Discharging of Cooling Water" as outlined in the MOEE publication *Deriving Received-Water Based, Point-Source Effluent Requirements for Ontario Waters* (1994).
Thallium	0.3 μg/L (Interim PWQO)[b]
Toluene	0.8 μg/L (Interim PWQO)[10]
Tributyl phosphate	0.6 μg/L (Interim PWQO)[a]
Tributyltin	0.000005 μg/L (Interim PWQO)[b]
Triethyl lead	0.4 μg/L (Interim PWQO)[7]
Tungsten	30 μg/L (Interim PWQO)[a]
Turbidity	PWQO[1]: · Suspended matter should not be added to surface water in concentrations that will change the natural Secchi disc reading by more than 10 percent.
Uranium	5 μg/L (Interim PWQO)[a]
Vanadium	7 μg/L (Interim PWQO)[b]
Water clarity	PWQO[1]: · The water in swimming areas should be sufficiently clear to estimate depth or to see submerged swimmers who may require assistance. To achieve this degree of safety, water clarity should be such that, if the bottom of the bathing area is not visible, the water should have a Secchi disc transparency of at least 1.2 m.
Zinc	30 μg/L (PWQO)[1]
Zinc (revised)	20 μg/L (Interim PWQO)[b]
Zirconium	4 μg/L (Interim PWQO)[a]

Table 12.IA. Percent NH_3 in aqueous ammonia solutions for 0-30 C and pH 6-10. All PWQOs are for the protection of aquatic life, unless otherwise noted. The PWQOs are based on the total concentration of an unfiltered sample, unless otherwise noted. The "CAS No." is the number assigned by the American Chemical Society's Chemical Abstracts Services.

	pH								
Temp °C	**6**	**6.5**	**7**	**7.5**	**8**	**8.5**	**9**	**9.5**	**10**
0	0.0083	0.026	0.083	0.26	0.82	2.6	7.6	21	45
1	0.009	0.028	0.09	0.28	0.89	2.8	8.3	22	47
2	0.0098	0.031	0.098	0.31	0.97	3	8.9	24	49
3	0.011	0.034	0.11	0.34	1.1	3.3	9.6	25	52
4	0.012	0.036	0.12	0.36	1.1	3.5	10	27	54
5	0.013	0.04	0.13	0.39	1.2	3.8	11	28	56
6	0.014	0.043	0.14	0.43	1.3	4.1	12	30	58
7	0.015	0.046	0.15	0.46	1.5	4.4	13	32	60
8	0.016	0.05	0.16	0.5	1.6	4.8	14	34	61
9	0.017	0.054	0.17	0.54	1.7	5.2	15	35	63
10	0.019	0.059	0.19	0.59	1.8	5.6	16	37	65
11	0.02	0.064	0.2	0.63	2	6	17	39	67
12	0.022	0.069	0.22	0.68	2.1	6.4	18	41	69
13	0.024	0.074	0.24	0.74	2.3	6.9	19	43	70
14	0.025	0.08	0.25	0.8	2.5	7.4	20	45	72
15	0.027	0.087	0.27	0.86	2.7	8	22	46	73
16	0.03	0.093	0.29	0.93	2.9	8.5	23	48	75
17	0.032	0.1	0.32	1	3.1	9.1	24	50	76
18	0.034	0.11	0.34	1.1	3.3	9.8	26	52	77
19	0.037	0.11	0.37	1.2	3.6	11	27	54	79
20	0.04	0.13	0.4	1.2	3.8	11	28	56	80
21	0.043	0.14	0.43	1.3	4.1	12	30	58	81
22	0.046	0.15	0.46	1.4	4.4	13	32	59	82
23	0.049	0.16	0.49	1.5	4.7	14	33	61	83
24	0.053	0.17	0.53	1.7	5	14	35	63	84
25	0.057	0.18	0.57	1.8	5.4	15	36	64	85
26	0.061	0.19	0.61	1.9	5.8	16	38	66	86
27	0.065	0.21	0.65	2	6.2	17	40	67	87
28	0.07	0.22	0.7	2.2	6.6	18	41	69	88
29	0.075	0.24	0.75	2.3	7	19	43	70	88
30	0.081	0.25	0.8	2.5	7.5	20	45	72	89

Table 12.IA is taken from Emerson *et al.* 1975[11] but percentages are rounded to two significant figures. The equations given by Emerson *et al.* may be used to interpolate values between those given in the table:

$$f = 1/(10^{pKa-pH} + 1), \text{ where } f \text{ is the fraction of } NH_3$$

$$pKa = 0.09018 + 2729.92/T, \text{ where T = ambient water temperature in Kelvin (K = C + 273.16)}$$

Results should be converted to percent and rounded to two significant figures. Extrapolations should not be made beyond the ranges of the table.

Note: Under certain temperature and pH conditions, the total ammonia criteria for the protection of aquatic life may be less stringent than the criteria for other beneficial uses (e.g. public water supply).

Table 12.IB. Dissolved oxygen concentrations for cold water biota and warm water biota.

Temperature	Dissolved Oxygen Concentration			
	Cold Water Biota		Warm Water Biota	
C	% Saturation	mg/L	% Saturation	mg/L
0	54	8	47	7
5	54	7	47	6
10	54	6	47	5
15	54	6	47	5
20	57	5	47	4
25	63	5	48	4

References for Appendix 12.I:

1. MOE. 1979. Rationale for the Establishment of Ontario's Water Quality Objectives. 236pp.
2. MOE. 1984. Scientific Criteria Document for Standard Development - Chlorinated Benzenes in the Aquatic Environment. 197 pp.
3. MOE. 1984. Scientific Criteria Document for Standard Development - Chlorinated

Phenol in the Aquatic Environment. 180 pp.
4. MOE. 1988. Scientific Criteria Document for the Development of Provincial Water Quality Objectives and Guidelines - Aluminum. 81 pp.
5. MOE. 1988. Scientific Criteria Document for the Development of Provincial Water Quality Objectives and Guidelines - Resin Acids. 60 pp.
6. MOEE. 1993. Scientific Criteria Document for the Development of Provincial Water Quality Guidelines for Chlorinated Ethanes & Chlorinated Ethylenes. 111 pp.
7. MOEE. 1994. Scientific Criteria Document for the Development of Provincial Water Quality Guidelines for Alkylleads. 50 pp.
8. MOEE. 1994. Ontario Drinking Water objectives (Revised 1994). 68 pp.
9. MOEE. 1994. Scientific Criteria Document for the Development of Provincial Water Quality Guidelines for Ethylbenzene. 32 pp.
10. MOEE. 1994. Scientific Criteria Document for the Development of Provincial Water Quality Guidelines for Toluene. 43 pp.
11. Emerson, K., F.C. Russo, R.E. Lund and R.V. Thurston. 1975 Aqueous ammonia equilibrium calculations: Effects of pH and temperature. J. Fish. Res. Board Can. 32: 2379-2383

APPENDIX 12.II

Tolerance Values for Applying to Rapid Bioassessment Protocols Using Macroinvertebrates

Definitions of feeding habits are at end of Appendix 12-II (? = undetermined)

Taxa	Tolerance Value	Feeding Habit
COELENTERATA		
HYDROZOA		
HYDROIDA		
Hydridae		
Hydra sp.	5	prd
NEMERTEA		
ENOPLA		
HOPLONEMERTINI		
Prostomatidae		
Prostoma rubrum	8	prd
PLATYHELMINTHES		
TURBELLARIA		
TRICLADIDA		
Planariidae - all species	6	prd
ANNELIDA		
POLYCHAETA		
SABELLIDA		
Sabellidae		
Manayunkia speciosa	6	c-g
OLIGOCHAETA		
LUMBRICINA		
? *Lumbricina*	8	c-g
LUMBRICULIDA		
Lumbriculidae - all species	8	c-g
TUBIFICIDA		
Enchytraeidae - all species	10	c-g
Tubificidae		
Aulodrilus spp.	8	c-g
Branchiura sowerbyi	10	c-g
Ilyodrilus templetoni	10	c-g
Isochaetides freyi	10	c-g
Limnodrilus spp.	10	c-g
Peloscolex sp.	10	c-g
Quistadrilus multisetosus	10	c-g
Spirosperma ferox	10	c-g
Tubifex tubifex	10	c-g
? Tubificidae w/capillary setae	10	c-g
? Tubificidae w/o capillary setae	10	c-g
Naididae		
Amphichaeta americana?	6	c-g
Arcteonais lomondi	6	c-g
Chaetogaster spp.	6	prd
Dero spp.	10	c-g
Haemonais waldvogeli	8	c-g
Nais barbata	8	c-g
Nais behningi	6	c-g
Nais bretscheri	6	c-g
Nais communis	8	c-g
Nais elinguis	10	c-g
Nais pardalis	8	c-g
Nais simplex	6	c-g
Nais variabilis	10	c-g
Nais sp.	8	c-g
Ophidonais serpentina	6	c-g
Pristina aequiseta	8	c-g
Pristina breviseta	8	c-g
Pristina leidyi	8	c-g
Pristina menoni	8	c-g
Pristina sp.	8	c-g
Pristinella jenkinae	10	c-g
Pristinella osborni	10	c-g
Pristina/Pristinella spp.	10	c-g
Ripistes parasita	8	c-f
Slavina appendiculata	6	c-g
Specaria josinae	6	c-g
Stylaria lacustris	8	c-g
Vejdovskyella comata	4	c-g
Vejdovskyella intermedia	7	c-g
Vejdovskyella sp.	4	c-g
HIRUDINEA		
RHYNCHOBDELLIDA		
Glossiphoniidae		

Batracobdella phalera	7	prd
Helobdella spp.	7	prd
Glossiphoniidae		
Placobdella montifera	7	prd
? Hirudinea	7	prd
APHANONEURA		
AEOLOSOMATIDA		
Aeolosomatidae		
Aeolosoma spp.	8	c-f
? Aeolosomatidae	8	c-f
BRANCHIOBDELLIDA		
BRANCHIOBDELLIDA		
Branchiobdellidae		
Branchiobdella sp.	6	c-g
? Branchiobdellidae	6	c-g
MOLLUSCA		
GASTROPODA		
BASOMMATOPHORA		
Physidae - all species	8	c-g
Lymnaeidae - all species	6	c-g
Planorbidae - all species	6	scr
Ancylidae		
Ferrissia rivularis	6	scr
MESOGASTROPODA		
Viviparidae		
Campeloma decisa	6	scr
Viviparus georgianus	6	scr
Pleuroceridae		
Elimia livescens	6	scr
Elimia virginica	6	scr
Elimia sp.	6	scr
Pleurocera acuta	6	scr
? Pleuroceridae	6	scr
Hydrobiidae		
Amnicola integra	5	scr
Amnicola limosa	5	scr
Amnicola lustrica	5	scr
Amnicola sp.	5	scr
Bithynia tentaculata	8	scr
Pomatiopsis lapidaria	8	scr
Probythinella lacustris	8	scr
? Hydrobiidae	8	scr
Valvatidae - all species	8	scr
PELECYPODA		
UNIONIDA		
Unionidae		
Anodonta cataracta	6	c-f
Anodonta implicata	6	c-f
Elliptio complanatus	8	c-f
Lampsilis radiata radiata	6	c-f
VENEROIDEA		
Dreisseniidae		
Dreissena polymorpha	8	c-f
Sphaeriidae		
Musculium spp.	6	c-f
Pisidium amnicum	6	c-f
Pisidium casertanum	6	c-f
Pisidium compressum	6	c-f
Pisidium variabile	6	c-f
Pisidium sp.	6	c-f
Sphaerium corneum	6	c-f
Sphaerium striatinum	6	c-f
Sphaerium sp.	6	c-f
? Sphaeriidae	6	c-f
ARTHROPODA		
CRUSTACEA		
ISOPODA		
Anthuridae		
Cyathura polita	5	c-g
Idoteidae		
Chiridotea almyra	5	c-g
Edotea sp.	5	c-g
Asellidae - all species	8	c-g
Lirceus sp.	8	c-g
AMPHIPODA		
Gammaridae		
Gammarus fasciatus	6	c-g
Gammarus pseudolimnaeus	4	c-g
Gammarus tigrinus	6	c-g
Gammarus sp.	6	c-g
Oedicerotidae		
Monoculodes edwardsi	5	c-g
Talitridae		
Hyalella azteca	8	c-g
CUMACEA		
Almyracuma proximoculi	5	c-g

DECAPODA		
Cambaridae		
Cambarus spp.	6	c-g
Orconectes spp.	6	c-g
ARACHNOIDEA		
HYDRACARINA - all species	6	prd
INSECTA		
COLLEMBOLA		
Isotomidae		
Isotomurus palustris	5	c-g
EPHEMEROPTERA		
Siphlonuridae		
Ameletus ludens	0	c-g
Ameletus sp.	0	c-g
Isonychia bicolor	2	c-g
Isonychia obscura	2	c-g
Siphlonurus sp.	7	c-g
Baetidae		
Acentrella sp.	4	scr
Baetis amplus	6	c-g
Baetis brunneicolor	4	c-g
Baetis flavistriga	4	c-g
Baetis intercalaris	6	c-g
Baetis macdunnoughi	5	c-g
Baetis pluto	6	c-g
Baetis propinquus	6	c-g
Baetis pygmaeus	4	c-g
Baetis tricaudatus	6	c-g
Baetis sp.	6	c-g
Callibaetis sp.	9	c-g
Centroptilum sp.	2	c-g
Cloeon sp.	4	c-g
Heterocloeon curiosum	2	scr
? Baetidae	6	c-g
Heptageniidae		
Cinygmula subaequalis	2	scr
Epeorus (*Iron*) sp.	0	scr
Heptagenia culacantha	2	scr
Heptageniidae		
Heptagenia flavescens	4	scr
Heptagenia marginalis	4	scr
Heptagenia pulla gr.	4	scr
Heptagenia sp.	4	scr
Leucrocuta sp.	1	scr
Nixe (*Nixe*) sp.	2	scr
Rithrogena sp.	0	c-g
Stenacron interpunctatus	7	scr
Stenonema exiguum	5	scr
Stenonema femoratum	7	scr
Stenonema integrum	4	scr
Stenonema ithaca	3	scr
Stenonema mediopunctatum	3	scr
Stenonema meririvulanum	2	scr
Stenonema modestum	1	scr
Stenonema pulchellum	3	scr
Stenonema terminatum	4	scr
Stenonema vicarium	2	scr
Stenonema sp.	3	scr
? Heptageniidae	3	scr
Leptophlebiidae		
Choroterpes sp.	2	c-g
Habrophlebia vibrans	4	c-g
Habrophlebia sp.	4	c-g
Habrophlebiodes sp.	6	scr
Leptophlebia sp.	4	c-g
Paraleptophlebia guttata	1	c-g
Paraleptophlebia mollis	1	c-g
Paraleptophlebia sp.	1	c-g
? Leptophlebiidae	4	c-g
Ephemerellidae		
Attenella attenuata	1	c-g
Attenella margarita	1	c-g
Dannella simplex	2	c-g
Dannella sp.	2	c-g
Drunella spp.	0	c-g
Ephemerella aurivillii	0	c-g
Ephemerella dorothea	1	c-g
Ephemerella excrucians?	1	c-g
Ephemerella invaria	1	c-g
Ephemerella needhami	1	c-g
Ephemerellidae		
Ephemerella rotunda	1	c-g
Ephemerella subvaria	1	c-g
Ephemerella sp.	1	c-g
Eurylophella funeralis	0	c-g
Eurylophella temporalis	5	c-g

Eurylophella verisimilis	2	c-g
Eurylophella sp.	2	c-g
Serratella deficiens	2	c-g
Serratella serrata	2	c-g
Serratella serratoides	2	c-g
Serratella sordida	2	c-g
Serratella sp.	2	c-g
? Ephemerellidae	2	c-g
Tricorythidae		
Tricorythodes sp.	4	c-g
Caenidae		
Brachycercus sp.	3	c-g
Caenis sp.	7	c-g
Baetiscidae		
Baetisca sp.	4	c-g
Potamanthidae		
Potamanthus verticis	4	c-g
Potamanthus sp.	4	c-g
Ephemeridae		
Ephemera guttulata	2	c-g
Ephemera sp.	2	c-g
Hexagenia sp.	6	c-g
Polymitarcyidae		
Ephoron leukon?	2	c-g
ODONATA		
Gomphidae		
Gomphus sp.	5	prd
Lanthus sp.	5	prd
Ophiogomphus sp.	1	prd
Stylurus sp.	4	prd
? Gomphidae	4	prd
Aeschnidae		
Basiaeschna janata	6	prd
Boyeria sp.	2	prd
Cordulegasteridae		
Cordulegaster sp.	3	prd
Libellulidae		
Macromia sp.	2	prd
Neurocordulia sp.	2	prd
Calopterygidae		
Calopteryx sp.	6	prd
? Calopterygidae	6	prd
Agrionidae		
Hetaerina sp.	6	prd
? Agrionidae	6	prd
Coenagrionidae		
Argia sp.	6	prd
Enallagma sp.	8	prd
Ischnura spp.	9	prd
? Coenagrionidae	8	prd
HEMIPTERA		
Corixidae		
Hesperocorixa sp.	5	prd
? Corixidae	5	prd
PLECOPTERA		
Capniidae		
Allocapnia vivipara	3	shr
Allocapnia sp.	3	shr
Paracapnia sp.	1	shr
? Capniidae	3	shr
Leuctridae - All species	0	shr
Nemouridae		
Amphinemura delosa	3	shr
Amphinemura nigritta	3	shr
Amphinemura wui	3	shr
Nemoura sp.	1	shr
Ostrocerca sp.	2	shr
Shipsa rotunda	2	shr
? Nemouridae	2	shr
Taeniopterygidae		
Strophopteryx fasciata	3	shr
Taeniopteryx spp.	2	shr
Perlidae		
Acroneuria spp.	0	prd
Agnetina capitata	2	prd
Agnetina flavescens	2	prd
Agnetina sp.	2	prd
Claasenia? sp.	3	prd
Perlidae		
Neoperla sp.	3	prd
Paragnetina immarginata	1	prd
Paragnetina media	1	prd
Perlesta placida	5	prd
? Perlidae	3	prd
Peltoperlidae		
Tallaperla sp.	0	shr

Chloroperlidae		
Alloperla sp.	0	c-g
Haploperla brevis	1	prd
Rasvena terna	0	c-g
Suwallia sp.	0	prd
Sweltsa sp.	0	prd
? Chloroperlidae	0	prd
Perlodidae		
Cultus decisus	2	prd
Helopicus subvarians	2	prd
Isogenoides hansoni	0	prd
Isoperla holochlora	2	prd
Isoperla namata	2	prd
Isoperla sp.	2	prd
Malirekus iroquois	2	prd
? Perlodidae	2	prd
Pteronarcidae - All species	0	shr
COLEOPTERA		
Haliplidae		
Haliplus sp.	5	shr
Peltodytes sp.	5	shr
Dytiscidae		
Agabetes sp.	5	prd
Agabus sp.	5	prd
Hydroporous sp.	5	prd
Laccophilus sp.	5	prd
? Dytiscidae	5	prd
Gyrinidae		
Dineutus sp.	4	prd
Hydrophilidae		
Berosus sp.	5	prd
Helochares sp.	5	prd
Helophorus sp.	5	shr
Hydrobius sp.	5	prd
Hydrophilidae		
Laccobius sp.	5	prd
Psephenidae		
Ectopria nervosa	5	scr
Ectopria sp.	5	scr
Psephenus herricki	4	scr
Psephenus sp.	4	scr
Dryopidae		
Helichus sp.	5	scr
Scirtidae		
? Scirtidae	5	scr
Elmidae		
Ancyronyx variegatus	5	c-g
Dubiraphia bivittata	6	c-g
Dubiraphia quadrinotata	5	c-g
Dubiraphia vittata	6	c-g
Dubiraphia sp.	6	c-g
Macronychus glabratus	5	shr
Optioservus spp.	4	scr
Oulimnius latiusculus	4	scr
Promoresia elegans	2	scr
Promoresia tardella	2	scr
Promoresia sp.	2	scr
Stenelmis spp.	5	scr
? Elmidae	5	scr
MEGALOPTERA		
Corydalidae		
Chauliodes sp.	4	prd
Corydalus cornutus	4	prd
Nigronia serricornis	0	prd
Sialidae		
Sialis sp.	4	prd
NEUROPTERA		
Sisyridae		
Climacia areolaris	5	prd
TRICHOPTERA		
Philopotamidae		
Chimarra aterrima?	4	c-f
Chimarra socia	4	c-f
Chimarra obscura?	4	c-f
Chimarra sp.	4	c-f
Dolophilodes sp.	0	c-f
Psychomyiidae		
Lype diversa	2	scr
Psychomyia flavida	2	c-g
Polycentropodidae		
Cyrnellus fraternus	8	c-f
Cyrnellus sp. 2	8	c-f
Neureclipsis bimaculata	7	c-f
Neureclipsis sp.	7	c-f
Nyctiophylax celta	5	prd
Nyctiophylax moestus	5	prd

Phylocentropus sp.	5	c-f
Polycentropus remotus	6	prd
Polycentropus sp.	6	prd
Hydropsychidae		
Arctopsyche sp.	1	c-f
Cheumatopsyche sp.	5	c-f
Diplectrona sp.	0	c-f
Hydropsyche betteni	6	c-f
Hydropsyche bronta	6	c-f
Hydropsyche nr. depravata	6	c-f
Hydropsyche dicantha	2	c-f
Hydropsyche leonardi	0	c-f
Hydropsyche morosa	6	c-f
Hydropsyche orris	5	c-f
Hydropsyche phalerata	1	c-f
Hydropsyche recurvata	4	c-f
Hydropsyche scalaris	2	c-f
Hydropsyche separata	4	c-f
Hydropsyche slossonae	4	c-f
Hydropsyche sparna	6	c-f
Hydropsyche valanis	6	c-f
Hydropsyche venularis	4	c-f
Hydropsyche sp.	4	c-f
Macrostemum carolina	3	c-f
Macrostemum zebratum	3	c-f
Macrostemum sp.	3	c-f
Parapsyche sp.	0	c-f
Potamyia sp.	5	c-f
? Hydropsychidae	5	c-f
Rhyacophilidae		
Rhyacophila carolina gr.	1	prd
Rhyacophila carpenteri?	1	prd
Rhyacophilidae		
Rhyacophila fuscula	0	prd
Rhyacophila glaberrima	1	prd
Rhyacophila melita	1	prd
Rhyacophila nigrita	1	prd
Rhyacophila sp.	1	prd
Glossosomatidae		
Agapetus sp.	0	scr
Glossosoma sp.	0	scr
Protoptila sp.	1	scr
Hydroptilidae		
Agraylea sp.	8	c-g
Alisotrichia sp.	6	scr
Hydroptila spp.	6	scr
Ithytrichia sp.	4	scr
Leucotrichia sp.	6	scr
Mayatrichia ayama	6	scr
Neotrichia sp.	2	scr
Orthotrichia sp.	6	shr
Oxyethira sp.	3	c-g
Palaeagapetus celsus	4	shr
Palaeagapetus sp.	1	shr
Phryganeidae		
Oligostomis ocelligera	2	prd
Ptilostomis sp.	5	shr
Brachycentridae		
Adicrophleps hitchcocki	2	shr
Brachycentrus appalachia	0	c-f
Brachycentrus incanus	0	c-f
Brachycentrus lateralis	1	c-f
Brachycentrus numerosus	1	c-f
Brachycentrus solomoni	1	c-f
Micrasema sp. 1	2	shr
Micrasema sp. 2	2	shr
Micrasema sp. 3	2	shr
? Brachycentridae	2	shr
Limnephilidae		
Apatania sp.	3	scr
Goera sp.	3	scr
Hesperophylax designatus	3	shr
Hydatophylax sp.	2	shr
Nemotaulius sp.	3	scr
Limnephilidae		
Neophylax concinnus	3	scr
Neophylax fuscus	3	scr
Neophylax sp.	3	scr
Platycentropus sp.	4	shr
Pseudostenophylax sp.	0	shr
Psychoglypha sp.	0	c-g
Pycnopsyche sp.	4	shr
? Limnephilidae	4	shr
Lepidostomatidae		
Lepidostoma sp.	1	shr
Odontoceridae		

Psilotreta sp.	0	scr
Molannidae		
Molanna sp.	6	scr
Helicopsychidae		
Helicopsyche borealis	3	scr
Helicopsyche sp.	3	scr
Leptoceridae		
Ceraclea punctata	3	c-g
Ceraclea sp.	3	c-g
Mystacides sepulchralis	4	c-g
Mystacides sp.	4	c-g
Nectopsyche sp.	3	shr
Oecetis avara	5	prd
Oecetis cinerascens	5	prd
Oecetis inconspicua	5	prd
Oecetis sp.	5	prd
Setodes sp.	2	c-g
Triaenodes sp.	6	shr
? Leptoceridae	4	prd
LEPIDOPTERA		
Arctiidae		
Estigmene sp.	5	shr
Nepticulidae		
? Nepticulidae	5	shr
Pyralidae		
Acentria sp.	5	shr
Nymphula sp.	7	shr
Parapoynx sp.	5	shr
Petrophila sp.	5	scr
Pyralidae		
? Lepidoptera	5	shr
DIPTERA		
Tipulidae		
Antocha spp.	*3*	*c-g*
Dicranota sp.	3	prd
Helius sp.	4	c-g
Hexatoma sp. 1	2	prd
Hexatoma sp. 2	2	prd
Hexatoma sp.	2	prd
Limonia sp.	6	shr
Pilaria sp.	7	prd
Tipula sp.	4	shr
? Tipulidae	4	shr
Psychodidae		
Pericoma sp.	4	c-g
? Psychodidae	10	c-g
Ptychopteridae		
Bittacomorpha clavipes	9	c-g
Blephariceridae		
? Blephariceridae	0	scr
Dixidae		
Dixa sp.	1	c-f
Chaoboridae		
Chaoborus punctipennis	8	prd
Ceratopogonidae		
Bezzia sp. 1	6	prd
Bezzia sp. 2	6	prd
Culicoides? sp.	10	prd
Forcipomyia sp.	6	scr
Probezzia sp. 1	6	prd
Probezzia sp. 2	6	prd
Sphaeromais longipennis	6	prd
? Ceratopogonidae	6	prd
Simuliidae		
Cnephia mutata	2	c-f
Prosimulium hirtipes	2	c-f
Prosimulium magnum	1	c-f
Prosimulium rhizophorum	2	c-f
Simulium aureum	7	c-f
Simulium decorum	7	c-f
Simulium fibrinflatum	6	c-f
Simulium gouldingi	3	c-f
Simulium jenningsi	4	c-f
Simulium latipes	4	c-f
Simulium parnassum	7	c-f
Simulium pictipes	4	c-f
Simulium rugglesi	5	c-f
Simulium tuberosum	4	c-f
Simulium venustum	5	c-f
Simulium vittatum	7	c-f
Simulium sp.	5	c-f
Tabanidae		
Chrysops sp.	5	c-g
Tabanus sp.	5	prd
? Tabanidae	5	prd
Rhagionidae		

Atherix sp.	4	prd
Empididae		
Chelifera sp.	6	prd
Clinocera sp.	6	prd
Hemerodromia sp.	6	prd
Wiedemannia? sp.	6	prd
Dolichopodidae		
? Dolichopodidae	4	prd
Ephydridae		
Hydrella sp.	6	shr
Muscidae		
? Muscidae	6	prd
Anthomyiidae		
? Anthomyiidae	6	prd
Chironomidae		
Tanypodinae		
Ablabesmyia spp.	8	prd
Clinotanypus pinguis	8	prd
Coelotanypus scapularis	4	prd
Conchapelopia spp.	6	prd
Guttipelopia guttipennis	5	prd
Hayesomyia senata	6	prd
Helopelopia cornuticaudata	6	prd
Hudsonimyia karelena	2	prd
Hudsonimyia parrishi	2	prd
Labrundinia pilosella	7	prd
Labrundinia nr. *virescens*	7	prd
Larsia canadensis	6	prd
Natarsia sp. A	8	prd
Natarsia baltimoreus	8	prd
Nilotanypus fimbriatus	8	prd
Nilotanypus sp.	*6*	*prd*
Paramerina sp.	6	prd
Pentaneura inconspicua?	6	prd
Procladius bellus	9	prd
Procladius sublettei	9	prd
Psectrotanypus dyari	10	prd
Rheopelopia perda?	4	prd
Rheopelopia sp. 2	4	prd
Rheopelopia sp. 3	4	prd
Tanypus punctipennis	10	prd
Tanypus stellatus	10	prd
Telopelopia okoboji	8	prd
Thienemannimyia gr. spp.	6	prd
Thienemannimyia norena	6	prd
Trissopelopia ogemawi	4	prd
Zavrelimyia sinuosa	8	prd
Zavrelimyia sp.	8	prd
? Tanypodinae	7	prd
Podonominae		
Paraboreochlus sp.	1	c-g
Diamesinae		
Diamesa spp.	5	c-g
Pagastia sp. A	1	c-g
Potthastia gaedii	2	c-g
Potthastia longimana	2	c-g
Pseudokiefferiella sp.	1	c-g
Sympotthastia sp.	2	c-g
? Diamesinae	2	c-g
Prodiamesinae		
Monodiamesa dipectinata	7	c-g
Prodiamesa sp. 1	3	c-g
Prodiamesa sp. 2	3	c-g
Orthocladiinae		
Acricotopus sp.	10	c-g
Brillia flavifrons	5	shr
Brillia parva	5	shr
Brillia sera	5	shr
Brillia sp.	5	shr
Cardiocladius albiplumus	5	prd
Cardiocladius obscurus	5	prd
Chaetocladius vitellinus gr.	6	c-g
Corynoneura celeripes	4	c-g
Corynoneura taris	4	c-g
Corynoneura sp.	4	c-g
Cricotopus bicinctus	7	c-g
Cricotopus nr. *cylindraceus*	7	shr
Cricotopus elegans	7	shr
Cricotopus festivellus gr.	7	c-g
Cricotopus intersectus gr.	7	shr
Cricotopus reversus gr.	7	shr
Cricotopus sylvestris gr.	7	shr
Cricotopus tremulus gr.	7	shr
Cricotopus triannulatus	7	shr
Cricotopus trifascia gr.	6	shr
Cricotopus vierriensis	7	shr

Diplocladius sp.	8	c-g
Epoicocladius sp.	4	c-g
Eukiefferiella brehmi gr.	4	c-g
Eukiefferiella brevicalcar gr.	4	c-g
Eukiefferiella claripennis gr.	8	c-g
Eukiefferiella coerulescens gr.	4	c-g
Eukiefferiella devonica gr.	4	c-g
Eukiefferiella gracei gr.	4	c-g
Eukiefferiella pseudomontana gr.	8	c-g
Heterotrissocladius marcidus gr.	4	c-g
Hydrobaenus pilipes	8	c-g
Krenosmittia sp.	1	c-g
Limnophyes sp.	8	c-g
Lopescladius sp.	4	c-g
Nanocladius (*Plecopteracoluthus*) sp.	3	c-g
Chironomidae		
Orthocladiinae		
Nanocladius spp.	3	c-g
Orthocladius (*Eudactylocladius*) sp.	6	c-g
Orthocladius (*Euorthoclad.*) I spp.	6	c-g
O. (*Euorthoclad.*) *rivulorum*	6	c-g
Orhtocladius annectens	6	c-g
Orthocladius carlatus	6	c-g
Orthocladius curtiseta	6	c-g
Orthocladius nr. *dentifer*	6	c-g
Orthocladius obumbratus	6	c-g
Orthocladius nr. *robacki*	6	c-g
Orthocladius trigonolabis	6	c-g
O. (*Symposiocladius*) *lignicola*	5	c-g
Orthocladius sp.	6	c-g
Parachaetocladius sp.	2	c-g
Paracricotopus sp.	4	c-g
Parakiefferiella triquetra gr.	4	c-g
Parakiefferiella sp.	4	c-g
Parametriocnemus lundbecki	5	c-g
Paraphaenocladius sp.	4	c-g
Paratrichocladius sp.	5	shr
Psectrocladius dilatatus gr.	8	c-g
Psectrocladius nigrus	8	c-g
Psectrocladius psilopterus gr.	8	c-g
Psectrocladius sordidellus gr.	8	c-g
Psectrocladius vernalis	8	c-g
Rheocricotopus robacki	6	c-g
Rheocricotopus tuberculatus	6	c-g

Rheocricotopus sp. 2	6	c-g
Rheocricotopus sp. 4	6	c-g
Synorthocladius nr. *semivirens*	6	c-g
Thienemanniella nr. fusca	6	c-g
Thienemanniella xena?	6	c-g
Thienemanniella sp.	6	c-g
Trissocladius sp.	5	c-g
Tvetenia bavarica gr.	5	c-g
Tvetenia vitracies	5	c-g
Unniella multivirga	4	c-g
Zalutschia zalutschicola	4	shr
? Orthocladiinae	5	c-g
Chironominae		
Axarus festivus gr.	6	c-g
Chironomus decorus gr.	10	c-g
Chironomus riparius gr.	10	c-g
Chironomus sp.	10	c-g
Cladopelma sp.	9	c-g
Cryptochironomus fulvus	8	prd
C. ponderosus	8	prd
Cryptotendipes spp.	6	c-g
Demicryptochironomus sp. 1	8	c-g
D. cuneatus	8	c-g
Dicrotendipes spp.	8	c-g
Einfeldia sp.	9	c-g
Endochironomus spp.	10	shr
Glyptotendipes lobiferus	10	shr
Glyptotendipes sp. 2	10	shr
Goeldichironomus sp.	8	c-g
Harnischia curtilamellata	8	c-g
Microchironomus sp.	8	c-g
Microtendipes rydalensis gr.	6	c-f
Microtendipes pedellus gr.	6	c-f
Nilothauma babiyi	2	c-g
Parachironomus spp.	10	prd
Paracladopelma nais	7	c-g
Paralauterborniella sp.	8	c-g
P. nigrohalteris	8	c-g
Paratendipes albimanus	6	c-g
Phaenopsectra dyari?	7	scr
Phaenopsectra flavipes	7	scr
Phaenopsectra sp.	7	scr
Polypedilum spp.	6	shr
Pseudochironomus sp. 1	5	c-g

Pseudochironomus sp. 2	5	c-g
Pseudochironomus sp. 3	5	c-g
Sergentia?	5	c-g
Stelechomyia sp.	7	c-g
Stenochironomus spp.	5	c-g
Stictochironomus sp.	9	c-g
Tribelos fuscicorne	5	c-g
Tribelos jucundum	5	c-g
Xenochironomus nr. *rogersi*	0	prd
Xenochironomus xenolabis	0	prd
? Chironomini	6	c-g
Cladotanytarsus spp.	7	c-f
Constempellina sp. 1	4	c-g
Constempellina sp. 2	4	c-g
Microspectra spp.	7	c-g
Paratanytarsus confusus	6	c-f
Paratanytarsus dimorphis	6	c-f
Rheotanytarsus exiguus gr.	6	c-f
R. distinctissimus gr.	6	c-f
Stempellina spp.	2	c-g
Stempellina spp.	4	c-g
Sublettea coffmani	4	c-f
Tanytarsus spp.	6	c-f
Zavrelia gr. spp.	4	c-f
? Tanytarsini	4	c-f

c-f = collector filterer
c-g = collector gatherer
prd = predator
scr = scraper
shr = shredder

CHAPTER 13

WATER POLLUTION, WATER POLLUTION CONTROL, AND WATER TREATMENT

Why Read This Chapter?

This chapter examines different kinds of water pollution, including artificial eutrophication, metals, pesticides, acid precipitation, thermal enrichment, sedimentation and radiation. Emphasis is placed upon the effects of the most toxic of elements and compounds within each class of pollutant on aquatic organisms and humans. The identification of pollutants and assessment of their impacts on aquatic organisms are discussed in Chapters 11 and 12, respectively.

This chapter also examines the fundamentals of water treatment and water pollution control. The term water treatment refers to the purification of water before it is used by the public for consumption. The term water pollution control refers to the processes which remove pollutants from water after it has been used by industries (e.g. effluents of factories) and humans (e.g. domestic waste).

By the end of this chapter you will know:

☞ 1. The characteristic changes as lakes eutrophy
☞ 2. The different kinds of metals that are toxic to aquatic life
☞ 3. The difference between "hard" and "soft" metals
☞ 4. Differences/similarities between essential, trace, heavy, micro- and macro- elements
☞ 5. The metals that are essential to humans
☞ 6. Five factors that affect metal toxicity
☞ 7. The toxic effects of mercury, cadmium, lead and arsenic to organisms
☞ 8. The sources of metal pollutants
☞ 9. The five types of metal formations
☞ 10. Cation exchange and adsorption mechanisms in soils

☞ 11. The levels of metal enrichment in algae
☞ 12. The levels of metal enrichment in macrophytes
☞ 13. The levels of metal enrichment in zooplankton
☞ 14. The levels of metal enrichment in benthos
☞ 15. The levels of metal enrichment in fish
☞ 16. Some common pesticides in the environment
☞ 17. Some common pesticides used in forestry
☞ 18. Some common pesticides used for controlling aquatic weeds, algae, snails and fish
☞ 19. The factors that affect movement of pesticides into aquatic environments
☞ 20. The meaning of bioconcentration and factors affecting it
☞ 21. The meaning of biomagnification and factors affecting it
☞ 22. How toxicants enter the aquatic environment
☞ 23. The effect of pesticides on bacteria and fungi
☞ 24. The effect of pesticides on algae
☞ 25. The effect of pesticides on invertebrates
☞ 26. The effect of pesticides on amphibians, reptiles and fish
☞ 27. The indirect effect of pesticides on the aquatic environment
☞ 28. The sources of acid precipitation
☞ 29. The pH scale
☞ 30. How geology affects the pH of lakes and streams
☞ 31. The meaning and mechanism of spring pH depression events
☞ 32. The effects of acidification on bacteria
☞ 33. The effects of acidification on algae
☞ 34. The effects of acidification on macrophytes
☞ 35. The effects of acidification on zooplankton
☞ 36. The effects of acidification on benthos
☞ 37. The effects of acidification on fish
☞ 38. Other kinds of atmospheric pollutants
☞ 39. How to combat lake and stream acidification
☞ 40. The sources and effects of thermal pollution on aquatic organisms
☞ 41. The natural sources of sediment loading
☞ 42. Three anthropogenic sources of sediment loading
☞ 43. Four main kinds of radiation
☞ 44. The units of radiation measurement
☞ 45. The effects of radiation on aquatic organisms
☞ 46. The kinds of oil pollution and their effects on aquatic organisms
☞ 47. The purpose and main components of a water pollution control plant (WPCP)
☞ 48. The processes involved in primary treatment of a WPCP
☞ 49. The processes involved in secondary treatment of a WPCP, including:
  ☞ a. The activated sewage sludge process
  ☞ b. The trickling filter process
☞ 50. The processes involved in tertiary treatment of a WPCP
☞ 51. The purpose and operation of a water treatment plant (WTP)

ARTIFICIAL EUTROPHICATION

Chapters 4 to 9 describe in great detail some of the physical, chemical and
1 ☞ biological changes that occur as lakes naturally evolve from the oligotrophic state to the
eutrophic state. Some of the most characteristic changes that occur as lakes eutrophy are:

1. An increase in nutrient (total phosphorous and total nitrogen) levels
2. An increase in suspended solids levels (especially phytoplankton) and colour (colourless to green, brown, etc.)
3. An increase in the mean epilimnetic temperature
4. A decrease in Secchi depth
5. An increase in the volume of the epilimnion and a decrease in the volume of the hypolimnion
6. Loss of dissolved oxygen in hypolimnetic waters during summer thermal stratification
7. Associated changes in levels of oxidized forms of carbon, nitrogen, sulphur, iron, manganese and silica to their reduced states, many of which are toxic (e.g. methane gas, nitrite, ammonia, hydrogen sulphide)
8. Decreases in diversity of organisms at all trophic levels, including autotrophs, zooplankton, submersed macrophytes, benthic macroinvertebrates and fish.
9. Increases in the biomass and productivity of species tolerant of organic enrichment at all trophic levels
10. Accumulation of organic material on the lake floor leading to a decrease in mean depth of the lake
11. Lake bottom profile changes from a U-shape to a steep V-shape, then a shallow V-shape
12. Eventual filling of the lake with organic matter and a transition from the aquatic state to a semi-aquatic state and eventually to a terrestrial state

While these changes occur naturally over geological time, human intervention may greatly accelerate the eutrophication process, but the processes remain the same. Figure 13.1 illustrates most of the changes described above. The diagnosis of the degree of organic enrichment is the same for lakes that eutrophy artificially (called ***cultural eutrophication***) as for lakes that eutrophy naturally.

The criteria used for assessing trophic status are developed in Chapters 4 to 9 and are summarized in Chapter 12 and need not be repeated here. The use of Secchi depth, in particular, has been recommended for assessing trophic status because it is highly correlated with nutrient and chlorophyll *a* levels and it is relatively easy to make and easy to use. Use of hypolimnetic dissolved oxygen concentration/saturation and spring turnover levels of total phosphorous and total nitrogen provide greater resolution of trophic status than does the Secchi depth, but they also require fairly labourious chemical analyses and some expertise. Similarly, indicator organisms are the integrators of environmental conditions and ultimately represent the overall impact of water quality on

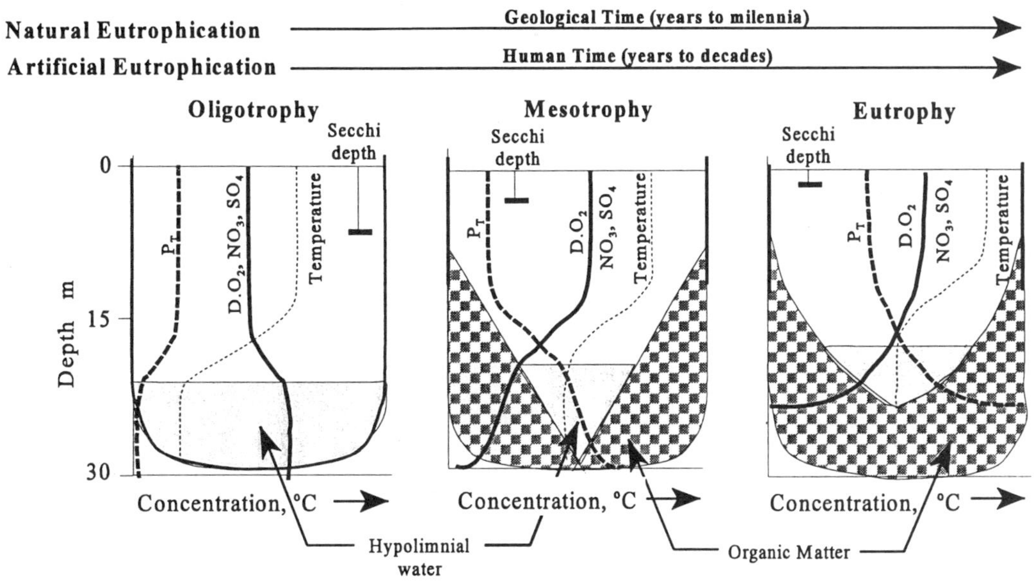

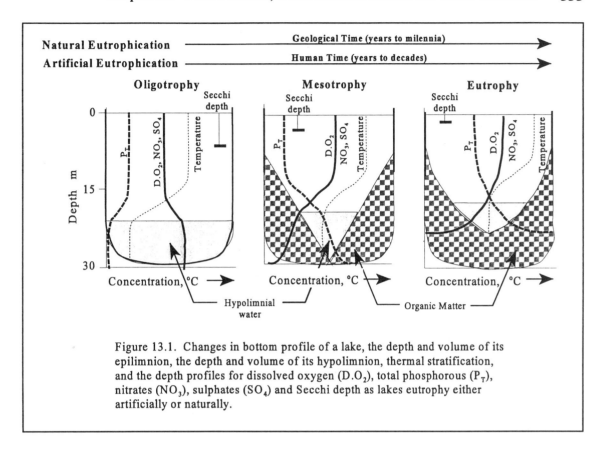

Figure 13.1. Changes in bottom profile of a lake, the depth and volume of its epilimnion, the depth and volume of its hypolimnion, thermal stratification, and the depth profiles for dissolved oxygen ($D.O_2$), total phosphorous (P_T), nitrates (NO_3), sulphates (SO_4) and Secchi depth as lakes eutrophy either artificially or naturally.

aquatic communities. But they too require specialized equipment and time to collect and some expertise for identification.

METAL POLLUTION

For an in-depth discussion of metal pollution in aquatic ecosystems, refer to
2☞ Forstner and Wittmann (1983)[1]; most of the following description is from their text. They cite several references to support their arguments, but most are omitted here for brevity. As in Forstner and Wittmann (1983), this section begins with brief summaries of the classifications of elements and metals and of trace elements essential to human life. The sources of metals, their toxicities, factors that affect their toxicity and the hazards of certain metals are then described.

The long form of the ***periodic table*** is shown in Table 13.1. It arranges elements into seven rows called ***periods***, or series, and sixteen columns called ***groups***, or ***families***.

---

[1]Forstner, U. and G. T. W. Wittmann. 1983. Metal pollution in aquatic environments. Springer-Verlag, New York. 486 pp.

Table 13.1. Long form of the periodic table of the elements (symbols) with the atomic number (1-103) above and the electronegativity value below. The concentrations of metals in the earth's crust are shown in different colours, the legend being given at the bottom of the table. Non-metals are shown without colour (boxes with 5% shading). Some elements, like silicon and selenium occur as non-metals and metals. The elements are listed according to their periods (or series, 1-7) and their group (or family, 1A - VIIA). The inert gases are shown in group 0, the last column. See text for explanation of the electronegativity value. Adapted from Forstner and Wittmann (1983).

Period	0	IIA	IIIB			IVB	VB	VIB	VIIB	VIII			IB	IIB	IIIA	IVA	VA	VIA	VIIA	0
1	1 H 2.1																		1 H	2 He -
2	3 Li 1.0	4 Be 1.5													5 B 2.0	6 C 2.5	7 N 3.0	8 O 3.5	9 F 4.0	10 Ne -
3	11 Na 0.9	12 Mg 1.2													13 Al 1.5	14 Si 1.8	15 P 2.1	16 S 2.5	17 Cl 3.0	18 Ar -
4	19 K 0.8	20 Ca 1.0	21 Sc 1.3			22 Ti 1.5	23 V 1.6	24 Cr 1.5	25 Mn 1.5	26 Fe 1.8	27 Co 1.8	28 Ni 1.8	29 Cu 1.8	30 Zn 1.6	31 Ga 1.6	32 Ge 1.8	33 As 2.0	34 Se 2.4	35 Br 2.8	36 Kr -
5	37 Rb 0.8	38 Sr 1.0	39 Y 1.2			40 Zr 1.4	41 Nb 1.6	42 Mo 1.8	43 Tc 1.9	44 Ru 2.2	45 Rh 2.2	46 Pd 2.2	47 Ag 1.9	48 Cd 1.7	49 In 1.7	50 Sn 1.8	51 Sb 1.9	52 Te 2.1	53 I 2.5	54 Xe -
6	55 Cs 0.7	56 Ba 0.9	57 La 1.1	58 Ce	71 Lu 1.2	72 Hf 1.3	73 Ta 1.5	74 W 1.7	75 Re 1.9	76 Os 2.2	77 Ir 2.2	78 Pt 2.2	79 Au 2.4	80 Hg 1.9	81 Tl 1.8	82 Pb [illegible]	83 Bi 1.9	84 Po 2.0	85 At 2.2	86 Rn -
7	87 Fr	88 Ra	89 Ac	90 Th	103 Lw															

Metal levels in earth's crust	> 1000	100-1000	10-100	1-10	0.1-1.0	<0.1	
	> 1000 ppm	100-1000 ppm	10-100 ppm	1-10 ppm	0.1-1.0 ppm	<0.1 ppm	Non-metals

Elements with similar electron configurations are ordered according to increasing atomic number. The ***periodic law*** states that most elements have closely related properties and "*these similar properties recur in a periodic manner with increase in atomic number*". Some of the physical and chemical properties which reveal periodicity are: (i) electrical and thermal conductivity; (ii) density; (iii) atomic and ionic radii; (iv) electronegativity; and (v) oxidation numbers.

Elements within the same group generally resemble one another closely. The properties of elements within a given period vary gradually from a highly metallic (electropositive) character at the left side to a highly non-metallic (electronegative) character on the right side of the row. Metallic elements are good conductors of electricity, their electrical resistance being directly proportional to the absolute temperature. They also have high thermal conductivity high density and are malleable. However, some non-metallic elements possess some of these traits but none are good electrical conductors making metals unique in this physical property. Seven of the ten most abundant elements in the earth's crust are metals: aluminum (at 7.5 weight percent); iron (at 4.7 weight percent); calcium (at 3.4 weight percent); sodium (at 2.6 weight percent); potassium (at 2.4 weight percent); magnesium (at 1.9 weight percent); and titanium (at 0.6 weight percent).

Boron, silicon, germanium, arsenic and tellurium have low electrical conductivity that decreases with increasing temperatures. These elements occur in the periodic table between metals and non-metals and hence are termed ***metalloids***, or half metals. The term ***heavy metal*** refers to metals that have densities exceeding 5.0 g/cm^3. Most of these have atomic numbers > 20 and atomic weights > 40. ***Trace elements*** are elements that have a natural occurrence < 0.1%.

Elements (mostly metals) that release electrons are usually referred to as being ***electropositive*** and those (mostly non-metals) that accept electrons are called ***electronegative***. The term ***electronegativity*** (EN) is a measure of the power of an atom to attract electrons to itself in a covalent bond. The EN of the elements are given in Table 13.1. The EN value tends to follow the periodic law within main groups, where within a given group (column) there is a general trend of decreasing EN with increasing atomic number. Conversely, within a series (row) the EN tends to increase gradually from left to right (metal to non-metal). The oxidation number of an element is a direct outcome of the EN concept and coincides numerically with the ionic charges of ions. The valence of an element is related to its oxidation state and represents its combining capacity.

Metals are classified according to their reactivity as electron-pair acceptors toward electron-pair donors, generalized by the equation:

$$A + :B \rightarrow A:B$$

The species, A:B, is referred to as an ***ion pair***, a ***metal complex***, a ***coordination
3☞ compound*** or a ***donor-acceptor complex***. The stability of the species, A:B, can be explained by classifying the acceptors and donors into "hard" and "soft" categories

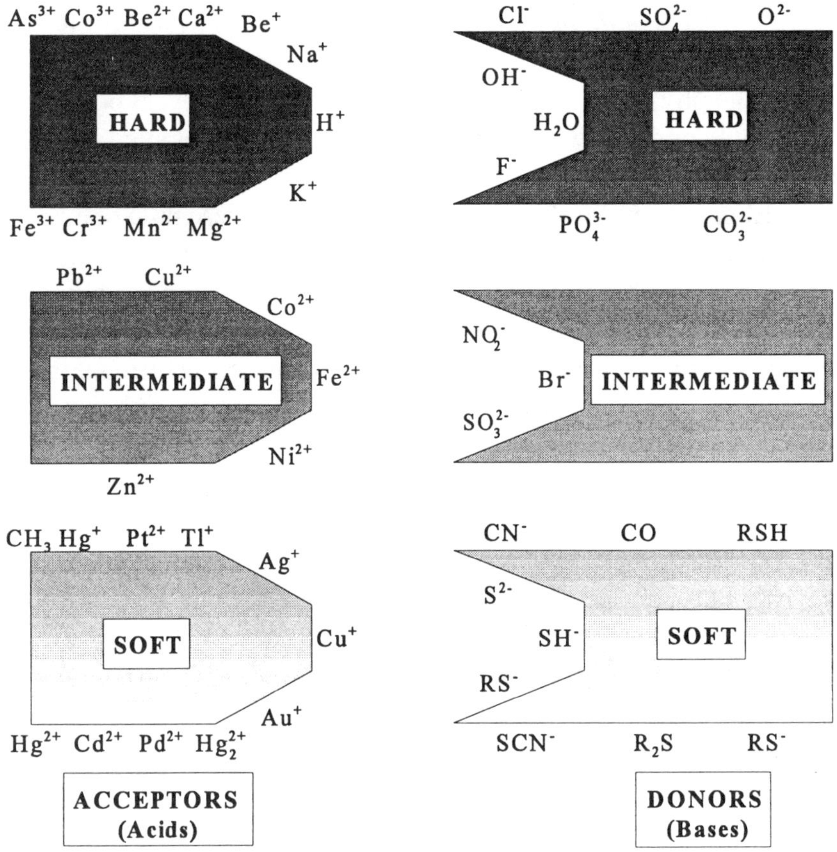

Fig. 13.2. Classification of acceptors and donors . R = alkyl or aryl.

(Pearson, 1968)[2]. In general, hard acceptors prefer to bind to hard donors and soft acceptors prefer to bind to soft donors to form stable compounds (Fig. 13.2). Electronegativity and electron mobility are the main criteria for classifying metals as hard or soft. Hard acceptors tend to have low mobility, low electronegativity and a large density with a positive charge (i.e. high oxidation state and small radius), while the reverse is true for soft acceptors. A hard donor has low mobility, high electronegativity and a high negative charge density, whereas the converse is true for soft donors.

Electron pair acceptors are generally classified as acids and electron pair donors are called bases. The ***preferential bonding between hard and soft metal species*** is known as the ***HSAB principle*** (hard and soft acids and bases). The principle is easily demonstrated in nature. For example, some metals occur in the earth's crust as ores of oxides or carbonates, others as ores of sulfides. Hard acids, such as Mg^{2+}, Ca^{2+}, and Al^{3+}

---

[2]Pearson, R. 1968. Hard and soft acids and bases, Part I, Fundamental principles. J. Chem. Educ. 45: 581-587.

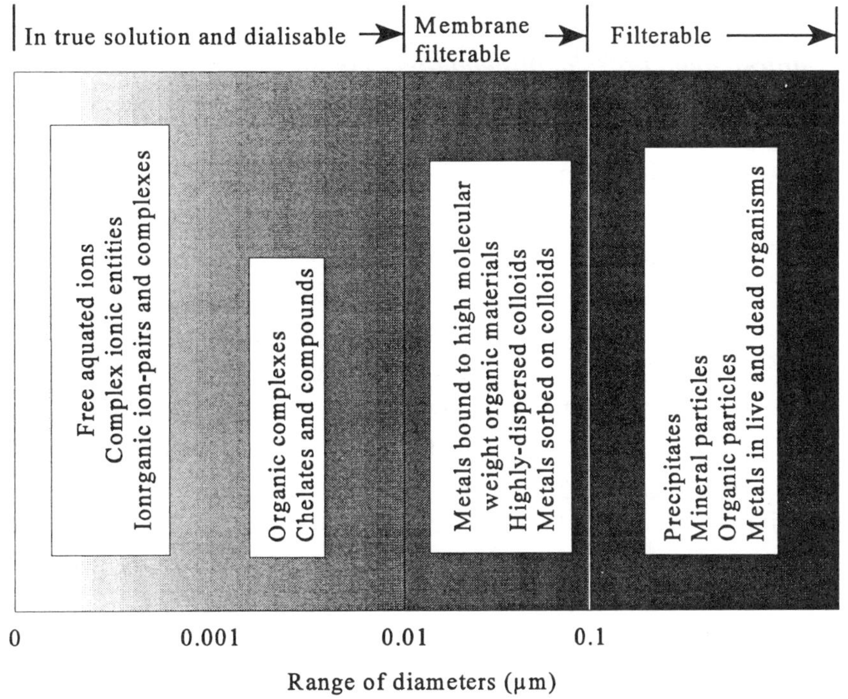

Figure 13.3. Types of metal species in water in relation to their particle size (diameter) and solubility.

form strong bonds with hard bases, especially CO_3^{2-} (e.g. limestone, $CaCO_3$), PO_4^{3-} (e.g. apatite, $CaPO_4$), and SO_4^{2-} (e.g. gypsum, $CaSO_4$). Conversely, soft acids such as Hg_2^{2+}, or Hg^{2+}, or Pb^{2+} are attracted to soft bases such as S^{2-}. According to the HSAB principle, weak bonds are formed from hard acid-soft base combinations and soft acid-hard base combinations and will be leached and carried away in runoff.

The toxicity of a metal depends not only on its total concentration but especially on its species. Metal speciation is based on the particle size fraction present (Fig. 13.3). In general, the larger the particle, the less toxic the species. Soluble metals are separated from particulate metals by filtration through a 0.45 μm pore size membrane filter. The group of soluble metal ions is separated on the basis of:

1. Simple aquated metal ions
2. Metal ions complexed by inorganic anions (e.g. $CuCO^3$)
3. Metal ions complexed by organic ligands such as amino, fulvic and humic acids.

Essential Trace Elements

Although the terms "***trace metal***", "***trace organic***", "***heavy metal***", "***microelements***", "***macroelements***", etc. have specific meanings, they are treated here as
4☞ synonymous with the term, ***trace elements***. The role of heavy metal ions in living systems follows the pattern of abundance and availability of the same metals occurring in nature (Wood 1974)[3].

An element is essential when: (i) it is consistently shown to be present in all healthy living tissues within a zoological family whereby tissue concentrations from species to species should not vary by a wide range; (ii) deficiency symptoms are noted with depletion or removal and disappear when the elements are returned to the tissue; (iii) the deficiency symptoms should be attributed to a distinct biochemical defect at the molecular level (Forstner and Wittmann 1983, cited earlier).

Table 13.2 lists some of the elements that are essential to human life. Sodium, potassium, magnesium and calcium are highly concentrated in body fluids and widely
5 ☞ distributed throughout the human body and thus cannot be considered trace elements. All four elements are often referred to as "***macroelements***". Most other essential elements are transition metals (e.g. Fe, Cu, Mo, Co, Zn) and are required for human growth, development, achievement and reproduction. The following is a summary of the biological roles of the ten most abundant essential elements in the human body.

Sodium and Potassium: Most mobile cations, involved in metabolic processes and nerve impulse conduction via the brain.

Magnesium and Calcium: These are the next most mobile cations, magnesium being more mobile than calcium. Magnesium ions participate in cell functions, nerve impulse transmission, muscle contractions, metabolic functions and are a structural complex of nucleic acids. Calcium ions have an affinity for oxygen-containing ligands, resulting in the presence of crystalline salts (e.g. phosphate, oxalate) in the blood. In bone and teeth formation, Ca^{++} is deposited as hydroxyl-apatite ($Ca_5(PO_4)_3(OH)$), making calcium the most abundant element in the human body.

Trace Metals: Most essential elements are transition metals that function primarily as catalysts, inducing or enhancing enzyme activity. In highly specific ***metalloenzymes*** the metal is firmly associated with the protein and usually constitutes the active center of the living cell, catalyzing a specific reaction or type of reaction. Hence, even trace concentrations of such metals can have a powerful, direct influence on biological functions within the human body.

Essential metals may have one of two separate functions:

Redox reactions- Transition metals with stable oxidation states may be

---

[3]Wood, J. M. Biological cycles for toxic elements in the environment. Science 183: 1049-1052.

Table 13.2. Essential minerals (listed alphabetically) in the diet of humans and their functions, symptoms of deficiency and excess and the safe level in water for drinking. Safe levels are based on provincial water quality objectives for recreational use and protection of aquatic life in Ontario waters. NC = Noncritical. *Copper safe level is based on hardness level: ≤ 1 if 0-20 mg $CaCO_3$/L; 5 if > 20 mg $CaCO_3$/L .

Mineral	Functions in body	Symptoms of deficiency	Symptoms of excess	[Safe] µg/L
Calcium (Ca)	Bone, tooth formation; nerve, muscle function; blood clotting	Loss of bone mass; stunted growth	None; high Ca intake needed	NC
Chlorine (Cl)	Acid-base; in gastric juices	Reduced appetite; cramps	> 2 mg/L undrinkable	2
Chromium (Cr)	Needed in glucose, energy metabolism	Impaired glucose metabolism	Causes lung tumours; skin sensitizer; CrIII salts less toxic than Cr IV compounds	<100
Cobalt (Co)	Component of vitamin B-12	B-12 deficiency	Death	<0.6
Copper (Cu)	Component of iron metabolism enzymes	Anemia; bone, cardio-vascular changes	Death	1-5*
Fluorine (F)	Maintains tooth, bone structure	Tooth decay increases	Undrinkable	<0.2
Iodine (I)	Component of thyroid hormones	Enlarges thyroid (goiter)	Undrinkable	<100
Iron (Fe)	Electron carrier in energy metabolism; component of haemoglobin	Anemia; weakness; reduced immunity	Water undrinkable when Fe in excess	<300
Magnesium (Mg)	Component of some amino acids	Nervous system disorders	None	NC
Manganese (Mn)	Component of some enzymes; glucose utilization	Abnormal bone, cartilage	Water undrinkable when Mn in excess	NC
Molybdenum (Mo)	Component of some enzymes	Impairs excretion of nitrogenous wastes	Acute or chronic effects unreported	<10
Phosphorous (P)	Bone, tooth, nucleotide formation; acid-base regulation	Loss of bone minerals; weakness	None, but leads to taste, odour problems	≤10
Potassium (K)	Acid-base, water balance; nerve function	Paralysis; muscular weakness	None	NC
Selenium (Se)	Component of enzymes; acts with vitamin E	Muscular pain	Death	<100
Sodium (Na)	Acid-base, water balance; nerve function	Reduced appetite; cramps	None	NC
Sulphur (S)	Component of some proteins	Symptoms of protein deficiency	Water undrinkable	NC
Zinc (Zn)	Component of some digestive enzymes and proteins	Impaired reproduction, growth, immunity; hair loss	May lead to increased heart disease	<20

involved in the electron transfer process, especially Fe(II)/Fe(III), Cu(I)/Cu(II) and Mo(V)/Mo(VI) systems.

Direct control of reaction mechanisms - Cobalt and zinc are both potent enzyme activators.

***Manganese*:** The second most abundant metal in nature (iron is first). Its chemistry resembles that of Mg^{++}, preferring weaker donors such as phosphate and carboxylate groupings to form stable bonds. Manganese is an enzyme activator and is involved in glucose utilization.

***Iron*:** The most abundant transition metal, essential in the formation of haemoglobin in blood.

***Cobalt*:** A component of vitamin B_{12}, which is a Co(III) complex, to form haemoglobin. Cobalt is a potent enzyme activator throughout the human body.

***Copper*:** An important constituent in the formation of haemocyanin, a blood pigment of snails and clams. Cu(I) is found in enzymes capable of carrying oxygen and is required in the formation of haemoglobin.

***Zinc*:** One of the most abundant essential elements in the human body. It is about 100 times more abundant than copper and is found in all mammals. Zn(II) is a potent enzyme activator.

***Molybdenum*:** Involved in the electron transfer processes and in nitrogen fixation.

Metal Toxicity

The deficiency or absence of an essential element could lead to the death of an
6☞ animal. Excessive supplies (>40 - 200 times the correct nutritional supply) can be toxic and also lead to the death of an organism. The two scenarios are depicted in Fig. 13.4.

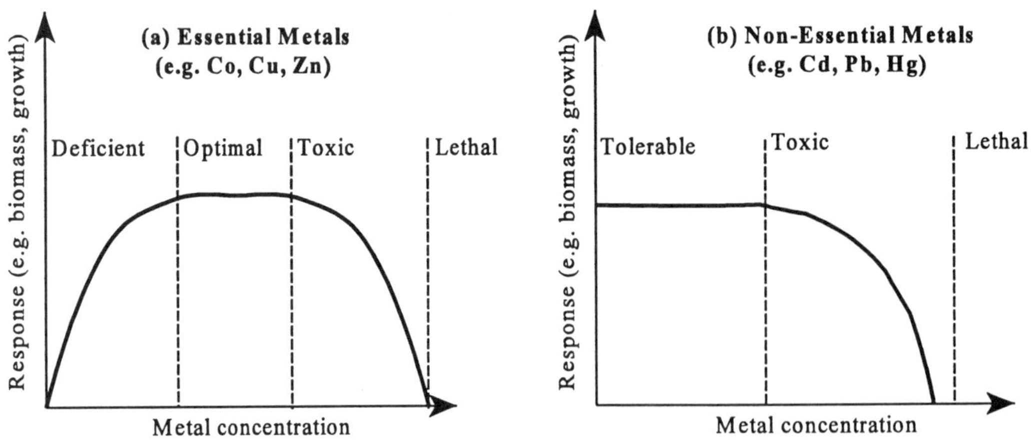

Figure 13.4. Deficiency and excessive supplies of (a) essential and (b) non-essential metals.

Elements can be classified according to their toxicity and availability, as (i) very toxic and relatively available, (ii) toxic but very insoluble or very rare, or (iii) non-critical (Table 13.3). Several factors can affect the toxicity of a metal in solution, including: (i) the form of the metal in water; (ii) the presence of other metals; (iii) the physiological response in relation to the physical and chemical characteristics of the water; (iv) the condition of the organism; (v) the behavioural response of the organism.

Table 13.3. Classification of elements according to toxicity and availability (Modified from Forstner and Wittmann (1983, cited earlier).

Very toxic, relatively accessible		**Toxic but very insoluble or very rare**		**Non-Critical**	
Be	Ag	Ti	Ga	Na	S
Co	Cd	Hf	La	K	Cl
Ni	Pt	Zr	Os	Mg	Br
Cu	Au	W	Rh	Ca	F
Zn	Hg	Nb	Ir	H	Li
Sn	Tl	Ta	Ru	O	Rb
As	Pb	Re	Ba	N	Sr
Se	Sb			C	Al
Te	Bi			P	Si
Pd				Fe	

(i) ***Form of metal in water***. Metals may exist in inorganic or organic forms, each of which may be soluble or particulate. Of the soluble phase, the metal may exist as an ion, a complex ion, a chelated ion or as a molecule. The particulate phase may be colloidal, precipitated or adsorbed.

(ii) ***Mixtures of other metals/toxicants***. Throughout the 1900s we experienced a variety of toxicants and contaminants that have threatened the very existence of human-kind. First it was eutrophication, then acid precipitation, dioxins, radiation, ozone depletion, and global warming. In retrospect, eutrophication has, arguably, been among the least of our worries. There is no question that threats to our aquatic environment have increased over time. Among the most enigmatic problems is the coincidental production of a plethora of metals and pesticides that have created a potpourri of potencies and interactions that few have been able to comprehend. So far, scientists have proposed relatively simple ***toxicological interactions*** in which exposure to two or three chemicals results in a biological response that is quantitatively different from that expected from the response of each of the chemicals alone. The toxicity of a chemical to an organism is affected by biotransformation, interaction between chemicals or modification of potency by the ambient chemical milieu. In the simplest of mixtures, exposure to two chemicals simultaneously may produce a response that is additive to the individual responses, or one that is greater than or less than expected from addition of their individual responses. The identification of additive, less-than-additive, antagonistic effects, etc. that occur with mixtures of chemicals is described and discussed in Chapter 11 (see Fig. 11.8).

(iii) ***Physiological response in relation to the physical and chemical characteristics of the water***. It is well known that temperature affects the metabolic rates of organisms. Generally, there is a doubling of the metabolic rate with a 10 °C rise in temperature. Temperature also affects the solubility of metal compounds, the solubility increasing with increasing temperature.

The solubility of many metal compounds is also affected by pH and water hardness. The degree of water hardness influences the toxicity by forming insoluble carbonates or by adsorption on calcium carbonate. For example, the acute toxicity of Co^{2+}, Ni^{2+}, Cd^{2+}, Cr^{6+}, and Cu^{2+} decreases with increasing hardness for *Daphnia*. In hard water, the concentrations of heavy metals needed to reach the level of lethal dosage is greater and in soft water it is lower. Apparently, magnesium and calcium ions compete with heavy metal ions for active sites in fish tissues, making the metals less available and less toxic.

All aquatic organisms have an optimal level of dissolved oxygen for growth and survival. Departures from the optimum level puts stress on the organism and lowers its ability to tolerate additions of metals and other toxicants to the water.

Light is important for photosynthesis in plants, but it is the red and infra-red light that causes heating of water. Many animals are photo-positive (move toward light), others are photo-negative (avoid light) and their physiological response may be indirectly affected by light conditions. Light also indirectly affects the oxygen levels by altering the rates of photosynthesis. Hence, it is often difficult to separate the effects of light, temperature and oxygen levels as factors affecting the toxicity of metals.

The amount of dissolved materials, (or conductivity and salinity) in the water, affects the response of animals to toxic levels of metals. All organisms have an optimal level of dissolved materials needed to survive, grow and reproduce. A decrease in the level of dissolved materials (e.g. low conductivity) has a similar effect as an increase in salinity by lowering the ability of fresh water animals to tolerate increases in metal levels. The effects are similar to a deficiency or oversupply of essential elements (Fig. 13.4).

(iv) ***Condition of the organism***. By condition is meant the organism's: (i) sex - sometimes sex is not a factor, males and females responding in a similar fashion; (ii) stage in life history (egg, larva, etc.) - generally larvae and juveniles are more sensitive than adults; (iii) changes in life cycle (e.g. moulting, reproduction) - generally, organisms are most sensitive during moulting and reproductive periods; (iv) age and size - generally tolerance increases with size and age, until near their life span when tolerance to metals decreases; (v) starvation - starved animals are more sensitive than sated animals; (vi) activity - highly active animals are more sensitive than quiescent ones; (vii) additional protection (e.g. shell, brood sac) - ovoviviparous animals are protected by body fluids of the parent; (viii) adaptation to metals - all animals have an ability to adapt to increases in metal levels and previous exposure increases their tolerance to the metal.

(v) ***Behavioural response of the organism***. The presence of certain metals, or increases in their concentrations, may alter the behaviour of an animal. Poisoning by methylmercury compounds causes neurological damage that alters the behaviour of animals, as in the case of Minamata disease discussed below.

Toxic Effects of Some Non-Essential Metals

The effects of deficiencies and oversupply of essential elements was discussed
7 ☞ earlier. In general, the effects of excessive amounts of non-essential elements is much less severe than the toxic effects of non-essential elements, particularly mercury, cadmium, lead and arsenic.

Mercury: The most infamous catastrophe associated with metal poisoning is the outbreak of ***Minamata disease*** in people eating fish from Minamata Bay, a freshwater lake in southwestern Japan, in the mid 1950s. People who ate fish and shellfish progressively suffered from a weakening of muscles, loss of vision, impairment of cerebral functions and eventually paralysis, or even death. Extensive research showed that the disease was caused by consumption of fish laden with methylmercury. The source of methylmercury was traced to the effluent from manufacturers of PVC plastics. Other outbreaks occurred in the mid 1960s in people eating fish and shellfish from a river, in Niigata, into which methylmercury was released from the effluent of an electrical industrial plant. These were the first cases of mercury poisoning in the aquatic environment. Other reports of Minamata disease have been associated with mercurial fungicides by paper mills (Sweden, 1960s) and wheat seed treated with mercurial fungicide (Iraq, 1972).

Cadmium: Next to mercury, cadmium is the most toxic metal. Cadmium poisoning in the aquatic environment is less widespread than mercury poisoning, but the toxic effects are as severe. Again, Japan is the source of the earliest report of cadmium poisoning. The source of cadmium was a zinc mine that released its effluent into the Jintsu River. The amounts of cadmium released were sublethal but bioaccumulation through the consumption of rice contaminated by sludge spread over rice fields over a 5- to 10- year period induced a painful disease of a "rheumatic nature". The disease became known as the "***itai-itai***" (meaning ouch-

ouch) disease that was always associated with cries of pain from patients with skeletal deformities. Some inflicted patients had as much as a 30 cm decrease in body heights! The disease culminated in the death of many patients.

Lead: Lead poisoning has been recorded for over 2000 years, the earliest record being "***plumbism disease***" among the Greeks who consumed water, wine and food stored in earthenware. Lead poisoning is still very common today, especially among children in slums and older urban areas where the plumbing, paints, putty and plaster still contain considerable amounts of lead. Lead poisoning is a major cause of brain damage, mental deficiency and serious behavioural problems. Plumbing that carries soft water (< 60 mg $CaCO_3$/L) contributes more lead to the water supply than does hard water (> 200 mg $CaCO_3$/L). Lead poisoning over a long term can lead to kidney infections called "***nephritis***", a common disease of "moonshiners" who consumed whiskey made in stills with lead-soldered tubing.

Arsenic: Arsenic poisoning was first reported in 1955 in Japan. Children consuming dry milk powder, to which arsenic contaminated sodium phosphate had been added as a stabilizer, exhibited mental disturbances and changes in EEGs (electroencephalograms). About 130 out of 12,000 cases were fatal. Other symptoms were exhibited by children inhaling fumes from a coal power plant in Czechoslovakia. The coal contained about 1000-1500 g Ar/ton and affected children developed respiratory problems and hearing loss. Other disorders associated with skin poisoning are skin cancer, hyperpigmentation and hyperkeratosis.

Sources of Metal Pollution

Forstner and Wittmann (1983, cited earlier) summarize sources of metal pollution
8☞ into five categories: (i) geologic weathering; (ii) industrial processing of ores and metals; (iii) use of metal and metal components; (iv) leaching of metals from garbage and solid waste dumps; (v) animals and humans. Fig. 13.5 illustrates the typical pathways and cycling of metals from natural and anthropogenic sources in the environment.

Geological Weathering

Multi-element analysis of rocks from geological surveys for ore deposits and oil have resulted in fairly detailed geochemical maps for large regions in several countries, including Canada. Geological weathering of mineralized zones is the source of baseline or background levels of most elements. However, mining the ores to retrieve and process the minerals results in elevated levels of metals through discharge of effluents, disposal of tailings and smelting operations which results in atmospheric pollution.

Mining Effluents

The effects of mine effluents on aquatic biota have been known since the mid

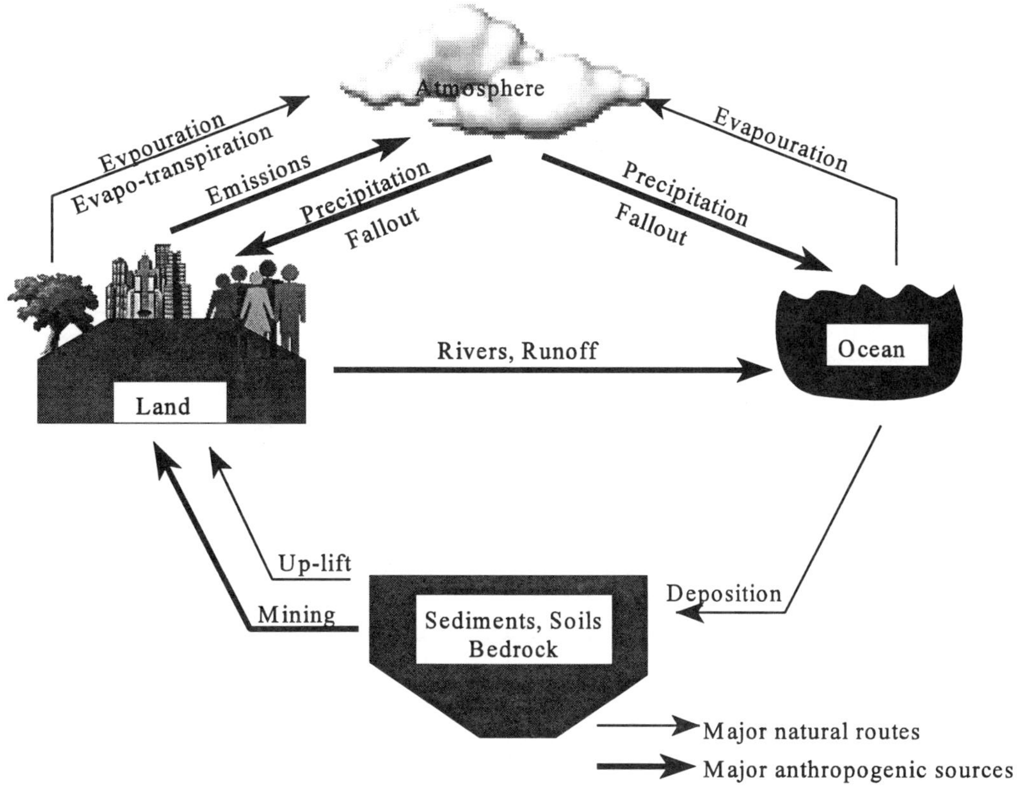

Figure 13.5. Sources and cycling of metals in the environment.

1800s. The mine effluent, or ***slimes***, is generally a potpourri of extremely poisonous metals. Erosion and dissolution of mine spoilage heaps and waste dumps and surface runoff from mine sites are also major sources of metals. Waste rock dumps and mine tailings often contain high concentrations of sulphides and/or sulpho salts which are associated with most ore and coal bodies. Exposure of pyrite (FeS_2) and of other sulphide minerals to atmospheric oxygen causes oxidation of the sulphide to produce sulphate, creating highly acidic conditions. The increased acidity leads to dissolution reactions of a variety of metals, including aluminum from alumino-silicates.

Waters receiving seepage from gold/uranium slimes and dams can be enriched by 4 orders of magnitude. For example, levels of dissolved manganese, cobalt and nickel are known to exceed normal surface water values by a factor > 10,000 for each metal. Iron, chromium, zinc and sulphate concentrations are frequently magnified 1000 times, and lead and cadmium levels can be enriched 100-fold.

Industrial Effluents

Table 13.4 lists major industries and the metals used and released in effluents by them. The industries are listed according to the diversity of metals employed in the manufacture of products, with the least diverse listed lastly. The metals are listed in decreasing order of use, with chromium listed first because it is used by all 12 industries. The multipurpose use of heavy metals usually causes major problems in definitive identification of the source metal that poisons receiving waters.

Table 13.4. Metals used by major industries.

Industry	Cr	Zn	Cd	Hg	Pb	Cu	Fe	Ni	Sn	Mn
Fertilizers	✝	✝	✝	✝	✝	✝	✝	✝		✝
Basic steel works foundries	✝	✝	✝	✝	✝	✝	✝	✝	✝	
Organic chemicals, petrochemicals	✝	✝	✝	✝	✝		✝		✝	
Alkalis, chlorine, inorganic chemicals	✝	✝	✝	✝	✝		✝		✝	
Petroleum refining	✝	✝	✝		✝	✝	✝	✝		
Basic non-ferrous metalworks, foundries	✝	✝	✝	✝	✝	✝				
Pulp, paper mills, paperboard, building paper, board mills	✝	✝		✝	✝	✝		✝		
Motor vehicles, aircraft plating, finishing	✝		✝			✝		✝		
Flat glass, cement, asbestos products, etc.	✝									
Textile mill products	✝									
Leather tanning, finishing	✝									
Steam generation power plants	✝									

Domestic Effluents and Urban Stormwater Runoff

Domestic effluents are characterized by enrichment with copper, lead, zinc and cadmium, mostly due to corrosion of the water supply system. Compared to contributions from crustal rocks, sewage sludge has, on average, 50 times more Cd. The ratio of other metals in sewage compared to crustal rock is: 40:1 for Zn; 30:1 for Pb; 15:1 for Cu; and 1.2:1 for Cr. Silver is known to be have a ratio of about 200:1 in effluents carrying photochemical wastes.

Metal Inputs from Rural Areas

Commercial forest, grassland and cultivated agricultural areas are among the largest potential non-point sources of metals. Soil cultivation greatly enhances soil erosion. Runoff from cultivated fields is a major source of sediment loading in streams. Animal wastes, fertilizers and pesticides in eroded soils are almost impossible to control in the runoff. The application of phosphate-laden fertilizers often causes excessive algal growth in receiving waters.

Dredged sediments from lakes and sewage sludge from water pollution control plants are often used as a plant nutrient source and soil conditioner. However, lake sediments and sewage sludge often contain high levels of metals that have sedimented out of solution and are a potential source of metals in agricultural runoff and drainage.

Atmospheric Sources

Metal enrichment in the atmosphere can be very high, especially in industrialized areas. Table 13.5 gives the enrichment factor (ratio of an elements concentration in air to that in the earth's crust) for several elements. Enrichment factors near unity (e.g. cobalt) indicate crustal weathering as the atmospheric source for the particular element. Most elements, especially, copper, zinc, arsenic, silver, mercury, cadmium, lead, selenium and antimony are enriched in atmospheric particles by one to four orders of magnitude. The enrichment factor for vanadium can be explained by the burning of heavy fuel oil which contains high levels of the element.

Table 13.5. Enrichment of metals in atmospheric particles (United States urban air) relative to the earth's crust

Element	Enrichment Factor	Element	Enrichment Factor
Aluminum	0.5	Arsenic	310
Cobalt	2	Silver	830
Chromium	11	Mercury	1100
Nickel	12	Cadmium	1900
Vanadium	42	Lead	2300
Copper	83	Selenium	2500
Zinc	270	Antimony	2800

Types of Metal Associations in Sediments

9☞ According to Forstner and Wittmann (1983, cited earlier) metals in sediments can originate from one of five natural sources:

Lithogenous Formations are weathered minerals from bedrock or rock debris from the river bed. The material is very resistant to change. The silicate minerals, feldspar and quartz, have very low metal contents. Metal bonding occurs predominantly in inert positions on clay particles. Erosion of refuse dumps is included in this class of minerals.

Lithologic components within a watershed are subject to weathering processes that consume H ions. Since CO_2 in equilibrium with H_2O results in carbonic acid (H_2CO_3) which has a pH near 5.6, the process is often referred to as ***carbonic acid weathering***. The acidification process is an acceleration of the weathering process in that H ion concentrations are greater, reaction rates are faster and surface waters receive elevated inputs. In reactions involving aluminosilicate minerals (e.g. $Al_2Si_{12}O_8$), H^+ react by either adsorption or dissolution mechanisms. ***Adsorption reactions*** with clay minerals, such as aluminosilicates, are a two-fold process. First, at pH less than 7 but greater than or equal to 4.5, surface exchange reactions occur whereby surface-adsorbed cations (e.g. Ca^{2+}, Mg^{2+}, Na^+, K^+) are exchanged for H^+. Secondly, at pH < 4.5 when clay surfaces have been depleted of exchangeable cations, inward diffusion by H^+ into the aluminosilicate mineral and outward diffusion of such elements as Al^{3+}, Cu^{2+}, Pb^{2+}, Zn^{2+}, U^{6+}, Cd^{2+} and As^{2+} occurs. ***Dissolution reactions*** involving aluminosilicates also occur. At pH < 4.5, H^+ ions are consumed by various dissolution reactions to yield major cations (as above), in addition to Al^{3+} and S^{2+} ions. Also, $Al(OH)_3$, which may form on surfaces of particles, may undergo dissolution to yield Al^{3+}. Therefore, by adsorption exchange and dissolution reactions, the export of these cations from a watershed dominated by aluminosilicate minerals may be increased.

Hydrogenous Formations are physico-chemical changes in the water that produce particles, precipitation products and/or substances with adsorbed minerals. Hydrogenous metal enrichment may occur anywhere between the outfall and the receiving body, including the sea.

Biogenous Formations include all biological remains, organic and inorganic, like siliceous and calcareous shells. Metal enrichment by this method also may occur anywhere between the site of death of the organism and the receiving waters, including the sea.

Atmogenous Formations are metal enrichments resulting from atmospheric fallout, in dissolved or solid form.

Cosmogenous Formations are extra-terrestrial particles produced upon impact of a meteor on earth.

Despite dissimilar origins, metals have characteristic types of bonding that can be categorized as: (i) adsorptive bonding; (ii) co-precipitation by hydrous iron and manganese oxides; (iii) complexation by organic molecules; (iv) incorporation in crystalline minerals. Many metals precipitate out of solution through physical sorption processes, chemical sorption processes (e.g. exchange of H^+ in fixed positions) or co-precipitation that results when the solubility product is exceeded in an area of the water course. These precipitation mechanisms are pH-dependent for hydroxides and oxides of Fe and Mn, ***bitumen*** (substances extractable with organic solvents such as benzene, ether and chloroform), lipids, humic materials, residual organics and calcium carbonate. The hydrous oxides of Al, Fe and Mn, especially the redox sensitive species of Fe-and Mn-hydroxides and -oxides under oxidizing conditions, constitute significant *sinks* of heavy metals in aquatic ecosystems, as discussed in Chapter 5. ***Humic materials*** include ***humin*** (high molecular mass, largely insoluble in aqueous solutions), ***humic acids*** (middle molecular size range, very complex but soluble under certain conditions), ***fulvic acids*** (less condensed humic substances, occur in a dissolved state) and ***yellow organic acids*** (low molecular mass, found in swamp waters, represent last stage of humic matter degradation). The bonding strength of heavy metals to humic or fulvic acids follows the sequence: $Hg^{2+} > Cu^{2+} > Pb^{2+} > Zn^{2+} > Ni^{2+} > Co^{2+} > Mn^{2+}$.

For heavy metals of hydroxides, carbonates and sulphides, precipitation occurs mainly as a result of exceeding the solubility product in the area of the water course. Metal hydroxides in amorphous form, or as a very fine crystalline precipitate, are derived from strongly over-saturated solutions. Complete dissolution is achieved only at pH levels < 4.0. Heavy metal sulphides are practically insoluble at neutral pH. Fe, Mn and Cd sulphides are very soluble in HCl, but Ni and Co sulphides dissolve with more difficulty. Cu, Pb and Hg are only soluble in oxidizing acids, especially HNO_3. The solubility of carbonates in aqueous solution, as discussed in Chapter 5, is highly dependent on the partial pressure of carbon dioxide.

Cation Exchange and Adsorption

Many sediment-forming materials with a large surface area, especially clays, iron hydroxides, amorphous silicic acids and organic substances, are capable of sorbing cations from solution and releasing equivalent amounts of other cations into the solution
10☞ by ***cation exchange***. The sum of exchangeable cations, including H^+, constitutes the ***exchange capacity*** and is expressed in milli-equivalents (meq)/100 g of material.

The amount of cation exchange is based entirely on the sorptive properties of negatively charged anionic sites in: (i) $SiOH^-$, $AlOH_2^-$ and AlOH-groups in clay minerals; (ii) FeOH-groups in iron hydroxides; (iii) carboxyl and phenolic OH-groups in organic substances towards positively charged cations. There is a preferential adsorption of specific cations and release of equivalent charges associated with other species. The accumulation of heavy metal ions at the solid-liquid interface as a result of intermolecular forces is termed ***adsorption***. With clay minerals, the exchange capacity increases greatly in the order kaolinite (3-15 meq/100 g) < chlorite (20-50meq/100 g) ≤ illite (10-40 meq/100 g) < montmorillonite (8-120 meq/100 g). The increase corresponds to the reduction in particle size and the related increase of surface area of all particles.

The affinity of cations towards exchangers is governed by: (i) ***valence and hydration effects***, the affinity increasing with increasing oxidation number (i.e. $Me^{+} < Me^{2+} < Me^{3+}$ ....) and with a decrease in the diameter of the hydrated cations (hence, alkali and alkaline earth metals have higher charge densities (e.g. Ba < Sr < Ca < Mg < Cs < Rb < K < Na < Li)); (ii) ***concentration of solution***, the number of exchanged cations increasing with increasing concentration of solution; (iii) ***cation exchange on organic and inorganic substances***, organic substances possessing a high degree of selectivity for divalent ions relative to monovalent ions. The affinity of heavy metals is greater than that of alkaline earth and alkali ions, such that:

$$Pb > Cu > Ni > C > Zn > Mn > Ba > Ca > Mg > NH_4 > K > Na;$$

(iv) ***Reactions involving hydrolized cations***, whereby cation exchange capacity increases with increasing hydrolysis of the exchanging cations, hence $CuOH^{+}$ and $FeOH^{2+}$ are sorbed in preference to Cu^{2+} and Fe^{3+}; (v) ***specific reactions between inorganic exchangers and cations***, such that it is impossible to establish an order of affinities that is generally applicable to the individual heavy metals based solely on ionic charge and radius. For example, the different properties of the exchange sites in the lattice, the influence of electrostatic field strength and the cation sorption process influence the affinity of K^{+} and NH_4^{+}.

Heavy Metal Enrichment in Aquatic Biota

Autotrophic Microorganisms:

Phytoplankton: Although numerous studies have shown little or no correlation between total phytoplankton biomass and metal levels in the ambient water, levels of
11☞ metals in some planktonic algae species do reflect the average metal content in water. However, the kind and level of metal bioaccumulated differ among species. Lead has been shown to be preferentially enriched in *Cladophora glomerata*, an attached green algae in rivers. The algal species has been used as a bio-indicator of Cd, Cu, Hg, Ni and Zn in water.

Macrophytes:

As with algae, metal levels can differ widely among marophytic species. Hence, use of macrophytes as bio-indicators should be restricted to one species within a river
12☞ system. The levels of metals in roots, leaves and stems is highly variable for different metals but, in general, highest metal levels are found in the leaves, the lowest in the roots. Species which have been used as bio-indicators of metal levels in water are *Potamogeton pectinatus*, *Potamogeton crispus*, *Callitriche palustris* and *Vallisneria americana*. All species studied showed above average, heavy metal enrichments, whereby the point source of heavy metals could be readily identified, especially in lotic

systems.

Heterotrophic Organisms:

Enrichment of metals in plants occurs through the soluble phase, but in heterotrophs, enrichment can occur by respiration (via gills or body surface), adsorption onto the body surface and/or from ingestion of solid and dissolved materials. The uptake of metals from food is a major source, but the amounts accumulated vary according to feeding habits. The functional feeding groups recognized are: grazers or scrapers (also sometimes known as herbivores) of algae (e.g. gastropods, some crustaceans and insects); collectors which are subdivided into filter feeders (e.g. rotifers, cladocerans, bivalves, blackflies) and gatherers, or sediment and detritus feeders (e.g. isopods and amphipods, many insects); predators or carnivores (e.g. predaceous zooplankton, many insects, crayfish, fish). Usually, there is no food chain enrichment in polluted waters with heavy metals as seen for organic pesticides. For example, one would expect biomagnification of metals to be ranked in the order: water < grazers < filter feeders < gatherers < predators. Instead, one observes enrichment of Cd and Pb in the food chain to have the following ranking: water < fish < sediments < aquatic invertebrates (with filter feeders (e.g. Sphaeriidae) < gatherers (e.g. Chironomidae, Tubificidae) < predators (e.g. crayfish, dragon fly nymphs) < grazers (e.g snails and Psychodidae) < sediments (Fig. 13.6). If fish and other carnivores in the food chain have elevated levels of metals, they must receive the major body burden from the water. This suggests that metal speciation may dictate not only acute toxicity but the availability of the metals as well.

Zooplankton:

The zooplankton are composed of grazing forms (e.g. *Daphnia*, *Bosmina*, *Ceriodaphnia*) and carnivorous forms (e.g. *Mysis*, *Polyphemus*, *Leptodora*). Although
13☞ most of the zooplankton studies that examined metal uptake have been in marine environments, there appears to be no clear effect of feeding type on metal levels in zooplankton species. The lack of any relationship between feeding types and metal levels may be due to the wide seasonal and spatial variations in zooplankton specie's composition and biomass. Near-shore plankton tend to have higher levels than off-shore plankton. In spite of these variations, several studies indicate that the levels of metals in many species of plankton correlate well with the levels in the water. Yan *et al* (1990)[4] showed that Cd levels in the crustacean zooplankton were positively correlated with aqueous Cd levels and negatively correlated with water levels of Ca, organic carbon and total phosphorous in 33 non-acidified south-central Ontario lakes. Factors other than Ca can affect the relationship between Cd levels in zooplankton and Cd levels in water. For

---

[4]Yan, N. D., G. L. Mackie and P. J. Dillon. 1990. Cadmium concentrations in crustacean zooplankton of acidified and nonacidified Canadian shield lakes. Environm. Sci. Technol. 24: 1367-1372.

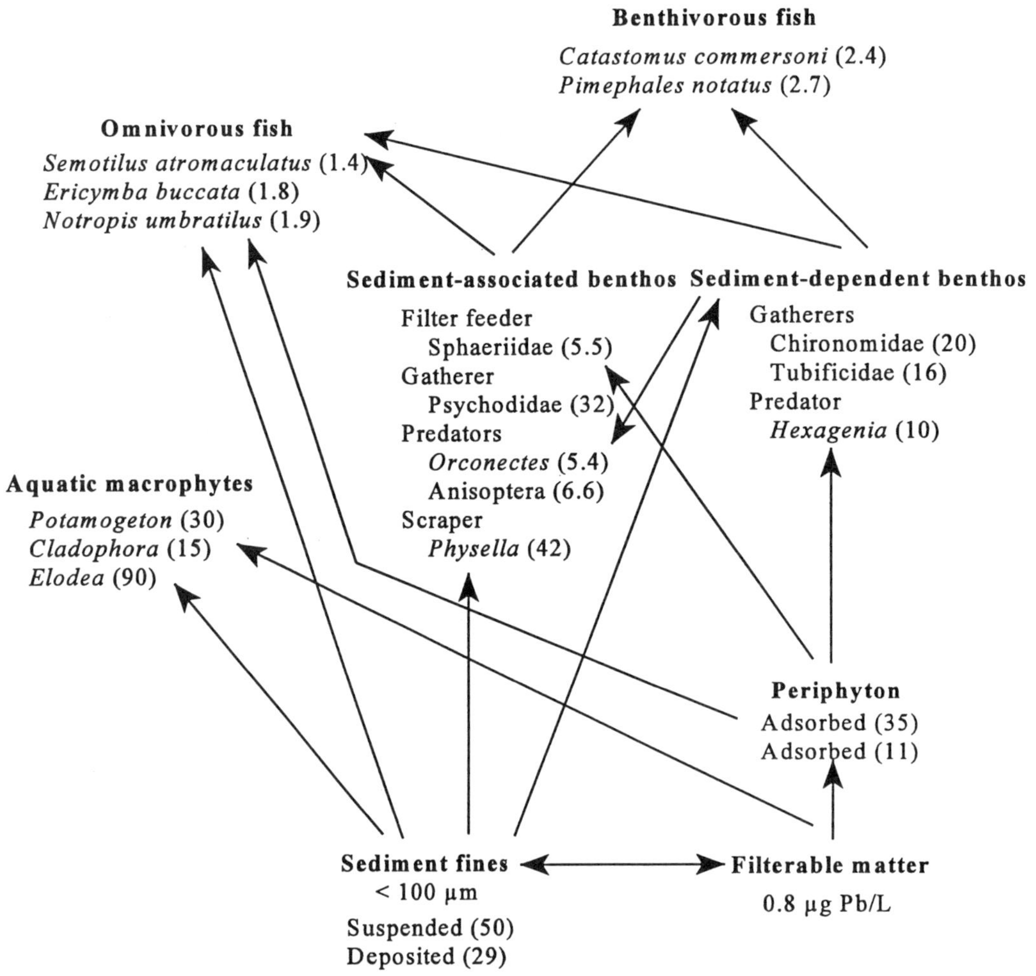

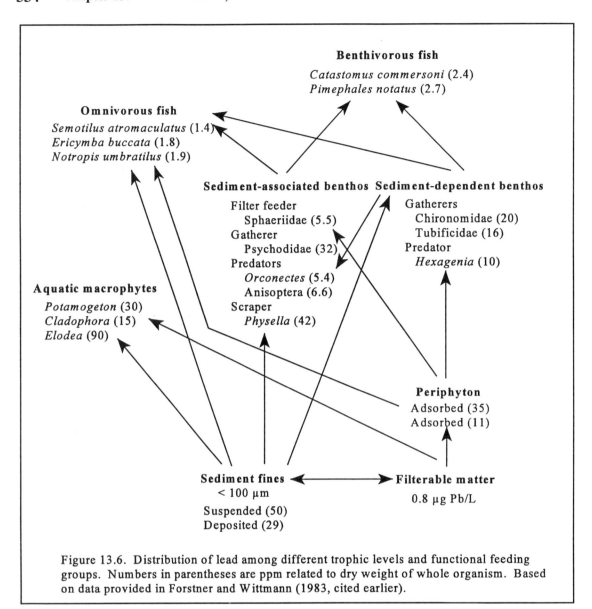

Figure 13.6. Distribution of lead among different trophic levels and functional feeding groups. Numbers in parentheses are ppm related to dry weight of whole organism. Based on data provided in Forstner and Wittmann (1983, cited earlier).

example, studies by Yan and Mackie (1990)[5] show a positive correlation between Cd levels in the cladoceran, *Holopedium gibberum*, and aqueous Cd levels, but a negative correlation with clutch size. Seasonal succession of species seems to be the principal cause of temporal variability in metal levels (Zn, Cd, Fe, Mn, Mg, Al, Ca, Cu, Ti, Sr, Ba, Ni) in the zooplankton of Canadian shield lakes in central Ontario, but variability among lakes is attributable to three factors, lake location (relative to smelters), acidity and

---

[5]Yan, N. D. and G. L. Mackie. 1990. Control of cadmium levels in *Holopedium gibberum* (Crustacea, Cladocera) in Canadian shield lakes. Environm. Toxicol. Chem. 9: 895-908.

community composition (Yan *et al* 1989)[6]. There is also some evidence of biomagnification of metals in zooplankton from lower trophic levels.

Benthic invertebrates

Marine bivalves have been studied extensively, partly because of their importance in human diets, but also because they are filter feeders and represent a link between the
14☞ sediments and water. Although there are differences in metal levels among species, many (e.g. zinc levels in the scallop, *Chlamys opercularis*, cadmium and zinc in the blue mussel, *Mytilus edulis*) are excellent indicators of metal levels in the water, even on a seasonal basis. Generally, body burdens of metals in bivalves are lowest during the spawning period and highest during periods of active growth and metabolism. The contents of metals in the organs of bivalves can vary by several orders of magnitude. For example, ^{58}Co varies from about 10^5 cpm (counts per minute)/g wet weight in the byssus to ~50,000 cpm/g in the liver, 10^4 cpm/g in the shell, 5,000 cpm/g in the gills, to about 10^3 cpm/g wet weight in adductor muscle and mantle tissue of *Mytilus edulis*. Metal levels in these tissues remain relatively constant after about 10 days of uptake from the water.

Studies of freshwater invertebrates as indicators of heavy metals are few. Studies on the isopod, *Caecidotea* (= *Asellus*), indicate that non-essential metals, like lead and cadmium, are enriched to much greater extent than are the essential metals, like zinc and copper, with maximum enrichment factors for the metals being 6.3 for Cd, 6.1 for Pb, 2.3 for Cu and 1.7 for Zn (Forstner and Wittmann 1983, cited earlier). Cd concentrations in the amphipod, *Hyalella azteca*, are correlated with total aqueous Cd, water hardness and dissolved organic carbon content of the water (Stephenson and Mackie 1988)[7]. They suggested that Cd ions compete with Ca ions for uptake sites at the gill surfaces of the species. They also found that increased dissolved organic carbon levels lead to lower Cd levels in *H. azteca*, suggesting that high levels of DOC may complex free Cd ions and reduce their concentration in solution. These data, and others, suggest that some freshwater benthic invertebrates are suitable indicators of heavy metal pollution. Also, the acute toxicity levels of Cd, Pb and Al appear to be more a function of the species than

---

[6] Yan, N. D., G. L. Mackie and D. Boomer. 1989. Chemical and biological correlates of metal levels in crustacean zooplankton from Canadian shield lakes: a multivariate analysis. Sci. Total Environm. 87/88 419-438.

[7] Stephenson, M. and G. L. Mackie. 1988. Multivariate analysis of correlations between environmental parameters and cadmium concentrations in *Hyalella azteca* (Crustacea: Amphipoda) from central Ontario lakes. Can. J. Fish. Aquat. Sci. 45: 1705-1710.

of their functional feeding behaviour (Mackie 1989)[8].

Fish

Much of the data on metal uptake by fish is for mercury which bioaccumulates in muscle tissue. There is an age-dependent Hg enrichment in several species, including
15☞ largemeouth bass (*Micropterous salmoides*), pike (*Esox lucius*) and lake whitefish (*Coregonus clupeaformis*). The levels of Hg in these species are generally dependent on the basic level present in the environment. Other metals, like Cd, Pb, Ni, Cd, As, Cu, Zn and Mn, do not show strong correlations between aqueous levels and fish muscle-tissue levels. There is some evidence of food chain enrichment but, in general, fish should not be used as a monitoring system for heavy metal pollution.

PESTICIDES

Herbicide use exceeds the usage of insecticides, fungicides and specialty products combined, in most years. Based on 1995 and 1996 data (Fig.13.7), herbicide use was

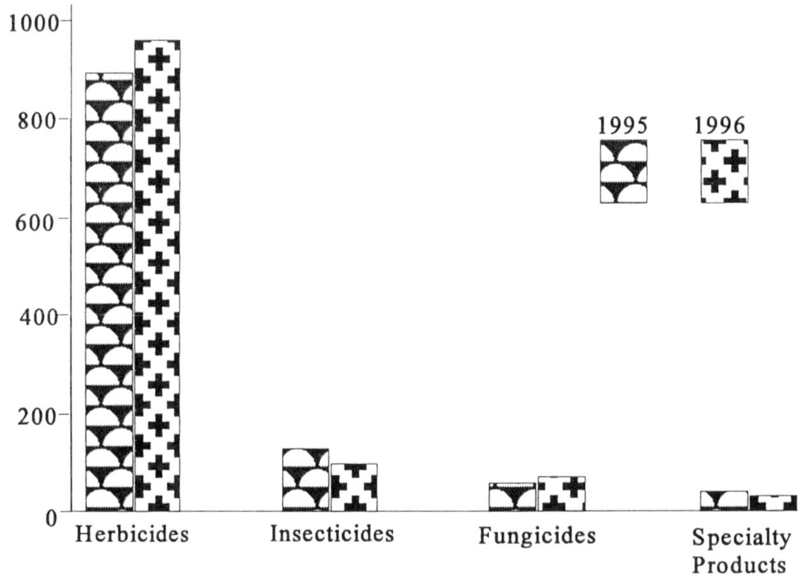

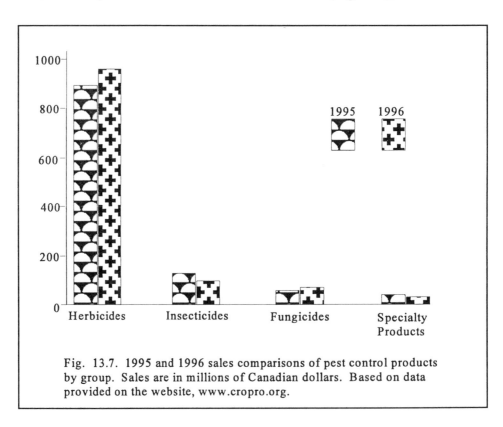

Fig. 13.7. 1995 and 1996 sales comparisons of pest control products by group. Sales are in millions of Canadian dollars. Based on data provided on the website, www.cropro.org.

---

[8]Mackie, G. L. 1989. Tolerances of five benthic invertebartes to hydrogen ions and metals (Cd, Pb, Al). Arch. Environm. Toxicol. 18: 215-223.

about an order of magnitude greater than any of the three pest control groups. Pesticides are used mostly to control noxious weeds and insects in the agricultural and forestry industries.

This section describes the most common and lethal organic compounds that are found today and in the most recent past. Among these are persistent organic pollutants, such as DDT, PCBs (polychlorinated biphenyls) and PAHs (polycyclic aromatic
16 hydrocarbons). DDT is an insecticide that exists in the form of different isomers (molecules with the same combination of atoms, but with some of the atoms, e.g. Cl, in different positions) (Fig.13.8). PCBs are produced by substituting chlorine atoms for some or all hydrogen atoms (Fig. 13.8). PAHs are produced by incomplete combustion of organic material, such as creosote (Fig. 13.8).

Pesticides in Forestry

Canada's forest resources cover 3.42 x 10^6 km^2, or about one-third of the total area of the country (Gustafson, 1988)[9]. Herbicides are applied in forestry during
17 replanting of forest stocks. They are used for site preparation to rid the site of any unwanted vegetation, to remove overtopping hardwoods and shrubs that compete with the growing conifer trees and for general control of weeds. Some common ***herbicides*** utilized are glyphosate , hexazinone, 2,4-D and simazine (Table 13.6).

Table 13.6. Relative use of different herbicides for forest management in Canada.

Herbicide	Ha treated	Percent
Glyphosate	176,536	81
2.4-D	29,094	13
Hexazinone	3,972	2
Simazine	8,224	4

The major insect pests of forests in Canada are the Western and Eastern spruce budworms, the jack pine budworm, the mountain pine beetle, the eastern hemlock looper, the forest tent caterpillar and the Gypsy moth. The Eastern tent caterpillar is a sporadic problem. The major ***insecticides*** for the control of insect pests are ***fenitrothion***, ***aminocarb*** and ***Bacillus thuringiensis***. The latter, also know as ***Bt***, is the only insecticide currently approved for aerial application to crown forests in Ontario and Quebec.

---

[9] Gustafson, D.I. 1988. Groundwater Ubiquity Score: A simple method for assessing pesticide leachability. Environmental Toxicology and Chemistry, 8: 339-357

In addition to these insecticides, a new nuclear *Polyhedrosis* virus, specific to the gypsy moth, has recently been successfully tested in Ontario and will shortly be marketed under the trade name "***Disparvirus***". ***DDT***, ***phosphamidon***, ***fenitrothion***, and ***matacil*** have all been heavily used in the past for insect pest control in Canada.

Volumes of pesticide use in forestry are very low compared to pesticide use in Canadian agriculture. Only 1.8% of the amount of pesticides used in agriculture was used in forestry in 1980 Campbell (1980)[10].

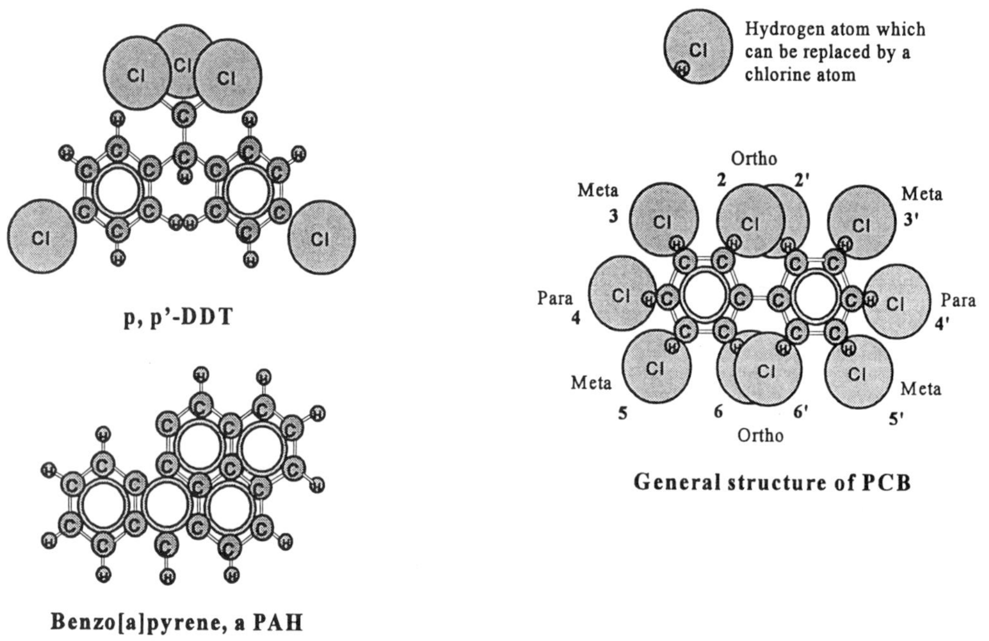

Figure 13.8. Molecular structure of some common persistent organic pollutants. DDT is an insecticide that exists in the form of different isomers (molecules with the same combination of atoms, but with some of the atoms, e.g. Cl, in different positions), this isomer being p,p' - DDT. Polychlorinated biphenyls (PCB)are produced by substituting chlorine atoms for some or all hydrogen atoms. The numbers are used for designating the positions of the chlorine atoms in the names of the individual congeners. For example, if there are Cl atoms at positions 3 and 3', 4 and 4' and 5 and 5' (total 6 Cl atoms) of a biphenyl molecule, the congener is named 3,3',4,4', 5,5' - hexachlorobiphenyl. Benzo[a]pyrene is an example of a polycyclic aromatic hydrocarbon (PAH) that is produced by incomplete combustion of organic material such as creosote.

---

[10]R. A. Campbell. 1980. Herbicide use for forest management in Canada: Where we are and where we are going. Forestry Chronicle, 66: 355-360.

Other Pesticides

TFM: TFM is a ***piscicide*** used to control sea lamprey in the Great Lakes. It acts on the respiratory system producing severe hemorrhaging in the capillaries. Since the sea lamprey breeds in the bottom mud of streams, TFM is formulated with sand to make it
18☞ sink rapidly. This serves to concentrate the material close to the target organism and reduce exposure to non-target species. TFM is often used in conjunction with ***clonitralid*** (see below) which acts as a ***synergist*** (increases activity of pesticide) and reduces costs. TFM is relatively water soluble and tends not to bioaccumulate in the environment. However, problems may be caused by incorrect application rates. It has a small margin of safety and clonitralid is non-selectively toxic to fish.

Clonitralid (Niclosamide): Clonitralid is used to control aquatic snails. It has a comparatively low toxicity to mammals. In is used in with TFM as a synergist to control sea lamprey in the Great Lakes. It is not bioaccumulated in the environment, however, it is toxic to fish and has a low margin of safety.

s-Triazine Herbicides: Atrazine, Hexazinone and Simazine - Atrazine is one of the most widely used herbicides to control pests on corn in Canada. It can be applied three different times in one season with little risk of crop injury. Along with persistence, movement of atrazine in the environment has been a major concern. Atrazine is the most common herbicide found in water. The greatest losses of the herbicide are to groundwater and surface waters in clayey watersheds. In surface waters, atrazine concentrations peak in early summer. In early spring, during large volumes of runoff, total loads in the watersheds are similar to summer levels. In small streams and tributaries, the peak concentrations are observed in June. However, in larger rivers, peaks occur in June-July. In large watersheds, contamination by atrazine ranges from 0.5 to 2% of the amount that was applied. Atrazine has been shown to affect algae, other aquatic plants and fish.

Hexazinone's mode of action is to block the electron flow in photosynthesis. It is used in broadcast applications (e.g. ***aerial spraying***) to control competing vegetation in forestry. Hexazinone is very water soluble, highly mobile and non-volatile (will not evaporate). Therefore, it has a very high potential for ***off-site movement*** to an aquatic system. Hexazinone has been documented to adversely affect primary producers and other organisms in the food chain, including zooplankton and aquatic invertebrates.

Simazine is the least water soluble of the triazine herbicides. It has many applications, including use to control insects on a wide variety of crops and in the control of aquatic weeds.

Rotenone: Rotenone reduces both oxygen uptake and phosphorylation (ATP formation) in mitochondria. It is a fish poison but also has high toxicity to aphids, lepidopterous larvae and mites. Insects and fish metabolize rotenone into other toxic metabolites. It is non-persistent in the environment, with a half-life of <1 day. Rotenone is also used by fisheries biologists to monitor sizes of fish populations. It can be detoxified with potassium permanganate, which is pumped into the stream below the sampling station where it oxidizes any remaining rotenone to non-toxic products.

Aldicarb: Aldicarb is an inhibitor of AChE (acetyl cholinesterase) which acts as a relay for signals within the nervous system and from the sensory receptors to the nervous system. This causes over stimulation of the nervous system. Aldicarb is highly toxic to mammals, with an LD_{50} of approximately 1 mg/kg. Aldicarb does not ***biomagnify***. However, once it is in groundwater and soil it is very persistent. For this reason it was withdrawn as a widely used pesticide for controlling potato pests in 1990.

Carbaryl: Carbaryl is used as a wide spectrum foliar spray in agriculture, in the home and garden. It also controls tree pests, such as the gypsy moth and tent caterpillar. It is an inhibitor of AChE and has a low toxicity to fish (250 μg/L in *Salmo salar,* 1,100 μg/L in *Onchorynchus mykiss* and 13,200 μg/L in *Carassius auratus).* However, it is relatively toxic to aquatic invertebrates, with LC_{50} values in the 1-20 μg/L range. Carbaryl is not persistent or mobile in the environment. It is relatively insoluble in water (40 mg/L), has a high K_{oc} and therefore, has a low mobility in soil. The half-life in high-pH water is very short but it is more stable in water of pH < 6.5. The half-life in water is between 1 and 5 days.

Organophosphorous Insecticides

This group of insecticides is the most diverse and widely used, with at least 100 different compounds in current use or being registered. They are also AChE inhibitors, however, organophosphorous insecticides block AChE for much longer periods than carbaryl pesticides (12 hours compared to 30-60 minutes). Examples of aquatically significant organophosphorous insecticides are ***diazinon*** and fenitrothion.

DDT & related compounds

DDT: Dichlorodiphenyl trichloroethane (DDT) was the first so-called modern insecticide. The site of action of DDT and its analogues (methoxychlor, dicofol) is the nerve axon where it affects the sodium or potassium gate or ATPase. This causes repetitive firing of the nerves in the affected organism producing tremors, convulsions and possible death. The LD_{50} of DDT on aerial insects is 2.8 mg/kg for oral and dermal application, while it is lethal to some species of fish at 0.001 mg/L! DDT is extremely ***persistent*** and has a high lipid solubility allowing it to bioaccumulate. ***Methoxychlor*** is also highly toxic to fish. The water solubility of DDT is 0.0012 mg/L and of methoxychlor, 0.69 mg/l. Due to DDT's potential for extreme environmental impacts it was banned in Canada in the 1970s and in USA, Europe, Japan and many other countries.

Chlorinated Cyclodienes

Aldrin, Dieldrin and Endrin: Dieldrin has been used as a soil insecticide against termites and ants as well as in baits. Endrin was used as a broad spectrum foliar insecticide and to control mice in orchards. The mode of action is to de-inhibit the nervous system by acting on post-synaptic receptors. The insecticides are highly persistent. Dieldrin is highly lipophilic and readily biomagnifies in the food chain. Chlorinated cyclodiens are banned or otherwise restricted in Canada and other countries. Endrin has never been used in Canada.

Endosulfan: Endosulfan is highly toxic to fish and used as a piscicide. It has also been widely used in agriculture against lepidopterous larvae, beetles and aphids. Its mode of action is in the nerve axon. Endosulfan has not caused many persistence problems and it does not bioaccumulate to a great extent. If used incorrectly it can cause poisonings in non-target fish.

Lindane: Lindane is a chlorinated aliphatic insecticide. Its mode of action is at synaptic receptors where it acts to de-inhibit the central nervous system. Its current use in Canada is mainly as a seed treatment in combination with fungicides. It is very toxic to fish and has been found in sediments of rivers and lakes in Ontario, suggesting a movement of the insecticide. The source of the residues may be atmospheric fallout because lindane and its isomers have a relatively high volatility. In anaerobic ecosystems, such as flooded soils and lake sediments, there is rapid degradation by anaerobic bacteria.

Pesticides Affecting Growth and Development and Cell Membranes

Bacillus Thuringiensis (Bt): The bacterium produces a sporangium containing a polypeptide endotoxin which is lethal to a number of insects when ingested. Bt is the only insecticide currently allowed for use on forests in Ontario and Quebec. It is not persistent and is almost non-toxic to non-insects. Bt is rapidly broken down in the environment and is among the most selective and environmentally friendly insecticides.

Diquat: Diquat is a 2-heterocyclic cationic compound that is referred to as ***bipyridylium***. It contains positively charged nitrogen atoms with four bonds in its structure. Diquat acts as a non-selective contact herbicide with essentially no activity in soil. It is used primarily as a seed crop desiccant and as an ***aquatic herbicide***. In water, it is absorbed by or adsorbed to aquatic plants. Long-term persistence in bottom sediments occurs where it is deactivated by the soil and organic material.

Plant hormone mimics

2,4-D and ***Silvex:*** 2,4-D is a phenoxy alkanoic acid and is one of the most important herbicides for the control of annual and perennial broadleaf weeds in cereals and for weed control in lawns, pastures, and aquatic macrophytes. The compounds are auxin mimics and affect RNA and protein synthesis in plants. In water they are degraded by microbes and photo-decomposition. They do not bioaccumulate in the environment, however, ***non-target toxicity*** has resulted from spray drift to aquatic plants.

Mitotic Inhibitors

MBC: MBC acts by binding to tubulin proteins and interfering with spindle fibres during cell division. It is not considered to be persistent or biomagnified in the environment. Very little leaching or run-off has been observed despite its half-life of three to six months. MBC is used to control Dutch Elm Disease.

Some Pesticides Affecting Metabolic Reactions

Urea herbicides (diuron, monuron): The mode of action is the inhibition of part of the synthetic pathway of the branched-chain amino acids. They are relatively non-toxic to aquatic animals but they may affect algae.

Glyphosate : Glyphosate is an important non-selective organophosphorous herbicide used in agriculture and forestry. Its mode of action is the inhibition of the enzyme 5-endolpyruvyl shikimate-3-P synthetase, an essential enzyme on the pathway to the synthesis of the aromatic amino acids. Glyophosate is toxic to most plants and environmental effects have been as non-target toxicity through accidental over spray.

Pesticides and their Effects on Aquatic Ecosystems

Most pesticides are not added intentionally to aquatic environments. Contamination is a result of the individual properties of a pesticide and usually occurs through a variety of mechanisms. The pesticides that have the greatest impact on aquatic environments have properties that make them significantly damaging (toxicity, persistence) or their particular usage causes a large amount of the pesticide to contaminate aquatic systems. For example, aerially-applied pesticides used in forestry may pose a significant threat to aquatic ecosystems because they are transported through the air and deposited on nearby lakes and rivers.

Movement of Pesticides to Aquatic Systems

Pesticides move from their target site to aquatic ecosystems through the land, air and water. The properties that affect the probability of this movement are a function of
19☞ the pesticide's ***toxicity, mobility, volume of use***, and ***persistence***. These factors together constitute the ***environmental risk of a pesticide***.

Toxicity and volume of use are fairly straight forward. Essentially, if the pesticide is toxic to aquatic organisms, but degrades very quickly and therefore does not reach the groundwater or enter a lake or river from runoff, then its impacts to the aquatic environment are minimal. However, if the pesticide is highly persistent and mobile, the chance of a greater impact on the aquatic environment is increased.

A large factor affecting movement of the pesticide is its fate in the soil at the target site. The pesticide can remain within the soil at the target site or it can degrade or move, its movement being influenced by a number of factors, listed in Table 13.7.

Table 13.7. Factors affecting the movement of a pesticide through the soil and its type of degradation.

Movement	Degradation
Adsorption to soil (like pins to a magnet; different from	Microbiological activity
Volume of use	Chemical stability
Uptake by plants and animals	Biodegradation
Leachability	Photodegradation
Erodibility of soil	

The risk of movement can be very significant when a pesticide moves through or over the soil because it can potentially pass through the watershed and into a lake or river. Also, if the pesticide is moving off its target site, its level of efficacy is decreased, therefore, it must be applied more frequently or in a higher volume of use. This results in even larger quantities of pesticides reaching the aquatic ecosystem.

The risk of movement of a pesticide can be rated by its K_{oc}, a measure of the average mobility of a chemical. It indicates the strength of attachment of the pesticide to the soil and therefore whether or not the pesticide will move with surface water runoff. Table 13.8 shows the risk level for different K_{oc} values.

Table 13.8. Risk levels for pesticides based on K_{oc}, water solubility and half-life.

Property	High Risk	Low Risk
K_{oc} (adsorbtion coefficient of pesticide on organic carbon in the soil, cc/g)	<500	>1000
Water Solubility (ppm)	>30	<1
Half life ($t_{1/2}$ days)	>21	<5

The ***groundwater ubiquity Score (GUS)*** calculates the potential of leaching of the pesticide into groundwater. The technique uses two relatively easily measured parameters of K_{oc} and the half-life of the pesticide in the soil.

$$GUS = \log_{10}(t_{1/2soil}) \times (4 - \log_{10}(K_{oc}))$$

A GUS value > 1.8 suggests a low potential for a pesticide to contaminate groundwater; a value of 1.8 - 2.8 indicates a moderate potential to contaminate; values >2.8 suggest a high potential for contamination.

All properties discussed above are dependent on the soil and vegetation type of the area. The potential impacts of a pesticide are also influenced by the presence of a riparian buffer zone (see Chapter 2) around the water body of concern.

Pesticides also reach aquatic ecosystems by transport through air or ***atmospheric fallout,*** discussed below. The rate and amount of movement of pesticides in the air is dependent on the vapour pressure of the pesticide.

Effects of Pesticides in Water

Once the pesticide reaches the aquatic environment, its effects are determined by the chemical, physical and biological properties of the water. Variables such as hardness and salinity can alter the effect of individual pesticides. For example, salinity decreases the toxicity of TFM. pH can also affect the stability of the pesticides. For example, pH influences the hydrolysis of organophosphorous and carbamate insecticides; the compounds are stable at low pH's (5-7) but are rapidly hydrolyzed at high pH's (7-10). In contrast, triazine herbicides are the most stable at pH 7.

An increase in temperature generally results in increases in rates of chemical reactions, solubility, adsorption, volatilization and biological degradation. In general, pesticides persist longer at lower temperatures. The sediment can also act as a sink for low water-soluble and persistent pesticides, such as DDT. Aquatic life will also aid in the metabolic breakdown of pesticides with biological degradation by microbes, plants and animals.

Bioconcentration and Biomagnification of Pesticides

Bioaccumulation, or bioconcentration, of a pesticide is defined as the movement of a chemical from the surrounding medium into an organism. It is a process that, like
20☞ other environmental impacts of a pesticide, is determined by a few major properties of the substance. The most important is the ***lipid solubility***, or lack of water solubility. It is expressed numerically as K_{ow}, or the octanol/water coefficient, which represents the ratio of the solubility of the pesticide in water to a lipid-type material, ***octanol.***

$$K_{OW} = [C]_{Octanol}/[C]_{Water}$$

K_{ow} is a good predictor of bioaccumulation, K_{ow} values < 500 indicating no environmental accumulation of the pesticide and K_{ow} values > 1000 indicating a high degree of bioaccumulation.

Other important factors are the affinity of the substance to sediments, the volatility of the substance and its rate of hydrolysis or biodegradation in various compartments of the ecosystem. In addition to the properties of the pesticide that cause it to bioaccumulate, the presence of a sink site in an organism must also exist. The most important sink site is the lipids in which organisms store energy. Lipids are non-polar and, in contrast to water and compounds which are lipid soluble, will tend to accumulate there.

21☞ Since biomagnification is dependent upon bioaccumulation at each trophic level, the two processes are related. ***Ecological magnification*** (***EM***) and a ***Bioactivity Index*** (***BI***) are useful measures of biomagnification. The EM is a combination of the uptake of the compound by the organism itself and the intake of the compound from the lower trophic levels, as the following equation shows:

$$\text{EM} = \frac{\text{Concentration of compound in organism}}{\text{Concentration of compound in water}}$$

The BI indicates how well an organism can metabolize a compound. It can be applied to any metabolite of the pesticide as well. It is important to include metabolites because they be more toxic that the parent pesticide (e.g. an extremely toxic metabolite of DDT is DDE). A BI value >0.1 indicates the pesticide is accumulating in the organism.

$$\text{BI} = \frac{\text{Concentration of polar metabolite in organism}}{\text{Concentration of non-polar metabolite in organism}}$$

Two major factors which affect biomagnification are the K_{ow} and the extent of biodegradation of the pesticide in the environment. If the pesticide is extremely lipophilic, yet is degraded very quickly, there is no bioaccumulation or subsequent biomagnification in the aquatic ecosystems. Other factors that affect biomagnification are:

- The lipid content of the organisms. There is a good correlation between fat content and the degree of accumulation of non-polar pesticides.
- Concentration of the pesticide in the environment. High pesticide levels increase the uptake.
- Dilution of pesticide into other organisms. The amount of available pesticide decreases because of the large number organisms bioaccumulating the pesticide- i.e. ***biological dilution***.
- Duration of exposure. Once a pesticide is removed from an area, the body burden of the pesticide decreases.
- Movement of the pesticide to a new environment. This reduces the exposure of the ecosystem to the pesticide and therefore, reduces the risk of bioaccumulation.

The above discussion provides a brief idea of different properties of pesticides and why they can have an affect on aquatic environments. Different families of pesticides have a wide diversity of chemical, physical and biological properties. They are just as diverse as the aquatic environments. It is important to assess each pesticide and its impact individually while taking into account the unique nature of the aquatic environment being studied.

Pesticide Inputs

Unintentional applications

Atmospheric Fallout: The atmosphere is a major source of pesticides in water,
22☞ particularly in the ocean because of the huge surface area. Small particles of pesticides can be transported very long distances and are deposited in water through fallout and precipitation. The area of the Great Lakes is about 250,000 km^2 and they receive a large amount of pesticide residues suspended or dissolved in rainfall. For example, in the 1970s, DDT residues and its metabolites (DDE) averaged about 5 ug/L in rainfall, which translates into approximately 6,000 kg/y of DDT falling onto the Great Lakes, not taking into account other local inputs! Particle drift from the air is a more important source of ***local contamination*** of fresh waters. The severity of the impact depends on the application techniques used to apply the pesticides, with ***aerial spraying*** having the greatest impact.

Soil Erosion: Pesticides bound to soil particles are transported into water. Erosion affects mostly the upper layers of the soil, which are also the most exposed to pesticides. Approximately 4 x 10^{12} kg of soil is lost to erosion each year in the USA.. Of this, 2 x 10^9 kg is lost into Lakes Michigan and Erie, and 37% of this originates from agricultural land! Representative of this problem is the presence of DDT and its metabolites in the sediments of Lake Erie to a depth of 12 cm.

Industrial: Sources of pesticides entering aquatic ecosystems include effluent from factories, sewage, illegal disposal of industrial wastes and accidental spills. One example is the extreme poisonings of fish in the river Rhine when a barge containing endosulfan sank, contaminating the river at toxic levels.

Intentional Applications of Pesticides

In addition to the unintentional additions, there are also pesticides used specifically against aquatic organisms, even to control some of the exotic species that we have discussed. For example, products called Clamtrol, Bulab 6002, and other polyquaternary ammonia and benzothiazole compounds are being used to control zebra mussels. Organochlorine and organophosphorous insecticides are used to control mosquitoes, blackflies and biting midges. Herbicides, such as diquat, 2,4 -D esters and atrazine, are used to kill macrophytes, and dichlorbenil and $CuSO_4$ are used to control algae. Rotenone, organochlorine and organophophorous insecticides and TFM (Nitrophenol piscicide) are used as piscicides. Molluscicides, such as Bayluscide, Frescon and niclosamide are used to control terrestrial gastropods of quarantine significance and marine biofouling bivalves.

Background pesticide residue levels in waters are normally in the order of pg/L while LC50 values for most aquatic organisms are in the ug/L range. However, there are direct and indirect effects of pesticides below the lethal levels.

Direct Effects of Pesticides on Aquatic Ecosystems

Fungi and Bacteria:

Little is known about the effects of fungicides on non-target fungi. It is assumed
23☞ that there are little to no direct effects. However, observed increases in numbers of
bacteria and fungi have been attributed to pesticides serving as food sources or providing
increases in available organic matter (from the vegetation killed by the pesticide).

Algae:

Pesticides whose mode of action is photosynthetic inhibition are the greatest
threat to algae. In contrast, applications of specific insecticides and herbicides have
24☞ resulted in algae blooms (which are not wanted either!). Examples are DDT, methyl
parathion, dichlobenil, and diquat. EC_{50} values for some algal pesticides are:

Urea herbicides (diuron, monuron)	10 - 100 ug/L
Altrazine	20 - 40 ug/L
MBC (benzimidazole fungicide)	340 ug/L
TFM	1,900 - 15,000 ug/L
2,4-D and 2,4,5-T	50,000 - 100,000 ug
Diquat	200,000 ug/L

***Aquatic invertebrates*:**

25☞ Pesticides have a diverse range of effects on invertebrates, ranging from death to
temporary displacement depending on the mode of action and the life stage of the species
during exposure.

Amphibians and reptiles:

Little is known of the impact of pesticides on these organisms. Studies done with
26☞ DDT have shown some sublethal effects of 0.5 ppm DDT on tadpoles. Residues of
more persistent pesticides have been found in amphibians and snakes.

***Fish*:**

In areas where forest spraying of insecticides has occurred, extensive mortality of fish has been observed. Generally smaller and younger fish are more susceptible to pesticides and different species have different rates of ***detoxification***. Pesticides can also have many ***sublethal effects*,** including behavioural changes, such as abnormal schooling behaviour (e.g. DDT and diazinon) and aggressiveness, changes in respiration, a reduction in cholinesterase activity, reduction in stamina, abnormal pathology in blood, brain, spinal cord, liver, heart (e.g. methoxychlor at 0.01 - 0.04 ppm resulted in abnormal pathology) and effects on reproduction. Table 13.9 shows the toxicities of four compounds to three species of fish.

Table 13.9. 96-h LC_{50} values for four insecticides on three species of fish.

	Toxicity (LC_{50}, ug/L)			
	DDT	**Lindane**	**Matacil**	**Carbaryl**
Rainbow Trout	7	27	170	4300
Coho Salmon	4	41	100	800
Channel Catfish	16	44	9000	15800

Bioaccumulation of pesticides in fish is an important factor for some toxicants, especially mercury and DDT. Fish that have very high levels of contaminants stored in their body fat are then consumed by people, birds and other animals higher up in the food chain leading to various health effects.

Indirect Effects of Pesticides on Aquatic Ecosystems

Indirect effects of pesticides are the most likely impact to expect in an aquatic environment. They include changes in the physical and chemical environment of the
27☞ lake or river. An example of this is creation of a ***low oxygen level*** due to the decomposition of organic materials provided by plants killed by pesticides and subsequent mortality or absence of oxygen sensitive species (e.g. 2,4-D against milfoil, silvex against alligator weed and ponds treated with diuron). Another example is the herbicide, dichlobenil, used against macrophytes, which causes a short-term decrease in oxygen level, although phytoplankton blooms are able to restore oxygen levels within 1 month.

Changes in ***competition*** for food is also observed. Herbicides increase the organic matter content in aquatic systems resulting in a rise in microbial populations. Herbicides also remove macrophytes which have a number of wide spread implications on aquatic ecosystems. Forest herbicides have been observed to kill phytoplankton through their photosysthetic mode of action which then affects the entire food chain, resulting in a ***bottom up effect*** (Chapter 7). This can have a profound effect on the oxygen content of the water and on the food supply of zooplankton, filter feeders, and grazers. Insecticides may also have an impact on food webs by influencing ***predator/prey relationships*** and competition, resulting in an increase in abundance of resistant species.

One example of direct and indirect effects of a pesticide is the spraying of forests with DDT in the 1950s to protect them against the attack of spruce budworm. The effects of spraying was seen dramatically in Canada's Atlantic salmon yields. By 1954 direct effects were evident, with a 100% kill of under-yearlings (0-1 year old) and 50 - 70% mortality in parr (greater than 1 yr old) a few days after spraying. Indirect effects also occurred, with a depletion of the salmon's food supply (aquatic invertebrates).

ACID PRECIPITATION

On a pH scale, one normally thinks of acidity as pH < 7.0. However,
28☞ precipitation (snow or rain) in equilibrium with carbon dioxide (i.e. carbonic acid) has a

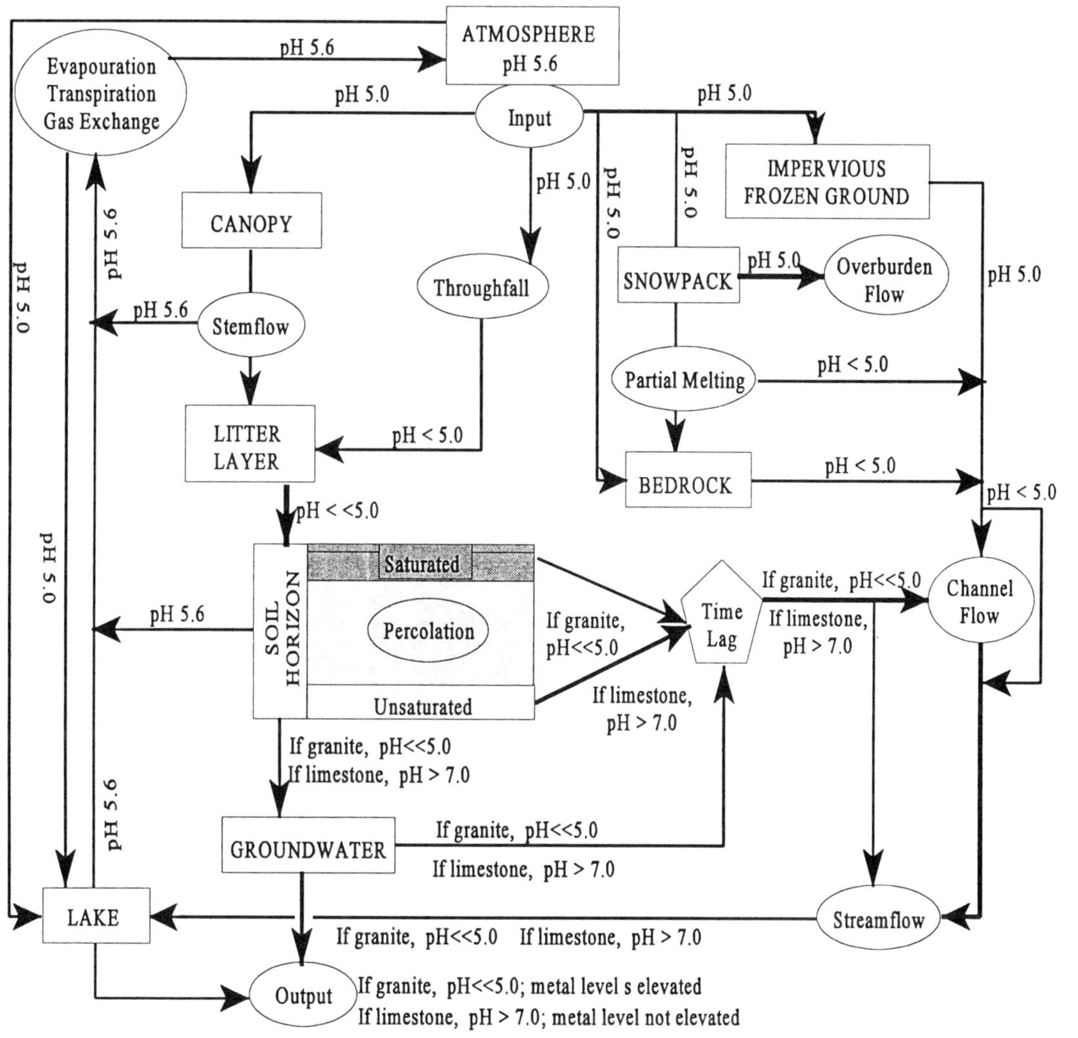

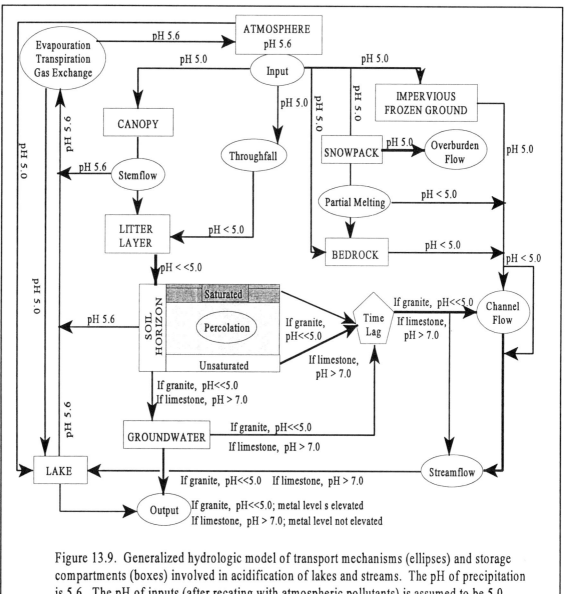

Figure 13.9. Generalized hydrologic model of transport mechanisms (ellipses) and storage compartments (boxes) involved in acidification of lakes and streams. The pH of precipitation is 5.6. The pH of inputs (after recating with atmospheric pollutants) is assumed to be 5.0.

sulphur and nitrogen compounds, the reactions that occur are:

$$H_2O + CO_2 \rightleftharpoons H_2CO_3 \text{ (carbonic acid)}$$

Such a condition exists only in areas that have little or no industrial development. In highly populated areas, and areas with industries that burn fossil fuels containing sulphur, elemental sulphur is oxidized to sulphur dioxide (SO_2). The sulphur dioxide is oxidized by ozone (O_3) in the atmosphere to form sulphite (SO_3) which dissolves readily in water droplets to form sulphuric acid. The reactions are:

$$SO_2 + O_3 \rightleftharpoons SO_3 + O_2$$
$$SO_3 + H_2O \rightleftharpoons H_2SO_4$$

Major producers of sulphur dioxide are power plants (public and industrial utilities) that combust coal containing sulphur; copper, lead and zinc smelters; steel plants; and urban areas where coal and fuel oil containing sulphur is used for heating. Automobile exhausts are a minor source of sulphurous compounds. Natural sources include volcanic eruptions, fresh and salt waters and biological reactions involving sulphur bacteria. In central Ontario, Canada, the mean pH and alkalinity of lakes are a function of the total S deposition rate, the total S loading being in excess of 1.0 $g/m^2/yr$.

Oxides of nitrogen are also created by the burning of nitrogen-rich fossil fuels at high temperatures. Sources (Fig. 13.9) include power plants, automobile exhaust, smelting operations and space heating. Elemental nitrogen (N) is oxidized readily in the atmosphere to nitrogen dioxide (NO_2), which dissolves in rain droplets to form nitric acid (NO_3), according to the following reactions:

$$N + O_2 \rightleftharpoons NO_2$$
$$3NO_2 + H_2O \rightleftharpoons 2HNO_3 + NO$$

However, the major source of nitrogen is not anthropogenic; it is from natural biological reactions involved in nitrification, denitrification and nitrogen fixation of the nitrogen cycle. See Chapter 5 for a review of the nitrogen cycle.

Clearly, inputs of sulphuric and nitric acid can depress the pH of precipitation well below 5.6. Usually a pH < 5.5 is taken to be indicative of artificially-induced acidification. The pH levels of precipitation can be depressed to as low as 3.5 in heavily industrialized areas, but winds can blow atmospheric pollutants well beyond their sources. Hence, even poorly populated unindustrialized areas downwind of industries, may experience similar levels of acid precipitation. The direction of prevailing winds can be shown by drawing a "***windrose***" diagram (Fig. 13.10). The windrose shown in Fig. 13.10 is for the Sudbury area and indicates a strong

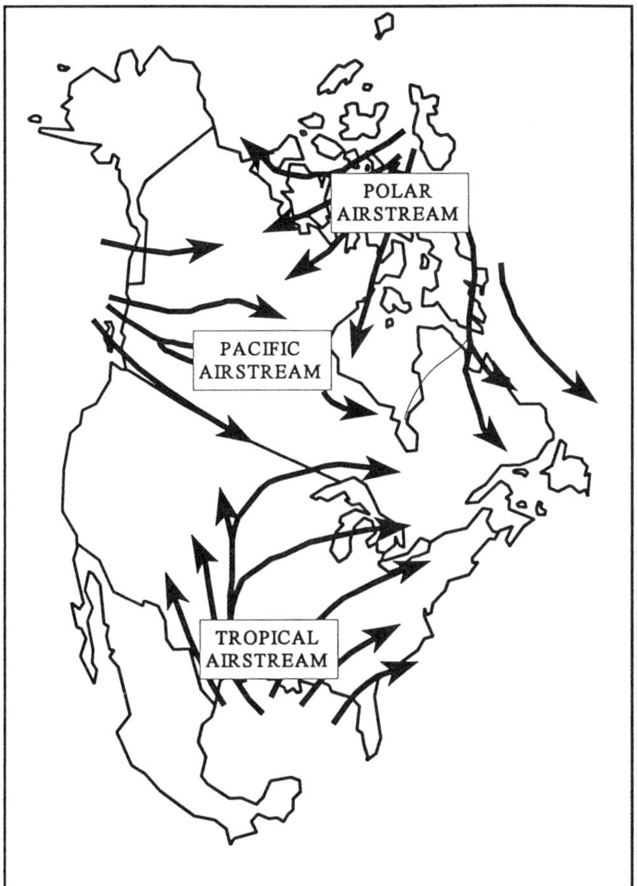

Figure 13.11. Resultant surface wind flow across North America during the summer.

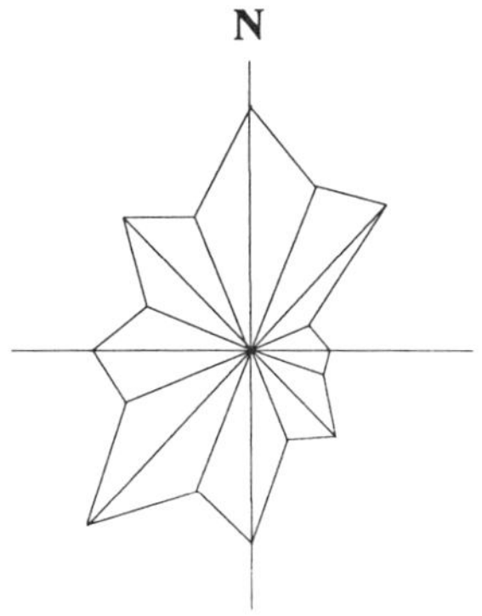

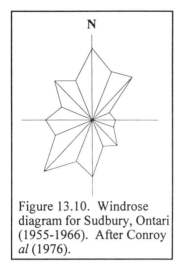

Figure 13.10. Windrose diagram for Sudbury, Ontari (1955-1966). After Conroy *al* (1976).

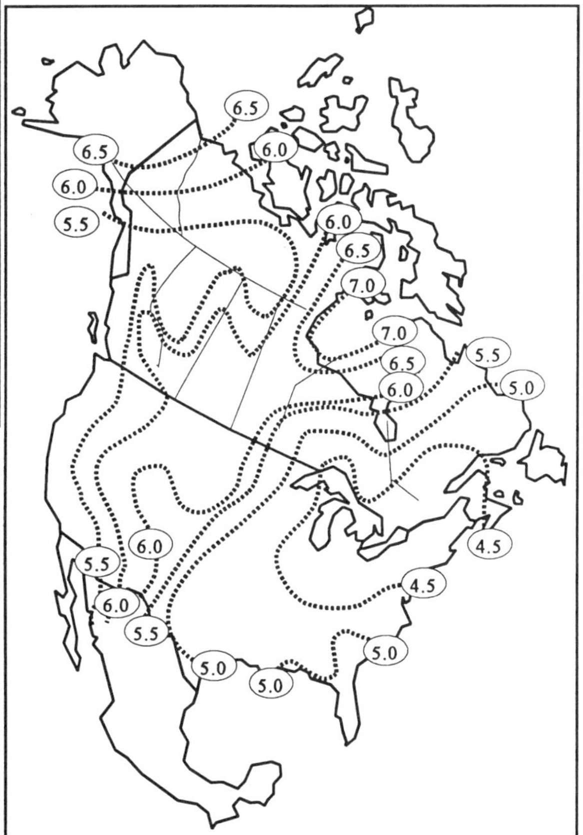

Figure 13.12. Mean pH isopleths for North America during 1976-1979 period.

tendency toward northeast-southwest winds during the period 1955-1966 and (Conroy *et al.* 1976)[11]. The prevailing winds (Fig. 13.11) explain the variations in pH that occur throughout North America (Fig. 13.12). Note that the lowest pH isopleths circumscribe highly industrial areas.

---

[11]Conroy, N, K. Hawley, W. Keller and C. Lafrance. 1976. Influences of the atmosphere on lakes in the Sudbury area. J. Intern. Assoc. Great Lakes Res. 2: 146-165.

The pH Scale

So far we have discussed only the changes in pH that occur in the atmosphere. Once the precipitation lands on trees and/or soil, profound changes in pH begin to occur (Fig. 13.9). The direction of these changes (i.e. a decrease or increase in pH) depends on the chemistry of the leaf litter and the soil. Before describing these changes, a review of the pH scale is in order.

pH is a logarithmic measure of the hydrogen ion concentration on a scale
29☞ ranging from 0 to 14 (Fig. 13.13). A solution that has equal amounts of hydrogen ions (H^+) and hydroxyl ions (OH) has a pH of 7, which is midway on the scale. The greater the hydrogen ion content, or acidity, the lower the pH. A change of one pH unit downward implies a tenfold increase in acidity (or a tenfold increase in hydrogen ion content); a change of two is a hundredfold; and a change of three is a thousandfold. Or, if the pH is 5, it is ten times more acidic than a pH of 6; a pH of 3 is a hundred times more acidic than a pH of 5.

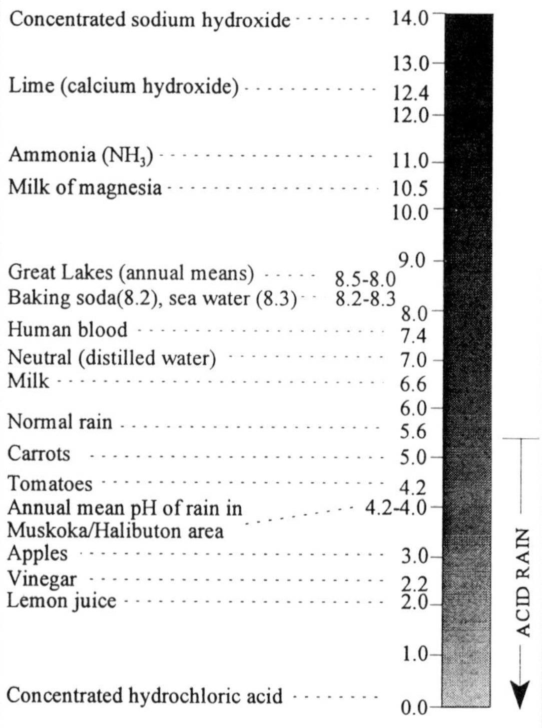

Figure 13.13. The pH acid scale, showing pH levels of common foods and substances.

Geochemical Influences

We learned in Chapter 5 the role of alkalinity in buffering additions of hydrogen
30☞ ions. Much of this buffering, however, may occur before the water(rain) reaches the streams and lakes. Soils and bedrocks containing calcium carbonate "consume" hydrogen ions by forming bicarbonates:

$$CaCO_3 + H^+ \rightleftharpoons Ca(HCO_3)_2$$

Once bicarbonates are formed, they too consume hydrogen ions:

$$Ca(HCO_3)_2 + 2H^+ \rightleftharpoons 2Ca^{++} + 2H_2CO_3$$

While carbonic acid is produced in the equilibrium process, the hydrogen ions dissociate much less strongly than most other acids:

$$H_2CO_3 \rightleftharpoons H^+ + HCO_3$$

These same reactions occur in lakes and streams that have carbonate and/or bicarbonate alkalinity. The end result is that the pH of precipitation (5.6 or less) is neutralized and the pH of the receiving waters (usually >8.0) remains basically unchanged.

Fig. 13.14 shows the geological areas of Ontario that have little to no buffering capacity and are sensitive to acid rain. An estimated 181,450 lakes occur in Ontario, and about 48,468 (27%) are considered sensitive to acid deposition. Lakes with a total alkalinity less than 0.3 meq/L (= 15 mg $CaCO_3$/L), pH < 6.0 and conductivity < 35 μS/cm (at 25 °C) are considered sensitive[12].

In areas where the soil and bedrock (e.g. granite, quartz) lack carbonates, the minerals present can slowly neutralize acid rain, but more **commonly**, the pH is depressed

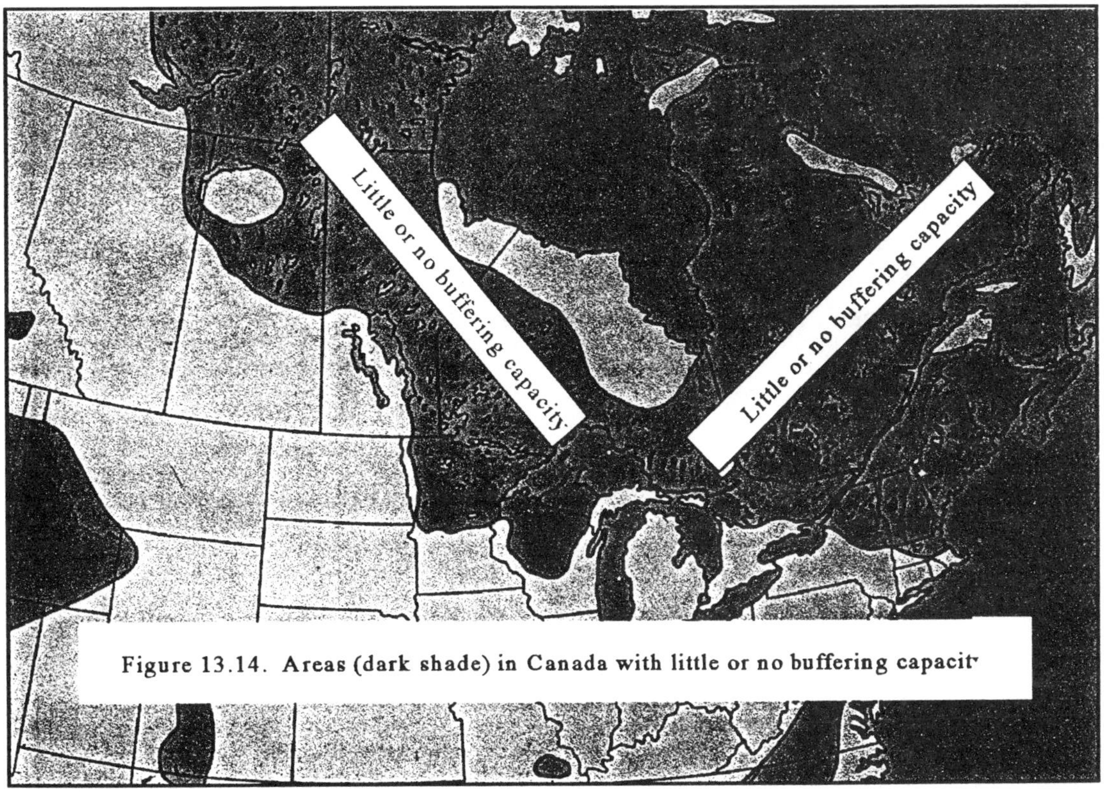

Figure 13.14. Areas (dark shade) in Canada with little or no buffering capacit

[12] Data are from a report, "Determination of the susceptibility to acidification of poorly buffered surface waters", Limnology and Toxicity Section, Ontario Ministry of the Environment, Rexdale, Ontario, March, 1979.

even lower than 5.6. Leaf litter is composed mostly of organic carbon which, when decomposed by bacteria, is converted to carbon dioxide. This process, and the respiration of bacteria, increase CO_2 levels well above that present in the atmosphere. As the concentration of CO_2 increases, the concentration of carbonic acid also increases, depressing the pH well below 5.6. The pH of concentrated carbonic acid is about 3.8. Without carbonates in the soil or bedrock, the pH remains depressed as water infiltrates and percolates through the soils and groundwater and enters streams and lakes in the watershed (Fig. 13.15).

Surface runoff during spring snow-melt cause severe pH depressions in streams lacking any buffering ability. During melt events through the winter, hydrogen ions
31 migrate to the base of the snow pack and become highly concentrated there. In the spring, when the ground is still frozen and the melt water, with its elevated levels of hydrogen ions, enters streams causing a "***spring pH depression***" in those lacking any buffering capacity (Fig. 13.15). The duration of the pH depression period varies from one to three weeks, but this varies from stream to stream and one year to the next. Because of the short duration of most spring pH depression periods, and the severe pH drop at the end of the period, the event causes severe chemical "shock" on all aquatic organisms. Hard water streams have sufficient amounts of alkalinity to buffer the additions of hydrogen ions, but soft water streams do not. Once pH levels approach 4.0, many metals, especially aluminum, become soluble and toxic and accompany the lethal levels of hydrogen ions into the receiving waters. In general, hydrogen ions become toxic to fish when pH levels fall below 5.5.

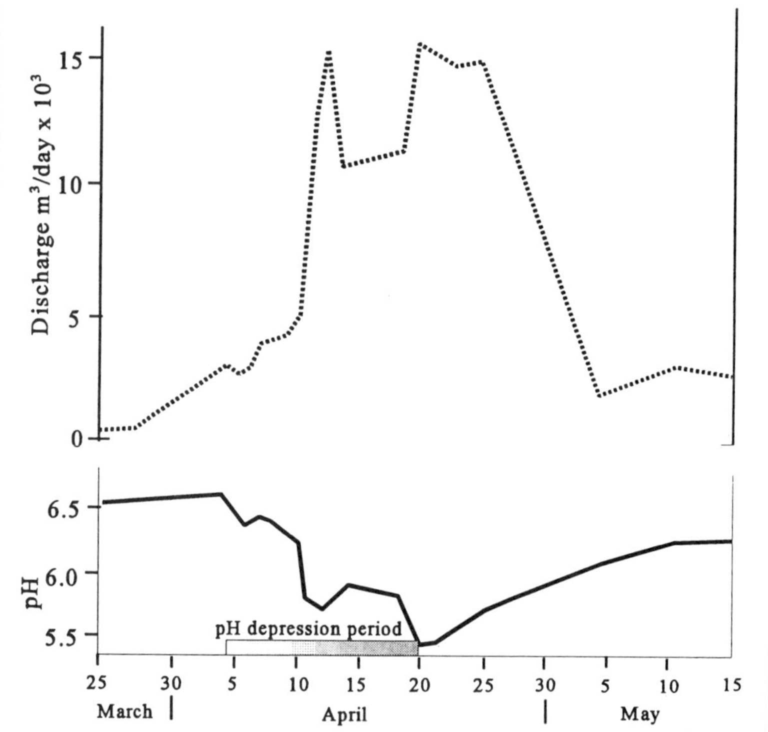

Figure 13.15. "Spring pH depression" of a stream.

Biological Effects of Lake/Stream Acidification

Decomposition Processes

In circumneutral lakes, allochthonous litter is decomposed rapidly by bacteria. But in acidified lakes there is a rapid accumulation of organic material, indicating
32 depressed metabolic activities of bacterial populations. Below pH 5.0, oxidation of ammonia compounds ceases, nitrification is reduced and bacterial cell counts drop. Only fungi are present below pH 4.0. Neutralization of acidified lakes with calcium hydroxide or lime, is followed closely by dramatic increases in microbial activity in sediments.

Algae

In general, the diversity and productivity of phytoplankton decrease with decreasing pH, due mostly to lower nutrient supplies. Exceptions are *acidophiles*
33 (species that prefer acidic waters), such as the diatom *Tabellaria flocculosa* and the blue-green algae *Mougeotia sp.*, which greatly increase in biomass down to about pH 4.0. Some dinoflaggellates (e.g. *Peridinium*) and chrysophytes (e.g. *Dinobryon*) also increase in numbers. However, once the pH falls below 4.0, the productivity of even acidophilic algae decrease. Associated with the decrease in algal diversity is an increase in water clarity, a process in acidic lakes often referred to as "***oligotrophication***" (i.e. lower nutrient supplies causes reduced algal productivity which increases water transparency).

Within the benthic assemblage, acidic lakes have three types of algal communities: (i) ***blue-green mats***; (ii) ***green mats***; and (iii) ***epiphytic or periphytic green clouds***[13] (Stokes, 1980). These are described in Chapter 12 (see bullet No. 8). Many are associated with macrophytes that are acid tolerant.

The significant increases in sizes of algal mats or clouds with decreasing pH is a common phenomenon in many Ontario lakes, especially in the Sudbury, Parry Sound, Muskoka and Haliburton areas, Quebec lakes and New York lakes.

Macrophytes

The macrophytes most commonly supporting the algal mats and clouds discussed above are *Lobelia dortmanni, Isoetes echinospora, Utricularia* spp., and some species of
34 *Potamogeton*. *Lobelia dortmanni* is the most tolerant macrophytic species but even it disappears below pH 4.0. The moss, *Sphagnum*, a plant characteristic of bogs, can become dominant in acidifying lakes. There are at least three explanations for the moss to dominate. (i) At the pH of acidic lakes, all the available inorganic carbon is in the form of CO_2 or H_2CO_3 Unlike most macrophytes, *Sphagnum* is not able to use HCO_3, and when HCO_3 is converted into CO_2 and H_2CO_3, all macrophytes, except *Sphagnum*,

---

[13] Stokes, P. M. 1980. Benthic algal communities in acidic lakes. *In:* Singer, R. (Ed.). Effects of acid precipitation on benthos. Proceedings of a Sypmosium, Colgate, University, Hamilton, N.Y., Aug. 8-9, 1980, pp. 119-138.

disappear. (ii) *Sphagnum* has an ion exchange capacity that results in the withdrawal of base cations, such as Ca, from solution, thus reducing their availability to other plants. (iii) The mats of *Sphagnum* become so dense that they are not a suitable substrate for other plants to grow on.

The roots of macrophytes pass oxygen to the sediments and substrates, with good macrophytic growth exhibiting good oxygen supplies. But in acid lakes, oxidation processes in sediments (e.g. nitrification) are greatly reduced, partly because of reduced macrophytic growth which generates the oxygen supply.

Zooplankton

In Canadian Shield lakes, the cladocerans *Holopedium gibberum*, *Daphnia galeata mendotae*, *Daphnia dubia* and *Daphnia pulex* collectively contribute an average
35☞ of 53% of the biomass of herbivorous zooplankton (Yan *et al* 1988)[14]. In these lakes, a positive relationship exists between pH and biomass of all four *Daphnia* species. However, Yan *et al* (1988) found that the *Daphnia*-pH pattern could not be ascribed to hydrogen ion toxicity. Some of the *Daphnia* mortality was hypothesized to be due to greater predation by zooplanktivores such as *Chaoborus,* which dominated many of the more acidic lakes. They also found that *Holopedium*, which is much more tolerant of acidification than *Daphnia* spp., had a negative relationship between its biomass and that of *D. galeata mendotae* or *D. pulex.*

Benthic Invertebrates

Much of the research on physiological responses of aquatic animals to low pH have been with invertebrates and fish. All freshwater organisms adjust the ion content of
36☞ their blood by osmoregulation. Sodium is one of the most abundant cations in freshwater organisms and its regulation, in particular, is adversely affected at low pH. Water moves by osmosis into cells of freshwater animals because the cellular salt content is much greater than the salt content in freshwater. Since the kidney is less than 100% efficient at resorbing salts, some salt is released into the urine and then to the water. Consequently, many freshwater animals, including fish, most crustaceans and some insect larvae, must actively remove salt from the water to maintain their high internal concentrations. Sodium is actively exchanged for either H^+ or NH_4^+, while Cl^- is exchanged for HCO_3^- (Havas, 1980)[15].

---

[14]Yan, N. D., W. Keller, J. R. Pitblado and G. L. Mackie. *Daphnia-Holopedium* relationships in Canadian Shield lakes ranging in acidity. Verh. Internat. Verein. Limnol. 23: 252-257.

[15]Havas, M. 1980. Physiological response of aquatic animals to low pH. In: Singer, R. (Ed.). Effects of acid precipitation on benthos. Proceedings of a Symposim, Colgate University,

Havas (1980) examined the effect of pH on survival and salt regulation in four species of crustaceans and three species of insect larvae. All species were collected from an alkaline pond (pH 8.2), except the blood worm *Chironomus riparius*, which was collected from an acidic pond (pH 2.8). The crustaceans were much more sensitive to low pH than the insect larvae, with mortality of the former occurring rapidly below pH 4.5. A significant loss (~ 50%) of Na and Cl occurred below pH 4.0 in *Daphnia middendorffiana*. The most sensitive insect was the chironomid *Orthocladius consobrinus*, which maintained constant Na levels from pH 8.2 to 4.5, and the mortality rate increased below pH 3.5, paralleling the decrease in internal Na levels. The most tolerant species was *C. riparius*, which maintained high internal sodium levels over the entire pH range tested (8.2 to 2.8). All species of *Chironomus* have anal papillae, which are the major site of sodium uptake and become enlarged in acid waters. Similar results were obtained for the waterbug, *Corixa punctata*, which was collected from an acidic pond (pH 4.0).

The effects of lake acidification on littoral benthic macroinvertebrate assemblages were nicely demonstrated by Stephenson *et al* (1993)[16]. They found that the benthic macroinvertebrate assemblage structure of central Ontario lakes could be predicted from a knowledge of the lake area and sensitivity of the lake to acidification. The benthic macroinvertebrate assemblages of small or calcium-poor and acidic lakes and streams experienced losses of gastropods (e.g. *Amnicola, Gyraulus*), oligochaetes (e.g. *Stylaria*), mayflies (e.g. *Stenonema, Ephemera*), some Trichoptera (e.g. *Mystacides, Helicopysche, Lepidostoma, Ceraclea*), some Coleoptera (e.g. *Stenelmis, Dubiraphia, Hygrotus*) and some Amphipoda, especially *Hyalella*. Simultaneously, a suite of taxa including some leeches (e.g. *Helobdella stagnalis*), some Odonata (e.g. *Boyeria, Enallagma, Libellula*), some Trichoptera (e.g. *Oecetis, Limnephilus*) and some chironomids (e.g. *Tanytarsus, Dicrotendipes, Microtendipes, Tribelos, Chironomus, Micropsectra,* and *Psectrocladius)* exhibited increases in numbers in acidic lakes and streams.

The changes in littoral benthic macroinvertebrate assemblages described above are typical for most fertile habitats. However, Arnold *et al* (1980)[17] provide evidence that low pH in infertile streams does not significantly affect: (i) biomass, production, or species diversity of periphyton, but may affect species composition; nor (ii) decomposition rates of leaves. However, the paucity of periphyton was accompanied by

---

Hamilton, N.Y., Aug. 8-9, 1980, pp. 49-66.

[16]Stephenson, M, G. Mierle, R., R. A. Reid and G. L. Mackie. Effects of experimental and cultural lake acidification on littoral benthic macroinvertebrate assemblages. Can. J. Fish. Aquat. Sci. 51: 1147-1161.

[17]Arnold, D. E., P. M. Bender, A. B. Hale and R. W. Light. Studies on infertile, acidic Pennsylvania streams and their benthic communities. *In:* Singer, R. (Ed.). Effects of acid precipitation on benthos. Proceedings of a Sypmosium, Colgate, University, Hamilton, N.Y., Aug. 8-9, 1980, pp. 15-34.

a decrease in diversity and numbers of insect grazers. Arnold *et al* (1980) attributed low alkalinities, rather than pH per se, as the main determinant of biological effects observed.

Studies on burrowing invertebrates, such as oligochaetes, some chironomids (e.g. *Chironomus*) and some mayflies (e.g. *Hexagenia, Ephemera*), have shown that the burrowing activities affect nutrient cycling by altering the transformations and fluxes of sulphur and nitrogen compounds. The dynamics of nitrogen and sulphur cycling is directly affected by the acidification process. Excretion of ammonia by burrowing organisms, together with the burrowing activities that help to mix lower with upper sediment layers, have been shown to cause: (i) an increase in redox potential (Eh); (ii) a corresponding decrease in pH; (iii) increases in oxidized forms of sulphur (e.g. sulphate) compounds in the sediments.

Finally, pH depressions such as those that occur during spring runoff in poorly buffered streams, elicit an immediate increase in the drift rate of benthic macroinvertebrates. Pratt and Hall (1980)[18] experimentally acidified a stream to pH 4.0 and found the increased drift leaving the acidified reach was more diverse in terms of taxa (orders), trophic functional groups and behavioural groups, and less diverse at the generic level than the drift entering the stream section. Megaloptera and some Plecoptera and Trichoptera were among the few orders that did not increase in drift density as the pH was lowered. In fact, it has been shown that the amount of emergence increases with decreasing pH for Megaloptera and Plecoptera. The functional groups that increased in drift density as the pH was lowered were gatherers, filter feeders, scrapers (grazers) and some predators. Peculiar behaviours are mainfest in several ways: (i) Aggressive behaviour is exhibited by the caddisfly, *Polycentrops flavomaculatus*, where fighting and killing occurs at pH 4.3 - 4.0 but not at pH 4.8 - 6.2; (ii) Avoidance behaviour occurs with the amphipod *Gammarus pulex*, which avoids waters with pH below 6.0, and the isopod *Caecidotea* (*Asellus*) *aquaticus*, which leaves the water, explaining why this species can tolerate high acidities; (iii) The mayfly *Baetis* does not lay eggs in water whose pH is less than 6.0.

Acidification to pH 5.7 or less also can cause decreased emergence (except in Megaloptera and Plecoptera) and growth of the insect drift. In general, sensitivity to low pH decreases with increasing size and age, eggs being more sensitive and showing greater mortality than adults at a given pH. Many invertebrates can tolerate short periods of acidification, but acclimation to low pH increases an organism's tolerance.

As a group, the Mollusca are the most sensitive to acidification because of their requirement for calcium carbonate in shell formation. Most snails die at ~pH 6.2, but *Amnicola limosa* and some fingernail clams (e.g. *Pisidium casertanum*) can survive (but do not grow well) down to pH 5.5. Unionidae clams are very sensitive and begin dying below pH 7.0; those that do survive appear to sacrifice some calcium in the shell to

---

[18]Pratt, J. M. and R. J. hall. 1980. Acute effects of stream acidification on the diversity of macroinvertebrate drift. *In:* Singer, R. (Ed.). Effects of acid precipitation on benthos. Proceedings of a Sypmosium, Colgate, University, Hamilton, N.Y., Aug. 8-9, 1980. Pp. 77-96.

maintain their calcium demands. However, eventually the calcium demands cannot be met and the clams succumb.

Of the arthropods, the Crustacea are most sensitive. The disappearance of the ubiquitous amphipod, *Hyalella azteca*, generally indicates that the lake has acidified because it is absent or depauperate below pH 5.6. Arthropods are particularly sensitive to high H^+ ion concentration during moulting events, when the permeability to water and ions is greatly increased. The crayfish, *Orconectes virilis*, is able to survive at pH 4.0 for short periods, but eventually dies due to interference with the uptake of calcium, which is needed in order to moult. In general, air-breathing insects (e.g. most Hemiptera, some Diptera and some Coleoptera) are more tolerant of low pH than are gilled forms.

Fish

The biological effects on fish of low pH waters are summarized in Fig. 13.16. In
37☞ addition to these are physiological effects very similar to those described above for benthic invertebrates.

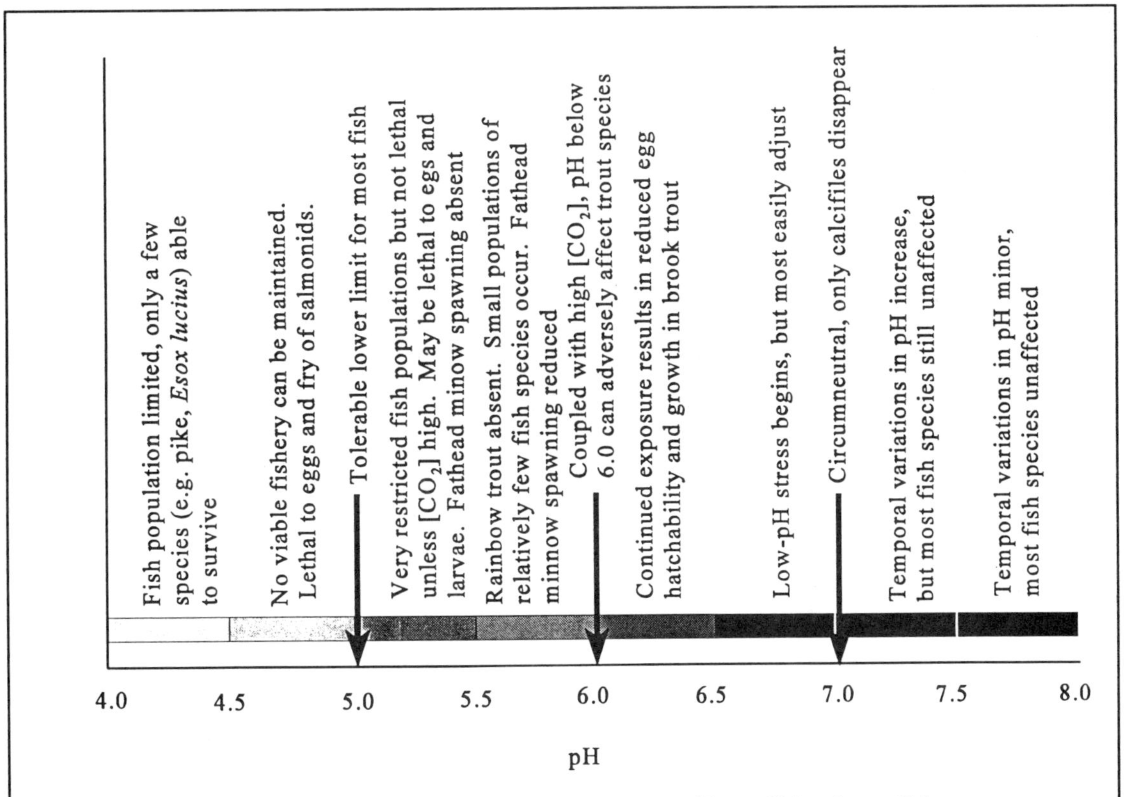

Figure 13.16. Summary of effects of low pH levels on fish.

Other Pollutants of the Atmosphere

While sulphur and nitrogen are the dominant pollutants in the atmosphere, the
38☞ fallout contains numerous other contaminants. The fallout may be in one of four forms; precipitation (rain, snow), gas, aerosols (mist) and/or solid particles (dust). Associated with each may be compounds of sulphur, nitrogen and carbon, all of which are discussed above, and VOCs (volatile organic compounds), CFCs (chlorofluorocarbons) and metals.

About 90% of the earth's ozone occurs in the stratosphere, the rest in the lower troposphere. Ozone is vital for absorbing the more harmful wavelengths of ultraviolet radiation from the sun, but high concentrations can cause a decrease in pulmonary functions of animals and damage vegetation. Ozone is not directly emitted as a pollutant by human activity. It forms as a secondary pollutant when sunlight penetrates air that is contaminated with VOCs and nitrogen oxides. High concentrations of ozone build up when a high air temperature is accompanied by direct sunlight, when VOCs and nitrogen oxide are present and when air mass movement stagnates for several days.

Ozone depletion in the stratosphere has been linked to emissions of CFCs, halons and other compounds. CFCs are non-toxic and long-lasting. They are used as refrigerants, foaming agents, solvents and propellant gases for antiperspirants and hair sprays. Many of these uses have since been banned because of the accumulation of CFCs in the atmosphere. CFCs do not accumulate at ground level or in water supplies. Halons contain carbon, fluorine, bromine and occasionally chlorine. Bromine, in particular, is a powerful catalyst for ozone destruction. Halons have three to six times more ozone-destroying potential than CFCs.

While VOCs and CFCs have no direct impact on aquatic systems, indirectly they allow greater amounts of UV light to penetrate the atmosphere and water columns. Development of fish and amphibian embryos have been linked to high UV exposure. Table 13.10 shows the levels of atmospheric enrichment of several metals (X). The EF (enrichment factor) is calculated as:

$$EF = ([X]/St)_{\text{atmosphere}}/([X]/St)_{\text{earth's crust}}$$

where
St = standard concentration against which all metals are compared in the atmosphere and earth's crust.

For a given element, X, an EF value near unity indicates crustal weathering as atmospheric source for that particular element (Forstner and Wittmann 1983, cited earlier). The data indicate that cobalt and, to a lesser extent chromium and nickel, in atmospheric particulates originate from the earth's crust. Other metals in aeolian particulates originate from anthropogenic sources. Once the metals enter the water, their solubilities and toxicities are greatly affected by pH, as discussed earlier under, "Metals".

Table 13.10. Enrichment of metals in atmospheric particles in United States urban air relative to the earth's crust. Adapted from Forstner and Wittmann (1983).

Metal (X)	EF	Metal (X)	EF
Iron (St)	1	Arsenic	310
Aluminum	1	Silver	830
Cobalt	2	Mercury	1,100
Chromium	11	Cadmium	1,900
Nickel	12	Lead	2,300
Vanadium	42	Selenium	2,500
Copper	83	Antimony	2,800
Zinc	270		

Combating Lake Acidification

The foregoing discussion reveals that lakes and streams acidify for two reasons: (i) acidic precipitation (pH < 5.6); (ii) poor buffering capacity of the receiving waters.
39☞ Therefore, in order to prevent acid precipitation, one must reduce the amounts of sulphur dioxide and nitrous oxide emitted into the atmosphere and/or artificially increase the buffering capacity of lakes and streams.

Reducing Sulphur Emissions

In North America the United States is the main source of sulphur dioxide. In 1979 the United States produced 29.7 million metric tonnes of sulphur dioxide, compared to 5.3 million metric tonnes in Canada. Of the annual 5.3 million metric tonnes, 2 million metric tonnes were produced by Sudbury smelters.

There are four potential solutions to reducing levels of sulphur dioxide nitrogen oxides in the atmosphere: (1) use low-sulphur fuels; (2) remove sulphur from fuels or sulphur-bearing ores prior to combustion (e.g. by coal washing and "scrubbing"); (3) flue gas desulphurization (FGD) which may even provide saleable sulphuric acid as a byproduct; (4) use an alternative energy source. There are disadvantages for each alternative. Switching to low sulphur coal from the western United States would be disastrous to mid-eastern state economies because of the loss of the coal industry. There are high costs involved will all sulphur removal alternatives, particularly to industries that would realize no monetary gains from clean water, although the environmental gains are obvious. FGD systems are expensive and the amount of sulphuric acid produced would flood the market, at which point the economy stagnates, again resulting in loss of jobs.

Use of nuclear power to generate electricity would solve the problem, but there are safety and environmental concerns about shipping and storing nuclear wastes.

Nevertheless, thanks to international co-operation, significant reductions in sulphur emissions have occurred over the past decade using one or more of the above alternatives. Lower demands of ferrous metals have also resulted in reduced emissions of sulphur dioxide. Studies are currently evaluating the results of the reduced emissions, but so far it seems that organisms, including fish, are again colonizing many acidic lakes and streams. Liming of several lakes has helped to accelerate their "recovery" rate.

Stream and Lake Neutralization

Calcium hydroxide ($Ca(OH)_2$) and calcium carbonate ($CaCO_3$) are the agents of choice for "liming" lakes and streams. They are inexpensive, safe and easy to handle and provide the chemical species that naturally occur in watersheds. Calcium hydroxide is added to raise the pH:

$$Ca(OH)_2 + H^+ \rightleftharpoons Ca(OH)^+ + H_2O$$
$$Ca(OH) + H^+ \rightleftharpoons Ca^{++} + H_2O$$

Calcium carbonate is added simultaneously to provide a residual buffering capacity:

(1) Dissolution: $CaCO_3 + H^+ \rightleftharpoons Ca^{2+} + HCO_3^-$
(2) Equilibrium: $HCO_3^- + H^+ \rightleftharpoons H_2CO_3$
(3) CO_2 exsolution: $H_2CO_3 \rightleftharpoons H_2O_+ CO_2$

The dissolution reaction controls the rate of neutralization at $pH < 5.0$. The carbon dioxide exsolution rate dictates the rate of liming at higher pHs. The amount of neutralization agents required is determined by titration.

Either the parent streams or the lake itself can be limed. There are advantages and disadvantages to both. By liming the streams, both the stream and lake ecosystems are neutralized, but liming must be done annually or a slowly-dissolving agent (e.g. stone-size limestone) must be used. However, the surfaces of the stones become rapidly covered by biofilm and the dissolution rate declines rapidly over time and the stone must be turned over to expose new surfaces. Other problems include: annual flow variations which affects the pH level (inversely); large seasonal variations, especially during the spring pH depression period; variations in precipitation, which affects both discharge and pH levels.

Stream Liming

Use of Silos: To address the above concerns, silos containing a lime slurry are used to provide lime at a rate and concentration electronically determined by the discharge level. An agitator in the silo ensures constant viscosity of the slurry and prevents sedimentation in the silo and the pipelines that carry the slurry to the stream. The resulting pH is monitored electronically downstream and the output is fed back to automatically adjust the discharge rate of the slurry. Potential problems with this method include: (i) designing systems to cope with hydrogen ion variations of three to four orders of magnitude; (ii) breakdowns and power losses during critical wither months; (iii) many streams are remote, poorly accessible and not close to a power supply.

Use of Crushed Limestone in River Beds: Most of the experiments to date have used crushed limestone in streams with pH between 4.8 and 5.0. The pH was increased by 0.0 units in the winter to 1.6 units during the summer. About 25% of the limestone was covered and deactivated (by chemical and biological coatings) and had to be replaced every year. There was also a tendency for some of the limestone to be covered by sediments or displaced downstream.

Lake Liming

Whole-lake neutralization seems to be the most widely used method for trying to rehabilitate acidified lakes. Powdered lime or slurries are spread by boats or airplanes over the surface of the lake. In many cases, the pH and alkalinity have been restored to pre-acidification levels. While variations occur from one lake to the next, in general, the response of the aquatic plant and animal communities to the increased pH levels appear to be much slower. In summary, relative to pre-acidification:

(1) No significant changes in chlorophyll *a* levels occur but there is a shift in dominance from blue-green species to chrysophyte species after liming.

(2) The macrophyte community is not affected and few or no plants colonize the lake after liming.

(3) Zooplankton show an increase in diversity but recovery is not complete. In many cases, changes in zooplankton diversity and abundance have been attributed to corresponding changes in abundance of planktivorous fish populations.

(4) Benthic communities appear to show some response to liming, but no consistent changes are seen from one limed lake to the next. However, profuse benthic filamentous algae growths can be eliminated by liming.

(5) Nutrient enrichment by internal or external loading can greatly affect the kinds of responses to lake liming.

By liming lakes, which have much greater retention times than their parent streams, liming only has to be done as frequently as the retention time of the lake. However, the type of lime and the method of application greatly affect the neutralization rate. Slurries and powders provide almost immediate amelioration, while limestone requires several weeks to have an effect. Liming of the lake by spreading on the lake surface provides a faster neutralization response than does liming of the parent streams.

THERMAL POLLUTION (CALEFACTION)

The term, ***calefaction***, is used to describe the thermal enrichment of waters by artificial means. Any increase in temperature beyond the natural ambient temperature can
40☞ be considered as calefaction, but organisms are not usually affected until a change (normally expressed as ΔT) of 10 °C occurs.

Electricity generating plants operate through the thermodynamic process known as the ***Rankine cycle***, in which high-pressure steam is produced in boilers and then expanded through turbines, which convert thermal energy into mechanical energy (Fig. 13.17). In fossil fuel plants, water is pumped from the source (e.g. large lake or river) into the condenser. The water is passed into feedwater heaters and pressurized, from whence it passes to the boiler to be converted to steam. The steam is superheated and expanded through the turbine, creating mechanical energy to drive the turbine and generator. The resulting low-pressure steam is passed to the condenser where heat is removed by the much cooler source water circulating through the condenser.

Nuclear power plants use less energy to generate power because they operate at lower temperatures and pressures than fossil fuel plants. However, the Rankine cycle is less efficient and a large amount of waste heat is still discharged.

The effluent (heated cooling-water) is warmer (but ΔT is less than 10 °C) than the influent but is mostly free of contaminants. Some contaminants, such as chlorine (often added to control biofouling by algae and zebra mussels), may be in the influent water but high water temperatures usually reduce the amounts in the effluent to undetectable levels.

Warm water has less dissolved oxygen than cool water, but oxygen levels are quickly restored as the effluent water mixes with the source water. However, the greater the ΔT, the lower the dissolved oxygen levels. If the water has a pH above 8.2 and a total

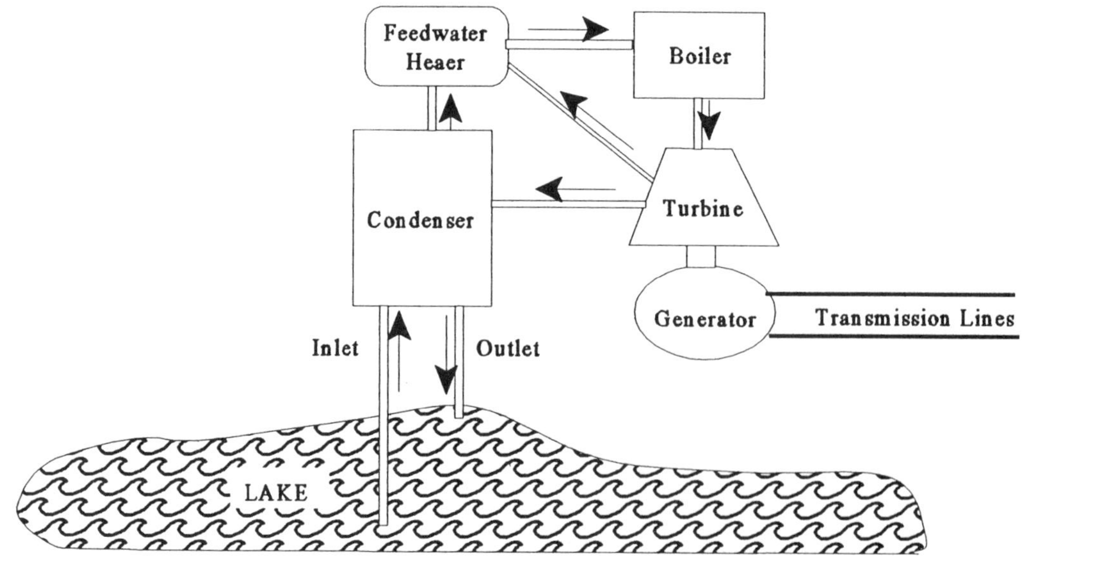

Figure 13.17. Basic components of the Rankin power conversion cycle.

hardness > 200 mg $CaCO_3$/L, the elevated temperatures also cause calcium carbonate to precipitate as scale on the walls of pipelines and cooling towers. Although the pipes and towers must be descaled periodically, the most significant impact is release of water with a lower alkalinity than in the influent.

Also, the influent water often contains considerable amounts of phytoplankton and zooplankton which perish when the water is heated. This material is released in the effluent as particulate organic matter and is often accompanied by increased levels of ammonia and organic nitrogen. However, these too are quickly oxidized into nitrates in the receiving waters.

SEDIMENT LOADING

High suspended solids loads in the water can be attributed to natural events, such as erosion by flowing waters over soft sediments and/or by spate events, and to anthropogenic inputs, including pulp and paper wastes (e.g. fibers), dredging operations and roadway and pipeline construction. Offshore waste disposal is not common in freshwater ecosystems but is common in oceans.

Natural Events

High turbidity usually accompanies floods, but even slow-flowing rivers with muddy or silty bottoms will be turbid. The quality of the material suspended is related
41 directly to the surface layers of the sediments being resuspended. Typically, high suspended solids loads interfere with photosynthesis and feeding mechanisms of zooplankton. Hence, reaches of rivers that are turbid tend to be less productive than upstream reaches that are less turbid. This is in direct contrast to the river continuum concept which predicts that productivity should increase downstream.

Anthropogenic Events

42 In contrast to natural events which tend to be of a recurring nature, anthropogenic events may be continuous to short-term or even "pulse events".

Pulp and Paper

The most common type of suspended material is wood fiber that is carried in the effluent. The fiber is carried downstream for a short distance and then deposited on the bottom. Wood fiber is refractory in nature and decomposes very slowly because there are few bacteria that can decompose woody elements. Hence, the fibers tend to accumulate on the bottom very quickly.

Most pulp and paper industries now have greatly improved technologies for reducing the suspended load of wood fibers. Nevertheless, many rivers, such as the Ottawa River, still have thick deposits of wood fibers and wood chips that were deposited over the years.

Dredging

Dredging operations are done occasionally to restore depths in shipping channels and artificial impoundments. Hence, the effects (discussed below) tend to be of short-term duration. If the sediments have pollutants, the waters may be temporarily contaminated. The degree of contamination, and the toxicity associated with the contaminants, depend on the rate of transformation from a precipitated state to a dissolved and often more toxic state.

Roadway/Pipeline Construction

The effects of roadway and pipeline construction are also of a short-term nature. However, of greater potential impact is the highway stormwater runoff into lakes and streams. The impact of the runoff potentially embraces nearly those from all types of pollution discussed in this chapter, especially suspended solid, thermal, metal and organic pollution and acid precipitation.

The main sources of highway pollutants are: (i) the vehicles and their products such as tires, brakes, sump oil and vehicle exhausts; (ii) roadway deicing materials; (iii) constituents of asphalt, tars, and bitumens; and (iv) precipitation. The pollutants can be categorized as dust, temperature, heavy metals, organic contaminants and petroleum products.

The main classes of potential impact are:

- habitat destruction, including changed hydrology
- flooding, especially immediately upstream of crossing
- groundwater contamination, especially shallow aquifers
- long- and medium-term physical environmental changes (e.g. changed fluvology)
- aquatic ecosystem changes (e.g. due to suspended solids loading and temperature changes)
- food chain effects, especially if contaminants bioaccumulate and biomagnify
- interruption of movement of aquatic organisms, especially migratory fish species
- reduced beneficial human uses of receiving waters (e.g. loss of angling)
- reduced aesthetic value

In general, high discharges of highway stormwater runoff is hot and turbid, has a high concentration of particulates and may be salty. It is also likely to comprise a broth of many other toxics that could have synergistic, additive or antagonistic effects on fish and benthic invertebrates. The degree of impact is probably associated with traffic density, average temperatures, geomorphology, geochemistry, hydrology, road salt applications and stage of the evolving impacted ecosystem.

RADIATION

The nucleus of an atom contains positively charged protons and neutral (electrically) neutrons. Electrons, which orbit the nucleus, carry a negative charge equal to the positive charge on a proton. Atoms of the same chemical element may vary in the number of neutrons they have and are called ***isotopes***. Some isotopes are unstable and they seek stability by releasing particles or electromagnetic rays, collectively called radiation.

Radiation is energy produced at one location and transmitted to another. Of most concern to humans and aquatic life is ***ionizing radiation***, because it has sufficient energy to ionize atoms and molecules. The most common propagators of ionizing radiation are helium nuclei, electrons and protons, gamma radiation, neutrons and protons. The natural sources of radiation are cosmic rays from outer space (10%), gamma radiation from rocks and soils (14%), radon and thoron gas in buildings (52%) and radiation accumulated in tissues from food and drink (12%).

There are four main types of ionizing radiation, with different powers of penetration:

1. ***Alpha radiation***, consists of particles made of two protons and two neutrons. The
43☞ radiation has very little penetrating power and lose its energy in a short distance (Fig. 13.18). However, they are strongly ionizing and if inhaled or ingested, can cause severe internal damage.

2. ***Beta radiation***, has greater penetrating power than alpha radiation, but they can be stopped by water, glass, concrete or metal. However, like alpha particles, beta particles can cause internal damage if swallowed or inhaled. Beta particles may be electrons (negative charge) or positrons (positive charge), depending on whether a neutron spontaneously changes into a proton, or a proton into neutron, in an unstable nucleus.

3. ***Gamma radiation***, is short-wave electromagnetic radiation, lacking mass and charge, but capable of penetrating many materials, but not lead (Fig. 13.18).

4. ***Neutron radiation***, found only in nuclear reactors, is composed of unstable, heavy

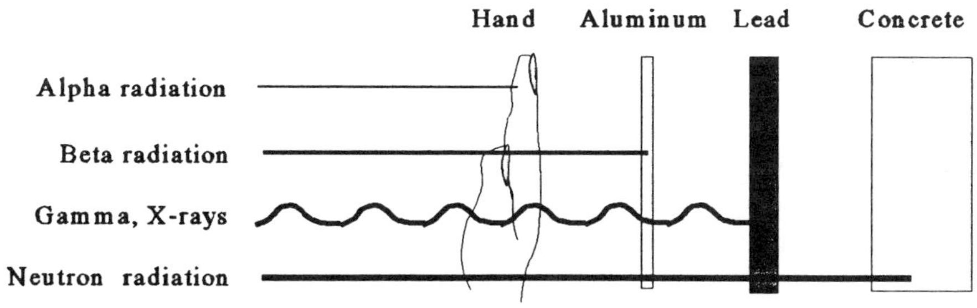

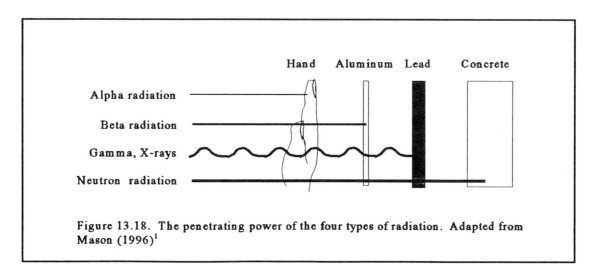

Figure 13.18. The penetrating power of the four types of radiation. Adapted from Mason (1996)[1]

nuclei which have a large excess of neutrons and the nucleus breaks into two fragments, by ***spontaneous fission***. The free neutrons produced in the fission process are highly penetrating.

In nuclear reactors, ***nuclear fission*** occurs when the uranium nucleus is hit with a neutron, the nucleus thence splitting into two equal pieces. The total mass of the fragments is slightly less than that of the original nucleus. This reduction in mass (m) appears as energy (E) according to Einstein's equation, $E = mc^2$, where c is the velocity of light (= 2.998×10^8 m/sec). The fission process releases two or three neutrons which can split other atoms, releasing more neutrons in chain reactions. The energy of the fission reaction is converted into heat as the fragments collide with one another and are slowed down after collision. The heat generated is passed to a heat exchanger or steam generator, where water is boiled and sent to turbines that generate power. The steam emerging from the turbine must be condensed, using cooling water drawn from a lake or river, before it is released back to he environment.

The nuclear fuel cycle involves several steps:

(i) the uranium ore is mined and milled
(ii) the uranium is then enriched to increase the concentration of U-235
(iii) the enriched uranium is fabricated into fuel rods
(iv) the fuel is then used in the reactor
(v) the spent fuel is reprocessed
(vi) finally, the wastes are disposed of.

Nuclear power has certain advantages and disadvantages relative to its alternatives (e.g. fossil fuel plants). Historically, nuclear power has produced less air pollutants (e.g. sulphur dioxide), has been cheaper than fossil fuel power, has had less impact on land and water use and has presented fewer occupational hazards and minor radiation exposures compared to those from natural radiation sources (Mason 1996)[19]. However, nuclear power has the potential to lead to environmental disasters. The most significant radiation effects in the past have occurred at the fuel reprocessing step. Also, nuclear power has greater safeguard problems and produces highly toxic wastes. To this day, disposal of radioactive wastes is a controversial issue.

Units of Radioactivity Measurement

Initially, "***rad***" and "***rem***" were the standard units for measuring radiation. But
44☞ these terms have been replaced by "***Gray***" and "***Sievert***", respectively. The following are definitions for these four units, and those currently in use for radiation work:

Past terminology

Rad is the acronym for ***radiation absorbed dose.*** 1 rad is the amount that leads to absorption of 1000 ergs/g of absorbing material.

---

[19]Mason. C. F. 1996. Biology of freshwater pollution. Longman. 3rd edition.

Rem is the acronym for ***roentgen equivalent man.*** 1 rem is the amount of radiation that produces the same biological damage in humans as 1 curie (Ci) of gamma radiation.

New terminology

Becquerel (Bq): A measure of radioactivity as the frequency with which radioactive disintegrations of the nucleus take place in a substance. 1 Bq = 1 disintegration/sec. The becquerel replaces the older unit, curie (Ci). 1 Bq = 27.03 pCi.

Gray (Gy): A measure of the amount of radioactivity absorbed by a tissue or organism. 1 Gy is the amount of radiation needed to cause 1 kg of tissue to absorb 1 joule of energy. The gray replaces the rad, where 1 Gy = 100 rad.

Sievert (Sv): An arbitrary unit which accounts for the fact that different kinds of radiation do different amounts of damage to living tissues and organisms for the same energy. Neutrons and alpha particles have about ten times the effect of beta or gamma particles for the same number of grays. The sievert replaces the rem, where 1 Sv =100 rem

The biological effects of ionizing radiation may be somatic (affecting the organisms health, as by induction of cancer) or genetic (hazardous to descendants of offspring, such as by mutations). Table 13.11 gives the amount of short-term doses (e.g. nuclear energy accidents or warfare) of radiation needed to cause different levels of damage in humans and mammals:

Table 13.11. Amount of short-term doses of radiation needed to cause different levels of damage in humans and mammals:

Dose (Gy)	Effect
1	Threshold level for observable effects
1-5	Slight temporary blood changes, such as deficiency of lymphocytes
5-10	Failure to generate blood cells
11-50	Intestinal failure, vomiting, diarrhoea
> 50	Convulsions, lethargy, tremor, central nervous system syndrome

Effect of Radiation on Aquatic Organisms

The effect of radiation on aquatic organisms is poorly described, largely because of the excellent safety record of nuclear power plants. The cooling waters, that are
45☞ eventually discharged back to the environment, inevitably receive some radiation. The following effects are in Mason (1996):

- Fish experimentally exposed to 0.6-3.6 mGy/day from sediments produced larger than average broods but with a higher incidence of abnormal embryos.

- Chironomid midge larvae living in same stream sediments showed increased chromosomal aberrations, but no effect on their numbers.
- Whitefish below a nuclear power station outfall with 5.6 mBq/ml in the effluent bioaccumulated 9 Bq/g, with concentration factors rising to as high as 5000. A person who ate 200 whitefish meals per year would receive an estimated 3 mSv/yr, or 20% of the annual limit.
- After the Chernobyl accident (April 26, 1986) in north Kiev, Ukraine, levels of ^{137}Cs in fish collected from Swedish waters had levels as high as 18,700 Bq/kg in trout (*Salmo trutta*), 14,200 Bq/kg in perch (*Perca fluviatilis*), 6,280 Bq/kg in rainbow trout (*Oncorhynchus mykiss*), 4,690 Bq/kg in pike (*Esox lucius*) and 3,130 Bq/kg in whitefish (*Coregonus lavaretus*). Biomagnification explained most of the differences observed in radiation body burdens of fish. No biological effects were reported for any of the species.
- To date, at current levels of radiation being discharged into the receiving waters, there appears to be no measurable effects of radiation on the biota *in situ*.

OIL AND PETROLEUM BI-PRODUCTS

Oil spills are mush less common in freshwater environments than in marine
46 environments. Hence, most of what we know about oil pollution is based on mishaps in the oceans. Perhaps the most infamous of these is the spill of about 35,000 tonnes of crude oil from *Exxon Valdez* when it ran aground in Prince William Sound, Alaska in the spring of 1989. Hundreds of thousands of seabirds, fish and otters were killed by the crude oil. The fate and effect of oil spills in freshwater systems can be just as devastating and are reviewed by Green and Trett (1989)[20].

Crude oil consists of hydrocarbons with up to 26 atoms per molecule. The three major types of hydrocarbons are ***aromatics*** (e.g. toluene, benzene, naphthalene), ***alkanes*** (e.g. butane, propane, ethane) and ***cyclohexanes*** (e.g. naphthenes). Compounds of sulphur, nitrogen and metals may be associated with the hydrocarbons. Some of the most toxic compounds are from oil products (e.g. crankcase oil, brake oil, grease, asphalt) and include ***PAHs*** (polycyclic aromatic hydrocarbons), ***PCBs*** (polychlorinated biphenyls) and metals, especially lead, aluminum, cadmium, copper, mercury and zinc. Indeed, many of the oil toxicants can be found in urban stormwater runoff (Latimer et al. 1990)[21]., not just

---

[20]Green, J. and M. W. Trett. (Eds.). 1989. The fate and effect of oil in fresh-water. Applied Science Publishers, London.

[21]Latimer, J. S., E. J. Hoffman, G. Hoffman, J. L. Fasching and J. G. Quinn. 1990. Sources of petroleum hydrocarbons in urban runoff. Water, Air, Soil Pollution 52: 1-21.

oil spills from tanker ships and trucks.

Effects on Biota

The effects of oil vary greatly among organisms, from one of stimulation to one of extreme toxicity and from direct effects to indirect effects, depending on the type of hydrocarbon.

Bacteria

After a spill, some of the oil that is deposited on the surface is dispersed by physical and chemical processes or is lost by evapouration and photochemical oxidation processes, the rest remains as hydrocarbons. Many heterotrophic bacteria and fungi are able to utilize the hydrocarbons and their growth can be stimulated. Almost 100% of crude oil can be degraded by microbes. If any hydrocarbons remain, they are represented mostly by asphaltenes, then heterocyclic compounds and lastly, aliphatics which tend to be the most innocuous components of oil.

Algae

Some algae are very sensitive to oil pollution but their death and the subsequent release of nutrients stimulates the growth and production of the more tolerant species of algae, especially blue-green algae. Photosynthesis by algae may be affected if a film of oil is present to absorb some of the PAR (photosynthetically active radiation, see Chapter 4) and/or by altering the permeability of cell membranes.

Macrophytes

Floating macrophytes, such as duckweed (*Lemna* spp.), are generally tolerant of crude oils but are much more sensitive to synthetic oils, particularly those that release aromatic hydrocarbons (Green and Trett 1989). Only when thick layers of crude oil are present is photosynthesis of submersed forms potentially affected, for the same reasons discussed above for algae. Emergent forms of macrophytes are less affected by surface films of oil because they rely on radiation in the atmosphere.

Plankton, Nekton and Neuston

The effects of oil pollution on neuston (floating organisms) and plankton and nekton (swimming organisms) are quite different. Floating oil affects the respiration, photosynthesis or feeding of organisms at the surface. Vertebrates, such as birds and otters, whose coats or feathers are covered with oil, lose buoyancy and insulation. Ingestion of oil may be directly toxic or indirectly affect natality rate by altering the development rate of eggs, or inducing ***teratogenesis*** (abnormality produced before birth), ***mutagenesis*** (alteration of genetic material) or ***carcinogenesis*** (increase in incidence of neoplasms, i.e. new growths).

Many fish are able to depurate themselves of toxic hydrocarbons by rapid metabolism which results in the excretion of foreign compounds. The metabolic capacity to remove hydrocarbons is due largely to "***mixed-function oxidase (MFO)***" systems and ***cytochrome P-450***, an isozyme contained within the liver. Any increase in concentrations of PAHs and PCBs, in particular, are followed immediately by an increase in the activity of MFO enzymes. The P-450 isozyme activates ***procarcinogens*** (chemicals, e.g. PAHs, that must be biotransformed into direct-acting carcinogens) in animals and man.

Benthic Macroinvertebrates

In general, phenols are more toxic to benthic invertebrates than are azarenes and PAHs (Green and Trett 1989). Snails appear to be more tolerant of hydrocarbons than arthropods, although functional feeding behaviour and/or their position in the water column may affect sensitivities. For example, many snails are pulmonates (e.g. *Physella gyrina*), capable of respiring atmospheric oxygen and can avoid exposure to, or the toxic effects of, aqueous phases of hydrocarbons by moving to or above the surface of the water to graze. The sensitivities among arthropods can be quite variable, with amphipods (e.g. *Gammarus minus*) being much more sensitive than midge fly larvae (e.g. *Chironomus tentans*). Both species are gatherers, but *Chironomus* burrows and gathers its food from the sediments, whereas *Gammarus* gathers food at the mud-water interface. Perhaps bacterial metabolism within the sediments "detoxifies" the hydrocarbons so that burrowing forms receive lower exposure concentrations than do those at the mud-water interface.

WATER POLLUTION CONTROL

Water pollution control herein refers to the processes which remove wastes from the sewage supplies of municipalities, or pollutants from the waste streams of industries
47☞ and factories. However, water pollution control in the strict sense would also include in-stream aeration dams to augment dissolved oxygen levels and use of phosphate-free biodegradable detergents to lower phosphorous loading levels.

In most cases, all processes of pollution control for municipal sewage and waste treatment are enclosed within ***water pollution control plants*** (***WPCP***), an example of which is shown in Fig. 13.20. The sources of water for the WPCP are domestic waste (i.e. toilets, sinks), roadway runoff, industries without their own water pollution control processes and colleges and universities. Large industries, colleges and universities may have some pre-treatment before sending the waste to the WPCP.

Waste, upon entering the WPCP, is first passed by a low lift ***Archimedes screw pump*** to a rack or screen to remove coarse floating material which may then be burned or buried (Fig. 13.20). The screened waste then enters a ***grit chambre*** where coarse settleable grit is removed and then disposed of as landfill material. Some WPCP have a ***comminutor*** which shreds and breaks coarse solids into small particles. The gritty waste is then sent for primary treatment.

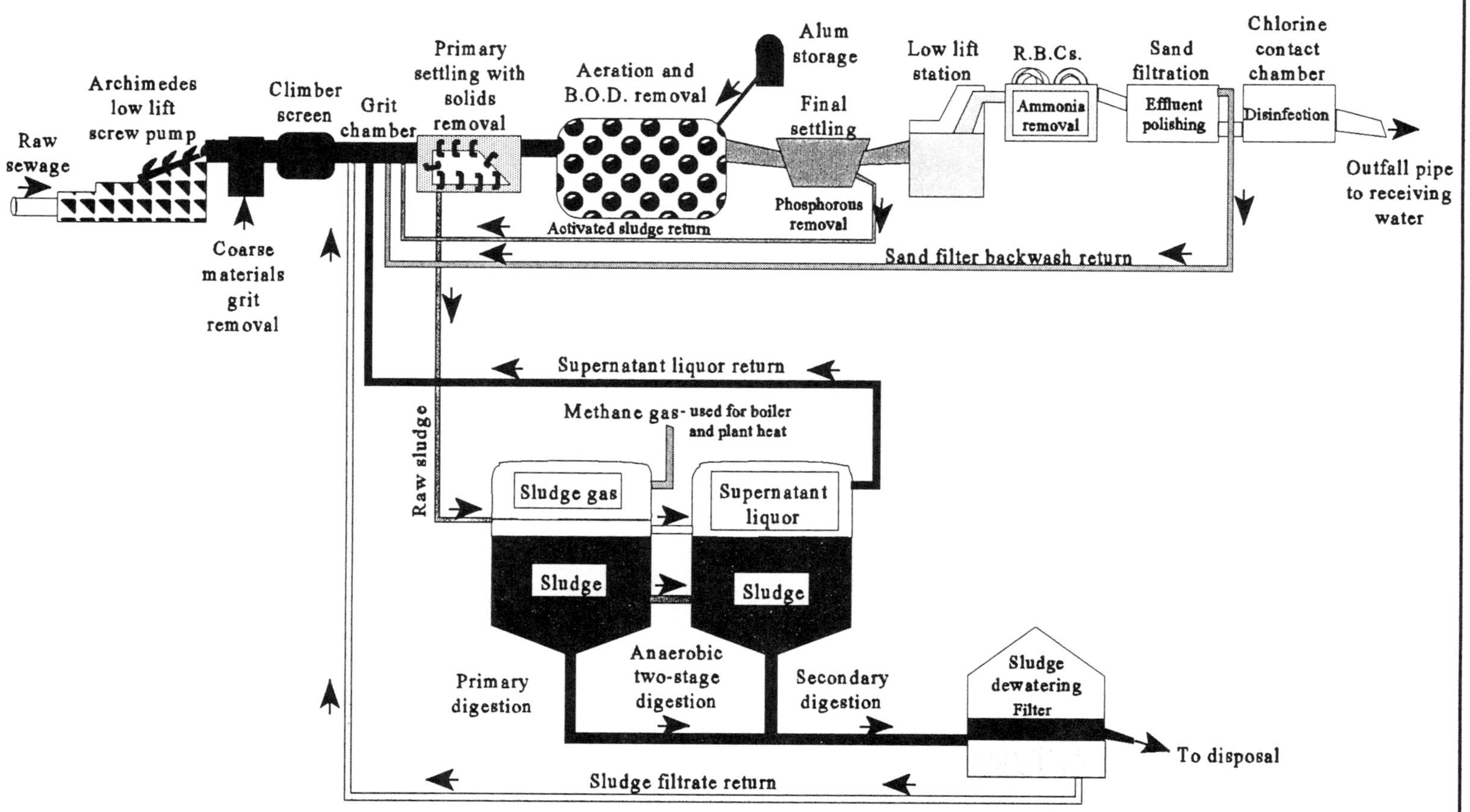

Figure 13.20. Typical components of a modern sewage treatment plant with tertiary treatment. Primary treatment involves solids removal in a primary settling tank. Secondary treatment involves aeration with B.O.D. removal followed by an activated sludge process that removes suspended solids in a final settling tank. Alum (aluminum sulphate) is usually added at this point to remove phosphorous. Trickling filter systems are often used instead of activated sludge in the secondary treatment process. Tertiary treatment involves removal of ammonia and other nitrogenous wastes using rotating biological contactors (RBCs) or other proceses. The effluent from the tertiary treatment (or secondary if tertiary is not used) is polished using sand filtration. Disinfection of the polished effluent is performed using sodium hypochlorite in a chlorine contact chamber. The receiving water is typically a river.

Primary Treatment

Primary treatment is a ***settling*** process that occurs in large rectangular or circular basins. Waste flow and turbulence is reduced and the waste water is detained long
48 enough for settleable solids to collect at the bottom. Only about 35% of oxygen-demanding organic materials are removed in the primary treatment process. The partly clarified water passes from primary treatment, a physical process, to secondary treatment, a biological process.

The sludge on the bottom is mechanically pushed into a sludge drain and sent for further treatment to large tanks where anaerobic bacterial digestion reduces the solid waste into liquid waste that can be returned to the primary treatment tanks. The anaerobic digestion process, which requires about 15 days, produces ***methane*** that can be used to heat the WPCP. Any remaining solids are piled in sheds and water is removed by air-drying. Alternatively, the solids can be sent to a large vat with a rotating drum inside. The wall of the rotating drum has perforations and is lined by a cloth. The wet solids adhere to the cloth walls of the drum and the water is removed by ***vacuum filtration***. The dried solids are then scraped off the cloth drum wall by knives and sent to be piled somewhere in the yard. Since most solids are relatively inert and rich in organic matter they are used as fertilizer or soil conditioner in municipal parks and gardens. If the solids do contain large amounts of contaminants they can be incinerated.

Secondary Treatment

49 Secondary treatment is an aerobic biological process. There are two alternatives to secondary treatment processes, "***activated sludge***" and the "***trickling filter***". The clarified flows from the activated sludge process or from the trickling filter process may be reduced by up to 80 to 90 percent. While most of the microbial community is also removed, or is dead by the time the secondary treatment is completed, chlorination of the effluent ensures that potentially hazardous microbes are killed before entering the receiving water. However, most large towns and cities now remove any remaining and resistant dissolved organic materials with tertiary treatment. Hence, instead of chlorinating and sending the flow to a receiving water, the secondary treatment effluent is sent directly for tertiary treatment.

Activated Sludge Process:

The clarified stream from the primary settling basin is mixed with a suspension of living and active (hence activated) micro-organisms, protists and metazoans in an
49a aerated tank. The dissolved organic matter (DOM) acts as nutrients for the growth and reproduction of the activated sludge population. The DOM is removed from the water by the organisms and incorporated into the biological matter of the sludge itself. The sludge is retained and activated upon for 6 to 8 hours before passing to a thickening tank. The resultant suspension is rich in solids and poor in dissolved organic matter. The solids

are allowed to settle out in the thickening tank, the upper clarified liquid being sent on for tertiary treatment (if present in the WPCP). The bottom suspension is often (but not always) separated into two flows, one that returns most of the suspension back to the activated sludge tank for re-processing, and the other that sends the suspension back to join the solids from primary settling for anaerobic sludge digestion, or for disposal.

The activated sludge process just described is the conventional process (Fig. 13.21a), but there are a number of variations, the configuration being determined in part by the population size of the community and costs. Most of the variations center around the aeration process, some plants using ***step aeration*** (Fig. 13.21b) where the activated sludge is recycled through several aeration steps, or ***re-aeration*** (Fig. 13.21c) where the activated sludge is returned for re-aeration until a clarified effluent is obtained. Whatever configuration is used, the activated sludge plant components may be replicated several times, depending on the amount of sewage that requires treatment. For example, the City of Guelph has three plants, each with a primary clarifier and a secondary clarifier (activated sludge treatment).

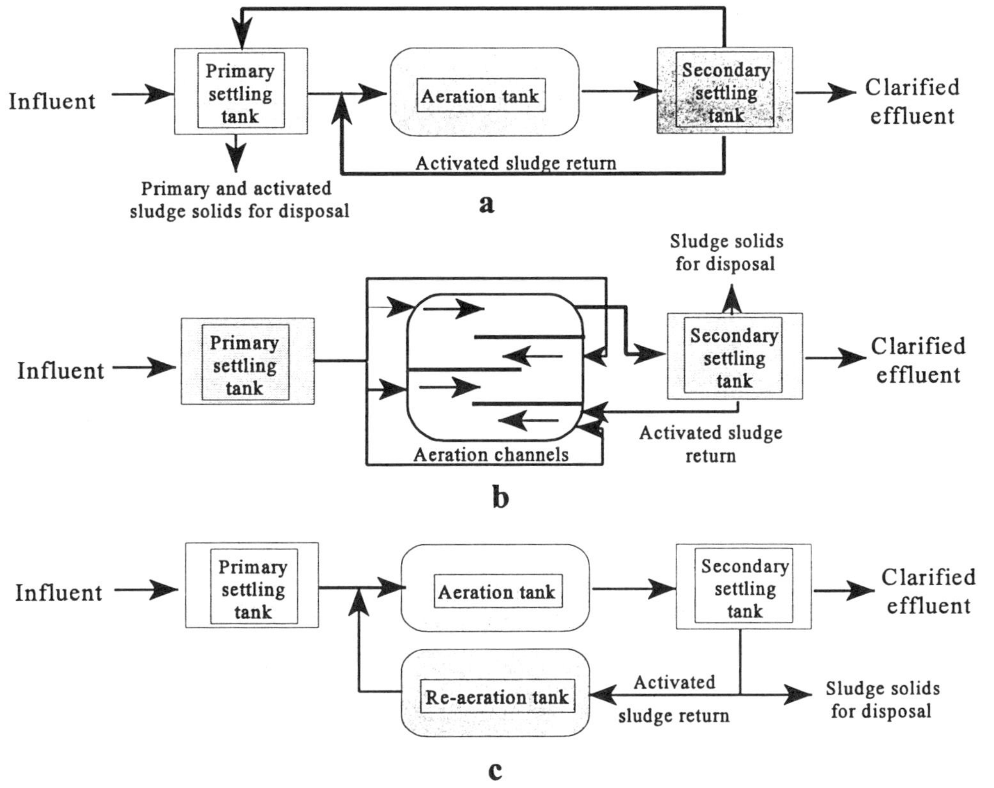

Figure 13.21. Variations in configuration of activated sludge treatment components: (a) conventional, (b) step aeration, (c) re-aeration.

Trickling Filter:

49b ☞ The filter is usually a circular concrete basin, filled with crushed stone to assure sufficient contact with air needed for aerobic digestion of the sludge. A biological film, consisting of similar kinds of organisms as found in activated sludge, covers the surface of the stones. Flow from the primary settler is distributed over the stone bed and trickles down and through the biological film to the basin outlet. The resulting flow is reduced in dissolved organics which are captured by the biological film as a nutrient source. However, some solids remain and they, with any slime masses that slough off the stones, are passed to a settling tank which separates the solid suspension from the effluent of the filter. The suspension of solids joins the suspended solids from the primary settler for anaerobic digestion or disposal.

As in activated sewage sludge systems, there are several configurations possible for trickling filters. Either one or two stage configurations is used (Fig. 13.22). The configuration used depends mostly on the population size of the community.

Tertiary Treatment

Tertiary treatment removes not only the remaining dissolved organics but phosphates and nitrogenous compounds that otherwise would enhance eutrophication in
50 ☞ lakes and rivers. It is essential that the volume of the receiving waters is large enough to dilute the effluent. Otherwise, an upstream reservoir is needed to increase the discharge of the receiving waters during low flow periods. The increase in flow is known as ***low flow augmentation*** and is practiced in many communities with artificial impoundments.

While primary treatment is a physical process and secondary treatment is a biological process, tertiary treatment is chemical for organic carbon and phosphorous removal and biological for nitrogenous compounds removal. In modern sewage treatment plants, most tertiary treatment processes remove at least the remaining dissolved organic carbon and phosphorous. The dissolved organics are removed by sending the clarified effluent from secondary settling tanks to tall towers (7-10 m high) containing activated carbon. The organic molecules adsorb to the surface of the carbon and are taken out of solution. The process is the same as described below for water treatment plants. Towers of activated carbon are typical in most water treatment plants.

Phosphorous removal occurs in activated sludge tanks or the secondary settling tanks (if trickling filter process is used). A solution of either alum (hydrated aluminum sulphate) or iron salts (e.g. hydrated ferrous sulphate) or lime ($CaCO_3$) is added to the sludge. The coagulant precipitates metallic phosphates which are retrieved as part of the suspension of solids and disposed of. Removal of 80 to 90 percent of phosphorous is usually achieved.

Removal of nitrogenous compounds as part of the tertiary process is not common. In some municipalities (e.g. City of Guelph) nitrogenous compounds are removed by sending the water to ***rotating biological contactors*** (***RBCs***). The RBCs employ a denitrification process whereby bacteria reduce nitrates to nitrogen gas.

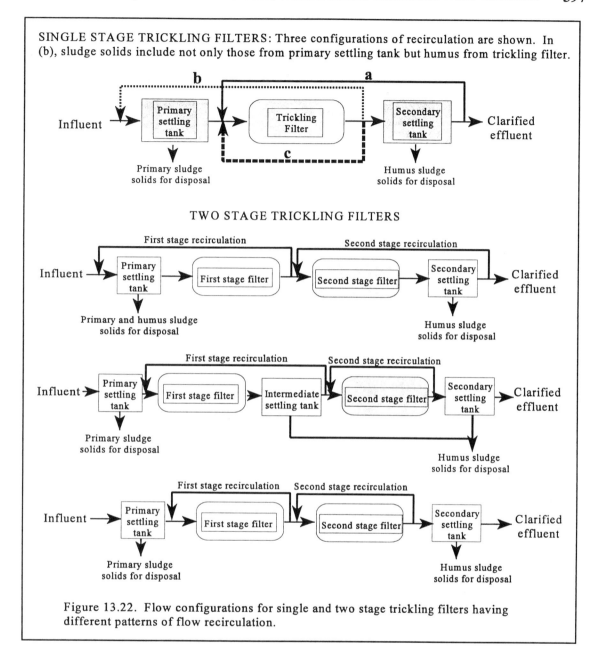

SINGLE STAGE TRICKLING FILTERS: Three configurations of recirculation are shown. In (b), sludge solids include not only those from primary settling tank but humus from trickling filter.

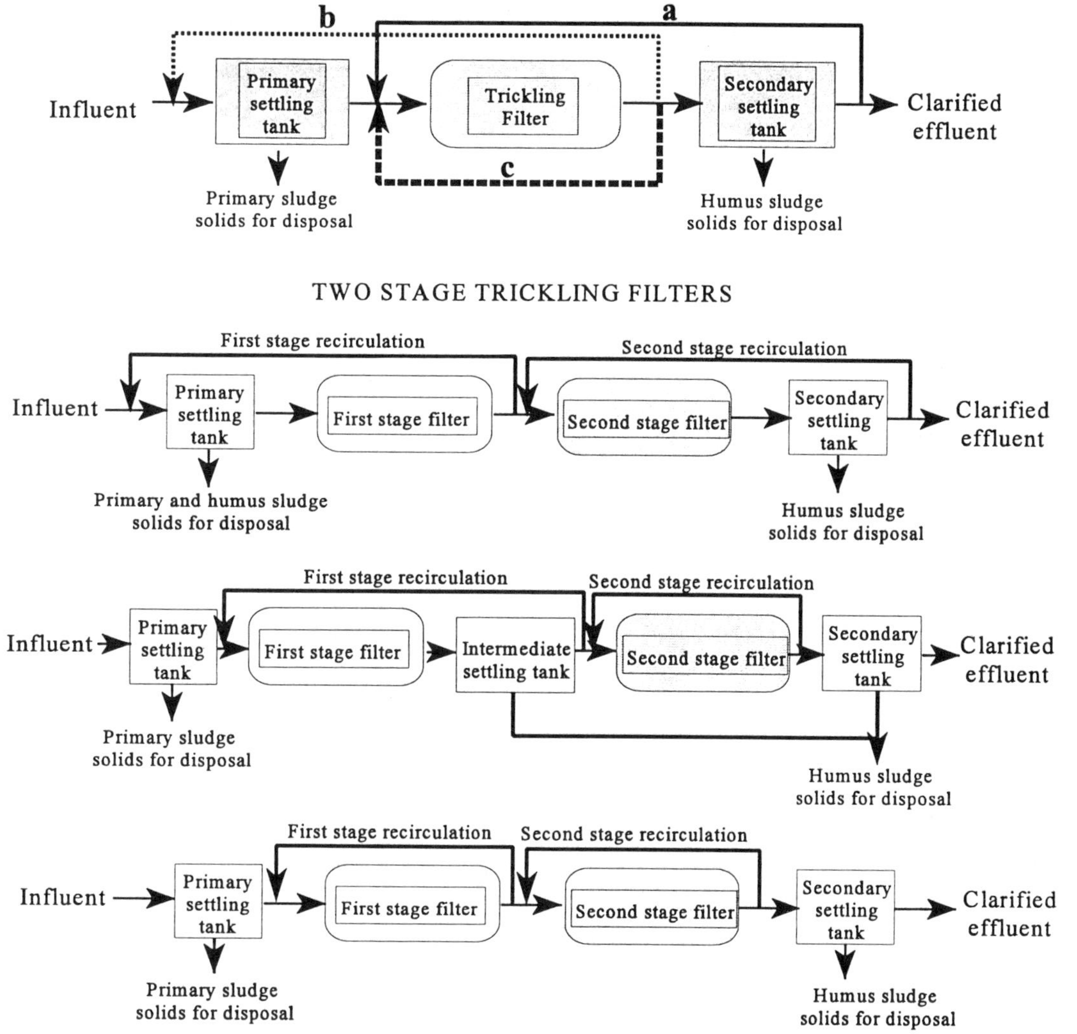

Figure 13.22. Flow configurations for single and two stage trickling filters having different patterns of flow recirculation.

Sand Filtration and Chlorination

The effluent from tertiary treatment (or secondary treatment if tertiary is absent) is often sent to sand filters which remove any suspended solids that remain after removal of organic carbon, phosphorous and/or nitrogenous compounds. The process is often referred to as ***polishing***. Since most of the suspended material is very fine, the sand filter needs to be back washed periodically, the backwash being returned to primary clarifiers. The clear water is then sent to a chlorine contact chamber for final disinfection before releasing the polished effluent to the receiving water.

WATER TREATMENT PLANTS

Water treatment describes processes which are used to purify water for human consumption (e.g. drinking, cooking, diluting). The water is delivered from the source
51 ☞ (e.g. well, lake, river, reservoir) by a pipeline (***water main***) to the treatment plant. All the processes of water treatment for municipalities are enclosed within ***water treatment plants*** (***WTP***), an example of which is shown in Fig. 13.23.

The processes involved in water treatment are:

1. If the source is a reservoir, lake or river, water is aerated by spraying jets of water into the air in a large basin filled with fountains. This removes carbon dioxide, hydrogen sulphide and methane gases that may be in the water.
2. The aerated water is then mixed with ***activated carbon*** in tall towers to remove some of the colloidal and organic material that are responsible for giving water its colour. The colloids and organic materials adsorb to the surfaces of activated carbon particles. The carbon particles, and their adsorbed materials, are removed in a sedimentation basin. Eventually, the adsorptive capacity of the carbon particles is exhausted and water flow is directed to another tower. While the second tower is in use, the carbon particles in the first tower are regenerated by heating to near 925 °C (1700 °F). The newly reactivated carbon is reused after the adsorptive capacity of particles in the second tower is exhausted. Five to eight percent of the carbon particles are destroyed (disintegrate) and must be replaced after each regeneration.
3. The partly clarified water is sent to a ***"coagulation"*** process which proceeds in several stages:
   i. The aerated water is mixed with ***alum*** (hydrated aluminum sulphate) or ***copperas*** (hydrated ferrous sulphate) or other coagulants. Colloids, such as clay particles and protein, and other compounds with a negative charge (which repels the particles from one another and keeps them in suspension) are attracted to and de-stabilized by the positively charged Al and Fe ions. In addition, insoluble oxides of Al and Fe also form positively charged colloids. Under normal conditions (e.g. 15 - 18 °C, 10 - 30 mg alum/L, mesotrophic waters), the coagulation process is very rapid, occurring in less than 5 min. In summary, coagulation is the agglomeration of particles by charge reduction.
   ii. The de-stabilized particles form a precipitant called "***floc***". The process of floc formation is called ***"flocculation"*** and is completed in about 15 to 20 minutes. In contrast to coagulation, flocculation is the agglomeration by inter-particle bridging.
   iii. The flocculated material is settled out in a sedimentation basin. Sometimes preliminary chlorination is applied either immediately before coagulation or immediately before sedimentation in order to kill viruses, bacteria and algae, which settle out in the sedimentation process.
4. The water leaving the sedimentation basin enters a ***"rapid sand filter"*** where most

of the remaining colloids, viruses, bacteria, and algae are removed. The rapid sand filter consists of layers sand, graded from the smallest at the top to largest at the bottom, followed by gravel. The spaces between the sand grains soon become clogged with particles and the filters must be periodically back washed with clean, filtered water to remove the clogged material. The backwash water can be recycled through the filter or sent directly to the drain as waste sludge.

5. The filtered or polished water is given a final disinfection using ozone or chlorine (as hypochlorous acid) prior to distribution. Water softening (or lowering or removal of water hardness) may be included before chlorination and distribution

Many variations in water treatment are possible. For example, the coagulation step may be omitted if cold, clear well water is used. Industries and utilities that use water simply as a coolant may also omit the coagulation step and put more emphasis on water softening to prevent scaling in boilers. Water softening is required especially if water hardness exceeds 200 mg $CaCO_3$/L.

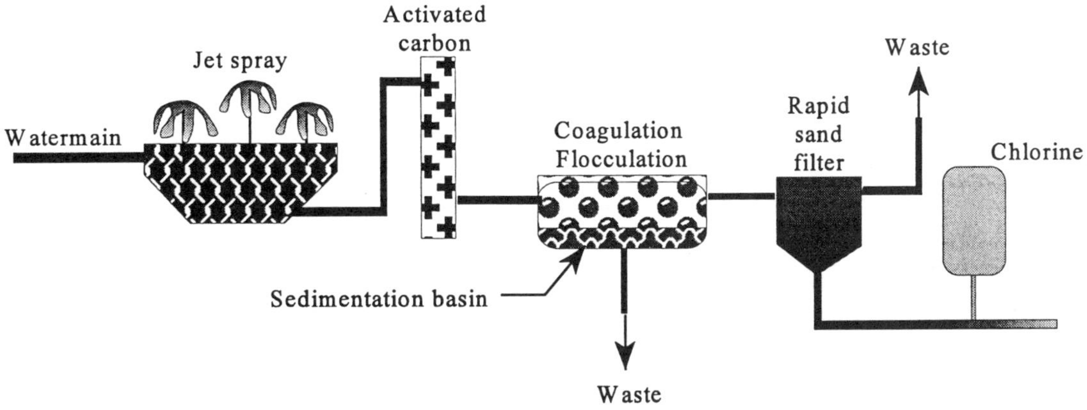

Figure 13.23. Typical components of a water treatment plant.

CHAPTER 14

BIODIVERSITY - THE WEAK AND THE STRONG

Why Read This Chapter?

All aquatic environments undergo natural changes over time and the organisms must either adapt or be replaced by those whose ecological tolerances and requirements best fit the new environment. In the past, human actions have been trivial when considered against the dominant processes of nature. But the continued intervention by humans has altered many of the fundamental processes of the aquatic environments, and indeed even the planet. This chapter examines two of these processes, the changes (mostly losses) in biodiversity and the introduction of new species.

The loss of biodiversity is due mainly to the inability of genetically and ecologically weaker species to adapt to changing environments and to compete with the new colonizers. Part of the new colonizers are exotic species, but not all of them are strong either. The new invaders must compete with the best of the native species. The strongest of the alien species make the headlines because most are opportunists and generally wreak havoc on the ecosystem, and even the economy. This chapter examines those factors that determine the weaknesses and the strengths of certain species and their potential for elimination, subordination or domination in aquatic environments.

The chapter begins by examining the different kinds of diversity. Then traits of generalists and opportunists are examined and rated for their potential to lead to the elimination of a species or to its domination. Then the alien species, especially those that have invaded the Great Lakes, their origins, dispersal mechanisms and impacts are reviewed. This is followed by an examination of some of the physical, chemical, mechanical and biological methods being used to control the zebra mussel in North America. The chapter ends with an examination of a new concept called biomimicry, where humans employ an array of natural designs and processes to solve human problems.

In this chapter you will learn:

☞ 1. What is meant by genetic diversity, genotype and phenotype
☞ 2. What is meant by microevolution and its causes
☞ 3. What is meant by species diversity

☞ 4. The importance of sentinel species as indicators of environmental quality
☞ 5. The importance of target species as indicators of environmental quality
☞ 6. The importance of keystone species in aquatic habitats
☞ 7. The importance of umbrella species in aquatic habitats
☞ 8. How habitat loss affects changes in biodiversity
☞ 9. How habitat fragmentation affects changes in biodiversity
☞ 10. How different kinds of pollution affect changes in biodiversity
☞ 11. How exotic species affect changes in biodiversity
☞ 12. Different kinds of life history traits that animals use to deal with changing environments
☞ 13. Som common traits of endangered species
☞ 14. Typical reproductive features of endangered species
☞ 15. The importance of a species' tolerances and requirements in adapting to habitats
☞ 16. The most effective kinds of mechanisms for dispersing a species
☞ 17. How habitat loss affects changes in biodiversity
☞ 18. The basic features of the Canadian Species at Risk Act (SARA)
☞ 19. The role of COSEWIC (Committee on the Status of Endangered Wildlife in Canada)
☞ 20. The definitions of different status designations and terms used by COSEWIC
☞ 21. Which aquatic species are currently designated as threatened, endangered or extirpated
☞ 22. The definitions of terms and designations used for global rankings
☞ 23. The definitions of terms and designations used for national rankings
☞ 24. The definitions of terms and designations used for provincial rankings
☞ 25. The criteria used in determining the conservation status of a species
☞ 26. How to determine when a change in community composition is occurring
☞ 27. The definition and characteristics of an exotic species
☞ 28. The variety of dispersal mechanisms used by invading species
☞ 29. Most of the aquatic species that have been introduced to the Great Lakes
☞ 30. The potential kinds of harmful ecological and socio-economic impacts caused by invaders
☞ 31. The potential kinds of beneficial impacts of invading species
☞ 32. A case history of the an exotic species - the zebra mussel
☞ 33. The biology of the zebra mussel
☞ 34. The negative impact of the zebra mussel
☞ 35. The beneficial impact of the zebra mussel
☞ 36. The chemical control of the zebra mussel
☞ 37. The physical control of zebra mussels
☞ 38. The mechanical control of zebra mussels
☞ 39. The biological control of zebra mussels
☞ 40. Some invasive plants and their impacts, including:
  ☞ a. Water hyacinth
  ☞ b. Water lettuce
  ☞ c. Water spangles

☞ d. Eurasian milfoil
☞ e. Common reed
☞ f. Purple loosestrife
☞ 41 Some facts about some of the most recent invaders to the Great Lakes, including:
☞ a. Round and tubenose gobies
☞ b. European ruffe
☞ c. Blueback herring
☞ d. Quagga mussel
☞ e. New Zealand snail
☞ f. Spiny water flea
☞ g. European gammarid
☞ 42. The meaning of biomimicry and ten lessons to be learned from its application

BIODIVERSITY

Diversity is the range of variation or differences among a set of entities. Biodiversity is commonly measured in terms of genetic diversity and/or species diversity. Ultimately, biodiversity depends to a large extent on the diversity within a habitat; the more diverse the habitat, the greater the diversity of organisms that can live there.

Genetic Diversity

1☞ Genetic diversity represents the heritable variation within and between populations of organisms. Species that are best adapted to existing environmental conditions will prevail. However, there is some degree of genetic variation among individuals in all populations. This variation can only be detected by biochemical methods. For example, electrophoresis can be used to examine genetic variation in populations.

The genetic makeup of an organism is known as a ***genotype***. Each genotype has an expressed, or physical trait (e.g. mayfly body colour), known as a ***phenotype***. The genotypic ratio may be different than the phenotypic ratio. For example, if a male mayfly with a dark brown exoskeleton (e.g. has alleles BB) mates with a light coloured female mayfly (e.g. has alleles bb), all the progeny will be dark brown, but the genotype will be Bb for all of them. When two individuals with Bb cross, their progeny will have a phenotypic ratio of 3 with dark brown and 1 with a light coloured exoskeleton, but the genotypic ratio will be 1 BB (dark brown), 2 Bb (dark brown) and 1 bb (light). In a head water stream environment, the dark coloured mayflies would probably be favoured (selected for) because the stream is shaded, lending the phenotypic variation as cryptic colouration that protects the mayflies from predators. Also, dark colours absorb more heat than light colours and body temperatures would rise faster, enabling earlier growth and reproduction than in light coloured mayflies. However, in shallower and exposed lower reaches, the light coloured mayflies would be favoured (selected for).

Each population has a ***gene pool***, the total collection of genes at any one time. The gene pool is a reservoir from which members of the next generation within the population derive their genes. The allele frequencies within the population gene pool will remain relatively constant, or in equilibrium, over time, unless acted upon by some agent. This equilibrium state is known as the ***Hardy-Weinberg equilibrium***. Five conditions are needed for Hardy-Weinberg equilibrium: (i) a very large population; (ii) an isolated population (no migration of individuals and their genes into or out of the population); (iii) all individuals are equal in reproductive success; (iv) mutations do not alter the gene pool; (v) mating is random.

In many cases, however, ***microevolution*** occurs. This is evolution on the smallest scale. Using our mayfly colour example, the higher frequencies of the light coloured mayfly downstream would "evolve" (be selected for) only after a number of generations.

There are five known potential causes of microevolution. (i) ***Genetic drift***, that is a change in the gene pool of a small population due to chance. This is most likely to
2☞ occur in populations with 100 or fewer individuals. In some instances, large populations can be reduced by a catastrophic event (e.g. habitat alteration by pipeline construction across a stream), resulting in a ***bottleneck effect***. The event causes death of large numbers of individuals unselectively, leaving a small population that is unlikely to have the same genetic makeup as the original population. The most prevalent alleles will be those that were present in the highest frequency in the original population; individuals with rare alleles would likely be destroyed and removed from the gene pool. Another mechanism that could occur through genetic drift is the colonization of a new habitat by a small number of individuals, leading to a ***founder effect***. The colonizers would bring with them only a small number of alleles that do not fully represent the parent population.

(ii) ***Gene flow*** is the gain or loss of alleles from a population by the movement of fertile individuals or gametes to another population. Gene flow tends to reduce genetic differences between populations. Today, air and sea travel has resulted in introductions of species from populations that were once geographically isolated but now are allowed to interbreed with North American populations.

(iii) ***Mutation*** is a change in an organism's DNA that creates a new allele. Not all mutations are harmful. Some are beneficial and result in a new gene pool or even species. However, mutations in a given gene are rare events, occurring once per gene locus per 10^5 or 10^6 gametes. Over the short term, mutation does not have much effect on a single generation. However, over the long term, mutation is vital to evolution because it is the only force that generates new alleles.

(iv) ***Nonrandom mating***. As implied, this is selective mating that results in a departure from the Hardy-Weinberg equilibrium requirements. This probably happens frequently in benthic populations due to the proximity of individuals. In lakes, it is unlikely that isopods on one rock mate with isopods on another rock. In streams, non-random mating is reduced in drifting organisms and instead enhances gene flow.

(v) ***Natural selection***, or differential success in reproduction, is a factor that is most likely to result in adaptive changes in a gene pool. Our example of different coloured mayflies is a form of natural selection. The conditions in the ambient media favour selection of dark bodies in dark, cold, upstream environments and light bodies in bright, warm, downstream environments. Adaptations to life in stream environments (e.g. streamlined bodies, flattened bodies, marginal contact, attachment claws, etc. - Chapter 8) are all a result of natural selection. One would think that natural selection would result in genetically uniform populations. However, most populations are heterozygous (carry both dominant and recessive alleles) and many recessive alleles are not displayed (in diploid organisms) because they are "hidden" or protected from natural selection and only the dominant allele influences the phenotype. For a recessive allele to be selected and expressed as the dominant phenotype, all individuals have to be homozygous (have identical alleles). Hence, in our mayfly example, only drifting mayflies that are heterozygous are likely to effect colour change in the downstream population but natural selection will "weed them out" as long as habitat conditions remain unchanged.

Before leaving the topic of genetic diversity, we should examine genetic variability of endangered and exotic species. ***Endangered species*** probably have low genetic variability. As populations are reduced, mostly through catastrophic events caused by humans (e.g. habitat alteration), their gene pool diversity also declines. Species particularly at risk are those with homozygous recessive alleles that underwent natural selection for the environmental conditions that existed prior to the catastrophic event. Such populations have no way to alter the gene pool for the new habitat conditions unless heterozygous individuals were introduced to the population. However, endangered species, by definition, are those with only a few populations still in existence. In most cases, the populations are in the same ecoregion and probably have similar gene pools. Endangered species also tend to display poor dispersal capabilities and are unlikely to migrate from one gene pool to another.

Exotic species, however, are the "weeds" of their taxonomic group. Weeds generally have great genetic variability. They also tend to have a variety of dispersal mechanisms and can migrate to and alter several other gene pools and therefore adapt easily to a variety of environmental conditions.

Species Diversity

Species diversity has two important variables, the number of species and their patterns of relative abundance. That is, given two areas, each with same number of
3☞ species, the one with an even distribution of individuals will be more diverse than one that has an unequal distribution. Hence, polluted habitats with a large number of individuals of some species and few of others, is less diverse than a pristine habitat with an equal number of a lot of species.

Generally, the greater the diversity of habitats, the greater the species diversity. However, stability of habitat types is also important; habitats in which environmental conditions have persisted for long periods of time allow species to evolve and specialize for a variety of microhabitats or food resources. This is known as the ***stability-time hypothesis***. It contrasts with the ***disturbance or cropper hypothesis*** which suggests that organisms are generalists (croppers) and increase in numbers until they reach the limit of a particular resource (e.g. food or space) that is in least supply. Then competition occurs and results in some species being eliminated. There is a third hypothesis, known as the ***area hypothesis***; the larger the area, the greater the number of species. Probably all three hypotheses apply to freshwater environments. Large lakes, like the Great Lakes, have large areas to accommodate a large number of habitats (e.g. littoral, sublittoral and profundal zones), each of which has a large number of microhabitats (e.g. different plant species, substrates, physical conditions), again each of which is utilized by generalists (e.g. herbivory, carnivory), specialists (e.g. insectivores, molluscivores, parasites) and opportunists (e.g. invaders). Generally, until humans intervene, larger habitats are more stable than smaller habitats. But stability may be relative; even temporary aquatic habitats have some stability because they are without water every year for about the same period and vernal species have evolved life history traits that correspond to this predictable variation. However, the diversity of life in temporary ponds (small and unstable) is generally less than in permanent ponds (larger and more stable) which in turn have less diversity than lakes (which are even larger and, arguably, more stable).

Chapter 8 alludes to benthic diversity as a spiders web, where each space between the longitudinal radii (spokes) represents a habitat and each cross-thread between the ribs represents a niche that is occupied by a species performing a specific function in the habitat. The same applies to any community of organisms and habitats. The strength of the web depends upon the numbers of ribs (habitats) and cross-threads (microhabitats with a certain species assemblage). Stability is measured by the "sag" in the web. Webs with only a few spokes (habitats) and cross-threads sag more than those with several habitats and microhabitats. If a spoke or cross-thread is removed (by pollution for example), the web sags a bit. The community is more stable and flexible as long as habitat diversity and niche diversity remains high. Hence, webs (aquatic environments) that are closely "knit" (or environments that have that have numerous habitats with an abundance of niches) can resist a lot of changes before the web sags (or the ecosystem exhibits a recognizable change in community structure).

Species diversity can be measured in a variety of ways. ***Species richness*** measures the absolute number of different species in a community without reference to the numbers of individuals in each. A ***species diversity index***, however, incorporates the number of individuals in each of the species present in a community. There are several diversity indices, including Shannons index (Shannon, 1948)[1], Simpson's index, 1949)[2],

---

[1]Shannon, C. E. 1949. A mathematical theory of communication. J. Bell Syst. Tech. 27: 379-423, 623-656.

[2]Simpson, E. H. 1949. Measurement of diversity. Nature 136: 688.

Margalef's index (Margalef, 1951)[3] and Menhinick's index (Menhinick, 1964)[4]. A variation of Simpson's index is given in the section, "Species Diversity", in Chapter 12 as a method for assessing impacts in streams.

Indicator Species Concept

Indicator species in each of the trophic levels has already been examined in depth. The term is applied to describe those species that respond to a specific set and levels of environmental conditions or variables. It is important to note that the mere presence or absence of a species has little or no indicator value. It is the relative abundance of the species in relation to other species that matters. Indicator species are used to monitor changes in the environment, such as the state of habitat conditions, appearance and levels of contaminants or population trends of other species.

There are several types of indicator species, including sentinel species, target species, keystone species and umbrella species. Examples of each are given below.

Sentinel Species

These are indicator species believed to be among the most sensitive to particular contaminants, chemicals, toxins or diseases in their environments. For example, changes
4☞ in diatom species assemblages or the diversity of mayflies (Ephemeroptera), stoneflies (Plecoptera) or caddisflies (Trichoptera) have been used as early warning indicators of changes in water quality. Sentinel species are selected because they are sensitive to environmental stress and should give an early warning to humans that adverse conditions exist.

Target Species

Groups of species are often considered more useful than single species because they provide more "integrated" information about a system. Groups of target species
5☞ have been used to monitor environmental conditions, such as pollution, habitat modification or fragmentation or environmental heterogeneity. They are less used than sentinel species because extensive, time-consuming surveys and multiple correlations are required to identify useful groups of target species. Different groups may be needed for different ecoregions and different environmental factors. The various biotic indices described in Chapter 12 employ the target species' concept.

---

[3]Margalef, R. 1951. Diversidad de especies en las comunidades naturales. Publicaciones del Institutio de Biologia Aplcada Barcelona 6: 59-72.

[4]Menhinick, E. F. 1964. A comparison of some species-individuals diversity indices applied to samples of field insects. Ecology 45: 859-861.

Keystone Species

Some species have the ability to not only affect the continued existence (at least in large numbers) of its competitors, or its prey, but to influence the structure of the entire
6☞ community of organisms. These are known as keystone species. Most of the examples cited in the literature are predators, like the starfish, *Pisaster*, which feeds on bivalves and prevents monopolization of space by other species. However, it seems that some opportunists, that are not predators, like the filter feeding zebra mussel, *Dreissena polymorpha*, also qualify. Soon after its arrival in Lake St. Clair, the exotic bivalve attached to native bivalves and stripped the water of planktonic foods, resulting in the starvation and death of all species of native clams, and altered both benthic and fish community structure. In the feeding process, zebra mussels utilized only a portion of the filtered food and released the rest as pseudofaeces that settled and covered much of the substrate. Hence, the zebra mussel not only eliminated its competitors, but its prolific production of pseudofaeces altered the entire structure of the benthic and planktonic communities. Planktivorous fish populations were also impacted.

Keystone species could be important for conservation because their removal from a system would cause radical changes in community structure. Perhaps 50 years from now, after (or if ever) we perceive the zebra mussel as "native", it could be considered an example of "conservation recovery" since, if the mussel was removed, it would cause radical changes in community structure. Some may argue that by getting rid of the mussel we could reverse the process and return to the original community structure.

Umbrella Species

The umbrella species concept has been applied to terrestrial organisms and is rarely, if ever, used in aquatic systems. They are species whose presence in an area
7☞ provides protection for most other species in the area. Umbrella species have been used to select areas to protect as nature reserves to maximize conservation of species diversity. An umbrella species may also be a keystone species, but each is a unique concept. When an umbrella species is preserved it allows many other species to be preserved. The absence of an umbrella species infers the absence of others, but the absence of others is not caused by the removal of the umbrella species, unless that umbrella is also a keystone species.

There are two types of umbrella species, a single large-bodied species that requires a large amount of food energy to survive and therefore has a large range to acquire this energy (Wilcox 1980)[5]; and a ***speciose group***, or "***focal taxa***", with a scattered distribution (Ryti 1992)[6]. Terrestrial examples of each are: Conserving a single

---

[5] Wilcox, B. A. 1980. Insular ecology and conservation. In: Soule, M. E. and B. A. Wilcox (Eds.). Conservation biology: an evolutionary-ecological perspective. Sinauer Assoc. Inc. Massachusetts. Pp. 95-117.

[6] Ryti, R. T. 1992. Effect of focal taxon on the selection of nature reserves. Ecological Applications 2: 404-410.

large species like the cougar may be useful for conserving most other species in a home range; conserving plant or bird groups may preserve other species on reserves. The concept has yet to be applied to aquatic environments, but there are some fish species and groups that have potential to be umbrella species or umbrella groups. For example, it is conceivable that a large lake trout population is a good umbrella for conserving a cold water fisheries, or oligotrophic species of plants, invertebrates and fish. It is also conceivable that a feeding guild of fishes, such as planktivores (e.g. rainbow smelt, alewife, gizzard shad), may be good focal taxa to conserve a fisheries or many other plant and invertebrate species in a lake.

Factors Affecting Changes in Biodiversity

Several biotic factors, such as genetic diversity, can affect biodiversity. Over the short term (e.g. one to ten years), biodiversity can fluctuate as different gene pools are randomly selected through short term changes in the environment. But over the long term (e.g. decades to centuries), biodiversity may decline due to both direct and indirect factors, such as habitat loss, habitat fragmentation, over exploitation, pollution and the introduction of exotic species. The effects of natural factors on biodiversity must first be eked from the anthropogenic factors listed. In lotic systems, the variables are variations in stream order size and the corresponding changes that occur downstream, as revealed in the river continuum concept (Chapter 2). These include substrate types, water velocity, depth and width of streams and sediment loads. All these physical attributes vary in relation to stream order, from coarse substrates (boulders, rocks, etc.) in clear, cold, well oxygenated water and narrow widths and shallow waters of low order streams (e.g. 1-3) to fine sediments (e.g. gravel, sand, silt, etc.) in more turbid, warmer, less oxygenated water and wider and deeper waters of higher stream orders (e.g. 4-6). In lentic systems, temporal variations occur due to natural eutrophication processes. If the magnitude of changes due to natural factors are known, we should be able to determine the magnitude of effects of habitat loss, habitat fragmentation, pollution and the introduction of exotic species.

Habitat Loss

Total destruction of many wetlands has resulted in the direct loss of many species of birds and fish, merely by removal of their "homes". In most cases, the wetlands have
8☞ been replaced by dirt, concrete and/or pavement, in the centre of which are new communities, high-rise buildings or structures which attract and foster only the "weed species" (pigeons, crows, etc.) that are already part of the regional species assemblage. At the very least, any aquatic species, especially rare and endangered ones, have been annihilated on the local scale. But replacement of wetlands by inert materials is not restricted to a local scale; it is a regional (affecting ecoregional and watershed diversity), continental (affecting continental, e.g. North American, diversity) and even a global (affecting global diversity) phenomenon.

Only the direct effects have been considered so far. Marshes and swamps, in particular, are biological filters that cleanse and remove suspended particles, pollutants and nutrients before they are released to lakes and rivers (Chapter 2). By removing the wetlands, one indirectly affects the quality and, therefore, the diversity of the receiving waters. Only the more resistant of the species survive, the weak are annihilated.

Habitat Fragmentation

In some cases, only a part of the wetland or aquatic ecosystem is destroyed, resulting in their fragmentation. One example of habitat fragmentation (in some cases, destruction) is the construction of numerous dirt roads and concrete or asphalt highways
9☞ across rivulets, creeks, streams and rivers within the same watershed. In fact, numerous stream orders can be crossed by the same highway, as occurs on the Grand River in southern Ontario. Scientific studies have shown that the diversity of native mussels (Unionidae) has declined by about 44% (from 32 living species to 18) in the Grand River watershed over the past 60 years. Of the 32 species, 12 are either extirpated or in imminent danger of being extirpated from the Grand River drainage basin. None of the 12 species has been collected alive from the Grand River drainage in the last 60 years. Another 12 are extremely rare to uncommon and are vulnerable. Only 25% (8 species) are relatively common in the Grand River drainage. Unfortunately, accompanying the direct effects of habitat fragmentation by highway construction are effects due to pollution from highway stormwater runoff (e.g. road salt, tars, oils, gasolines, metals, rubber tire derivatives, etc.), pollution from agriculture (nutrients, herbicides, pesticides, etc.), silt loadings and elevations in water temperature.

Another example of habitat fragmentation is the creation of artificial impoundments. The building of dams imposes a lentic habitat within a lotic system. The aquatic communities must suddenly adjust to the changes in physical, chemical and biological attributes of riverine systems to those of lacustrine systems. Some species are adapted to a lotic existence and perish when a lentic system is imposed upon them. Others, mostly highly tolerant forms like chironomids and tubificid worms, exploit the new habitat and explode in biomass. In many cases, the dams are built on streams of stream order 3, 4 or 5. In this scenario, species diversity usually (i.e. it depends on depth and size of impoundment) declines, but if it does not, the species assemblage certainly changes from one dominated by lotic shredders, filter feeders, grazers and predators to lentic herbivores, filter feeders and predators, with corresponding changes in the fish community.

Pollution

Agriculture

In some watersheds, runoff from agricultural lands accounts for almost all the discharge into major tributaries. For example, over 95% of the Grand River watershed is
10☞ agricultural land, with less than 5% containing forest cover. Mixed farming is common throughout the drainage but livestock breeding predominates in the upper and middle parts of the watershed while dairy farming predominates in the lower regions. The direct

impact of cattle crossings on survival and diversity of aquatic species is obvious and can account for significant changes observed in benthic and fish community structure.

Indirect effects of agriculture are also apparent. Direct inputs of manure by cattle results in heavy growths of blue-green surface algae (e.g. *Microcystis, Aphanizomenon*), attached green algae (e.g. *Cladophora*) and submerged macrophytes (e.g. *Myriophyllum*). In some cases, plant biomass is so dense that stream flow is reduced to stagnant pools and eddies, containing soft, black mud that emits a strong hydrogen sulfide smell. Organic enrichment is further enhanced by the spreading of inorganic (e.g. phosphates) and organic (e.g. manure) nutrients on the fields, some of which ends up in the runoff during heavy precipitation events.

Compounding the problem of organic enrichment is the input of pesticides from agricultural runoff. Herbicides are applied almost annually, after the crops have begun to grow, and enter the stream indirectly as runoff or directly by going too close and accidentally spreading the pesticide into the stream.

Urban and Highway Runoff

Aquatic communities downstream of many municipalities change due to the effects of urban storm water runoff and solid waste disposal. Storm water runoff has similar constituents as highway runoff, with road salt, tars, oils, gasolines, metals and
10☞ rubber tire derivatives entering streams as a broth of contaminants. Runoff from asphalt also has a significantly higher water temperature, often resulting in greater than a 10 °C increase in stream temperature immediately below the runoff. Silt loads are also high so that the runoff is hot and turbid. Much of the effluent results in increases in sediment concentrations of total hydrocarbons, aromatic hydrocarbons (both in lubricating oils and fuels), and heavy metals (e.g. lead in fuel (in the past), copper in brake linings, zinc and cadmium in tires, and chromium and copper in de-icing salts). The toxicity of this "broth" of contaminants in the water and sediments is complex, with synergistic and antagonistic effects. Benthic and fish communities respond accordingly, but over the long term, fish are eliminated and the benthos are dominated by pollution tolerant forms, like tubificid worms and chironomid larvae. Salt concentrations exceeding 1,000 mg Cl / L are accompanied by increases in drift rates of benthic macroinvertebrates, decreases in biomass and diversity of algae and increases in bacterial counts. A recent review and study of ecosystem impacts from highway runoff is provided by Maltby *et al* (1995)[7]. In general, the type and size of the receiving water, the potential for dispersion, the size of the surrounding catchment area and the biological diversity of the ecosystem are some of the factors determining the importance of runoff effects.

Solid waste disposal also impacts the aesthetic value of the stream. It is not an uncommon sight to see rubber tires, grocery carts, automobile parts, plastics, etc. wedged

---

[7]Maltby, L., D. M. Forrow, A. B. A. Boxall, P. Calow and C. I. Beeton. 1995. The effects of motorway runoff on freshwater ecosystems: 1. Field study. Environ. Tooxicol. Chem. 14: 1079-1092. Ibid. 2: Identifying major toxicants. Environ.. Tooxicol. Chem. 14: 1093-1101.

into the bottom or hung up on rocks in the middle of a stream.

Exotic Species

The impact of exotic species on native fauna and species diversity is well documented and discussed in more detail below. However, not all exotic species result
11 ☞ in a decline of species diversity; some add to the diversity.

Two of the most notorious exotic species are the lamprey and the zebra mussel. The sea lamprey caused the decline of native lake trout populations in the Great Lakes about 60 years ago. Enormous populations of sea lamprey developed in Lakes Huron, Michigan and Superior in the early 1940s and quickly decimated their lake trout and whitefish populations. Before the lamprey, the combined annual U. S. and Canadian catches in these lakes was about 15 million pounds. By the early 1960s the catch had declined to about 300 thousand pounds.

The zebra mussel, and its ally, the quagga mussel, were responsible for the near total loss of native freshwater pearly mussels in Lake St. Clair, western basin of Lake Erie and the St. Lawrence River. Scientists predicted that native mussels that carried, on average, a mass of zebra mussels equal to or greater than their own mass, would be extirpated. They were right! The impact of zebra and quagga mussels in river and stream environments, however, should be less severe, provided there is no impoundment in the river system. Because the exotic mussels have planktonic larvae (called veligers), stream currents will carry the larvae downstream from their point of introduction (see Chapter 8). The distance that veligers are carried downstream will be equal to the water velocity multiplied by the development time. Even using an average water velocity of 0.1 m/sec and a development time of 20 days, the veligers would be carried about 173 km before they settle on the bottom. Stream populations can be expected to survive only one generation (about 2 years), unless the populations are rejuvenated by an upstream source or by a re-introduction at the site. Impoundments with retention times greater than 20-30 days will give the veligers enough time to develop and settle; the impoundment populations will then "seed" populations downstream of the impoundments on a annual basis. ***Therefore, it is vital to prevent the introduction of zebra mussels into impoundments.***

Species' Strengths and Weaknesses

Before examining different concepts of endangered and exotic species, it is necessary to review the strategies used by different species in different ecosystems. The following is summarized from descriptions in Stearns (1976)[8].

Species are thought to adopt one of three life history strategies in order to live in stable and unstable environments: (i) ***r*-selection**; (ii) ***K*-selection**; or (iii) **bet-hedging**.

---

[8]Stearns, S. C. 1976. Life-history tactics: A review of the ideas. Quart. Rev. Biol. 51: 3-47.

Stable environments are those in climates that are relatively constant and/or predictable, as in tropical climates. Unstable and/or unpredictable environments are characteristic of variable climates, like many in the temperate zone. Stable environments are
12☞ characterized by species with a *K*-strategy, while fluctuating environments are characterized by species with an *r*-strategy. Advocates of *r*- and *K*-selection deal with models in which fecundity and mortality schedules fluctuate. Bet-hedging is advocated when fluctuations in these life history traits occur. Table 14.1 is a summary of Stearn's (1976) correlates of r- and K-selection.

Table 14.1. Some strategies of r- and K-selection traits.

Traits	r-selection	K-selection
Mortality	Density-independent; high adult mortality	Density-dependent; high juvenile mortality
Population size	Variable in time, no equilibrium; recolonization frequently needed; usually below carrying capacity	Constant in time and equilibrium; recolonization rarely needed; at or near carrying capacity
Competition	Often lax	Usually keen
Selection favours:	1. Rapid development 2. High intrinsic rate of increase 3. Early reproduction 4. High resource thresholds 5. Small body size 6. Semelparity 7. Increased birth rate	1. Slow development 2. Competitive 3. Delayed reproduction 4. Low resource threshold 5. Large body size 6. Iteroparity 7. Decreased death rate
Life span	Short, < 1 year	Long, > 1 year
Leads to:	Productivity	Efficiency
Relative apportionment of energy to reproduction:	Relatively large	Relatively small
1. Mass of young/parent/ brood	Larger	Smaller
2. Mass of young/parent/lifetime	Larger	Smaller
3. Size of young	Smaller	Larger
4. Parental care	Less	More

Most of the correlations in Table 14.1 are self-explanatory. In fluctuating environments, such as temporary ponds, species must develop quickly and reproduce before the pond dries. A high resource threshold permits the organisms to utilize the abundance of resources that suddenly become available when the pond fills in the spring. Small body size enables the organisms to aestivate in the mud. Because adult mortality is usually high and life spans are short, semelparity, or "***big bang***" reproduction (a bunch of eggs or young are released and then the parent dies) is favoured in unstable environments.

Organisms in stable environments, on the other hand, can afford to grow slowly, put more energy in growth and development early and then apportion energy to reproduction. With environmental stability, species have evolved long life spans which leads to larger body sizes, reproduction once a year (after a long period reaching sexual maturity) and low adult mortality until they reach senescence (a case where being big helps) .

While r- and K-selection explain observed life history traits in most environments, a mix of traits is seen in others. For example, juvenile mortality is often higher in unstable environments than in stable environments and parental care is seen in many species (e.g. fingernail clams) in temporary ponds. Such species have adopted a "bet-hedging" strategy which has resulted in reversal of r- and K-selection tactics (Table 14.2). Note that the descriptions used are subjective and relative to the groups under consideration. For example, within the freshwater bivalve families, Unionidae (commonly called freshwater pearly mussels) and Dreissenidae (e.g. zebra mussels), dreissenids are opportunists because they live for only 2 to 3 years, they have high fecundities (up to 1 million eggs per female) but high larval and juvenile mortality (up to 99%), very rapid growth rates (attain 90% of their growth in 4 months) and utilize as many as 23 different dispersal mechanisms (it took zebra mussels only eight years to disperse from Lake St. Clair, where they first appeared in North America, to the Gulf of Mexico!). Unionids, on the other hand, live for 10-20 years (some up to 100 years), have high fecundities and high juvenile mortality (opportunistic traits), very slow growth rates (continue growing throughout life) and very poor dispersal mechanisms (they produce

Table 14.2. Contrasting predictions of r- and K-selection and bet-hedging Modified from Stearn 1976).

Bet-hedging Predictions When Adult Mortality is Variable	
Stable Environments	**Fluctuating Environments**
Slow development, late maturity	Rapid development, early maturity
Iteroparity	Semelparity
Smaller reproductive effort	Larger reproductive effort
Fewer young	More young
Long life span	Short life span
Bet-hedging Predictions When Juvenile Mortality is Variable	
Early maturity	Late maturity
Iteroparity	Semelparity
Larger reproductive effort	Smaller reproductive effort
Fewer broods but more young per brood	More broods but fewer young per brood
Short life span	Long life span

glochidia larvae that require certain species of fish to develop and disperse them (e.g. water currents). In fact there are no exotic species of Unionidae in North America!. Clearly, unionids are generalists relative to dreissenids. Yet, if one compares species within the family Unionidae, some species (e.g. those with a North American distribution and live for only 10 years) would be considered opportunists relative to other unionid species (e.g. those that have very restricted distributions and live for 50-100 years).

ENDANGERED SPECIES

All aquatic ecosystems change in morphological, physical and chemical characteristics over time. For example, Chapter 2 describes how streams constantly erode new paths that are accompanied by corresponding changes in physical and chemical attributes. Eutrophication is a natural process but eons of time are required to change an oligotrophic lake into a eutrophic one. While the ecosystem changes slowly, organisms can adapt slowly to the conditions present. But if the rate of change is suddenly increased, only those organisms with life history traits that can accommodate the increasing rate of change, or an unstable environment, will prevail. Species that have incorporated life history traits adapted for a stable environment will succumb. Hence, species that have incorporated K-strategy traits will probably perish before those with r-strategy traits. But many of the strategies are relative, for example, slower vs faster, larger vs smaller, more vs less, etc. Is there a specific rate, size or quantity that separates an endangered species from a "weed" species? Probably not because each species' strengths and weaknesses depends on the combinations, kinds and magnitudes of stressors present. But there are some general principles to follow.

The various kinds of life history strategies employed by organisms have already been examined. Now consider which attributes would contribute to the decline and, perhaps, the disappearance of a species. In other words, what are the potential weaknesses of endangered species? One way is to rank the different traits, as done in Table 14.3, from those that give a species a competitive "edge" to those that would contribute to its probable extinction (i.e. on a global scale). Table 14.3 summarizes the ranked orders for several traits, the rationales being provided below. Benthic invertebrates and fish are given as examples for each trait.

Life Span and Size

The gene pool of species with a short life span will change faster than for species with a long life span. If the rates of change in environmental quality and conditions
13☞ increase, the genotypes and phenotypes selected will probably be from species with short life spans. Moreover, most species with short life spans become reproductively mature at an earlier age than species with long life spans. For example, some species of unionid clams live close to 100 years and do not begin reproducing until their tenth year of life. Most unionid species with life spans shorter than 10 years begin reproducing during or

Table 14.3. Ranking of life history traits, ecological tolerances and requirements and dispersal potential that potentially could lead to extinction of a species under rapidly changing aquatic (freshwater) conditions.

Life History Traits	Rank (from survivorship to extinction)
1. Life span	a, Short-lived > b, long-lived
2. Sex	a, Hermaphroditism > b, Separate sexes
3. Number of generations per year	a, multivoltinism > b, bivoltinism > c, univoltinism
4. Life time number of generations	a, iteroparity > b, semelparity
5. Parental care	a, oviparity > b, ovoviviparity > c, viviparity > d, parasitism
6. Fecundity	a, High numbers > b, low numbers of eggs
7. Natality	a, High survival > b, low survival rate of embryos
8. Development duration and rate	a, Short development time, fast rate > b, long development time, slow rate
9. Age at sexual maturity	a, Early sexual maturation > b, late sexual maturation
10. Size	a, Small > b, Large
11. Ecological tolerances and requirements	a, hardy and tolerant (e.g. eutrophic indicators) > b, moderately hardy and tolerant (e.g mesotrophic indicators) > c, very sensitive, requiring pristine condition (e.g. oligotrophic indicators)
12. Dispersal potential	a, Small size, has evolved dispersal mechanism(s), uses a variety of passive and active dispersal mechanisms > b, large size, has evolved few if any dispersal mechanisms, relies mostly on passive mechanisms

immediately after their first year of life and contribute to the gene pool at a rate of one to two orders of magnitude faster than those living 100 years. Lake sturgeon (*Acipenser fulvescens*), with typical maximum ages of 50 to 80 years (depending on sex), not only sexually mature at a very late age (8 - 13 years) but spawning takes place even later, at 12-19 years for males and 14-23 years for females. With such a life history trait, the species is doomed to extinction if humans do not change their habits of building dams or polluting the rivers that are part of the sturgeon's migratory patterns.

Another correlate of life span is size; long-lived species are generally larger than short-lived species. Size affects not only a species' potential for dispersal, as discussed below, but its reproductive potential, larger species generally producing more eggs than smaller species.

Reproductive Potential

14☞ Several aspects of reproductive potential need to be considered here: (i) the species' sexual state (e.g. separate sex or hermaphrodite); (ii) its egg-laying habit (e.g.

oviparous, ovoviviparous, viviparous); (iii) its fecundity (no of eggs produced); (iv) its natality (number of eggs surviving); (v) its annual frequency of egg-laying habits (e.g. univoltine, bivoltine, multivoltine); (vi) its life-time frequency of egg-laying habits (e.g. semelparity, iteroparity).

Hermaphroditism reduces the risk of a species being eliminated during periods when it is difficult to find a mate. Parthenogenesis would also allow a species to reproduce when mates are difficult to find. Hence, a species is more likely to become endangered if a dioecious habit is employed. Oviparity (egg laying with young hatching from egg) is more common than ovoviviparity (brooding of eggs and young, with birth of miniature adults) in freshwater animals. Viviparity is mostly absent in freshwater animals. Ovoviviparity seems to be more common in hermaphrodites than in dioecious animals. Snails of the family Viviparidae are ovoviviparous (in spite of family name) and dioecious, but most species are also capable of parthenogenetic reproduction. The brooding behaviour is usually accompanied by few, small-sized young with a high survival rate (i.e. high natality rate), while oviparity often results in enormous numbers of eggs, many of which perish during development, but those that do survive mature into good competitors. Hermaphrodites alsc tend to have shorter life spans and higher frequencies of reproductive events per year (e.g. bivoltinism, trivoltinism or multivoltinism) but they have fewer reproductive events in their life time (i.e. are semelparous) than do most dioecious species. In summary, species that are dioecious, semelparous, univoltine, ovoviviparous and have low fecundities and/or natalities are more likely to become extinct than species that are hermphroditic, iteroparous, multivoltine, oviparous and have high fecundities and/or natalities.

Some of the above traits are mutually exclusive. For example, ovoviviparity is rarely seen in dioecious aquatic animals but it is favoured over oviparity for overland dispersal because the young are protected by the parent. However, the reproductive tactics can be variable, with both oviparous and ovoviviparous forms displaying a range of fecundities and natalities and voltinism. Hence, based on reproductive traits alone, the potential for a species' dispersal ability can be ranked from high (e.g. ovoviviparity, high fecundity, high natality, mulivoltine), to moderate (e.g. ovoviviparity, low fecundity, high natality, bivoltine) or low (e.g. oviparity, low fecundity, low natality, univoltine).

Species Tolerances and Requirements

The physiological and ecological tolerances and requirements describe the "hardiness" of a species. The more hardy a species is, the greater its ability to adapt to
15☞ quickly changing environments. "Weed" species are not likely to become endangered or extinct. They are widely distributed and if pollution or intentional destruction by humans eradicates them in one part of the country, other populations will perpetuate and perhaps even reintroduce the species. If humans alter the rate of change in habitat quality, pollution (or eutrophic) indicator species have less potential to become extinct than do clean water (or oligotrophic) indicator species. For example, of the fingernail clams, the

arctic-alpine clam, *Pisidium conventus,* is more likely to become extinct than the ubiquitous pea clam, *Pisidium casertanum*; or deep water sculpins (*Myoxocephalus thompsoni (=quadricornis*) would likely have a faster extinction rate than the stream- and lake-dwelling slimy sculpin (*Cottus cognatus*). Support for this argument can be seen in Table 14.5 where many, if not most, of the fish species that are listed as endangered or threatened are cold-water species adapted to oligotrophic or pristine conditions.

Dispersal Potential

This attribute determines the range and numbers of populations that can be established by a species. A specie's dispersal potential is dictated, in large part, by the
16 ☞ factors previously discussed. For example, if the species does not have wide physiological and ecological tolerances and requirements, if it is too large to disperse, or if it does not have reproductive traits conducive to its dispersal, the species is destined to isolation (i.e. has a very small range) and probable extinction. Each of these is elaborated upon below, but the various dispersal mechanisms are reviewed when the traits of exotic species are examined (see Table 14.8).

There are two basic types of dispersal mechanisms, ***passive*** and ***active dispersal***. Passive dispersal is hitch-hiking a ride using abiotic (e.g. water currents, wind, ships,
17 ☞ boats, etc.) or biotic (e.g. birds, insects, mammals) vectors. Most of the active dispersal mechanisms are natural; most of the passive dispersal mechanisms are anthropogenic. Active dispersal is dispersal using the swimming (e.g fish) or flying (e.g. adult insects with aquatic larval stages) abilities of the organism.

Most widely-distributed species probably have evolved with dispersal ability as part of the natural selection process. For example, the ephippium of cladocerans is considered an adaptation for overland dispersal by wind or animal vectors because the enclosed eggs are able to survive long periods out of the water. However, dispersal is somewhat random and the ephippium may end up in a harsher environment than whence it came. Because dispersal is somewhat random, species with wide ecological tolerances are likely to be dispersed more widely than species with narrow ecological tolerances. Since humans seem to be continually altering aquatic habitats on a global scale, the species most destined to extinction are those specialized to live within very narrow ranges of physiological and ecological tolerances and requirements.

The size of a species is important because it determines, in part, the potential for a species to spread great distances. In general, small species disperse greater distances than large species. For example, fingernail clams (Family Sphaeriidae) have a greater global distribution than freshwater pearly mussels (Family Unionidae). There appears to be two reasons for this trend. Firstly, big is more noticeable than small and the dispersal agent is more likely to "dump" a large hitch-hiker sooner than a small one. Secondly, big also means heavy and a vector would have to spend more energy, and therefore risk its own life, to transport itself and its baggage great distances. Perhaps the only exception to the size rule is intentional introductions by humans. In this instance, large is an advantage if the introduction is for food (or sport). Coho salmon and carp are good examples of intentional introductions of large species. The mystery snails, *Cipangopaludina malleatus chinensis* and *C. japonicus* are also examples of large species introduced for escargot.

Table 14.3 summarizes and ranks the different life history ecological and dispersal traits that occur in freshwater organisms. Different combinations of traits would lead to different rates of extinction. For example, a species with a combination of 1a, 2a, 3a, 4a, 5a, 6a, 7a, 8a, 9a, 10a, 11a and 12a traits would survive longer than a species with 1b, 2b, 3b, 4b, 5b, 6b, 7b, 8b, 9b, 10b, 11b and 12b traits, which in turn would survive longer than a species with 1b, 2b, 3c, 4b, 5d, 6b, 7b, 8b, 9b, 10b, 11c and 12b traits. Support for the latter ranking can be seen in the four families of freshwater bivalves. Two families (Unionidae and Sphaeriidae) are native to North America and two (Dreissenidae and Corbiculidae) are introduced. The Unionidae (pearly mussels) are long-lived (up to 100 years), dioecious, univoltine and iteroparous. They produce parasitic larvae called glochidia that require a fish to complete development of most of their organ systems. Some unionids are very host specific, requiring only a certain fish species to parasitize. They have very high fecundities (about 1 to 2 millicn eggs are produced) but very low survival of young (< 0.0007%), because most glochidia do not find a fish host and those that do may perish because the fish is preyed upon. However, the glochidia have a relatively short (about 30 days as a parasite) development time. Some adults attain sexual maturation after 1 to 5 years but many require up to 10 years to mature.

Unionids are the largest of the freshwater bivalves. As a group, they are very sensitive to changing environmental conditions, with habitat alteration being the most commonly cited cause of the high species extinction rate. Of the two native families of bivalves, the Unionidae have 72% of the total number of species (~300) listed as either extinct, endangered, threatened or of special concern and only 24% are currently stable. In fact, only the Unionidae are not represented by exotic species in North America.

The Sphaeriidae (fingernail clams) are short-lived (1 to 2 years), hermaphroditic, univoltine to bivoltine, semelparous to iteroparous and ovoviviparous, brooding their larvae for 2 to 5 weeks. Most sphaeriids have low fecundities (5 -50 young per parent) but high natalities and short development times (most are ready for birth in 2 to 5 weeks). Adults are sexually mature shortly after birth. The sphaeriids are the smallest of the freshwater bivalves, with some species growing only to about 1.5 mm in shell length. Of the 40 or so species of Sphaeriidae, only one (*Pisidium ultramontanum*) is of special concern. They also have good dispersal powers, with 5 species introduced to North America from Eurasia. Some species (e.g. *Pisidium casertanum*) are very tolerant of organic enrichment, others are very sensitive (e.g. the oligotrophic indicators, *Pisidium conventus* and *Sphaerium nitidum*).

The Corbiculidae (Asian clams) and Dreissenidae (zebra and quagga mussels) are the most tolerant and prolific of the four families of bivalves. The corbiculids are short-lived (2 to 3 years), monoecious to dioecious, univoltine to bivoltine (some populations have continuous breeding for 3 to 4 months) and are iteroparous. They brood up to ten thousand larvae for 5 to 6 weeks (some for 2 to 3 months) and then release them to a planktonic existence for 3 to 5 days. Apparently most of the larvae survive and settle. The juveniles grow quickly into moderately large (4 to 7 cm) adults which attain relatively early sexual maturity. Asian clams are extremely tolerant, even being used to clarify sewage, and excellent competitors, known to displace unionids and sphaeriids.

The Dreissenidae are short-lived (1 to 2 years on average), dioecious, uni- to bivoltine and iteroparous. Zebra and quagga mussels have extremely high fecundities (~ one million eggs per female), the eggs developing into planktonic larvae that have a short development time (2 to 4 weeks). However, the larvae have a very low survival rate (~ 1%) because many do not find an appropriate substrate to settle on. The adults have an early sexual maturity (about 8 weeks or 8 mm in shell length) and grow to only 3 to 4 cm shell length on average. Zebra and quagga mussels are very tolerant, able to survive oxygen levels down to 2 mg/L, tolerate salinities up to 8 ppt, and grow best under mesotrophic to eutrophic conditions. They are also excellent competitors, known to have displaced entire unionid communities in some lakes (e.g. Lake St. Clair and the western basin of Lake Erie).

The Canadian Endangered Species Protection Act

The Government of Canada first considered the establishment of a Canadian Endangered Species Protection Act (CESPA)[9] in 1996 and would have provided the federal component with a national commitment to protect and conserve endangered species. The goal of the Act, and of its programs that flowed from it, was to prevent wild
18☞ Canadian species from becoming extinct as a consequence of human activities and to recover species where possible and economically feasible. The Act would have protected all native, wild, non-domesticated species, subspecies and geographically distinct populations within Canada and Canadian waters. CESPA was tabled October 31, 1996, but died on the order paper when a federal election was called in April 1997). However, the Government of Canada now has a new Bill, C-33, that is commonly referred to as the Canadian Species at Risk Act, or SARA[10], that is currently tabled and should be proclaimed by spring 2001. Most provinces have their own stand-alone Endangered Species Acts. SARA has a similar goal, purpose and scope as CESPA.

---

[9]The Canadian Endangered Species Protection Act: A legislative proposal. 1995. Minister of Supply and Services Canada. Available from Minister of the Environment, House of Commons, Room 509-S, Centre Block, Ottawa, Ontario K1A 0A6.

[10] The House of Commons of Canada, Bill C-33. An Act respecting the protection of wildlife species at risk in Canada. First reading, April 11, 2000. Available from: Public Works and Government Services Canada - Publishing, Ottawa, Ontario K1A 0S9.

The status of wild species, subspecies and separate populations in Canada is determined by COSEWIC (Committee on the Status of Endangered Wildlife in
19☞ Canada), legislatively established by SARA. COSEWIC is composed of representatives from each provincial and territorial government wildlife agency, four federal agencies (Canadian Museum of Nature, Canadian Parks Service, Canadian Wildlife Service and Department of Fisheries and Oceans) and three national conservation organizations (Canadian Nature Federation, Canadian Wildlife Federation and World Wildlife Fund, all referred to as NGOs (non-government organizations)), as well as experts (chairs and co-chairs) for each of the taxa listed in Table 14.4. The committee meets annually in the spring and fall to consider status reports of candidate species. With the best available information, COSEWIC's main functions are to:

(a) assess the status of each wildlife species considered to be at risk, identify existing and potential threats to the species and:

(i) classify the species as extinct, extirpated, endangered, threatened or of special concern,

(ii) indicate that COSEWIC does not have sufficient data to classify the species, or

(iii) indicate that the species is not currently at risk.

(b) rank the urgency of assessing particular wildlife species;

(c) reassess the status of species at risk and, if necessary, reclassify or declassify them;

(d) develop and periodically review criteria for assessing the status of wildlife species and for classifying them and recommend the criteria to the Minister and the Canadian Endangered Species Council (CESC);

(e) provide advice to the Minister and CESC and perform any other functions that the Minister, after consultation with that council, may assign.

All native fish, amphibians, reptiles, plants and animals are included, with freshwater and marine molluscs and Lepidoptera added to COSEWIC's mandate in 1995. Besides fish and molluscs, no other freshwater group is currently included in COSEWIC's mandate. Table 14.4 summarizes COSEWIC's status designations as of April 1999.

Table 14.4. Status designations of the Committee on the Status of Endangered Wildlife in Canada, April 1999. Mammals are divided into terrestrial and marine, with marine species given in parentheses; for amphibians & reptiles, reptiles are given in parentheses; for molluscs and Lepidoptera, Lepidoptera are given in parentheses.

Category	Birds	Mammals	Fish	Amphibians & Reptiles	Molluscs & Lepidoptera	Vascular Plants	Lichens	Total
Extinct	3	1 (1)	6	0 (0)	1 (0)	0	0	12
Extirpated	2	2 (2)	2	0 (1)	1 (3)	2	0	15
Endangered	18	6 (6)	4	2 (4)	4 (1)	40	1	86
Threatened	7	5 (5)	18	1 (6)	2 (0)	30	0	75
Vulnerable	22	19 (8)	42	9 (8)	0 (1)	39	3	151
Total	**52**	**33 (22)**	**72**	**12 (19)**	**8 (5)**	**111**	**4**	**339**

Definitions for Conservation Status and Rankings

20☞ COSEWIC's definition of terms and "risk" categories are as follows:

Wildlife: All wild life, including wild mammals, birds, reptiles, amphibians, fishes, invertebrates, plants, fungi, algae, bacteria, and other wild organisms.

Species: Any indigenous species, subspecies, variety or geographically defined population of wild fauna and flora which have been regularly occurring in Canada for more than 50 years.

Peripheral Species: Any terrestrial or freshwater species with an historical range in Canada confined to within 50 km of the national border, and a spatial distribution within Canada less than 10% of its global distribution, or a marine species, with a Canadian distribution less than 0% of the global distribution. COSEWIC's mandate does not include peripheral species.

Regularity of Occurrence: Species which occur regularly in Canada, excluding vagrants.

Non-resident or Migratory Species: Any species which, although not a full-time resident in Canada, meets other eligibility criteria and requires a habitat in Canada for a key life history stage.

Extinct (X): A species that no longer exists in the world.

Critically Endangered (CE): This designation is not currently in use but is used in global designations. It includes those species with a 50% risk of extinction in 10 years or 3 generations.

Extirpated (XT): A species no longer existing in the wild in Canada, but occurring elsewhere.

Endangered (E): A species facing imminent extirpation or extinction.

Special Concern (SC): Species which are particularly sensitive to human activities or natural events but are not endangered or threatened species. This designation was made in 2000

Threatened (T): A species likely to become endangered if limiting factors are not reversed. Designation used up to 1999, discontinued in 2000.

Vulnerable (V): Applies to a species of special concern because of characteristics that make it particularly sensitive to human activities or natural events. Vulnerable was used up to 1999 and discontinued in 2000.

Data Deficient (DD): A species for which there is inadequate information to make a direct, or indirect, assessment of its risk of extinction. The designation replaced "Indeterminate" in 2000.

Not at Risk (NAR): A species that has been evaluated and found to be not at risk.

Indeterminate (I): A species for which there is insufficient scientific information to support status designation. The term has been replaced by "Data Deficient".

Table 14.5 lists the species of freshwater organisms currently designated with a conservation status. The year of listing is also given. The list is based on information
21☞ in COSEWIC's April 1999 update of Canadian species at risk.

The terms and definitions used at the national level differ from those used at the global, continental and provincial levels. The following are the terminologies and definitions used at each level.

Global Rankings

These are assigned by a consensus of the network of natural heritage programs, scientific experts, and The Nature Conservancy to designate a rarity rank based on the
22☞ range-wide status of a species or subspecies. Ranks for the list are provided to the National Heritage Information Centre (NHIC), a joint venture of the Ontario Ministry of Natural Resources, Nature Conservancy Canada, Natural Heritage League and The Nature Conservancy.

G1 Extremely rare; usually five or fewer occurrences in the overall range or very few remaining individuals; or because of some factor(s) making it especially vulnerable to extinction.

G2 Very rare; usually between 5 and 20 occurrences in the overall range or with many individuals in fewer occurrences; or because of some factor(s) making it vulnerable to extinction.

G3 Rare to uncommon; usually between 20 and 100 occurrences; may have fewer occurrences, but with a large number of individuals in some populations; may be

Table 14.5. Species of fish, macrophytes and molluscs that are listed as either extinct, endangered, threatened or vulnerable in Canada. The species listed by COSEWIC as "not at risk" are not listed here.

Species	Province	Status
Fish:		
Acadian whitefish (*Coregonus huntsmani*)	NS	Endangered/1984
Aurora trout (*Salvelinus fontinalis timagamiensis*)	ON	Endangered/1987
Banded killifish (*Fundulus diaphanus*)	NF	Vulnerable/1989
Banff longnose dace (*Rhinichthyes cataractae smithi*)	AB	Extinct/1987
Benthic Hadley Lake stickleback (*Gasterosteus* sp.)	BC	Endangered/1999
Benthic Paxton Lake stickleback (*Gasteroteus* sp.)	BC	Threatened/1999
Benthic Vananda Lake stickleback (*Gasteroteus* sp.)	BC	Threatened/1999
Bigmouth buffalo (*Ictiobus cyprinellus*)	MB, ON, SK	Vulnerable/1989
Bigmouth shiner (*Notropis dorsalis*)	MB	Vulnerable/1985
Black buffalo (*Ictiobus niger*)	ON	Vulnerable/1989
Black redhorse (*Moxostome duquesnei*)	ON	Threatened/1988
Blackfin cisco (*Coregonus nigripinnis*)	ON	Threatened/1988
Blackstripe topminnow (*Funduls notatus*)	ON	Vulnerable/1985
Blue walleye (*Stizostedion vitreum glaucum*)	ON	Extinct/ 1985
Bridle shiner (*Notropis bifrenatus*)	ON	Vulnerable/1999
Brindled madtom (*Noturus miurus*)	ON	Vulnerable/1985
Central stoneroller (*Campostoma anomalum*)	ON	Vulnerable/1985
Channel darter (*Percina copelandi*)	ON, QC	Threatened/1993
Chestnut lamprey (*Ichthyomyzon castaneus*)	MB, SK	Vulnerable/1991
Copper redhorse (*Moxostoma hubbsi*)	QC	Threatened/1987
Cultus pygmy sculpin (*Cottus* sp.)	BC	Vulnerable/1997
Deepwater cisco (*Coregonus johannae*)	ON	Extinct/1988
Easter sand darter (*Ammocrypta pellucida*)	ON, QC	Threatened/1994
Enos Lake Stickleback (*Gasterosteus* sp.)	BC	Threatened/1988
Giant stickleback (*Gasterosteus* sp.)	BC	Vulnerable/1980
Gravel chub (*Erimystax x-punctatus*)	ON	Extirpated/1987

Great Lakes deepwater sculpin (*Myoxocephalus thompsoni*)	AB/MB/NT/ON/QC/SK	Threatened/1987
Green sturgeon (*Acipenser medirostris*)	BC	Vulnerable/1987
Greenside darter (*Etheostoma blennoides*)	ON	Vulnerable/1990
Kiyi (*Coregonus kiyi*)	ON	Vulnerable/1988
Lake chubsucker (*Erimyzon sucetta*)	On	Vulnerable/1994
Lake lamprey (*Lampetra macrostoma*)	BC	Vulnerable/1998
Lake Simcoe whitefish (*Coregonus clupeaformis*)	ON	Threatened/1987
Lake Utopia dwarf smelt (*Osmerus* sp.)	NB	Threatened/1998
Limnetic Hadley Lake stickleback (*Gasterosteus* sp.)	BC	Endangered/1999
Limnetic Paxton Lake stickleback (*Gasteroteus* sp.)	BC	Threatened/1999
Limnetic Vananda Lake stickleback (*Gasteroteus* sp.)	BC	Threatened/1999
Longjaw cisco (*Coregonus alpenae*)	ON	Extinct/1985
Margined madtom (*Noturus insignis*)	ON, QC	Threatened/1989
Morrison Creek lamprey (*Lampetra richardsoni*)	BC	Threatened/1999
Nooksack dace (*Rhinichthyes* sp.)	BC	Endangered/1996
Northern brook lamprey (*Ichthyomyzon fossor*)	MB, ON, QC	Vulnerable/1991
Northern madtom (*Noturus stigmosus*)	ON	Vulnerable/1998
Orangespotted sunfish (*Lepomis humilis*)	ON	Vulnerable/1989
Paddlefish (*Polydon spathula*)	ON	Extirpated/1991
Pugnose minnow (*Opsopoedus emilae*)	ON	Vulnerable/1985
Pugnose shiner (*Notropis anogenus*)	ON	Vulnerable/1985
Redbreast sunfish (*Lepomis auritus*)	NB	Vulnerable/1989
Redside dace (*Clinostomus elongatus*)	ON	Vulnerable/1987
River redhorse (*Moxostoma carinatum*)	ON, QC	Vulnerable/1987
Rosyface shiner (*Notropis rubellus*)	MB	Vulnerable/1994
Salish sucker (*Catastomus* sp.)	BC	Endangered/1986
Shorthead sculpin (*Cottus confusus*)	BC	Threatened/1984
Shortjaw cisco (*Coregonus zenithicus*)	AB/MB/NT/ON/SK	Threatened/1987
Shortnose cisco (*Coregonus reighardti*)	ON	Threatened/1987
Shortnose sturgeon (*Acipenser brevirostrum*)	NB	Vulnerable/1980
Silver chub (*Macrhybopsis storeriana*)	MB, ON	Vulnerable/1985

Silver shiner (*Notropis photogenis*)	ON	Vulnerable/1987
Speckled dace (*Rhinichthys osculus*)	BC	Vulnerable/1980
Spotted gar (*Lepisosteus oculatus*)	ON	Vulnerable/1994
Spotted sucker (*Minytrema melanops*)	ON	Vulnerable/1994
Spring cisco (*Coregonus* sp.)	QC	Vulnerable/1992
Squanga whitefish (*Coregonus* sp.)	YT	Vulnerable/1987
Umatilla dace (*Rhinichthyes umatilla*)	BC	Vulnerable/1988
Warmouth (*Lepomis gulosus*)	ON	Vulnerable/1994
Western silvery minnow (*Hybognathus argyritus)*	AB	Vulnerable/1997
White sturgeon (*Acipenser transmontanus*)	BC	Vulnerable/1990
Molluscs:		
Banff Springs snail (*Physella johnsoni*)	AB	Threatened/1997
Dwarf wedgemussel (*Alasmidonta heterodon*)	NB	Extirpated/1999
Eelgrass limpet (*Lottia alveus*)	NF, NS, QC	Extinct/1996
Hotwater physa (*Physella wrighti*)	BC	Endangered/1998
Northern abalone (*Haliotis kamtschatkana*)	BC	Threatened/1999
Northern riffleshell (*Epioblasma torulosa rangiana*)	ON	Endangered/1999
Rayed bean (*Villosa fabalis*)	ON	Endangered/1999
Wavy-rayed lampmussel (*Lampsilis fasciola*)	ON	Endangered/1999
Macrophytes:		
Bolander's quillwort (*Isoetes bolanderi*)	AB	Vulnerable/1985
Engelmann's quillwort (*Isoetes engelmannii*)	ON	Endangered/1992
Hill's pondweed (*Potamogeton hillii*)	ON	Vulnerable/1986
Water-plantain (*Ranunculus alismaefolius*)	BC	Endangered/1996

susceptible to large-scale disturbances.

G4 Common; usually more than 100 occurrences; usually not susceptible to immediate threats.

G5 Very common; demonstrably secure under present conditions.

T Denotes that the rank applies to a subspecies or variety

North American Conservation Status Rankings

23☞ ***R*** Rare species; any indigenous species that, because of its biological characteristics, or because it occurs at the fringe of its range, or for some other reason, exists in low numbers or in very restricted areas but it is not a threatened species.

T Threatened species; any indigenous species that is likely to become endangered if the factors affecting its vulnerability do not become reversed.

E Endangered species; any indigenous species whose existence is threatened with immediate extinction through all or a significant portion of its range, owing to the action of man.

ET Extirpated species; any indigenous species no longer existing in the wild in North America but existing elsewhere.

EX Extinct species; any species formerly indigenous to North America but no longer existing anywhere in the world.

Provincial Rankings

Provincial ranks are used by the NHIC to set protection priorities for rare species and natural communities but the ranks are not legal designations. The most
24☞ important factors considered in assigning provincial ranks are the total number of known extant sites in Ontario, and the degree to which they are potentially or actively threatened with destruction. Other criteria include the number of known populations considered to be securely protected, the size of the various populations, and the ability of the taxon to persist at its known sites.

The ranks are used by COSSARO (Committee on the Status of Species at Risk in Ontario), a committee established in 1995, and has a mandate to list species of flora and fauna as endangered, threatened or vulnerable. To date, less than 50 species (flora and fauna) have been listed.

S1 Extremely rare in Ontario; usually five or fewer occurrences in the province or very few remaining individuals; often especially vulnerable to extirpation.

S2 Very rare in Ontario; usually between 5 and 20 occurrences in the province; or with many individuals in fewer occurrences; often susceptible to extirpation.

S3 Rare to uncommon in Ontario; usually between 20 and 100 occurrences in the province; may have fewer occurrences, but with a large number of individuals in some populations; may be susceptible to large-scale disturbances.

S4 Common in Ontario; usually more than 100 occurrences; usually not susceptible to immediate threats.

S5 Very common in Ontario; demonstrably secure under present conditions.

SH Historical; of only historical occurrence in the province (no occurrences verified in the past 20 years), but with expectation that it may still be extant.

SX Apparently extirpated from Ontario, with little likelihood of rediscovery. Typically not seen in the province for many decades, despite searches at known historic sites.

Conservation Status Assessment Criteria

COSEWIC has a list of criteria for assessing the conservation status of any
25☞ species. The criteria follow the concepts of The Nature Conservancy (TNC) whose criteria are widely used in the USA with whom Canada shares a large number of species. TNC criteria are used in other countries as well. One advantage of the TNC system is that global ranks (see above) have been established by TNC for a large number of species. However, COSEWIC has now adapted international criteria, namely those of the International Union for the Conservation of Nature and Natural Resources (IUCN). The IUCN criteria differ from COSEWIC's new criteria in three areas: IUCN has a "Critically Endangered" status, COSEWIC has "Endangered" only; COSEWIC uses "Special Concern", while IUCN uses "Vulnerable" for the same definition; the criteria for each status is necessarily different because the areas covered by each are very different (i.e. Canada's area is much smaller than the global area). The IUCN (1996)[11] criteria for assessing risk of extinction are given in Table 14.6. A review of Table 14.6 reveals that IUCN's (and COSEWIC's) main criteria for determining the status category of a species are population numbers and sizes and their rates of change. This information would be available in any status report written for COSEWIC because one may use ***any*** of the A to E criteria.

In accordance with SARA (see above), COSEWIC maintains three lists: (i) species at risk (Extinct, Endangered, Threatened or Special Concern); (ii) species examined and designated in the not at risk category; and (iii) species examined and designated in the indeterminate category because of insufficient scientific information. COSEWIC provides advice regarding recovery priorities for listed species. This advice

---

[11]IUCN. 1996. 1996 red list of threatened animals. IUCN, Gland, Switzerland. Pp. Introduction 1-70 + 368 + Annex 1-10. (Also cited as Baillie, J. and B. Groombridge. 1996).

would assist jurisdictions to determine which species and activities to address first. For taxa not listed yet, scientists may provide advice to COSEWIC and subsequently seek permission to determine the conservation status of one or more species. COSEWIC then commissions the preparation of a ***status report*** for priority candidate species. The approved format for status reports is:

1. Technical Summary
2. Status and Summary of Reasons
3. Table of Contents
4. Executive Summary
5. Species Information
    - Name, Classification
    - Description
6. Distribution
    - Global Range
    - Canadian Range
7. Habitat
    - Definition
    - Trends
    - Protection/Ownership
8. General Biology
    - General
    - Reproduction
    - Survival
    - Physiology
    - Movements/Dispersal
    - Nutrition and interspecific interactions
    - Behaviour/Adaptability
9. Population Sizes and Trends
10. Limiting Factors and Threats
11. Special Significance of the Species
12. Evaluation and Proposed Status
    - Existing Protection or Other Status
    - Assessment of Status and Recommendation
13. Acknowledgments

Estimating Changes in Community Composition

When examining changes over time in a community due to loss of species, one
26☞ can determine the degree of similarity in community composition between time periods using Pappantoniou and Dale's (1982)[12] ***Community Overlap Index*** (C_λ) as follows:

$$C_\lambda = 2(\sum_{i=1}^{s} x_i\, y_i) \,/\, \sum_{i=1}^{s} x_i^2 + \sum_{i=1}^{s} y_i^2)$$

where,
x_i and y_i are the proportions that species i represents in communities (or time periods) x and y, and s is the total number of species. An index value of 1.0 indicates identical compositions. Community composition not only indicates which species are present, but how common or rare they are relative to one another. It is a more sensitive indicator of biological integrity than is species richness, a count of the numbers of species present.

Suppose, for example, that a community had 15 species in 1900 (= x) and 10

---

[12]Pappantoniou, A. and G. Dale. 1982. Comparative food habits of two minnow species: blacknose dace, *Rhinichthys atraculatus* and longnose dace, *Rhinichthys cataractae*. J. Freshwater Ecology 1: 361-364.

Table 14.6. Summary of new IUCN categories and criteria for assessing risk of extinction.

Use any of the A - E criteria	Critically Endangered	Endangered	Vulnerable
A. Declining population Population decline rate at least using either: 1. Population reduction observed, estimated, inferred or suspected in the past or:	80% in 10 years or 3 generations	50% in 10 years or 3 generations	20% in 10 years or 3 generations
2. Population decline projected or suspected in future based on A. Direct observation B. An index of abundance appropriate for the taxon C. A decline in area of occupancy, extent of occurrence and /or quality of habitat D. Actual or potential levels of exploitation E. Effects of introduced taxa, hybridization, pathogens, pollutants, competitors or parasites			
B. Small distribution and decline or fluctuation Either extent of occurrence or area of occupancy and 2 of the following 3 (1-3):	< 100 km^2 < 10 km^2	< 5,000 km^2 < 500 km^2	< 20,000 km^2 < 2,000 km^2
1. Either severely fragmented: (isolated subpopulations with a reduced probability of recolonization, if once extinct) or known to exist at a number of locations.	1	≤ 5	≤ 10
2. Continuing decline in any of the following: A. Extent of occurrence; B. Area of occupancy; C. Area, extent and/or quality of habitat; D. Number of locations or subpopulations; E. Number of mature individuals	Any rate	Any rate	Any rate
3. Fluctuating in any of the following: A. Extent of occurrence; B. Area of occupancy; C. Area, extent and/or quality of habitat; D. Number of locations or subpopulations; E. Number of mature individuals	> 1 order of magnitude	> 1 order of magnitude	> 1 order of magnitude
C. Small population size and decline Number of mature individuals and of the following 2 (1-2):	< 250	< 2,500	< 10,000
1. Rapid decline rate	25% in 3 years or 1 generation	20% in 5 years or 2 generations	10% in 10 years or 3 generations
2. Continuing decline and either: A. Fragmented; or B. All individuals in a single population	Any rate all subpops ≤ 5	Any rate all subpops ≤ 250	Any rate all subpops ≤ 1,000
D. Very small or restricted, either:			
1. Number of mature individuals or	< 50	< 250	< 1,000
2. Population is susceptible	(Not applicable)	(Not applicable)	area of occupancy < 100 km^2 or # of locations < 5
E. Quantitative analysis, indicating probability of extinction in the wild to be at least	50% in 10 years or 3 generations	20% in 20 years or 5 generations	10% in 100 years

species in 1997 (= y), with the numbers of living specimens in each species as shown in Table 14.7. The community overlap index, C_λ, for the two time periods is 0.234/0.320 = 0.73, indicating that the 1997 community has changed by about 27% since 1900.

Table 14.7. Calculations for estimating community overlap index (C_λ) for a hypothetical community of unionid bivalves between 1900 and 1997.

Species (#)	1900 (= x)	1997 (= y)	$\sum x_i y_i$	$\sum x^2 y^2$
a	10	2	0.003	0.004
b	9	2	0.003	0.003
c	8	2	0.002	0.003
d	7	0	0.000	0.002
e	2	0	0.000	0.000
f	14	5	0.010	0.010
g	3	0	0.000	0.000
h	21	9	0.027	0.027
i	18	4	0.010	0.014
j	5	0	0.000	0.001
k	14	28	0.056	0.118
l	12	5	0.009	0.009
m	32	25	0.113	0.125
n	8	2	0.002	0.003
o	5	0	0.000	0.001
Totals	168	84	0.234	0.320

EXOTIC SPECIES

What we Should Know About a Newly Introduced Species

As much as possible should be known a new species, but at least the following
27☞ ten attributes need consideration:

(i) Species characteristics - Give morphological features that are diagnostic of the species, with a key to separate it from closely related native species. Most of the characteristics can be taken from published taxonomic keys.

(ii) Mode of life and habitat - Describe how and where the organism lives, particularly if it has potential to exploit a poorly used habitat. For example, before the zebra mussel arrived, there was no native bivalve species that attached to hard surfaces so the habitat was "uncontested" by native bivalves. The zebra mussel displaced all other attached forms of fauna, including caddisflies, snails and sponges. A search through the literature of the specie's country of origin will reveal the kinds of habitats that the new exotic species will potentially invade.

(iii) Biology - Only unique aspects of the organism's functional anatomy need to be described. Emphasis should be placed on those features that may be potentially targeted for controlling the spread of the organism. For example, the dreissenids are the only species with a byssal apparatus that is used to attach the animal to a firm substrate. Describing the byssus, its rate of production and attachment properties lead to the development of coatings that resist attachment by zebra and quagga mussels. A study of the allometry of growth is also useful because it helps one to determine the organism's shape and size at different life stages. This provides knowledge of the sizes of screens needed to keep an organism out of a facility that draws water for cooling or for manufacturing purposes.

(iv) Reproductive potential and life cycle - Knowledge of: (i) the specie's sexual state (e.g. separate sex or hermaphrodite); (ii) its egg-laying habit (e.g. oviparous, ovoviviparous, viviparous); (iii) its fecundity (number of eggs produced); (iv) its natality (number of eggs surviving); (v) its annual frequency of egg-laying habits (e.g. univoltine, bivoltine, multivoltine); (vi) its life-time frequency of egg-laying habits (e.g. semelparity, iteroparity) is essential for predicting the impact potential of a species. For example, although the numbers are highly variable, ovoviviparous forms are about an order of magnitude less fecund than oviparous forms (e.g. 10:100). Even though many ovoviviparous forms are parthenogenetic, reducing the risk of having to find a mate, the fecundities are still relatively low. Even ovipositing, oviparous forms have low natalities relative to planktonic, oviparous forms. Ovipositing, oviparous forms are about three to four orders of magnitude less fecund than planktonic oviparous forms (e.g. 100:100,000 to 100:1,000,000). Species that have planktonic larval stages have particular biofouling potential for two reasons: (i) they usually produce large numbers of eggs; (ii) the developing (planktonic) larvae can enter a facility through the water intake by the millions and then grow and reproduce to establish biofouling populations inside the facility.

Knowledge of the organism's size at sexual maturity and the period(s) of gametogenesis requires histological examination of gonads in different size classes. The information may help develop control options that focus on delaying sexual maturity or inhibit gametogenesis. For example, ***pheromones*** (biochemicals that are specie's specific and affect only the species producing them) can be applied at times when gametogenesis is occurring.

(v) Food and feeding habits - Knowledge of the organisms functional feeding habits (e.g. shredder, gatherer, filter feeder, grazer, predator, parasite) or guilds (e.g. herbivore, benthivore, planktivore, piscivore, omnivore), and the food(s) it eats provides information on potential competition and impacts on native species. Indirect effects may also be observed, such as the production by adult zebra mussels of large amounts of pseudofaeces and their accumulation on the bottom. This accumulation has caused changes in the composition of benthic macroinvertebrate assemblages in many parts of the Great Lakes.

(vi) Adaptations - Any morphological, behavioural and/or physiological adaptations that explain, at least in part, the species' success in a particular habitat should be described. For example, describe any adaptations that might explain a species' success for: (i) avoiding desiccation or surviving prolonged periods of exposure; (ii) living an infaunal existence in the soft sediments; (iii) living an epifaunal existence on firm substrates; (iv) tolerating high turbidities characteristic of high order streams/rivers; (v) being eurythermous over their normal temperature range (e.g. tropical eurytherms and temperate eurytherms); or (vi) dealing with short periods of anoxia or low oxygen tensions. There are distinct differences in the abilities of exotic species to tolerate anoxia but few, if any, can survive prolonged anoxic conditions.

Knowing the physiological tolerances and requirements of the organism will provide information on the potential spread and continental limits of distribution of the species. Since most species require calcium for bone growth or development of exoskeletons, knowledge of the lower threshold calcium level for growth and reproduction will enable predictions of its ability to live in low-calcium waters. Similarly, all species have thermal thresholds for growth and reproduction, and upper and lower thermal tolerance levels. This information will not only enable predictions for an organism's ability to survive and grow in cold (i.e. its northern limit of distribution) and warm temperate waters (i.e. its southern limit of distribution), but have potential as a control option (e.g. using thermal back-flushing or recycling warm to hot water to kill biofoulers already inside a facility).

(vii) Predators - Knowing the predators of a species may be of use as a potential biological control option. Fish and waterfowl, and occasionally aquatic mammals (muskrat, otters), are the top predators in most aquatic systems. For example, predation by waterfowl is sufficient to control zebra mussels in some lakes in the Netherlands. The waterfowl species do not migrate and feed on zebra mussels all year. But in North America, waterfowl have little impact on mussel populations because most species migrate and are able to feed on zebra mussels only during the summer months. Very few native fish species have mouth parts adapted for crushing mollusc shells and those that do (e.g. freshwater drum) are so few in numbers that the fish has had little (noticeable) impact on zebra mussel densities.

If the new exotic species is a predator itself, it may have potential for controlling densities of other exotic species. For example, round gobies have mouth parts adapted for crushing shells of molluscs and are known to feed on zebra mussels. However, the potential impact of gobies on zebra and quagga mussels is presently unknown.

(viii) Parasites - The parasites of freshwater organisms fall under five groups: Protista, especially ciliates; Digenea; Aspidogastrea; Nematoda; and Arthropoda, especially hydrachnids (parasitic mites). A study of the parasites of the new exotic species may have potential control value, especially if the parasite is host specific.

(ix) Dispersal mechanisms - The dispersal mechanisms discussed for endangered species apply to exotic species as well. However, exotic species typically have an array of mechanisms and vectors. Table 14.8 lists numerous mechanisms available to organisms for dispersal over short or long distances. The table shows the potential of each mechanism for dispersal within a region (e.g. by "leap-frogging from lake to lake to eventually disperse throughout a province or state), a continent (e.g. intra-continental, or within North America or within Eurasia) or inter-continentally (e.g. from Eurasia to North America). The list is based mainly on dispersal mechanisms used by zebra mussels (Carlton 1993)[13] but it is not necessarily restricted to them.

Of the natural mechanisms, external transport (e.g. feet and feathers) is generally a more effective dispersal mechanism than internal transport via the digestive tract. Large insects are able to disperse only small organisms. Fish and semi-aquatic vertebrates (e.g. amphibians, reptiles) are able to disperse organisms only within a watershed. Lake currents disperse only plankton or planktonic stages. River currents can disperse organisms only downstream of their introduction. Water spouts are probably of little dispersal value unless they are large and "spill over" into nearby and adjacent water bodies, and then only (mostly?) planktonic organisms can be dispersed this way.

The anthropogenic mechanisms are split between intentional (or deliberate) and non-intentional (or accidental) releases because the former usually involves dispersal of large organisms for their food or sport value. Most exotic fish species have both food and sport value, but molluscs such as *Cipangopaludina chinensis malleatus* and *C. japonicus*

---

[13]Carlton, J. T. 1993. Dispersal mechanisms of the zebra mussel (*Dreissena polymorpha*). In: Nalepa, T. F. and D. W. Schloesser (Eds.). Zebra mussels: Biology, impact, and control. Lewis Publishers, Boca Raton, FL. pp. 667-697.

Table 14.8. Dispersal mechanisms available to organisms for short-range (e.g. regional = ⌂) mid-range
28☞ (e.g. intra-continental = ☎) or long-range (e.g. inter-continental = ✈) transport.

Dispersal Mechanisms	Potential
Natural Mechanisms	
By insects, birds or mammals	⌂☎
By fish or semi-aquatic vertebrates	⌂
Currents	⌂☎
Water spouts (planktonic stages only)	⌂
Wind	⌂☎
Unintentional Anthropogenic Mechanisms	
Interiors (e.g. ballast tanks) or exteriors (e.g anchor holds) of ocean vessels	☎✈
Interiors (e.g. fish wells) or exteriors (e.g. hulls) of ships and crafts of rivers and lakes	⌂☎
Canals (irrigation and vessels)	⌂☎
Navigation and marker buoys and floats	⌂
Marina and boatyard equipment	⌂
Fisheries equipment (e.g. cages, nets, bait buckets)	⌂
Amphibious and fire-fighting planes	⌂☎
Fire-truck water	⌂
Commercial products (e.g. logs, aesthetic and medicinal plants)	⌂☎✈
Aquarium releases	⌂☎✈
Recreational equipment (e.g. floating docks)	⌂
Litter (e.g. tires)	⌂
Scientific research	⌂☎✈
Intentional Anthropogenic Mechanisms	
Food	⌂☎✈
Sport	⌂☎✈

were introduced by orientals purely for their food value as escargot. Mills *et al* (1993)[14] have attributed deliberate releases to 11 of 139 exotic species in the Great Lakes, most of these being fish. Of the 144 species introduced into the Great Lakes (at least 5 more have been introduced since 1993), 81 have originated in Eurasia and were introduced by ballast water exchange, the main inter-continental release mechanism. About 32 species originated from somewhere in North America (e.g. southern U. S., Pacific, Mississippi drainage, and Atlantic and Pacific sources, Fig. 14.1). Mills *et al* (1993) attribute aquarium releases to some species in the Great Lakes, claiming that many people released their "pets" without any intention of establishing self-sustaining populations. Similarly, the accidental escape of cultivated and ornamental plants have occurred since the times when early settlers brought over plants for medicinal (e.g. bittersweet), gastronomical (e.g. water cress) and ornamental (e.g. yellow flag) purposes. Aquatic plants account for 59 of the 144 species introduced to the Great Lakes.

Figure 14.1 shows a timeline of introductions of four groups of organisms (plants, algae, invertebrates, fish) in the Great Lakes. The completion of the Welland Canal in 1829, the locks at Sault Ste. Marie in 1855 and the St. Lawrence River Canal system in 1847 opened up all the Great Lakes to shipping activities from distant drainage basins in

Figure 14.1. A timeline of introductions for each of four groups of organisms.

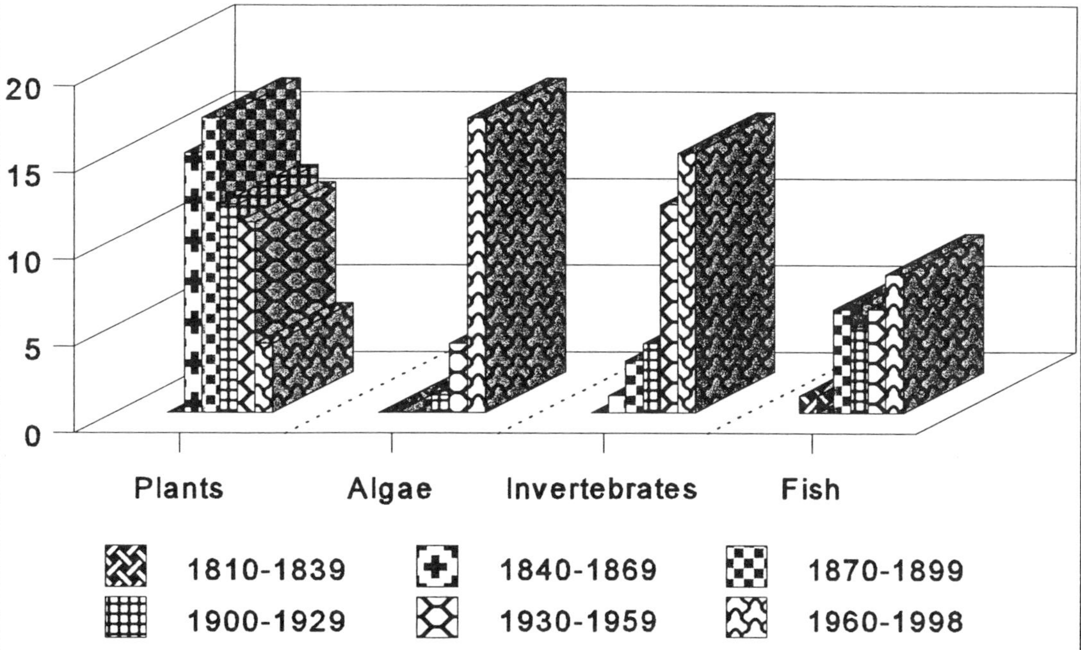

[14]Mills, E. L., J. H. Leach, J. T. Carlton and C. L. Secor. (1993). Exotic species in the Great Lakes: A history of biotic crises and anthropogenic introductions. J. Great Lakes Res. 19: 1-54.

North America and from European continents. Ballast water was in use by the 1880s and there began a dramatic increase in releases of exotic species (Fig. 14.1). Because water quality was deteriorating on all continents, due to the industrial revolution, the only species that could survive the 10- to 14-day voyage across the ocean were the highly tolerant forms. With increased environmental awareness and development of faster ships by the 1950s and improved water quality beginning in the 1960s, more species were able to survive the voyage, resulting in about 30% (44 of 144) of the releases in the last 30 years. The total number of species for each time period are: 1810-1839 = 1; 1840-1869 =16; 1870-1899 = 26; 1900-1929 = 22; 1930-1959 = 33; 1960-1998 = 44.

29☞ Tables 14.8 to 14.12 list the species that have been introduced into the Great Lakes. They include most (many marsh plants are omitted) of those listed by Mills *et al* (1993) and others, including the amphipod *Echinogammarus ischnus*, the snail, *Potamopyrgus antipodarum* and the Blueback herring, *Alosa aestivalis*, that were introduced after 1993.

30☞ ***(x) Impact potential*** - All ecological and/or socio-economic impact(s) ascribed to a species should be provided. The potential impacts are of three general types: (a) harmful ecological and/or socio-economic impacts; (b) potential beneficial ecological and/or socio-economic impacts (uses); (c) little or no potential ecological or socio-economic impact. The extent of the impacts will depend not only on the species that invade a habitat but the quality of the habitat visa viz the tolerances and requirements of the invading species. Below are numerous types of impacts that have been recorded for different exotic species.

Ecological Impacts

(1) ***Reduced seston levels/increased water transparency:*** If the exotic species is a filter feeder, the filtration of suspended materials (seston) from the water may result in reduced phytoplankton biomass and increased water clarity. This may have a negative impact on planktivorous fish. Conversely, beneficial impacts may potentially accrue from increased water transparency. For example, macrophyte and benthic algae biomass and diversity may increase and/or SCUBA divers may benefit from the increased water clarity.

(2) ***Loss of endemic species:*** Competition for food and/or space may result in the elimination of closely related species and/or species with the same functional feeding habits.

(3) ***Alteration of benthic community structure and abundance:*** A direct result of item (2) or an indirect result through bottom-up or top-down effects. A bottom-up effect would be an increase in the diversity and biomass of algae and macrophytes followed by changes in the benthic invertebrate community, with grazers and herbivores showing a greater contribution to the total diversity of functional feeding groups. A top-down effect would be an increase in the numbers of planktivorous fish species followed by a decline in zooplankton and an increase in autotroph numbers and biomass.

Table 14.8. Origin and date of first sighting of non-indigenous bacteria, protists and algae of the Great Lakes. Species are listed alphabetically. (Modified from Mills *et al* (1993)[15]).

Taxon	Species	Common Name	Date	Origin
Bacteria	*Aeromonas salmonitica*	Causes furunculosis disease	<1902	Unknown
Protozoa	*Glugea hertwigi*	Microsporidean parasite	1960	Eurasia
Protozoa	*Myxobolus cerebralis*	Salmonid whirling disease	1968	Unknown
Bacillariophyta	*Actinocyclus normanii f. subsalsa*	diatom	1938	Eurasia
Bacillariophyta	*Bangia atropurpurea*	diatom	1964	Widespread
Bacillariophyta	*Biddulphia laevis*	diatom	1978	Widespread
Bacillariophyta	*Chaetoceros hohnii*	diatom	1978	unknown
Bacillariophyta	*Chroodactylon ramosum*	diatom	1964	Atlantic
Bacillariophyta	*Cyclotella atomus*	diatom	1964	Widespread
Bacillariophyta	*Cyclotella pseudostelligera*	diatom	1946	Widespread
Bacillariophyta	*Cyclotella cryptica*	diatom	1964	Widespread
Bacillariophyta	*Cyclotella woltereki*	diatom	1964	Widespread
Bacillariophyta	*Diatoma ehrenbergii*	diatom	1930s	Widespread
Chlorophyta	*Enteromorpha prolifera*	Green alga	1979	Atlantic
Chlorophyta	*Enteromorpha intestinalis*	Green alga	1926	Atlantic
Chrysophyta	*Hymenomonas roseola*	Green alga	1975	Eurasia
Chlorophyta	*Nitellopsis obtusa*	Green alga	1983	Eurasia
Bacillariophyta	*Skeletonema subsalsum*	diatom	1973	Eurasia
Bacillariophyta	*Skeletonema potamos*	diatom	1963	Widespread
Bacillariophyta	*Sphacelaria fluviatilis*	diatom	1975	Asia
Bacillariophyta	*Sphacelaria lacustris*	diatom	1975	Unknown
Bacillariophyta	*Stephanodiscus binderanus*	diatom	1938	Eurasia
Bacillariophyta	*Stephanodiscus subtilus*	diatom	1946	Eurasia
Bacillariophyta	*Thalassioosira weissflogii*	diatom	1962	Widespread
Bacillariophyta	*Thalassiosira lacustris*	diatom	<1978	Widespread
Bacillariophyta	*Thalassiosira guillardii*	diatom	1973	Widespread
Bacillariophyta	*Thalassiosira pseudonana*	diatom	1973	Widespread

[15]Mills, E. D., J. H. Leach, J. T. Carlton and C. L. Secor. 1993. Exotic species in the Great Lakes: A history of biotic crises and anthropogenic introductions. J. Great Lakes Res. 19: 1-54.

Table 14.9. Origin and date of first sighting of non-indigenous submersed and emergent macrophytes of the Great Lakes. Species are listed alphabetically in each group. (Modified from Mills *et al* (1993)

Submersed Species	Common Name	Date	Origin
Cabomba caroliniana	Fanwort	1935	Southern U.S.
Hydrocharis morus-ranae	European frogbit	1972	Eurasia
Marsilea quadrifolia	European water clover	<1925	Eurasia
Myriophyllum spicatum	Eurasian milfoil	1952	Eurasia
Najas marina	spiny naiad	1864	Eurasia
Najas minor	minor naiad	1932	Eurasia
Nasturtium aquaticum	Water cress	1847	Eurasia
Nymphoides peltata	Yellow floating heart	1930	Eurasia
Potamogeton crispus	Curly pondweed	1879	Eurasia
Trapa natans	Water chestnut	<1959	Eurasia
Emergent Species	**Common Name**	**Date**	**Origin**
Butomus umbellatus	Flowering rush	<1930	Eurasia
Carex acutiformis	Swamp sedge	1951	Eurasia
Conium maculatum	Poison hemlock	<1843	Eurasia
Glyceria maxima	Reed sweet-grass	1940	Eurasia
Iris pseudacorus	Yellow flag	1886	Eurasia
Juncus compressus	Flattened rush	<1895	Eurasia
Juncus inflexus	Rush	1922	Eurasia
Lysimachia vulgaris	Garden loosestrife	1913	Eurasia
Lysimachia nummularia	Moneywort	1882	Eurasia
Lythrum salicaria	Purple loosestrife	1869	Eurasia
Mentha gentilis	Creeping whorled mint	1915	Eurasia
Mentha piperita	Peppermint	<1843	Eurasia
Mentha spicata	Spearmint	<1843	Eurasia
Mysotis scorpioides	True forget-me-not	1886	Eurasia
Polygonum caespitosum v. longisetum	Bristly lady's thumb	1960	Asia
Polygonum persicaria	Lady's thumb	<1843	Eurasia
Rorippa sylvestris	Creeping yellow cress	1884	Eurasia
Rumex longifolius	Yard dock	1901	Eurasia
Rumex obtusifolius	Bitter dock	<1840	Eurasia
Solanum dulcamara	Bittersweet nightshade	<1843	Eurasia
Sparganium glomeratum	Bur reed	1936	Eurasia
Typha angustifolia	Narrow-leaved cattail	1880s	Eurasia
Veronica beccabunga	European brooklime	1915	Eurasia

Table 14.10. Origin and date of first sighting of non-indigenous invertebrates of the Great Lakes. Species are listed alphabetically in each group. (Modified from Mills *et al* (1993).

Mollusc Species	Common Name	Date	Origin
Bithynia tentaculata	Faucet snail	1871	Eurasia
Cipangopaludina japonica	Japanese mystery snail	1940s	Asia
Cipangopaludina chinensis malleatus	Oriental mystery snail	1931	Asia
Corbicula fluminea	Asian clam	1980	Asia
Dreissena polymorpha	Zebra mussel	1985	Eurasia
Dreissena bugensis	Quagga mussel	1990	Eurasia
Elimia virginica	Virginia snail	1860	Atlantic
Gillia altilis	Hydrobiid snail	1918	Atlantic
Lasmigona subviridis	Green floater	<1959	Atlantic
Pisidium supinum?	Hump-backed pea clam	1950s	Eurasia
Pisidium amnicum	Greater European pill clam	1897	Eurasia
Pisidium henslowanum?	Henslow's pea clam	1950s	Eurasia
Potamopyrgus antipodarum	New Zealand snail	1991	England
Radix auricularia	European ear snail	1901	Eurasia
Sphaerium corneum	European fingernail clam	1952	Eurasia
Valvata piscinalis	European valve snail	1897	Eurasia
Viviparus georgianus	Banded mystery snail	<1906	Mississippi
Crustaceans	**Common Name**	**Date**	**Origin**
Argulus japonicus	Parasitic copepod	< 1988	Asia
Bythotrephes cederstroemi	Spiny water flea	1984	Eurasia
Echinogammarus ischnus	Eurasian amphipod	1996	Eurasia
Eubosmina coregoni	Water flea	1966	Eurasia
Eurytemora affinis	Calanoid copepod	1958	Widespread
Gammarus fasciatus	Gammarid amphipod	<1940	Atlantic
Orconectes rusticus	Rusty crayfish	1960	United States
Skistodiaptomus pallidus	Calanoid copepod	1967	Mississippi
Oligochaetes	**Common Name**	**Date**	**Origin**
Branchiura sowyerbi	Gilled tubificid	1951	Asia
Phallodrilus aquaedulis	Tubificid oligochaete	1983	Eurasia
Ripistes parasitica	Naidid oligochaete	1980	Eurasia
Platyhelmnithes	**Common Name**	**Date**	**Origin**
Dugesia polychroa	Flatworm	1968	Eurasia
Hydrozoa	**Common Name**	**Date**	**Origin**
Cordylophora caspia	Hydroid	1956	Unknown
Cordylophora macronatum	Hydroid	1997	Eurasia
Craspedacusta sowyerbi	Freshwater jellyfish	1933	Eurasia
Insecta	**Common Name**	**Date**	**Orgigin**
Acentropus niveus	Aquatic moth	1950	Eurasia
Tanysphyrus lemnae	Aquatic weevil	<1943	Eurasia

Table 14.11. Origin and date of first sighting of non-indigenous fish species of the Great Lakes. Species are listed alphabetically. (Modified from Mills *et al* (1993)

Fish Species	Common Name	Date	Origin
Alosa aestivalis	Blueback herring	1994	Atlantic
Alosa pseudoharengus	Alewife	1873	Atlantic
Apeltes quadracus	Fourspine stickleback	1986	Atlantic
Carassius auratus	Goldfish	<1878	Asia
Cyprinus carpio	Common carp	1879	Asia
Enneacanthus gloriosus	Bluespotted sunfish	1971	Atlantic
Gambusia affinis	Western mosquitofish	1923	Misissippi
Gymnocephalus cernuus	Ruffe	1986	Eurasia
Lepomis humilis	Orangespotted sunfish	1929	Mississippi
Lepomis microlophus	Redear sunfish	1928	Southern U.S.
Misgurnus anguillicaudatus	Oriental weatherfish	1939	Asia
Morone americana	White perch	1950	Atlantic
Neogobius melanostomus	Round goby	1990	Eurasia
Notropis buchanani	Ghost shiner	1979	Mississippi
Noturus insignis	Margined madtom	1928	Atlantic
Oncorhynchus gorbuscha	Pink salmon	1956	Pacific
Oncorhynchus kisutch	Coho salmon	1933	Pacific
Oncorhynchus mykiss	Rainbow trout	1876	Pacific
Oncorhynchus nerka	Kokanee	1950	Pacific
Oncorhynchus tshawytscha	Chinook salmon	1873	Pacific
Osmerus mordax	Rainbow smelt	1912	Atlantic
Petromyzon marinus	Sea lamprey	1830s	Atlantic
Phenacobius mirabilis	Suckermouth minnow	1950	Mississippi
Proterorhinus marmoratus	Tubenose goby	1990	Eurasia
Salmo trutta	Brown trout	1883	Eurasia
Scardinius erythrophthalmus	Rudd	1989	Eurasia

(4) ***Increased potential as vectors of parasites whose definitive hosts are commercially important species of fish, waterfowl or humans:*** This has been demonstrated by the introduction of rainbow smelts which brought with it the protozoan parasite, *Glugea hertwigi*. The parasite caused high mortalities of the rainbow smelt in the 1960s and 1970s but infestation levels have not been as high since the 1970s.

Navigational and Vessel Impacts
(5) ***Biofouling of hulls of sailing vessels:*** Biofouling includes any attached organism (e.g. algae, sponges, zebra mussels) that affect the sailing efficiency or aesthetic value of the vessel.
(6) ***Biofouling of navigation buoys:*** Extensive biofouling (e.g. by zebra mussels) could sink the buoys deeper in the water than normal.

Private Property Impacts
(7) ***Formation of shoals of debris on beaches that detract from the beach's recreational and aesthetic value:*** Ice scouring in the winter months or winter-kill may result in the piling up moribund bodies or parts of organisms (e.g. weeds, shells of molluscs, fish).
(8) ***Biofouling of cottage plumbing and intake structures:*** Large numbers or biomasses of organisms (e.g. weeds, zebra mussels) have been shown to plug or restrict flow through foot valves and/or screens protecting the plumbing lines.

Beneficial Impacts of Exotic Species

There is usually some benefit derived from even nuisance organisms. The
31☞ following are some examples of beneficial impacts, most of them being of socio-economic value.

(1) ***Manufacturers of Products for Control of the Species:*** If the exotic species is a nuisance, entrepreneurs quickly respond by manufacturing products (e.g. chemicals, filters, electronics, etc.) to control the organism. One example is the appearance in Canada and the United States of numerous kinds of devices (e.g. filters, magnets, coatings) for preventing biofouling of cottage water intake structures, boat hulls, outboard motors, docks, etc. Numerous options for controlling species in domestic and industrial facilities are given below and many were developed after the zebra mussel arrived in North America.
(2) ***Manufacturers of Food for Domesticated Animals or Agricultural Products:*** All organisms have some organic (e.g. proteins) and inorganic (e.g. calcium) components that can be ground up or pelletized and used as soil fertilizer, soil conditioning or food for domesticated animals.
(3) ***Food for Human Consumption, Display or Sport Value:*** Many organisms (e.g. water cress, escargo) have been introduced (deliberately or unintentionally) for their gastronomical value, medicinal value (e.g. bittersweet), ornamental value (e.g. yellow flag), trinkets (e.g. broaches, pins, ear rings made of zebra mussel shells) or recreational value (e.g. sport fisheries).
(4) ***Indirect Recreational Value:*** Occasionally some benefits can be derived from alteration of the habitat, such as increased water clarity by filter feeding exotic specie; this has benefited SCUBA divers who now have increased water clarity in the Great Lakes..

Species With Little or No Apparent Ecological Impact

It seems that most introduced species have had little or no obvious impacts on the ecology of the ecosystems into which they were introduced. This appears to apply to molluscs in particular. Of the 22 species introduced to North America, 14 have had no documented impacts, although the effects may be insidious and below the level of detection. Most freshwater molluscs are intermediate hosts of numerous species of trematodes whose definitive hosts are fish, waterfowl or humans, but no studies have assessed the added impact of introduced molluscs infected with their trematodes on native definitive hosts. It is not anticipated that the exotic species of molluscs will add significantly to the infestation of definitive hosts that are already infected by trematodes of native species of molluscs. All exotic species will have to compete with their native counterparts for food and space but very few (*Bythinia tentaculata* may be an exception) are known to have replaced native species; rather they seem to have added to the diversity of molluscan fauna.

(xi) Control measures - Control measures need to be described only for biofouling species.

CASE HISTORY FOR BIOFOULING - North America and the Great Lakes

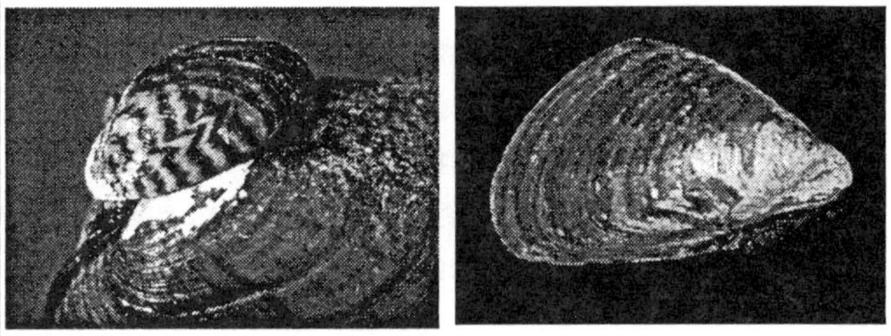

a b

Figure 14.2. (a) Zebra mussel attached to a native clam, *Lampsilis siliquoidea*. The zebra mussel specimen was the first one reported to the Ontario Ministry of Environment and Energy in 1988. (b) Quagga mussel.

Zebra and Quagga Mussels (Fig. 14.2): 32☞ Because of the enormous impact of the zebra and quagga mussels on industrial and domestic pipelines and structures, and on cottagers and body contact recreation, the two species are used as a worst-case scenario. The case history follows the recommendations described above. The descriptions for the biology, impact and control of the species are summaries of those given in Mackie (1998)[16].

Dreissena polymorpha (zebra mussel, Fig. 14.2a) and *Dreissena bugensis* (quagga mussel, Fig. 14.2b)

---

[16]Mackie, G. L. 1998. Mollusca: Ballast water introductions; Aquarium trade introductions; Food value introductions. In: Claudi, R. and J. Leach (Eds.). Non-indigenous freshwater organisms in North America: Their biology and impact. In press.

The zebra mussel was found and described for the first time in the northern part of the Caspian Sea and in the Ural River. In the early 19th, century the geographical range of *D. polymorpha* was dramatically extended: to Hungary in 1794, then a rapid invasion of Britain, first in Cambridgeshire in the 1820's, London in 1824, Yorkshire 1831-33, Forth and Clyde Canals in 1833, and in the Union Canal near Edinburgh in 1834, with numerous other records from 1835 onwards (Morton 1993). The invasion was just as rapid in Germany (e.g. to Rotterdam in 1827, Hamburg in 1830 and Copenhagen in 1840). The invasion of the zebra mussel throughout the USSR was just as startling; from the Davina River Basin in 1845, through the Mariinsk Canal system to western Europe, and more recently (1960's) through the Oginski and Moscow Canals. Extension of its range in the USSR is still occurring at a rapid rate. The species appeared in the Scandinavias in the 1940's and has since appeared in numerous Swiss lakes. It is now in Italy, Finland, Ireland, and the Iranian coast of the Caspian Sea. Similar trends and rates of spread occurred for the quagga mussel in Europe, but not in North America.

The zebra mussel was first introduced into Lake St. Clair in 1986. The most probable entry mechanism was from an oceangoing vessel that discharged its ballast into Lake Erie before entering the Detroit River. Fishing vessels then apparently dispersed the mussel into Lake St. Clair. The species spread to all the Great lakes within a period of two years. By 1991 it had been dispersed from the Great Lakes drainage systems to the Mississippi River and within two years migrated downstream to its current southern limit at New Orleans. The species now occupies most of the major river systems in North America, including the St. Lawrence, Mississippi, Missouri, Ohio, Illinois, Tennessee, Cumberland and Hudson Rivers.

The quagga mussel was first found in Lake Ontario in 1989 and may have been introduced at the same time as the zebra mussel but, because it can live in deeper water than the zebra mussel, the species may not have been discovered until later. Its ecological and socio-economic impacts appear to be similar to the zebra mussel.

Species characteristics: The zebra mussel shell typically has yellow, brown and black stripes. A few have no stripes. The byssal opening on the ventral side is in the anterior
33☞ third of shell. Its ventral side is flat to concave.

Shell ornamentations are generally lacking in *D. bugensis* and its ventral surface is convex, with the byssal opening more distal in position (see Chapter 8 for additional descriptions).

Mode of life and habitat: Large freshwater lakes and rivers are the favourite habitats of dreissenids but they also do well in cooling ponds, quarries, and irrigation ponds of golf
33☞ courses. Zebra mussels are capable of living in brackish water or estuaries where the salinity does not exceed 8 to 12 ppt. The incipient lethal salinity for zebra mussel post-veligers is near 2 ppt and for adults (5-15 mm) it is between 2 and 4 ppt. Hence, the species is able to invade estuarine portions of the Hudson River where salinities range between 3 and 5 ppt.

The mode of life and habitat of the quagga mussel is similar to zebra mussels, with two major exceptions. The quagga mussel can live as an infaunal species in littoral and profundal sediments and it can live as an epifaunal or infaunal species in either the littoral or profundal zones of lakes. *Dreissena polymorpha* has an exclusively epifaunal habit in shallow waters 1 to 15 m). In the Great Lakes, the ratio of quagga to zebra mussels increases with increasing depth down to about 50 m. The greatest densities of both species in Lake Erie occur at about 12 m. Only quagga mussels occur in Ekman grab samples taken at the 50 m depth. It appears that quagga mussels are able to alter their shell allometry for an infaunal existence; they are thinner and lighter in weight in soft profundal sediments than on firm inshore substrates, whereas zebra mussels exhibit little or no changes in shell allometry on firm or soft substrates.

With the adaptability of quagga mussels to life in soft sediments at great depths, provided oxygen is near saturation, quagga mussels have the potential to invade more northern lakes than the zebra mussel. Although dreissenids tend to develop greater biomasses in flowing waters (e.g. pipelines) of moderate velocities (< 1.5 m/sec), where their filter feeding efficiency can be enhanced by a constant delivery of food to the siphons, the planktonic larvae will prevent them from establishing large and permanent populations in natural streams and rivers that lack impoundments, as discussed below.

Dispersal mechanisms: Carlton (1993, see reference earlier) describes twenty-three different mechanisms by which the mussels have been or can be dispersed in the larval
33☞ and adult stages. Their dispersal prowess is divided into three natural (e.g. water currents, birds and other animals) and twenty human-mediated mechanisms (e.g. ballast water, hulls of sailing vessels, etc.). The great diversity of dispersal mechanisms has impressed upon us the near impossibility of preventing the spread of zebra mussels once they have been introduced. "It is not a question of IF it will get here, it is a question of WHEN it will get here", is now a cliche for describing the dispersal powers of biofouling dreissenids.

While currents are an effective mechanism for dispersing planktonic larval stages, it is not effective for sustaining populations in streams. The byssal apparatus will help to maintain the position of the adults in the streams but the flow of water will carry the larvae well downstream of the parent population which will ultimately disappear, unless an upstream population can rejuvenate the colony. As stated earlier, the distance that the larvae are carried depends upon the velocity and the duration of the planktonic stage. Dreissenids introduced to a stream will survive one life span at best, unless they are dispersed again to the same site, which is not likely if there are no adult populations upstream, or if there is an impoundment to retain the planktonic larvae. Indeed, reservoirs will serve to provide breeding habitats for establishing and maintaining populations downstream. Once rivers become slow enough and currents are such that the position of the developing larvae can be maintained up to and including settlement, the populations can be self sustained by the adults attached to the bottom. Without mainstream reservoirs, dreissenids would not succeed (or at least would not be as great a pest) in most rivers in North America. Streams with an average velocity of 0.1 m/sec (= 8.64 km/day) would carry larvae with development times of 20 to 30 days, about 173 to

260 km downstream before settlement would occur. For a population to maintain its position within 100 m of itself, the average water velocity must be less than 0.00006 m/sec, a velocity exceeded in most streams, and possibly many reservoirs (based on Drift Distance = Average Development Time x Water Velocity; therefore, velocity = 100 m/21 days = 100 m/(21 days x 24 hrs/day x 60 min/hr x 60 sec/min) = 0.00006 m/sec). Only back eddies in pools of rivers, mainstream impoundments or reservoirs would provide such currents for the complete development of the larvae.

Adaptations: Claudi and Mackie (1994)[17] review some of the more important morphological adaptations and the physiological tolerances and requirements of the zebra
33☞ mussel in North America.

The shell morphology of the species allows it to exploit an epifaunal existence in fresh waters. The ventral surface is flat and, with help from the byssal apparatus, allows the species to be pulled tightly against the surface of the substrate. A triangular-shape shell (in end view), makes it difficult for predators to grasp and pull from a surface. The zebra mussel is the only freshwater bivalve that has this adaptation.

Other adaptations include abilities to survive periods of desiccation, turbidity and, to a lesser extent, hypoxia. Quagga and zebra mussels can survive 5 to 13 days in low humidity. They eventually succumb to accumulations of toxic anaeorobic end-products during desiccation but are know to survive out of water for up to at least 7 days under damp conditions (e.g. rain). The time to death of zebra mussels is rapid at freezing temperatures, occurring in less than 24 h at -3°C (= 27°F).

Of all the physiological parameters of importance to exotic molluscs, tolerances to turbidity is the least understood. Associated with increasing loads of suspended particles is increasing sedimentation rates, especially in lakes or streams with depositional current velocities. One effect of turbidity on native bivalves is a depression of the respiratory rate, with normal levels of turbidity having little or no effect on respiration rate. However, in *D. polymorpha*, the respiration rate is relatively depressed at low turbidities (5 NTU) and both zebra and quagga mussels can partially acclimate over a 4-week period to turbid water conditions (80 NTU) by adjusting their metabolic rate. Apparently there are ecophenotypic differences within populations of *D. polymorpha* in which populations in habitats characterized by high suspended loads show a marked increase in labial palp to gill area ratios.

Increased turbidity reduces light penetration resulting in lower primary production. Therefore, non-grazing and infaunal forms of exotic molluscs will be favoured over grazing and epifaunal forms. As a result, gastropods will be replaced by filter-feeding bivalves. Although filter-feeding rates may be depressed in waters with extremely high silt loads, much of the inorganic material (i.e. silt) is captured by mucous that is rejected as pseudofaeces. The onset of pseudofaeces production indicates overload of the filtration apparatus.

---

[17]Claudi, R. and G. L. Mackie. 1994. A practical manual for monitoring and control of zebra mussels. Lewis Publishers, Boca Raton, FL. 227 p.

The population filtration rate depends on individual clearance rates (volume of water cleared of particles per unit time), the percentage of mussels filtering, and the time of day, with filtering activity being about 6-9% higher at night than during the day. However, no diel change in clearance rates have been observed.

Studies in European lakes and canals indicate that *Dreissena* plays a significant role in processes of biological self-purification and improvement in water quality in aquatic systems. This is a consequence of the enormous numbers of mussels that usually prevail in aquatic systems and not of a high individual filtration capacity. As bivalves, the zebra mussel's filtration rate (10-100 ml/ind./hr) is intermediate between Unionidae (60-490 ml/ind./hr) and Sphaeriidae (0.6-8.3 ml/ind./hr). Scientists have shown that size, temperature, acclimation turbidity and ambient turbidity significantly affect oxygen consumption rates and that zebra mussels adjust their metabolic rate in response to chronic exposure turbidity.

In areas heavily populated by zebra mussels, large biomasses of pseudofaeces accumulate on the bottom, often causing a shift in energy from the pelagic zone to the benthic zone. This has implications on altering concepts in reservoir ecology, especially pulse stability and successional changes. Dreissenids may add to the duration and/or the magnitude of pulse stability because their population densities exhibit their own pulse stabilities; the filtering activities of enormous numbers of dreissenids is bound to affect the pulse stability of a reservoir. Similarly with succession, dreissenids dramatically accelerate the sedimentation rate of suspended particles and the rates of trophic changes that normally occur in new impoundments could be dramatically altered.

The thermal tolerances of biofouling dreissenids are well described. For zebra mussels the chronic lethal temperature is 34-37°C, while the acute lethal temperature ranges from about 33°C to 42.3°C. However, previous acclimation temperature greatly affects both the acute and chronic lethal temperatures. The tolerance times at different temperatures increases with increasing acclimation temperature and decreasing shell size, and decreases with increasing exposure temperature.

Nearly all thermal tolerance studies were conducted to determine the feasibility of using thermal stress to control biofoulers. Epilimnial temperatures in lakes and reservoirs are usually higher than temperatures in the parent streams, but the hypolimnial temperatures are colder than parent stream temperatures. In general, molluscs, like all other aquatic organisms in impoundments, adjust their upper thermal tolerance limits by acclimating to increasing summer temperatures as the summer progresses. Most molluscs have the ability to seek a preferred temperature and can usually find it in thermally stratified waters. This includes even the byssate forms because they can translocate by releasing themselves from their byssal attachment and resettling in a more suitable thermal regime, as long as the dissolved oxygen levels are appropriate. Dreissenids cannot tolerate even short periods of anoxia but can survive short periods of hypoxia, down to 2 mg/L for a few days.

Both zebra and quagga mussels appear to have the same requirements for calcium. Neither can survive in waters with calcium levels below 7 mg/L. Infestation levels of growth and reproduction occur in waters with calcium levels above 25 mg/L.

The quagga mussel can grow and reproduce at lower temperatures (< 9 °C) than the zebra mussel (8-10 °C for growth, 12-15 °C for reproduction) and therefore has a greater potential to infest lakes more northern in latitude. However, some scientists claim that quagga mussels can be more abundant at thermally-enriched sites than at sites unaffected by thermal discharges and they dispute claims that zebra mussels can tolerate higher temperatures than quagga mussels. It appears that quagga mussels prefer the relatively constant temperatures characteristic of deep water (e.g. 1-4 °C) than the highly variable temperatures found in shallow waters.

Apparently quagga mussels have a lower upper thermal limit and a greater instantaneous mortality rate across acclimation temperatures than do zebra mussels. However, there appears to be no difference in instantaneous temperatures required to cause 100% mortality in either species. Quagga mussels also have the same functional responses as zebra mussels to turbidity, acclimation turbidity and ambient turbidity and both respire normally in turbidities as high as 80 NTU when previously acclimated to high suspended loads.

Predators: The dreissenids have planktonic larval stages that are preyed upon mainly by crustacean zooplankton (e.g. *Cyclops*) and larval fish. The relative importance of these
33☞ prey groups to the total mortality of larval stages is unknown. Scientists have reported predation of zebra mussel larvae by the cnidarian, *Hydra americana*. These and other hydrozoans are often present in large numbers attached to mussels and capture food in currents generated by the siphons of the mussels.

Adult zebra mussels have a high nutritional value of the tissues, with 60.7% protein, 19.0% carbohydrate, 12.0% lipid and 5.9% ash and are consumed in large quantities by crayfish, fish, and waterfowl. The nutritional value of zebra mussels changes seasonally so prey selection by waterfowl may change accordingly. Predation by crayfish has been documented in Europe and North America. The extent of predation by crayfish on zebra mussels appears to depend upon at least the size classes of mussels present, on colonization of mussels by micro-organisms and a learning period for the crayfish.

The relative contribution of fish to total mortality by predation of dreissenids is unknown. Zebra mussels have been found in the stomachs of freshwater drum, white suckers, walleye, yellow perch, bass and a few others, some of which are exotic species themselves, such as the round (*Neogobius melanostomus*) and tubenose gobies (*Proterorhinus marmoratus*). The freshwater drum, *Aplodinotus grunniens*, and the round gobie have molariform pharyngeal teeth for crushing mollusc shells. The drum is most common in large lakes and rivers. Predation on zebra mussels increases as drum size increases, with large drum feeding almost exclusively on zebra mussels. Apparently, prey size is limited by the crushing power of the jaws. Scientists have observed size-selective predation in both young freshwater drum and yellow perch *(Perca flavescens*), but the predation intensity by large fish seems to be related more to the availability of mussels.

In North America, dreissenids form a major part of the diet of greater (*Aythya marila*) and lesser (*A. affinis*) scaup, buffleheads (*Bucephala albeola*), oldsquaws (*Clangula hyemalis*) and white-winged scoters (*Melanitta deglandi*). The distribution of greater and lesser scaup at Long Point, Lake Erie, appears to be influenced by the distribution of dreissenid beds. The impact of waterfowl on zebra mussel densities is variable, from little or no effect to significant decreases, up to 97% in zebra mussel biomass in some lakes in The Netherlands, especially in the winter months.

Parasites: Dreissenids are not common vectors of parasites. Protists and digeans are the most common parasites, with Nematoda observed sporadically in zebra mussels. Many
33 ☞ protists are common parasites but they do not seem to affect the numbers of zebra mussels. Trematodes are less common parasites of zebra mussels than are protists, the greatest infestation rate recorded being 10%. The most dangerous protists are ciliates of the family Ophryoglenidae which parasitize the digestive gland and may kill the mussel. The only other parasitic ciliate reported in dreissenids is *Concophthirus*, but adverse effects on the bivalve host are unknown.

Species of *Phyllodistomum* and *Bucephalus* are important parasites of dreissenids in lakes of The Netherlands. The infestation prevalence is usually about 1% and may go as high as 10%. The effects of parasites on *D. polymorpha* appear to be minimal, at least until high incidences of cercariae of *B. polymorphus* occur. The intensity of parasitism by *P. folium* is directly correlated to shell size of *D. polymorpha*, the maximum number recorded being 200 at a shell size of 24-28 mm.

In North America, scientists have reported a 2.9% prevalence of plagiorchiid metacercariae in mussels and 2.7% prevalence of adults and juvenile aspidogastrids in mussels from two sites in Lake Erie. The ciliate, *Ophryoglena*, occurred with 1.3% prevalence at a site in Lake St. Clair and 2.7 to 4.3% prevalence at two sites in Lake Erie. It appears that the mass development of zebra mussels may increase the infection rate in definitive hosts, especially fish and waterfowl. The oligochaete, *Chaetogaster limnaei*, and the chironomid *Paratanytarsus*, have also been reported as commensals in zebra and quagga mussels.

Sponges may also act as parasites. Lethal overgrowths of three sponge species, *Eunapius fragilis*, *Ephydatia muelleri* and *Spongilla lacustris* on zebra mussels can occur. The sponge colonies spread over the entire shell of the zebra mussel and smother their siphons. The sponges may also inhibit settlement of dreissenids and out- compete them for hard substrate.

Reproductive potential and life cycle: Zebra and quagga mussels are the most prolific mollusc species in freshwater, producing over one million eggs per female each year in
33 ☞ their two to three year life span. As in European populations, there are one or two spawning seasons per year. The first lasts about three months (early to mid-May to early to mid-August) and is comprised of several spawning events; the second occurs in August or September. If only one spawning season occurs, spawning peaks about the end of August, but several small spawning events may occur even into October in the Great Lakes.

The larvae pass through several developmental stages, including trochophore, D-shape, veliconcha, pediveliger, and plantigrade. However, up to 99% mortality of the larval stages may occur. Most of the mortality in zebra mussels probably occurs during the settling event if the plantigrade form does not find a suitable substrate on which to attach. In spite of this high mortality, the dreissenids are the most productive of all the exotic molluscs. Even *Corbicula* displays a lower fecundity than *Dreissena*, with 25,000 to 75,000 veligers produced in the lifetime of a single clam. Native species of Sphaeriidae have very low fecundities (e.g 1 - 40 eggs per adult) because they are brooders. Since only a small number of larvae can be brooded by any one parent, the number of larvae that are produced is rather small. Hence, external fertilization and development partly explains why *Dreissena* is much more prolific than native species of bivalves in North American surface waters.

Quagga mussels have a greater reproductive potential than the zebra mussel because they can reproduce at lower temperatures, as low as 7 °C and grow at lower temperatures (4 °C). The species is also capable of an infaunal existence, which means it can occupy a greater percentage of the bottom of lakes than can the zebra mussel.

***Food and feeding habits*:** Adult zebra mussels feed on (filter) phytoplankton, protists, bacteria and other microscopic plankton and zooplankton, including their own veligers.
33☞ Studies with stable isotopes of nitrogen and carbon have shown that zebra mussels utilize the entire seston resource, whereas other filter feeders, such as *Daphnia*, use distinct sources of organic carbon. The mussels can filter particles less than 1 µm and are capable of filtering a wide range of bacteria ranging in size from 1- 4 µm.

The filtration rate of dreissenids is affected by their shell size, turbidity, temperature and concentrations of certain sizes and kinds of algal cells (e.g. *Chlamydomonas*) and bacterial cells (See reviews in Neumann and Jenner 1992[18]). The filtration capability of *D. polymorpha* in relation to its role as a clarifier of water in an entire lake, epilimnion or littoral zone is well documented. Europeans are using the clarifying ability of zebra mussels to manage water quality. Zebra mussels are not only capable of removing suspended solids but also of significantly reducing biochemical oxygen demand and phosphorous levels in dilute (15%) activated sewage sludge (Mackie and Wright 1994)[19]. Although dreissenids are efficient water clarifiers, much of the filtered materials are deposited on the bottom as faeces and pseudofaeces. The size of faecal and pseudofaecal pellets varies with mussel size. The settling velocities and rate of

---

[18]Neumann, D. and H. A. Jenner (Eds). 1992. The zebra mussel, *Dreissena polymorpha.* Ecology, biological monitoring and first applications in the water quality management. Gustav Fischer, New York.

[19]MACKIE, G. L., C. A. WRIGHT. 1994. Biodeposition of seston and removal of phosphorous and biochemical oxygen demand from activated sewage sludge by *Dreissena polymorpha* (Bivlavia: Dreissenidae). Water Research 28: 1123-1130.

Fig. 14.3. Total economic impact of zebra mussels by water use (x $1000)

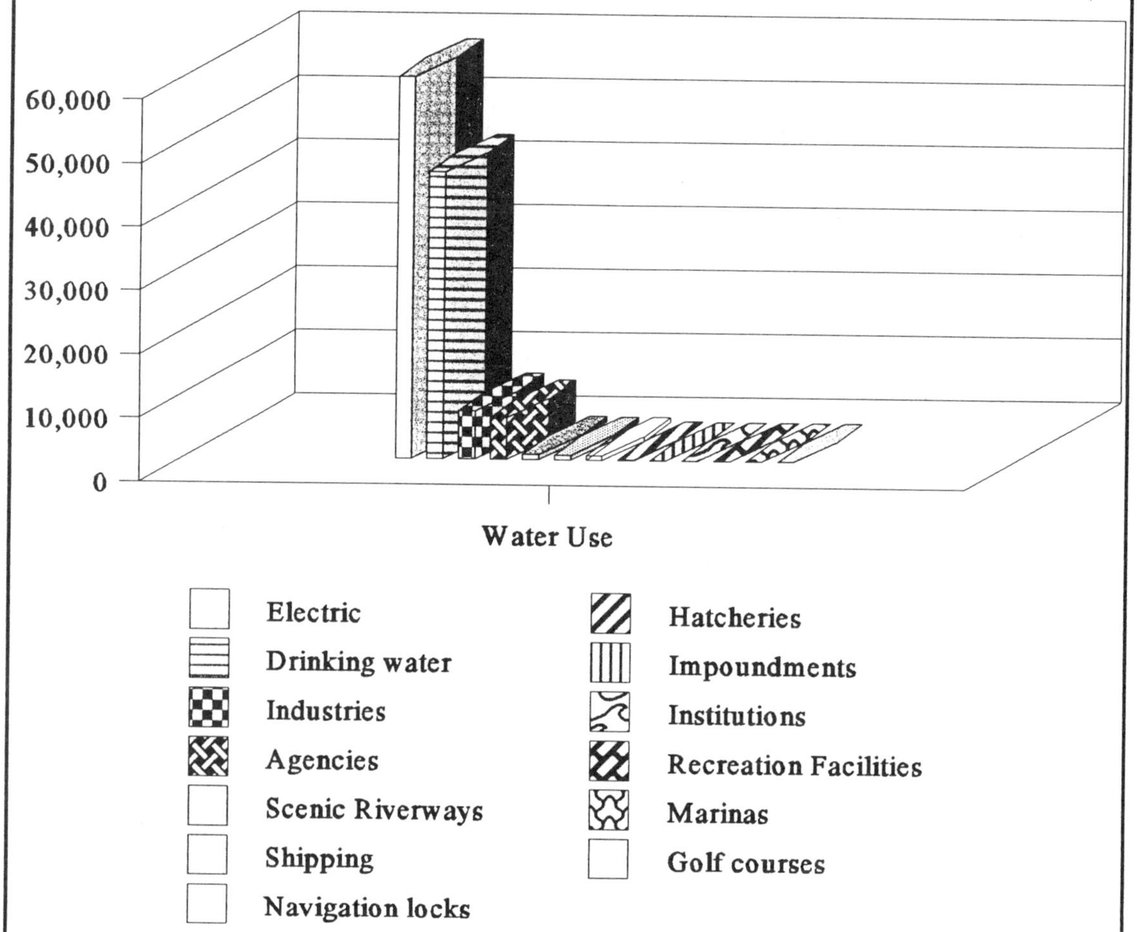

accumulation on the bottom greatly exceed normal sedimentation processes.

Impact potential: The ecological and socio-economic impacts are the severest of all
34 species introductions to date, with direct and indirect costs expected to approach 10 billion dollars (Fig. 14.3). The following impacts have been documented to date.

Industrial/Utility Impacts

(1) ***Industrial and domestic pipelines*:** Accumulations cause (i) reductions in the bore of pipes; (ii) reduced flow through the pipe due to friction loss (turbulent flow instead of laminar); (iii) electro-corrosion of steel or cast iron pipes; (iv) deposition of empty mussel shells at the pipe outlet; (v) tainting and possible contamination of the water upon death (especially when killed as part of a massive control program). All have been documented
34 in North American facilities.

(2) ***Underground irrigation systems*:** Some golf courses that draw water from the Great

Lakes have reported plugged irrigation systems.

34☞ **Fisheries and Wildlife and Other Biological Impacts**

(3) ***Fouling of fishing gear:*** Trap nets and pound nets that are left in the water for extended periods of time were commonly reported in Lake Erie.

(4) ***Potential loss of commercial fisheries:*** Fish species that feed on plankton will be affected through the filter feeding activities of the zebra mussels. This has not yet been proved for Great Lakes fisheries but commercial fishermen on Lake Erie and Lake St. Clair believe that impacts are only now being manifested as smaller catches of perch, smelt and walleye. This may be related, in part, to declines in some deepwater benthos. Apparently the recent declines in smelt populations are being enhanced by diversion of energy to large *D. bugensis* populations in eastern Lake Erie, both through induced changes to the biomass and composition of the plankton as well as reduced amphipod (*Diporeia hoyi*) populations. With the introduction of the zebra mussel, there has been a shift from a pelagic-dominated food web to a benthic-dominated food web. In some lakes (e.g. Lake St. Clair and Lake Ontario) there has been an increase in benthic biomass, especially in annelids, gastropods, amphipods, and crayfish.

(5) ***Elevated levels of contaminants:*** Some *Dreissena* have elevated levels of organic and inorganic contaminants which are passed up the food chain. Biomagnification has resulted in reproductive problems, such as reduced clutch sizes and high embryo mortality in ducks.

(6) ***Reduced seston levels/increased water transparency:*** The dense populations of zebra mussels in North American waters has resulted in the filtration of suspended materials (seston) from the water which has resulted in reduced phytoplankton biomass and nutrient levels and increased water clarity. The reduced levels of plankton are predicted to have a negative impact on planktivorous fish.

However, there have been some beneficial impacts. Macrophyte and benthic algae biomass and diversity has increased and SCUBA divers have benefited from the increased water clarity.

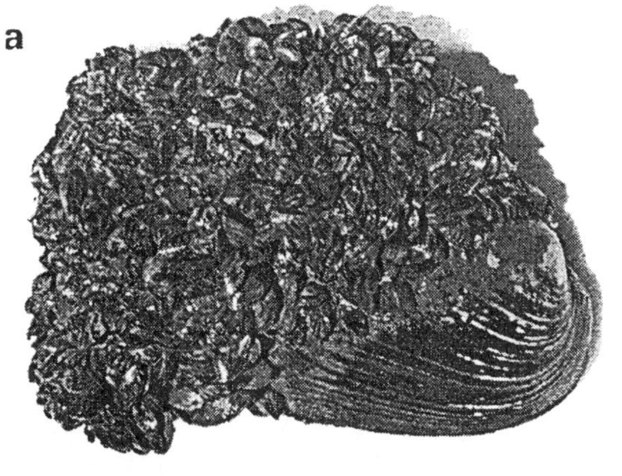

Figure 14.4. A native clam, *Potamilus alatus*, covered with about 5000 zebra mussels.

(7) ***Loss of endemic species of Unionidae:*** All native unionids on the Ontario shores of Lake St. Clair, where the mussel was first introduced, were eliminated by the zebra mussel. The mussels are drawn toward the unionid by their siphoning activity and attach to the exposed posterior end (Fig. 14.4). The mussels become so numerous that the unionid cannot upright itself after rising out of the sediments. Food and dissolved oxygen is stripped from the water by the

zebra mussels, causing the unionid to succumb to starvation and suffocation. Similar results have been documented for other North American lakes (mostly Great Lakes) and rivers. Significant declines in unionid densities begin to occur when infestation levels exceed 10/unionid (Fig. 14.4). The decline in unionid diversity may also be due, in part, to species-specific rates of starvation in unionids.

(8) ***Alteration of benthic community structure and abundance:*** Changes in biomass of benthic algae and macrophytes (see 6 above) were followed by changes in the benthic invertebrate community, with grazers and herbivores showing a greater contribution to the total diversity of functional feeding groups. Numerous studies document changes in community structure and abundance in benthic invertebrates, in part due to increased loadings of faeces, pseudofaeces and seston placed on the bottom by zebra mussel filtering activities, which ultimately may affect fish community structure and abundance.

(9) ***Increased potential as vectors of parasites whose definitive hosts are commercially important species of fish and/or waterfowl:*** This has not yet been demonstrated in North American waters.

Navigational and Vessel Impacts

(10) ***Encrusting the hulls of boating and sailing vessels:*** This is a common sight in the Great Lakes.

(11) ***Encrusting of navigation buoys to the point that the buoys sink deeper in the water than normal:*** This has also been demonstrated in the Great Lakes, especially for fishing markers/buoys.

Private Property Impacts

(12) ***Formation of shoals of shell debris on beaches that will detract from the beach's recreational and aesthetic value:*** This was a common sight on some large beaches in Lake Erie in 1990-1992.

(13) ***Fouling of cottage plumbing and intake structures:*** There are increasing numbers of reports that foot valves and/or screens protecting the plumbing lines are being fouled by zebra mussels.

Socio-economic impacts

34☞ In 1991 economists were predicting that the total economic costs to users of Great Lakes water and the ecological costs caused by the zebra mussel would amount to about 5 billion dollars over a 10-year period. Figure 14.3 shows the total zebra mussel economic impacts on water users in the United States and Canada (Ontario and Quebec only). The data are from O'Neill (1997)[20] and cover the period from 1989 to 1995. As of 1995, approximately $113,000,000 had been spent to combat the zebra mussel. These costs include research, training monitoring, mechanical repairs, planning, prevention, chemical and no-chemical mitigation, retrofitting and lost revenue due to shut-downs. However, the total annual zebra mussel-related costs have risen from an average of 8 million dollars in 1991 to 17.5 million dollars in 1995, and these costs will increase as the zebra mussel extends its range into the U. S.

Beneficial Impacts

(1) ***Manufacturers of Products for Control of the Species:*** Numerous chemical and physical methods have been developed for controlling zebra mussels in a variety of
35☞ industrial and domestic facilities (see Tables 14-13 and 14-14). Most molluscicides were developed specifically for control of nuisance bivalves like the Asian clam and the zebra and quagga mussels. Chlorine is, by far, the most common control option in present use, mainly because of its low operating and capital costs since most industries and utilities were using it previously to control algal and bacterial fouling.

(2) ***Manufacturers of Food for Domesticated Animals or Agricultural Products:*** Some experiments are underway to determine if crushed and powdered zebra mussel shells (and soft parts) will enhance productivity of some fruit crops, like apples.

(3) ***Food for Human Consumption, Display or Sport Value:*** A few entrepreneurs in Florida have used the more colourful zebra mussel shells to make ear-rings, broaches and paper weights (when embedded in plastic). Zebra mussels are not eatable, as many Ontario Ministry of Natural Resource personnel can attest to after a challenge for one of their annual picnics; apparently more than a dozen "dishes" were prepared, including entrees, soups and desserts, and not one was eatable.

(4) ***Indirect Recreational Value:*** Since the appearance of the zebra and quagga mussels, their filtering activities have almost tripled water clarity in Lake Ct. Clair and Lake Erie to the pleasure of SCUBA divers who are able to see objects (e.g. sunken ships) deeper in the water than ever before.

---

[20]O'Neill, C. R., Jr. 1997. Economic impact of zebra mussels - Results of the 1995 National zebra mussel Information Clearinghouse Study. Great Lakes Research Review, Vol. 3, No. 1, pp. 35-44.

Control

Claudi and Mackie (1994, cited earlier) review numerous methods that are categorized as chemical, physical, mechanical or biological control. Tables 14.12 to 14.16 summarize the various methods tested or being tried in North America and Europe.

There are two basic strategies used to control biofouling. A ***proactive strategy*** employs methods that prevents the initiation of biofouling. A ***reactive strategy*** allows some biofouling to occur, but control methods are applied just before nuisance growths and problems occur. The proactive strategy generally employs low, sublethal doses or levels all year, or most of the year, while the reactive strategy employs high, lethal doses or levels to a specific life stage for part of the year. For example, as a proactive strategy, chlorine may be applied at levels that do not kill veliger larvae but disturbs them enough to prevent the plantigrade larvae (settling stage) from attaching to the inside walls of pipes and wet wells. The chlorine would have to be applied from about June to November, while the larvae are in the water. As a reactive strategy, an industry may want to wait until November, after all larvae have settled, and then apply lethal doses (~2 mg/L at end of system) of chlorine until all attached adults are dead. While the costs may be lower for the reactive strategy, there are some disadvantages that are not usually encountered with a proactive strategy. Firstly, large numbers of dead (and often putrefying) shells must be discarded. Secondly, the effluent water usually has to be detoxified because such high levels of chemicals are needed to kill the adults. If chlorine is used, sodium metabisulphite must be used at the end of the pipe to dechlorinate the water before it enters the receiving waters.

There are three main areas that require protection in most industrial and domestic facilities; the intake structure, its delivery system and the facility itself. The intake
36☞ structure is often several kilometres off shore and has "trash bars" that keep large objects, like logs, from entering the intake. The trash bars and inside walls of the intake structure, in particular, require protection because mussels can occlude the spaces between the bars. To date, either coatings, such as silicon-based or copper-based barriers, or mechanical methods, such as scraping or hydro-blasting have been used. The delivery system includes the intake pipeline and the wet-well(s) that maintain a certain water level for traveling screens to remove plant growths, algae and fish from the intake water. The facility has several components that need protection but the components vary according to industrial needs. Nearly all have a fire protection system, an emergency system, a service water system (e.g. sand filtration, water clarification) and a circulating water system (e.g for cooling water that has been circulated at least once). Chemicals are most commonly used to treat the delivery system and facility components, but there is concern that some chemicals may be spilled or released into the water source (i.e. lake or river), and physical methods are being sought to replace chemical treatment.

Mitigation by Chemical Methods

36☞ Nearly all chemical methods can be used either proactively or reactively, the only differences being the concentrations used (e.g. low for proactive, high for reactive treatment) and the frequency of application (e.g. continual for proactive treatment, intermittent for reactive treatment. The cost of the chemical usually dictates which strategy will be used.

Chlorine is the most popular method used because it is relatively inexpensive and most of the capital costs of installing chlorine holding and dispensing equipment have already been made to control biofouling by bacteria and algae. About 90% of all the industries and utilities use chlorination to combat zebra mussel biofouling. Some water treatment utilities use ozone, which has high capital costs but low operating costs, and is very effective for controlling bacterial, algal, viral and macro-biofouling. Other chemicals used by water treatment utilities are potassium permanganate, hydrogen peroxide, and chlorine dioxide (Table 14.12), all of which are stronger oxidizers than chlorine and are more potent but more costly. Potassium permanganate cannot be used in hard water because it tends to turn the water pink. The main disadvantage with chlorine is its tendency to form trihalomethanes (THMs) that are well-known carcinogens.

Non-oxidizing chemicals (e.g. all the molluscides listed in Table 14.12) are very effective at killing zebra mussels but most are expensive and their use is limited to a reactive strategy. Also, in many cases, the chemical has to be detoxified at the end of the pipe before it is released to the receiving waters.

Apparently, some cottagers are using chlorine pucks that are contained within a holding device somewhere on the pipeline between the pump and the cottage. The puck
36☞ slowly dissolves as water flows through the device and delivers about 1 mg/L to prevent or kill mussels within the pipeline. However, no chemicals can be used to protect the intakes or other structures within the lake (e.g. foot valves) and physical or mechanical methods must be used instead

Mitigation Using Physical Methods

Proactive methods: Because the release of chemicals into the environment, even
37☞ at low to near undetectable levels, may have incipient effects on non-target organisms, the use of chemicals as molluscicides is strictly controlled and generally discouraged by environmental agencies. As a result, a variety of physical control technologies have been or are being developed. Because many industries have enormous flows (e.g. > 1,000,000 gpm), the physical method employed must be able to handle high flow rates. Of the methods listed in Table 14.13, the most commonly used is filtration, either by mechanical filtration or by filtration galleries. The filters must have a minimum pore size of $40\mu m$ to remove the eggs of zebra or quagga mussels. Because of the small pore size, frequent back-flushing is needed to expel the huge amounts of fine material that collects on the screen filters. Infiltration galleries are filters embedded in the bottom of the lake and use the sand bottom to filter the water. This is the only method needed to protect all areas of a facility (intake, delivery system and facility components). All other methods require at least two kinds of treatments, one for the trash bars and another for the delivery system and facility components.

Table 14.12. Chemical control options that have been or are being used by different industries and utilities in Canada and U. S. A. Options for nuclear (N), fuel (F) and hydro-electric (H) control are separated for electrical generating stations. A= approved for use as a zebra mussel control agent; A* = approved for clarifying water but kills veligers in the process; Exp. = mostly experimental, but is effective for waters low in alkalinity ; NA= not approved or limited approval for use as a zebra mussel control agent; x = used in Europe in fire protection systems and other closed loop systems. The chemicals are listed alphabetically.

Chemical Control Options	Status of Approval		Brewery	Cement Manuf.	Chemical Manuf.	Cottage	Electrical			Food Process.	Pulp & Paper	Steel Manuf.	Water treat't
	Can	U. S.					N	F	H				
Ammonium nitrate	NA	NA					x	x	x				
Bromine	NA	A			*		*	*	*				*
Chlorine - Gas	A	A			*					*	*		*
Chlorine dioxide	NA	A								*			*
Chloramine	NA	NA											
Sodium hypochlorite	A	A	*	*	*	*	*	*	*	*	*	*	*
Calcium hypochlorite	NA	NA											
Hydrogen peroxide	NA	A					*	*	*		*		*
Molluscicides													
Bayer's Bayluscide	NA	NA											
Betz's Clam-Trol	NA	A					*	*	*				
Buckman's Bulab 6002	NA	A					*	*	*				
Calgon's H1-30	NA	A					*	*	*				
Nalco Actibrom	NA	A					*	*	*				
Ozone	NA	A			*		*	*	*	*			*
Potassium salts	NA	A											
Potassium permanganate	NA	A											*
Some other methods													
Alum	A*	A*	*		*					*			*
Endod	Exp.	Exp.											
High [CO_2]	Exp.	Exp.				*							
Low [Ca], pH	Exp.	Exp.				*							*
Salinity	NA	A					*	*	*				
Sodium metabisulphite	A	A	*	*		*	*	*	*	*	*	0	

Cottagers are now being offered an array of mechanical filters to protect cottage water supply systems. However, cottagers should be aware that if the filter is placed between the pump and the cottage, the pump and its intake (e.g. foot valve) will need to be protected using another method (e.g. annual removal or shut down and removal of submerged components). A filter placed in the lake will protect the pump, but the filters will either have to be back-washed automatically or replaced at frequent intervals, the interval being shorter for turbid waters.

Water turbidity often dictates what physical method can be used. For example, U. V. light and fine filters cannot be used effectively in waters with high turbidities, but otherwise the methods are effective at killing or removing veliger larvae. Low levels (e.g. 4-8 v/cm) of electrical current (AC) applied continuously are very effective but of only recent development and have not gained popular use yet, but is being used to control settlement on trash bars and concrete walls of wet wells in some industries. Plasma pulse technology is a method that uses high levels of electrical energy to create an intense shockwave, or plasma channel, that will rupture and/or kill veligers suspended in the water column. Pulsed power is similar to lithotripsy, a technique used to destroy kidney stones and gall stones in humans, in that both use electrically-generated pulsed (e.g. about 6,000 v every 40 micro-seconds) acoustic shock waves. Cathodic protection relies on electrical current densities to repel veligers from steel surfaces. Field experiments have shown that 8-10 mA/ft^2 will prevent settlement on steel surfaces.

Antifouling coatings are effective at preventing zebra mussel attachment to structures exposed to lake water. They do not offer protection to facility components. The most successful antifoulant coatings to date are silicones and copper- or zinc-based paints. The silicone-based coatings must be applied in several different layers to a perfectly clean and dry (< 10% moisture) surface, making them costly (~ \$100/m^2 ± \$20/m^2), with a service life of ~5 years. The copper- and zinc-based paints are applied in a single layer, are more scratch resistant and apparently have a longer service life (8 - 10 years) and are less costly (~ \$80/m^2 ± \$20/m^2) than silicone coatings. However, the surfaces may have to be "reactivated" or abraded to expose fresh coating.

Even the use of different materials can help reduce amounts of attachment. Field experiments have shown that mussel attachment intensity in pipes tends to be least on copper (90% Cu) and greatest on plastics (e.g. ABS), in the following order of efficacy: copper < brass < galvanized iron < aluminum < acrylic < black steel < polyethylene < PVC < ABS. However, the experiments were done in standing waters and, as discussed below under reactive strategies, flow rates affects the amount of settlement in pipelines.

Water treatment plants and industries that clarify water use alum (aluminum sulphate) as a flocculant to remove suspended materials from the water. Field and laboratory experiments have shown that the levels of alum used in water treatment plants (10-30 mg/L) is not sufficient to kill veligers but the floc formed around the larvae does physically remove them. The addition of alum to water depresses the pH below 5 which is lethal to the veligers. Hence, the toxicity of hydrogen ions and the physical action of the alum combine to remove veligers in water treatment facilities.

Table 14.13. Physical control options that have been or are being used in different industries and utilities in Canada and the U. S. A. Options for nuclear (N), fuel (F) and hydro-electric (H) control are separated for electrical generating stations. A= Approved for use as a zebra mussel control method; A* = Desiccation may be used any time an industry shuts down and has to de-water for repairs. Flocculation by alum is effective at veliger removal in utilities that use it for water clarification; Gamma radiation is entirely experimental and intended only for nuclear power plants. Desiccation has been used in Europe.; Exp. = Plasma pulse technology is use of ultra-high pressure wave created by releasing high voltage (e.g. 6,000 V) of very short duration (e.g. 40 milliseconds). Centrifugation is effective experimentally but has not been tried in industries. The physical methods are listed alphabetically.

Physical Control Options	Status of Approval		Brewery	Cement Manuf.	Chemical Manuf.	Cottage	Electrical			Food Process.	Pulp & Paper	Steel Manuf.	Water treat't
	Can.	U. S.					N	F	H				
Acoustics, Plasma pulse	Exp.	Exp							*			*	*
Cathodic Protection	A	A	*		*		*	*	*		*	*	
Centrifugation	Exp.	Exp.											
Coatings	A	A	*	*	*	*	*	*	*	*	*	*	*
Desiccation	A*	A*											
Electroshocking	A	A	*	*	*	*	*	*	*	*	*	*	*
Filtration	A	A											
Infiltration gallery	A	A											*
Sand	A	A	*	*	*	*	*	*	*	*	*	*	*
Screen, mesh	A	A	*	*		*	*	*	*		*	*	
Flocculation	A*	A*	*		*					*			*
Flushing	A	A	*	*	*	*	*	*	*	*	*	*	*
Freezing	A	A				*		*	*				
Heat Treatment	A	A			*	*	*	*	*			*	
Irradiation (Gamma)	A*	A*											
Pressure	A	A											
Ultraviolet Light	A	A											

Reactive Strategies: Numerous studies have shown that the amount of attachment varies with water velocity. Mussel attachment rates tend to increase from 0 to about 0.5 m/sec but above 0.5 m/sec, mussel settlement intensity decreases with increasing water velocity, until 1.5 m/sec when no settlement occurs at all and at 2.0 m/sec, any attached mussels can be flushed off the walls. Hence, flushing of pipes with flows exceeding 2.0 m/sec usually clears pipelines. While increased flushing rates can be used to clear cottage pipelines, few, if any, cottage systems are designed for such fast flow rates. To determine water velocity in a cottage pipeline, divide the flow rate of water (ft^3/sec or m^3/sec) delivered by a pump by the cross-sectional area of the pipeline. For example, if a cottage has a pipeline with 1.5 in (3.8 cm) diameter (cross-sectional area = 1.77 cm^2) and a one-half horse-power jet pump with a 3 m water lift to supply a flow of 0.85 L/sec ($\approx$14.4 U.S gal/min), a velocity of 4.8 m/sec can be achieved and used to clear the line of mussels.

Thermal shock, or rapid increases in water temperature, are very effective at removing attached mussels. However, the difference in temperature between ambient and the upper lethal temperature and the acclimation temperature affect the rate of mortality. Using data provided in Claudi and Mackie (1994), Table 14.14 provides an estimate of the time needed to kill 100% of the attached mussels at different acclimation and lethal temperatures. Only those industries (e.g. hydro-generating utilities) that are able to recirculate water that has been heated to temperatures required to kill attached mussels can use "thermal back-flushing" as a mitigation strategy. To kill mussels attached to foot valves, or other removable structures, cottagers need only to immerse the structure in boiling water for 5 to 10 minutes to kill all the adults.

Table 14.14. Times (hours) required to kill 100% of zebra mussels at different acclimation temperatures (e.g. temperature of lake water) and various upper lethal temperatures (34 - 38° C).

Lethal temperature °C	Acclimation Temperature °C			
	5	10	20	30
34	6.98	6.6	11.45	—
35	4.05	3.85	4.52	8.75
36	3.48	1.78	3.37	4.35
37	1.93	0.87	2.1	2.55
38	—	—	1.1	1.3

Desiccation, freezing and oxygen deprivation (a chemical method) are effective but not practical for controlling zebra mussel infestations at most industries and utilities. Prolonged shut-downs (about 2 weeks) are required for desiccation and freezing to induce

100% mortality. Oxygen deprivation can be achieved by adding oxygen scavenging chemicals, like sodium metabisulphite, but then the facility has to be shut down so the water can stagnate and lose oxygen. Zebra mussels are relatively intolerant of hypoxia or anoxia, exhibiting 100% mortality within 6 days at 17-18 °C, 4 days at 20-21 °C and 3 days at 23-24 °C. However, anoxia tolerance increases with increasing acclimation temperature. Industries usually prefer to use well-oxygenated water because anoxic water exacerbates corrosion problems.

Mitigation Using Mechanical Methods

Use of explosives to remove attached mussels in pipelines has been shown experimentally to be effective, but is not practical or encouraged for clearing industrial or
38☞ domestic pipelines. Disposable substrates are large plates or coupons made of material that attracts zebra mussels and encourages settlement before the mussels reach more sensitive components of a facility. However, it is labour-intensive, requiring constant monitoring and removal of the disposable substrate.

The remaining mechanical control options (Table 14.15) are commonly used in many industries and utilities. Hydroblasting involves the use of a powerful jet of water to blast the mussels off pumps and walls of wet-wells after de-watering. A pressure of 3,000 psi is recommended to remove heavy build-ups of mussels.

Thermal shock blasting involves the use of carbon dioxide pellets and is an alternative to hydroblasting. The method was developed for removing organic deposits from heat exchangers. Liquid CO_2 is pumped into a pelletizer and the resulting dry-ice pellets are introduced into a pressurized air line and directed out a blasting nozzle onto a surface to be cleaned. The high velocity, rapid expansion (~ 700x) and resultant fragmentation of pellets, and low blast temperature (-84 °C) of the fragments on impact causes embrittlement of the deposit material and easy removal.

Pigging is the use of cylindrical plugs (called pigs) that are placed inside a pipe and then forced, by water pressure, along the length of the pipe. The plug has a diameter equal to that of the inside diameter of a pipe and scrapes the attached mussels off the walls of the pipe as it passes along the pipe.

RCVs, or remote controlled vehicles, is a safe method of either visualizing infestation intensities within pipes, using a remote-controlled camera on the RCV, or even removing some attached mussels, using remote-control scrapers on the RCV. Alternatively, SCUBA divers can enter the pipeline (if it's large enough) and either photograph or scrape mussels off the walls of the pipe. However, this is a dangerous practice and many divers have lost their lives, mainly due to negligence.

Mitigation Using Biological Controls

The biological control options listed in Table 14.16 and 4.17 are operating all the time and serve to help regulate population levels of zebra and quagga mussels in water
39☞ bodies, including forebays of industrial and domestic facilities. However, none have proven to eliminate zebra or quagga mussels in any water body in North America. The only potential biological control option for eliminating zebra and quagga mussels in lakes

Table 14.15. Mechanical control options that have been or are being used in Canada and U. S. A. A = approved for use as a zebra mussel control method. Disposable substrates, as a control option, has been used in Europe but not in North America. The mechanical methods are listed alphabetically.

Mechanical Control Options	Status of Approval		Brewery	Cement Manuf.	Chemical Manuf.	Cottage	Electrical			Food Process.	Pulp & Paper	Steel Manuf.	Water Treat't
	Can.	U. S.					N	F	H				
Blasting	A	A											
Disposable substrates	A	A											
Hydroblasting	A	A	*	*	*		*	*	*	*	*	*	*
Pigging	A	A	*	*	*		*	*	*	*	*	*	*
Robotics	A	A					*	*	*				
Scraping	A	A	*	*	*	*	*	*	*	*	*	*	*
Screens/nets	A	A	*	*	*	*	*	*	*	*	*	*	*
Thermal shock blasting	A	A			*	*	*	*	*			*	

Table 14.16. Biological control options that have been or could be used in Canada and U. S. A.

Biological Control Options	Status of Approval		Brewery	Cement Manuf.	Chemical Manuf.	Cottage	Electrical			Food Process.	Pulp & Paper	Steel Manuf.	Water Treat't
	Can.	U.S.					N	F	H				
Bacteria	NA	NA											
Bird Predation	A	A				*							*
Fish Predation	A	A	*	*	*	*	*	*	*	*	*	*	*
Parasites	NA	NA											
Natural	A	A	*	*	*	*	*	*	*	*	*	*	*

A = Approved for use as a zebra mussel control method
NA = No bacteria or parasites have been approved for use as a method of control

Table 14.17. Species of organisms that serve to limit population sizes of zebra mussels in North America. The list is a summary of those given by Malloy *et al* (1997)[21]. Species are listed alphabetically for each group.

Species	Common name	Comments
Fish Predators		
Acipencer fulvescens	Lake sturgeon	Feed on juveniles or adults
Acipenser brevirostrum	Shortnose sturgeon	Feed on juveniles or adults
Alosa pseudoharrengus	Alewife	Feed mainly on planktonic larvae
Alosa aestivalis	Blueback herring	Feed mainly on planktonic larvae
Anguilla rostrata	American eel	Feed on juveniles or adults
Aplodinotus grunniens	Freshwater drum	Feed on juveniles or adults
Catostomus commerconi	White sucker	Feed on juveniles or adults
Coregonus clupeaformis	Lake whitefish	Feed on juveniles or adults
Cyprinus carpio	Common carp	Feed on juveniles or adults
Dorosoma cepedianum	Gizzard shad	Feed mainly on planktonic larvae
Ictalurus nebulosus	Brown bullhead	Feed on juveniles or adults
Ictalurus punctata	Channel catfish	Feed on juveniles or adults
Ictiobus niger	Black buffalo	Feed on juveniles or adults
Ictiobus bubalus	Smallmouth buffalo	Feed on juveniles or adults
Lepomis gibbosus	Pumpkinseed	Feed on juveniles or adults
Lepomis auritus	Redbreast sunfish	Feed on juveniles or adults
Lepomis microlophus	Redear sunfish	Feed on juveniles or adults
Lepomis macrochirus	Bluegill	Feed on juveniles or adults
Morone chrysops	White bass	Feed on juveniles or adults
Morone americana	White perch	Feed on larvae and adults
Moxostoma valenciennesi	Greater redhorse	Feed on juveniles or adults
Moxostoma carinatum	River redhorse	Feed on juveniles or adults
Neogobius melanostomus	Round goby	Feed on juveniles or adults
Osmerus mordax	Rainbow smelt	Feed mainly on planktonic larvae
Perca flavescens	Yellow perch	Feed on juveniles or adults
Stizostedion vitreum	Walleye	Feed on juveniles or adults

[21]Malloy, D. P., A. Y. Karatayev, L. E. Burlakova, D. P. Kurandina and F. Laruelle. 1997. Natural enemies of zebra mussels: predators, parasites, and ecological competitors. Rev. Fish. Sci. 5(1): 27-97.

Table 14.17 Continued

Species	Common name	Comments
Waterfowl Predators		
Aythya affinis	Lesser scaup	Diving duck, 11-21 mm mussels preferred
Aythya americana	Redhead	Diving duck
Aythya collaris	Ring-necked duck	Diving duck
Aythya marila	Greater scaup	Diving duck, 11-13 mm mussels preferred
Aythya valisnineria	Canvasback	Diving duck
Bucephala albeola	Bufflehead	Diving duck, 11-13 mm mussels preferred
Bucephala clangula	Goldeneye	Diving duck, 8-12 mm mussels consumed
Calidris alpina	Dunlin	Shorebird
Calidris maritima	Purple sandpiper	Shorebird
Charadrius vociferus	Killdeer	Shorebird
Clangula hyemalis	Oldsquaw	Diving duck
Euphagus carolinus	Rusty blackbird	Passeriform
Fulica americana	American coot	Diving rail
Larus argentatus	Herring gull	Gull
Larus ridibundus	Blackheaded gull	Gull
Lophodytes cucullatus	Hooded merganser	Diving bird
Melanitta fusca	White-winged scoter	Diving duck
Melanitta nigra	Black scoter	Diving duck
Melanitta perspicillata	Surf scoter	Diving duck
Sturnus vulgaris	European starling	Passeriform
Other Predators		
Callinectes sapidus	Blue crab	Feeds on mussels in Hudson River
Cambarus robustus	Crayfish	Based on laboratory feeding trials
Glossiphonia complanata	Leech	Based on European studies
Graptemys geographica	Map turtle	Consumes 4-32 m mussels
Hydra americana	Hydra	Feeds on planktonic larvae
Mesocyclops	Copepod	Feeds on planktonic larvae
Ondatra zibethicus	Muskrat	Based on European studies; 3% of diet are unionids
Orconectes propinquus	Crayfish	Prefers 3-5 mm, feeds on 3-14 mm mussels
Orconectes virilis	Crayfish	Based on laboratory feeding trials

Table 14.17 Continued.

Species	Common name	Comments
Parasites		
Aspidogaster conchiola	Aspidogaster	Only about 2.7% of mussels have ≤ 2 worms
Bucephalus polymorphus	Bucephalid trematode	Mussel is first intermediate host
Chaetogaster limnaei	Oligochaete	Symbiont, lives in mantle cavity of mussels
Echinoparyphium recurvatum	Echinostmatid trematode	Mussel is second intermediate host
Paratanytarsus	Parasitic chironomid	Probably a commensal relationship
Phyllodistomum folium	Gorgoderid trematode	Mussel is first intermediate host
Unionicola	Parasitic mite	Lives in mantle cavity of bivalves

are microbes (bacteria or virus) or microbial products. For example, *Bacillus thuringiensis* (*B.t.*) produces a toxin that is contained within a crystal in each bacterial cell. The toxin is non-toxic to non-target species. Insecticides based on *B.t.* are very effective against a wide variety of leaf-eating caterpillars and are used on vegetable crops. Numerous bacteria have been screened for toxicity of a crystal specific for zebra and quagga mussels, but none have been found so far.

Table 14.17 lists several species of fish and waterfowl that are known to prey on zebra and quagga mussels and parasites. Waterfowl are known to reduce population densities of zebra mussels by 97% in the Netherlands, but the waterfowl do not migrate and the lakes in which this level of predation occurs do not freeze over. Hence, waterfowl can feed on zebra mussels all year. The impact of waterfowl on zebra mussels in lakes at lower latitudes (i.e. where waterfowl are present all year and lakes do not freeze) of North America have yet to be determined.

Several species of fish feed on the planktonic veligers (Table 14.17). Most of them are planktivores as adults. *Morone americana* feds on both veligers and adults. The majority of the fish species feed on juvenile and adult mussels. Only the drum and round goby have mouth parts adapted for crushing mollusc shells.

Most of the waterfowl are diving birds, with three shore birds, three gulls and two passeriforms also feeding on adult mussels (Table 14.17). Predation of mussels by waterfowl is limited to the diving depths of the birds, mainly 4-6 m maximum. The dive time of the ducks decreases with increasing density of mussels. Tufted ducks sieve mussels (< 16 mm long) in a water-suction flow generated by repeated tongue movements. Longer mussels are picked up individually. Dramatic increases in waterfowl population sizes and alterations in the timing and routes of migration have ben attributed to the invasion of zebra mussels in the Great Lakes.

Other aquatic organisms also feed on zebra and quagga mussels, including crayfish, crabs, turtles and muskrats. About 3% of the muskrat's diet is pearly mussels (Unionidae) and they probably feed heavily on zebra mussels in some areas..

The parasites are mostly trematodes that use zebra mussels as the first or second intermediate host for the development of sporocysts, rediae and/or cercariae. A few mites, chironomids and oligochaetes live in the mantle cavity of zebra and quagga mussels (and other bivalves), but most have a commensal relationship with the host.

Some Invasive Plants and Their Impacts

40☞ Many plants have been introduced for aesthetic reasons, mostly as cover for fish ponds. These include the water hyacinth, *Eichhornia crassipes*, the water lettuce, *Pistia stratiotes*, and *Salvinia minima*, all of which are tropical and cannot survive naturally in temperate climates where water freezes (hence are not listed in Table 14.9). The majority of exotic plant species are emergent forms, like the common reed, *Phragmites australis*, and purple loosestrife, *Lythrum salicaria*. The most nuisance submersed species is Eurasian milfoil, *Myriophyllum spicatum*. All six species are described below.

Pistia stratiotes, Water lettuce (Fig. 14. 5)

40a☞ The plants are free-floating and new plantlets remain attached by root stalks. The leaves are sessile in a rosette, entire, about 25 cm long, strongly ribbed, densely pubescent and appearing gray-green. Both male and female flowers occur on a short stalk in the axils of leaves. Fruits are green berries with numerous seeds that are dispersed by water.

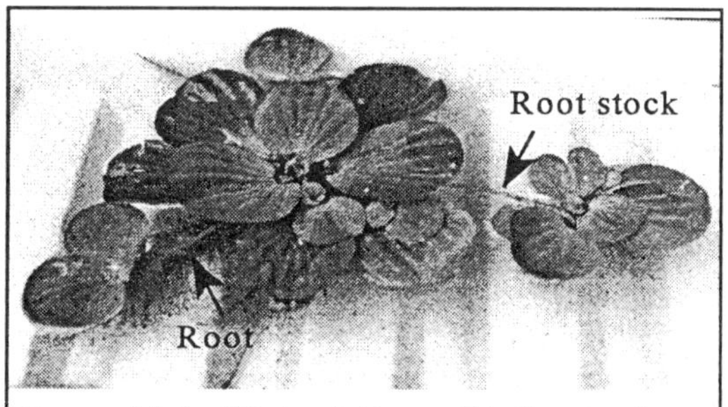

Figure 14.5. Water lettuce, *Pistia stratiotes*.

In its native range, water lettuce grows in lakes and in low velocity streams and canals. The species forms dense mats and clogs waterways. A weevil and a moth, are used for biological control of the species, causing significant reductions in population sizes.

Eichhornia crassipes*, *Water hyacinth (Fig. 14.6)

40b☞ Water hyacinth floats on the surface of the water. The individual plants consist of several thickened, leathery leaves in rosettes and are connected by root stalks. Prominent, black roots hang from each rosette. The leaf petiole is usually inflated, spongy, and can grow to 20 cm in length. The leaf blades are 2 to 15 cm long and 2 to 10 cm wide and broadly elliptic. The leaf bases tend to be heart-shaped. The plant produces a spike with

Figure 14.6. Water hyacinth, *Eichhornia crassipes*

several light-blue to bluish-purple flowers that have a yellow blotch. The fruit is a many seeded capsule that serves to disperse the species.

Eichhornia crassipes is native to South America, probably Brazil. Because of its showy flowers, water hyacinth is sold as an ornamental for small fish ponds and sometimes escapes, or is intentionally introduced, into larger water bodies such as lakes and reservoirs.

In its native range, water hyacinth grows in ponds, canals, freshwater and coastal marshes, lakes and back water sloughs and oxbows along rivers. Reproduction occurs primarily by vegetative means (runners or root stocks) that allows the plant to quickly colonize large areas in relatively short periods of time. Water hyacinth can survive periods of drought as seeds that remain dormant until re-flooding occurs. Wind and water currents are the main mechanisms used to distribute plants within a water body.

Water hyacinth is a nuisance species in its native range, clogging canals and water intakes and restricting navigation along rivers and lakes. It can also negatively impact water quality (especially oxygen levels by preventing light from penetrating the surface) and exclude native vegetation.

***Salvinia minima*, Water spangles** (Fig. 14.7)

The leaves (actually fronds) of *S. minima* are small, 1 to 4 cm long, in whorls of 3 along a slender stem
40c☞ (rhizome) and are of two kinds. Two of the 3 leaves from each node float on the water, are elliptical in shape and about 1 to 1.5 cm long, and nearly as wide. The upper leaf surface has numerous stiff hairs that are divided distally into 4 branches and prevent the leaves from becoming wet. The third leaf is dissected into numerous root-like segments and hangs down in the water.

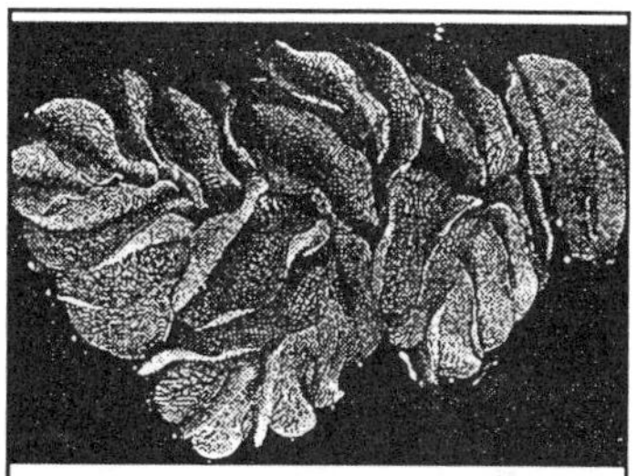
Figure 14.7. Water spangles *Salvinia minima*.

Salvinia minima is native to Central America and Mexico. It is now established in Florida and along the Gulf Coastal Plain from Georgia to Louisiana. Apparently the species *Salvinia auriculata*, *Salvinia natans* and *Salvinia rotundifolia* are commonly misapplied to *Salvinia minima* in the United States. *Salvinia molesta* and *S. auriculata*) are commonly grown as aquarium plants and potentially could be threats if they escape from cultivation.

Plants of *Salvinia* reproduce by vegetative fragments and, occasionally, by the production of spores borne in small round structures called ***sporocarps***. In their native ranges, the various species of *Salvinia* grow in still waters of ponds, small lakes, canals and slow streams. If *S. minima* is cultivated in aquaria and decorative pools in temperate climates, the plants must be brought inside during freeze up.

Some species of *Salvinia* have tremendous growth rates and can cover large bodies of water in a short time to shade out desirable submersed species. *Salvinia molesta* is considered one of the world's worst weeds. In its native range, *Salvinia* clogs irrigation systems, restricts navigation, negatively impacts fisheries, interferes with power production and clogs water intakes.

***Myriophyllum spicatum*, Eurasian milfoil** (Fig. 14.8)

40d☞ Plants of Eurasian milfoil are rooted and submersed, except for a short (3 to 8 cm) emergent flowering spike. Primary stems are generally branched, forming a dense canopy on the water surface. Leaves are whorled, with 4 or 5 (rarely) leaves per node. Each leaf is pinnately dissected into narrow, linear segments and clings to the stem above each node when removed from the water. The flowers are also whorled, in spikes, with the pistillate flowers at the lower nodes of the spike and staminate flowers at the upper nodes. The stem below the inflorescence is curved, lays parallel with the water surface and is about twice the diameter of the lower stem. This feature separates Eurasian milfoil from native species which are not thickened below the inflorescence. *Myriophyllum spicatum* is native to Europe, Asia, and North Africa. It became established in the United States between 1900 and 1940's.

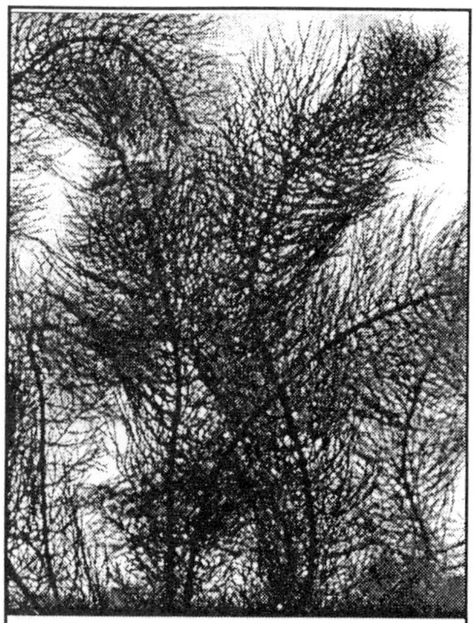

Figure 14.8. Eurasian milfoil, *Myriophyllum spicatum*.

Eurasian milfoil is a highly invasive and aggressive species that colonizes reservoirs, lakes, ponds, streams, small rivers and brackish waters of estuaries and bays. Stems of Eurasian milfoil near the water surface branch profusely and usually form a dense canopy that reduces light availability for other submersed species. *Myriophyllum spicatum* dies back to propagating root crowns during the winter months and spreads primarily by asexual means. Long range dispersal is primarily by fragmentation, the fragments being transported over long distances by water currents, boat trailers or even waterfowl. Once established, individual plants may extend a few meters by the production of stolons. Seeds are considered to be of minor importance in the dispersal of *M. spicatum*.

The most notable nuisance aspects of Eurasian milfoil are: (i) shading out and out-competing desirable native species to form monospecific colonies over large areas; (ii) dense mats and colonies restricting swimming, boating and bank fishing; (iii) negatively impacting any aesthetic appeal; (iv) fragments and floating mats clogging water intakes at power generation facilities and potable water intakes; (v) dense stands providing habitat for mosquitoes, even increasing populations of some species.

***Phragmites australis,* Common reed** (Fig. 14.9)

40e☞ Common reed is a tall (up to 4.5 m), coarse (5 to 15 mm thick), perennial with stout rhizomes that are deep seated in the substrate. The stems are leafy throughout, the sheaths are overlapping with a large, dense, terminal panicle. The leaves are serrated, 1 to 6 cm broad, up to 6 dm long, tapering to long-attenuate tips and stiff.

Figure 14.9. Common reed, *Phragmites australis*

The ligule at the base of each leaf is a ring of dense, short, stiff hairs. The panicle is terminal, plume-like, tawny to purplish or silvery, 15 to 50 cm long, 2 dm broad and with many branches.

Phragmites australis is widespread in temperate and tropical regions of the world and is native to some states in the United States (Gould 1968). The species is found in marshes and in shallow water along the shoreline of lakes and in ponds, swamps, ditches, streams, canals, rivers and estuaries. Large quantities of seed may be produced, but in many cases, very few are viable. The seed will not germinate in more than about 5 cm of water. Once established, *Phragmites* spreads by rhizomes and stolons, often forming dense, monospecific colonies along shorelines and shallow water areas. Rhizomes may grow up to about 2 m per year and be as long as 20 m.

Dense colonies of common reed impede water flow, interfere with recreational activities, such as fishing, and restrict one's view from shoreline areas. *Phragmites* is often cultivated for the production of fiber and is sometimes used in constructing wetlands for the removal of nutrients and pollutants.

***Lythrum salicaria*, Purple loosestrife** (Fig. 14.10)

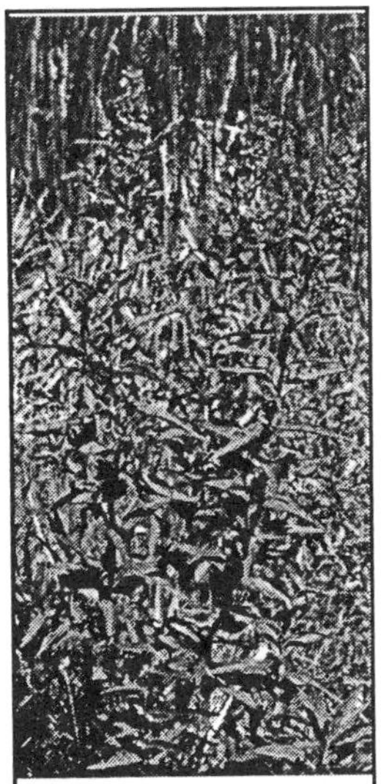

Figure 14.10. Purple loosestrife, *Lythrum salicaria*.

Plants are erect, emergent, much-branched perennials,
40f☞ growing to 2 m tall. The stems are 4-angled. The leaves are opposite or whorled, sessile, mostly longer than the internode above, 2 to 10 cm long and 0.5 to 1.5 cm wide. Flowers are whorled in a showy, terminal spike-like inflorescence. The flowers have 6 petals that are rose-purple and up to 10 mm long. There are usually 12 stamens.

Lythrum salicaria is native to Eurasia and is now naturalized over large areas of the North America. It was introduced into coastal areas of the northeastern United States in the early 1800's. Due to its attractive flowers, it was planted as an ornamental garden species but escaped from cultivation to come to occupy at least 40 states and most of Canada.

Purple loosestrife grows anywhere with a source of water, such as in marshes, along pond, lake and river margins, canals, wet meadows, prairies and ditches. The plant spreads asexually, from a strong root stock, and sexually, a single plant producing as many as 2.5 million seeds annually.

Populations of purple loosestrife may spread so aggressively that native vegetation becomes suppressed, altering the structure and function of wetlands and reducing the value of wetlands for wildlife. Purple loosestrife may also impede the flow of water in irrigation canals and degrade the quality of pasture and hay. It is estimated that 200,000 ha of wetlands in the United States are degraded annually through invasion of purple loosestrife.

Most Recent Invasive Animals and Their Potential Impacts

41☞ The catastrophic impacts of the lamprey and the zebra mussel are well known, but what can we expect from the more recent introductions, that is, those that have been introduced since 1990. The newest invaders are the round and tubenose gobies, the European ruffe, the blueback herring, the quagga mussel, the New Zealand snail, the spiny waterflea and the European gammarid.

41a☞ **Round goby, *Neogobius melanostomus*, and tubenose goby, *Proterorhinus marmoratus*** (Fig. 14.11)

The round and tubenose gobies were first discovered in the St. Clair River in 1990. Both species are benthivores. The round goby is more aggressive than the tubenose goby. Both species are native to the Black or Caspian Seas, but the tubenose goby is endangered in its native range.

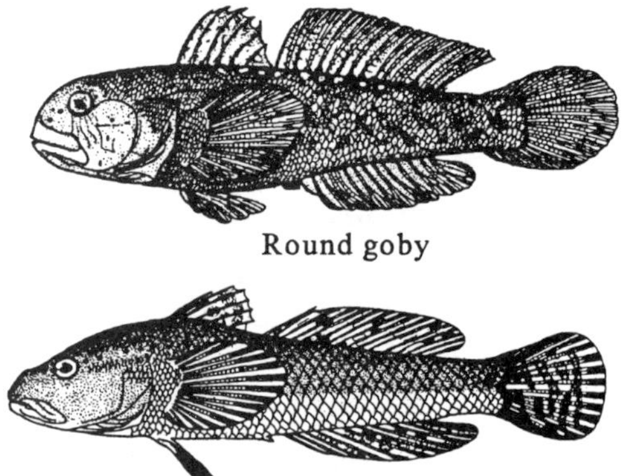

Figure 14.11. Round and tubenose gobies.

The round goby posesses several characteristics of invasive species: (i) tolerance of a wide range of environmental conditions; (ii) a broad diet; (iii) aggressive behaviour; (iv) high fecundity; (v) iteroparity; (vi) nest guarding by males; (vii) a small body size (relative to other fish, but large relative to fish with similar benthic habits, e.g. sculpins). Round gobies larger than 10-cm long feed almost exclusively on zebra mussels. At smaller sizes, they feed on the same benthic foods as mottled sculpins (*Cottus bairdi*), darters (*Etheostoma* spp), logperch *(Percina caprodes)* and some minnows (*Notropis* spp.). The round gobie's aggressive nature drives mottled sculpins from their prime feeding, shelter and spawning areas in shallow (3-5 m) rivers. In deep lakes, it is expected (hoped?!) that the mottled sculpin will out-compete the round goby. Like the freshwater drum, the round goby has robust molariform teeth used to crush the shells of molluscs, including those of zebra and quagga mussels.

The round goby lives mostly in inshore waters from 20 m to 60 m deep, on gravel, shell or sand bottoms. Spawning typically occurs throughout the spring and summer. Its prolific reproductive nature is attributed to several repeated spawnings, up to 6 times every 18 - 20 days. Peak spawning occurs at 15 °C. The life span is typically 3 to 4 years, occasionally exceeding 4 years. Round gobies are preyed upon by bass, walleye and trout.

In Europe and Asia, the tubenose goby lives in fresh and slightly brackish, shallow waters of rivers and estuaries. It is found under rocks in weedy areas, feeding on benthic invertebrates. Spawning occurs in late spring and summer, with eggs (up to 6,100) laid under stones and mollusc shells.

European ruffe, *Gymnocephalus cernuus* (Fig. 14.12)

41b ☞ The European ruffe was first discovered in 1986 in plankton hauls from the St. Louis River, the western-most tributary of Lake Superior. It was fist discovered in Lake Superior near Thunderbay, Ontario, in 1991, but by this time it had become the second most abundant fish species in the St. Louis River. The European ruffe is tolerant of a wide range of environmental conditions, typically occurring in slow-flowing, often turbid rivers. It has a high fecundity, aggressive behaviour and is a voracious carnivore (of benthic invertebrates, zooplankton and fish eggs), attributes that could lead to significant reductions in some native fish stocks, particularly those whose eggs it preys upon.

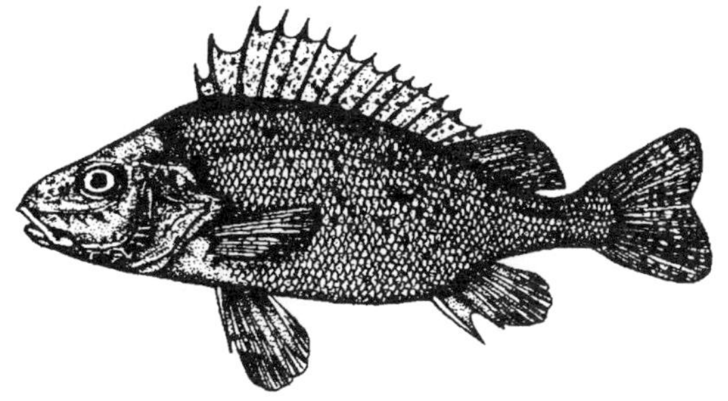

Figure 14.12. The European ruffe.

Spawning occurs in spring and summer, on a hard bottom of clay, sand or gravel. Up to 200,000 eggs are laid. European ruffe have an extended spawning period, longer than most native percids (e.g. yellow perch), a trait that ensures that some eggs meet favourable environmental conditions. This trait, its predation on eggs of fish, and its potential ubiquitous nature, have raised concerns of its potential serious effects on Great Lakes fisheries and the difficulty of eradicating the species.

Blueback herring, *Alosa aestivalis*
(Fig. 14.13)

41c☞ The blueback herring was first found in Lake Ontario in 1995. MacNeill (1998)[22] suggests that the species was introduced from the Hudson River through the inland canal system that connects the river to Lake Ontario. Like its sibling species, the alewife (*Alosa pseudoharrengus*), the

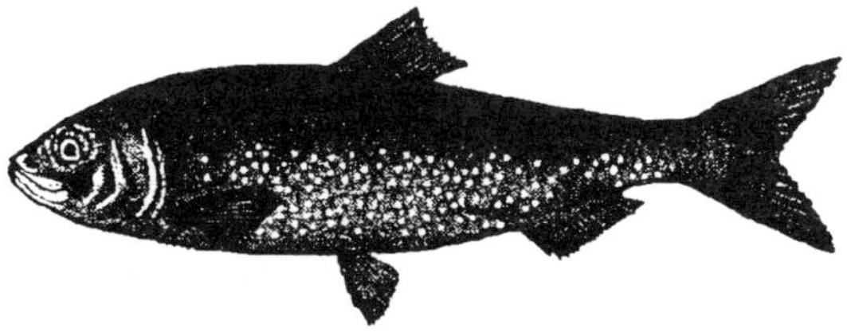

Figure 14.13. The blueback herring.

---

[22]MacNeill, D. 1998. The discovery of blueback herring, *Alosa aestivlis*, in the Great Lakes raises speculations on its colonization potential and species impacts. Great Lakes Research Review, Vol. 3, No. 2, pp. 8-12.

blueback herring is anadromous, dwelling in the sea as adults and spawning in fresh waters. However, the two species have different ecological requirements, with slightly different temperature, depth and food preferences, the alewife preferring colder, deeper water and feeding on larger benthic zooplankton than the blueback herring which feeds on small zooplankton, including veligers of the zebra and quagga mussels. Potential impacts include competition with, and declines in, native planktivorous fish species populations.

Quagga mussel, *Dreissena bugensis* (Fig. 14.2b)

The species was first found in Lake Ontario in 1989 and may have been introduced at the same time as the zebra mussel but, because it can live in deeper water
41d☞ than the zebra mussel, the species may not have been discovered until later. Its ecological and socio-economic impacts appear to be similar to the zebra mussel.

Shell ornamentations are somewhat similar to the zebra mussel but less colourful, the ventral surface is convex and the byssal opening is more distal in position. The mode of life and habitat of quagga mussels is similar to zebra mussels, with two major exceptions. The quagga mussel has an ability to live an infaunal existence, in deeper waters, and an epifaunal existence in shallower waters (zebra mussels grow best on hard substrates in warm, shallow water). In the Great Lakes, the ratio of quagga mussels to zebra mussels increases with increasing depth, down to about 150 m. It appears that quagga mussels are able to alter their shell allometry for an infaunal existence. They are thinner and lighter in weight in soft, profundal sediments than on firm, inshore substrates, whereas zebra mussels exhibit little or no changes in shell allometry on firm or soft substrates.

With the adaptability of quagga mussels to life in soft sediments at great depths, provided oxygen is near saturation, quagga mussels have the potential to invade more northern lakes than can the zebra mussel. Also, the species can grow and reproduce at lower temperatures (< 9 °C) than the zebra mussel (8-10°C for growth, 12-15°C for reproduction) and therefore has a greater potential to infest lakes that are more northern in latitude. However, several have observed quagga mussels being more abundant at thermally-enriched sites than at sites unaffected by thermal discharges. Perhaps quagga mussels prefer the relatively constant temperatures characteristic of deep water (e.g. 1-4 °C) than the highly variable temperatures found in shallow waters.

Although dreissenids tend to develop greater biomasses in flowing waters (e.g. pipelines) of moderate velocities (< 1.5 m/sec), where their filter feeding efficiency can be enhanced by a constant delivery of food to the siphons, the planktonic larvae will prevent them from establishing large and permanent populations in natural streams and rivers that lack impoundments, as discussed earlier.

The quagga mussel has the same impact potential as the zebra mussel, but because it can live as an infaunal species in profundal sediments, its ecological impacts will be extended to deeper water and to lakes further north in latitude than will the zebra mussel.

New Zealand snail, *Potamopyrgus antipodarum* (Fig. 14.14)

The first North American record of the New Zealand snail, *Potamopyrgus*
41e☞ *antipodarum*, is Green River, Colarado in 1990. It was first found in the Great Lakes in Lake Ontario and the Saint Lawrence River in 1991 but the specimens were archived for a few years before the introduction was discovered (Zaranko *et al* 1997[23]). The species is widespread throughout Europe, including Britain, Denmark, Switzerland, Poland, Hungary and Slovakia. It is also widely distributed in south eastern Australia and Tasmania where it was introduced in the middle of the last century.

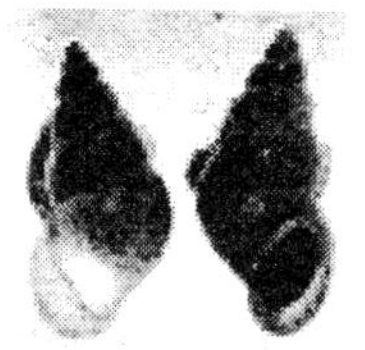

Figure 14.14. *Potamopyrgus antipodarum* showing rear and aperture views of shell.

In Lake Ontario *P. antipodarum* lives on silty sand, but in Britain and New Zealand it occurs in lowland rivers, stony streams, creeks, ponds, lakes, springs and estuaries where salinities can range up to $26^{o}/_{oo}$. It is found only in permanent waters with a wide range in calcium content, on soft and hard substrates and amongst vegetation. In Australia, the species appears to live almost entirely in, or close to, disturbed habitats, such as those that have been altered by urban development or by agricultural or forestry activities (Mackie 1999[24]). Apparently, the snail cannot tolerate water temperatures greater than 28 °C.

The species spreads passively by river and current transport, by attaching to extremities on birds and fish, and by internal transport in the digestive tract of fish as a mechanism. Part of its effectiveness at passive dispersal is its ability to survive in estuarine habitats. The snails have been known to breed in freshwater tanks and reservoirs in the Sydney area of Australia and have been distributed through water pipes to emerge from domestic taps.

Natural enemies include fish and waterfowl, perhaps some amphibians. The species is ovoviviparous, but populations may be entirely parthenogenic or contain varying proportions of sexually functional males. Sexual reproduction probably occurs to some extent in many populations because males constitute up to 20% in some European populations, 9% in some Australian populations, but usually 0 to <3% in most populations. In Lake Ontario only females usually occur.

The species has not been present long enough in North America to make any evaluations on its positive or negative impact potential. Zaranko *et al* (1997) suggest that the biofouling potential is low, its most serious threat being competition with native

---

[23]Zaranko, D. T., D. G. Farara and F. G. Thompson. 1997. Another exotic mollusc in the Great Lakes: the New Zealand native, *Potamopyrgus antipodarum* (Gray 1843) (Gastropoda: Hydrobiidae). Canad. J. Fisher. Aquat. Sci. 54: 809-814.

[24]Mackie, G. L. 1999. Ballast water introductions. *In:* R. Claudi and J. Leach. 1999. Non-indigenous freshwater organisms: Vectors, Biology and Impact. CRC Press, Boca Raton, FL.

gastropods. However, in Australia, *P. antipodarum* has been recorded as blocking water pipes and meters in Australia. There is no evidence to suggest that the species requires immediate control in the Great Lakes.

Spiny water flea, *Bythotrephes cederstroemi* (Fig. 14.15)

41f☞ *Bythotrephes* (pronounced bith-o-treh-feez) *cederstroemi* entered the Great Lakes in the 1980s, from the ballast water of a European ship. The species is native to northern Europe, including the British isles, Scandinavia and the Soviet Union. The first living specimen in the North America was found in Lake Huron in December 1984. *Bythotrephes* spread to Lakes Erie and Ontario in 1985, Lake Michigan by 1986 and Lake Superior by 1987.

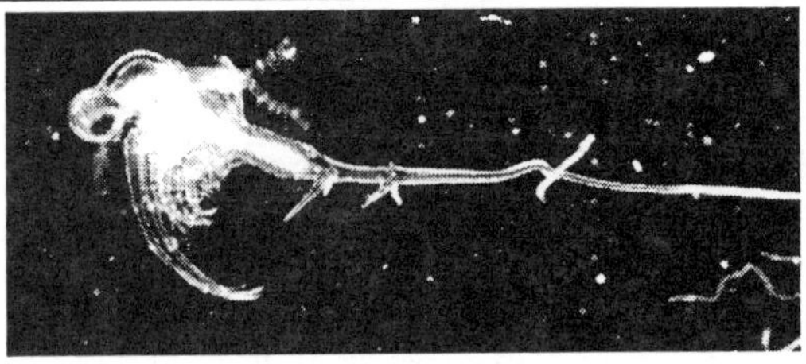

Figure 14.15. *Bythotrephes cederstroemi*. Photograph provided by Jonathan Witt.

The species is easily recognized by its long, spiny tail (Fig. 14.15). The tail is disproportionately long, often comprising over seventy percent of the animal's total length, and contains one to four pairs of thorn-like barbs. The numbers of barbs can be used to determine the approximate age of the animal, juveniles possessing just one pair. The head has a single large eye, filled with black pigment, and a pair of sickle-shaped mandibles, or jaws that are used to pierce and shred its prey. Just behind the head is a pair of swimming antennae, that propel the animal through the water. The animal has four pairs of legs, the first pair being much longer than the others, and used for catching prey; the other pairs of limbs are designed for grasping prey as they are being consumed.

Bythotrephes has a very rapid reproduction rate. Like other cladocerans, reproductive females carry their eggs (1 to 10) in a brood pouch. Most of the time, female *Bythotrephes* reproduces asexually by parthenogenesis. The new females are genetic replicas, or clones, of the mother. During the summer, when the surface water of the lake is warm, new generations are produced by parthenogenesis in less than two weeks. The sex of the offspring is determined mostly by environmental factors. When food becomes limiting, or when the lake cools in the fall, males begin to appear. When adult females detect declining environmental quality, they respond by producing male rather than female offspring. The males are able to mate with surviving females to produce resting eggs. The resting eggs are first carried as orange-brown spheres in the female brood pouch. The eggs are later released and fall to the bottom of the lake where they over-winter. In spring, or early summer, the eggs hatch into juvenile females that begin parthenogenetic reproduction again. See Chapter 7 for details on reproduction in Cladocera.

Spiny water fleas eat other common zooplankton, especially *Daphnia*. However, *Daphnia*, are an important food item for small, juvenile fish and thus compete directly
41g☞ with young fish for food. The rapid population growth enables *Bythotrephes* to exploit

the food supply to the eventual detriment of the fish. Although *Bythotrephes* is preyed upon by fish, its spiny tail discourages feeding by most small fish. The species is rarely found in stomachs of fish less than 5 cm (2 inches) in length. Apparently the growth rates and survival of young fish can be adversely affected by the presence of *Bythotrephes,* owing mostly to competition for food. In general, food availability decreases with increasing abundance of *Bythotrephes*.

European gammarid, *Echinogammarus ischnus* (Fig. 14.16)

Like the gobies and the zebra and quagga mussels, the European gammarid is a euryhaline amphipod native to the Caspian-Black Sea system (Witt et al 1997)[25]. Laboratory bioassays have shown that the incipient lethal level of salinity for the European gammarid is about 19 ppt (at 21 °C), and it can survive in fresh water with as little as 3.8 mg Ca/L. Its ability to live under a wide range of environmental conditions suggests that it will invade a large number of lakes throughout North America. The species has the potential to spread as rapidly as it has in Europe, through natural and man-made watercourses.

Figure 14.16. *Echinogammarus ischnus*. Photograph provided by Jonathan Witt.

The specie's range will probably overlap that of the zebra mussel with which it is commonly associated, feeding on organisms and material in the pseudofaeces of the zebra mussel. Currently, the European gammarid is found only in the Great Lakes (Lake Erie and Lake St. Clair as of 1998). Its presence in the Great Lakes is predicted to result in the displacement of native amphipods and emulate the effects of the North American euryhaline species, *Gammarus tigrinis*, which displaced *Gammarus duebeni*, *Gammarus zaddachi* and *Gammarus pulex* in several lakes in The Netherlands. In Lake St. Clair, the European gammarid is found in the same habitats as *Gammarus fasciatus* and *Hyalella azteca*. In 1995, when the *E. ischnus* was first discovered, native amphipod species dominated the habitats. But in 1998, *E. ischnus* was the dominant species, suggesting gradual competitive displacement of *G. fasciatus* and *H. azteca*.

BIOMIMICRY

42☞ Biomimicry is derived from the Greek words for life (i.e. bios) and imitation (i.e.

---

[25]Witt, J. D. S., P. D. N. Hebert and W. B. Morton. 1997. *Echinogammarus ischnus*: another crustacean invader in the Laurentian Great Lakes. Can. J. Fish. Aq. Sci. 54: 264-268.

mimesis). Benyus (1998)[26] describes the term as "innovation inspired by nature". She examined an array of natural designs and processes and how we have used them to solve human problems, for example, a solar cell inspired by a leaf. She asks numerous questions that are answered by learning from nature, questions like, "How should we grow our food?", "How should we make our materials?" and "How should we power ourselves, heal ourselves, store what we learn?".

Although Benyus (1998) uses terrestrial examples, like "farming the way nature farms", there are also numerous aquatic examples of biomimicry. Benyus (1998) describes the process that *Mytilus edulis* (blue mussel) uses to produce a byssus when attaching to a surface: first it cleans, or primes the surface (explaining why we prime surfaces before painting them); then it applies the primer that is water proof and has cohesive (glue molecules sticking to one another) and adhesive (binds to a surface) properties, like the paints that humans have now developed.

Oyster and mussel farming also is based on observations made on natural oyster beds and mussel beds. The production of cultured pearls is based on observations made on the production of natural pearl oysters. Many of our health foods, such as *Chlorella* pellets/capsules and watercress (*Nasturtium*), are cultured in environmentally-controlled ponds that provide the same optimal growth requirements observed in nature.

Benyus (1998) gives us ten lessons that she learned from nature (quotations are hers):

1. Use waste as a resource (e.g. by recycling materials in the sediments and water column).
2. Diversify and cooperate to fully use the habitat (e.g. all stable habitats are diverse and have a variety of functional feeding behaviours at all trophic levels, hence little is wasted).
3. Gather and use energy efficiently. Not everything needed by industry can be recycled, and even in natural systems, only nutrients and minerals can be recycled. For example, energy that is converted into heat by doing work is unavailable to do more work. But in nature, very little energy is wasted.
4. Optimize rather than maximize. Once throughput (e.g. biomass accumulation) and recruitment has been maximized, nature puts more emphasis on optimizing resources to ensure that one or two offspring survive.
5. Use materials sparingly. "Organisms build for durability, but they don't overbuild". They build only what is needed, with a bare minimum of materials and effort (e.g honeycombs provide maximum space with minimum structural material).
6. Don't foul the nests. If you are going to sleep, eat and breathe in a facility, keep it clean. Nature has ways of eliminating any toxic materials (e.g. ammonia) produced by organisms (e.g. by oxidizing them and making nitrate, a nutrient, in

---

[26]Benyus, J. M. 1998. Biomimicry. Quill, William Morrow & Company, Inc, New York, NY.

the nitrogen cycle). Organisms don't pollute their environment, people do.

7. Do not draw down resources. "Organisms in a mature ecosystem live on harvestable interest, not principle". The best predators and grazers are those that don't completely eliminate their prey and plant crops. "Don't use nonrenewable resources faster than you can develop substitutes" and "don't use renewable resources faster than they can regenerate themselves".
8. Remain in balance with the biosphere. The biosphere is a closed system, with no materials being exported or imported. Within the biosphere are numerous cycles (e.g. nitrogen cycle) that have subcycles (e.g. terrestrial, aquatic nitrogen cycles). "The stocks of carbon, nitrogen, sulphur and phosphorous stay pretty much the same, even though they are actively traded among organisms".
9. Run on information. All trophic levels have some kind of feedback information. Both negative and positive feedback are required to keep ecosystems balanced. "Don't use nonrenewable resources faster than you can develop substitutes" and "don't use renewable resources faster than they regenerate themselves". Successful body designs and behaviours are high in information content.
10. "Shop locally". Organisms obtain their products locally. By shopping locally they become experts in their own back yard. Predators know where their prey are and which prey provides the most energy for the least amount of effort. "Nature does not commute to work", a lesson that is almost totally ignored by humans.

CHAPTER 15

GLOSSARY

For scientific and common names of fish and macrophytes, see end of glossary

A

Active dispersal
is dispersal using the swimming (e.g fish) or flying (e.g. adult insects with aquatic larval stages) abilities of the organism.

Acute toxicity
of a substance is the lethal amount when it is administered in a single dose; coming quickly to a crisis, usually within 96 h.

Additive effect
occurs when the combined effect of two chemicals is equal to the sum of the effects of the individual chemicals applied alone (e.g. 1 + 1 = 2).

Advection
process by which solutes are transported by the bulk of flowing fluid such as the flowing ground water; horizontal transfer of heat by large-scale motions of the atmosphere.

Aeolian lakes
are formed by wind action by the toppling of trees and subsequent piling up of debris and soil across rivers, which over time accumulates enough organic and inorganic materials to effectively dam the river.

Aestivate
is a type of inactivity or dormancy during times of drought, especially in temperate and tropical regions; in lungfish, some snails, and many reptiles and insects.

Agitation aggregation
is one of the two mechanisms that converts DOM to POM; tiny plate-like aggregates of organic composition form when air is bubbled through filtered water. Aggregates occur because some compounds, such as soluble fatty acids, proteins, and humic materials, readily adhere to bubble surfaces.

Algivores
organisms that feed on algae.

Alkali lakes
lakes in arid regions; the pH of the water is greater than 12.

Alkalinity
is a measure of the ability of the water to resist a change in pH. Water resists changes in pH due to the presence of anions of carbonates, bicarbonates and hydroxides.

Allochthonous
material generated outside a particular habitat but brought into that habitat, such as debris brought into a lake by a river.

Amictic
refers to a female rotifer which produces only diploid (2N) eggs incapable of being fertilized by a male; refers to the egg that this female rotifer produces.

Ammocoete larvae
is the larval stage of lampreys found in streams; blind and toothless; usually in U-shaped tubes; stage lasts for three to seven years.

Ammonification
is a complex of processes involved in the bacterial conversion of proteins to ammonia (NH_3)and other compounds.

Amoeboid
1. refers to organisms that move or feed by forming pseudopods - false feet 2. refers to an *Amoeba* or a related protozoan. 3. refers to an organism having movement resembling that of an *Amoeba.*

Amphipoda
a large order of the class Malacostraca (in the subphylum Crustacea) including a great

diversity of species in over 100 families; examples are the sideswimmers, sand hoppers, etc.; body compressed; first thoracic segment fused with head; no true carapace; mostly scavengers; most spp. marine, burrowing or moving about on the bottom and debris; some parasitic, a few in fresh waters, and several terrestrial; over 6000 spp.

Ampitheater
: stepwise arrangement of lakes on mountain

Anadromous
: refers to an organism that lives in the sea as an adult and returns to fresh water to spawn in the spring.

Annelida
: phylum consisting of elongated, segmented worms; includes earthworms, leeches, and many kinds of marine forms; also referred to as the True Worms.

Antagonism
: occurs when two chemicals applied together interfere with each other's potency

Aphotic zone
: is the portion of the lake where incident light level is less than 1% (i.e. total darkness).

Application factor (AF)
: a factor used to predict safe concentrations of a contaminant under field conditions. It is intended to provide an estimate of the relationship between a contaminants chronic and acute toxicity and is calculated as the ratio of the MATC to LC_{50} values, or AF = MATC / LC_{50}

Arthropoda
: a vast phylum containing crustaceans, insects, spiders and their relatives, centipedes, millipedes, etc.; also referred to as the arthropods

Astroblemes
: are "star scars" created by meteors upon impact with the Earth and create huge depressions that fill with water to create *meteoric crater lakes*.

Autochthonous
: native substances; generated within a particular habitat; indigenous.

Auxotrophic
: refers to species that require vitamins for growth. The vitamins required are vitamin B_{12}, thiamin and biotin.

B

Bacillariophyta (Bacillariophyceae)
: phylum of algae also referred to as the *diatoms*, a class of the division Chrysophyta; unicellular and eukaryotic autotrophs with a siliceous frustrule (a box-like cell wall made of silica); mostly planktonic.

Back eddies
: pools along a stream that are created behind rocks and snags where the whirls around but delay its downstream passage.

Basaltic minerals
: consist of fine-grained igneous rock, black or greenish black in colour, and rich in iron, magnesium and calcium.

Basin characteristics or Basin features
: Features that describe the shape, slope, area, length, number of stream segments, etc. See Chapter 2.

Bifurcation ratio (BR)
: the ratio of the number of stream segments in an order relative to the number of stream segments in the next highest order; $BR = n_i/n_{i+1}$ where, n_i = number of stream segments in order i and n_{i+1} is the number of stream segments in the next highest order.

Bathymetry
: the depth characteristics of a lake.

Behavioural drift
: is of two types, passive behavioural drift and active behavioural drift. Passive behavioural drift is an innate behaviour usually with some *diel* periodicity associated with it. Active behavioural drift involves some specific movements that will move them downstream or upstream.

Benthic
: refers to collectively, all those animals and plants living on the bottom of a lake or sea, from the waters edge to the greatest depths (benthos); bottom in a pelagic area.

Bet-hedging
: models of life history evolution considering instances in which selection in temporarily variable environments results in the fixation of phenotypes with lower variance in fitness which are not necessarily those with highest expected fitness.

Binary fission

asexual reproduction in animals by a simple division of the body into two parts.

Bioaccumulation (or Bioconcentration)
is the accumulation of contaminants in body tissues; occurs when contaminants such as metals and halogenated hydrocarbons, which cannot be excreted, remain in the body in an unchanged state and are continually added during the life of the organism.

Bioassay
is the establishment of the character and strength of a drug or other substance by measuring its effect upon a living laboratory animal; The bioassay may be an acute, sublethal or chronic toxicity test.

Bioavailability
refers to the amount of chemical entering the bloodstream or tissue fluids to become available to the cells in the body.

Biochemical oxygen demand
in the case of organic wastes, the difference in oxygen concentration in a sample before and after bacterial digestion has taken place; provides a direct measure of the oxygen utilization in bacterial degradation of the sample.

Biocides
a chemical substance that kills or limits living organisms.

Biodiversity
is commonly measured in terms of genetic diversity and/or species diversity.

Biofouling
Fouling by organisms, such as bacteria, algae, zebra mussels, of surfaces, water, or anything useable by other organisms, including humans.

Biomagnification
is the progressively increasing amounts of contaminants in body tissues of animals up the food chain. For example, carnivores will contain more than benthivores, which will contain more than planktivores, which will contain more than autotrophs.

Biomass
is the total weight of organisms per unit area at any given moment in time.

Biotransformation
transfer of a compound to modify its potency.

Bivalvia
class of invertebrates in the phylum Mollusca including clams, oysters and mussels.

Bivoltine (Bivoltinism)
reproduction in which two generations per year are produced.

Bogs
are a type of *wetland* characterized by a low species diversity, few higher plants and abundance of sphagnum moss (that is acidic bogs); they depend on mineral-poor rainwater; they are dominated by acidophilic mosses and sedges.

Bottleneck effect
occurs when an event causes death of large numbers of individuals non-selectively, leaving a small population that is unlikely to have the same genetic makeup as the original population. The most prevalent alleles will be those that were present in the highest frequency in the original population; individuals with rare alleles would likely be destroyed and removed from the gene pool.

Bryozoa
phylum which is also referred to as the Moss Animals; small tufted or branched marine and freshwater organisms

Budding
type of asexual reproduction involving the appearance of new animals as outgrowths from an older individual and their subsequent growth and maturation.

Byssal threads
are used by zebra mussels to attach to substrate; they are secreted by the byssal gland in the base of the foot.

C

Calcareous water
is water that contains calcium as carbonate

Calderas
are basins formed by the subsistence of the low rim or roof around the volcanic cone after the magma has been ejected.

Capillary wave
also known as a Faraday wave, it develops when the wind blows over the ocean surface and the pressure and stress deform the ocean surface into small rounded waves with the wavelengths less than 11.74cm; the dominant force working to destroy them and smooth the ocean surface is surface tension; have rounded crests and V-shape troughs.

Cartometer
is an instrument for measuring the perimeter of lakes or the shoreline.

Catadromous
refers to fishes which spend most of their life in fresh water but migrate into salt water to spawn.

Catastrophic drift
is a movement of organisms as a result of flooding, spring runoff, ice or any other changes in physical conditions of the water. It is especially common in the winter and spring.

Charophyta
a phylum of algae also referred to as the stoneworts. The charophytes are really macrophytes because all are large (up to 0.4 m in height), submersed plants in the littoral or sublittoral zone.

Chemo-organotrophic
refers to heterotrophs (primary consumers) that use organic chemicals to derive for their energy; they require organic carbon compounds to provide carbon chains for biosynthesis and energy to drive biosynthesis

Chironomidae (Chironomids)
a family also referred to as the non-biting midges; a speciose order Diptera (true flies).

Chlorophyta
a phylum of algae also referred to as the chlorophytes or green algae that are mostly confined to freshwater and terrestrial environments with only 10% being marine.

Chronic toxicity testing
is toxicity testing that is continuous over a long time, usually over the life span of the individual.

Chrysophyta
a phylum of algae also referred to as the yellow-green and golden-brown algae; known as silicoflagellates, characterized by a star-shaped internal skeleton made of silica or cellulose. Some divide this phylum into two groups, the Xanthophyceae (yellow-green algae) and Chrysophyceae (golden-brown algae).

Cirque lake
are formed by glacial forces. Such lakes evolve in rocky cliff basins where the slow downhill movement of ice and the continual freezing and thawing activity erodes and fractures the rock; they have *amphitheater* shapes

Cladoceran
is an Order of crustaceans also referred to as the Water Fleas.

Clinograde profiles
are oxygen profiles that exhibit 100% saturation in the epilimnion and greatly reduced levels in the hypolimnion

Cnidaria
phylum of invertebrate including hydras, jellyfish, sea anemones, and corals.

Coastal lake
are formed by shoreline forces when sand bars form across bays.

Cohorts
a group of equal-aged individuals whose survivorship and fecundity are to be followed throughout the life span of the entire group; a generation.

Cold stenotherms
organisms having a very narrow thermal tolerance and preferring cooler temperatures characteristic of the hypolimnion and metalimnion.

Commensals
one of two of the members of commensalism relationship; close association between two different spp. whereby one member of the association derives an advantage and the other member has neither an advantage nor a disadvantage

Compensation depth
depth at which the rate of photosynthesis equals the rate of respiration.

Constant drift
occurs at all times at low levels and, like *catastrophic drift*, is represented by most taxa in the stream. The reasons for constant drift are not known

Convection
the transmission of heat by the movement of the heated particles.

Copepoda
is a subclass in the class Crustacea; copepods; body more or less cylindrical; head fused with one or more thoracic segments to form a cephalothorax; with five or six free thoracic segments, each with a pair of swimming legs; three to five abdominal segments; usually less than 10 mm; many spp. microscopic; marine and freshwater spp.; ~ 6300 spp.

Coprozoic zone
is a zone where the organisms only live in faeces; consists only of anaerobic bacteria or, *Boda edax*, a zooflagellate that occurs in stagnant water.

Corbiculids
are Asian bivalve molluscs; exotic spp. in North America.

Coriolis forces
on air it arises as a result of the Earth's rotation. It deflects air to the right (left) of its intended path in the Northern (Southern) Hemisphere due to the frictional drag of the Earth's surface on the atmosphere above it.

Coulee lakes
are formed at the base of the volcanoes by lava forming walls to contain pools or basins of water. Lava flowing across a river may also dam the river and form a small lake upstream.

COSEWIC
Committee on the Status of Endangered Wildlife in Canada; primary mandate is to develop a national listing of Canadian species at risk based on the best scientific evidence available; for more info: http://blizzard.cc.mcgill/Redpath/cosewic.htm

Crater lakes
are formed by extra-terrestrial forces, such as meteors, whose impact with earth create huge depressions that fill with water.

Critically Endangered (COSEWIC)
includes those species with a 50% risk of extinction in 10 years or 3 generations; this term is not currently in use but is recommended for use.

Cryogenic, or Thaw lakes
are formed by glacial forces; they occur when frozen ground or permafrost thaws, leaving the ice melt to fill the depression left thawing

Cryptophyta
phylum of algae also referred to as the cryptomonads which have two flagella and lack a skeleton.

Cultural Eutrophication
is organic enrichment of a lake induced by human activity.

Cumulative toxicity
is brought about, or increased in strength, by successive additions of the substance.

Cyanophyta (Cyanobacteria)
a phylum of algae also referred to as the blue-green algae; specialized group of photosynthetic bacteria; among the most primitive plant-like organisms.

Cyclomorphosis
is a seasonal change in morphology.

Cysts
is a resistant resting stage formed by many different organisms, especially as a response to adverse environmental conditions.

D

Dalton's Law
states that the pressure of each component of gas in a mixture is proportional to its concentration in the mixture, the total pressure of all gases being equal to the sum of its components. e.g. when concentration of carbon dioxide is high, the concentration of oxygen in the water will be low.

Deep water waves
are waves whose depth is greater than one-half the average wavelength and who are not affected by the bottom.

Delta
occur when rivers, streams, tidal inlets enter quiet waters of a lake or ocean and deposit large amounts of sediments; typically triangular in shape and formed at the mouth of a river.

Deltaic lakes
are long narrow lakes found within many deltas. Deltaic lakes formed adjacent to oceans receive saltwater at high tides and are brackish.

Dendritic pattern
branched pattern that emulates nerve dendrites.

Denitrification
is an **anaerobic bacterial process** where nitrates, ammonia, and nitrites are reduced to release N_2 gas as the end product
$NO_3 \rightarrow NO_2 \rightarrow N_2O \rightarrow N_2$

Density
the number per unit as an area; the ratio of the mass of an object to its volume.

Deposit feeding (or Pedal Feeding)
is a method of feeding of bivalves where they use the cilia on their foot to take up detritus, algae, bacteria and other food deposited on the sediments.

Desiccation
the loss of water or drying out of aquatic organisms when they are out of the water.

Desmid
is one of a group of *green algae* in which the cells are composed of mirror-image halves; their reproduction is by union of amoeboid gametes.

Detritus
inorganic or organic debris.

Detritivores
an animal that feeds on a decaying organic material; a variety of dead plant materials, bacteria and algae trapped in the surface film. See *Gatherers*.

Detritus pool
is the organic source of carbon which consists of *dissolved organic material* (DOM) and *particulate organic material* (POM).

Diapause
a period of arrested growth and development; often used to survive harsh environmental conditions.

Diatom
is the common name for the phylum of algae, Bacillariophyta.

Centrate forms
circular with radial symmetry.

Pennate forms (Diatom)
elongate with bilateral symmetry.

Diel cycle
referring to the 24 hour day. The evening period is known as the *nocturnal* part of the diel cycle and the daytime period is known as *diurnal* part of the diel cycle.

Dinoflagellate
see *Pyrrophyta.*

Direct toxicity
results from the toxic agent acting more or less directly at the site(s) of action in and/or on the organism.

Discharge features

Current velocity
is the downstream water movement in the stream channel.

Cross-sectional area of a stream

Discharge (Q)
is the volume of water passing through a cross-sectional area of the stream channel per unit of time, or Q = Av.

Hydrographs
are plots of discharge against time.

Mean channel depth
is the sum of the channel depths divided by the number of measurements plus 1.

Rapids
a part of a stream where the current is moving with a greater swiftness than usual and where the water surface is broken by obstructions, but without a sufficient break in slope to form a waterfall, as where the water descends over a series of small slopes. It commonly results from a sudden steepening of the stream gradient, from the presence of a restricted channel, or from the unequal resistance of the successive rocks traversed by the stream.

Riffles
movements of water in a river characterized by shallow, turbulent water passing through or over stones or gravel of a fairly uniform size.

Dissected leaves
are leaves that are divided into many slender segments.

Distal
situated away from the base or point of attachment.

Divergence
horizontal movements of water or air in different directions from a common zone, the *divergence zone*, which is the region along which crustal plates move apart and new lithospheric material solidifies from rising volcanic magma.

Dune lakes
are formed by wind action in arid regions that piles up pieces of eroded or broken rocks or re-distributes sand, resulting in the formation of lake basins. Most are temporary.

Dystrophic
a brown-water lake with a very low lime content and a very high humous content, often characterized by a severe poverty of nutrients; a lake with an extreme case of oligotrophy and in a certain sense in the final stage of the trophic series.

E

Ecophenotypic
morphological traits of an organism in response to environmental factor(s)

Ecoregion

an area related by similar climate, physiography, oceanography, hydrology, vegetation and wildlife potential

Ecotone
is the transition or interdigitated area between two adjacent communities.

Edaphic
referring to soil characteristics, such as water content, pH, texture and nutrient availability, that influence the type and quantity of vegetation in an area.

Effective concentration (EC)
a term used when an effect other than death is being used as the end point for toxicity testing. For example, respiratory stress in fish, valve gaping in bivalves, byssal detachment of byssate mussels. Again, 50% percent mortality is usually used for the end point and an EC_{50} is given for the result.

Ekman transport
the net movement of water, 90° from the wind direction, in the upper 100 m of water.

Ekman grabs
a benthic sampler designed for taking samples on soft muck, mud, or fine, peaty material; utilizes a messenger to release the two-pin jaw release mechanism.

Emergent plants
are those which are erect and most of the plant is above water. If the emergent part is extensive, it usually floats on the surface because the stem has no stiffening tissues to keep the emergent part upright.

Endangered
any indigenous species facing imminent extirpation or extinction from a specified area. See *Species.*

Epeirogeny
an endogenic (originating within the earth) process that affects large parts of the continents and oceans, primarily through upward or downward movements, and produces plateaus and basins; uplifting.

Ephemeral pools
pools of water that have brief aquatic lives of a few weeks to months; most are < 1 m deep

Ephemeroptera
is an order of insects which includes the mayflies; delicate, soft-bodied insects with vestigial mouth parts in the adult stage; wings membranous and held vertically when at rest; forewings much larger than hind wings; tip of abdomen with two or three long cerci and tarsi with 1 claw each.

Ephippium
special postero-dorsal part of the carapace of certain female Cladocera; thick-walled and contains one to several eggs (usually fertilized); the whole device is separated from the body and then it may sink or float, and the contained eggs are highly resistant to adverse environmental conditions.

Epibenthos
the benthos that lives on top of the sediments, rocks, logs or plants. The prefix, "epi", simply means attached to the surface or living freely upon the surface.

Epifauna
epibenthos organisms attached to animals (e.g. zebra mussels often attach to clams, crustaceans, or snails).

Epilimnion
the warm, oxygen rich, upper layer of a lake.

Epilithic
periphytic organisms (ex. algae) that live or grow on rocks or stone surfaces.

Epiphytic
organisms that live attached on plants.

Epiphyte
a plant that is not rooted in the bottom but rather uses other plants as a substrate without penetrating into them and without withdrawing nutrient substances from them (psuedoparasites).

Episammic
organisms which are living upon or moving through mud or sand.

Estuaries
a semi-enclosed, tidal body of saline water with free connection to the sea; are places where the river meets the sea and combines some river features with ocean features.

Euglenophyta
a phylum of algae also referred to as the Euglenas including some chlorophyllous genera and some colourless ones.

Eutrophic
is a classification of a lake meaning that the lake is highly enriched ("eu" meaning true; "troph" meaning food or nutrients, hence truly enriched). Such lakes are often shallow (usually v-shaped basins) and warm and have

anoxic (no oxygen) waters in deepest portion.

Eutrophication
the process of organic enrichment of a body of water as a result of increased nutrient loading and the subsequent increase in autotrophic production. See *Cultural Eutrophication.*

Eusaprobity
contains nontoxic sewage and industrial waste

Katharobity
drinking water quality

Limnosaprobity
clean to polluted surface and underground water

Transsaprobity
has toxic wastes, inaccessible to bacteria

Evorsion
the erosional action of river eddies.

Extinct
a species that no longer exists in the world. See *Species.*

Extinction coefficient
is an expression of the exponential decrease in light intensity with increasing depth in water; or the fraction of light absorbed per metre of water. The higher the extinction coefficient, the lower the transmission of light, or the less transparent the water.

Extirpated (COSEWIC)
any indigenous species no longer existing in the wild in Canada but existing elsewhere.

Extracellular products of photosynthesis (ECPP)
exudates released from autotrophs.

F

False karsts
solution lakes that form within sandstone.

Feeding guilds
a group of populations that utilizes a gradient of resources in a similar way.

Fens
are a type of wetland that are alkaline (higher pH) and species rich with aquatic macrophytes and mosses; typically supplied by mineral rich groundwater; characteristically underlain by peat deposits.

Ferrel cells
circulating cells which form a the mid-latitudes of a rotating planet to balance the transport by the Hadley and Polar cells, the other two giant cells of atmospheric circulation in each hemisphere; their air motion is opposite to planetary motion; forms the southwesterly prevailing westerlies at the surface; formed when air rises at 55 degrees latitude.

Fetch
the distance that wind can blow over water without interruption by land.

Filter feeders
consumes FPOM (fine particulate organic matter) suspended in the water column. Filter feeders dominate in lower reaches after all the CPOM has been reduced to FPOM.

Fjord lakes
are formed by glacier forces in mountainous regions along glaciated coast lines. Because of the proximity of fjord lakes to oceans and seas they are often saline.

Floating plants
live at the surface but some species (e.g. water lilies) have long stems with roots that anchor the plant to the substrate. Other floating species (e.g. duckweed) have roots, but little or no stem and the roots float immediately below the leaves at the surface.

Flocculant
a substance capable of aggregating into small lumps, especially colloids and those in soils.

Floodplains
are the vegetated (usually) areas at the edge of streams. Some have grasses; others have shrubs or trees; landward edge of the floodplain is usually taken as twice the distance of the highest-water mark.

Fluviatile lake
are lakes that are generated by the activities of flowing waters.

Fork lengths
is a type of measurement for fish where tip of snout to notch in tail is measured.

Founder effect
population started by a small number of colonists, which contain only a small and often biased sample of genetic variation of the parent population; it may result in a markedly different new population.

Fragmentation
the mechanical breaking of an animal into two or more parts, each of which regenerates into a complete new animal.

Frequency
the inverse of a period (in terms of waves).

Functional feeding group
refers to consumer groups based on their morpho-behavioural mechanisms of food acquisition; they are, *predators*, *shredders*, *scrapers*, *collectors*, *filter feeders*, *gatherers* (*deposit, detritus feeders*) and *parasites.*

Fusiform
spindle-shaped, tapering at each end.

G

Gametogenesis
complex of processes by which oogonia become ova and spermatiogonia become sperm.

Gastropoda
is one of the six classes of the phylum Mollusca; snails, slugs, limpets, whelks, nudibranchs, etc.; usually with a asymmetrical spiral one-piece shell lined with mantle and containing much of the viscera; marine, fresh-water, and terrestrial spp., 80 000 spp.

Gatherers
consume FPOM (fine particulate organic matter) laying on the bottom (also called *detritivores or deposit feeders).*

Gene flow
is the gain or loss of alleles from a population by the movement of fertile individuals or gametes to another population.

Generalists
r-strategists that are capable of exploiting new environments successfully.

Genetic diversity
represents the heritable variation within and between populations of organisms.

Genetic drift
is the colonization of a new habitat by a small number of individuals, leading to a *founder effect.* The colonizers would bring with them only a small number of alleles that do not fully represent the parent population.

Geostrophic effects
are due to the rotation of the earth.

Geostrophic currents
are the water movements, caused by the rotation of the earth *(Coriolis forces)*. They are found in oceans and large lakes. Geostrophic currents also affect weather patterns.

Glochidia (glochidium)
is a bivalve larval stage characteristic of fresh-water mussels; upon release from the parent it must come in contact with a fish, then quickly encysts in gills, fins, and other superficial tissues; after several weeks to a year, the cyst ruptures and the small juvenile mussel is released and falls to the substrate.

Graben lakes
are the most common type of tectonic basin. They are formed as a result of large depressed areas located between adjacent faults, where the fault block is untilted and forms a flat bottom in a trough.

Gravity waves
short surface waves with a wave length greater than $2 \times \pi = 6.28$ cm.

Grazers
herbivores that consume plants, like algae and macrophytes that are *periphyton* or *phytoplankton.*

Gross production
refers to observed changes in weight within a unit area (= biomass), plus all losses (predatory and non-predatory), divided by the time interval.

Groundwater
water that fills up the pores and hollows within the earth in a larger and mostly subsurface water table; closely connected with lakes and streams; shallow ground water is in the overburden (glacial sand and gravel) and deep ground water is in the limestome bedrock; ground water moves in the direction of the hydraulic head ie. Higher water table to lower water table.

H

Hardy-Weinberg equilibrium
is the principle that shuffling of genes that occurs during sexual reproduction, by itself, cannot change the overall generic makeup of a population.

Heavy metals
are those metallic elements with atomic weights greater than 40 (or atomic number greater than 20). These exclude alkali earth metals (e.g. Ca, Mg), alkali metals (e.g. Na,

K) and uranium.

Henry's Law
states that the concentration of a solution of a gas which has reached equilibrium is proportional to the partial pressure at which the gas is supplied.

Hermaphroditic
organisms having male and female gonads.

Heterocysts
specialized cells produced by *cyanobacteria* (blue-green algae) that are the sites of nitrogen fixation.

Holozoic
refers to organisms, usually protists, that ingest solid particles of food, such as bacteria, algae, other protists and debris.

Homeostatic
tendency towards stability in normal internal fluids and metabolic conditions; steady internal state; ecological stability.

Humic acids
large-molecule organic acids that dissolve in water.

Hydraulic residence time
the time required to refill an empty lake with its natural inflow (see *Retention time*).

Hyetograph
plots precipitation intensity (e.g. mm/hr) on the y-axis against discharge (e.g. m^3 /sec) on the x-axis; they are used to determine the frequency and size of flood events.

Hypereutrophic
pertaining to a lake or other body of water characterized by excessive nutrient concentrations such as nitrogen and phosphorous and resulting high productivity; often shallow, with algal blooms and periods of oxygen deficiency; generally undesirable for drinking water and other needs;

Hypolimnetic oxygen minimum profile
also known as the *negative heterograde oxygen curve*, exhibits the minimum concentration of oxygen in the *metalimnion*. Such profiles arise when there is an accumulation of organic material in the *metalimnion* which through oxidative metabolism by bacteria cause a decrease in the oxygen content.

Hypolimnion
a lower cold layer of a lake that lies below the *metalimnion*.

Hypoxia
is a state of having too little oxygen in the tissues for normal metabolism.

I

Impoundment
body of water such as a pond, confined by a dam, dike, floodgate or other barrier; used to collect and store water for future use; in terms of water quality, it is generally an artificial collection and storage area for water or wastewater confined by a damn, dike, floodgate or other barrier.

Incipient lethal level (ILL)
is the concentration at which acute toxicity ceases. Organisms will eventually die above this value and live indefinitely below this value. It is usually taken as the concentration at which 50% of the population can survive for an indefinite time.

Indicator organisms
that become numerically dominant only under a specific set of environmental conditions; used to develop biological indices that assess environmental impacts of organic and inorganic contaminants; always appears in large numbers or biomass under enriched (eutrophic) or organically polluted waters and poorly enriched (oligotrophic) or pristine waters.

Indirect toxicity
results from the influence of changes in the physical (e.g. decreased water clarity for plants), chemical (e.g. dissolved ion content) and/or biological (e.g. quality and quantity of food) environment.

Infaunal
the animals that live in the sediments; they obtain their food and dissolved oxygen primarily from interstitial water held between the sediment particles.

Instar
growth stage between two successive moults.

Intensity
(of light) is the number of ***quanta*** or ***photons*** passing through a unit of given area.

Interflow
the pattern that water from an input makes if the stream temperatures are similar to surface lake temperatures.

Internal seiches

also known as internal gravity waves which occur at the thermocline and can only be formed under stratified conditions; they are not apparent at the lake's surface and are normally larger than surface waves and may be as much as 10 m high. *(see Seiches)*.

Inverted orthograde profile
mirror image of orthograde profile.

Isopoda
an order of the class Malacostraca (in the subphylum Crustacea); includes pillbugs, sow bugs, and wood lice; first thoracic segment fused with head; abdomen short, some or all segments fused; 4000 spp described; in bottom debris of marine and fresh waters i.e. benthic; some terrestrial spp but mostly marine.

Iteroparity (Iteroparous)
is an egg laying habit where the species spawns more than once during the course of its life.

K

***K*-selection**
is selection under carrying capacity conditions and high level of competition.

Karsts
are lakes formed within limestone bedrock.

Kelvin wave
is a long propagating internal gravity wave that produces a significant current in the hypolimnion of large lakes. Kelvin waves are produced from internal seiches when the Coriolis force induces a counter-clockwise rotation in the northern hemisphere (on the plain of the internal seiche) and a clockwise rotation in the southern hemisphere. In Kelvin waves, the thermocline not only rocks back and forth, but it tilts and rotates to the side in a back and forth fashion. This sets up wave amplitudes that increase exponentially toward the shore.

Kettle lakes
are glacial lakes and form when blocks of ice trapped in glacial debris, called *moraine*, melt and fill the depression created by the moraine being deposited around the melting ice block. Such lakes usually have steep sides.

Keystone Species
are species which have the ability to not only affect the continued existence (at least in large numbers) of its competitors, or its prey, but to affect the structure of the entire community of organisms.

L

Labial palp
is one of two thin, fleshy palps on either side of the mouth region in many bivalve mollusks; function in separating inedible from edible material 2. is a one to four segmented sensory appendage on each side of a typical insect labium.

Labile
material easily broken down to CO_2 and C

Lacustrine systems
refers to lakes or living in lakes.

Lake whitening
occurs when phytoplankton use up so much dissolved carbon dioxide in the lake water that a fine precipitate of calcium carbonate is produced; the lake water appears cloudy.

Laminar Flow
occurs when water particles slide smoothly past each other; in parallel layers or sheets; a non-turbulent flow; occurs in lakes at depths less than 5 m and where water current velocity is less than 1 cm/sec.

Langmuir circulation
are vertical, helical currents in the upper layers of lakes. They are generated by wind, which sets the water particles into motion in their cycloid paths, and the *Coriolis force,* which converts the cycloid paths into horizontal helices. Because of friction, adjacent cycloid paths rotate in opposite directions, causing the helices to rotate in opposite directions. This creates alternating zones of convergence and divergence.

Lentic
refers to standing water. Lentic ecosystems include lakes, ponds, puddles and, to some extent, estuaries.

Lethal toxicity
causing death by direct action.

Lethal dose (LD)
a dosage (e.g. applied orally or hypodermally) that causes death directly.

Lethal time (LT)
is the time to mortality. Usually taken when 50% of the organisms are dead and is expressed as LT_{50}.

Lethal concentration (LC)
the concentration that causes death to a certain percentage of organisms. Usually 50% mortality is sought after and the lethal concentration is expressed as LC_{50}.

Levee lake
are formed by fluviatile forces when the deposition of silt at the sides of a river isolates a body of water from the mainstream. Levee lakes are often flushed or destroyed by *spring spates*.

Limnetic zone
the open water zone of fresh waters. Limnological terms, definitions, and concepts for more info: http://www.cfn.cs.dal.ca/Science/SWCS/glossary.html

Lithophils
are fish that lay eggs in interstices (crevices, cracks) of sand, gravel, or cobble.

Littoral zone
is that zone predominated by *emergent* and floating leaved aquatic plants.

Local effects
occur at the primary site of contact, for example, skin or gill inflammation.

Long wave
occurs when the wave length is greater than 20 times the water depth and the cycloid movements are transformed into to and fro sloshing motions that extend to lake bottom.

Lowest Effect Concentration (LOEC)
is the concentration at which some significant deleterious effect is observed.

M

Maars
are lakes formed by volcanic forces and are small depressions with diameters less than 2 km; they are nearly circular in shape and can be extremely deep in relation to their small surface area.

Macroinvertebrates
organisms larger than meiobenthos (small benthic organisms) and which are retained on course sieves with a mesh size ≥2 mm.

Macrophytes
are large aquatic plants more commonly referred to as weeds, sedges, grasses, reeds, bullrushes, etc.

Mantle
thin extension of the dorsal body wall of molluscs underlying the shell and responsible for its secretion.

Marshes
a type of *wetland* characterized by emergent aquatic macrophytes, namely reeds, grasses and sedges and submerged or floating macrophytes, such as pondweeds and waterlilies; they are also distinguished by the absence of trees and shrubs, the accumulation of peat and an external source of water; pH 5.1-7.0.

Maximum acceptable toxicant concentration
(MATC) is the concentration of a toxic substance that does not harm the productivity and use of the receiving water.

Maximum fetch
the maximum distance the wind may blow without being interrupted by land.

Mean flow
is a type of lake water movement. It has a specified direction and *advects* substances around the lake.

Megalopterans
is a small order of insects which includes the dobsonflies, alderflies, fishflies and sialids; medium to large spp. with complete metamorphosis; biting mouth parts and long antennae; wings large, similar; larvae aquatic, carnivorous, and with segmented lateral abdominal gills.

Mesotrophic
are those lakes that are moderately enriched, between oligotrophy and eutrophy

Metalimnetic oxygen maximum
also known as the *positive heterograde oxygen curve*. There is an increase in the oxygen concentration in the metalimnion. Generally, the oxygen maximum is attributed to an accumulation of cold, *stenotherm* (have a physiological requirement for water with a cold but narrow temperature range) algae, mostly *Oscillatoria*, a filamentous, blue-green form. Photosynthesis by this algae contributes to the oxygen maximum layer.

Metalimnion
is the intermediate layer in lakes between the epilimnion and the hypolimnion. The metalimnion is the zone of very rapid changes in temperature, more than 1°C per metre advance in the water column.

Metalloids
are non-metallic elements, with metallic properties, like silicon and arsenic.

Microbial loop
refers to the recycling of organic matter (DOC) in the food web. Dissolved organic matter and small suspended particles provide food for bacteria, heterotrophic microflagellates, and ciliates, which in turn are fed upon by larger zooplankton and small fish. It includes heterotrophic microbes, including bacteria, ciliates, flagellates and some sarcodines.

Microevolution
evolution on its smallest scale that is caused by 1. Genetic drift 2. Gene flow 3. Mutation 4. Nonrandom mating 5. Natural selection

Mixotrophic
utilize carbon both from carbon dioxide used in photosynthesis and from organic compounds metabolized in the dark.

Molecular diffusion
causes water motions on small scale. Occurs when water is heated and in ionic solutions.

Mollusca
is the largest macroinvertebrate phylum including clams, oysters, squids, octopods and snails. They thrive in both marine and fresh water. They are represented in freshwater by only two classes, Gastropoda (the snails and limpets) and Bivalvia (the clams and mussels).

Molluscicide
a pesticide used against molluscs.

Monogononta
is one of the two classes of rotifers; with a single ovary.

Moraine
glacial debris.

Moribund
in a dying state.

Morphometry
the method of measuring and analyzing the physical dimensions of a lake; a function of underwater contour lines, the shape of the lake, and its geologic origin when characterizing or describing a lake.

Multivoltine
More than two generations per year.

Mutation
new, abrupt genetic feature in any individual produced by changes in the individual genes or chromosomes.

Myxophycean
Blue-green algae. see *Cyanophyta.*

N

Naiad
common name applied to: (1) nymphal stage of insects, especially dragonflies and damselflies; (2) any memeber of the plant genus, *Najas*; (3) any member of the family of clams belonging to Unionidae.

Natality
the production of new individuals; it can be represented as either a crude birthrate or a specific birthrate; pertains to the number of eggs surviving of an organism.

Nekton
is the open water community of strong swimmers. They are capable of swimming against strong currents and their position in the water column is dictated by themselves, not by wind-driven water currents.

Nematoda
a phylum also referred to as the true roundworms; body slender, cylindrical, often tapered near ends, and covered with a cuticle; no segmentation, cilia, or circular muscle fibers; with a true digestive tract and pseudocoeloem; sexes separate, marine, fresh-water, terrestrial, and parasites of plants and animals; about 13,000 spp.

Neonates
refers to newly-hatched young

Net production
is the difference between gross production and the losses, divided by the time interval.

Neuston
are collectively, the organisms that associate with just the surface film of water in aquatic environments, taking advantage of the surface tension of water.
Epineuston
they live on the surface film.
Hyponeuston

organisms that live just below surface film.

Pleuston
is the community of macroorganisms floating on the surface of the sea.

Nitrification
is an aerobic bacterial process where ammonium is first oxidized to nitrite by *Nitrosomonas* and then the nitrite is oxidized to nitrate by *Nitrobacter.*

$NH_4^+ \rightarrow N_2O \rightarrow NO_2 \rightarrow NO_3$

OR is the conversion of organic nitrogen compounds into ammonium compounds, nitrites, and nitrates by bacteria in soil and aquatic environments.

No Effect Concentration (NOEC)
the concentration of a contaminant at which no significant deleterious effect is observed.

Nomogram
is used to determine the oxygen saturation values for water when temperature and dissolved oxygen content are known

Nonrandom mating
is selective mating that results in a departure from the *Hardy-Weinberg equilibrium* requirements.

Nymphs
all the immature stages from hatching to the adult in aquatic insects; also called *naiads*

O

Ocher
$Fe(OH)_3$, a rust-colored sediment layer.

Odonata
is the order of insects which includes the damselflies, dragonflies, darning needles, snake doctors, and mosquito hawks; head large, with biting mouth parts and very large compound eyes; wings long, narrow, transparent, and net-veined; more than 5000 spp.

Oligochaetes
is one of the classes in the Phylum Annelida; earthworms; definite head; they live in moist soil, fresh water and marine habitats; 2700 spp.

Oligotrophic
is a trophic classification of lakes having little or few nutrients, ("oligo" meaning poor; "troph" meaning food or nutrients, hence poor nutrient levels). They are generally deep (usually U-shaped basins), cold lakes with close to 100% oxygen saturation at all depths.

Oligotrophication
is the reversal of *eutrophication.*

Operculum
a gill cover; a flap-like outer protective covering to a chamber-like structure i.e. the gills of fish, certain immature amphibians, and a few aquatic invertebrate. (pl. opercula).

Opportunists
organisms able to exploit temporary habitats or conditions; an r-strategist; invaders; they are characteristically, hardy, aggressive, prolific breeders and capable of dispersing.

Orthograde profiles
are oxygen profiles which exhibit 75% to 100% saturation at all depths

Overflow
occurs often during the spring when the spring melt water's density flowing into lakes from parent streams is usually lower than that of the lake (light water on top of heavy water.

Oviparous
refers to females that release eggs from which the young later hatch out; birds, most insects, and many aquatic invertebrates.

Oviposition
the reproductive strategy of organisms by which they lay eggs.

Ovoviparous
refers to bringing forth a live egg in which the embryo is borne by the mother but nourished by the yolk sac; least common pattern of reproduction; some mollusc, insect, fish, amphibians and reptile spp.

Ovoviviparous
females that lay eggs which hatch within the bod of the female and are released as free-living offspring.

Oxbow lakes
are formed by fluviatile forces and are formed on large meandering rivers if one of the loops is cut off by silt deposition.

Oxygen deficit
is the amount of oxygen required to reach the saturation level.; it is an expression of the loss of oxygen.

P

Parthenogenetic

is a type of unisexual reproduction involving the production of young by females which are not fertilized by males; a common method of reproduction in rotifers, cladocerans, aphids, bees, ants and wasps; depending on the sp., a parthenogenetic egg may be haploid or diploid.

Partuition
bringing forth young, especially in mammals; delivery; birth; depends on hormone balance.

Passive dispersal
'hitch-hiking' a ride using abiotic (e.g. water currents, wind, ships, boats, etc.) or biotic (e.g. birds, insects, mammals) vectors.

Paternoster lakes
are formed by glacial forces; a chain of *cirque lakes*, much like a string of beads.

Pedal Feeding (or Deposit Feeding)
is a method of feeding of bivalves where cilia on their foot are used to take up detritus, algae, bacteria and other food deposited on the sediments.

Pelagic zone
the open water zone of lakes and oceans; referring to the water column away from the bottom.

Percent saturation
the amount of a substance that is dissolved in a solution compared to the maximum amount that could be dissolved in it. e.g.. the amount of oxygen (mg/L) dissolved in the water divided by the amount of oxygen that the water can hold at the observed temperature.

Periostracum
is the proteinaceous outer layer of mollusc shells secreted by the mantle edge.

Peripheral Species (COSEWIC's definition)
any terrestrial or freshwater species with an historical range in Canada confined to within 50 km of the national border, and a spatial distribution within Canada less than 10% of its global distribution, or a marine species, with a Canadian distribution less than 0% of the global distribution. COSEWIC's mandate does not include peripheral species.

Periphyton
are algae that grow on a variety of submerged substrates, such as rocks, plants or debris, in lakes and streams. They are commonly referred to as attached algae but the term periphyton means, "on plants", or the community of algae growing on plants.

Permanent hardness
is the total concentration of cations of Ca, Mg, and Fe combined with non-carbonates (e.g. SO_4, Cl, PO_4, etc.); they cannot be removed by boiling.

Phaeophyta
phylum of algae also referred to as the brown algae varying from olive green to dark brown due to the presence of yellow pigments.

Pheromones
biochemicals secreted by an organism that are species specific and affect only individuals of species that produce them.

Photons
a quantum of light energy.

Photosynthetically active radiation (PAR)
light that is useable by most aquatic plants; range of PAR is between the wave lengths of 400 to 700 nm since only half of the underwater spectrum of transmitted light can be used by algae for photosynthesis.

Phytoplankton
the plant members of the plankton community which are suspended or free-floating in the water column of aquatic habitats and the primary producers of the oceans; includes algae and fungi.

Phytotoxic effects
toxicity due to metabolites of algae.

Piedmont lakes
are formed by glacial forces in mountainous regions, they are found at the foot of mountains in alpine areas.

Piercers
organisms that have specialized mouth parts to pierce plant or animal cells and suck out the cell or body contents.

Pisidiids
a type of fingernail clam of the genus, *Pisidium*; molluscs that are shallow-burrowing suspension and filter feeders with the respiratory margin oriented at or near the sediment/water interface; inhabit ephermeral or permanent lentic (quiet-water) environments.

Planimeters
used for estimating areas of odd-shaped patterns and designed for measuring surface areas.

Planimetry
estimation of areas using a planimeter

Plankton
are collectively, all those organisms suspended in the water of an aquatic habitat which are not independent of the currents and other water movements; most such organisms are microscopic and commonly include bacteria, algae, protozoans, rotifers, larvae, and small crustaceans.
Macroplankton
are plankton with a size range of > 2000 μm.
Mesoplankton
are plankton with a size range of 200 - 2000 μm.
Microplankton
are plankton with a size range of 20 - 200 μm.
Nanoplankton
are plankton with a size range 2 - 20 μm.
Picoplankton
are plankton with a size range 0.2 - 2 μm.

Platyhelminthes
a phylum also referred to as the flatworms; consisting of elongated, dorsoventrally flattened worms, and including free-living turbellarians, parasitic flukes, and parasitic tapeworms; usually hermaphroditic; about 10,000 spp.

Plecoptera
an order of soft-bodied insects; includes the stoneflies and salmonflies; moderate to large size; wings large and membranous but flight is weak; antennae long in adults; chewing mouth parts, often absent in adult; two long cerci at tip of abdomen and tarci each with 2 claws in nymphs; gradual metamorphosis; adults aerial but all spp. have a series of fresh-water naiad stages which are abundant in streams.

Plunging breakers
a *breaker wave* where the forward face of the wave becomes convex and the crest curls over and collapses underneath to complete a vortex.

Poincaré Waves
caused by the Coriolis effect and occur in the open water of large lakes. They are transformed into Kelvin waves when encountering shore boundaries. Poincaré waves undulate in a standing wave pattern both across and along the modal channel or basin producing a cellular pattern of wave-associated currents.

Polar cells
created from air circulation patterns when air from the poles descends; are at the 60° latitudes (northern and southern hemispheres).

Porifera
a Phylum consisting of the sponges; they have no organs, only tissues; 4500 spp. are marine and a few (the Spongillidae) being found in fresh waters.

Potentiation
is also called a " less-than-additive effect"; occurs when one chemical has a toxic effect only if applied with another chemical

.Pothole lakes
formed by the erosional action of river eddies is known as *evorsion.*

Prehensile
refers to an appendage that is adapted for grasping or seizing, especially by curling or wrapping around.

Probit analysis
standard method for determining a LC_{50} value. Probits are a way of expressing standard deviations for a sigmoid curve with a normal distribution when percent mortality is plotted against the concentration of a toxicant on ordinary arithmetic graph paper.

Production (Primary)
the amount or weight (biomass) of organic matter synthesized by organisms from inorganic substances, such as carbon dioxide, by autotrophs (photosynthesizers) per unit of time in a defined volume of water; also called *gross primary production* and includes any losses (e.g. due to injury, respiration, excretion, death, grazing) that occurred during that period.

Profundal zone
the plantless part of the lake, where there is too little light to support the growth of plants and the bottom is usually muddy and very soft.

Progressive waves
see *travelling surface waves.*

Pseudofaeces
undigested food that has been packaged up in mucous and deposited on the bottom of the lake.

Pupa
is the inactive stage between larva and adult

in some insects (holometabolous).

Putrefaction
the decomposition of proteins by bacteria and yeasts; characterized by the production of foul-smelling substances, such as hydrogen sulfide, ammonia, and mercaptans.

Pyrrophyta
a division of algae also referred to as the *dinoflagellates* which are a large group of planktonic, unicellular organisms important in the economy of the oceans and characterized by two unequal flagella.

Q

Quantum
an elemental unit of energy.

R

r-selection
selection under low population densities; favours high reproductive rates under conditions of low competition.

RAHOD
Relative Areal Hypolimnial Oxygen Deficit; the sum of the difference in oxygen contents for each layer between the beginning and end of the stratified period; **Oligotrophic lakes have RAHOD values < 0.017 mg O_2 lost cm^{-2} day^{-1} and eutrophic lakes have > 0.033 mg O_2 lost cm^{-2} day^{-1}.** These values reflect the amount of respiration by hypolimnetic bacteria in decomposing all the *moribund phytoplankton* that once grew in the *epilimnion* during the stratified period.

Redox potential
reduction/oxidation potential, indicated with the symbol Eh, is the electrical voltage that occurs between two electrodes, one made of hydrogen and the other made of platinum. The change in oxidation state of many metallic ions is defined by the redox potential. At neutral pH and 25°C oxygenated lake water has a redox potential of about +500 millivolts.

Refractory
resistant to decomposition.

Relative hypolimnial oxygen deficit
it accounts for oxygen deficits in layers below the hypolimnetic surface.

Retention time
the time required to refill an empty lake with its natural inflow and flush itself out; may be somewhat different from the *hydraulic residence time*, since sedimentation and recycling take place within the lake.

Reynolds Number
(R_e) is the ratio of inertial forces to viscous forces, or the stirring energy divided by the viscosity of the water.

Rhodophyta
phylum of marine algae also referred to as the red algae; characterized by the presence of red pigments that mask chlorophyll *a*.

Rift lakes
formed by tectonic forces; they occur with a single fault event, the depression brought about by tilting (Fig 3-2).

Riparian zone
the edge adjacent to the stream usually taken as the vegetated area 30 m from the water's edge plus another 30 m to allow for wind-fallen trees/shrubs, or a total of 60 m from the water's edge; "*buffer strips*" that absorb (or buffer) chemical and physical extremes.

Ripples
see *Capillary waves.*

River continuum concept
physical, chemical and biological attributes of a stream that change as a continuum of gradients as stream order size increases from its headwaters to its mouth. The structure and function of the benthic invertebrate community, from headwaters to river mouth, is strongly regulated by the gradient of allochthonous and autochthonous organic matter.

S

Safe level or concentration
the maximum concentration of a toxic compound that has no observable effect on a species after exposure over at least one generation.

Saprophytic
use organic substances (dead plant and animal matter) for growth.

Saprozoic nutrition
a type of nutrition involving the absorption of dissolved salts and simple organic materials (e.g. DOC) from the water and

Scrapers
functional feeding group that feeds on algae that is encrusted on surfaces of rocks, macrophytes, logs, etc.

Secchi disc
a disc, usually 20 cm in diameter, with black and white quadrats. It is used to calculate *Secchi depth.*

Secchi depth
a measure of the relative depth of light penetration. It is used to predict the *trophic status* of a lake.

Seiches
the periodic exposure and drying period of shallow *littoral zones*. They are long standing waves with rhythmic motion and wavelengths of the same order as the length of the lake basin. The water surface at opposite ends of the lake oscillate up and down, like a teeter totter, about a line of no vertical motion at the node.

Selfpurification
the natural process of "cleaning" that a lake's ecosystem preforms; wind often stirs up the water column allowing the water to splash over rocks and aerating the stream water; the opposite to *Eutrophication.*; below is the sequence of changes that occur in a polluted river from septic to pristine condition in the Saprobien system.
Antisaprobity
Metasaprobity (septic zone, highest coliform counts)
Polysaprobity (recovery has begun)
α-mesosaprobity
β-mesosaprobity
Oligosprobity
Xenosaprobity
Katharobity (the most pristine)

Semelparity
an egg laying habit of a species where the organism spawns only once in its life time.

Senescence
in fish, a period of old age covering the time period when the growth of the mature fish has become exceptionally slow or even ceased; production of gametes is severely reduced; can last for several years in some species or a few days or weeks in others; in terms of lakes, it is the aging of lakes eventually resulting in the formation of wetland area.

Sentinel Species
are *indicator species* believed to be among the most sensitive to particular contaminants, chemicals, toxins or diseases in their environments. Sentinel species are selected because they are sensitive to environmental stress and should give an early warning to· humans that adverse conditions exist.

Seston
collectively, all the living and dead suspended microscopic particulate matter in aquatic habitats.

Shallow water wave
See *Long wave.*

Shoaling
a non-polarized, non-directional aggregation of fish. Shoals function to protect individual members, since the presence of a large number of fishes tend to confuse a predator, and it increases both the chances of reproductive success and efficiency at finding food.

Shoreline lakes
formed by the activities of waves on the shoreline of some lakes; they are fairly shallow, usually temporary pools, typically found adjacent to large bodies of water, such as seas or large lakes, that possess some irregularity or indentation along the shoreline.

Shoreline development
measures the degree of dissectedness of the shoreline; reflects the degree of shoreline irregularity; is equal to the length of the shoreline divided by two times the square root of pie A. A being the area of the lake and pie being 3.1416.

Shoreline length
is the length of the shore line perimeter derived from an outline map.

Short waves
See *Deep water waves.*

Shredders
reduce large plants (*CPOM* or *coarse particulate organic matter*) to small bits, or *FPOM* (*fine particulate organic matter*). They dominate in the head waters of streams where there is plenty of leaf litter and other kinds of CPOM.

Simuliids
a family of small humpbacked flies in the Order Diptera; buffalo gnats, black

flies; larvae and pupae occur only on rocks, etc. in swift streams, sometimes in enormous numbers.

Siphons
a tubular fold of the molluscan mantle used to direct water to and from the mantle cavity.

Solar constant quantifies the amount of energy emitted from the sun. It varies somewhat but is presently about 1.94 calories per square centimetre per minute.

Solar radiation
the heat and light that is emitted from the sun; it provides the heat that contributes to the world's wind patterns.

Solution lakes
are created in areas where there are deposits of soluble rock that slowly dissolves by purcolating water; they are usually very circular and conically-shaped; commonly referred to as sinks, dolines, or swallow holes.

Species
a group of organisms which interbreed and which are reproductively isolated from all other such groups; the fundamental category of classification, intended to designate a single kind.

Species richness
measures the absolute number of different species in a community without reference to the numbers of individuals in each.

Specific gravity
the amount of heat required to raise the temperature of 1 gram of water 1°C.

Spilling breakers
is a breaker wave where the crest collapses forward spilling downward over the front of the wave

Spring spates
heavy rain downpours in the springtime.

Spring turnover
part of the annual temperature cycles of lakes when the entire water column is homothermous.

Standard lengths
is a type of measurement for fish where the tip of snout to tip of tail is measured.

Standing waves
see *Seiches.*

Stenotherm
an organism that has a physiological requirement for water with a cold but narrow temperature range.

Sublethal effects
are the effects of a chemical or toxin without causing death; generally divided into three classes: biochemical and physiological; behavioural; and histological.

Sublethal level or toxicity
Below the level of a toxicant that directly causes death.

Submergent plants
are those which remain submerged for most of the year, although the tips may penetrate the surface near the end of the summer. Most submerged species have well developed roots and are firmly anchored in the bottom. Other submerged species lack roots and merely sit or float on the bottom.

Subsampling
sampling within a sample group or breaking up the sample group into smaller subunits.

Substratum
the substrate.

Subwatersheds
a watershed that forms part of the area of a larger watershed; any tributary or tributary network of a river will have one that lies within that river's watershed.

Summer Thermal Stratification
is part of the annual temperature cycles of lakes when the heating and mixing processes continue and eventually result in the formation of three thermal layers in the lake.

Surface seiches
is the periodic exposure and drying period of shallow littoral zones due to wind-induced tilting of the surface waters and of the *thermocline.*

Swamps
a type of wetland that is dominated by trees and shrubs, although a variety of aquatic macrophytes grow in the more open sunlit areas; generally little peat is accumulated and they usually have an external source of water (ground and river waters) which can either be rich or poor in nutrients.

Synergistic effect
also called a more-than-additive effect; occurs when the combined effect of two chemicals is much greater than the sum of effects of the individual chemicals applied alone.

Systemic effects
occur when the toxicant gains access to the internal medium of the organism, such as through the circulatory system. They require absorption and distribution of the toxicant to a remote site from the original contact or entry site.

T

T-samplers
quantitative stream sampler shaped like a "T".

Target Species
are used to monitor environmental conditions, such as pollution, habitat modification or fragmentation, or environmental heterogeneity. Groups of species are often considered more useful than single species because they provide more "integrated" information about a system.

Taxa
the plural of taxon; any formal taxonomic unit or category of organisms; e.g. species., genus, family, order, etc.

Temperature inversion
a condition different (inverted) from the normal one where warm water usually sits upon cold water. Temperature inversion is the typical pattern in winter months.

Temperature polygons
graphs used to determine the acclimation and lethal temperatures of fish.

Thermocline
see *Metalimnion.*

Thienemann's Principles
three principles that explain the diversity of benthic species in correspondence to seasonal and spatial variations in habitat.

Threatened
any indigenous species that is likely to become endangered if the factors affecting its vulnerability do not become reversed.

Time-independent threshold value
see *ILL*

Topographic map
a map that represents the surface features of a region, as hills, rivers, cities, etc.

Toxic
refers to substance when it is an agent that can produce an adverse effect in an organism sufficient to cause damage to its structure and functioning or terminating in death.

Trace metals
elements that are absolutely essential to life because they are essential to the function of specific enzymes although present naturally in minuscule amounts; includes alkali earth metals (e.g. Ca, Mg), alkali metals (e.g. Na, K) uranium, As, Cd, Cu, Ni, Pb, Se, Mn, Fe, Mo, V, I, Zn and heavy metals.

Travelling surface waves
the waves that are seen on the surface of lakes. They are created by the friction of the wind blowing over the lake surface. Their energy is confined primarily to the surface layer of the lake.

Trematodes
one of the three classes of the Phylum Platyhelminthes; the flukes; all parasitic; body covered with a resistant cuticle; one or more muscular suckers on the external surface; several thousand spp.

Trichoptera
an order of insects which includes the caddis flies; adults are soft-bodied, with two pairs of wings coated with hairs and scales; complete metamorphosis; mouth parts feebly biting; antennae and legs long; larvae aquatic, most spp. living in distinctive movable or fixed elongated cases in running or standing water; cases constructed of sand grains, gravel, or bits of vegetation cemented together with a salivary secretion; about 5000 spp.

Trochophore
a generalized type of minute, translucent, free-swimming larva found in several invertebrate groups.

Trophic status
enrichment status of a lake. Lakes are classified as either *oligotrophic*, *mesotrophic* or *eutrophic*, depending on the amount of nutrients contained.

Trophogenic zone
the region of a lake where photosynthetic production occurs; in terms of biological productivity, it is one of two different regions in a lake, the other being the tropholytic zone or region of breakdown.

Tubificids (Tubificidae)
one of the ten families of the class Oligochaeta, the aquatic earthworms.

Turbellarians

one of the three classes of the Phylum *Platyhelminthes*; free living flatworms.

Turbulent flow
random, chaotic movement of water particles around each other or any object in the water and is sufficient to create eddies;water motion that stirs up the water column and prevents stratification (separation into layers). Flow often proceeds from *laminar* to turbulent, the onset of turbulence being predicted from *Reynolds Number*, R_e.

U

Umbrella Species
are *indicator species* whose presence in an area provides protection for most other species in the area; is applied to terrestrial organisms and is rarely, if ever, used in aquatic systems.

Underflow
when cooler, heavier water flows under warmer, lighter water during the summer; occurs because stream temperatures are often cooler than the surface temperatures of lakes; also occurs with turbid waters where heavy, turbid water flows under lighter, clearer water.

V

Veligers
free-swimming larva of most marine Gastropoda, Scaphopoda and Pelecypoda; has the beginning of a foot, mantle, shell, etc.

Vernal pools
seasonal wetlands.

Viscosity
the internal resistance of a substance to being fluid, caused by molecular attraction.

Viscous drag
caused because the viscosity of water decreases with increasing temperature. Animals must work harder to swim in cold water than in warm water, or cold water flows more slowly than warm water.

Vulnerable
a species of special concern because of characteristics that make it particularly sensitive to human activities or natural events. This term may change to "Lower Risk" if a subcommittee recommendation of *COSEWIC* is passed.

W

Watershed
an area of land that intercepts and drains precipitation and collects water for a particular stream or other water body; a drainage basin; a catchment.

Wave period
the time required for the passage of two crests or two troughs past a point.

Wave length
1. Distance between two adjacent crests or troughs. 2. defines the colour of the light beam; abbreviated by lamda (λ).

Wave height
the vertical distance between a crest and a trough.

Wave amplitude
is half the wave height.

Weather
is the state of the atmosphere at a specific time and place.

Weberian ossicles
a series of four small bones forming the Weberian organ in certain fishes; conveys pressure changes to the inner ear and plays a role in reception and production of sound.

Wetland
an area of land where water is at, near or above the land surface; found at the edges of rivers and lakes; classified as either a swamp, bog, marsh or fen.

Wildlife (*COSEWIC*)
all wild life, including wild mammals, birds, reptiles, amphibians, fishes, invertebrates, plants, fungi, algae, bacteria, and one-celled protists.

Winter thermal stratification
part of the annual temperature cycle of lakes when a low density surface layer (under the ice) at 0 °C and sits on top of a dense (actually the densest) layer near 4 °C.

Z

Zenith sun
when the sun is directly overhead.

Zoophagous
animals which feed on animal material.

Zooplankton
collectively, all those **animals** suspended in the water of an aquatic habitat which are not independent of currents and water movements; most such organisms are microscopic and commonly include protozonans, rotifers and small crustaceans. Also may include invertebrates that migrate from bottom, like *Chaoborus* and *Mysis*.

Water Words Dictionary

for more definitions: http://www.state.nv.us/cnr/ndwp/dict-1/WORD_H.htm

Limnological terms

http://www.cfn.cs.dal.ca/Science/SWCS/glossary.html

COSEWIC: for more information:

http://blizzard.cc.mcgill/Redpath/cosewic.htm

List of macrophytes arranged alphabetically by scientific names

Scientific Name	Common Name
Bryophyta	Liverworts and mosses
Butomus umbellatus	Flowering rush
Cabomba caroliniana	Fanwort
Callitriche intermedia	Water starwort
Carex acutiformis	Swamp sedge
Carex disticha	Eurasian sedge
Ceratophyllum demersum	Hornwort or coontail
Conium maculatum	Poison hemlock
Eichhornia crassipes	Water hyacinth
Elodea (Anacharis)	Water weed
Elodea canadensis	Waterweed
Equisetum	Horsetail or scouring rush
Eriocaulon	Buttonwort
Glyceria fluitans	Manna grass
Glyceria maxima	Reed sweet-grass
Heteranthera dubia	Water star grass
Hydrocharis morus-ranae	European frogbit
Iris pseudacorus	Yellow flag
Isoetes bolanderi	Bolander's quillwort
Isoetes echinospora	Quillwort
Isoetes engelmannii	Engelmann's quillwort
Isoetes lacustris	Quillwort
Juncus compressus	Flattened rush
Juncus inflexus	Rush
Lemna minor	Duckweed
Lemna trisulca	Star Duckweed
Lobelia dortmanna	Water lobelia
Ludwigia	False loosestrife
Lysimachia nummularia	Moneywort

Lysimachia vulgaris	Garden loosestrife
Lythrum salicaria	Purple loosestrife
Marsilea	Aquatic fern or pepperwort
Marsilea quadrifolia	European water clover
Megalodonta beckii	Water Marigold
Mentha gentilis	Creeping whorled mint
Mentha piperita	Peppermint
Mentha spicata	Spearmint
Myriophyllum alterniflorum	Water milfoil
Myriophyllum spicatum	Eurasian milfoil
Mysotis scorpioides	True forget-me-not
Najas flexilis	Bushy pondweed, or naiad
Najas marina	Spiny naiad
Najas minor	European naiad, or minor naiad
Najas quadulepenis	Bushy pondweed
Nasturtium aquaticum	Water cress
Nasturtium officinale	Water cress
Nuphar alba	Yellow Water Lily
Nymphaea alba	Water lily
Nymphoides peltata	Yellow floating heart
Phragmites	Reed grass, or cane reed
Pistia stratiotes	Water lettuce
Polygonum	Smartweed or knotweed
Polygonum amphibium	Buckwheat
Polygonum caespitosum v. longisetum	Bristly lady's thumb
Polygonum persicaria	Lady's thumb
Pontederia	Pickerel weed
Potamogeton amplifolius	Large leaf Pondweed, Muskie Weed, or Bass weed.
Potamogeton berchtoldii	Pondweed
Potamogeton crispus	Curly pondweed
Potamogeton densus	Pondweed

Potamogeton filiformis	Pondweed
Potamogeton foliosus	Leafy Pondweed
Potamogeton friesii	Pondweed
Potamogeton gramineus	Variable Pondweed
Potamogeton hillii	Hill's pondweed
Potamogeton lucens	Pondweed
Potamogeton natans	Floating brownleaf
Potamogeton nodosus	Pondweed
Potamogeton obtusifolius	Pondweed
Potamogeton pectinatus	Sago pondweed
Potamogeton perfoliatus	Pondweed
Potamogeton perfoliatus	Pondweed
Potamogeton polygonifolius	Pondweed
Potamogeton praelongus	Whitestem Pondweed, or Muskie Weed
Potamogeton pusillus	Pondweed
Potamogeton richardsonii	Clasping -leaf Pondweed, or Bass weed
Potamogeton robbinsii	Robins' pondweed
Potamogeton zosteriformis	Flat-stemmed pondweed
Ranunculus	Buttercups or crowfoot
Ranunculus alismaefolius	Water-plantain
Ranunculus aquatilis	Water crowfoot
Ranunculus circinatus	Buttercup
Ranunculus longirostris	Stiff Water Crowfoot
Riccia and Ricciocarpus	Thallose Liverworts
Rorippa (Nasturtium) aquaticum	March cress
Rorippa sylvestris	Creeping yellow cress
Rumex longifolius	Yard dock
Rumex obtusifolius	Bitter dock
Sagittaria	Arrowhead, Duck Potato
Salvinia rotundifolia	Water fern, or floating moss
Scirpus	Bullrush

Solanum dulcamara	Bittersweet nightshade
Sparganium angustifolium	Floating leaf bur reed
Sparganium glomeratum	Bur reed
Sphagnum antipyretica	Moss
Sphagnum fontinalis	Moss
Trapa natans	Water chestnut
Typha angustifolia	Narrow-leaved cattail
Utricularia	Bladderwort
Vallisneria americana	Eel grass, wild celery
Veronica beccabunga	European brooklime
Zannichellia palustris	Horned Pondweed
Zizania	Wild rice

List of macrophytes arranged alphabetically by common names

Common Name	Scientific Name
Aquatic fern or pepperwort	*Marsilea*
Arrowhead, Duck Potato	*Sagittaria*
Bitter dock	*Rumex obtusifolius*
Bittersweet nightshade	*Solanum dulcamara*
Bladderwort	*Utricularia*
Bolander's quillwort	*Isoetes bolanderi*
Bristly lady's thumb	*Polygonum caespitosum v. longisetum*
Buckwheat	*Polygonum amphibium*
Bullrush	*Scirpus*
Bur reed	*Sparganium glomeratum*
Bushy pondweed, or naiad	*Najas flexilis*
Bushy pondweed	*Najas quadulepenis*
Buttercup	*Ranunculus circinatus*
Buttercups or crowfoot	*Ranunculus*
Buttonwort	*Eriocaulon*

Clasping -leaf Pondweed, or Bass weed	*Potamogeton richardsonii*
Creeping whorled mint	*Mentha gentilis*
Creeping yellow cress	*Rorippa sylvestris*
Curly pondweed	*Potamogeton crispus*
Duckweed	*Lemna minor*
Eel grass	*Vallisneria americana*
Engelmann's quillwort	*Isoetes engelmannii*
Eurasian milfoil	*Myriophyllum spicatum*
Eurasian sedge	*Carex disticha*
European brooklime	*Veronica beccabunga*
European frogbit	*Hydrocharis morus-ranae*
European naiad, or minor naiad	*Najas minor*
European water clover	*Marsilea quadrifolia*
False loosestrife	*Ludwigia*
Fanwort	*Cabomba caroliniana*
Flat-stemmed pondweed	*Potamogeton zosteriformis*
Flattened rush	*Juncus compressus*
Floating brownleaf	*Potamogeton natans*
Floating leaf bur reed	*Sparganium angustifolium*
Flowering rush	*Butomus umbellatus*
Garden loosestrife	*Lysimachia vulgaris*
Hill's pondweed	*Potamogeton hillii*
Horned Pondweed	*Zannichellia palustris*
Hornwort or coontail	*Ceratophyllum demersum*
Horsetail or scouring rush	*Equisetum*
Lady's thumb	*Polygonum persicaria*
Large leaf Pondweed, Muskie Weed, or Bass weed.	*Potamogeton amplifolius*
Leafy Pondweed	*Potamogeton foliosus*
Liverworts and mosses	*Bryophyta*
Manna grass	*Glyceria fluitans*
March cress	*Rorippa (Nasturtium) aquaticum*

Moneywort	*Lysimachia nummularia*
Moss	*Sphagnum antipyretica*
Moss	*Sphagnum fontinalis*
Narrow-leaved cattail	*Typha angustifolia*
Peppermint	*Mentha piperita*
Pickerel weed	*Pontederia*
Poison hemlock	*Conium maculatum*
Pondweed	*Potamogeton filiformis, amplifolius*
Pondweed	*Potamogeton perfoliatus*
Pondweed	*Potamogeton berchtoldii*
Pondweed	*Potamogeton friesii*
Pondweed	*Potamogeton perfoliatus*
Pondweed	*Potamogeton obtusifolius*
Pondweed	*Potamogeton densus*
Pondweed	*Potamogeton pusillus*
Pondweed	*Potamogeton polygonifolius*
Pondweed	*Potamogeton nodosus*
Pondweed	*Potamogeton lucens*
Purple loosestrife	*Lythrum salicaria*
Quillwort	*Isoetes lacustris*
Quillwort	*Isoetes echinospora*
Reed grass, or cane reed	*Phragmites*
Reed sweet-grass	*Glyceria maxima*
Robins' pondweed	*Potamogeton robbinsii*
Rush	*Juncus inflexus*
Sago pondweed	*Potamogeton pectinatus*
Scouring rush	*Equisetum* spp.
Smartweed or knotweed	*Polygonum*
Spearmint	*Mentha spicata*
Spiny naiad	*Najas marina*
Star Duckweed	*Lemna trisulca*

Stiff Water Crowfoot	*Ranunculus longirostris*
Swamp sedge	*Carex acutiformis*
Thallose Liverworts	*Riccia and Ricciocarpus*
True forget-me-not	*Mysotis scorpioides*
Variable Pondweed	*Potamogeton gramineus*
Water chestnut	*Trapa natans*
Water cress	*Nasturtium officinale*
Water cress	*Nasturtium aquaticum*
Water crowfoot	*Ranunculus aquatilis*
Water fern, or floating moss	*Salvinia rotundifolia*
Water hyacinth	*Eichhornia crassipes*
Water lettuce	*Pistia stratiotes*
Water lily	*Nymphaea alba*
Water lobelia	*Lobelia dortmanna*
Water Marigold	*Megalodonta beckii*
Water milfoil	*Myriophyllum alterniflorum*
Water star grass	*Heteranthera dubia*
Water starwort	*Callitriche intermedia*
Water weed	*Elodea (Anacharis)*
Water-plantain	*Ranunculus alismaefolius*
Waterweed	*Elodea canadensis*
Whitestem Pondweed, or Muskie Weed	*Potamogeton praelongus*
Wild celery	*Vallisneria americana*
Wild rice	*Zizania*
Yard dock	*Rumex longifolius*
Yellow flag	*Iris pseudacorus*
Yellow floating heart	*Nymphoides peltata*
Yellow Water Lily	*Nuphar alba*

List of fish species arranged alphabetically by scientific name

Scientific Name	Common Name
Acipencer fulvescens	Lake sturgeon
Acipenser medirostris	Green sturgeon
Acipenser brevirostrum	Shortnose sturgeon
Acipenser transmontanus	White sturgeon
Alosa aestivalis	Blueback herring
Alosa pseudoharrengus	Alewife
Ambloplites rupestris	Rock bass
Amia calva	Bowfin
Ammocrypta pellucida	Eastern sand darter
Anguilla rostrata	American eel
Catostomus commersoni	white sucker
Apeltes quadracus	Fourspine stickleback
Aplodinotes grunniens	Freshwater drum
Campostoma anomalum	Central stoneroller
Carassius auratus	Goldfish
Catastomus sp.	Salish sucker
Catastomus commerconi	White sucker
Chrosomus eos	Northern redbelly dace
Clinostomus elongatus	Redside dace
Clupeidae	Herring Family
Coregonus clupeaformis	Lake whitefish
Coregonus johannae	Deepwater cisco
Coregonus alpenae	Longjaw cisco
Coregonus nigripinnis	Blackfin cisco
Coregonus reighardti	Shortnose cisco
Coregonus artedii	Lake herring
Coregonus sp.	Squanga whitefish
Coregonus sp.	Spring cisco

Coregonus kiyi	Kiyi
Coregonus huntsmani	Acadian whitefish
Cottidae	Sculpin Family
Cottus confusus	Shorthead sculpin
Cottus bairdi	Mottled sculpins
Cottus cognatus	Slimy sculpin
Couesius plumbeus	Lake chub
Culea inconstans	Brook stickleback
Cyprinidae	Minnow or Carp Family
Cyprinus carpio	Common carp
Dorosoma cepedianum	Gizzard shad
Enneacanthus gloriosus	Bluespotted sunfish
Erimystax x-punctatus	Gravel chub
Erimyzon sucetta	Lake chubsucker
Esocidae	Pike Family
Esox lucius	Northern pike
Esox masquinongy	Muskellunge
Esox americanus vermiculatus	Grass pickerel
Etheostoma nigrum	Johnny darter
Etheostoma caeruleum	Rainbow darters
Etheostoma flabellare	Fantail darters
Etheostoma blennoides	Greenside darter
Funduls notatus	Blackstripe topminnow
Fundulus diaphanus	Banded killifish
Gambusia affinis	Western mosquitofish
Gasterosteus sp.	Stickleback
Gymnocephalus cernuus	Ruffe
Hiodontidae	Mooneye Family
Hypentelium nigricans	Northern hogsucker
Ichthyomyzon castaneus	Chestnut lamprey

Ichthyomyzon fossor	Northern brook lamprey
Ictalurus punctatus	Channel catfish
Ictalurus nebulosus	Brown bullhead
Ictiobus cyprinellus	Bigmouth buffalo
Ictiobus bubalus	Smallmouth buffalo
Ictiobus niger	Black buffalo
Lampetra lamottei	American brook lamprey
Lampetra macrostoma	Lake lamprey
Lepisosteus oculatus	Spotted gar
Lepisosteus osseus	Longnose gar
Lepomis gulosus	Warmouth
Lepomis gibbosus	Pumpkinseed
Lepomis cyanellus	Green sunfish
Lepomis humilis	Orangespotted sunfish
Lepomis microlophus	Redear sunfish
Lepomis auritus	Redbreast sunfish
Lepomis macrochirus	Bluegill
Lota lota	Burbot
Lucioperca lucioperca	Pikeperch
Macrhybopsis storeriana	Silver chub
Micropterus salmoides	Largemouth bass
Micropterus dolomieu	Smallmouth bass
Minytrema melanops	Spotted sucker
Misgurnus anguillicaudatus	Oriental weatherfish
Morone americana	White perch
Morone chrysops	White bass
Moxostoma hubbsi	Copper redhorse
Moxostoma carinatum	River redhorse
Moxostoma valenciennesi	Greater redhorse
Moxostome duquesnei	Black redhorse

Myoxocephalus quadricornis	Deep water sculpin
Myoxocephalus thompsoni	Great Lakes deepwater sculpin
Neogobius melanostomus	Round goby
Nocomis biguttatus	Hornyhead chub
Notemigonus crysoleucas	Golden shiner
Notropis atherinoides	Emerald shiner
Notropis cornutus	Common shiner
Notropis rubellus	Rosyface shiner
Notropis spilopterus	Spotfin shiners
Notropis anogenus	Pugnose shiner
Notropis umbratilis	Redfin shiner
Notropis dorsalis	Bigmouth shiner
Notropis stramineus	Sand shiners
Notropis photogenis	Silver shiner
Notropis buchanani	Ghost shiner
Notropis hudsonius	Spottail shiner
Noturus insignis	Margined madtom
Noturus miurus	Brindled madtom
Oncorhynchus kisutch	Coho salmon
Oncorhynchus gorbuscha	Pink salmon
Oncorhynchus mykiss	Rainbow trout, steelhead trout
Oncorhynchus tshawytscha	Chinook salmon
Oncorhynchus nerka	Kokanee
Opsopoedus emilae	Pugnose minnow
Osmeridae	Smelt Family
Osmerus mordax	Rainbow smelt
Perca flavescens	Yellow perch
Percina caprodes	Logperch
Percina copelandi	Channel darter
Petromyzon marinus	Sea lamprey

Phenacobius mirabilis	Suckermouth minnow
Pimephales promelas	Fathead minnow
Pimephales notatus	Bluntnose minnow
Pimephales promelas	Fathead minnow
Polydon spathula	Paddlefish
Proterorhinus marmoratus	Tubenose goby
Rhinichthys atratulus	Blacknose dace
Rhinichthys cataractae smithi	Banff longnose dace
Rhinichthys umatilla	Umatilla dace
Rhinichthys cataractae	Longnose dace
Rhinichthys osculus	Speckled dace
Salmo salar	Atlantic Salmon
Salmo trutta	Brown trout
Salvelinus nemaycush	Lake trout
Salvelinus fontinalis	Brook trout
Salvelinus timagamiensis	Aurora trout
Salvelinus alpinus	Arctic char
Scardinius erythrophthalmus	Rudd
Semotilus atromaculatus	Creek chub
Stizostedion canadense	Sauger
Stizostedion glaucum	Blue walleye
Stizostedion vitreum	Yellow walleye
Tinca tinca	Tench
Umbra limi	Central mudminnow

List of fish species arranged alphabetically by common name

Common Name	Scientific Name
Acadian whitefish	*Coregonus huntsmani*
Alewife	*Alosa pseudoharrengus*
American brook lamprey	*Lampetra lamottei*
American eel	*Anguilla rostrata*
Arctic char	*Salvelinus alpinus*
Atlantic Salmon	*Salmo salar*
Aurora trout	*Salvelinus timagamiensis*
Banded killifish	*Fundulus diaphanus*
Banff longnose dace	*Rhinichthys cataractae smithi*
Bigmouth buffalo	*Ictiobus cyprinellus*
Bigmouth shiner	*Notropis dorsalis*
Black buffalo	*Ictiobus niger*
Black redhorse	*Moxostome duquesnei*
Blackfin cisco	*Coregonus nigripinnis*
Blacknose dace	*Rhinichthys atratulus*
Blackstripe topminnow	*Funduls notatus*
Blue walleye	*Stizostedion glaucum*
Blueback herring	*Alosa aestivalis*
Bluegill	*Lepomis macrochirus*
Bluespotted sunfish	*Enneacanthus gloriosus*
Bluntnose minnow	*Pimephales notatus*
Bowfin	*Amia calva*
Brindled madtom	*Noturus miurus*
Brook trout	*Salvelinus fontinalis*

Brook stickleback	*Culea inconstans*
Brook trout	*Salvelinus fontinalis*
Brown bullhead	*Ictalurus nebulosus*
Brown trout	*Salmo trutta*
Burbot	*Lota lota*
Central mudminnow	*Umbra limi*
Central stoneroller	*Campostoma anomalum*
Channel catfish	*Ictalurus punctatus*
Channel darter	*Percina copelandi*
Chestnut lamprey	*Ichthyomyzon castaneus*
Chinook salmon	*Oncorhynchus tshawytscha*
Coho salmon	*Oncorhynchus kisutch*
Common carp	*Cyprinus carpio*
Common shiner	*Notropis cornutus*
Copper redhorse	*Moxostoma hubbsi*
Creek chub	*Semotilus atromaculatus*
Deep water sculpin	*Myoxocephalus quadricornis*
Deepwater cisco	*Coregonus johannae*
Eastern sand darter	*Ammocrypta pellucida*
Emerald shiner	*Notropis atherinoides*
Enos Lake Stickleback	*Gasterosteus* sp.
Fantail darters	*Etheostoma flabellare*
Fathead minnow	*Pimephales promelas*
Fathead minnow	*Pimephales promelas*
Fourspine stickleback	*Apeltes quadracus*
Freshwater drum	*Aplodinotes grunniens*

Ghost shiner	*Notropis buchanani*
Giant stickleback	*Gasterosteus* sp.
Gizzard shad	*Dorosoma cepedianum*
Golden shiner	*Notemigonus crysoleucas*
Goldfish	*Carassius auratus*
Grass pickerel	*Esox americanus vermiculatus*
Gravel chub	*Erimystax x-punctatus*
Great Lakes deepwater sculpin	*Myoxocephalus thompsoni*
Greater redhorse	*Moxostoma valenciennesi*
Green sturgeon	*Acipenser medirostris*
Green sunfish	*Lepomis cyanellus*
Greenside darter	*Etheostoma blennoides*
Herring Family	Clupeidae
Hornyhead chub	*Nocomis biguttatus*
Johnny darter	*Etheostoma nigrum*
Kiyi	*Coregonus kiyi*
Kokanee	*Oncorhynchus nerka*
Lake chub	*Couesius* plumbeus
Lake chubsucker	*Erimyzon sucetta*
Lake herring	*Coregonus artedii*
Lake lamprey	*Lampetra macrostoma*
Lake sturgeon	*Acipencer fulvescens*
Lake trout	*Salvelinus nemaycush*
Lake whitefish	*Coregonus clupeaformis*
Largemouth bass	*Micropterus salmoides*
Logperch	*Percina caprodes*

Longjaw cisco	*Coregonus alpenae*
Longnose dace	*Rhinichthys cataractae*
Longnose gar	*Lepisosteus osseus*
Margined madtom	*Noturus insignis*
Minnow or Carp Family	Cyprinidae
Mooneye Family	Hiodontidae
Mottled sculpins	*Cottus bairdi*
Muskellunge	*Esox masquinongy*
Nooksack dace	*Rhinichthys* sp.
Northern brook lamprey	*Ichthyomyzon fossor*
Northern hogsucker	*Hypentelium nigricans*
Northern pike	*Esox lucius*
Northern redbelly dace	*Chrosomus eos*
Orangespotted sunfish	*Lepomis humilis*
Oriental weatherfish	*Misgurnus anguillicaudatus*
Paddlefish	*Polydon spathula*
Pike Family	Esocidae
Pikeperch	*Lucioperca lucioperca*
Pink salmon	*Oncorhynchus gorbuscha*
Pugnose minnow	*Opsopoedus emilae*
Pugnose shiner	*Notropis anogenus*
Pumpkinseed	*Lepomis gibbosus*
Rainbow darters	*Etheostoma caeruleum*
Rainbow smelt	*Osmerus mordax*
Rainbow trout, steelhead trout	*Oncorhynchus mykiss*
Redbreast sunfish	*Lepomis auritus*

Redear sunfish	*Lepomis microlophus*
Redfin shiner	*Notropis umbratilis*
Redside dace	*Clinostomus elongatus*
River redhorse	*Moxostoma carinatum*
Rock bass	*Ambloplites rupestris*
Rosyface shiner	*Notropis rubellus*
Round goby	*Neogobius melanostomus*
Rudd	*Scardinius erythrophthalmus*
Ruffe	*Gymnocephalus cernuus*
Salish sucker	*Catastomus* sp.
Sand shiners	*Notropis stramineus*
Sauger	*Stizostedion canadense*
Sculpin Family	Cottidae
Sea lamprey	*Petromyzon marinus*
Shorthead sculpin	*Cottus confusus*
Shortnose cisco	*Coregonus reighardti*
Shortnose sturgeon	*Acipenser brevirostrum*
Silver chub	*Macrhybopsis storeriana*
Silver shiner	*Notropis photogenis*
Slimy sculpin	*Cottus cognatus*
Smallmouth bass	*Micropterus dolomieu*
Smallmouth buffalo	*Ictiobus bubalus*
Smelt Family	Osmeridae
Speckled dace	*Rhinichthys osculus*
Spotfin shiners	*Notropis spilopterus*
Spottail shiner	*Notropis hudsonius*

Spotted gar	*Lepisosteus oculatus*
Spotted sucker	*Minytrema melanops*
Spring cisco	*Coregonus* sp.
Squanga whitefish	*Coregonus* sp.
Suckermouth minnow	*Phenacobius mirabilis*
Tench	*Tinca tinca*
Tubenose goby	*Proterorhinus marmoratus*
Umatilla dace	*Rhinichthys umatilla*
Warmouth	*Lepomis gulosus*
Western mosquitofish	*Gambusia affinis*
White sucker	*Catostomus commersoni*
White bass	*Morone chrysops*
White perch	*Morone americana*
White sturgeon	*Acipenser transmontanus*
Yellow perch	*Perca flavescens*
Yellow walleye	*Stizostedion vitreum*